T0142288

Advances in Intelligent Systems and Computing

Volume 990

The series "Advances in Intelligent Systems and Computing" contains publications on theory, applications, and design methods of Intelligent Systems and Intelligent Computing. Virtually all disciplines such as engineering, natural sciences, computer and information science, ICT, economics, business, e-commerce, environment, healthcare, life science are covered. The list of topics spans all the areas of modern intelligent systems and computing such as: computational intelligence, soft computing including neural networks, fuzzy systems, evolutionary computing and the fusion of these paradigms, social intelligence, ambient intelligence, computational neuroscience, artificial life, virtual worlds and society, cognitive science and systems, Perception and Vision, DNA and immune based systems, self-organizing and adaptive systems, e-Learning and teaching, human-centered and human-centric computing, recommender systems, intelligent control, robotics and mechatronics including human-machine teaming, knowledge-based paradigms, learning paradigms, machine ethics, intelligent data analysis, knowledge management, intelligent agents, intelligent decision making and support, intelligent network security, trust management, interactive entertainment, Web intelligence and multimedia.

The publications within "Advances in Intelligent Systems and Computing" are primarily proceedings of important conferences, symposia and congresses. They cover significant recent developments in the field, both of a foundational and applicable character. An important characteristic feature of the series is the short publication time and world-wide distribution. This permits a rapid and broad dissemination of research results.

**** Indexing: The books of this series are submitted to ISI Proceedings, EI-Compendex, DBLP, SCOPUS, Google Scholar and Springerlink ****

More information about this series at http://www.springer.com/series/11156

Himansu Sekhar Behera · Janmenjoy Nayak ·
Bighnaraj Naik · Danilo Pelusi
Editors

Computational Intelligence in Data Mining

Proceedings of the International Conference on ICCIDM 2018

Editors
Himansu Sekhar Behera
Department of Information Technology
Veer Surendra Sai University of Technology
Burla, Sambalpur, Odisha, India

Bighnaraj Naik
Department of Computer Application
Veer Surendra Sai University of Technology
Burla, Sambalpur, Odisha, India

Janmenjoy Nayak
Department of Computer Science
and Engineering
Sri Sivani College of Engineering
Srikakulam, Andhra Pradesh, India

Danilo Pelusi
Faculty of Communication Sciences
University of Teramo
Teramo, Italy

ISSN 2194-5357 ISSN 2194-5365 (electronic)
Advances in Intelligent Systems and Computing
ISBN 978-981-13-8675-6 ISBN 978-981-13-8676-3 (eBook)
https://doi.org/10.1007/978-981-13-8676-3

This Springer imprint is published by the registered company Springer Nature Singapore Pte Ltd.
The registered company address is: 152 Beach Road, #21-01/04 Gateway East, Singapore 189721, Singapore

ICCIDM Committee

Chief Patron & President

Prof. Atal Chaudhuri, Vice Chancellor, VSSUT

Co-Patron

Prof. P. K. Hota, Dean CDCE, VSSUT, Burla

Honorary Advisory Chair

Prof. S. K. Pal, Sr. Member IEEE, LFIEEE, FIAPR, FIFSA, FNA, FASc, FNASc, FNAE, Distinguished Scientist and Former Director, Indian Statistical Institute, India
Prof. V. E. Balas, Sr. Member IEEE, Aurel Vlaicu University, Romania

Honorary General Chair

Prof. B. Majhi, Director, IIIT Kanchipuram, Tamil Nadu, India
Prof. B. K. Panigrahi, Indian Institute of Technology (IIT), Delhi, India

General Chair

Dr. Himansu Sekhar Behera, Associate Professor, Department of IT, Veer Surendra Sai University of Technology (VSSUT), Burla, Odisha, India

Program Chair

Dr. D. P. Mohapatra, NIT Rourkela, Odisha, India
Dr. J. C. Bansal, South Asian University, New Delhi, India
Dr. A. K. Das, IIEST, Shibpur, WB

Organizing Chair

Dr. Bighnaraj Naik, Department of Computer Application, Veer Surendra Sai University of Technology (VSSUT), Burla, Odisha, India
Dr. Janmenjoy Nayak, Sri Sivani College of Engineering, Srikakulam, Andhra Pradesh, India

International Advisory Committee

Prof. A. Abraham, Machine Intelligence Research Labs, USA
Prof. Dungki Min, Konkuk University, Republic of Korea
Prof. Francesco Marcelloni, University of Pisa, Italy
Prof. Francisco Herrera, University of Granada, Spain
Prof. A. Adamatzky, Unconventional Computing Centre UWE, UK
Prof. H. P. Proença, University of Beira Interior, Portugal
Prof. P. Mohapatra, University of California
Prof. S. Naik, University of Waterloo, Canada
Prof. George A. Tsihrintzis, University of Piraeus, Greece
Prof. Richard Le, Latrob University, AUS
Prof. Khalid Saeed, A.U.S.T, Poland
Prof. Yew-Soon Ong, Singapore
Prof. Andrey V. Savchenko, N.R.U.H.S.E, Russia
Prof. P. Mitra, P.S. University, USA
Prof. D. Sharma, University of Canberra, Australia
Prof. Istvan Erlich, University of Duisburg-Essen, Germany
Prof. Michele Nappi, University of Salerno, Italy

Prof. Somesh Jha, University of Wisconsin, USA
Prof. Sushil Jajodia, George Mason University, USA
Prof. S. Auephanwiriyakul, Chiang Mai University, Thailand
Prof. Carlos A. Coello Coello, Mexico
Prof. M. Crochemore, University de Marne-la-Vallée, France
Prof. T. Erlebach, University of Leicester, Leicester, UK
Prof. T. Baeck, Universiteit Leiden, Leiden, The Netherlands
Prof. J. Biamonte, ISI Foundation, Torino, Italy
Prof. C. S. Calude, University of Auckland, New Zealand
Prof. P. Degano, Università di Pisa, Pisa, Italy
Prof. M. Murugappan, University of Malaysia
Prof. Raouf Boutaba, University of Waterloo, Canada
Prof. Kenji Suzuki, University of Chicago
Prof. Raj Jain, WU, USA
Prof. D. Al-Jumeily, Liverpool J. Moores University, UK
Prof. M. S. Obaidat, Monmouth University, USA
Prof. P. N. Suganthan, NTU, Singapore
Prof. Biju Issac, Teesside University, UK
Prof. Brijesh Verma, C.Q.U, Australia
Prof. Ouri E. Wolfson, University of Illinois, USA
Prof. Klaus David, University of Kassel, Germany
Prof. M. Dash, NTU, Singapore
Prof. L. Kari, Western University, London, Canada
Prof. A. S. M. Sajeev, Australia
Prof. Tony Clark, M.S.U, UK
Prof. Sanjib ku. Panda, NUS, Singapore
Prof. R. C. Hansdah, IISC, Bangalore
Prof. G. Chakraborty, Iwate Pref. University
Prof. Atul Prakash, University of Michigan, USA
Prof. Sara Foresti, University of degli Studi di Milano, Italy
Prof. Pascal Lorenz, University of Haute Alsace, France
Prof. G. Ausiello, University di Roma "La Sapienza", Italy
Prof. X. Deng, University of Liverpool, England, UK
Prof. Z. Esik, University of Szeged, Szeged, Hungary
Prof. A. G. Barto, University of Massachusetts, USA
Prof. G. Brassard, University de Montréal, Montréal, Canada
Prof. L. Cardelli, Microsoft Research, England, UK
Prof. A. E. Eiben, VU University, The Netherlands
Prof. Patrick Siarry, Université de Paris, Paris
Prof. R. Herrera Lara, EEQ, Ecuador

National Advisory Committee

Sri A. K. Choudhury, Registrar (VSSUT, Burla)
Prof. R. P. Panda (VSSUT, Burla)
Prof. A. N. Nayak, Dean, SRIC (VSSUT, Burla)
Prof. D. Mishra, Dean, Students' Welfare (VSSUT, Burla)
Prof. P. K. Kar, Dean, Faculty and Planning (VSSUT, Burla)
Prof. P. K. Das, Dean, Academic Affairs (VSSUT, Burla)
Prof. S. K. Swain, Dean, PGS & R (VSSUT, Burla)
Prof. S. Panda, Coordinator TEQIP (VSSUT, Burla)
Prof. U. R. Jena (VSSUT, Burla)
Prof. D. K. Pratihar (IIT, Kharagpur)
Prof. K. Chandrasekaran (NIT, Karnataka)
Prof. S. G. Sanjeevi (NIT, Warangal)
Prof. G. Saniel (NIT, Durgapur)
Prof. B. B. Amberker (NIT, Warangal)
Prof. R. K. Agrawal (JNU, New Delhi)
Prof. U. Maulik (Jadavpur University)
Prof. Sonajharia Minz (JNU, New Delhi)
Prof. G. Jena (BVC Engineering College, JNTU, Kakinada, Andhra Pradesh)
Prof. C. R. Tripathy (VC, BPUT, Rourkela)
Prof. B. B. Pati (VSSUT, Burla)
Prof. B. B. Biswal (Director, NIT Meghalaya)
Prof. K. Shyamala (Osmania University, Hyderabad)
Prof. P. Sanyal (WBUT, Kolkata)
Prof. G. Panda (IIT, BBSR)
Prof. B. B. Choudhury (ISI, Kolkata)
Prof. K. K. Shukla (IIT, BHU)
Prof. G. C. Nandy (IIIT, Allahabad)
Prof. R. C. Hansdah (IISC, Bangalore)
Prof. S. K. Basu (BHU, India)
Prof. J. V. R. Murthy (JNTU, Kakinada)
Prof. DVLN Somayajulu (NIT, Warangal)
Prof. G. K. Nayak (IIIT, BBSR)
Prof. P. P. Choudhury (ISI, Kolkata)
Prof. Amiya Ku. Rath (VSSUT, Burla)
Dr. R. Mohanty, HOD CSE (VSSUT, Burla)
Prof. S. Bhattacharjee (NIT, Surat)

Technical Committee

Dr. Adel M. Alimi, REGIM-Lab., ENIS, University of Sfax, Tunisia
Dr. Chaomin Luo, University of Detroit Mercy, Detroit, Michigan, USA
Dr. Istvan Erlich, Department of EE and IT, University of Duisburg-Essen, Germany
Dr. Tzyh Jong Tarn, Washington University in St. Louis, USA
Dr. Simon X. Yang, University of Guelph, Canada
Dr. Raffaele Di Gregorio, University of Ferrara, Italy
Dr. Kun Ma, Shandong Provincial Key Laboratory of Network Based Intelligent Computing, University of Jinan, China
Dr. Azah Kamilah Muda, Faculty of ICT, Universiti Teknikal Malaysia Melaka, Malaysia
Dr. Biju Issac, Teesside University, Middlesbrough, England, UK
Dr. Bijan Shirinzadeh, Monash University, Australia
Dr. Enver Tatlicioglu, Izmir Institute of Technology, Turkey
Dr. Hajime Asama, The University of Tokyo, Japan
Dr. N. P. Padhy, Department of EE, IIT, Roorkee, India
Dr. Ch. Satyanarayana, Department of Computer Science and Engineering, JNTU Kakinada, India
Dr. M. R. Kabat, Department of CSE, VSSUT, India
Dr. A. K. Barisal, HOD, Department of EE and EEE, VSSUT, Burla
Dr. M. Murugappan, University of Malaysia
Dr. Danilo Pelusi, University of Teramo, Italy
Dr. L. Sumalatha, Department of Computer Science and Engineering, JNTU Kakinada, India
Dr. K. N. Rao, Andhra University, Visakhapatnam, India
Dr. S. Das, Indian Statistical Institute, Kolkata, India
Dr. D. P. Mohapatra, Head, Department of CSE, N.I.T, Rourkela, India
Dr. A. K. Turuk, Department of CSE, NIT, RKL, India
Dr. M. P. Singh, Department of CSE, NIT, Patna, India
Dr. R. Behera, Department of EE, IIT, Patna, India
Dr. P. Kumar, Department of CSE, NIT, Patna, India
Dr. A. Das, Department of CSE, IIEST, WB, India
Dr. J. P. Singh, Department of CSE, NIT, Patna, India
Dr. M. Patra, Berhampur University, Odisha, India
Dr. A. Deepak, Department of CSE, NIT, Patna, India
Dr. D. Dash, Department of CSE, NIT, Patna, India
Dr. Bhanu Prakash Kolla, K. L. Deemed to be University, Vijayawada, Andhra Pradesh

Publicity Chair

Prof. P. C. Swain, VSSUT, Burla
Dr. P. K. Sahu, VSSUT, Burla
Dr. G. T. Chandrasekhar, SSCE, Srikakulam, Andhra Pradesh
Dr. Apurba Sarkar, IIEST, Shibpur, WB, India
Dr. Soumen Kumar Pati, St. Thomas' College of Engineering and Technology, West Bengal, India

Sponsorship Chair

Dr. Satyabrata Das, VSSUT, Burla
Mr. Sanjaya Kumar Panda, VSSUT, Burla

Registration Chair

Dr. Sucheta. Panda, VSSUT, Burla
Ms. Gargi Bhattacharjee, VSSUT, Burla

Web Chair

Dr. Manas Ranjan Senapati, VSSUT, Burla

Publication Chair

Dr. Janmenjoy Nayak, Sri Sivani College of Engineering, Srikakulam, Andhra Pradesh, India
Dr. Bighnaraj Naik, Department of Computer Application, VSSUT, Burla, Odisha, India

Organizing Committee

Mr. D. C. Rao, VSSUT, Burla
Mr. Kishore Kumar Sahu, VSSUT, Burla

Mr. Sanjib Nayak, VSSUT, Burla
Mrs. Sasmita Behera, VSSUT, Burla
Mrs. Sasmita Acharya, VSSUT, Burla
Dr. Santosh Majhi, VSSUT, Burla
Dr. Pradipta Kumar Das, VSSUT, Burla
Dr. Kshiramani Naik, VSSUT, Burla
Mr. Sujaya Kumar Sathua, VSSUT, Burla
Mr. Atul Vikas Lakra, VSSUT, Burla
Mr. Gyanaranjan Shial, VSSUT, Burla
Mr. Suresh Kumar Srichandan, VSSUT, Burla
Mrs. E. Oram, VSSUT, Burla
Mr. S. Mohapatra, VSSUT, Burla

Organizing Secretary

Mr. D. P. Kanungo, Department of IT, VSSUT, Burla, Odisha, India

Technical Reviewers

Dr. Asit Ku. Das, Indian Institute of Engineering Science and Technology, Shibpur, Kolkata
Dr. Soumen Pati, St. Thomas College of Engineering, Kolkata
Dr. D. Shyam Prasad, CVR College of Engineering, Hyderabad
Mrs. D. Priyanka, Sri Sivani College of Engineering, Srikakulam, Andhra Pradesh
Dr. Santosh Ku Sahoo, CVR College of Engineering, Hyderabad
Dr. Kola Bhanu Prakash, K L University, Vijayawada
Dr. Sarat Ch. Nayak, CMR College of Engineering and Technology (Autonomous), Hyderabad
Dr. Sunanda Das, Neotia Institute of Technology, Kolkata
Mr. Himansu Das, Kalinga Institute of Industrial Technology (KIIT), Bhubaneswar, Odisha
Dr. G. T. Chandra Sekhar, Sri Sivani College of Engineering, Srikakulam, Andhra Pradesh
Dr. V. Ramesh, CMR College of Engineering and Technology (Autonomous), Hyderabad
Mr. Ch. Jagan Mohan Rao, Veer Surendra Sai University of Technology, Burla, Odisha
Mr. Jagannadh Panigrahi, Sri Sivani College of Engineering, Srikakulam, Andhra Pradesh
Dr. B. Manmadha Kumar, Aditya Institute of Technology and Management, Andhra Pradesh

Dr. Narendra B. Mustare, CVR College of Engineering, Hyderabad
Dr. P. Vittal, Aditya Institute of Technology and Management, Andhra Pradesh
Mr. P. Suresh Kumar, Dr. Lankapalli Bullayya College, Visakhapatnam, Andhra Pradesh
Dr. Ram Chandra Barik, Vikash Engineering College, Bargarh, Odisha
Dr. K. Dhayalini, K L University, Vijayawada
Dr. Shaik Nazeer, Bapatla Engineering College (Autonomous), Guntur District, Bapatla, Andhra Pradesh
Dr. K. V. D. Kiran, K L University, Vaddeswaram, Guntur, Andhra Pradesh
Dr. P. Pradeepa, Jain University, Bangalore
Dr. S. Nageswari, Alagappa Chettiar Government College of Engineering and Technology, Karaikudi
Dr. R. Arulmurugan, S R Engineering College, Warangal
Dr. K. C. Bhuyan, College of Engineering and Technology, Bhubaneswar, Odisha
Mr. Dillip Khamari, Vikash Institute of Technology, Bargarh, Odisha
Dr. Meenakhi Rout, Kalinga Institute of Industrial Technology (KIIT), Bhubaneswar, Odisha
Dr. Suneeta Mohanty, Kalinga Institute of Industrial Technology (KIIT), Bhubaneswar, Odisha
Mr. Suresh Chandra Moharana, Kalinga Institute of Industrial Technology (KIIT), Bhubaneswar, Odisha
Dr. P. M. K. Prasad, GVP College of Engineering for Women Visakhapatnam, Andhra Pradesh, India
Dr. Manoj Kumar Mishra, Kalinga Institute of Industrial Technology (KIIT), Bhubaneswar, Odisha
Mr. Debasis Mohapatra, Parala Maharaja Engineering College (Govt. Odisha), Berhampur, Odisha
Mr. Rashmi Ranjan Sahoo, Parala Maharaja Engineering College (Govt. Odisha), Berhampur, Odisha
Dr. S. K. Padhy, Veer Surendra Sai University of Technology, Burla, Odisha
D. K. Behera, Trident Academy of Technology, Bhubaneswar, Odisha
Dr. M. Mishra, ITER, Siksha 'O' Anusandhan (Deemed to be University), Bhubaneswar
Mr. Umashankar Ghugar, Berhampur University, Berhampur, Odisha
Dr. S. K. Panda, Veer Surendra Sai University of Technology, Burla, Odisha
Dr. Sourav Kumar Bhoi, Parala Maharaja Engineering College (Govt. of Odisha), Berhampur
Mr. Siba Prasada Tripathy, NIST, Berhampur, Odisha, India
Mr. Purnendu Mishra, Vikash Institute of Technology, Bargarh, Odisha
Mr. Sibarama Panigrahi, Sambalpur University, Institute of Information Technology, Sambalpur, Odisha
Dr. Manas Senapati, Veer Surendra Sai University of Technology, Burla, Odisha
Dr. Chittaranjan Pradhan, Kalinga Institute of Industrial Technology, Bhubaneswar, Odisha
Dr. Sucheta Panda, Veer Surendra Sai University of Technology, Burla, Odisha

Dr. M. Marimuthu, Coimbatore Institute of Technology, Coimbatore
Ms. Santwana Sagnika, Kalinga Institute of Industrial Technology, Bhubaneswar, Odisha
Aditya Hota, Veer Surendra Sai University of Technology, Burla, Odisha
Dr. Santosh Kumar Pani, Veer Surendra Sai University of Technology, Burla, Odisha
Dr. Prakash Bhat, Senior Engineer at Siemens PLM Software Inc., Houston, Texas, USA
Dr. Ajay Kumar Jena, Kalinga Institute of Industrial Technology (KIIT), Bhubaneswar
Dr. S. K. Majhi, Veer Surendra Sai University of Technology, Burla, Odisha
Dr. Nageswara Rao, Velagapudi Ramakrishna Siddhartha Engineering College
Dr. M. Nazma B. J. Naskar, Kalinga Institute of Industrial Technology (KIIT), Bhubaneswar, Odisha
Mr. R. Paul, University of South Florida, Tampa
Dr. S. Albert Alexander, Kongu Engineering College, Perundurai, Tamil Nadu
Mr. Shyam Prasad Devulapalli, JNTU Hyderabad, Hyderabad
Dr. M. R. Narasinga Rao, K L University, Vaddeswaram, Andhra Pradesh
Dr. B. Satyanarayana, CMR College of Engineering and Technology (Autonomous), Hyderabad
Dr. G. Syam Prasad, K L University, Vaddeswaram, Andhra Pradesh
Dr. A. Abdul Rasheed, CMR Institute of Technology, Bengaluru
Mr. Pinaki Sankar Chatterjee, Kalinga Institute of Industrial Technology (KIIT), Bhubaneswar
Dr. P. V. R. D. Prasada Rao, K L University, Vaddeswaram, Andhra Pradesh
Dr. Satyabrata Das, Veer Surendra Sai University of Technology, Burla, Odisha
Dr. Subrat Nayak, Siksha 'O' Anusandhan (Deemed to be University), Bhubaneswar, Odisha
Dr. Himansu Sekhar Behera, Veer Surendra Sai University of Technology, Burla, Odisha
Radhamohan Pattnayak, Godavari Institute of Engineering and Technology, Rajahmundry, Andhra Pradesh
Dr. Banchhanidhi Dash, JIGIGA University, Ethiopia
Mrs. P. Sruthi, CMR College of Engineering and Technology (Autonomous), Hyderabad
Mr. Jay Sarraf, Kalinga Institute of Industrial Technology, Bhubaneswar, Odisha
Dr. K. Suvarna Vani, Velagapudi Ramakrishna Siddhartha Engineering College, Vijayawada, Andhra Pradesh
Dr. P. Vidya Sagar, Velagapudi Ramakrishna Siddhartha Engineering College, Vijayawada, Andhra Pradesh
Dr. M. Panda, Utkal University, Bhubaneswar, Odisha
Dr. M. Patra, Berhampur University, Berhampur, Odisha
Dr. R. Umamaheswari, Velammal Engineering College, Chennai, India
Dr. K. L. S. Soujanya, CMR College of Engineering and Technology (Autonomous), Hyderabad

Mrs. A. Poongodai, CMR College of Engineering and Technology (Autonomous), Hyderabad
Mr. A. Vivekanad, CMR College of Engineering and Technology (Autonomous), Hyderabad
Mr. Susil Kumar Mohanty, University of Hyderabad, Hyderabad, Telengana
Dr. Rajesh Kumar, JNTU, Hyderabad
Dr. Rajesh Kumar Tripathy, BITS, Pilani
Dr. Koteswara Rao, JNTU, Hyderabad
Dr. Sanjay Kumar Sahoo, Lovely Professional University, Punjab
Dr. S. Harivardhangini, JNTU, Hyderabad
Dr. Shampa Sengupta, MCKV Engineering College, West Bengal
Dr. Md. Dilsad Ansari, CMR College of Engineering and Technology (Autonomous), Hyderabad
Dr. Santanu Phadikar, MAKAUT, West Bengal
Mr. Arka Ghosh, IIEST, Shibpur
Dr. Vijaya Kumar Koppula, CMR College of Engineering and Technology (Autonomous), Hyderabad
Dr. Samit Biswas, IIEST, Shibpur, West Bengal
Dr. Apurba Sarkar, IIEST, Shibpur, West Bengal
Mrs. Sarita Mohapatra, Siksha 'O' Anusandhan (Deemed to be University), Bhubaneswar, Odisha
Mrs. Subhra Swetanisha, Trident Academy of Technology, Bhubaneswar
Dr. P. Raman, GMR Institute of Technology, Andhra Pradesh
Ms. Yasmin Ghazaala, St. Thomas Engineering College, West Bengal
Mrs. Soumi Dutta, IEM, West Bengal
Dr. Amit Paul, St. Thomas Engineering College, West Bengal
Dr. K. Ramakrishna, CMR College of Engineering and Technology (Autonomous), Hyderabad
Dr. Arijit Ghoshal, St. Thomas Engineering College, West Bengal
Amiya Halder, St. Thomas Engineering College, West Bengal
Dr. Ramesh Prusty, Veer Surendra Sai University of Technology, Burla, Odisha
Mr. Karthik Jilla, CMR College of Engineering and Technology (Autonomous), Hyderabad
Ms. D. Mishra, College of Engineering and Technology, Bhubaneswar, Odisha

Preface

The expansion of data, structured and unstructured, is creating a complex analytical scenario for both industries and academic institutions. The challenge is to structure and review these data (information) for effective decision-making. The current solution is computational intelligence tools which use adaptive mechanisms for the understanding of data in a complex and altering environment.

The fifth international conference on computational intelligence in data mining (ICCIDM 2018), organized by Veer Surendra Sai University of Technology (VSSUT), Burla, Sambalpur, Odisha, India, during 15–16 December 2018, has given a platform for research community to ponder over these challenging problems and find a solution.

The conference has received more than 250 submissions from researchers working in the field of data mining and computational intelligence. After a thorough review, knowledgeable subject experts from different countries have selected 66 best quality papers for presentation and publication. With an acceptance rate of 26.4%, the papers present the latest research findings in the subject and discuss new and interesting applications.

The papers discuss current research findings in the fields of soft computing, data mining, computational intelligence, green computing, big data analysis and nature-inspired computing.

Hope the content of the proceeding will be useful for readers in their field of study.

Burla, Sambalpur, India	Himansu Sekhar Behera
Srikakulam, India	Janmenjoy Nayak
Burla, Sambalpur, India	Bighnaraj Naik
Teramo, Italy	Danilo Pelusi

Acknowledgements

The conference has attracted more than 250 researchers globally to participate and share their research findings in the form of quality research papers. Thanks to all the participants for their contribution.

The editors highly appreciate the organizing committee for their active involvement from day one and during each stage of the conference. We express our heartfelt gratitude to the international and national advisory members and technical programme committee. The team is highly obliged and thankful to the reviewers for their remarkable effort in providing their valuable comments and finalizing the content of the proceedings.

We also want to thank the volunteers for their untiring effort to make this event successful.

Last but not least, we appreciate the help rendered by the editorial team of Springer Nature in publishing this proceeding in AISC book series.

About the Conference

After four triumphant interpretations of ICCIDM, the international conference on "Computational Intelligence in Data Mining" (ICCIDM) has accomplished to its fifth consequential version with a high endeavor. ICCIDM 2018 has been entrenched itself as one of the prominent and leading conference which will assist the progress of cross cooperation across the different provincial research communities in India along with other international provincial research partners and programs. Such vigorous deliberations and brainstorming conferences among national and international research communities is requisite to address many challenges, applications of computational intelligence in the area of engineering, science and technology and new trends are proceeding by each passing moment. The 2018 version of ICCIDM is an appropriate platform for researchers, scientists, scholars, academicians and practitioners to contribute their inventive thoughts and research discussion on the confronts, state-of-the-art developments and unsettled open troubles in the area of data mining and computational intelligence.

The conference aims to:

- Administrate a sneak examination into the strength and weakness of trending relevance and research results in the field of computational intelligence and data mining.
- Augment the exchange of thoughts and consistency among the different computational intelligence schemes.
- Develop the significance and utilization of data mining application fields for end-users and neophyte user application.
- Endorse peculiar high eminent research verdicts and innovative results to the challenging problems in intelligent computing.
- Contribute research endorsements for upcoming assessment reports.

By the end of the conference, we anticipate that the participants will enhance their knowledge by latest perceptions and examinations on existing research topics from leading academician, researchers and scientists around the globe, contribute their own thoughts on significant research topics like computational intelligence and data mining as well as to networks and conspire with their international compeer.

Contents

About the Editors

Dr. Himansu Sekhar Behera is an Associate Professor & Head of the Department at the Department of Information Technology, VSSUT, Burla, Odisha, India. He has published more than 80 research papers in various international journals and conference proceedings, edited 11 books, and serves on the Editorial/Reviewer Board of various international journals. His research interests include Data Mining and Intelligent Computing. He is associated with various educational and research societies, e.g. the OITS, ISTE, SMIAENG, SMCSTA etc.

Dr. Janmenjoy Nayak is an Associate Professor & Head of the Department at the Department of Computer Science & Engineering, Sri Sivani College of Engineering, Srikakulam, India. He was awarded an INSPIRE Fellowship by the DST, Government of India to pursue his doctoral research at the VSSUT, Burla, India. He has published more than 50 research papers in various reputed peer-reviewed journals and conference proceedings. He is the recipient of the Best Researcher Award from the JNTU, Kakinada (2018) and the Young Faculty in Engineering Award from Venus International Foundation (2017). With more than eight years of teaching and research experience, his research interests include Data Mining, Nature Inspired Algorithms and Soft Computing. He is an active reviewer for various reputed peer-reviewed journals, e.g. IEEE, IET, Elsevier, Springer and Inderscience journals. Further, Dr. Nayak is a life member of several societies, e.g. the IEEE, CSI India, IAENG (Hong Kong) etc.

Dr. Bighnaraj Naik is an Assistant Professor at the Department of Computer Application, Veer Surendra Sai University of Technology, Burla, India. He has published more than 50 research papers in prominent journals and conference proceedings. With more than eight years of teaching experience in the field of Computer Science and Information Technology, Dr. Naik's research interests include Data Mining and Soft Computing. He is a life member of several societies, e.g. the IEEE, IAENG (Hong Kong) etc.

Dr. Danilo Pelusi is an Associate Professor at the Faculty of Communication Sciences, University of Teramo, Italy, where he also received his Ph.D. in Computational Astrophysics in 2006. His current research interests include Information Theory, Fuzzy Logic, Neural Networks and Evolutionary Algorithms. He is an Associate Editor of IEEE Transactions on Emerging Topics in Computational Intelligence and IEEE Access, and has served as a guest editor for several Inderscience and Springer journals.

Cluster Validity Using Modified Fuzzy Silhouette Index on Large Dynamic Data Set

Chatti Subbalakshmi, Rishi Sayal and H. S. Saini

Abstract Cluster validity index is applied to evaluate clustering results. It can be performed based on different measures and it can accomplish at data point level or cluster center level. In distance-based clustering methods, silhouette is an efficient point to point index measure, which defines relation based on compactness and separation distances. To validate fuzzy partitions, fuzzy silhouette index is used by applying defuzzification process. One of the applications of cluster validity is finding an optimal number of clusters in distance-based methods. As data size increases, point-wise index measure calculation takes more execution time. Hence, we proposed approaches to reduce time complexity by modifying fuzzy silhouette index at center to center and center to mean levels. All these methods are applied to find the right number of cluster and they are giving correct value in minimum execution time. All work is implemented in Matlab and effective results are given by our proposed methods.

Keywords Cluster validity · Silhouette width · Fuzzy clustering

1 Introduction

In Customer Relation Management (CRM), Social Networks, Web user profile tracking, medical data analysis and many other data analytics applications, data clustering has been used in data mining [1]. Clustering defines the relations between the data points by partition data into clusters. The quality of relations is determined with putting similar objects in the same group and dissimilar objects in distinct partitions.

C. Subbalakshmi (✉) · R. Sayal · H. S. Saini
Guru Nanak Institutions Technical Campus, Hyderabad 501506, Telangana, India
e-mail: subbalakshmichatti@gmail.com

R. Sayal
e-mail: ad.rs@gniindia.org

H. S. Saini
e-mail: md@gniindia.org

H. S. Behera et al. (eds.), *Computational Intelligence in Data Mining*, Advances in Intelligent Systems and Computing 990,
https://doi.org/10.1007/978-981-13-8676-3_1

Cluster results are evaluated by applying cluster validity indices. Many types of clustering methods are being introduced for support of application data bases. The basic methods are distance-based clustering, model-based clustering and density-based clustering. All these algorithms require certain input values which depend on data set and they are defined prior to the execution of algorithms. For example, for distance-based clustering algorithms [2, 3], we need to input two minimum parameters, i.e. data set (D) and number of clusters (k) and k value always depends on data set characteristics. Using cluster validity indices, we can find optimal number of clusters. In literature, many cluster validity measures have been proposed.

The complexity of cluster analysis on static data sets is less due to no need for change in input parameters as data does not change. But, in recent years, most of the web applications are generating dynamic large amount of data. And it always data changes time to time depends on application requirements. Due to this nature, data may be uncertain, incomplete and overlapped. To do clustering on overlapped and uncertain data, fuzzy distance-based algorithms are suitable [4] due to simple and efficiency. To perform data analysis, complete clustering process needs to be repeated by updating input values (cluster number) when data changes. It increases the complexity of time and space for large data bases. Hence, it requires a new approach for reduction of time complexity for large data bases with dynamic nature.

In this paper, we took the problem clustering fuzzy partitions on large dynamic data set. For that, to validate and to find optimal cluster number, we selected fuzzy silhouette index approach. Basic fuzzy silhouette index is applied on point to point level, where it takes more execution time as data size increasing and changing. Hence, we proposed an approach to reduce the time complexity in this paper. We modified the fuzzy silhouette index by incorporating center to point level and center to mean of clusters in calculating compactness and separation distances. The more effective results given by our proposed methods compare to basic method.

The paper is organized as follows: related work is given in Sect. 2, proposed methods are presented in Sect. 3 and results and conclusion are given in Sects. 4 and 5.

2 Related Work

The cluster validity is evaluating the clustering quality of given partitions by applying cluster validity index. For validating the distance-based clustering, the separation and compactness measures are applied on clusters. The partition clustering methods come under distance-based clustering, where finding the number of clusters is a major problem of cluster validity [5]. Therefore, right number of clusters gives good clustering and developing a validity measure to indicate the right number of clusters is a major research problem in the literature.

Fuzzy clustering can be used for highly complex data, partial membership and incomplete data, where measures are not partitioned into mutually exclusive clusters. The fuzzy approach, objects are overlapping between the clusters and many measures are defined for cluster quality. The basic fuzzy cluster indices are given in following subsections.

2.1 *Fuzzy Cluster Validity Indices*

The criteria for evaluating fuzzy partitions are framed on membership value (U) of cluster point or distance relation between cluster centers. At indices values are minimum or maximum, cluster structure is well defined. Some of the fuzzy partition cluster validity indices are given below.

2.1.1 Least Squared Error (SE) Index

It is the sum of squared error function within the clusters [6]. Its value is minimum when the cluster structure is perfect. It is used as an objective function in fuzzy c-means algorithm.

$$SE(x, u, c) = \sum_{i=1}^{n} \sum_{j=1}^{c} u_{ij}^{m} \left\| x_i - c_j \right\|^2 \tag{1}$$

where x_i is data point, c_j is jth cluster center and u_{ij} is membership value of ith point in jth cluster. The fuzzy c-means algorithms terminate at minimum SE value, which indicates the well fuzzy partitioned.

2.1.2 Partition Coefficient

Bezdek [7] defined the partition coefficient (PC index), which measures the amount of overlapping or fuzziness in partitions. It is defined in Eq. 1, based on object membership values and it gets boundary values between 1/c and 1. The index has its minimum at the fuzziest possible into c clusters.

$$PC(U) = \frac{1}{n} \sum_{i=1}^{n} \sum_{j=1}^{c} u_{ij}^{2} \tag{2}$$

This measure also used for selection of the number of clusters c, on given membership matrix U of clusters. The PC value is minimum at right number of clusters but it will not hold the relationships between data points such as separation and compactness distances.

2.1.3 Xie-Beni Index

In [8, 9], defined fuzzy cluster validity index by introducing the geometric relations between the data points in their measure. The total variation of data is defined as compactness and minimum distance between clusters is defined as separation of clusters.

$$XB(X, U, V) = \frac{\pi}{d_{min}^2} = \frac{\sum_{i=1}^{c}\sum_{j=1}^{n} u_{ij}^2 d^2(c_i, x_j)}{n \ \ min_{1\le i\ne h\le c} d^2(c_i, c_j)} \tag{3}$$

Xie-Beni index uses minimum center to center distance, it does well where clusters become more compact but not well separated. And it has more consistency due to it takes only squared distances in measures.

2.1.4 Partition Entropy Index (PE)

It was defined in [10], data relation is computed using logarithmic on membership values of cluster points as

$$PE(u) = -\frac{1}{n}\sum_{i=1}^{n}\sum_{j=1}^{c} u_{ij} log_b(u_{ij}) \tag{4}$$

where b is logarithmic base value and index value is minimum at right cluster structure.

2.1.5 Modified Partition Coefficient Index (MPE)

Modification of Partition Coefficient (PC) was defined in [11, 12], it reduces the monotonic tendency and its value is maximum at best clustering on given data set

$$MPC = 1 - \frac{c}{c-1}(1 - PC) \tag{5}$$

where c is the number of clusters.

2.1.6 Weighted Inter-intra Index (Wint)

The weighted inter-intra (Wint) measure is proposed in [13], it defines the relation compactness and separation of data as

$$Wint = \left(1 - \frac{2c}{n}\right)(1 - \frac{\sum_i \frac{1}{n-|c_i| \sum_{j \neq i} inter(c_i, c_j)}}{\sum_i \frac{2}{|c_i|-1} intra(c_i)}) \tag{6}$$

where intra(c_i) is compactness distance of ith cluster and inter (c_i, c_j) is separation distance between the clusters c_i and c_j. the index value is maximum when given data is clustered perfectly.

2.2 *Fuzzy Silhouette Width (Point-Wise)*

Silhouette is an efficient point-wise measure of cluster validity defined in [12]. Basically, it is defined in view of crisp/hard partitions. The silhouette width *S* of data point x_t defined as

$$S_t = S(x_t) = \frac{b_t - a_t}{\max\{a_t, b_t\}} \tag{7}$$

where compactness distance calculated as

$$a(x_t) = \frac{\sum_{i=1, i \neq t}^{n} d(x_i, x_t).u_i}{\sum_{i=1}^{n} u_i} \tag{8}$$

Separation distance is calculated as

$$b(x_t) = \min\{\frac{\sum_{i=1, i \neq t}^{n} d(x_i, x_t).u_i}{\sum_{i=1}^{n} u_i} \tag{9}$$

where a_t, b_t is compactness and separation distance of point x in cluster t. The S value ranges between 0 and 1. At right cluster number, it gives the largest average cluster silhouette value. Basically, S(t) is proposed to apply on hard partitions. By introducing defuzzification step in process, it can be applied to fuzzy partitions [13].

For each point (x_t), silhouette width S is weighted (w_i) by the difference between the two largest cluster membership values of it and its value calculated as

$$w(x_t) = u_{pt} - u_{qt} \tag{10}$$

The extended fuzzy point silhouette width $S(x_t)$ is computed by Eq. 3

$$EFS(x_t) = \frac{\sum_{i=1}^{n} w_i S_t}{\sum_{i=1}^{n} w_i} \tag{11}$$

It is well-performed in detecting areas of high density and fuzzy c-means [4] produces partitions in which the points near the area of cluster centers assuming larger weights and points in overlapping areas smaller weights. Smaller weight points can produce risk in cluster quality assessment. Therefore, point-wise cluster evaluation by silhouette is well suited for hard partition.

3 Proposed Method

Cluster validity index can be used for two purposes, to evaluate results of clustering algorithm and to find right cluster number at maximum validity. The process of calculating good fitted cluster numbers is shown in Fig. 1.

We executed the point-wise fuzzy silhouette index as defined in Eq. 4 on two-dimensional synthetic data set and two application data set. The results are given in Table 1. From results, number of calls on distance function increasing as data size is increasing. Hence, total execution time also proportionally increasing.

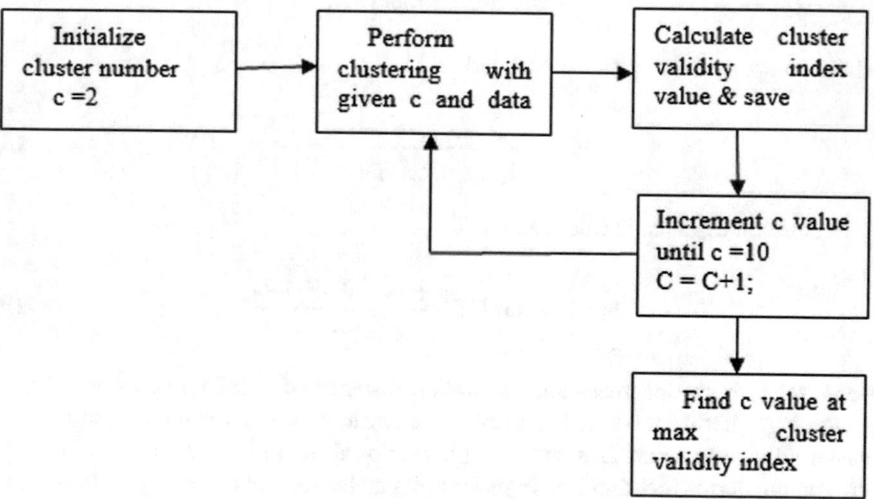

Fig. 1 Steps to find optimal number of clusters

Table 1 Input parameters and their values of proposed algorithm

Input parameter	Value
Maximum number of clusters	10
Maximum iterations	100
Fuzzification coefficient (m)	2

3.1 *Modified Fuzzy Silhouette Indices*

In modified extended fuzzy silhouette index, we proposed two indices at center to point level and center to center level. The main aim of these indices is to reduce the number of calls on distance method in calculating compactness and separation relations. Hence, it reduces the total execution time. As per that, we defined compactness and separation distance for assessment of cluster quality which takes less execution time.

Center to point level:
Here, the relation between cluster center and each data points in same cluster is defined as compactness of cluster. The relation between cluster center to data points in other cluster is called separation of cluster. The compactness *a(t)* is calculated as average distance between cluster center (c_t) and each data point (x_t) in the same clustert. The separation distance *b(t)* is defined as minimum average distance between cluster center *(c_t)* and each point in other clusters.

The compactness distance *a(t)* is computed to

$$a(c_t) = \frac{\sum_{j=1, j\neq t}^{n} u_{tj} d(c_t, x_i)}{\sum_{j=1, j\neq t}^{n} u_{tj}} \tag{14}$$

The separation distance *b(t)* is computed to

$$b(c_t) = min_{i\neq t}\{\frac{\sum_{j=1, j\neq t}^{n} u_{tj} d\left(c_t, x_j\right)}{\sum_{j=1, j\neq t}^{n} u_{tj}} \tag{15}$$

The silhouette index is calculated using Eq. 1 and extended weighted silhouette is computed Eq. 2.

Center to center level:
In this level, we defined relations on data points in the cluster based on cluster centers and mean of total cluster centers. The compactness *a(t)* is calculated as average distance between cluster center (c_t) and each data point (x_t) in the same cluster *t*. The separation distance *b(t)* is defined as minimum average distance between cluster center *(c_t)* and other cluster centers (c_j).

The compactness distance *a(t)* is computed to

$$a(c_t) = \frac{\sum_{j=1, j\neq t}^{n} u_{tj} d(c_t, x_j)}{\sum_{j=1, j\neq t}^{n} u_{tj}} \tag{16}$$

The separation distance *b(t)* is computed to

$$b(c_t) = min_{i\neq t}\{\frac{\sum_{i=1}^{n} d(c_i, \bar{c}).n_i}{n} \tag{17}$$

The silhouette index is calculated using Eq. 1 and extended weighted silhouette is computed Eq. 2.

Here, we proposed an algorithm for finding the right number of clusters using our modified fuzzy silhouette index is given below.

4 Execution Setup and Results

In this section, execution setup, data set description and result analysis are presented.

4.1 Execution Setup

The proposed algorithms are implemented and executed on Intel(R) Dual-Core CPU @ 3.07 GHz, Matlab version 2012. In all proposed methods, fuzzy partitions are generated using fuzzy c-means basic fuzzy clustering algorithm. The input parameter required for this algorithm are listed in Table 1.

4.2 Data Sets

The proposed methods are executed on both artificial data set and two application data set. The synthetic data set was created using random data generation method in MATLAB. Breast cancer data (WBCD) and wholesale customer data (WCD) sets are from the UCI public repository. The data sets description is given Table 2.

4.3 Results Analysis

The point-wise fuzzy silhouette values for all data sets are given Table 6 and results are giving that, in calculating compactness and separation distance, we were calling pdist2 method in Mat lab to calculate point to point Euclidean distance. But, as

Table 2 Data set description used in executions

Synthetic data		
Data set	Dimensions	Instances
data1000	2	1000
data2000	2	2000
data3000	2	3000
data4000	2	4000
data5000	2	5000
Application data set		
WBCD	9	683
WCD	8	440

data points are increasing total execution time and number of calls for pdist2 also increasing (Tables 3, 4 and 5).

This complexity further reduced using center-point fuzzy silhouette and results are given Table 5. The execution time is reduced by 50% of point-wise calculations and by giving same number of right c value. Further, execution time and call to pdist2 are reduced by using center-mean calculations which are given in Table 6. The point–point and center to point measures not finding subclusters, but using center to mean calculations is finding subclusters.

5 Conclusions

Cluster analysis has been using many web-based applications, where generated is changes frequently. Therefore, there is need of refreshing clustering process when there is a change and validating clustering results. It reflects in finding the input parameters of algorithms. We have considered one of the distance-based clustering on fuzzy partitions, fuzzy c-means which require number of clusters prior to execution. Fuzzy silhouette cluster validity index is basic approach at point to point-wise, but it takes more time as data size increase. Hence, we proposed random sampling method and modified fuzzy silhouette index in the view of reducing execution time by producing correct cluster validity.

Table 3 Results of finding optimal number of clusters on customer data set

Iterations	c = 2	c = 3	c = 4	c = 5	c = 6	c = 7	c = 8	c = 9	c = 10	Optimal	Total time (s)
	Finding on sample data size = 168										
First	**0.38877**	0.385527	0.379181	0.38553	0.344055	0.355501	0.337385	0.286223	0.309415	2	6.455
Second	**0.371433**	0.367097	0.298029	0.293325	0.277539	0.355105	0.354874	0.343498	0.285881	2	6.926
Third	**0.35945**	0.373231	0.367485	0.355793	0.331695	0.303576	0.279072	0.260451	0.296422	2	6.907
Fourth	**0.453469**	0.435189	0.437915	0.384194	0.370764	0.329526	0.331033	0.380341	0.389346	2	6.546
Fifth	**0.372805**	0.354444	0.276909	0.305168	0.239359	0.239015	0.219307	0.271908	0.308471	2	6.655
Average	0.390683	0.379687	0.355024	0.340705	0.316506	0.31588	0.300287	0.303306	0.311848	2	**6.6978**
	Finding on total data size = 440										
First	**0.370823**	0.365415	0.325748	0.295871	0.262239	0.250908	0.263596	0.282714	0.239766	2	11.607
Second	**0.370823**	0.365415	0.325748	0.295856	0.262674	0.268794	0.263593	0.252431	0.240321	2	10.439
Third	**0.370823**	0.365415	0.325748	0.295868	0.263506	0.250555	0.26359	0.257661	0.243761	2	11.776
Fourth	**0.370823**	0.365415	0.325748	0.295871	0.269526	0.250893	0.263611	0.252383	0.240633	2	11.555
Fifth	**0.370823**	0.365415	0.325748	0.295871	0.265049	0.268561	0.263435	0.260764	0.239014	2	11.378
Average	**0.370823**	**0.365415**	**0.325748**	**0.295867**	**0.264599**	**0.257942**	**0.263565**	**0.261191**	**0.240699**	**2**	**11.351**

Table 4 Results of point-wise fuzzy silhouette index on data sets

Data size	c = 1	c = 3	c = 4	c = 5	c = 6	c = 7	c = 8	c = 9	c = 10	Right c	Total time (s)	No of pdist	Pdist time (s)
Point-wise fuzzy silhouette width on synthetic data													
1000	0.4118	0.4829	**0.5029**	0.4912	0.4481	0.4594	0.4745	0.4751	0.4786	4	**17.678**	54000	10.546
2000	0.4229	0.4729	**0.4946**	0.4605	0.46555	0.462552	0.4776191	0.47981	0.47706	4	**34.573**	108000	21.896
3000	0.4124	0.468	**04851**	0.4525	0.47702	0.469422	0.4801764	0.48758	0.48367	4	**49.323**	162000	33.335
4000	0.4129	0.4737	**0.4982**	0.4794	0.45433	0.456677	0.4665116	0.47152	0.46103	4	**66.943**	216000	45.822
5000	0.4156	0.471	**0.5004**	0.4604	0.44734	0.452377	0.4777732	0.48186	0.46307	4	**80.567**	270000	58.646
Application data set													
WBC	**0.56786**	0.4039	0.28396	0.23781	0.20745	0.148771	0.1390604	0.15915	0.16855	2	11.937	37638	7.504
WCD	**0.37082**	**0.3654**	0.32575	0.29583	0.26263	0.263522	0.2636035	0.28111	0.24736	2,3	7.376	23760	4.644

Table 5 Results of center-point fuzzy modified silhouette width on data sets

Data set (x, y)	c = 2	c = 3	c = 4	c = 5	c = 6	c = 7	c = 8	c = 9	c = 10	c	Total time (s)	Nopdist	Pdist time (s)
Data = 1000	0.2563	0.338	**0.361**	0.2884	0.268	0.2837	0.297042	0.2918	0.2715	2	2.881	384	0.093
Data = 2000	0.2625	0.325	**0.347**	0.293	0.2666	0.2803	0.303477	0.2987	0.2809	2	4.687	384	0.092
Data = 3000	0.2506	0.323	**0.3399**	0.2812	0.2748	0.2935	0.302006	0.3093	0.2838	2	5.897	384	0.092
Data = 4000	0.2541	0.329	**0.3591**	0.3033	0.2694	0.2631	0.285488	0.2873	0.2705	2	7.397	384	0.061
Data = 5000	0.2487	0.327	**0.361**	0.3092	0.2674	0.2704	0.303093	0.3021	0.2732	2	8.078	384	0.108
Cancer	**0.5124**	0.263	0.178	0.0814	0.0726	0.0683	0.035683	0.0351	0.037	2	2.138	384	0.093
Customer	**0.2522**	**0.26**	0.204	0.1837	0.1435	0.1269	0.136577	0.1377	0.1133	2,3	1.637	384	0.078

Table 6 Results of center-mean fuzzy modified silhouette width on data sets

Data set (x, y)	c = 2	c = 3	c = 4	c = 5	c = 6	c = 7	c = 8	c = 9	c = 10	c	Total time (s)	Nc of pdist	Pdist time (s)
Data = 1000	−0.0875501	0.1611615	0.3050583	0.2979159	0.3292666	0.3321804	**0.3653813**	0.3542083	0.3542083	8	**4.963**	108	0.062
Data = 2000	−0.088562	0.1030352	0.2931982	0.2683945	0.2520886	0.3321222	**0.379828**	0.3589728	0.3569945	8	**6.265**	108	0.048
Data = 3000	−0.1022257	0.1329412	0.2878234	0.2863575	0.335119	0.2345365	**0.3695076**	0.3104281	0.3330753	8	**8.065**	108	0 062
Data = 4000	−0.1027756	0.1065246	0.3024385	0.3274001	0.3268314	0.3056472	**0.3595745**	0.2772211	0.3511922	8	**7.394**	108	0.047
Data = 5000	−0.0953743	0.1127269	0.3073142	0.2738046	0.3005258	0.3201339	**0.3780988**	0.2945727	0.3570849	8	**8.067**	108	0 063
Customer	−0.10914	0.1651655	0.2251644	**0.3062306**	0.2177685	0.2957776	0.2213157	0.2752939	0.2177271	5	1.849	108	0.063
Cancer	**0.4665016**	**0.2724246**	0.2547241	0.1783553	0.1292219	0.1312726	0.1191513	0.1058377	0.1058121	2	2.305	108	0.094

References

1. Hartigan, J.A.: Clustering Algorithms. John Wiley & Sons, Inc., New York (1975)
2. Hartigan, J.A., Wong, M.A.: Algorithm AS 136: a k-means clustering algorithm. J. Roy. Stat. Soc. Ser. **C28**(1), 100–108 (1979)
3. Kaufman, L. and Rousseau, P.J. (1987), Clustering by means of Medoids, in Statistical Data Analysis Based on the L_1–Norm and Related Methods, edited by Y. Dodge, North-Holland, 405–416
4. Park, H.S., Jun, C.H.: A simple and fast algorithm for K-medoids clustering. Exp. Syst. Appl. **36**(2), 3336–3341 (2009)
5. Crespo, F., Weber, R.: A methodology for dynamic data mining based on fuzzy clustering. Fuzzy Sets Syst. **150**(2), 1 (2005)
6. Nock, R., Nielsen, F.: On weighting clustering. IEEE Trans. Pattern Anal. Mach. Intell. **28**(8), 1–13 (2006)
7. Bezdek, J.C.: Cluster validity with fuzzy sets 58–73 (1973)
8. Bezdek, James C. (1981). Pattern Recognition with Fuzzy Objective Function Algorithms. ISBN 0-306-40671-3
9. Xie, X.L., Beni, G.: A validity measure for fuzzy clustering. IEEE Trans. Pattern Anal. Mach. Intell. **13**(8), 841–847 (1991)
10. Pal, N.R., Bezdek, J.C.: On cluster validity for the fuzzy c-means model. IEEE Trans. Fuzzy Syst. **3**(3), 370–379 (1995)
11. Bezdek, J.C., Coray, C., Gunderson, R., Watson, J.: Detection and characteristics of cluster substructure and linear structure: Fuzzy c-lines. SIAM J. Appl. Math. **40**(2), 339–357
12. Rousseau, P.J.: Silhouettes: a graphical aid to the interpretation and validation of cluster analysis. Comput. Appl. Math. **20**, 53–65 (1987)
13. Peters, G., Weber, R., Nowatzke, R.: Dynamic rough clustering and its applications. J. Appl. Soft Comput. **12**(2012), 3193–3207 (2012)

RRF-BD: Ranger Random Forest Algorithm for Big Data Classification

G. Madhukar Rao, Dharavath Ramesh and Abhishek Kumar

Abstract In the current era, data are growing with a faster rate in terms of exponential form where these data create a major challenge for suitable classification to classify the statistical data. The relevance of this topic is extraction of data, insights, mining of information from the dataset with an efficient and faster manner has attracted attention towards the best classification strategy. This paper presents a Ranger Random forest (RRF) algorithm for high-dimensional data classification. Random Forest (RF) has been treated as a most popular ensemble technique of classification due to its measure variable importance, out-of-bag error, proximities, etc. To make the classification constraint possible, in this paper, we use three different datasets in order to accommodate the runtime and memory utilization effectively with the same efficiency as given by the traditional random forest. We also depict the improvements of Random Forest in terms of computational time and memory without affecting the efficiency of the traditional Random Forest. Experimental results show that the proposed RRF outperforms with others in terms of memory utilization and computation time.

Keywords Random forest · Ranger random forest · Machine learning · Classification

G. M. Rao · D. Ramesh (✉) · A. Kumar
Department of Computer Science and Engineering, Indian Institute of Technology (ISM), Dhanbad 826004, Jharkhand, India
e-mail: drramesh@iitism.ac.in

G. M. Rao
e-mail: madhukar.iitism@gmail.com

A. Kumar
e-mail: abhishek.cse.iitism@gmail.com

H. S. Behera et al. (eds.), *Computational Intelligence in Data Mining*, Advances in Intelligent Systems and Computing 990,
https://doi.org/10.1007/978-981-13-8676-3_2

1 Introduction

Random Forest is one of the most used classification algorithms in many applications like medical disease classification, land classification, genome-wide association studies, etc. Random Forest based on decision trees together with aggregation and bootstrap ideas were introduced by Breiman [1]. This is a powerful nonparametric statistical method, which allows considering regression problem as well as two class and multiclass classification problem in a single and versatile framework. With the growing ensemble decision trees and vote for the dominant class resulted in the momentous improvements in classification accuracy. Usually, random vectors are generated in order to build these ensembles and manage the gain of each tree in the ensemble manner. Bagging is one of the examples of this, where the random selection (without replacement) of features from dataset has been made to grow every tree. In this strategy, random vector performs random split selection, where the split is selected randomly at each node among the best *K* splits. Breiman develops new training sets from the original training set by randomizing the outputs [1]. Selecting the training set from a random set of weights of the samples in the training set is another approach to perform the random selection. Based on the "*random subspace*" method, a number of works have been presented, which randomly selects a subset of features used to build each tree. For kth tree, the trivial element in all of these procedures, a random vector Θk is generated. This vector is not dependent on the past random vectors ($\Theta_1, \ldots, \Theta k_{-1}$) but has the same dissemination. Using the training set and Θ_k a tree is growing, resulting in a classifier $h(\text{x}, \Theta_k)$, where x represents an input vector. For example, the random vector Θ in bagging is developed as the counts in *N* boxes resulting from *N* darts thrown at the boxes at random, where *N* is number of samples (rows in the dataset) in the training set. The symbol Θ consists of a number between 1 and K which is not dependent on random split selection. The dimensionality and nature of Θ depend on its use in development of tree. Finally, voting for the most prominent class after many numbers of trees are generated. We call these procedures as random forests.

1.1 Ranger Random Forest (RRF)

The implementations of RF have their own advantages and disadvantages. RF has been originally implemented by Breiman and Cutler. Random Forest in R implemented by Liaw and Wiener is feature-rich and mostly used, but it has been found that this could not be optimized for the use with high-dimensional dataset [2]. RRF is basically implemented in C++ and utilizes standard libraries only. For example, for a random number generation, random library is used and can be applied on all parallel processing platforms. At the end, an optimized implementation of RF named "Random Jungle" is generated for inspecting the data with many features, i.e. high-dimensional data [3]. This Random Jungle package can only accessible as a C++

application with library dependencies, and it is not handy to python, R or any other statistical programming language.

Hence, therefore, "**RAN**dom forest **GE**ne**R**ator", a new software package "ranger" in R has been implemented by Marvin [1]. The main goal in ranger was to build a platform independent and modular framework for the analysis with RF for data which has huge features. Second, the package should be available in R and ease to use with a run time to compute result not larger than the Random Jungle. Another achievement of this approach is that this machine learning method utilizes the memory and consumes lesser memory to compute the result than the traditional RF.

2 Related Works

In the literature, a good number of works have been compiled in medical diagnosis with random forest like diagnosis and prognostic of breast [4]. In this work, the authors used feature selection methodology to select important features and then considered RF classifier for diagnosing and prognosticating breast cancer [4]. In order to classify Wisconsin Breast Cancer Prognostic Dataset and Diagnosis Dataset, the authors maintained important features and removed repeated features based on their weights. They obtained around 99.8% classification efficiency on average and 100% in the best case on these datasets. The authors of [5] proposed a random forest classifier (RFC) method for lymph disease diagnosis. Targeting to make important features to be selective, the first step is to choose, the important features by a feature selection algorithm on the lymph diseases dataset for dimensionality reduction. There are some popular dimensionality reduction algorithms such as principal component analysis (PCA), genetic algorithm (GA), Fisher, Relief-F, sequential backward floating search (SBFS), and the sequential forward floating search (SFFS) have been applied for suitable classifications.

In the second step, after applying feature selection RFC model has been constructed for classification. In this step, gained feature subsets from the first step are then dealt with the RFC for accurate classification. On this lymph diseases dataset, it was found that GA-RFC gave the maximum classification efficiency of 92.2%. The authors of [6] used the random forest classifier (RFC) for a complex area to classify land cover. Evaluation of land cover classification was based on different parameters such as sensitivity to data set size, noise, and mapping accuracy. Thematic Mapper data "*Landsat-5*" has been taken in European summer and the spring was used with extra variables borrowed from a model "digital terrain". On this land cover dataset, they obtained reasonable land cover classification result with total efficiency of 92% and with Kappa index 0.92. Different kappa values obtained in RF used for noise addition values and data reduction more than 20 and 50%, respectively. With the help of a machine learning model, the authors of [7], correlated the accelerometers worn on the hip and wrist and added value of heart rate (HR) to predict Electrical energy expenditure (EE) and physical activity (PA).

In [8], Quanlong Feng acquired optical imagery by a mini-unmanned aerial vehicle (UAV). They used optical imagery for monitoring the severe urban waterlogging in Yuyao city, China. They derived the texture features from the gray-level co-occurrence matrix. Texture features were added so that the separability between several ground objects could be increased. For the extraction of flooded areas in the textural feature space, an RFC was used. The efficiency of the proposed machine learning method was measured by using confusion matrix of RFC. The results of the RFC are expressed in the following manner; (i) RF showed satisfactory achievement with an overall efficiency of 87.3% and with a Kappa index of 0.746 in urban flood mapping, (ii) involvement of texture features increased classification efficiency in a good manner, and (iii) Random Forest showed a similar performance to SVM and outperformed artificial neural network and maximum likelihood. In recent years, another efficient algorithm named fast compressive tracking (FCT) has been proposed. On the other hand, it increases the difficulty in dealing with pose variation, occlusion, and appearance changes, etc.

The authors of [9] proposed a machine learning method which was an improved and fast compressive tracking based online random forest (FCT-ORF) for powerful visual tracking. This performs better than the fast compressive tracking algorithms in the area of pose variation, barrier, and presence changes. Wang and Wan proposed a machine learning approach named as incremental extremely random forest (IERF) [10]. The aim of this method was to deal with the online learning classification of streaming data (i.e., mainly on short streaming data). Mursalin and Zhang proposed an RF classifier along with improved correlation-based feature selection method (ICFS) [11]. This method was used to show an analysis approach for the epileptic seizure detection from the electroencephalogram (EEG) signal. In this method, first, the most important features of the time domain, frequency domain, and entropy-based features are selected using ICFS and in the second step ensemble RF classifier learned the set of features which have been selected by ICFS. For multiclass disease classification problem, the authors of [12] introduced an improved-RFC (Random Forest Classifier) method. This method is the combination of RF machine learning algorithm, an instance filter method, and an attribute evaluator method which achieves the best performance of the Random Forest algorithm. The methodology designed by Patti et al. [13], explores the recognition of sleep spindles using the RFC which is well-known to do not over fit data as in the other supervised classifiers. In [14], the authors have proposed an application of two data mining classifiers for the classification security and assessment of a problem in multiclass security. On the other hand, RF also has been used for Big Data research in its own variants. In [15], the authors used RF in their big data research and proposed five variants of RF. They have used big dataset which had millions of instances and found that every variant of RF has their own importance in terms of OOB error, efficiency, and runtime.

3 Proposed Work

In this paper, a variant of RF named Ranger Random Forest (RRF) has been proposed for the classification of high dimensional data (i.e. structured big data). RRF is the fastest implementation of random forest. RRF classifier gives the result faster than the random forest without affecting the original accuracy. RRF is also applicable for high dimensional dataset, where RF can't apply. RRF supports classification, regression as well as survival tree. Classification and regression forest has been implemented in the original Random Forest [16] and Survival Forest [17]. The algorithmic structure of the RRF is depicted in the below manner.

Random Forest (RF) Algorithm

Usually, there exist two phases in the RR algorithm. In the first phase, there is a creation of random forest and in a second phase, the data was used by the created random forest for the prediction. The working of Random Forest is shown below.

$\textbf{Step}_1$: Randomly choose "*k*" features out of "*m*" features from the dataset such that $k << m$.
$\textbf{Step}_2$: Compute the best split for node "*d*" from selected "*k*" features in step_1.
$\textbf{Step}_3$: Split the node "*d*" into child node again by computing best split.
$\textbf{Step}_4$: Repeat steps 1–3 till "*l*" nodes will be arrived.
$\textbf{Step}_5$: Create "*n*" number of trees by repeating steps 1–4 "*n*" times to grow forest.

Ranger Random Forest (RRF) Algorithm

This algorithm contains three stages of operations as mentioned below.

$\textbf{Stage}_1$: Build a random forest of training data.
$\textbf{Stage}_2$: The weights of all training examples are calculated for every example of test data by counting the number of trees, where terminal node having both examples.
$\textbf{Stage}_3$: Build weighted random forest for the training data for every test example, using the weights calculated in Stage_2. Estimate the result of the test example as usual.

The ranger uses most of the random forest features and new features were added which causes it to perform faster than the RF as well as memory efficient. For example, survival tree is now supported in addition to classification and regression trees (CART). Moreover, for multiclass classification problems, class probabilities can be measured as described by Kruppa et al. [3]. From the R package GenABEL ([18] Aulchenko et al.), objects can be loaded from genome-wide association studies

(GWAS) and evaluated directly for better analysis of data. There is an option given to the user to choose between modes optimized for memory efficiency or runtime for other data formats.

The Gini index measures the node impurity for the classification trees and with the measured response variance for regression trees. Decrease in node impurity is the split criteria to split the node for regression and classification of RF and log-rank test is the splitting criteria for the survival of RF. Build the trees as regression trees to calculate the probability. The node impurity or permutations are used to determine the variable importance. Prediction error is calculated by out-of-bag data in order to measure the mean square error of regression and misclassification frequency for classification. The prediction error of survival tree has been calculated as one minus the C index [19].

Ranger determines bottlenecks and optimizes the significant algorithms for different forms of input data. The most essential factor is splitting the node where all values of all *mtry* contestants features require to be decided as splitting contestants. There are two distinct algorithms have been used for splitting the nodes. First algorithm sorts the value of features earlier and accesses them by using their index. Second algorithm says that the raw values are fetched and then sorted them at the time of splitting. In runtime optimized mode, first algorithm is used for large nodes and for small nodes second algorithm is used. Only the second algorithm has been used in memory utilization mode. Splitting features in every node by drawing the *mtry* contestant was another bottleneck for many features and big*mtry* values. For sampling without replacement, Knuth algorithm [20] gives major improvement in terms of runtime and memory utilization. Memory efficiency achievement has been done by avoiding copies of the original data and uses simple data structure for saving node information.

4 Result Analysis

In this section, we compare the runtime of two machine learning models. We apply these models on both the small dataset as well as large datasets. We use three datasets such as CTG, Adult, and Forest cover type to validate the proposed algorithm. Cardiotocograms (CTGs) database having 23 numbers of attributes and 2,126 the number of instances. We use this dataset to check if there is a large difference in the result of accuracy or not. Both the models were successfully applied on this dataset and found that both are giving almost same accuracy. But, there found a difference between RF and RRF, where RF is taking more time by consuming more memory than the ranger in all the situations. The predicted results of Cardiotocography dataset are shown in Table 1.

As the results are shown in Table 1, it is concluded that RF takes more time as compared to RRF. On the other hand, RF consumes more memory than the ranger but OOB error of both the algorithm is almost same. So for large dataset, RF consumes more memory and takes time in terms of hours to compute the results. Hence ranger

Table 1 Comparison between RF and RRF for Cardiotocograms dataset

Algorithm used	No. of trees	mtry	Time elapsed (s)	Memory (Mb)	OOB error (%)
RF	500	8	3.691	20.2	5.35
Ranger	500	8	0.37	2.5	5.55
RF	800	8	5.859	22	5.15
Ranger	800	8	0.509	4	5.28
RF	1100	8	8.242	23.5	5.48
Ranger	1100	8	0.704	5.5	5.15

could be the best option for prediction. We also test both the algorithms on the dataset for different number of trees and found that RF takes large amount of time to compute the result but ranger takes lesser time to compute the result.

Cardiotocography is the monitoring technique which gives important and vital information on fetal status during antepartum and intrapartum periods. This dataset has some features which are measured as important by both RF and RRF. Some of them are, abnormal short term variability (ASTV), mean value of short term variable (MSTV), abnormal long term variability (ALTV). NSP feature has been used to classify fetal state. Based on these features we have classified fetal state whether it is normal, suspicious or pathological.

The comparison between growths in time with respect to the number of trees has been shown in the graph. Figure 1 shows that if we increase the number of trees, then runtime of RF increases faster, where the runtime of RRF increases linearly. While in Fig. 2, we compare the runtime of RF and RRF with respect to the number

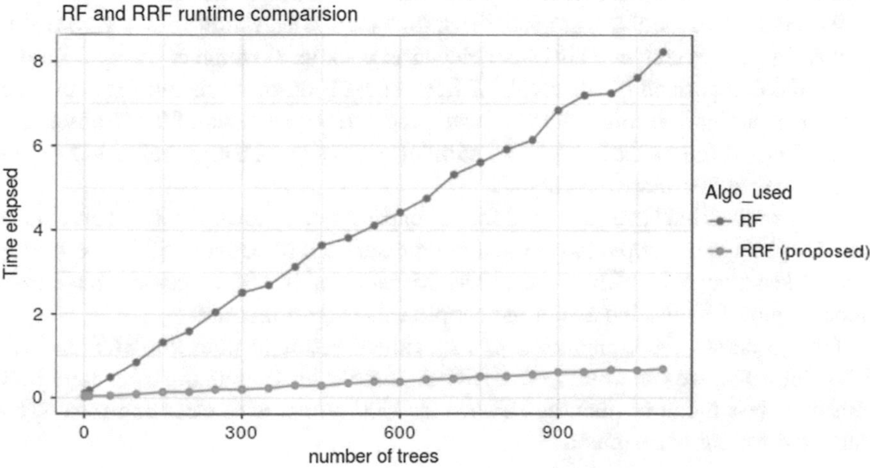

Fig. 1 Comparison between RF and ranger runtime with respect to number of trees

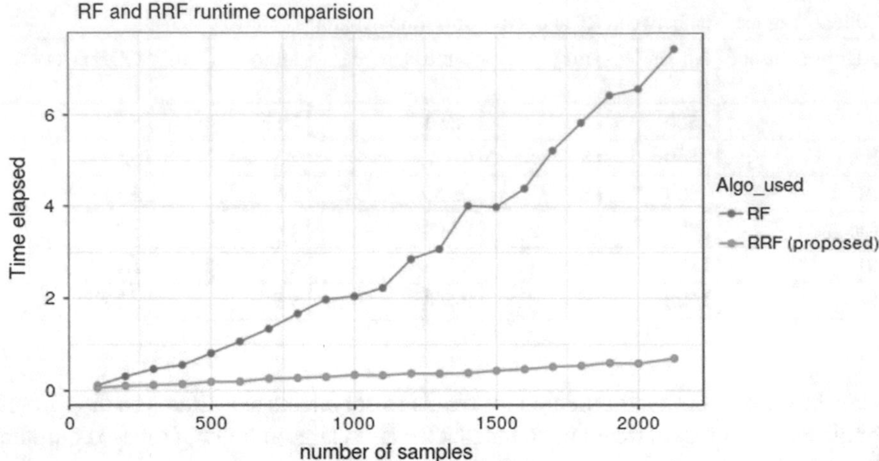

Fig. 2 Comparison between RF and ranger runtime with respect to number of samples

of sample of the Cardiotocograms dataset. In this case, it has been also observed that RRF takes less runtime to compute the result.

Another dataset which is used for experiment was Forest cover type dataset (cov-type). This dataset has 53 features (after data preprocessing), 14,498 instances and 7 cover types.

Forest covertype dataset has been used for seven class experiments. While using this dataset, it is found that both the algorithms RF and RRF successfully applied and calculated the result with the same accuracy. With this, RRF taken much lesser memory and gave the result faster than the RF. Figure 3 shows the scatter and bar plot of forest cover dataset which explains the distribution of the dataset.

We have drawn the above graph of the dataset feature Elevation vs Covertype using ggplot2 by [21], which explains what elevation belongs to which covertype. Table 2 shows the comparison of the result of RRF and RF on the forest cover data type. Both the machine learning methods show good performance with 86. 6% efficiency. We use both of the methods with 300 number of trees and 16 mtry value and get the following result as shown in Table 2.

Third dataset that have been used for our experiment is **Adult** dataset. This dataset contains the age of people having income greater than or equal to 50 K and having income less than 50 K. This dataset has 15 features, 14,998 instances and 2 class labels. Figure 4 shows the age of the people versus their income.

Table 3 shows the comparison of experiment result of both the RRF and RF. From this table, we can observe that RRF runtime is lesser than the RF, where RRF consumes less memory than the RF. We use 300 numbers of trees and *mtry* value equal to 4 for our experiment.

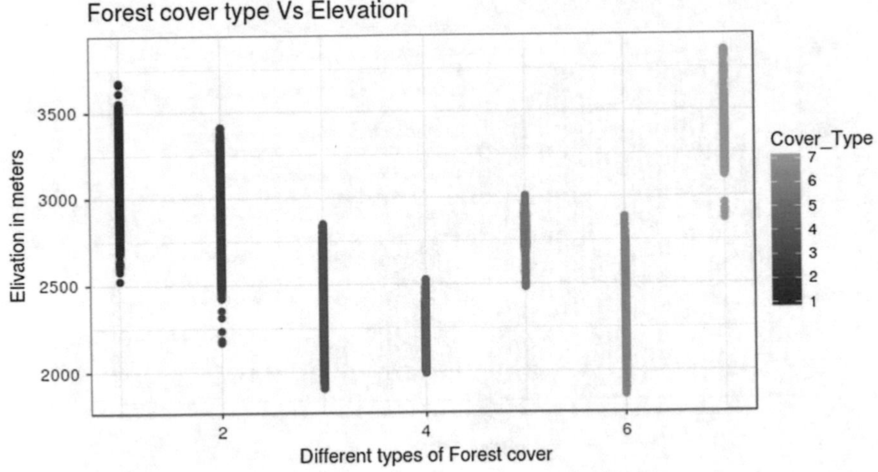

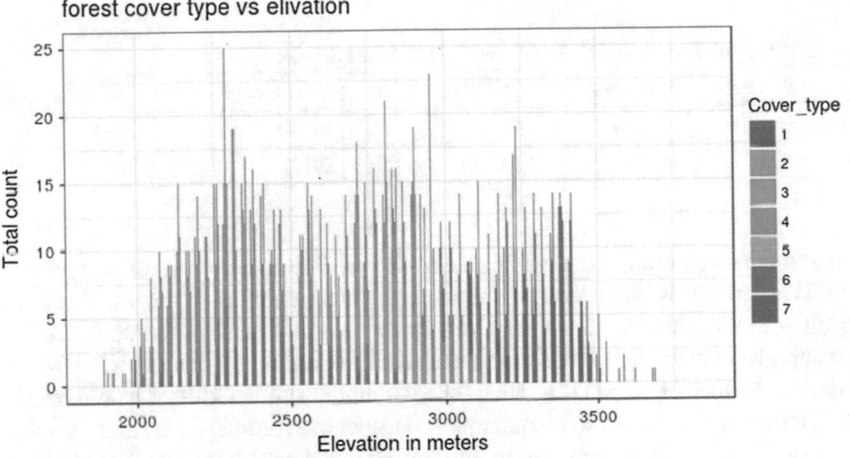

Fig. 3 Scatter plot (above) and bar plot (below) of Forest cover type dataset

Table 2 Comparison of RRF and RF for forest covertype dataset

Algorithm used	Runtime (s)	OOB error (%)	Used memory (Mb)	Accuracy (%)
RRF	6.19	13.22	25	86.59
RF	146.196	13.29	895.7	86.61

5 Conclusion and Future Scope

In this paper, in order to classify the data, a variant of RF named RRF has been presented. At the same time, both the algorithms have been validated on different datasets to compare their runtime and memory. From the results predicted, we con-

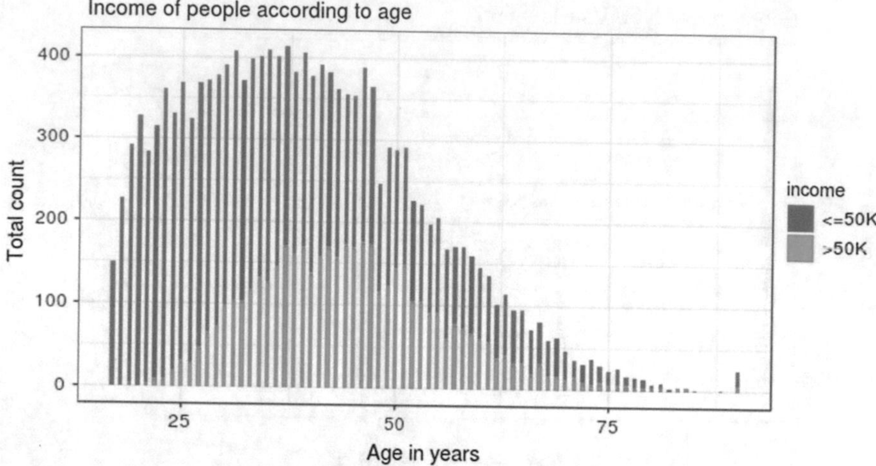

Fig. 4 Age of people versus income

Table 3 Comparison of result of RRF and RF for **Adult** dataset

Algorithm used	Runtime (s)	OOB error (%)	Used memory (Mb)	Accuracy (%)
RRF	1.747	13.1	25	86.9
RF	109.77	13.17	895.7	86.83

clude that ranger could be a better choice for prediction. Another important factor of this algorithm is that it utilizes the memory in an efficient manner and gives the result with the same efficiency as given by the original random forest. Since RF is not applicable to the data having large number of features, we can use RRF for those datasets. As this algorithm is applied for high number of features, it would be a best choice for the data that has the maximum number of features. As a future work, the proposed RRF methodology can be further extended for high dimensional big data and the same can be validated by using the spark framework to calculate the result in a faster manner as compared to parallel RF.

Acknowledgements This research work is supported by Indian Institute of Technology (ISM), Government of India. The authors would like to express their gratitude and heartiest thanks to the Department of Computer Science and Engineering, Indian Institute of Technology (ISM), Dhanbad, India for providing their research support.

References

1. Wright, M.N., Ziegler, A.: Ranger: a fast implementation of random forests for high dimensional data in C++ and R. arXiv:1508.04409 (2015)
2. Liaw, A., Wiener, M.: Classification and regression by randomForest. R News **2**(3), 18–22 (2002)
3. Kruppa, J., Liu, Y., Biau, G., Kohler, M., König, I.R., Malley, J.D., Ziegler, A.: Probability estimation with machine learning methods for dichotomous and multicategory outcome: Theory. Biom. J. **56**(4), 534–563 (2014)
4. Nguyen, C., Wang, Y., Nguyen, H.N.: Random forest classifier combined with feature selection for breast cancer diagnosis and prognostic. J. Biomed. Sci. Eng. **6**(05), 551 (2013)
5. Azar, A.T., Elshazly, H.I., Hassanien, A.E., Elkorany, A.M.: A random forest classifier for lymph diseases. Comput. Methods Programs Biomed. **113**(2), 465–473 (2014)
6. Rodriguez-Galiano, V.F., Ghimire, B., Rogan, J., Chica-Olmo, M., Rigol-Sanchez, J.P.: An assessment of the effectiveness of a random forest classifier for land-cover classification. ISPRS J. Photogramm. Remote. Sens. **67**, 93–104 (2012)
7. Ellis, K., Kerr, J., Godbole, S., Lanckriet, G., Wing, D., Marshall, S.: A random forest classifier for the prediction of energy expenditure and type of physical activity from wrist and hip accelerometers. Physiol. Meas. **35**(11), 2191 (2014)
8. Feng, Q., Liu, J., Gong, J.: Urban flood mapping based on unmanned aerial vehicle remote sensing and random forest classifier—A case of Yuyao. China. Water **7**(4), 1437–1455 (2015)
9. Xiong, J., Pan, J., Yang, J., Zhong, Z., Zou, R., Zhu, B.: An improved fast compressive tracking algorithm based on online random forest classifier. In: MATEC Web of Conferences, vol. 59. EDP Sciences (2016)
10. Wang, A.P., Wan, G.W., Cheng, Z.Q., Li, S.K.: Incremental learning extremely random forest classifier for online learning. RuanjianXuebao/J. Softw. **22**(9), 2059–2074 (2011)
11. Mursalin, M., Zhang, Y., Chen, Y., Chawla, N.V.: Automated epileptic seizure detection using improved correlation-based feature selection with random forest classifier. Neurocomputing **241**, 204–214 (2017)
12. Chaudhary, A., Kolhe, S., Kamal, R.: An improved random forest classifier for multi-class classification. Inf. Process. Agric. **3**(4), 215–222 (2016)
13. Patti, C.R., Shahrbabaki, S.S., Dissanayaka, C. Cvetkovic, D.: Application of random forest classifier for automatic sleep spindle detection. In: Biomedical Circuits and Systems Conference (BioCAS), 2015, IEEE, pp. 1–4. IEEE (2015)
14. Sekhar, P., Mohanty, S.: Classification and assessment of power system static security using decision tree and random forest classifiers. Int. J. Numer. Model. Electron. Netw. Devices Fields **29**(3), 465–474 (2016)
15. Genuer, R., Poggi, J.M., Tuleau-Malot, C., Villa-Vialaneix, N.: Random forests for big data. Big Data Res. **9**, 28–46 (2017)
16. Breiman, L.: Random forests. Mach. Learn. **45**(1), 5–32 (2001)
17. Ishwaran, H., Kogalur, U.B., Blackstone, E.H., Lauer, M.S.: Random survival forests. Ann. Appl. Stat. 841–860 (2008)
18. Aulchenko, Y.S., Ripke, S., Isaacs, A., van Duijn, C.M.: GenABEL: an R library for genome-wide association analysis. Bioinformatics **23**(10), 1294–1296 (2007)
19. Harrell Jr., F.E., Califf, R.M., Pryor, D.B., Lee, K.L., Rosati, R.A.: Evaluating the yield of medical tests. JAMA **247**(18), 2543–2546 (1982)
20. Epstein, J.M.: Agent-based computational models and generative social science. Complexity **4**(5), 41–60 (1999)
21. Wickham, H.: Positioning. ggplot2, pp. 115–137. Springer, New York, NY (2009)

Survey on Plagiarism Detection Systems and Their Comparison

Lovepreet, Vishal Gupta and Rohit Kumar

Abstract Plagiarism occurs when a person uses someone's work, ideas, words, expressions without giving the required attribution. Plagiarism is a common problem in fields like academia, Research papers, Publications, Patents, etc. In this paper, we deliberate the techniques for detecting the extrinsic plagiarism. These techniques are based on linguistic features, Semantic role labelling, vector space model, Fuzzy semantic string matching, and N-gram approach. And are tested on PAN plagiarism corpus 2009 and PAN plagiarism corpus 2011.

Keywords Plagiarism detection system · Extrinsic plagiarism detection · Semantic similarity · PAN-PC-09 · PAN-PC-11

1 Introduction

Plagiarism is abducting someone's work, ideas and passing them off as one's own. Person behind plagiarism is known as plagiarius. Plagiarius ask for the rights on the work that he stole from other authors. In the modern era of digitalization, due to tremendous amount of data available on the internet, people knowingly or unknowingly are involved in plagiarism. Plagiarism can be found in both academic and non-academic fields. Plagiarism sometimes takes place in the work when the person knowingly copies the sentences from other person's work or when he does paraphrasing or replace the words in the existing text with its synonyms.

Lovepreet (✉) · V. Gupta (✉) · R. Kumar
Department of Computer Science and Engineering, UIET,
Panjab University, Chandigarh 160014, India
e-mail: loveahuja12@yahoo.com

V. Gupta
e-mail: vishal_gupta100@yahoo.co.in

R. Kumar
e-mail: rklachotra@gmail.com

H. S. Behera et al. (eds.), *Computational Intelligence in Data Mining*, Advances in Intelligent Systems and Computing 990,
https://doi.org/10.1007/978-981-13-8676-3_3

It's very much difficult to check the plagiarism manually in a suspicious document for the given source documents. So there are many automatic plagiarism detection tools which can help us to avoid the plagiarism in our work, e.g. turnitin, PlagScan, iThenticate, etc. Some lexical databases like WordNet, Gene Ontology, and Transfer Standard are used in order to find the similarity between concepts without any intervention of human because of the reason that the computers cannot interpret the semantics.

Plagiarism detection systems can be of two types: 1. Intrinsic plagiarism detection system [1] and 2. Extrinsic plagiarism detection system [2]. Intrinsic plagiarism detection is a technique in which the given query document is compared with the source documents by the same author, in order to find out whether the suspicious document is scripted by the same author or not. In the intrinsic plagiarism detection, the main focus of the system is given on the author's writing style, paragraph structuring, section formulations, etc. On the other hand, extrinsic plagiarism detection system the query document is compared with the set of reference documents which can also be written by the other authors or a database.

The main focus of this paper will be on the techniques which are based on extrinsic plagiarism detection. Researchers proposed many techniques for detecting the extrinsic plagiarism. The techniques proposed by them are based on linguistic features, Semantic role labelling [3], vector space model [2, 4, 5], Fuzzy semantic string matching [6], N-gram approach [7], Singular value decomposition (SVD) [7], Latent Semantic Analysis (LSA), etc. We discuss the results of some majorly used techniques that used PAN-PC-09 and PAN-PC-11 corpora, on the basis of precision, F-measure, PlagDet score, and recall in further sections.

2 Related Work

This part discusses the plagiarism detection systems which were implemented with PAN plagiarism corpus 2009 and PAN plagiarism corpus 2011.

2.1 Plagiarism Detection Using Linguistic Features

In this approach [8], various linguistic feature functions are used to find relatedness among the sentences and it has potency in detecting extreme cases of plagiarism like an exact copy, rephrasing, and line modification. The proposed approach uses semantic knowledge bases such as WordNet.

The first step carried out in this system is pre-processing, which includes steps like Text segmentation, removing Stop-words, Stemming. Second step is comparing the relatedness between the sentences (source and suspicious) by varying weights of the functions (depth function, local density function, inverse path length function, and estimating depth function) and combining them. Semantic ($relscore_{sem}$ (Q, S))

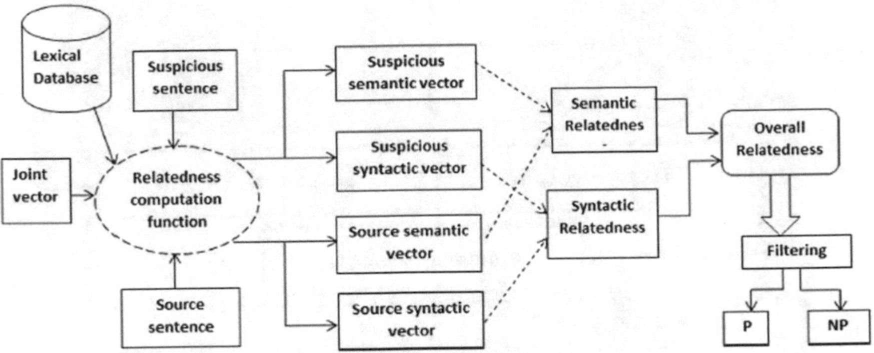

Fig. 1 Architecture of PDIS [8]

and syntactic relatedness ($relscore_{syn}$ (Q, S)) is calculated as shown is Fig. 1. Where Q is query sentence and S is source sentence. The final plagiarism score is calculated as:

$$relscore_{total}(Q, S) = \alpha.relscore_{sem}(Q, S) + (1 - \alpha).relscore_{syn}(Q, S) \quad (1)$$

If this score exceeds the threshold value, i.e. 0.45, then the sentence is considered as plagiarised. The results of the above said systems are: Precision: 0.96, Recall: 0.74, F-measure: 0.84.

2.2 *Extrinsic Plagiarism Detection Based on Semantic Role Labelling*

This paper [9] launches an External Plagiarism Detection System, which can detect divergent categories of plagiarism such as *verbatim copying*, rephrasing, and modification of sentences and changing of words. It employs the semantic and syntactic knowledge with the Semantic Role Labelling (SRL) technique. This technique helps us in avoiding the cases in which the relatedness in source and the query sentence is high and the sense (meaning) is poles apart.

This method has four steps first, the NLP techniques is applied to both original and suspicious documents. Second, to reduce the size of the collection, candidate retrieval stage is used. Third, the detailed comparison step is conducted to determine the plagiarism suspicion sections and their potential sources. The last step is the post-processing stage to select the sentences that are plagiarised. If this similarity exceeds the threshold value, i.e. 0.5, then the sentence is considered as plagiarised. The results of the system are: Precision: 0.92, Recall: 0.62, F-measure: 0.74, Plagdet: 0.74 (Fig. 2).

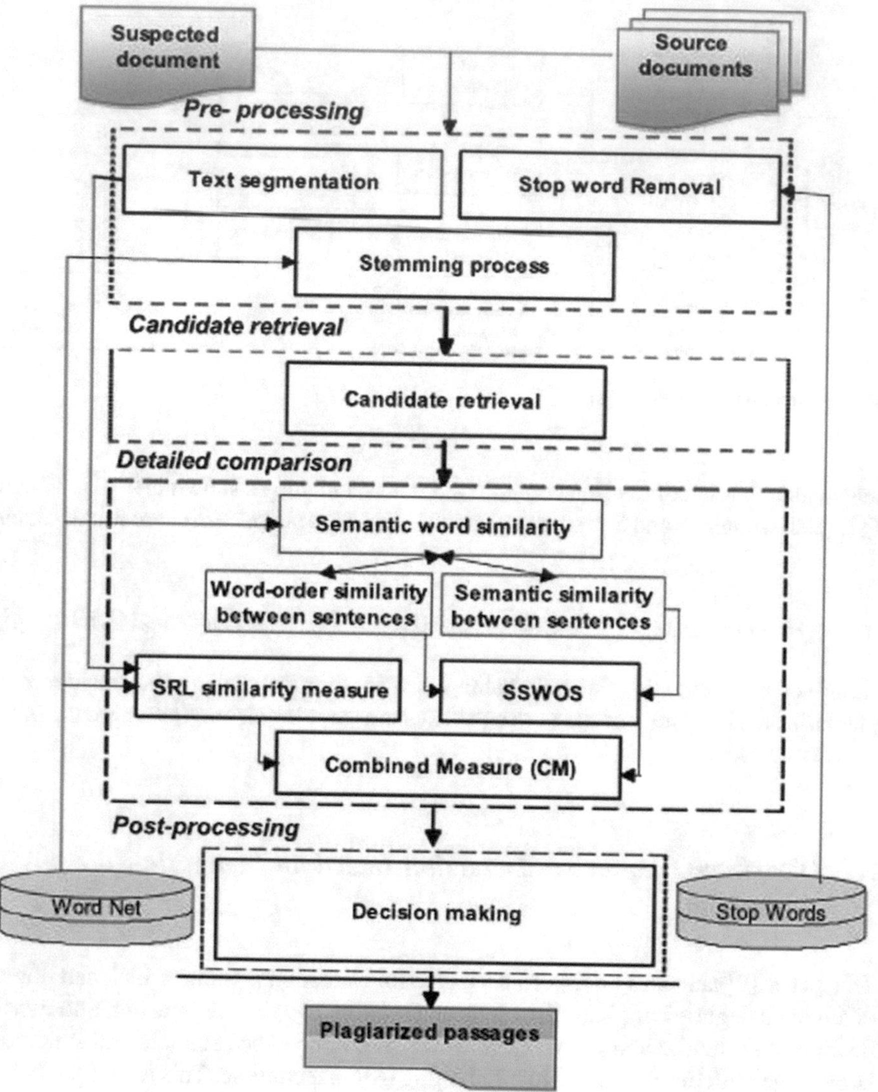

Fig. 2 Architecture of IEPDM [9]

2.3 *Plagiarism Detection Based on Linguistic Knowledge*

This approach [10] detects extrinsic plagiarism by the combination of semantic relations between terms and their occurrence pattern. Sentences with similar surface text may have different meanings but some techniques consider such text to be plagiarised giving unnecessary results. To avoid such false results and to bridge the lexical gaps for semantically similar contexts, method of plagiarism detection using linguistic knowledge is beneficial. It is also capable of detecting usual activities performed by plagiarists like copying the same text, rephrasing, and modification of word string and reforming the pattern of words in the word string.

This plagiarism detection technique comprises of three phases. First phase includes pre-processing of the documents, which cover the steps like segmentation, removing stop-words, and stemming of the terms. Second phase consists of calculating the similarity of sentences which are segmented in the previous stage. In this stage, if the calculated similarity (semantic and syntactic) overreaches the threshold, i.e. 0.6, then they are considered as plagiarised. The last step finally selects the sentence pair (in query doc. and source doc.) which are decided as plagiarised. The results of above system are: Precision: 0.90, Recall: 0.70, F-measure: 0.79, Plagdet: 0.79 (Fig. 3).

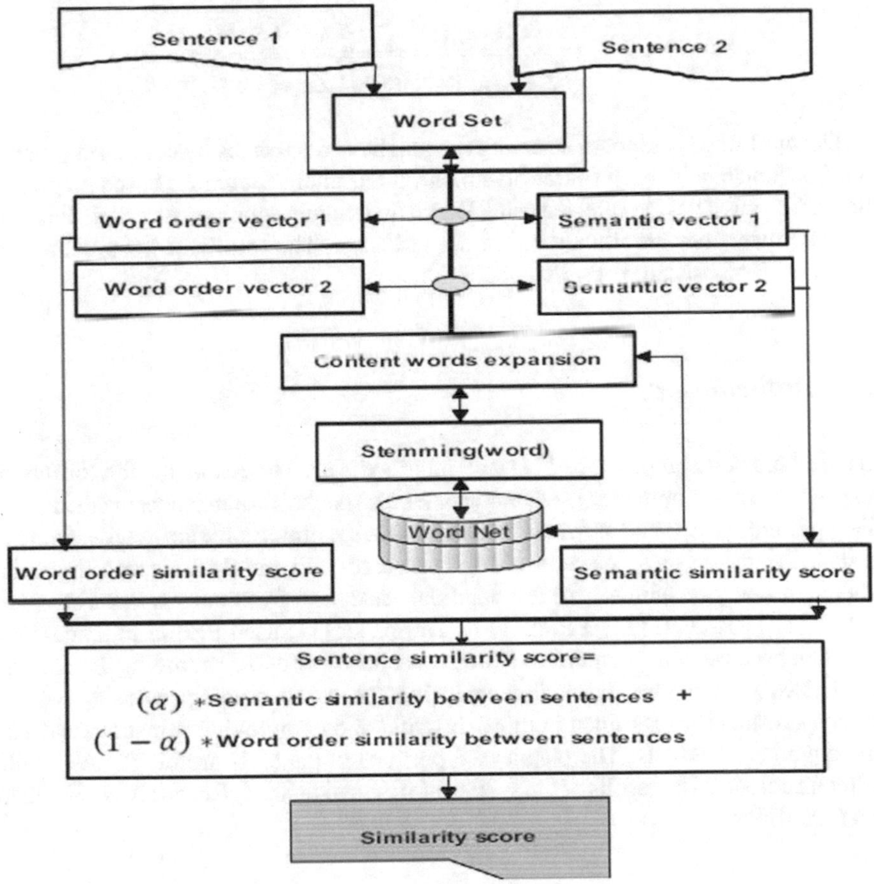

Fig. 3 Architecture of PDLK [10]

2.4 VSM Approach for Detecting Extrinsic Plagiarism in Text

Extrinsic plagiarism detection techniques are generally grounded on matching strings, Vector Space Modelling (VSM) and Fingerprinting. The VSM for detecting plagiarism is a standard method in Information Retrieval (IR). This technique [4] uses vectors to represent a given documents and calculates the relatedness between document-pairs by calculating the distance among the conforming vectors. The proposed method includes four stages. First stage includes pre-processing of documents which is done by tokenization, lemmatization, extracting Part-of-Speech (PoS) classes, named-entity (NE) classes, sentence numbers and character-offsets. The next stage includes candidate selection by using Vector Space Model. Here we call our text files as vectors. The similarity between text files d_q (query doc.) and d_s (source doc.) is calculated as

$$similarity_{d_q,d_s} = \frac{\sum_{t_\omega \in d_q \cup d_s} \left(tf - idf_{w,q} \cdot tf - idf_{w,s}\right)}{\sqrt{\sum_{t_\omega \in q \cup d_s} tf - idf_{w,q}^2 \cdot \sum_{t_\omega \in d_q \cup d_s} tf - idf_{w,s}^2}} \tag{2}$$

The third stage comprises selecting the plagiarised passages by comparing suspicious document with each source document. It this method which is based on graph, plagiarised passages are selected using Depth first search technique on graph. Finally, the false detections are filtered out in the last stage. The results of the system are: Precision: 0.66, Recall: 0.19, Plagdet: 0.29.

2.5 AuthentiCop

In the architecture proposed by [11] web interface is used to upload the file (different format files can be uploaded) and hand it over to Apache Tika that accepts divergent file types and extracts text and metadata. After this, stemming and stop-words removal is done. Then the pre-processed file is compared with the datasets and the web. There are many techniques for the candidate selection from corpus like Encoplot [12], FASTDOCODE [13], PPJoin [14]. This paper [11] used PPJoin for candidate selection because of the benefits of PPJoin over the other two. For finding the precise plagiarised portion, there is detailed analysis stage and then post-processing, where previous results are examined in more description by employing semantic analysis and alignment methods. The plagiarised portions in the texts are highlighted with different colours. The results of the system are: Precision: 0.76, Recall: 0.34, Plagdet: 0.4 (Fig. 4).

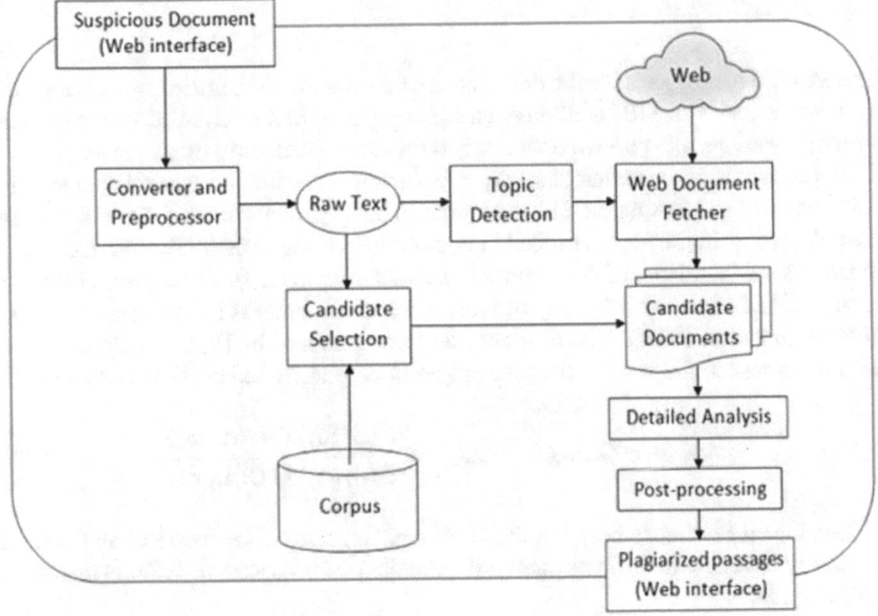

Fig. 4 Working of AuthentiCop [11]

2.6 *External Plagiarism Detection Technique Using Vector Space Model*

Rao et al. [2] proposed a system for detecting extrinsic plagiarism which is grounded on VSM. VSM here is used for the candidate selection. The candidate selection is done by generating the relatedness score among the query and source document from the source and computing their scalar product.

$$\cos\alpha = \frac{sd.sus}{||sd|| \cdot ||sus||} \tag{3}$$

Further the plagiarism is detected by N-gram technique by comparing words (excluding the stop-words) from the query document with the source document. At first 7 words are compared if they matched then 200 characters (25 words) are compared. Then by selecting the threshold, i.e. 0.5 it is decided whether the document is plagiarised or not. For the documents that are not in English, Google Language Identifier identifies the language and translate it to English with the help of Google translator API. The author of this paper [2] also discuss the technique for intrinsic plagiarism detection but we are discussing extrinsic plagiarism detection techniques only. The results of the system are: Precision: 0.45, Recall: 0.16, Plagdet: 0.2.

2.7 Semantic Role Labelling

The SRL [3] technique have the dominance of producing the semantic arguments for each statement. The SRL method detects copy paste, replacement of words having similar meanings and passive or active modification (structure) of sentence.

In proposed approach first of all pre-processing of the file is done and then the file is further processed using the SRL technique. The arguments that are drawn out from the pre-processed file are assembled in nodes conforming to their characteristics and the nodes are then labelled by Argument Label Group (ALG). Then abstract idea of each node is derived with the help of WordNet and this process is known as Semantic Term Annotation (STA). The abstract ideas are gathered in Topic Signature Node which further led the way to discover the required plagiarised portion of the text.

$$\text{similarity}\, C_i\left(ArgS_j, ArgS_k\right) = \frac{C\left(ArgS_j\right) \cap C(ArgS_k)}{C\left(ArgS_j\right) \cup C(ArgS_k)} \tag{4}$$

Calculated Similarity between the arguments is then summed up to find the total similarity. The results of the system are: Precision: 0.64, Recall: 0.82, F-measure: 0.72 (Fig. 5).

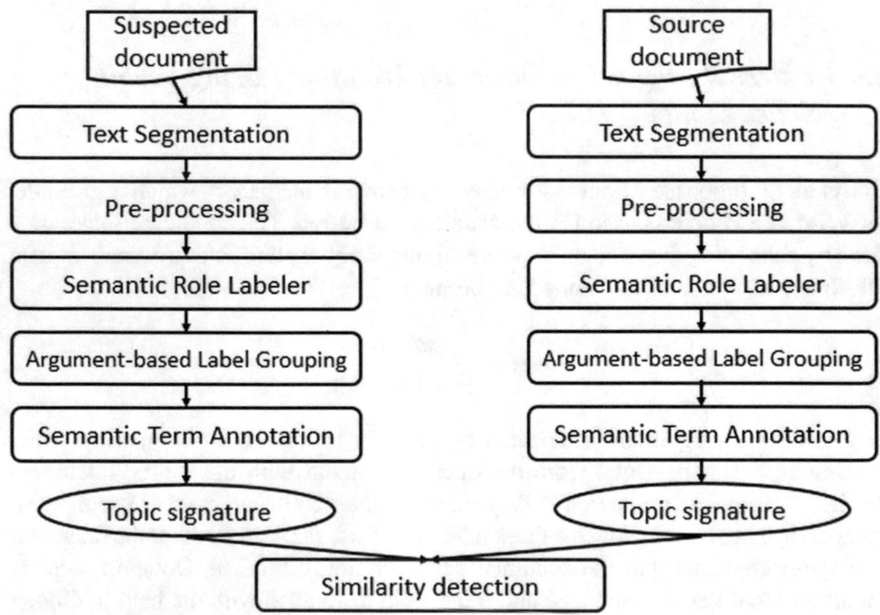

Fig. 5 Architecture of SRL technique [3]

2.8 Semantic Role Labelling Using Argument Weighting

This [15] is an improved method of the Semantic Role Labelling (SRL) [3] with the application weighting of an argument to learn their performance and influence in detecting plagiarism. The weighting of the arguments helped in selecting only the important arguments while computing the relatedness. Weight (X) to an argument is given by the formula:

$$X\ Arg_j = \frac{no.of\ \ args\, having\, the\, value\, in\, group\, j}{no.of\ \ entire\, documents\, in\, group\, j} \tag{5}$$

The results of the system are: Precision: 0.9, Recall: 0.83, F-measure: 0.86.

This technique was further improved by Paul and Jamal [16] using sentence ranking.

2.9 Extrinsic Plagiarism Detection Using Vector Space and Nearest Neighbour Search

Extrinsic plagiarism can be detected by considering passages as vectors in vector space. If the passage from the reference set is close enough to the passage of the suspicious set in vector space then it is considered to be plagiarised. This detection [5] is done via similarity threshold on nearest neighbour list. For the sentences having less than fifteen words, the threshold is 0.7. Sentence having maximum of 35 words the threshold is set to 0.5, for bigger sentences the threshold is set to 0.4.

$$Cosine\ similarity(p, q) = \frac{\langle p, q \rangle}{||p||\ ||q||} \tag{6}$$

In the final step, plagiarised sentence are merged together and marked as a continuous block. The results of the system are: Precision: 0.82, Recall: 0.48, F-measure: 0.60.

2.10 Fuzzy Technique for Detecting Plagiarism Based on Semantic String Similarity

Alzahrani and Salim [6] presents a fuzzy system for detecting plagiarism which is grounded on semantic string similarity. In this system, first step includes tokenization of documents, stop-words removal and stemming. Further the candidate documents are selected for the given query document, with the help of shingling and Jaccard measure. The fuzzy similarity between two lines is L_q (query) and L_x (candidate's line) is computed as Eqs. (7), (8), (9). If this similarity exceeds the threshold value, i.e. 0.65, then the sentence is considered as plagiarised.

$$\beta_{q,x} = 1 - \prod_{w_p \in L_x} \left(1 - S_{q,p}\right) \tag{7}$$

where w_p are the words in L_x and $S_{q,p}$ is a fuzzy similarity among the terms w_q and w_p. $S_{q,p}$ is computed as following:

$$S_{q,p} = \begin{cases} 1 \text{ } if \text{ } w_q \text{ } and \text{ } w_p are \text{ } same \\ 0.5 \quad if \text{ } w_p \text{ } is \text{ } synset \text{ } of \text{ } w_q \\ 0 \qquad otherwise \end{cases} \tag{8}$$

Similarity among (L_q and L_x) is calculated using the equation:

$$Similarity\left(\mathrm{L_q}, \mathrm{L}_x\right) = \left(\beta_{1,x} + \beta_{2,x} + \cdots + \beta_{q,x} + \cdots + \beta_{m,x}\right)/\mathrm{p} \tag{9}$$

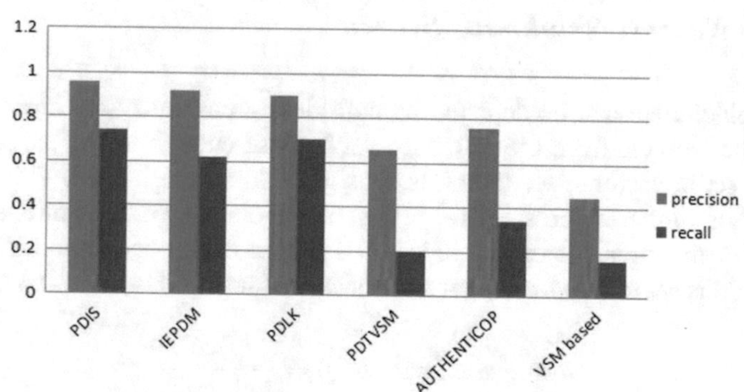

Fig. 6 Performance comparison of systems on PAN-PC-11 dataset

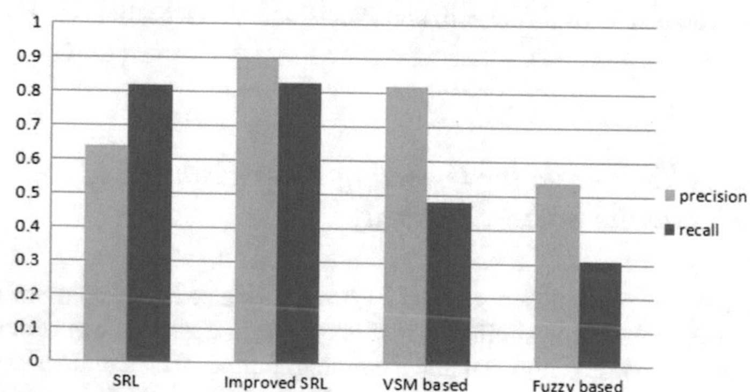

Fig. 7 Performance comparison of systems on PAN-PC-09 dataset

where *p* is the count of words in L_q.

The results of the system are: Precision: 0.54, Recall: 0.31, Plagdet: 0.13.

3 Analysis of Performance of the Extrinsic Plagiarism Detection Systems

This section discusses the accuracy of the systems detecting plagiarism on PAN plagiarism corpus 2011 in Fig. 6 and PAN plagiarism corpus 2009 in Fig. 7. Out of the six systems that used PAN-PC-11 dataset, PDIS gave the best precision value and the lowest precision was given by VSM based technique for extrinsic plagiarism detection. The systems, IEPDM and PDLK also gave the good precision value as

Table 1 Comparison of systems on PAN-PC-11 dataset

Name	Technique used	Results	Pros	Cons
PDIS [8]	Linguistic feature functions	Precision: 0.96 Recall: 0.74 F-measure: 0.84	Can detects sentence transformation, rewording, paraphrasing	To study the system by extensive weight variations to linguistic features
IEPDM [9]	SRL technique with semantic and syntactic information	Precision: 0.92 Recall: 0.62 F-measure: 0.74 Plagdet: 0.74	Can detect active and passive sentences	Scope of improvement in their candidate retrieval module
PDLK [10]	Linguistic feature functions	Precision: 0.90 Recall: 0.70 F-measure: 0.79 Plagdet: 0.79	Can differentiate between the meanings of two sentences	Can't detect passive and active sentences
Plagiarism detection in text using VSM (PDTVSM) [4]	Vector space model	Precision: 0.66 Recall: 0.19 Plagdet: 0.29	This is greedy technique that is not affected by variations in normal plagiarism detection	High obfuscation cases are not detected
AuthentiCop [11]	PPJoin	Precision: 0.76 Recall: 0.34 Plagdet: 0.4	Can detect highly obfuscated passages	To advance stemming support and semantic study modules
VSM based [2]	Vector space model	Precision: 0.45 Recall: 0.16 Plagdet: 0.2	Very fast candidate retrieval method	Parameter tuning is required

Table 2 Comparison of systems on PAN-PC-09 dataset

Name	Technique used	Results	Pros	Cons
SRL [3]	Semantic role labelling	Precision: 0.64 Recall: 0.82 F-measure: 0.72	Can detect passive and active sentences	Small dataset used
Improved SRL [15]	Semantic role labelling with argument weighting	Precision: 0.9 Recall: 0.83 F-measure: 0.86	Required importance to an argument is given	To improve the technique for larger dataset
VSM based [5]	Vector space model	Precision: 0.82 Recall: 0.48 F-measure: 0.60	Detects very near duplicates	Real nearest neighbour of vector can't be decided
Fuzzy based [6]	Fuzzy semantic string matching	Precision: 0.54 Recall: 0.31 Plagdet: 0.13	Can discover the parts that have similar meaning but different pattern	Less detection efficiency and more time consuming

compared to the other systems which have very low precision. PDIS gave the best recall among all.

Out of the four systems that used PAN-PC-09 dataset, improved SRL technique gave the best precision value and the lowest precision was given by Fussy based system. The recall value of improved SRL and SRL is much higher than VSM-based and fussy-based plagiarism detection systems (Tables 1 and 2).

4 Conclusion

In order to improve the quality of work, plagiarism should not be tolerated in any field either academic or non-academic. Extrinsic plagiarism detection systems are the main focus of this paper. Systems that are built using linguistic features and semantic role labelling are very much improved and have shown good results but the system based on vector space model and fuzzy are still to be improved especially their recall values. The other thing with some systems is that they are tested on small datasets and are to be tested of larger datasets.

References

1. Oberreuter, G., VeláSquez, J.D.: Text mining applied to plagiarism detection: the use of words for detecting deviations in the writing style. Expert Syst. Appl. **40**(9), 3756–3763 (2013)
2. Rao, S., Gupta, P., Singhal, K., Majumder, P.: External & Intrinsic Plagiarism Detection: VSM & Discourse Markers Based Approach Notebook for PAN at CLEF 2011 (2011)
3. Osman, A.H., Salim, N., Binwahlan, M.S., Twaha, S., Kumar, Y.J., Abuobieda, A.: Plagiarism detection scheme based on semantic role labeling. In: 2012 International Conference on Information Retrieval & Knowledge Management (CAMP), pp. 30–33. IEEE (2012)
4. Ekbal, A., Saha, S., Choudhary, G.: Plagiarism detection in text using vector space model. In: 2012 12th International Conference on Hybrid Intelligent Systems (HIS), pp. 366–371. IEEE (2012)
5. Zechner, M., Muhr, M., Kern, R., Granitzer, M.: External and intrinsic plagiarism detection using vector space models. Proc. SEPLN **32**, 47–55 (2009)
6. Alzahrani, S., Salim, N.: Fuzzy semantic-based string similarity for extrinsic plagiarism detection. Braschler Harman **1176**, 1–8 (2010)
7. Hussein, A.S.: Arabic document similarity analysis using n-grams and singular value decomposition. In: 2015 IEEE 9th International Conference on Research Challenges in Information Science (RCIS), pp. 445–455. IEEE (2015)
8. Sahi, M., Gupta, V.: A novel technique for detecting plagiarism in documents exploiting information sources. Cogn. Comput. **9**(6), 852–867 (2017)
9. Abdi, A., Shamsuddin, S.M., Idris, N., Alguliyev, R.M., Aliguliyev, R.M.: A linguistic treatment for automatic external plagiarism detection. Knowl. Based Syst. **135**, 135–146 (2017)
10. Abdi, A., Idris, N., Alguliyev, R.M., Aliguliyev, R.M.: PDLK: plagiarism detection using linguistic knowledge. Expert Syst. Appl. **42**(22), 8936–8946 (2015)
11. Buruiana, F.C., Scoica, A., Rebedea, T., Rughinis, R.: Automatic plagiarism detection system for specialized corpora. In: 2013 19th International Conference on Control Systems and Computer Science (CSCS), pp. 77–82. IEEE (2013)
12. Grozea, C., Popescu, M.: Who's the thief? Automatic detection of the direction of plagiarism. In: International Conference on Intelligent Text Processing and Computational Linguistics, pp. 700–710. Springer, Berlin, Heidelberg (2010)
13. Oberreuter, G., Ríos, S.A., Velásquez, J.D.: FASTDOCODE: finding approximated segments of n-grams for document copy detection lab report for PAN at CLEF 2010 (2010)
14. Xiao, C., Wang, W., Lin, X., Yu, J.X., Wang, G.: Efficient similarity joins for near-duplicate detection. ACM Trans. Database Syst. (TODS) **36**(3), 15 (2011)
15. Osman, A.H., Salim, N., Binwahlan, M.S., Alteeb, R., Abuobieda, A.: An improved plagiarism detection scheme based on semantic role labeling. Appl. Soft Comput. **12**(5), 1493–1502 (2012)
16. Paul, M., Jamal, S.: An improved SRL based plagiarism detection technique using sentence ranking. Procedia Comput. Sci. **46**, 223–230 (2015)

IoT in Home Automation

Prasad Suman Sourav, Mishra Jyoti Prakash and Mishra Sambit Kumar

Abstract The application of intelligence as well as representation of relevant facts toward home automation may sometimes be associated with application associated with sensing technology as well as controls with timings to embedded devices to enhance energy efficiency and security. The technology associated with this technique is somehow responsible toward many relevant areas including Internet of Things. Basically, the concept is based on multiuser and distributed architectures. In particular, it is intended to include the basic mechanism toward distributed system architecture in home automation. Applying this concept, a smart multi-agent object may be thought of to support the artificial intelligence framework and to focus on the associated resources with suitable implementation.

Keywords Control systems · Embedded system · Distributed system · Multi-agent · Sensor · Actuator · Artificial intelligence

1 Introduction

In general, the term Internet of Things may be thought of a complete network of objects having the capacity of sharing information, data and resources, and subsequent changes in the environment. It may be associated with different sensors, actuators, peripherals, and desired other information. While communicating with the devices, there may be slight interference by users toward operation, as every device may be connected to every other device. Smart Things usually may be con-

P. S. Sourav (✉)
Ajay Binay Institute of Technology, Cuttack 750314, Odisha, India
e-mail: prasadsuman800@rediffmail.com

M. J. Prakash (✉) · M. S. Kumar (✉)
Gandhi Institute for Education and Technology, Baniatangi, Khurdha 752060, Odisha, India
e-mail: jpmishra@gietbbsr.com

M. S. Kumar
e-mail: sambitmishra@gietbbsr.com

H. S. Behera et al. (eds.), *Computational Intelligence in Data Mining*, Advances in Intelligent Systems and Computing 990,
https://doi.org/10.1007/978-981-13-8676-3_4

sidered as a group of devices which can be monitored and controlled via central processors and web services. In this case, the users may also be able to initiate and control directly through the applications associated with smart objects. Basically, it may be linked with hub associated with smart objects, cloud, gateway for getting events and messages to or from the Cloud or providing the abstraction, and intelligence layers as well as the web services that support the presentation layer along with providing the presentation layer for smart things.

2 Review of Literature

Normann and Jamison [1] in their work have highlighted on hybrid imagination approach which may be useful for external changing conditions as empowerment and accordingly, may support the understanding of an experimenting approach. In simple term, it is the general term to explore and experiment to acknowledge the changes in external conditions. Wang [2] in their work have focused on performance-based, service-based, and system based definition. They have evaluated the performance associated with the service functions of communication as well as automation in different fields. Xiaojuan [3] in their work have focused on the gateway approach associated with smart home not only to control the associated devices but also to take care of illegal access from the external sources. Basil Hamed [4] in their work have focused on implementation as well as design part of control associated with home automation. In this case, the technology may be applied to control various activities with the help of remote control system. Basma et al. [5] in their work have focused on wireless sensor network as well as associated technologies. They have estimated the approximate cost of the system and proposed the technology for building automation system. Dixon et al. [6] in research have highlighted on architectural approach for IoT implemented as source linked with sensor and actuator network which may result in many systems-of-systems interference problems. Munir et al. [7] in their study have analyzed the activities associated with models of healthcare, living style along with safety measures. It may be sometimes challenging due to different behavioral aspects of human beings. Gross et al. [8] in their research have discussed on authentication and privacy by applying the desired security techniques. Practically, though the techniques adopted in this case are computationally expensive, for specific application to IoT devices these may be desirable. Accordingly based upon circumstances, the privacy may be presumed while associated with the IoT services. Uckelmann et al. [9] in their work have highlighted the requirements associated with different security levels along with initiating control toward unauthorized access of sensor data. Vermesan and Friess [10] in their work have discussed on needs toward IoT community primarily focusing on integrity of data, authentication, different encryption techniques, nature of interconnected devices, authentication as well as policy implementation, and management. Babar et al. [11] in their work have proposed the utility of lightweight cryptographic algorithms to provide data protection toward processing as well as storage capabilities. In this regard, they have thought for Datagram

Transport Layer Security (DTLS) to provide end-to-end security for the application layer. Hong et al. [12] in their research activity have focused on Smart Home Energy Management System to effectively implement most of the home appliances. The idea has been implemented to evaluate the performance of various appliances. Rani and Ahmed [13] in their work have discussed on services based on advanced solutions along with voluminous data rate. As observed, the data rate internet services may increase in the smart home environment. Accordingly, they have thought for applying the technologies in different dimensions in a smart home for proper differentiation. Al-Fuqaha et al. [14] in their analysis have discussed on improvement in the embedded architecture. They have analyzed that IoT may bring new light in technology toward decision making as well as machine learning. Khajenasiri et al. [15] in their research activity have focused on urban life including data sets collected from different devices associated with smart technology. The application associated with this technology may be linked to energy management. Hui et al. [16] in their work have discussed the automation system with Raspberry Pi utilizing IoT.

3 Necessity of Home Automation

In the present situation, the smart home may be very useful and effective to make life easier. In the present day, the modern home automation are linked with the internet. Accordingly, the domestic activities as well as the home appliances may be controlled without using a wired connection.

The minimum components may be required in the smart home Simulation are:

(i) Two Raspberry Pi 2 model B: The main processing and controlling unit of the system. One was used for the room model and the other for the surveillance car.
(ii) Servo Motor: It acts as the door lock.
(iii) Infrared (IR) sensor: Shows the current state of the front door, either opened or closed.
(iv) Web Camera: Acts as a surveillance camera for the room streaming images of room that are processed by the Raspberry Pi. It utilizes OpenCV's image processing to be able to detect objects in the room.
(v) Smoke Detector: It detects _re, ensuring the safety of the home.
(vi) Two H-Bridges: Each H-Bridge controls two motors, two are used to control the four DC motors of the device.
(vii) Wi-Fi Dongle: Attached to the Raspberry Pi through USB port to allow its connection to wireless internet instead of using Ethernet cable.

4 Artificial Intelligence in Smart Home

The concept of home automation is primarily based on the application of intelligence and knowledge representation. In the present scenario, it may not be commercially viable for the general application of home automation. Sometimes it may also be desirable to focus more on reasoning based system with proper methodologies. It has also been observed that the application of artificial intelligence to home automation sometimes may use technologies associated with a server over the internet. The prediction mechanisms along with the sensor inputs then may schedule services to the user. The proper implementation of user information by using sensors in home automation may be sometimes a big challenge and may be used t detect the changes in the state.

5 Problem Formulation with Basic Design Principle

In a practical situation, it may be necessary to analyze the methodologies which should support the research activities associated with the iterative process model. The smart objects and sensors associated with home automation may be used to exchange information along with predicted actions. The basic purpose, in this case, maybe to control desired functions to improve efficiency in the home automation. While associated with the process of exploring as well as implementation, the changes in external conditions may be viewed and accordingly based on the present knowledge level, the inference rules may be formed which may guide the next level processes and techniques. Sometimes common sense may be applied to find the desired path and required result. The main challenges like inflexibility, ownership, improper manageability and difficulty while achieving security may be observed while achieving security. The main objective of this work is to design the mechanism for home automation using intelligence and basic approaches of the Internet of Things. The principles of the Object Orientated Method may be thought of while designing the smart home simulator model using unified modeling language based class diagram. The encapsulation mechanism in the object-oriented approach may reduce coupling between objects and use the basic principles of generalization and polymorphism. By applying the desired technique, coupling may be essential toward the parametric values in the model. The techniques, as well as procedure associated with the implementation of home with artificial intelligence, may be modeled using the software ASP.NET and java script library. Similarly, jQuery may be used for building Ajax-based web applications. The problem with jQuery may be written primarily for desktop applications. For better efficiency, jQuery Mobile may be the alternative as compared to jQuery. Sometimes ASP.NET may be used for implementing scalable as well as web applications. It may be effective for various web applications as we as search engines. All the required sites referred to the jQuery and JavaScript file may be used to define metadata for the HTML document. The <meta name =

”viewport”/> meta tag may be used to control HTML content. The data attributes may as allow to add simple metadata to individual elements interpreted by the jQuery Mobile JavaScript functions.

6 Steps While Designing Storage Engine

Step 1: **MySQL provides support for many storage engines which manage storage and retrieval.**

Using MyISAM and InnoDB, search engine for required table may be created. The basic storage engine allows read-only tables. Most engines read data from files and feed to MYSQL. On the other hand, the idea is to retrieve data from APIs and feed it to MYSQL. It may allow conducting complex analysis using SQL on data retrieved from API"s. The idea is to write an engine for each set of related APIs.

Step 2: **The mySQL storage engine retrieves data from Gmail and feeds to mySQL.**

Step 3: **When user enters a select query on Email table, the control passes to Email search engine using predefined method**

The predefined method checks that table name is really “Email.” Then it checks that where clause parse tree is present or not. If yes, then where clause parse tree is traversed to be converted to IMAP search commands. Otherwise, all emails in the inbox are fed to MySQL.

Step 4: **Dynamically add access database columns at run time using vb.net**

Step 5: **Storing complex properties as text in database**

Step 6: **Database Setup**

Step 7: **Mapping columns to properties**

In order to map Tags and Owner columns, it may be needed separate properties for them. It may be also be required to add two internal _Tags and _Owner properties to map the columns. These two fields may contain text taken from database. It may be a utility class that is used to define properties in Owner property, it is not mapped to database table (Figs. 1, 2 and Tables 1, 2).

Fig. 1 Database allocation with associated cost

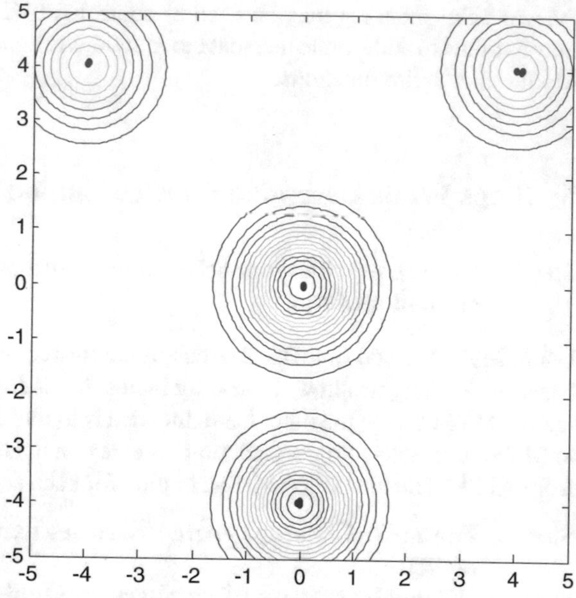

Fig. 2 Bound limits associated with servers with fitness values

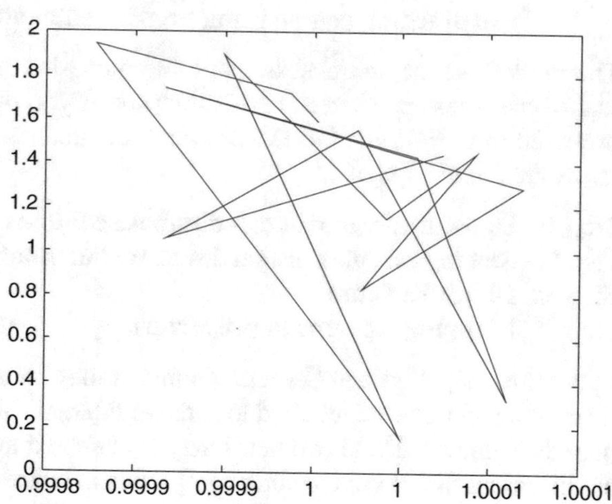

Table 1 Allocation of database with associated cost

Databases associated with server	Range of databases in server (x-direction)	Range of databases in server (y-direction)	Cost associated with the databases
20–30	−4.0066	4.0832	0.9931
30–40	4.0768	4.0238	0.9936
40–50	−0.0670	−3.9611	1.9880
50–60	0.0068	0.0153	1.9994
60–70	4.0356	3.9920	0.9987

Table 2 Bound limits associated with servers with fitness values

Range of queries	Bound limits associated with the servers	Fitness values
20–30	0.1–0.5	1
30–40	0.4–1	1.0001
40–50	1.2–1.4	0.9999
50–60	1.5–1.8	0.9999

7 Analyzing Databases Using Metaheuristic Approach

a. Initial assumption

No. of data centers = 20
No. of iterations = 500
Absorption coefficient, Ac = 1
Randomness = 0.5
Parametric coefficient of database servers associated with data centers, Pc = 0.2

b. Simple limits for the dimensional problems

No. of database servers linked with each data center, C = 15;
  Lower bound = zeros(1, C);
  Upper bound = 2*ones(1, C);
  Initial random guess

$$u0 = \text{Lower bound} + (\text{Upper bound} - \text{Lower bound}). * \text{rand}(1, C);$$
$$\text{Evaluate dimension, i.e., length}(u0);$$

c. Evaluate the performance and objective function of each database server

z = sum((x-1).^2); where x denotes no. of databases associated with each database server

d. Move all database servers to the better locations

[ns] = dsa_move(n,C,ns,ds,nso, randomness,nbest,,Ac, Pc, Lower bound, Upper bound);

8 Performance Evaluation

It has been observed that sometimes the heterogeneous wireless network may be responsible for the performance of the smart home entities. In this regard, it may affect the communication between sensors and sensor coordinators. To eradicate this interference as well as the problems associated with the entire system, the dedicated sensors may be considered. Each desired appliance should be connected with a sensor, as well as sensor coordinators. The coordinator may be responsible for generalization as well as aggregation of data linked with the sensors.

9 Discussion and Future Direction

There are so many ways to determine the basic structure of the IoT. Similarly, the architecture, as well as visualization associated with IoT applications, may provide the desired value to the whole system which will help to escalate more efficiently. The components associated with the system may use separate technologies to find the weaknesses and analyze integrity of the system.

10 Conclusion

In the present scenario, the Internet of Things state can make previously unknown but significant work in secured embedded computer devices. But sometimes, the security may obstruct toward initiation vectors point associated with the IoT devices. The data associated with IoT devices may be linked with higher exposure and it is required to foster on security point of view. The recent analysis, as well as challenges, should be initiated for better improvement linked with security objectives at early design stages and efficient and effective application of security standardized solutions at production stages.

References

1. Normann, A., Jamison, A.: Knowledge Making in Transition: On the Changing Contexts of Science and Technology. University of Pittsburgh Press, Pittsburgh (2011)
2. Wang, S.: Intelligent Buildings and Building Automation, 2nd edn. Spon press, New York (2010)
3. Xiaojuan, Z.: The strategy of smart home control system design based on wireless network. In: International Conference on Computer Engineering and Technology (ICCET), pp. 4-37–4-40 (2010)
4. Hamed, B.: Design & Implementation of Smart House Control Using LabVIEW. Int. J. Soft Comput. Eng. (IJSCE) ISSN **1**(6), 2231–2307 (2012)
5. El-Basioni, B.M.M., El-kader, S.M.A., Abdelmonim Fakhreldin, M.: Smart home design using wireless sensor network and biometric technologies. **2**(3) (2013)
6. Dixon, C., Mahajan, R., Agarwal, S., Brush, A., Lee, B., Saroiu, S., Bahl, P.: An Operating System for the Home. In: NSDI (2012)
7. Munir, S., Stankovic, J., Liang, C., Lin, S.: New cyber physical system challenges for human-in-the- loop control. In: 8th International Workshop on Feedback Computing (2013)

8. Gross, H., Holbl, M., Slamanig, D., Spreitzer, R.: Privacy-aware authentication in the internet of things. In: Cryptology and Network Security. Springer International Publishing, pp. 32–39 (2015)
9. Uckelmann, D., Harrison, M., Michahelles, F.: An architectural approach towards the future internet of things. Springer, Berlin (2011)
10. Vermesan, O., Friess, P.: Internet of Things: Converging Technologies for Smart Environments and Integrated Ecosystems. River Publishers, Aalborg (2013)
11. Babar, S., Stango, A., Prasad, N., Sen, J., Prasad, R.: Proposed embedded security framework for internet of things (iot). In: 2011 2nd International Conference on Wireless Communication, Vehicular Technology, Information Theory and Aerospace & Electronic Systems Technology (Wireless VITAE), pp. 1–5. IEEE (2011)
12. Hong, S.H., Yu, M., Huang, X.: A real-time demand response algorithm for heterogeneous devices in buildings and homes. Energy **80**, 123–132 (2015)
13. Rani, S., Ahmed, S.: Multi-hop Routing in Wireless Sensor Networks: An Overview, Taxonomy, and Research Challenges. Springer, Berlin (2016)
14. Al-Fuqaha, A., Guizani, M., Mohammadi, M., Aledhari, M., Ayyash, M.: Internet of things: a survey on enabling technologies, protocols, and applications. IEEE Commun. Surv. Tutor. **17**(4), 2347–2376 (2015)
15. Khajenasiri, I., Estebsari, A., Verhelst, M., Gielen, G.: A review on internet of things solutions for intelligent energy control in buildings for smart city applications. Energy Procedia **111**, 770–779 (2017)
16. Hui, T.K., Sherratt, R.S., Sánchez, D.D.: Major requirements for building smart homes in smart cities based on internet of things technologies. Futur. Gener. Comput. Syst. (2016)

Finding Correlation Between Twitter Influence Metrics and Centrality Measures for Detection of Influential Users

Kunal Chakma, Rohit Chakraborty and Shubhanshu Kumar Singh

Abstract With such a great ease of sharing thoughts with the whole world and ever-increasing delightful features, social media such as Twitter has been influencing many aspects of our lives such as product recommendations, movie reviews, political campaigns, game predictions, and so on. It can be found that often few individuals tend to be influential enough to drive a whole group or network towards a particular thought. Network Centrality is one of the most widely studied and used concepts in social network analysis. There are other Twitter-specific measures known as influence metrics which are often used for social network analysis. The work presented in this paper reports on the implementation of some centrality measures and Twitter-specific influence measures are applied on a Twitter network of professional wrestling. The paper also reports on the correlation among the different measures for detecting influential users.

1 Introduction

With ever-increasing users joining the social network everyday, social media analysis is greatly shaping political campaigns and marketing strategies of the business organizations. Even though the social media posts often do not come with authenticated proofs, yet people heavily rely on them. Out of many social media platforms such as Facebook, Twitter, Quora, Instagram, and so on, Twitter's uniqueness lies in its feature of short and public tweets which anyone can see, like or retweet. The whole Twitter network can be thought of as clusters of interconnected networks of followers and followees. In each such networks, often there exist some users who are

K. Chakma (✉) · R. Chakraborty · S. K. Singh
National Institute of Technology Agartala, Jirania, Tripura, India
e-mail: kchakma.cse@nita.ac.in

R. Chakraborty
e-mail: rohitcse1996@gmail.com

S. K. Singh
e-mail: mpsshubhanshu.123@gmail.com

H. S. Behera et al. (eds.), *Computational Intelligence in Data Mining*, Advances in Intelligent Systems and Computing 990,
https://doi.org/10.1007/978-981-13-8676-3_5

more reliable or influential than others. The reason for that influence is not always straightforward. In a complex network, it is truly a tough task for a person to find out the influencers. The work presented in this paper aims to analyze the features or patterns exhibited by the nodes (users) in a social network graph for the correct prediction of the influential users from various perspectives. The need to find the most influential user in a social network struck to our minds when we were pondering about the social media advertisements, specifically Facebook ads, and promotional tweets, and the cost incurred for the same. What if the ad goes to the most influential user of a small network and not to any other member of that network? If the ad proves to be relevant, the other users will also come to know. It is a natural human tendency to share whatever new we come across. Also, most often the recommendation from an influencer proves to be of more convincing than the advertisement banner. Then, we looked for the other benefits of targeting the most influential user in different domains such as political campaigns and predicting game outcomes. After this, we browsed through the already existing works in this domain and future research scopes. We have considered a small Twitter network comprising of WWE[1] superstars and fans as nodes with directed edges from followers to followees, forming a multiple directed graphs. Our network comprises an interconnected multiple directed graphs with 22,236 nodes and 38,758 edges. The aim is to analyze the influence of the nodes based on centrality measures [1] such as *indegree* and *closeness* along with Twitter metrics [2] such as *source of a tweet (celebrity, organization, etc.)*, *number of retweets*, *number of mentions*, and to study how different features affect the influence of a user. The rest of the paper is organized as follows. In Sect. 2, the related work is discussed. In Sect. 3, the data collection and nature of the network is discussed. Section 4 discusses the features adopted for implementation. The results and analysis are discussed in Sect. 5 which is followed by the concluding remarks in Sect. 6.

2 Related Work

In social network analysis, the definition of influential user is difficult to establish as new influence measures are emerging rapidly, each offering different measurement criteria. References [3, 4] present a study on several Twitter influence measures that exist so far in literature. Existing measures can be categorized into *activity measures* [2, 5–9], *popularity measures* [10–14] and *influence measures* [15–20]. Reference [19] states that influential users are the most able to spread information within the network. In [20], the authors have presented an in-depth comparison of three measures of influence: *indegree*, *retweets*, and *mentions*, using a large amount of data collected from Twitter. The authors in [20] observed that spawning retweets or mentions are not necessarily influenced by popular users who have high indegree. Further, [21] states that influential users in a social network are not neccessarily

[1] https://www.wwe.com.

influential in the real life. Reference [22] try to classify different types of centrality criteria. Reference [23] focuses on the location of influential users in a social network and calculation of influence based on the content generated. Reference [24] report on the development of a regression-based machine learning system for measuring influence of Twitter users. Reference [25] measured the correlation between five centralities such as (a) Degree Centrality, (b) Betweenness Centrality, (c) Closeness Centrality (d) Eccentricity Centrality and (e) Eigenvector Centrality. Reference [26] propose a novel partition of influence theories in three major categories with divergent performance.

Our work intends to find out the influence measures significant to the subject of application. The most influential celebrities for fitness products advertisements may not be as effective in spreading propagandas. Although the traditional rank-based approach gives a reasonable outcome of who is more influential. But, it is too general and non-specific. Often, a great influencer of a single small network may not be as such in a complex bigger network. Furthermore, let's say $a, b\ and\ c$ are three persons in a network where the influence order can be shown as $a > b > c$. However, a separate network of $a, c\ and\ d$ may exist cohesively where the influence between a and c may follow a different order, lets say, $c > d > a$. To analyze the influence in such a case, we require a more subjective network-based analysis.

3 Data Collection

We started with a list of WWE superstars[2] and their Twitter handles. Using official Twitter API's,[3] we fetched the followees of those superstars. This gave us an inter-connected multiple directed graph with 22,236 nodes and 38,758 edges. We didn't go for any further depth to have a reasonable complexity for visualization and analysis. Next, using the same APIs, we collected tweets of all the 22,236 users from April 1, 2018 to April 8, 2018. The most awaited annual event of WWE, *Wrestlemania 34*, was scheduled to take place on April 8, 2018 at New Orleans, Los Angeles. So, we got around 278,000 tweets. We needed a Twitter data set with data about the graphical structure of followers–followees and tweets across the network regarding same topic of discussion. The WWE Wrestlemania dataset we collected served both the purposes fairly well for our analysis. For the analysis to start, we had to manually pre-process the tweets to remove objectionable content and resolve the non-printable characters. We ignored the images and links as we concentrated on the textual content only.

---

[2]Obtained from https://wwe.com.

[3]https://developer.twitter.com/en/docs/api-reference-index.

4 Features

We used both Twitter metrics [2] based features (*number of retweets*, *mentions*, *top words*, *top hashtags*) as well as centrality measures [27] based features (*indegree centrality* and *closeness centrality*) to find the influence. We intend to find the correlation among each individual features to gain subjective insights of which factor is more responsible for determining influence depending on the application. Our aim is to find the patterns followed by the nodes with respect to the features.

4.1 Centrality Features

We intend to form a multiple directed graph from the list of edges we collected. After that, the nodes list is also to be taken into account in case any isolated node creeps in. The direction of the edges go from the followers to the followees. Hence, a high indegree will suggest the user has a large number of followers and a high outdegree will suggest the user follows a large number of people.

4.1.1 Calculation of Indegree and Closeness Centrallity

Degree is one of the most basic measures of network centrality [27]. It is based on the assumption that the node which has more number of direct neighbours is more central to the network. For a directed graph such as ours, there exist two types of degrees depending upon the direction of edges—*indegree* and *outdegree*. A large indegree of a node suggests that there are many nodes directed towards it. Hence, it can be regarded as a factor that depicts the influence of a node inside a network. Our proposed idea is to measure the indegree of each of the nodes and sort them into decreasing order to obtain top nodes based on indegree. In order to do that we first, found out the indegree centralities of each individual nodes. Then we picked up top 100 users with largest indegree centralities for further analysis. The steps are illustrated in Algorithm 1. The function *indegree_centrality* returns a dictionary with each individual nodes of the graph G as keys and their corresponding indegree centralities as the values.

The shortest distance of the destination nodes from a source node is taken into account for measuring closeness [27] centrality across a network. It is based on the assumption that the important/influential nodes are closer to all the other nodes. The closeness centrality is measured as

$$C_{closeness}(v_i) = \frac{1}{\frac{1}{N-1}\sum_{v_j \neq v_i} l_{i,j}} \quad (1)$$

where N is the total number of nodes and $l_{i,j}$ is node v_i's average shortest path length to other nodes. The smaller the average shortest path length, the higher the centrality for the node. Using Eq. 1 [28], for each node, we took the average distance of that node from all the other nodes and then normalized it with the total number of nodes to find its closeness centrality. Again, we took top 100 users with largest closeness centralities for further analysis. The steps are illustrated in Algorithm 1.

Algorithm 1 Finding Indegree and Closeness Centrality

```
begin
    Initialize G(nodes, edges)
    Declare dictionary I, C
    I = indegree_centrality(G)
    C = closeness_centrality(G)
    Sort(I.items(), key = x[1]|x ∈ I, descending)
    Sort(C.items(), key = y[1]|y ∈ C, descending)
    Out(I[1 : 100])
    Out(C[1 : 100])
end
```

4.2 *Metrics Based Features*

A *metric* is a simple mathematical expression that helps us to provide basic information about the social network in the form of a numerical value [3]. As suggested by Pal and Counts [2], metric is a set that involves *original tweets*, *replies*, *retweets*, *mentions* and *graph characteristics*.

4.2.1 Finding Top Words, Retweets, Mentions and Hashtags

The most popular words are to be found out by maintaining a global dictionary structure with key-value pairs, where the individual words are the keys and their corresponding counts in the whole corpus are the values. After scanning the entire corpus, the dictionary is sorted in descending order of the values. The corresponding keys give us the desired list of top words. The top words give an indication of the topics being discussed across the network. We maintained a dictionary W which kept record of all the words as keys with their corresponding counts as values. The output obtained was written to a file *Top_Words*. Altogether, we found out 461,884 distinct words. The most popular hashtags are found by the similar approach adopted for finding the most popular words. The most popular hashtags denote the most followed campaigns across the network. The analysis of the emergence might give us useful insights into finding the influence. The retweets in Twitter are preceded by 'RT:'. Although Twitter API's give the retweets count of each tweet, that is the global retweets count. For our analysis, we manually found the most retweeted tweets using regular expressions from our corpus, maintaining the counts in the similar dictionary structure as proposed above. In this manner, we get the exact retweets counts for each tweet in the local network we are considering. The most retweets received for a particular tweet is also a factor to measure influence across a network. Any user is mentioned in tweets in the format '@user_handle'. Here also, we intend to find the most mentions received in the local network we are considering. So, a similar approach to retweets received is adopted. The users getting more mentions clearly indicates that he is more influential. Algorithm 2 describes the steps followed for extracting top words, hashtags, retweets and mentions. Twitter APIs provide built-in functions for these features, but the counts of retweets and mentions obtained are global counts. As we have considered our own local network, we implemented our own functions to locally count the retweets and mentions received for each users. We maintained separate dictionaries for keeping counts of hashtags, retweets and mentions received.

Algorithm 2 Finding Top Words, Hashtags, Retweets and Mentions

```
begin
    Initialize dictionaries W, H, R, M
    For lines in Tweets
        s = lines.split()
        if s[0] == 'RT'
            R[s[1]]++
        For words in s
            if words[0] == '@'
                M[words]++
            else if words[0] == '#'
                H[words]++
            else
                W[words]++
    out(W, H, R, M)
end
```

Further, we intend to find whether the classification of Tweets with respect to its source such as celebrity, organization, and so on has any effect on influence or not. This comes from the general intuition that a tweet from a celebrity will of course reach more users than a tweet from any non-celebrity.

5 Results and Analysis

In this section, we discuss the results obtained after feature extraction. The top 100 words obtained from our corpus is depicted by the word cloud in Fig. 1a. From the visualization, it is evident that most of the discussions were related to Wrestling Entertainment, more specifically, the then-upcoming event of 'WWE Wrestlemania 34'. The events such as 'NXTTakeOver', 'WrestleMania', 'UFC223', 'WWE', 'wrestling', 'watching', and so on have found their place in our list of most frequently used words. The top hashtags depicted in Fig. 1b makes the domain of tweets even clearer. The top hashtags include '#WrestleMania', '#WWEHOF', '#WWE', '#UFC223', '#RAW', '#NewOrleans' and so on. This clearly correlates to the fact inferred from the top words. Users whose tweets have been retweeted most number of times is depicted in Fig. 1c. Figure 1d reveals the users receiving the most number of mentions. From the pie chart in Fig. 2a and b, there is not much difference in the proportion of celebrities and common people appearing in the list. This may come from the fact that reply to a tweet or referring someone happens more often in our local friend circle. The number of organizations involved is also of modest count. Indegree Centrality allows us to compare users directly based on the number of followers they have. In our local WWE network under consideration, the nodes having maximum indegree centralities are depicted in Fig. 1e. The most renowned WWE superstars such as TripleH, Stephanie McMahon, Mick Foley, John Cena, Cris Jericho, Charlotte, Finn Balor, Randy Orton, Becky Lynch, Sheamus, Natalya and so on are among the list. An important trend to notice here is that those superstars who have been there for more time have more followers. For example, Chris Jericho, Mick Foley are not active inside the ring. Still they enjoy more followers than the current WWE Champions. As depicted in Fig. 1e and f, most of the users having more indegree centrality also have more closeness centrality. The pie chart in Fig. 2c and d depicts the proportion of celebrities, organizations and common people in the list. Figure 2c and d show that indegree and closeness are almost equal for celebrities as well as for common users. However, there are some users such as Shinsuke Nakamura and Steve Austin who seem to be more prominent in case of closeness. Also, the percentage of celebrities sharing the top closeness centralities list is higher than the top indegree centralities list. Here, we can see a new name—YouTube. This comes from the fact that any organization focuses on many celebrities and wide range of topics instead of a single one. Therefore, the followers of each of them retweet the relevant tweets. Hence, the number of retweets reflect the cumulative retweets from the followers of many superstars. The pie chart in Fig. 2a and d depicts the most striking observation, i.e., the percentage of organizations having most retweets is 21% which is much higher than the percentage of organizations having top centrality measures which is 3%.

(a) Top words (b) Top hashtags

(c) Top retweets (d) Top mentions (e) Top indegree (f) Top closeness

Fig. 1 Word cloud depicting the top metrics

5.1 Correlation Among Features

Figure 3a represents the intersection of influential users with respect to different features. *I* stands for indegree centrality, *C* stands for closeness centrality, *R* stands for retweets, *M* stands for mentions. Each circle represents the set of top 100 users with respect to the corresponding features. Among top 100 influencers with respect to each category, there are only 22 common users who are present in all the individual top 100s list. Figure 3b depicts the correlation among the features. From this it can be seen that indegree centrality and closeness centrality are highly correlated (**0.83**) for our dataset. All other features have weak correlation among themselves. The number of mentions and retweets exhibit the least correlation (**0.0089**). A higher retweet count for a user does not necessarily mean that it will have higher mention count or vice versa. For example, the user 'StephMcMahon' has a retweet count of 48 with mention count of 94 whereas, the user 'BellaTwins' has retweet count of 32 and mention count of 185. Though the user 'BellaTwins' has lesser retweets than the user 'StephMcMahon', the former has higher mention counts than the later. Figure 4 shows how different influencing features vary for the top 20 influencers.

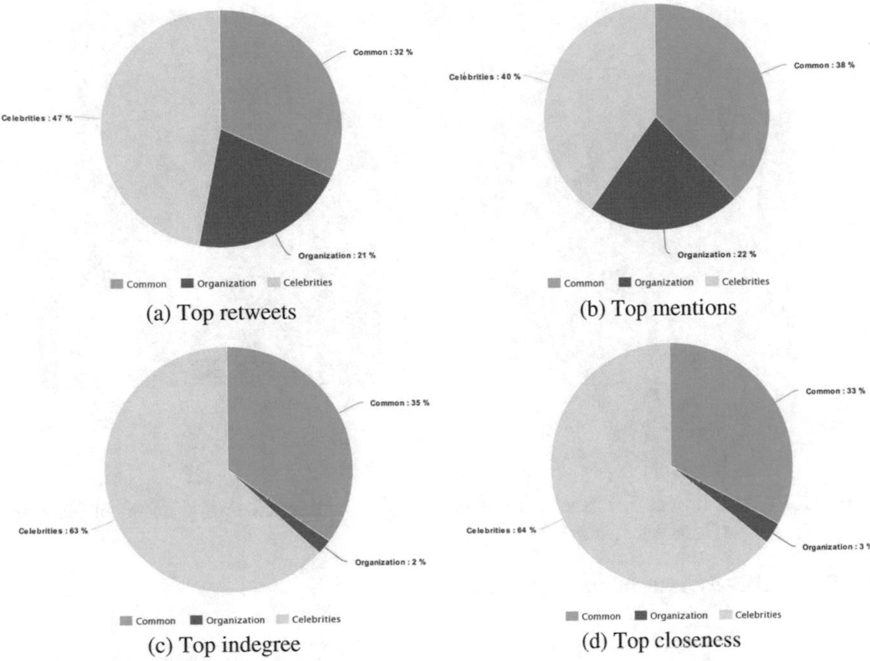

(a) Top retweets

(b) Top mentions

(c) Top indegree

(d) Top closeness

Fig. 2 Percentage distributions of metrics across common users, organizations and celebrities

The values have been scaled to obtain a comparative visualization. As clear from the graph, the features are not strongly correlated. Every feature has its own significance and utility. The indegree centrality values have been multiplied by 100000 and the closeness centrality values by 700 to bring all the features in the same range of 0–400. In the graph, it can be seen that after user15, there is a sharp decline in the closeness centralities. It can be inferred that only few influencers are closer to a large number of users in the network.

6 Conclusion and Future Direction of Work

In this paper, we studied and analyzed the measures for identifying influential users in a social network. We compared two sets of features: centrality measures based and metrics based. Under centrality based features, we measured the *indegree* and *closeness* centralities. For metrics based features, we measured the top occurring *words*, *retweets*, *mentions* and *hashtags*. Our study shows that celebrities have large indegree centralities and closeness centralities. This comes from the fact that they have a huge fan following. The celebrities who have been around for a long time have more closeness centralities. So, to reach out to a fan, they can be more effective than their newer counterparts. The mentions are equally likely to be received by the

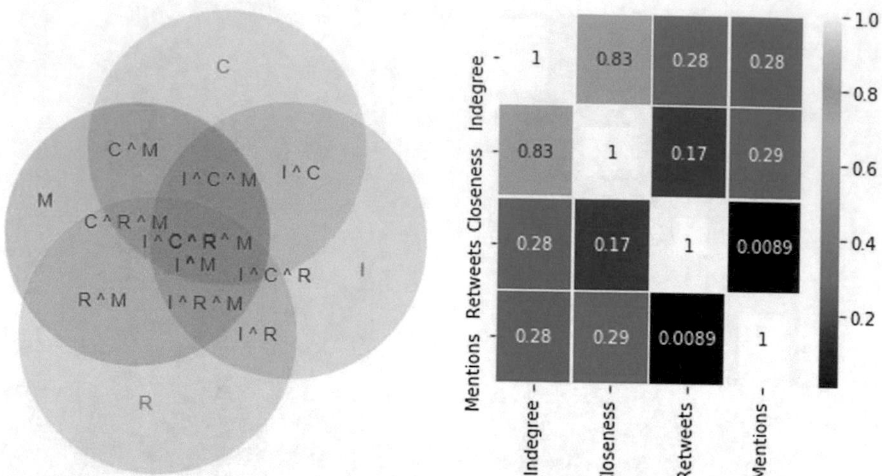

(a) Intersection of top 100 users with respect to different features (b) Correlation among different features

Fig. 3 Correlation among features for the top 100 users

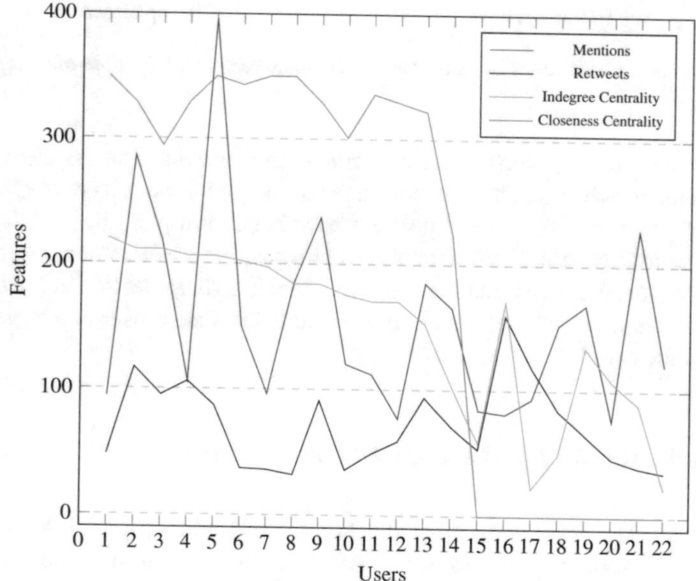

Fig. 4 Comparison of top influencers based on top features

celebrities and common people. It comes naturally from the fact that reply to a tweet or referring someone can happen more often in our local friend circle. From the above inferences, it is evident that finding influencers across a network is subjective. It depends on the motive behind the application of that influence. For marketing purposes, well-established organizations and renowned celebrities who have been there for a long time can be more influential. For promoting new trends or fashion, newer celebrities can play a major part. Organizations can be highly influencing in spreading propagandas and campaigns. The work can be extended by extensively studying the influence patterns of social networks according to the application as discussed before. Analyzing social media platforms other than Twitter such as Instagram, Facebook, Quora, etc. may also give useful outcomes. Using machine learning based predictive analysis, outcomes of matches, entertainment series endings, etc. can be predicted. Recommendation systems can be made smarter by offering recommendations from the social circle of a user.

References

1. Newman, M.E.J.: Networks: an introduction. Oxford University Press, Oxford (2010)
2. Pal, A., Counts, S.: Identifying topical authorities in microblogs. WSDM 45–54 (2011)
3. Riquelme, F., Gonzlez-Cantergiani, P.: Measuring user influence on Twitter: a survey. Inf. Process. Manag. **52**(5), 949–975 (2016)
4. Sun, J., Tang, J.: A survey of models and algorithms for social influence analysis. Social Network Data Analytics, pp. 177–214. Springer, Berlin (2011)
5. Noro, T., Ru, F., Xiao, F., Tokuda, T.: Twitter user rank using keyword search. Information Modelling and Knowledge Bases XXIV. Frontiers in Artificial Intelligence and Applications, vol. 251, pp. 31–40. IOS press, Amsterdam (2013)
6. Lee, C., Kwak, H., Park, H., Moon, S.B.: Finding influentials based on the temporal order of information adoption in Twitter. In: WWW, pp. 1137–1138 (2010)
7. Jabeur, L.B., Tamine, L., Boughanem, M.: Active microbloggers: identifying influencers, leaders and discussers in microblogging networks. In: SPIRE, pp. 111–117 (2012)
8. Romero, D.M., Galuba, W., Asur, S., Huberman, B.A.: Influence and passivity in social media. In: Machine Learning and Knowledge Discovery in Databases - European Conference, ECML PKDD, vol. 6913, pp. 18–33 (2011)
9. Zhang, B., Zhong, S., Wen, K., Li, R., Gu, X.: Finding high-influence microblog users with an improved PSO algorithm. IJMIC **18**(4), 349–356 (2013)
10. Nagmoti, R., Teredesai, A., De Cock, M.: Ranking approaches for microblog search. In: International Conference on Web Intelligence, WI 2010, pp. 153–157 (2010)
11. Aleahmad, A., Karisani, P., Rahgozar, M., Oroumchian, F.: OLFinder: finding opinion leaders in online social networks. J. Inf. Sci. 1–16 (2015)
12. Gayo-Avello, D.: Nepotistic relationships in Twitter and their impact on rank prestige algorithms. Inf. Process. Manag. **49**(6), 1250–1280 (2013)
13. Khrabrov, A., Cybenko, G.: Discovering influence in communication networks using dynamic graph analysis. In: PASSAT, pp. 288–294 (2010)
14. Srinivasan, M.S., Srinivasa, S., Thulasidasan, S.: Exploring celebrity dynamics on Twitter. In: Proceedings of the 5th IBM Collaborative Academia Research Exchange Workshop, I-CARE 2013, pp. 1–4 (2013)
15. Hajian, B., White, T.: Modelling influence in a social network: metrics and evaluation. In: PASSAT/SocialCom, pp. 497–500 (2011)

16. Jin, X., Wang, Y.: Research on social network structure and public opinions dissemination of micro-blog based on complex network analysis. J. Netw. **8**(7), 1543–1550 (2013)
17. Hirsch, J.E.: An index to quantify an individual's scientific research output that takes into account the effect of multiple coauthorship. Scientometrics **85**(3), 741–754 (2010)
18. Anger, I., Kittl, C.: Measuring influence on Twitter. In: Lindstaedt, S.N., Granitzer, M. (eds.) I-KNOW 2011, 11th International Conference on Knowledge Management and Knowledge Technologies
19. Morone, F., Makse, H.A.: Influence maximization in complex networks through optimal percolation. Nature **524**, 65–68 (2015)
20. Cha, M., Gummadi, K.P.: Measuring user influence in twitter: the million follower fallacy. In: Association for the Advancement of Artificial Intelligence, pp. 10–17 (2010)
21. Cossu, J.-V., Dugu, N., Labatut, V.: Detecting real-world influence through Twitter. In: Second European Network Intelligence Conference, ENIC, vol. 2015, pp. 83–90 (2015)
22. Probst, F., Grosswiele, L., Pfleger, R.: Who will lead and who will follow: identifying influential users in online social networks - a critical review and future research directions. Bus. Inf. Syst. Eng. **5**(3), 179–193 (2013)
23. Vogiatzis, D.: Influential users in social networks. Semantic Hyper/Multimedia Adaptation - Schemes and Applications. Studies in Computational Intelligence, vol. 418, pp. 271–295. Springer, Berlin (2013)
24. Nargundkar, A., Rao, Y.S.: InfluenceRank: a machine learning approach to measure influence of Twitter users. IEEE Explore, pp. 1–6 (2016). https://doi.org/10.1109/ICRTIT.2016.7569535
25. Batool, K., Niazi, M.A.: Towards a methodology for validation of centrality measures in complex networks (2014). https://doi.org/10.1371/journal.pone.0090283
26. Kardara, M., Papadakis, G., Papaoikonomou, A., Tserpes, K., Varvarigou, T.A.: Large-scale evaluation framework for local influence theories in Twitter. Inf. Process. Manag. **51**(1), 226–252 (2015)
27. Freeman, L.C.: Centrality in social networks conceptual clarification. Soc. Netw. **1**(3), 215–239 (1978) (Elsevier)
28. Bavelas, A.: A mathematical model for small group structures. Hum. Organ. **7**, 16–30 (1948)

QMAA: QoS and Mobility Aware ACO Based Opportunistic Routing Protocol for MANET

Aparna A. Junnarkar, Y. P. Singh and Vivek S. Deshpande

Abstract In ad hoc networks, the dynamics of mobile nodes and topology leads to the frequent links breaks and formation as mobile nodes move randomly across the network area. The frequent changes in network topology may suffer from the frequent task of link breaks and links building to transfer data from the source node destination node. The quality of service (QoS) performance of such networks mainly depends on how well the routes are stables and efficient in network. The tasks of route discovery and route formation and data transmission are the key factors of mobile ad hoc network (MANET) routing protocols. To improve the QoS performance of routing protocols, the state-of-art algorithms presented in recent past. However, designing scalable QoS aware routing protocol is still key problem. In this research paper, we investigated the novel routing algorithm based on optimization algorithm to select more stable and mobility aware routes to forwarded data from source to destination. The proposed routing protocol is based on optimization scheme called Ant Colony Optimization (ACO) with swarm intelligence approach. This proposed method used the RSSI measurements and energy metrics to determine the distance between two mobile nodes in order to select efficient path for communication. This new routing protocol is named as QoS Mobility Aware ACO (QMAA) Routing Protocol. The simulation results show that QMAA routing protocol achieved the significant performance improvement as compared to state-of-art routing protocols.

Keywords Ant colony optimization · Quality of service · Routing protocols · Ad hoc network · RSSI · Route selection

A. A. Junnarkar (✉) · Y. P. Singh · V. S. Deshpande
Computer Science & Engineering, Kalinga University, Raipur 492101, Chhattisgarh, India
e-mail: aparna.junnarkar@gmail.com

Y. P. Singh
e-mail: ypsingh10@rediffmail.com

V. S. Deshpande
e-mail: vsd.deshpande@gmail.com

H. S. Behera et al. (eds.), *Computational Intelligence in Data Mining*, Advances in Intelligent Systems and Computing 990,
https://doi.org/10.1007/978-981-13-8676-3_6

1 Introduction

At display, the MANETs has involved into refinement as like they substantial the conceivable to the transfiguring several components of our life, from general communication, ecological and military uses. Be that as it may, various specialized hindrance still remain behind and these are used to the assurance and improved before one can involve the filled capability of MANETs [1]. It is contended that segments in MANETs must be actualized adaptable and responsive as for altering into the network topology, load condition, node connectivity, and the end-to-end QoS viewpoints. The MANETs do control various limitations that are must to comprehend guarantee higher QoS [2–6]. Finally, in view of extending associating with nature of multimedia applications and totally expecting business organization of MANETs, the QoS prop up in MANETs has changed into a crucial thing [7]. This condition needs in all cases association between the middle focuses, both to set up the course and to remain the favorable circumstances required for streamlining perfect QoS [8]. Some sensible social requests have portrayed QoS as an arrangement of organization necessities to be refined by the correspondence network while performing transportation of a package stream from source hub to sink or to target hub [9]. The quality of service (QoS) routing is dissimilar from the resource reservation and they possess two distinctive responsibilities that can be either together or individual in QoS architectures. The QoS routing protocol is employed for finding an alternate path that could facilitate the QoS requirements, but it is the QoS signaling that reserves, even maintains, and releases resources across the network elements. The QoS signaling would function better in case it coordinates with QoS routing but majority of QoS routing methods are either very complex or very costly and hence leads to substantial overhead in MANET. The QoS is the entire performance of the communication or computer network, specifically the performance detected by the user of the network are data rate, average throughput, bandwidth efficiency, end-to-end delay, jitter, packet delivery rate, etc. These parameters used to define the QoS performance of MANET [10, 11].

The QoS performance of MANET can be improved by number of different ways such as optimal route selection, distance based, traffic based, load balancing based, etc. However, the mobility which is major cause of performance degradation is yet to efficiently address for MANET routing protocols. In this research paper, we investigated decent optimization and stream intelligent based approach for MANET routing to enhance the extendable and efficiency against the state-of-art technique. The proposed ACO based routing protocol for MANET is presented with goal of improving the overall efficiency of MANET routing. In Sect. 2, we presented the study of related works. In Sect. 3, proposed methodology and algorithm design are presented. In Sect. 4, results are discussed. Finally the Sect. 5 represents the conclusion and future work.

2 Related Work

This section presents the recent works reported on QoS efficient for MANET routing protocols based on different methodologies. Also we presented the load balancing methods presented in recent past.

In [1], another GA based routing scheme designed by authors. They designed GA algorithm for efficient route discovery process in order to deliver the optimal path as well as path maintenance to address infeasible chromosomes. They had evaluated the efficiency of both uni-path and multipath using the well known NS2 and Qualnet Software's.

In [2], author conducted the extensive performance analysis study among the ACO based routing method, DSR, DSDV, and AODV routing protocols. They had considered the delay, packet deliver rate, and packet arrival time, etc. for wireless sensor networks (WSNs).

In [3], author presented the hybrid optimization technique based on (ACO) as well as CS (Cuckoo Search) for the MANET routing performance improvement. They improved the AODV routing method by contributing to the hybrid optimization technique. There proposed method had shown the improved routing performance in simulation studies.

In [4], they proposed the swarm intelligence based routing method to search the efficient route for the data transmission from source station to destination station. They designed the scheduling method based on weight computations. Such scheduling based method improved the packets and queues allocation for data transmission.

In [5], they presented the evaluation of three routing method such as AntHocNet, DSR and AODV schemes using the NS2. They demonstrated the results of all three routing methods with scalability objectives. From their results, they claimed that AntHocNet had efficient routing performance as compared to both DSR and AODV routing methods with varying network scenarios. AntHocNet is the ACO based routing scheme.

In [6], author aimed to present the efficient technique to discover the optimized route among the source and destination nodes as well as detected the attacker's mobile nodes in MANET. Due to the misbehaving mobile nodes, routing protocols suffering from the worst QoS performance, they addressed this problem with their method. They designed the GA based algorithm for efficient detection of attacker nodes in network by modifying the existing DSR routing scheme. They conducted the simulation using the NS2 tool. They additionally proposed to use ACO method also for optimal path selection against the GA algorithm.

In [7], they contributed to this paper in two ways. First, they modified the existing AODV and DSDV routing schemes to address their limitations by utilizing the properties of ACO algorithm. Second, they conducted the comparative study among all three routing schemes for the QoS performance as well as fairness. Their simulation results claimed that their proposed routing scheme outperformed the existing DSDV and AODV methods.

In [8], author examined and evaluated the different ant colony optimization based routing schemes in MANET. They have evaluated the routing methods based on ACO in terms of delay, packet delivery ratio, etc. They motivated by modifying the AODV routing method using ACO for MANET routing performance optimization. They designed ACO based technique for latency reduction.

In [9], they introduced the method of adaptive steady load balancing gateway selection was simulated for MANET. Their method based on load balancing of path via the computation of path load as well as metrics of path queue length. Their method utilized the GA technique for the optimal selection of gateway. The GA fitness function used the fuzzy logic technique using the three different network parameters. The characteristics of their proposed routing protocol were gateway advertisement periodicity and range dynamically adjusted.

In [10], author presented the recent hybrid ACO based routing scheme. This technique generated the paths automatically and dynamically with mechanism of similar load dispersion in MANET. The neighborhood seek part of ACO was adjusted by them with a specific end goal to give the productive network QoS performance.

In [11], the authors Yoo et al. presented that, the load balancing strategy called SLBA implies basic load balancing approach. This strategy can without much of a stretch added to any existing routing protocols. This SLBA strategy decreases the movement fixation by empowering each node for dropping the RREQ packets or for surrendering the packet sending.

In [12], author Khamayseh et al. introduced novel MLR (Mobility and Load aware Routing) strategy for diminishing the effects of broadcasting issue. Flooding process is controlled by MLR strategy by confining messages of rebroadcast in light of less speed and less loaded mobile nodes. In this technique each mobile node takes decision on received RREQ message contingent upon number of parameters, for example, routing load, speed, etc.

In [13], authors Cheng et al. introduced an approach for composing issue of dynamic load-balanced clustering into the issue of dynamic optimization. To determine this issue, the author introduced the distinctive sorts of dynamic genetic algorithms in MANETs.

In [14], author Guodong et al. introduce the novel technique for load balancing with improved energy efficiency based on geographic routing protocol. This approach is defined as ELGR means energy-efficiency and load-balanced geographic routing. Researcher investigated that the ELGR techniques for lossy types of MANETs. For making the routing decision, this approach combining load balancing as well as energy efficiency together. This is only the first method in which for improving the network lifetime performance, method of link estimation is adopted for packet reporting rate. Then, load learning approach is used for sensing the current loads across the local network in order to balance the load.

In [15], author Meri et al. designed and proposed load balancing scheme based on table-driven routing approach which is able to balance local nodes load. The notion of proactive routing is utilized for this approach. After the building the routing tables, nodes will decide every message relay without having interaction with any node further.

However, the existing load balancing and QoS based solutions proven to be inefficient under the highly dynamic network conditions due to frequent and random mobility of mobile nodes. Therefore, our work is designed to alleviate the mobility aware stable routes problem in state-of-art solutions in this paper. Next section elaborates the proposed QoS aware routing protocol.

3 Qmaa

The QAMM stands for QoS and Mobility Aware ACO based routing protocol for MANET. The new ACO based method for MANET routing method proposed to select the healthy and QoS efficient route via source mobile node to the destination mobile node. The innovated QMAA routing protocol aimed to discover the route which can not only deliver the QoS efficiency but also minimize the energy consumption of mobile nodes. For QoS and energy control, this technique manages the transmission energy while sending data from one hop to next hop in current path. The first important step of proposed algorithm is collection of mobile nodes location information based on RSSI computations. In position based ACO method, the search of route is done if there are the data packets to forwards from source to destination node. In QMAA, when any mobile node changes their position or joins the network, it sends the Hello packet in network. Then using the RSSI value, every mobile station finds the distance with its neighbors (immediate) by utilizing the relationship among actual distance and RSSI value. Figure 1 is showing the flowchart of proposed QMAA algorithm.

The algorithm of QMAA is described in below steps:

Algorithm 1: QMAA

Input: Routing table, packet, source S, destination D, $E_{threshold}$, $P_{threshold}$
Output: Transmitted signals optimal path

1. Mobile node location estimation using RSSI
2. RT = routediscovery (S, D).
3. Extract RSSI based nodes location
4. Extract routes based location from S to D
    4.1 S broadcast RREQ to 1 hop neighbor
    4.2 Routing table entries are computed and updated for each 1 hop neighbor
    4.3 RREQ received node determines the routing table entry for node D.
    4.4 If (entry found)
    4.5 {
    4.6 Sent RREP to Node S
    4.7 } else
    4.8 {
    4.9 Go to step 4.4
    4.10 }
    4.11 RREQ packets forwarding from 1 hop neighbours to their 1 hop neighbours

4.12 Update and append the current entries in routing table.
4.13 Call to steps 4.2-4.3
4.14 If node S moves and shows itself as intermediate nodes 1 hope neighbor, then discard the RREQ packets with ID of S.
4.15 RREP packet is sent by node D to S for every RREQ packet from different intermediate nodes.

5. All extracted paths from S to D are stored into routing table
6. Start ACO based route selection process

6.1 Scan all routes using FANT (Forward ANT) to select the path with intermediate nodes those are satisfy the constraints of battery level and load conditions.
6.2 if (battery (node) > threshold && load (node) > threshold)
6.3 {
6.4 select node
6.5 }
6.6 S initiate route establishment via forwarding FANT to two possible routes.
6.7 Once FANT packet received at every next mobile node in route, then at each node the equations 1 to 6 (according to pheromone rule) are applied to check whether node is eligible or not for path selection process.
6.8 Update the pheromone values using EW (energy weight) and LW (load weight).
6.9 On receiving FANT packet at D node, BANT (backward ANT) generated at node D.
6.10 BANT traces the path recorded by FANT.

7. After receiving all FANTs by node D, it computes the optimal path using our evaluation criteria and rules specified.
8. If optimal path discovered, destination node puts all the possible entries in routing table as backup routes.
9. Node S starts sending data through selected path to node D.
10. Stop

Above algorithm uses the FANT packet which is having the format showing in Table 1.

where

SID = Source Address
E_{needed} = Energy required transferring the number of packets NP
EWcum = Cumulative Edge Weight
DID = Destination address.

The computations to perform as each node in path selection problem are done by below equations as used in Algorithm 1.

$$RE_{node} - E_{needed} > E_{threshold} \; then \; EW_{ij} = 1 \tag{1}$$

$$RE_{node} - E_{needed} = E_{threshold} \; then \; EW_{ij} = 0 \tag{2}$$

$$RE_{node} - E_{needed} < E_{threshold} \; then \; EW_{ij} = -1 \tag{3}$$

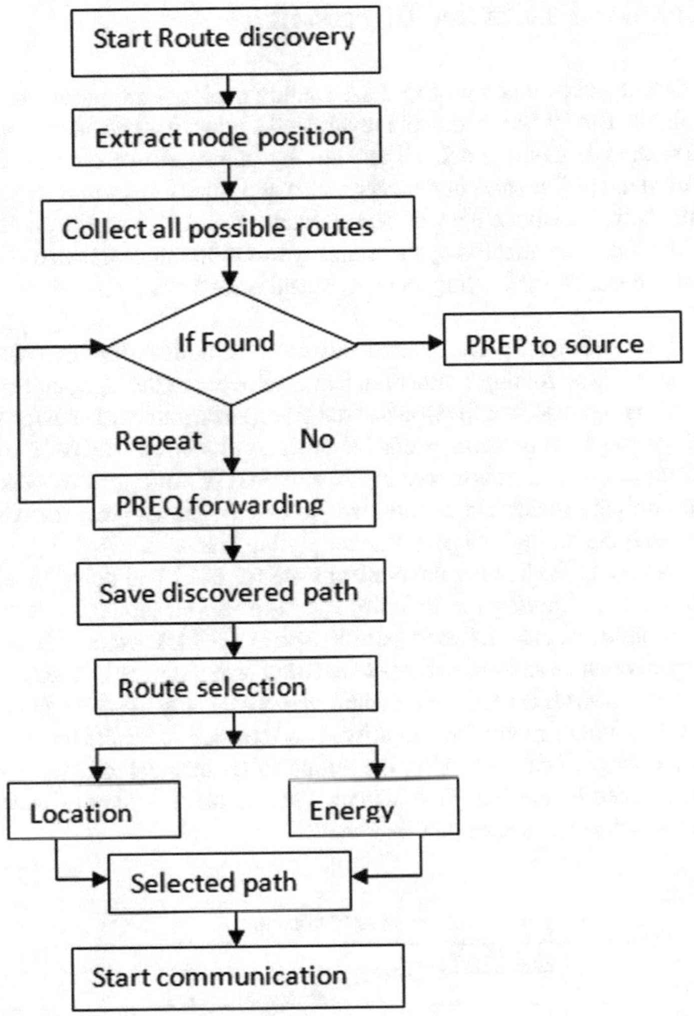

Fig. 1 Proposed algorithm flowchart

Table 1 FANT packet format

SID	E_{needed}	NP	EWcum	DID

$$NP_{otr} + NP > NP_{threshhold} \text{ then } LW_{ij} = -1 \quad (4)$$

$$NP_{otr} + NP = NP_{threshhold} \text{ then } LW_{ij} = 0 \quad (5)$$

$$NP_{otr} + NP < NP_{threshhold} \text{ then } LW_{ij} = 1 \quad (6)$$

4 Simulation Results and Discussion

The simulation and evaluation of QMAA routing protocol are performed by using NS3 simulator. The QMAA protocol is evaluated against the two state-of-art routing protocols such as AODV and ARA. There are four main performance metrics thus as throughput, delay; PDR (packet delivery ratio) and jitter are chosen to evaluate the QoS of all routing methods with respect to varying mobility speed. As the routing performance if mainly affected by the mobility of mobile nodes, hence we prepared the network scenarios for varying mobility speed from 5 to 25 m/s with total mobile users are 55.

Figure 2 is showing the performance analysis of methods AODV and ARA against the proposed QMAA routing protocol in terms of average throughput. For wireless networks, throughput defines the QoS and data rate performance. The ARA method is showing its throughput performance is better as compared to AODV. The proposed QMAA routing outperforming both ARA and AODV routing protocols in every scenario of mobility speed. The results also indicating that with respect to increasing mobility speed, the throughput performances going less.

Figures 3 and 4 are showing the performance for PDR and delay for all routing protocols. We have varying the mobility speed of mobile nodes by keeping total number of mobile nodes 55 for each mobility speed. QMAA protocol is showing the enhanced performance as differentiates to existing protocols for QoS efficiency. The performance of QMAA is showing the improvement by only 20–30% approximately as compared to AODV in varying mobility speed scenarios. Similar case is showing in Fig. 4 of jitter performance. Also, the notice about increasing delay and jitter as the mobility speed increasing. This indicates the mobility leads to frequent routes breaks due which performance gets degraded.

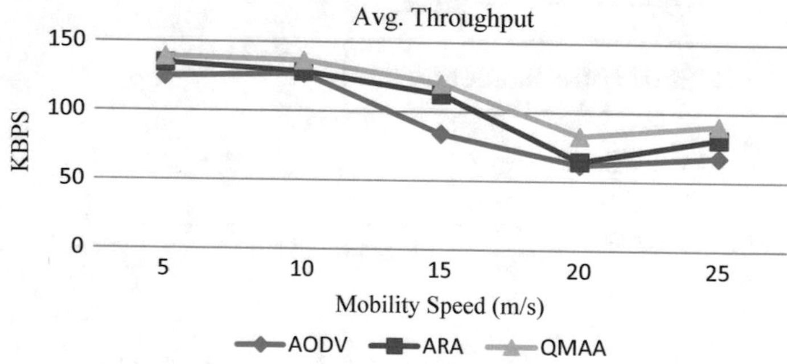

Fig. 2 The throughput evaluation for QMAA

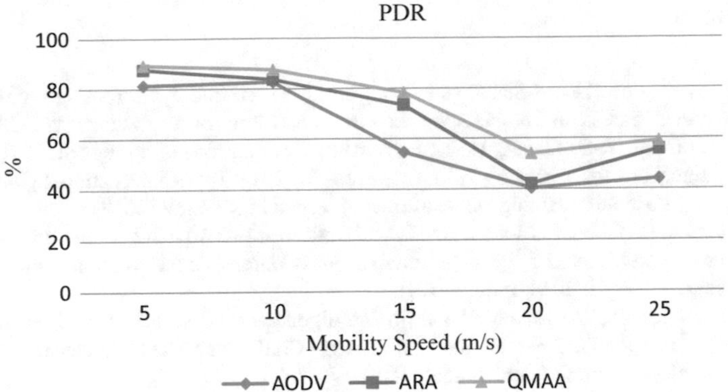

Fig. 3 The PDR evaluation for QMAA

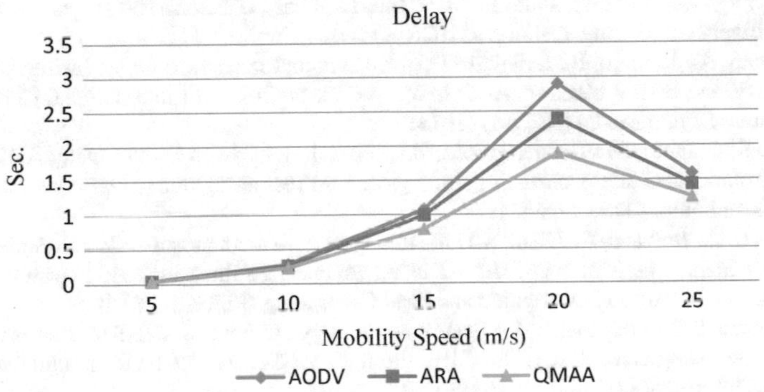

Fig. 4 The average end-to-end delay evaluation for QMAA

5 Conclusion and Future Work

This research paper proposed the novel optimization technique based path selection routing algorithm in our first contribution. We presented the extensive literature review on route optimization in MANET with respect to different parameters such as mobility speed, energy consumption, load, etc. The QoS enhancement techniques are based on routing protocol based algorithms. First, we are studies on the optimization based routing protocols for MANET and other wireless networks for QoS improvement. Then present novel QoS efficient ACO based routing algorithm using nodes location information to handle the mobility issues of MANET. The proposed QMAA routing protocol is designed route selection method using FANT and BANT with route selection rules. The simulation results with varying mobility speed scenarios claim the effectiveness of proposed QMAA against the state-of-art methods. For the future work, there is suggestion to work on reliability to enhance the QoS further under the presence of attackers in network.

References

1. Biradar, A., Thool, R.C.: Reliable genetic algorithm based intelligent routing for MANET. In: 2014 World Congress on Computer Applications and Information Systems (WCCAIS) (2014)
2. Lin, T.-L., Chen, Y.-S., Chang, H.-Y.: Performance evaluations of an ant colony optimization routing algorithm for wireless sensor networks. In: 2014 Tenth International Conference on Intelligent Information Hiding and Multimedia Signal Processing (2014)
3. Nancharaiah, B., Chandra Mohan, B.: Hybrid optimization using ant colony optimization and cuckoo search in MANET routing. In: International Conference on Communication and Signal Processing, 3–5 April 2014, India (2014)
4. Sharma, P., Kumar, R.: Enhanced swarm intelligence-based scheduler to improve QoS for MANETs. In: 2014 5th International Conference - Confluence The Next Generation Information Technology Summit (Confluence) (2014)
5. Taraka, N., Emani, A.: Routing in Ad Hoc networks using ant colony optimization. In: 2014 Fifth International Conference on Intelligent Systems, Modelling and Simulation (2014)
6. Varshney, T., Katiyar, A., Sharma, P.: Performance improvement of MANET under DSR protocol using swarm optimization. In: 2014 International Conference on Issues and Challenges in Intelligent Computing Techniques (ICICT) (2014)
7. Hartaman, A., Rahmat, B., Istikmal.: Performance and fairness analysis (using Jain's Index) of AODV and DSDV based on ACO in MANETs. In: 2015 4th International Conference on Interactive Digital Media (ICIDM) (2015)
8. Khatri, S.K., Dixit, M.: Reducing route discovery latency in MANETs using ACO. In: 2015 4th International Conference on Reliability, Infocom Technologies and Optimization (ICRITO) (Trends and Future Directions)
9. Zaman, R.U., Tayyaba, A., Khan, K.U.R., Reddy, A.V.: Enhancement of load balanced gateway selection in integrated Internet-MANET using genetic algorithm. In: 2016 Fourth International Conference on Parallel, Distributed and Grid Computing (PDGC) (2016)
10. Prabaharan, S.B., Ponnusamy, R.: Secure and energy efficient MANET routing incorporating trust values using Hybrid ACO. In: 2016 International Conference on Computer Communication and Informatics (ICCCI -2016) (2016)
11. Yoo, Y., Ahn, S., Agrawal, D.P.: Impact of a simple load balancing approach and an incentive-based scheme on MANET performance. J. Parallel Distrib. Comput. **70**(2), 71–83 (2010)
12. Khamayseh, Y., Obiedat, G., Yassin, M.B.: Mobility and load aware routing protocol for ad hoc networks. J. King Saud Univ.-Comput. Inf. Sci. **23**(2), 105–113 (2011)
13. Cheng H, Yang S, Cao J.: Dynamic genetic algorithms for the dynamic load balanced clustering problem in mobile Ad Hoc Networks. Exp. Syst. Appl. (2012)
14. Guodong, W., Gang, W., Jun, Z.: ELGR: an energy-efficiency and load-balanced geographic routing algorithm for lossy mobile Ad Hoc networks. J. Aeronaut. **23**(3), 334–340 (2010)
15. Mei, A., Piroso, N., Vavala, B.: Fine grained load balancing in multi-hop wireless networks. J. Parallel Distrib. Comput. **72**(4), 475–488 (2012)

An Effective Trajectory Planning for a Material Handling Robot Using PSO Algorithm

S. Pattanayak and B. B. Choudhury

Abstract This paper utilizes the potentiality of PSO algorithm to design and optimize the trajectory for a material handling robot. This approach is based on the behavior of fish schooling and birds flocking. Layout of the Institute machine shop is selected as an environment for determining the trajectory length. Total fifteen numbers of obstacles (machines) are acknowledged during this analysis. The selected approach not only delivers curtail path length but also generates a traffic-free trajectory. The material handling robot is not colliding with any machines during movement. The programming codes for the selected approach are written, compiled, and run through a software: MATLAB.

Keywords Material handling robot · Trajectory planning · Obstacle avoidance · Particle swarm optimization

1 Introduction

Requirement of endless supply of raw materials, tools, and other necessary accessories to a machining station, warehouse in the least possible time to avoid starving condition in a processing station. It limits the working of human beings in any industries. These limitations are possibly overcome by the development of a mobile material handling robot, which can be guided wirelessly through any electronic gadgets. Flexibility of material handling system directly related with the production rate which indirectly point out the level of automation of the industry. The time required

S. Pattanayak (✉)
Production Engineering Department, Indira Gandhi Institute of Technology, Sarang, Dhenkanal 759146, Odisha, India
e-mail: suvranshupattanayak@gmail.com

B. B. Choudhury
Mechanical Engineering Department, Indira Gandhi Institute of Technology, Sarang, Dhenkanal 759146, Odisha, India
e-mail: bbcigit@gmail.com

H. S. Behera et al. (eds.), *Computational Intelligence in Data Mining*, Advances in Intelligent Systems and Computing 990,
https://doi.org/10.1007/978-981-13-8676-3_7

for processing (machining time) a task is generally predetermined. The cycle time is only increases/decreases by the change in tool and raw material handling time. The time spends in tool and material handling can be saved by the use of material handling robot. It freely delivers and retrieves the needy things to and from the required machining/processing station, when there is no obstacle in its predetermined path. Thus path planning is an important concern for a material handling robot. Different algorithms are followed for the development of optimal path. Path planning is directed towards the determination of best possible shortest trajectory between starting and target point by averting meeting with machines. There are certain key factors upon which the trajectory planning procedure also depends, such as type of environment, number of obstacles, obstacle/machine position, trajectory length and travel time.

Tsuzuki et al. [1] proposed the Simulated Annealing (SA) approach for path planning of robot. This approach relies on graph searching which is associated with space decomposition. Here the trajectory is characterized by polyline, Bezier curve and spline interpolated curve. A conflict-free coordinated trajectory is designed by Chiddarwar et al. [2] for multiple robots by implementing a practically viable approach. The stated approach is a hybrid between A* algorithm and path modification sequence (PMS). A* algorithm is designed for attaining the collision-free path, whereas PMS is employed for obtaining the coordination among multiple robots. Husain et al. [3] applied Ant Colony Optimization (ACO) method to achieve the best motion planning sequence for the material handling robot. This study is directed towards robot path planning in warehouse for material handling. Augmented Reality based (RPAR-II) system is adopted by Fang et al. [4] for trajectory planning of robot. Here dynamic constraints are taken into consideration while determining the trajectory length. Li et al. [5] proposed fuzzy logic and obstacle avoidance algorithm for trajectory planning of mobile robot in dynamic environment. Shum et al. [6] established a comparison between Fast Marching Method (FMM) and Ordered Upwind Method (OUM) for path planning of autonomous vehicles. It is reported that OUM saves 20–70% computational time and more accurate than FMM. A* algorithm is designed by Jafri et al. [7] to generate the global optimum path for mobile robot in outdoor terrain. Aerial images are obtained either by Satellite or by aerial drones for navigation of the robot. A* algorithm was proposed by Emharraf et al. [8] to plan the travel trajectory for a mobile robot in indoor environment having stagnant obstacles. Han et al. [9] determined the feasible shortest optimal path between starting and target point by employing PSO algorithm, which is based on former and latter points (PI_FLP). Surrounding point set (SPS) are also considered during evaluation of optimum trajectory. Self-adaptive learning particle swarm optimization (SLPSO) is designed by Li et al. [10] using different learning strategies for searching an optimal trajectory. Three objectives are considered while defining the objective function such as: path length, degree of collision risk and smoothness of path. Bound violation handling scheme is further adopted for boosting the feasibility of generated path by restricting particle velocity and position.

Till-date few research works have been carried out on trajectory planning for a material handling robot. This work is directed towards the development of a material handling robot and then the diminished trajectory between starting and target point is estimated using an optimization approach. Intention of this work is, to develop a trajectory, which is crash free, smooth and having curtailed length. To achieve the above-stated objectives, PSO algorithm is implemented for generating the global optimum path. The work envelope selected here is a layout of our Institute machine shop, where different types of machines are positioned at different location. Hear each machine is treated as an obstacle during the simulation work.

2 Experimental Setup

A material handling robot is designed for calculating the trajectory length between origin and target point. The photographic view of the robot is presented in Fig. 1. It can moves freely by escaping from conflict with the obstacles/machines appears in its path, without any human guidance.

The environmental condition employed for accomplishing this work is listed in Table 1. It is a static environment, consists of fifteen numbers of obstacles (machine). The dimension and positioning of each machine member is presented in Table 2. While evaluating the trajectory length, the dimension and positioning of the machines member remain the same as the environment is static in nature. A view of the environment is shown in Fig. 2.

To simplify the programming for PSO approach the rectangular shaped obstacles are changed to circular one. Also extra clearance can be provided between machine member and robot by changing them to circular shapes. So the chances of colli-

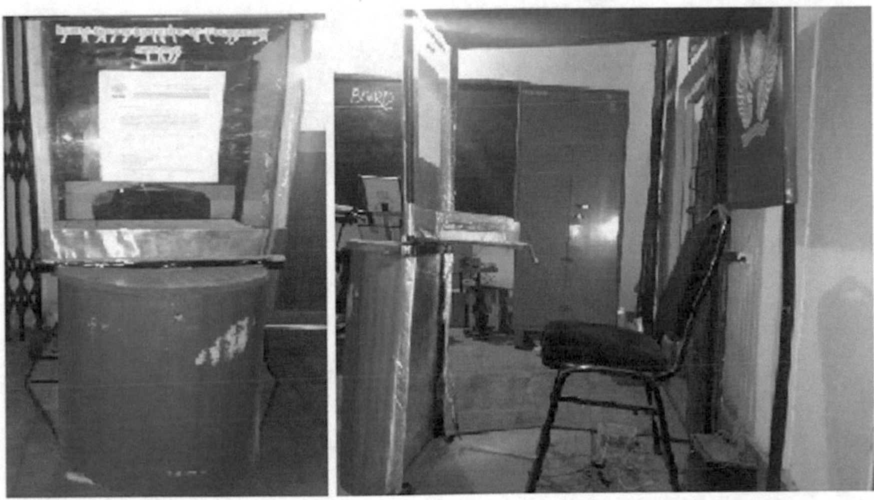

Fig. 1 Mobile material handling robot

Table 1 Environmental setup

Sl. no	Total number of machine member	Starting co-ordinates (meter)	Co-ordinates of target point (meter)
1	15	(0.30, 0.30)	(11, 32)

Table 2 Position and dimension of the machine members

Obstacle number	Obstacle name	X-position (meter)	Y-position (meter)	(Length × Width) of obstacle
1	Planner Machine	1.5	1.8	(3.1 × 2.4)
2	Power Hacksaw-1	1.5	9.1	(1.2 × 0.6)
3	Horizontal Milling Machine-1	1.5	12.1	(1.8 × 1.5)
4	Horizontal Milling Machine-2	1.5	16.7	(1.8 × 1.5)
5	Shaper Machine	1.5	22.8	(1.5 × 1.5)
6	Tool Cutting and Grinding	4.8	4.5	(1.2 × 1.2)
7	Capston Lathe	5.1	10.3	(1.8 × 0.9)
8	Vertical Milling Machine	5.6	13.7	(1.8 × 1.5)
9	Surface Grinding Machine	6.1	22.8	(2.4 × 1.2)
10	Tool Grinder	8.2	2.1	(2.4 × 1.2)
11	Lathe Machine-1	8.2	9.1	(2.4 × 0.9)
12	Lathe Machine-2	8.2	16.1	(2.4 × 0.9)
13	Power Hacksaw-2	8.2	20.1	(1.2 × 0.6)
14	Grinder	7.6	28.3	(0.9 × 0.6)
15	Radial Drilling Machine	2.4	28.9	(1.8 × 1.2)

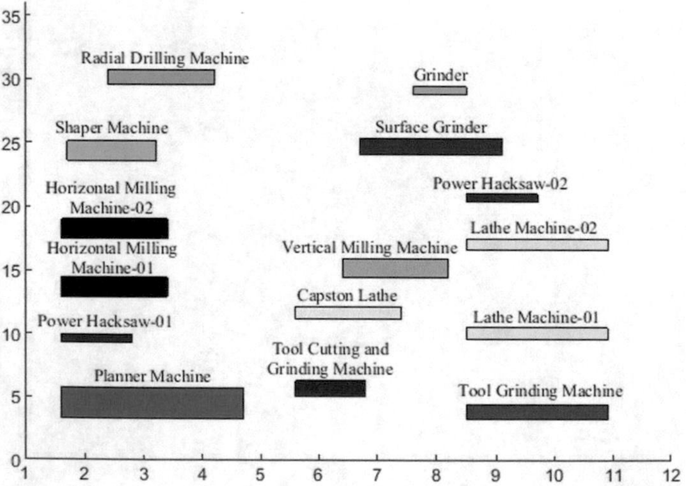

Fig. 2 Static view of the environment

Table 3 Color pattern of each machine member

Obstacle number	Obstacle name	Radius (meter)	Color pattern
1	Planner Machine	1.6	Blue
2	Power Hacksaw-1	0.7	Violet
3	Horizontal Milling Machine-1	1.06	Black
4	Horizontal Milling Machine-2	1.06	Black
5	Shaper Machine	0.9	Sky
6	Tool Cutting and Grinding	0.7	Brown
7	Capston Lathe	1.06	Silver
8	Vertical Milling Machine	1.06	Grey
9	Surface Grinding Machine	1.3	Red
10	Tool Grinder	1.3	Pink
11	Lathe Machine-1	1.3	Yellow
12	Lathe Machine-2	1.3	Yellow
13	Power Hacksaw-2	0.7	Violet
14	Grinder	0.7	Light Green
15	Radial Drilling Machine	1.06	Deep Green

sion are somehow avoided. The radius of the circular shaped obstacle is listed in Table 3. Different color patterns are used for each machine member for their easy identification. Table 3 shows the color pattern used for each machine member.

3 Algorithm

Programming codes are employed for defining the PSO and for achieving global optimum trajectory. PSO algorithm is fitting well, when the complete evidence of the machine member's position and dimensions are available. The position and velocity of the particles are updated after completion of each iteration by using the following mathematical formulae.

$$\begin{aligned} PP_{ij} &= PP_{ij} - 1 + PV_{ij} \\ PV_{ij} &= \left[W.\, PP_{ij} - 1 + C_1\, R_1\left(Pb_{ij} - PP_{ij} - 2\right) + C_2\, R_2\left(Gb_i - PP_{ij} - 2\right)\right] \end{aligned} \tag{1}$$

where 'PP_{ij}' is jth particle position in ith iteration, 'PV_{ij}' is jth particle velocity in ith iteration, 'W' is weight factor, 'C_1, C_2' are the Acceleration coefficient, 'R_1 and R_2' random no. amid 0 and 1, 'Pb_{ij}' jth particle best position in ith iteration, 'Gb_i' best position in the entire swarm.

Table 4 Inputs required for defining PSO

Sl. no	Fixed inputs name	Symbol	Parameter value
1	Swarm size	i	(10–10000)
2	No. of iterations	it	50
3	Acceleration coefficient	C_1, C_2	c1 = c2 = 1.5
4	Inertia weight	W	1

There are certain inputs (as shown in Table 4), whose values are constant during execution of PSO algorithm to estimate the 'Gb' value.

4 Results

After completion of each iteration, 'Eq. 1' is employed for revising the particle velocity and position. After that a searching is performed to notice whether the particles are residing in obstacles or not. If particle is within the obstacle, cancel the swarm, else measure the gbest and pbest value. Then termination condition is verified to know whether it is fulfilled or not. If the termination condition is fulfilled, then terminate the swarm, else again evaluate the fitness function. Total twenty tests are carried out to pinpoint the dwindle trajectory between (0.3, 0.3) meter and (11,

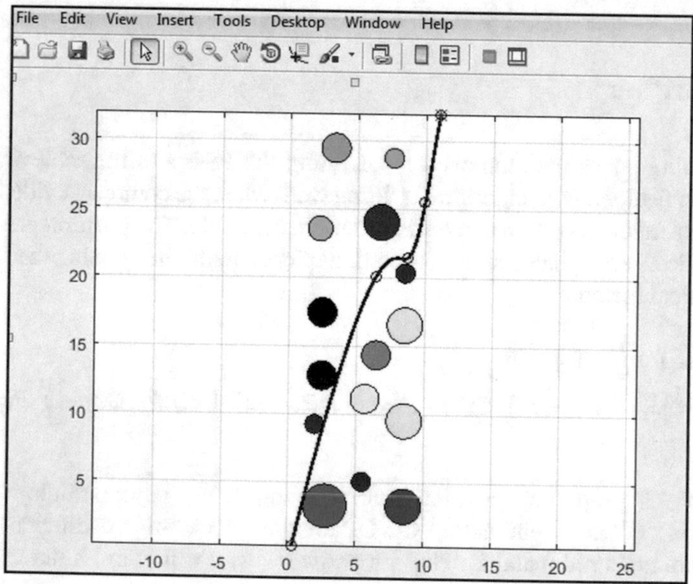

Fig. 3 Shortest trajectory using PSO

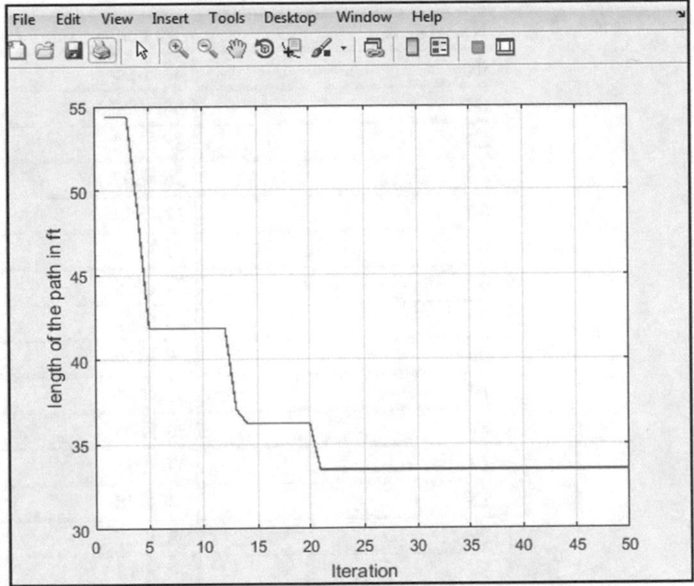

Fig. 4 Trajectory length in every iteration

32) meter. The diminished trajectory for the material handling robot is displayed in Fig. 3. The travel trajectory is represented by the black line and the trajectory length obtained after each iteration for the global optimum path is shown in Fig. 4. Different color pattern distinguishes among the machine members for easy identification. The trajectory length attained after each test is presented in Table 5.

The optimum trajectory length is found as 33.4937 meter between (0.3, 0.3) meter and (11, 32) meter. After getting the global optimum path the material handling robot is programmed to chase that path. Lastly, that global optimum trajectory is designed in the environment, so that material handling robot will follow the trajectory. The experimental value of trajectory length is determined. The experimental trajectory length is found as 34.2193 meter. The experimental value is slightly higher than the simulation result because, in actual practice, adequate clearance is provided between machine member and the robot to avoid crash.

5 Conclusion

A crash less global optimum trajectory is obtained by implementing PSO soft-computing approach in static environment for the material handling robot. This robot overcomes the limitations like endless supply and retrieval of raw part, tools, and other necessary accessories at lesser time, and avoids starving condition in the processing stations. The global optimum trajectory length is determined as 33.4937 meter.

Table 5 Trajectory length generated by PSO

No. of tests	Trajectory length (meter)
01	37.6795
02	35.9096
03	38.5542
04	36.8607
05	37.7584
06	37.3612
07	37.1509
08	35.5049
09	35.3990
10	37.8454
11	38.1764
12	37.7891
13	34.2378
14	38.2267
15	36.9314
16	34.6927
17	36.4592
18	33.6698
19	36.3158
20	33.4937

The experimental trajectory length is found as 34.2193 meter. The simulation result shows the applicability, effectiveness, and feasibility of the PSO approach. Also the PSO approach fulfils all the objectives of this work, i.e., collision-free trajectory, smoothness of path and diminished length.

References

1. Tsuzuki, M.S.G., Martins, T.C., Takase, F.K.: Robot path planning using simulated annealing. In: 12th IFAC/IFIP/IFORS/IEEE/IMS Symposium, Information Control Problems in Manufacturing, vol. 39, pp. 175–180. Elsevier Ltd. (2006)
2. Chiddarwar, S.S., Babu, N.R.: Conflict free coordinated path planning for multiple robots using a dynamic path modification sequence. Robot. Auton. Syst. **59**, 508–518 (2011) (Elsevier B.V)
3. Husain, A.M., Sohail, S.M., Narwane, V.S.: Path planning of material handling robot using Ant Colony Optimization (ACO) technique. Int. J. Eng. Res. Appl. **2**, 1698–1701 (2012)
4. Fang, H.C., Ong, S.K., Nee, A.Y.C.: Interactive robot trajectory planning and simulation using augmented reality. Robot. Comput. Integr. Manuf. **28**, 227–237 (2012) (Elsevier Ltd.)
5. Li, X., Choi, B.J.: Design of obstacle avoidance system for mobile robot using fuzzy logic systems. Int. J. Smart Home **7**, 321–328 (2013) (Science and Engineering Research Support Society)

6. Shum, A., Morris, K., Khajepour, A.: Direction-dependent optimal path planning for autonomous vehicles. Robot. Auton. Syst. **70**, 202–214 (2015) (Elsevier B.V)
7. Jafri, S.M.H., Kala, R.: Path planning of a mobile robot in outdoor Terrain. Adv. Intell. Syst. Comput. **385**, 187–195. Springer International Publishing Switzerland (2016)
8. Emharraf, M., Saber, M., Rahmoun, M., Azizi, M.: Online local path planning for mobile robot navigate in unknown indoor environment. In: Mediterranean Conference on Information & Communication Technologies, vol. 2, pp. 69–76. Springer International Publishing Switzerland (2016)
9. Han, J., Seo, Y.: Mobile robot path planning with surrounding point set and path improvement. Appl. Soft Comput. **57**, 35–47 (2017) (Elsevier B.V)
10. Li, G., Chou, W.: Path planning for mobile robot using self-adaptive learning particle swarm optimization. Sci. China Inf. Sci. **61**, 1–18 (2017) (Science China Press and Springer-Verlag, Berlin Heidelberg)

An Approach to Compress English Posts from Social Media Texts

Dwijen Rudrapal, Manaswita Datta, Ankita Datta, Puja Das and Subhankar Das

Abstract Compression of sentences in Facebook posts and Twitter is one of the important tasks for automatic summarization of social media text. The task can be formally defined as transformation of sentence into precise form by preserving the original meaning of the sentence. In this paper, we propose an approach for compressing sentences from Facebook English posts by dropping those words who contribute very less importance to the overall meaning of sentences. We develop one parallel corpus of Facebook English posts and corresponding compressed sentences for our research task. We also report evaluation result of our approach through experiments on develop dataset.

Keywords Sentence compression · Social media text · Machine learning · Facebook posts · Natural language processing

1 Introduction

Social media has received much attention to the researcher over time. Social network sites have become the major source for automatic dissimilation of important and current information about events taking place all over the world. User gets real-time

D. Rudrapal (✉) · M. Datta · A. Datta · P. Das · S. Das
National Institute of Technology, Agartala 799046, Tripura, India
e-mail: dwijen.rudrapal@gmail.com

M. Datta
e-mail: manaswitadatta@gmail.com

A. Datta
e-mail: ankitadatta@gmail.com

P. Das
e-mail: pujadas@gmail.com

S. Das
e-mail: sdasnitacs@gmail.com

H. S. Behera et al. (eds.), *Computational Intelligence in Data Mining*, Advances in Intelligent Systems and Computing 990,
https://doi.org/10.1007/978-981-13-8676-3_8

news, information and public opinion about every subject matter instantly. In short, social media is the information powerhouse to all kind of users.

When needed, a user can search for any particular topic and retrieve the relevant information as desired. But interpreting the information is quite challenging as there is no way that the most significant information can be comprehended by the users from such a sheer volume of data. To make it easy for the users to find out the information they are interested in, an automatic summarization system is required that can generate a condensed content.

Although various approaches are proposed for summarization of social media text, those methods use features such as term frequency [18], query relevance [13] and assign a score to each sentence. Then extract the sentences which maximize the score to compose the intended summary. Those approaches don't shorten the sentence which leads to redundant words, phrases and holds frequent non-significant words. Our work is motivated to solve this shortening problem.

Sentence compression is a sub-task for text summarization. Besides automatic summarization of text, sentence compression is required for various other Natural Language Processing (NLP) processes like semantic role labeling, factoid questioning. The state-of-the-art sentence compression methods are only for standardized texts. Standardized texts are those found in newspaper articles, documents. But social media text differs from the former. The coarse nature of social media text creates certain difficulties for the existing methods [10]. Variation of writing style among users create a lot of challenges like misspelled words, non-standard abbreviation, typological errors, chat slang, etc. Detection of sentence boundary is difficult because of inconsistent punctuation and unwanted capitalization. Emoticons, letter repetition "hiiii", lack of grammaticality are other concerns for any syntactic analysis of such kind of text. Henceforth, the existing methods fail to perform well on social media text.

According to Knight and Marcu [15], given an input source sentence of words $x = x_1, x_2, \ldots x_n$ a compression is formed by dropping any subset of words. Good compression are those which use fewer words than the source sentence, retain the most important information from the source sentence and remain grammatically correct. In this paper, we propose an approach to obtain the compressed version of a sentence by deletion of non-significant words or phrases. To the best of our knowledge, this is the first work on sentence compression for social media text.

The rest of the details are divided into the following sections: In Sect. 2, we discuss the promising research relevant to sentence compression. Section 3, describes the dataset preparation method followed by Sect. 4, stating our proposed approach. The experiment setup and result, result analysis is detailed in Sects. 5 and 6, respectively. Section 7, concludes our research work and provides direction for future work.

2 Related Work

Several approaches for sentence compression of standardized text, both machine learning based and rule-based have been proposed to date, some of the promising research works are discussed below.

Sentence compression is one of the most experimented text-to-text generation methods. Most of the compression method proposed in various work [14, 22] detect data redundancy and apply reduction to generate summaries. Some works [4, 11] used compressed forms of texts to generate small screens subtitles. Knight and Marcu proposed model formulated the task as a word or phrase deletion problem. They proposed two models for sentence compression: a probabilistic noisy channel model and a decision-based model to generate abstractive summaries by dropping a subset of words. The deletion problem was further improvised to address global dependency in deleting local words [4, 11].

Some sophisticated models were introduced to improve the accuracy further. Nguyen et al. [19] proposed two sentence reduction algorithms. One, the template-based learning model which learns lexical translation rules and another, a Hidden Markov model which learns sequence of lexical rules. Both the model performed better than the earlier research in terms of grammaticality and importance measure. Filippova and Strube [11] used rules for syntactic and semantic modification of parse tree to generate the compressed form. Cohn and Lapata [5] gave a compression method that extracts tree transduction rules from aligned, parsed texts. Then using a max-margin learning algorithm learns weights on transformations. Later, they [6] presented tree-to-tree transducer that is capable of transforming an input parse tree to a compressed parse tree based on Synchronous Tree Substitution Grammar. Lexicalized Markov Grammar was used by Galley and Mckeown [12] for sentence compression. Clarke and Lapata [3] used Integer Programming approach and encoded various decision variables and constraints in the model to guarantee the grammatical intactness of the reduced sentence. The unsupervised model proposed by Cordeiro et al. [8] used Inductive Logic programming (ILP). Banerjee et al. [2] used ILP for the generation of abstractive summaries by sentence compression. Most recently, Lu et al. [16] present an approach based on a Re-read Mechanism and Bayesian combination model for sentences compression.

All of them are designed for traditional texts. Our work is focused on social media texts, more precisely, Facebook English posts.

3 Dataset Preparation

In this section, we describe the methodology adopted for collecting posts, selecting the required posts, tokenization of the corpus, preparing the parallel corpus and tagging the tokens.

3.1 Facebook English Post Collection

We have collected a corpus of 1884 English posts from Facebook through Facebook API[1] related to UN election 2017 (CNN Facebook page[2]).

After a manual inspection, few posts were removed, being code-mixed or being non-English, as we restrict to English posts only. From these remaining posts, the first sentence of each was extracted since our task is related to sentence compression. Soon it was discovered that a proportion of the posts were identical. Those sentences were removed to make the posts unique in nature. Some sentences were also dropped because of the following reasons:

- Sentences contain unwanted symbols or noise in an intermediate position.
- Sentences didn't convey any meaning.
- Sentences seemed like a single sentence due to the lack of proper end marker.

Finally, 1200 unique and consistent sentences were manually selected for further processing.

3.2 Annotation Process and Challenges

The collected data were annotated using a semi-automatic technique to speed up the manual tagging. Tokenization is difficult for social media texts because of the noisy and informal nature. Ritter's tagger[3] was employed to tokenize and tag the data in the first place. From this 1200 sentences, a total of 16445 tokens are formed i.e. approximately 13.70 tokens for each sentence. The tagger also provided the POS tag, the chunk, and the named entity for each token.

However, the accuracy of Ritter's tagger for our corpus is only 85.3%. To avoid erroneous tagging, manually inspection of the whole corpus was done to correct wrong tags.

The reason behind wrong tokenization and tagging was mainly due to the fact that:

1. Lack of white space between words. For example, "Because he legitimately hasnt passed 1piece of legislation....", "1piece" was treated as a single entity even though two entities "1", "piece" are there.
2. Use of multiple periods between words. For example, "Yeah liberals love a nationalist military state and a narcissistic leader with a bad haircut....oh wait!", "haircut....oh" was considered a single token.
3. Sometimes word was tokenized into two due improper use of punctuation. For example, "Leftists should like N Korea, Gov't runs everything", "Gov't" was divided into two tokens; viz. "Gov" and "t".

[1]https://developers.facebook.com/.

[2]https://www.facebook.com/cnn/.

[3]https://github.com/aritter/twitternlp.

Table 1 Examples of sentence compression

Sentence 01:	Hahahah Todd just contradicted his own argument
Compressed form:	Todd contradicted his own argument.
Sentence 02:	Yeah liberals love a nationalist military state and a narcissistic leader with a bad haircut....oh wait!
Compressed form:	Liberals love a military state and a narcissistic leader!

For preparing the parallel compressed corpus, two annotators are involved in our work. Annotators are native English speakers. Both the annotators have annotated the whole corpus as per our defined annotation guidelines. To prepare gold standard dataset for our experiment, disagreed compression is resolved through discussion among the annotators. Following instructions are provided to the annotators for annotation task. The instructions are:

1. All the posts are related to the US election 2017.
2. Read the sentence and compress it by preserving its original meaning.
3. Restricted to delete words or phrase from the source sentence.
4. Restricted to insert, substitute or reorder the words of the sentence.
5. Not allowed to recreate two sentences from a single sentence when sentences are connected through conjunctions.
6. Sentences that cannot be shortened, keep unaltered.

Table 1 shows two examples of annotated sentences.

4 Proposed Approach

This section discusses the proposed approach to train a machine learning algorithm for compressing English sentences. We formulate some rules to drop words or phrases and accordingly features are derived to train classifiers.

4.1 Rules Formation and Features Selection

Since the parallel corpus is formed by deletion of a subset of words from source sentence, we frame a set of rules to find the viable pattern by which deletion was carried out. The rules are:

- Delete adjective occurring to the left of a noun. For example, in table I, sentence 02, the term “nationalist” is removed.
- Delete adverb occurring to the left of a verb. For example, in table I, sentence 01, the term “just” removed from the sentence.
- Delete the determiner.
- Name entity can be reduced to a single word specifically, reduced to last word of the noun phrase (e.g. the United States of America can be reduced to America).

Considering these set of rules, we select various features based on the lexical and syntactic nature of the sentences to generate the model for automatic compression. The features are:

- The current word (W).
- The previous word (PW).
- Previous to previous word (PPW).
- Next word (NW).
- Next to next word (NNW).
- The chunk (P): While analyzing the parallel corpus, it is observed that either the whole chunk was removed or reduced to its corresponding head word to obtain the compressed sentence.
- POS tag of W, PW, PPW, NW, NNW, P. POS tags resolve the ambiguity that exists between words [9].
- Named entity (NE): Named entities are nouns and hence kept intact for most of the sentences. These tokens are mostly not removed for compression.
- The headword of the chunk (H): The headword in a chunk is that word which is essential to the core meaning of the chunk. For e.g. in case of a noun phrase, the headword is likely to be the subject. The headword of each chunk was identified by following Collins work [7].

The former two lexical features are inspired from the work by Jurafsky et al. [21] where the relationship existing between the current word and its surrounding words are examined.

4.2 Classifier Selection

Choosing the right classifier is crucial for the performance of the experiment. For this experiment, a classifier that can detect whether a given word should be deleted or retained is needed. Since we are dealing with sentences extracted from Facebook, classifiers that work firmly for sequential tagging of data is required. The four classifiers used are namely, Sequential Minimal Optimization (SMO), Naive Bayes (NB), Logistic Regression (LR) and Random Forest (RF).

A NB classifier is based on Bayes theorem and considers all the features to be independent. Empirical study shows that NB [17] is quite effective for sequential tagging of data. SMO is related to both Support Vector Machines (SVM) and optimization

Table 2 Performance analysis of approaches

Evaluation metrics	Approaches			
	SMO	NB	LR	RF
Weighted average F-measure	**0.789**	0.770	0.700	0.759

algorithms and shown comparable performance in previous relevant works [20]. LR is selected because it performs well for correlated features. RF classifier averages multiple decision trees based on random samples from the dataset and proves to be effective for sequential data tagging.

5 Experiment Setup and Result

We used Weka[4] for implementing the machine learning algorithms. On the dataset, tenfold cross validation is performed. Table 2 describes the evaluation metric obtained after applying the algorithms on the dataset. Often, accuracy is used to measure how correct the prediction of a classifier is. However since we have an imbalanced dataset, it will not be helpful. We, therefore use F-measure (harmonic mean of precision and recall) to evaluate the performance of our proposed model. It can be clearly observed that SMO gives the highest weighted average F-measure (0.789) while LR (0.700) performed least.

6 Result Analysis

Table 3 lists two compression examples resulted by experimented algorithms. Although in the most cases the grammatical structure is not retained, still the intended meaning can be conveyed easily. SMO generated compressed sentence is very close to the manually compressed one. While other algorithms output is deviated in terms of compactness as well as readability. Further analysis showed that for most of the sentences, the length of the compressed version and machine-generated compressed version didn't differ much.

7 Conclusion and Future Work

We present a machine learning approach to train social media text corpus for obtaining compressed sentences. We detail on collecting, annotating the corpus using a semi-automatic technique and incorporate various features to determine whether a given

[4]https://www.cs.waikato.ac.nz/ml/weka/.

Table 3 Compressed output by trained models

Sentence 03:	I personally think that it's disrespectful to the n korean people to go there.
Gold form:	It's disrespectful to the n korean to go there.
SMO output:	It's disrespectful to the n korean people to go there.
NB output:	That it's disrespectful to the n korean people to go there.
LR output:	Think that it's is to the n korean people to go there.
RF output:	Personally think it's disrespectful to the n korean people to go there.
Sentence 04:	Normally i wouldn't be too happy about the govt telling me where i can and i can't go.
Gold form:	I wouldn't be happy about govt telling me where i can and i can't go.
SMO output:	I wouldn't be too happy about the govt telling me where i can and i can't go.
NB output:	I wouldn't be too happy about the govt telling me i can and i can't go.
LR output:	I wouldn't be too happy about the telling i can and i can't go.
RF output:	Normally wouldn't be too happy about the govt telling where i can and i can't go.

word should be deleted or retained. Drawing from the result obtained after applying four different machine learning algorithm (SMO, NB, LR, RF), SMO performed best with an F-measure of 0.789.

Being the first attempt at social media text compression, we want to further modify the model to improve the grammaticality of the compressed sentence. We want to increase the size of the corpus substantially to decrease the number of unknown and non-standard words. Owing to the extensive use of non-standard words, we want to employ lexical normalization [1].

References

1. Baldwin, T., Li, Y.: An in-depth analysis of the effect of text normalization in social media. In: Proceedings of the 2015 Conference of the North American Chapter of the Association for Computational Linguistics: Human Language Technologies, pp. 420–429 (2015)
2. Banerjee, S., Mitra, P., Sugiyama, K.: Multi-document abstractive summarization using ilp based multi-sentence compression. In: IJCAI, pp. 1208–1214 (2015)
3. Clarke, J., Lapata, M.: Constraint-based sentence compression an integer programming approach. In: Proceedings of the COLING/ACL on Main conference poster sessions, pp. 144–151. Association for Computational Linguistics (2006)
4. Clarke, J., Lapata, M.: Global inference for sentence compression: An integer linear programming approach. J. Artif. Intell. Res. **31**, 399–429 (2008)
5. Cohn, T., Lapata, M.: Sentence compression beyond word deletion. In: Proceedings of the 22nd International Conference on Computational Linguistics-Volume 1, pp. 137–144. Association for Computational Linguistics (2008)
6. Cohn, T., Lapata, M.: An abstractive approach to sentence compression. ACM Trans. Intell. Syst. Technol. (TIST) **4**(3), 41 (2013)
7. Collins, M.: Head-driven statistical models for natural language parsing. Comput. Linguist. **29**(4), 589–637 (2003)

8. Cordeiro, J., Dias, G., Brazdil, P.: Unsupervised induction of sentence compression rules. In: Proceedings of the 2009 Workshop on Language Generation and Summarisation, pp. 15–22. Association for Computational Linguistics (2009)
9. Jurafsky, D., Martin, J.H.: Part-of-speech tagging. In: Speech and Language Processing, 3rd edn, pp. 142–167 (2017). draft(ch. 10)
10. Eisenstein, J.: What to do about bad language on the internet. In: Proceedings of the 2013 Conference of the North American Chapter of the Association for Computational Linguistics: Human language technologies, pp. 359–369 (2013)
11. Filippova, K., Strube, M.: Dependency tree based sentence compression. In: Proceedings of the Fifth International Natural Language Generation Conference, pp. 25–32. Association for Computational Linguistics (2008)
12. Galley, M., McKeown, K.: Lexicalized markov grammars for sentence compression. In: Human Language Technologies 2007: The Conference of the North American Chapter of the Association for Computational Linguistics; Proceedings of the Main Conference, pp. 180–187 (2007)
13. Gupta, S., Nenkova, A., Jurafsky, D.: Measuring importance and query relevance in topic-focused multi-document summarization. In: Proceedings of the 45th Annual Meeting of the ACL on Interactive Poster and Demonstration Sessions, pp. 193–196. Association for Computational Linguistics (2007)
14. Jing, H.: Sentence reduction for automatic text summarization. In: Proceedings of the Sixth Conference on Applied Natural Language Processing, pp. 310–315. Association for Computational Linguistics (2000)
15. Knight, K., Marcu, D.: Summarization beyond sentence extraction: A probabilistic approach to sentence compression. Artif. Intell. **139**(1), 91–107 (2002)
16. Lu, Z., Liu, W., Zhou, Y., Hu, X., Wang, B.: An effective approach of sentence compression based on re-read mechanism and bayesian combination model. In: Chinese National Conference on Social Media Processing, pp. 129–140. Springer, Berlin (2017)
17. Najibullah, A.: Indonesian text summarization based on naïve bayes method. In: Proceeding of the International Seminar and Conference on Global Issues, vol. 1 (2015)
18. Nenkova, A., Vanderwende, L., McKeown, K.: A compositional context sensitive multi-document summarizer: exploring the factors that influence summarization. In: Proceedings of the 29th Annual International ACM SIGIR Conference on Research and Development in Information Retrieval, pp. 573–580. ACM (2006)
19. Nguyen, M.L., Horiguchi, S., Shimazu, A., Ho, B.T.: Example-based sentence reduction using the hidden markov model. ACM Trans. Asian Lang. Inf. Process. (TALIP) **3**(2), 146–158 (2004)
20. Platt, J.: Sequential minimal optimization: A fast algorithm for training support vector machines (1998)
21. Tseng, H., Jurafsky, D., Manning, C.: Morphological features help pos tagging of unknown words across language varieties. In: Proceedings of the Fourth SIGHAN Workshop on Chinese Language Processing (2005)
22. Unno, Y., Ninomiya, T., Miyao, Y., Tsujii, J.: Trimming cfg parse trees for sentence compression using machine learning approaches. In: Proceedings of the COLING/ACL on Main Conference Poster Sessions, pp. 850–857. Association for Computational Linguistics (2006)

A Deep Learning Approach to Predict Football Match Result

Dwijen Rudrapal, Sasank Boro, Jatin Srivastava and Shyamu Singh

Abstract Predicting a match result is a very challenging task and has its own features. Automatic prediction of a football match result is extensively studied in last two decades and provided the probabilities of outcomes of a scheduled match. In this paper we proposed a deep neural network based model to automatically predict result of a football match. The model is trained on selective features and evaluated through experiment results. We compared our proposed approach with the performance of feature-based classical machine learning algorithms. We also reported the challenges and situations where proposed system could not predict the outcome of a match.

Keywords Predictive modelling · Match result · Machine learning · Multi-layer perceptron · Deep learning

1 Introduction

Football is the most played and most watched sport in the world. Footballs governing body, the International Federation of Association Football (FIFA), estimated that at the turn of the 21st century there were approximately 250 million football players and over 1.3 billion people interested in football.[1] Popularity of this game created huge number of people's association. Therefore, prediction of match result in advance is very attractive to the experts and researchers. But it is very difficult to guess the result of a football match by experts or past statistics. There are many factors who can influence the match outcome like, skills, player combination, key players forms, teamwork, home advantage and many others. Prediction becomes more challenging when match played with extra time or substitute players or injuries.

[1] https://www.britannica.com/sports/football-soccer.

D. Rudrapal (✉) · S. Boro · J. Srivastava · S. Singh
National Institute of Technology, Agartala 799046, Tripura, India
e-mail: dwijen.rudrapal@gmail.com

H. S. Behera et al. (eds.), *Computational Intelligence in Data Mining*, Advances in Intelligent Systems and Computing 990,
https://doi.org/10.1007/978-981-13-8676-3_9

Football sports related data is now publicly available and grown interest in developing intelligent system to forecast the outcomes of matches. In last two decades, researchers proposed several state-of-the-art methodology to predict outcome using past and available data. Most of the previous research devise the prediction problem as a classification problem [1, 13]. The classifier predicts the result with one class among win, loss or draw class [9].

In this paper, we focus on deep neural network for this problem. Deep neural network technique has shown to be effective in various classification problems in many domains [4] including in this domain [5, 7].

The main contribution of this paper are as follows:

- We propose Multi-Layer Perceptron based prediction model to predict football match result. We experimented our model with classical machine learning based models and report our evaluation result.
- We propose new set of features to train our predictive model by detail analysis of features performance and their role on match outcome and from head-to-head match result.

The remainder of this paper is organized as follows. A summary of previous promising work on football predictions is presented in Sect. 2. In Sect. 3, the proposed approach including the dataset, is presented followed by experiment result in Sect. 4. We discuss error analysis of our approach in Sect. 5 and finally, the conclusion is in Sect. 6.

2 Related Work

In this section, we have discussed promising research work related to predictive model for football match result. The research for predicting the results of football matches outcome as started as early 1977 by [11]. Proposed a model to predict the match outcome using matrix of goal scoring distribution. Purucker proposed [10] an ANN model to predict results in the National Football League (NFL) using hamming, adaptive resonance theory (ART), self-organizing map (SOM) and back-propagation (BP). The approach uses 5 features and experimented on data of 8 round matches in the competition. The work extended in the approach proposed by Kahn [5] achieve greater accuracy by using large dataset of 208 matches in the 2003 season of National Football League (NFL). The work includes features like home team and away team indicator. The approach [8] presents a framework for football match result prediction based on two components: rules-based component and Bayesian network component. The performance of the system depends on sufficient expert knowledge of past data. McCabe and Trevathan predict result in their approach [7] for 4 different kinds of sports. Proposed approach targeted Rugby League, Australian Rules football, Rugby Union and EPL. A multi-layer perceptron model trained using BP and conjugative-gradient algorithms for predicting task.

Table 1 Statistics of dataset

Season count	Match count	Team count	Player count
15	11400	42	462

The system in the work [3] proposed a prediction model to generate forecasts of football match outcome for EPL matches using objective and subjective information of teams like team strength, team form, psychological impact and fatigue. Ben Ulmer and Matthew Fernandez used features like game day data and current team performance in the proposed work [14]. Albina Yezus [15] used both static and dynamic features in proposed approach. Static features include forms of the players and teams, concentration and motivation, while dynamic features include goal difference, score difference and history.

Tax and Joustra proposed a prediction system [12] for Dutch football competition based on past 13 years data to predict the match results. Koopman and Lit developed a statistical model in their work [6] for predicting football match results. The work used Bayesian Networks and assumes a bivariate Poisson distribution with intensity coefficients that change randomly over time. Recently Bunker and Thabtah [2] investigated machine learning approaches for sport result prediction and focused on the neural network for prediction problem. The work identifics the learning methodologies, data sources, evaluation and challenges proposed a prediction framework using machine learning approach as a learning strategy.

3 Proposed Methods for Football Match Prediction

3.1 Data Collection

English Premier League (EPL) is the most watched football league in the world with almost 4.7 billion viewers. Its competitiveness as well as its outcomes are random in nature. For example, in the season of 2010–11, the distribution of wins, losses and draws was 35.5%, 35.5% and 29% respectively. For our current research work, we concentrated on EPL matches. We collected different data like team statistics and squad, player and result from various online sources.[2,3,4,5] The dataset includes information of 11,400 matches from the season 2000–01 to the season 2015–2016. We analyzed what parameters or features are influential or highly influential in determining the outcome of a match. Detail statistics of the dataset is reported in Table 1.

[2]www.football-data.co.uk/.

[3]http://fifaindex.com/.

[4]https://www.skysports.com/.

[5]https://kaggle.com/.

Table 2 Features distribution details

Features class	Features count		
	Home team	Away team	Total
Team	9	9	18
Player	7	7	14
Head-to-head match	4	4	8
Total	20	20	40

3.2 Feature Selection

From literature survey of previous promising research work, we have drawn 40 features from our prepared dataset. Features are drawn from different class of statistics like team, player and head-to-head match statistics. Each match is classified into two classes, home match and away match. Through our analysis, we found that this classification plays crucial role to predict match result due to the advantages of venue, crowd support and known environment to the home team. So, all the features are calculated for home as well as away matches. Distribution of features are reported in Table 2. Details explanation of the features are discussed in following subsections.

Team Related Features

1. **Attack Point**: It describes the attacking ability of a team.
2. **Mid Field Point**: It describes the mid-field ability of a team.
3. **Defence Point**: It describes the defensive ability of a team.
4. **Team Rating**: Calculated considering the attack, mid-field and defence point.
5. **EPL Times Point**: It is the times that a team played EPL during the 18 seasons. If a team played all the 18 seasons they are given 4 points.
6. **EPL Streak Point**: It describes how consistent a team is playing in EPL.
7. **EPL Performance Point**: It describes the best performance of a team in EPL during these 18 seasons. A champion is given maximum of 4 points.
8. **EPL Point**: It is the average of EPL times Point, EPL Streak Point and EPL Performance Point.
9. **Standings Point**: It is the standing point of each team in whichever league they are. For example, point can vary from 1–4 according to their position in the points table.

Player Related Features

1. **Palyer Rating**: It describes the overall rating of a player considering parameters like aggression, tackle, shooting accuracy, dribbling and many more. This gives an overview of the players performance. 94 was the highest overall rating of a player in our dataset.

2. **Player Potential**: Player potential describes the maximum ability of a player. The player rating can have a maximum value equal to player potential when the player is at his best form.
3. **Market Value**: Market Value gives an idea of the form of player, how dependable he is and how many seasons still left in his career. The maximum value of this feature is 123 million in our corpus.
4. **Goal Keeper Point**: It describes the ability of the Goal keeper of each team.
5. **Defence Point**: It includes total of 11 defence parameters of a player like aggression, marking, positioning, sliding tackle, standing tackle. The average value of these parameters is taken and calculated with respect to 4.
6. **Mid-Field Point**: It includes total of 13 mid-field parameters of a player like acceleration, agility, ball control, crossing, dribbling.
7. **Striking Point**: It gives an idea of the striking or scoring ability of each player. Includes total of 19 striking parameters of a player like acceleration, curve, finishing, long shots, power shot.

Head-to-Head Match Related Features

1. **Team Points**: Team point is the average of the points achieved by a home or away team. Every win conceded 3 points, draw and lost conceded 1 and 0 points respectively.
2. **Form Points**: It describes the form of a team during head -to-head matches. Last 5 home and 5 away matches prior to the match is considered.
3. **Goal Difference**: It is the difference between goals scored and goals conceded divided by the total number of matches.
4. **Form Points Difference**: It is the difference between the home form point and away form point.

3.3 Classifier Selection

In this paper, we have proposed Multi-Layer Perceptron (MLP) for football match result prediction. In earlier similar research work [5, 10] MLP has shown comparable performance. We have also experimented with popular statistical machine learning algorithms like Support vector machine (SVM), Gaussian Naive Bayes and Random Forest to evaluate our approach.

4 Experiment and Result

Proposed MLP classifier having 10 hidden states trained on our dataset. We used holdout method to split our dataset into training set and testing set. So, 70% of the dataset is tagged as training dataset and rest 30% tagged as test dataset. In Table 3,

Table 3 Performance evaluation of proposed approach

Parameter	Accuracy	Sensitivity	Specificity	F1-score
Score	73.57	71.02	75.48	71.45

Table 4 Analysis of proposed approach

Algorithm	Accuracy (in %)	F1-score (in %)
MLP	73.57	71.45
SVM	58.77	50.07
Gaussian Naive Bayes	65.84	64.26
Random forrest	72.92	66.07

we report the accuracy as 73.57% on the test dataset. We also report sensitivity, specificity, precision, recall and F1-score of our proposed approach.

As sensitivity and specificity of our model does not differ so much, therefore we can conclude that our model has a decent accuracy. We have compared the accuracy of our model with promising statistical machine learning algorithms in this domain research like SVM, Gaussian Naive Bayes and Random Forest.

5 Error Analysis

In this section, we make an in-depth analysis of our model and report the situations where the model failed to predict match outcome (Table 4).

1. In our current research work, we are considering the features of only the starting 11 players. But sometimes substitute of a player also determines very important role in the match. Sometimes, playing 11 may also change in another match. So, these challenges can be resolved by considering all the players playing 11 and substitutes.
2. There are many hidden features like crowd support, unfairness of a referee, unpredictable nature of the game etc. In our current work we could not include those features. Inclusion of those features will improve our model.
3. As the features which we have considered are dynamic in nature, the value changes slowly and gradually with the passage of seasons. So, performance of our approach is mostly dependant on recent past matches.

6 Conclusion

In this work we have achieved a decent accuracy of predicting the outcome of a football match. Our model works very well considering recent past match statistics. We made detail analysis of features and drawn effective features to train our model. We have also shown that match wise features like home and away match features are effective to the model.

Our future scope of the work is to study the importance of the features like team's reserve player strength, coach rating and inclusion of those to make our model more robust.

References

1. Abdelhamid, N., Ayesh, A., Thabtah, F., Ahmadi, S., Hadi, W.: Mac: a multiclass associative classification algorithm. J. Inf. Knowl. Manag. **11**(02), 1250011 (2012)
2. Bunker, R.P., Thabtah, F.: A machine learning framework for sport result prediction. Appl. Comput. Inform. (2017)
3. Constantinou, A.C., Fenton, N.E., Neil, M.: PI-football: a Bayesian network model for forecasting association football match outcomes. Knowl.-Based Syst. **36**, 322–339 (2012)
4. Davoodi, E., Khanteymoori, A.R.: Horse racing prediction using artificial neural networks. Recent. Adv. Neural Netw. Fuzzy Syst. Evol. Comput. **2010**, 155–160 (2010)
5. Kahn, J.: Neural network prediction of NFL football games, pp. 9–15 (2003)
6. Koopman, S.J., Lit, R.: A dynamic bivariate poisson model for analysing and forecasting match results in the english premier league. J. R. Stat. Soc.: Ser. (Stat. Soc.) **178**(1), 167–186 (2015)
7. McCabe, A., Trevathan, J.: Artificial intelligence in sports prediction. In: 2008 Fifth International Conference on Information Technology: New Generations, ITNG 2008, pp. 1194–1197. IEEE (2008)
8. Min, B., Kim, J., Choe, C., Eom, H., McKay, R.B.: A compound framework for sports results prediction: a football case study. Knowl.-Based Syst. **21**(7), 551–562 (2008)
9. Prasetio, D., et al.: Predicting football match results with logistic regression. In: 2016 International Conference on Advanced Informatics: Concepts, Theory And Application (ICAICTA), pp. 1–5. IEEE (2016)
10. Purucker, M.C:: Neural network quarterbacking. IEEE Potentials **15**(3), 9–15 (1996)
11. Stefani, R.T.: Football and basketball predictions using least squares. IEEE Trans. Syst. Man Cybern. **7**, 117–121 (1977)
12. Tax, N., Joustra, Y.: Predicting the dutch football competition using public data: a machine learning approach. Trans. Knowl. Data Eng. **10**(10), 1–13 (2015)
13. Thabtah, F., Hammoud, S., Abdel-Jaber, H.: Parallel associative classification data mining frameworks based mapreduce. Parallel Process. Lett. **25**(02), 1550002 (2015)
14. Ulmer, B., Fernandez, M., Peterson, M.: Predicting Soccer Match Results in the English Premier League. Ph.D. thesis, Doctoral dissertation, Ph. D. dissertation, Stanford (2013)
15. Yezus, A.: Predicting outcome of soccer matches using machine learning. Saint-Petersburg University (2014)

Prediction of Breast Cancer Recurrence: A Machine Learning Approach

Kashish Goyal, Preeti Aggarwal and Mukesh Kumar

Abstract Breast Cancer is a major health problem and is one of the significant causes of death among women. Recurrence occurs when cancer returns after few years of treatment. To aid in the medical treatment and for clinical management, the early cancer diagnosis and prognosis have become the necessity in breast cancer recurrence. As the medical data is increasing with the advancement in medical technology, data mining facilitates to manage the data and provide useful medical progression and treatment of cancerous conditions. Various machine learning techniques can be used to support the doctors in effective and accurate decision making. In this paper, various classifiers have been tested for the prediction of type of breast cancer recurrence and the results show that neural networks outperform others. These findings will help the physicians in identifying the best features which can lead to breast recurrence.

Keywords Breast cancer · Recurrence-events · Non-recurrence-events · K-Means clustering · Classification models

1 Introduction

Breast Cancer is one of the deadly diseases and is the second major cause of death among women today. It is a malignant tumor that starts as non-cancerous but as the cells divide uncontrollably in the breast or nearby areas it becomes carcinogenic. As per Indian Council Medical Research (ICMR), breast cancer rate is expected to rise

K. Goyal (✉) · P. Aggarwal · M. Kumar
Department of Computer Science & Engineering, University Institute of Engineering and Technology, Panjab University, Chandigarh 160023, India
e-mail: kashishgoyal31@gmail.com

P. Aggarwal
e-mail: pree_agg@pu.ac.in

M. Kumar
e-mail: mukesh_rai9@pu.ac.in

H. S. Behera et al. (eds.), *Computational Intelligence in Data Mining*, Advances in Intelligent Systems and Computing 990,
https://doi.org/10.1007/978-981-13-8676-3_10

to 1.9 lakh by 2020 [1]. In some cases, breast cancer may return after few years of treatment. Nearly 40% of cases of all patients with breast cancer suffer recurrence [2]. The likelihood of recurrence is never zero but the risk is high in first 2–3 years. There are some symptoms which can be considered and taken care of if the recurrence occurs [3]. Breast cancer recurrence is of three types:

1. Local Recurrence (LR): In this, the cancer redevelops at the same part of the breast where it was diagnosed earlier.
2. Regional Recurrence (RR): The areas near the breast such as in lymph nodes, armpits or collar bones are affected.
3. Distant Metastasis (DM): When other parts of the body such as lungs, bones or brain, etc. and even sometimes the opposite breast are affected by cancer, it is called Distant Recurrence.

Normally during diagnosis, local recurrence and regional recurrence are interpreted as Locoregional recurrence (LLR) as they both show similar symptoms and a different diagnosis is considered for Metastatic recurrence [4, 5]. Using ML tools key features can be extracted from the complex datasets. Various data mining algorithms such as Support Vector Machine (SVM), Decision Tree (DT), Naive Bayes (NB) and Neural Networks which includes Generalized Regression Neural Network (GRNN) and Feed Forward Back Propagation Network (FFBPN) can be used for the prediction of type of breast cancer recurrence.

The paper is organized in the following manner. Section 2 presents the implementation part i.e. the detailed analysis of the dataset including the data cleaning preprocessing. Also it includes the details of the machine learning classifiers. The results and the overall comparison of the classifiers are shown in Sects. 3 and 4 include conclusion and future scope followed by references.

2 Methods and Materials

This section elaborately represents the methodology proposed by the work. The first section contains about the dataset followed by the pre-processing steps and machine learning used for the work and their architectures.

2.1 Dataset

The breast cancer recurrence dataset has been taken from UCI Machine Learning Repository available online [6]. It is provided by the University Medical Centre, Institute of Oncology, Ljubljana, Yugoslavia. It consists of 286 instances and 10 attributes (explained in 0, 1) which includes a Class attribute that decides the outcome of the breast cancer being recurrence and non-recurrence. The italicized terms in Table 1 are the standard terms used in UCI Machine Learning Repository.

Table 1 Breast cancer dataset

Attributes	Details	Values
Age	It determines the age when the primary tumor was detected	10–19, 20–29, 30–39, 40–49, 50–59, 60–69, 70–79, 80–89, 90–99
Menopause	The age when the menstruation cycle stops in women. Here the menopause status of the patient at the time of diagnosis is considered	lt40, ge40, premeno
tumor-size	It describes the size of the lump that is formed. The tumor size is measured in millimeter (mm)	0–4, 5–9, 10–14, 15–19, 20–24, 25–29, 30–34, 35–39, 40–44, 45–49, 50–54, 55–59
inv-nodes	It tells the number of axillary nodes that carry symptoms of breast cancer when the histological examination is done	0–2, 3–5, 6–8, 9–11, 12–14, 15–17, 18–20, 21–23, 24–26, 27–29, 30–32, 33–35, 36–39
node-caps	It tells if the tumor has diffused into the node capsule or not	yes, no
deg-malig	Range 1–3 the histological grade of the tumor i.e. the resemblance of tumor cells with normal cells	1, 2, 3
Breast	Breast cancer may occur in either breast	left, right
breast-quad	Breast can be divided into four quadrants considering nipple as the central point	left-up, left-low, right-up, right-low, central
Irradiation	Radiation therapy is a treatment to destroy cancer cells using high-energy x-rays	yes, no
Class	Output class depending upon reappearing symptoms of breast cancer in the patients after treatment	no-recurrence-events, recurrence-events

2.2 *Preprocessing of Data*

The downloaded data from UCI Machine Learning Repository was not in the required format which may lead to low classification accuracy. It had missing data and few duplicate rows. Also the dataset was not balanced. Hence following preprocessing techniques were employed to make the dataset compatible for classification. Table 2 presents the number of instances in the data set.

Table 2 No. of instances after preprocessing

Total no. of instances (Actual): 286(R = 85, NR = 201)	Total no. of instances (After Removing Duplicity): 272(R = 81, NR = 191)	
Total no. of instances to be processed = 272(R = 81, NR = 191)		
SMOTE(1)	Total no. of instances = 353	
k = 5	R	NR
N = 100	162	191
	Total no. of samples after removing duplicate rows = 327	
	R	NR
	136	191
SMOTE(2)	Total no. of instances = 463	
k = 5	R	NR
N = 100	272	191
	Total no. of samples after removing duplicate rows = 356	
	R	NR
	165	191
SMOTE(3)	Total no. of instances = 521	
k = 5	R	NR
N = 100	330	191
	Total no. of samples after removing duplicate rows = 369	
	R	NR
	178	191
Missing data elimination	Total no. of instances = 360	
	R	NR
	174	186

2.2.1 Synthetic Minority Oversampling Technique (SMOTE) and Removal of Duplicate Rows

SMOTE is an oversampling technique and is generally used when the data is highly imbalanced. It synthetically generates the new instances of minority class rather than traditional methods which simply replicates the minority class or eliminates the instances from majority class [7]. SMOTE uses interpolation method to create the instances of minor class and thus avoids over fitting. SMOTE considers two parameters:

- nearest neighbor (k),
- Percentage of the synthetic rows to be generated from the minor class (N).

Since the dataset is not balanced, SMOTE is applied multiple times to get the right proportion of recurrence and non -recurrence samples. On using this technique the duplicate instances are generated, so they are eliminated at each step to get the balanced data. Table 2 demonstrates the results of applying SMOTE and remaining dataset after removing duplicate instances.

2.2.2 Missing Data Elimination

The dataset now consists of 369 samples, but 9 rows have missing data, hence eliminated. The final no. of instances in the dataset along with their class division is presented in Table 2.

2.2.3 Data Normalization

A new range of data is formed from an existing range. It is generally done in data mining to get the better results for prediction and forecasting purposes [8]. For normalization, the values of the attributes like *age, tumor-size* and *inv-nodes* are taken as the values of lower limit and a fixed numeric value is chosen for the other attributes. Hence the final dataset after preprocessing is nominal in nature. Subsequently, it is converted into numeric form to be used for further processing.

2.2.4 K-Means Clustering

In this n objects are divided into k clusters based on certain attributes such that k < n. The main motive is to define k centroids to each of these clusters, and to fill these clusters with the nearest items to them. [9]. As discussed before recurrence breast cancer can be broadly classified as locoregional and distant recurrence so to differentiate between the two, the instances of Recurrence Breast Cancer are divided into two parts using K-Means Clustering [10, 11]. This clustering works well on numeric form of data. The pseudo code for K-Means is represented in Table 3.

Now the entire dataset is divided into three classes as Class 1 and Class 2 belong to Recurrence instances and Class 3 belongs to Non-Recurrence instances. The division of instances is described in Table 4.

Since there was lack of medical assistance so the division of the classes could not be justified, it can be probably said that Class 1 belongs to LocoRegional recurrence [12] and Class 2 belongs to Distant Metastasis [13, 14] and class 3 is kept unchanged, i.e., the Non-Recurrence Class.

Table 3 Pseudo code for K-Means

```
Begin
k=2; n=174;
for i=1 to 500,
        assign each of the n (no. of recurrence events)
      samples according to the nearest k
        // cluster assignment step
        for every value of k calculate the mean of the
      points in that cluster and assign the centroid to
      the mean
        // move centroid step
 Until no change in centroid
  return k1, k2 newly created clusters
End
```

Table 4 No. of instances after K-means clustering

Total No. of instances	Class 1 (Probably LocoRegional Recurrence)	Class 2 (Probably Distant Metastasis)	Class 3 Non-recurrence Class
360	71	103	186

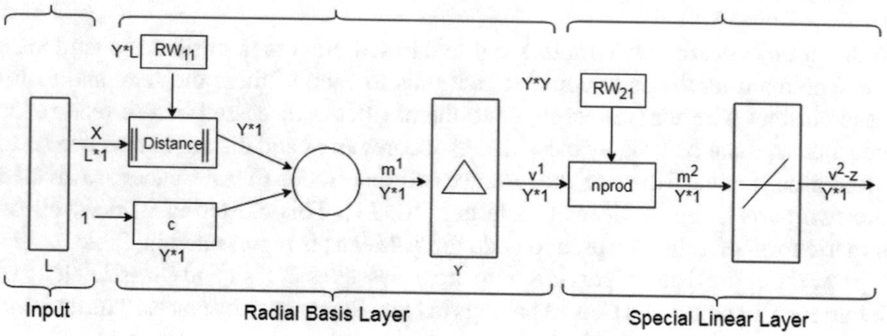

Fig. 1 Structure of GRNN

2.3 *Machine Learning Classifiers*

2.3.1 Generalized Regression Neural Network (GRNN)

GRNN is the variant of radial basis neural network which is used for function approximation. Kernel regression networks are used in this model. The main idea of this algorithm is to map the approximation function between the target and the input vector with the minimum error [15]. Figure 1 shows the basic architecture of GRNN.

It consists of two layers: radial basis layer and special linear layer. The no. of hidden neurons in radial basis layer is equal to the no. of inputs. Each neuron is given a net input which is the product of its bias and the weighted input [16]. Next is the linear layer which is connected to the radial basis layer. In this case, GRNN estimates the output (n) value based on the training inputs (m), i.e., the network calculates the probability density function of m and n. The output is determined by (1)

$$\hat{f} = \frac{\sum_{i=1}^{n} n^i exp\left(\frac{-Z_i^2}{2\sigma^2}\right)}{\sum_{i=1}^{n} exp\left(\frac{-Z_i^2}{2\sigma^2}\right)} \tag{1}$$

$$Z = \left(m - m^i\right)^T \left(m - m^i\right) \tag{2}$$

where $\hat{f}$ is the probability estimator,σ is the smoothing factor and Z is the Mean Squared Distance (refer (2)) between the input vector and training vector. σ is an important parameter of GRNN [17]. It is considered as the size of region of the neuron. Thus multiple values of σ must be tested to get the best smoothening factor.

2.3.2 Feed Forward Back Propagation Neural Network (FFBPN)

FFBPN is one of the algorithms in Artificial Neural Network (ANN) used very commonly and has large number of applications in engineering. In this network feed forward algorithm is used for neural processing and recalling patterns whereas the back propagation method is used to train the neural network [18, 19].

Each node in this is activated by the previous node and the weighted sum of the input is calculated as given in (3).

$$S_j = \sum_{i=1}^{n} w_{ji} y_i \tag{3}$$

where

n *no. of inputs*
w *weight of connection between node j and i*
y *input from node i*

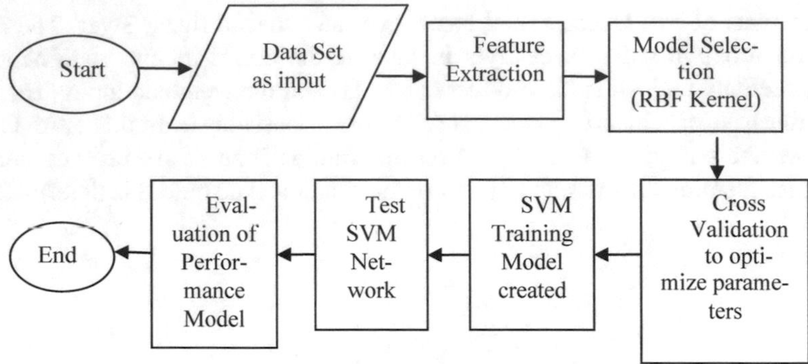

Fig. 2 Flowchart for SVM

Output of the node is calculated by applying the transfer function (T) to the weighted value S and is given by (4).

$$T_j = f(S_j) \tag{4}$$

2.3.3 Support Vector Machine (SVM)

SVM is a supervised learning technique which is used for non-linear complex functions. It can be used for both classification and regression purposes.. New input vectors are mapped into the same space and the prediction is done regarding their class and on which side of the hyper plane they exist [20]. There are Kernels used in SVM which check the similarity between the instances. The Kernel used here is radial basis function: exp (−gamma * |u – v|^2).

The resultant classifier formed is now generalized enough to be used for classification of new samples. Figure 2 represents the flow chart of the SVM.

2.3.4 Decision Tree (DT)

Decision Tree builds the models for classification and regression in the form of tree structure. Classification of the instance is done by starting at the root node of the tree, testing the attribute specified by this node, then going down the tree branch corresponding to the value of the attribute [21]. A new sample can be classified while traversing down the tree.

2.3.5 Naive Bayes (NB)

It is based on Bayes Theorem. In this, we assume that predictors are independent, i.e., knowing the value of one attribute does not tell the value of other attributes. Bayes' Theorem finds the probability of an event occurring given the probability of another event that has already occurred. It is stated mathematically as given in (5).

$$P(k/l) = \frac{P(l/k)P(k)}{P(l)} \quad (5)$$

where

$P(k/l)$ *Probability of the target class when predictor attribute is given.*
$P(k)$ *Prior probability of the class.*
$P(l/k)$ *Likelihood which is the probability of predictor when the class is given.*
$P(l)$ *Prior probability of the predictor.*

The outcome of the prediction is the class with highest posterior probability.

3 Results

The accuracy of recurrence classification on breast cancer data is calculated. The neural network classifiers like GRNN and FFBPN are used for classification in MATLAB on the numeric form of data with training and testing data ratio as 70% and 30%, respectively. The data is randomly selected for training and testing. The value of $\sigma = 1$ in case of GRNN. The output value calculated is rounded off to compare with the target values to calculate the accuracy of correctly classified instances. The accuracy of other classifiers such as SVM, DT and NB are calculated on the nominal form of data with 80% data split. The accuracy results of the classifiers are presented in Table 5. The highest accuracy of 85.18% is achieved by FFBPN.

Regression Plots of GRNN and FFBPN in Training and Testing phase are shown in Fig. 3. The data points are represented using small circles and the dotted line in the graph demonstrates the line at which the target value matches the output value, i.e.,

Table 5 Classification Accuracy of ML Classifiers

Classifiers	Correctly classified instances from testing data	Classification accuracy (%)
GRNN	90/108 (70% data split)	83.33
FFBPN	92/108 (70% data split)	85.18
SVM	56/72 (80% data split)	77.77
DT	51/72 (80% data split)	70.83
Naive Bayes	52/72 (80% data split)	72.22

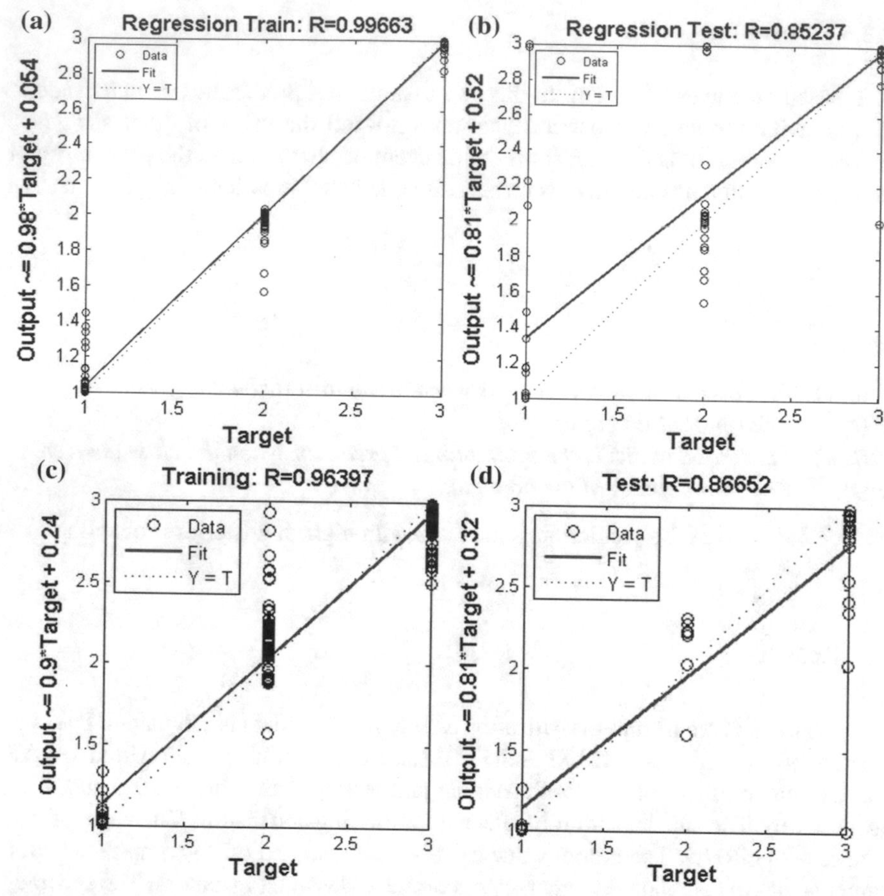

Fig. 3 Regression plot for training and testing for GRNN (**a**), (**b**) and FFBPN (**c**), (**d**)

it is the ideal line for the perfect fit. If the data points lie along the line of 45% then it can be said that data fits appropriately. At this line, the target values exactly match the output values. The blue line is the actual result of the data fit. The training plots for GRNN and FFBPN shows quite good results with values of R = 0.99 and 0.96 respectively. At the value of R = 1, the output line completely overlaps the fitting line. In the case of testing plot, the value of R = 0.85 for GRNN and R = 0.86 for FFBPN which are reasonable fine.

The classification accuracy of FFBPN (best results) is analyzed on the numeric form of data in MATLAB. The data is divided randomly for training (70%) and testing (30%). The parameters are chosen as follow:

- Training function: TRAINLM
- Number of layers: 2
- Number of neurons: 10

Table 7 Accuracy measures of different classifiers

Classifier	Sensitivity	Specificity	Precision	Recall	Class
SVM	0.941	0.927	0.800	0.941	Class1
	0.421	1	1.000	0.421	Class2
	0.889	0.667	0.727	0.889	Class3
Decision tree	1.000	0.891	0.739	1.000	Class1
	0.579	0.868	0.611	0.579	Class2
	0.639	0.778	0.742	0.639	Class3
Naive bayes	0.941	0.909	0.762	0.941	Class1
	0.632	0.868	0.632	0.632	Class2
	0.667	0.778	0.750	0.667	Class3
GRNN	0.782	0.945	0.782	0.782	Class1
	0.782	0.934	0.750	0.782	Class2
	0.871	0.868	0.885	0.871	Class3
FFBPN	0.833	0.968	0.833	0.833	Class1
	0.821	0.909	0.741	0.821	Class2
	0.871	0.902	0.915	0.871	Class3

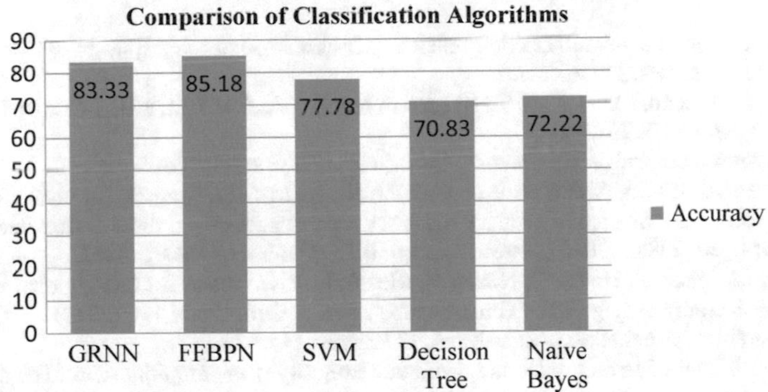

Fig. 4 Overall accuracy results of machine learning classifiers

- Transfer function: TANSIG.

The output values calculated are rounded off to compare with the target values to calculate the accuracy of correctly classified instances in training and testing phase.

The accuracy measures such as Sensitivity, Specificity, Precision and Recall [22] for all the classifiers are presented in Table 7.

Overall accuracy measure when compared among the machine learning classifiers is represented in graphical form in Fig. 4.

The results of recurrence classification are performed over the UCI Machine Learning Repository. It is seen from the table and the graph that best results are obtained for FFBPN. This justifies and validates our classification of data set.

4 Conclusion and Future Scope

In this paper, the type of breast cancer recurrence (only two classes) is predicted along with support to non-recurrence class. Popular classification algorithms have been used here and it can be concluded that neural network classifiers have performed better than other machine learning classifiers. In future, the accuracy can be increased by adding more features or by increasing the instances of the dataset. Also, the combination of existing classification techniques can be used to enhance the efficiency. Besides this, by verifying the output with medical professionals can increase the chances to identify the type of breast cancer recurrence adequately.

References

1. https://timesofindia.indiatimes.com/india/Cancer-cases-in-India-likely-to-soar-25-by-2020-ICMR/articleshow/52334632.cms
2. Gerber, B., Freund, M., Reimer, T.: Recurrent breast cancer **107**(6), 85–91 (2010). https://doi.org/10.3238/arztebl.2010.0085
3. https://www.medicinenet.com/breast_cancer_recurrence/article.htm
4. Witteveen, A., Kwast, A. B.G., Sonke, G.S., IJzerman, M.J., Siesling, S.: Survival after locoregional recurrence or second primary breast cancer: impact of the disease-free interval. Plos One **10**(4), e0120832 (2015). https://doi.org/10.1371/journal.pone.0120832
5. Hastings J, Iganej S, Huang C, Huang R, Slezak J.: Risk factors for locoregional recurrence after mastectomy in stage T1 N0 breast cancer. Am. J. Clin. Oncol 1–6 (2013)
6. https://archive.ics.uci.edu/ml/datasets/breast+cancer
7. Qazi, N.: Effect of feature selection, synthetic minority over-sampling (SMOTE) and under-sampling on class imbalance classification (2012). https://doi.org/10.1109/UKSim.116
8. Patro, S.G.K., Sahu, K.K.: Normalization: a preprocessing stage. Iarjset 20–22 (2015). https://doi.org/10.17148/IARJSET.2015.2305
9. Sharmila, S., Kumar, M.: An optimized farthest first clustering algorithm. In: Nirma University International Conference on Engineering, NUiCONE 2013, pp. 1–5 (2013). https://doi.org/10.1109/NUiCONE.2013.6780070
10. Sharma, R., Rani, A.: K-means clustering in spatial data mining using weka interface. In: International Conference on Advances in Communication and Computing Technologies (ICACCT), pp. 26–30 (2012)
11. Belciug, S., Gorunescu, F., Salem, A.B., Gorunescu, M.: Clustering-based approach for detecting breast cancer recurrence. In: 10th International Conference on Intelligent Systems Design and Applications (ISDA), pp. 533–538 (2010). https://doi.org/10.1109/ISDA.2010.5687211
12. Kwast, A., Groothuis-Oudshoorn, K., Grandjean, I., Ho, V., Voogd, A., Menke-Pluymers, M., et al.: Histological type is not an independent prognostic factor for the risk pattern of breast cancer recurrences. Breast Cancer Res. **135**, 271–80 (2012). https://doi.org/10.1007/s10549-012-2160-z, PMID: 22810087

13. Yadav, B., Sharma, S., Patel, F., Ghoshal, S., Kapoor, R.: Second primary in the contralateral breast after treatment of breast cancer. Radiother. Oncol. **86**, 171–6. PMID: 17961777 (2008)
14. Vichapat, V., Garmo, H., Holmberg, L., Fentiman, I., Tutt, A., Gillett, C.: et al. Prognosis of metachronous contralateral breast cancer: importance of stage, age and interval time between the two diagnoses. Breast Cancer Res. Treat. **130**, 609–18 (2011). https://doi.org/10.1007/s10549-011-1618-8, PMID: 21671018
15. Kumar, G., Malik, H.: Generalized regression neural network based wind speed prediction model for western region of India. Procedia Comput. Sci. **93**(September), 26–32 (2016). https://doi.org/10.1016/j.procs.07.177
16. Sun, G., Hoff, S., Zelle, B., Nelson, M.: Development and comparison of backpropagation and generalized regression neural network models to predict diurnal and seasonal gas and PM 10 concentrations and emissions from swine buildings, **0300**(08) (2008)
17. Cigizoglu, H.K.: Generalized regression neural network in monthly flow forecasting. Civil Eng. Environ. Syst. **22**(2), 71–84 (2005). https://doi.org/10.1080/10286600500126256
18. Cross, A.J., Rohrer, G.A., Brown-Brandl, T.M., Cassady, J.P., Keel, B.N.: Feed-forward and generalised regression neural networks in modelling feeding behaviour of pigs in the grow-finish phase. Biosyst. Eng. 1–10 (2018). https://doi.org/10.1016/j.biosystemseng.2018.02.005
19. Manickam, R.: Back propagation neural network for prediction of some shell moulding parameters. Period. Polytech. Mech. Eng. **60**(4), 203–208 (2016). https://doi.org/10.3311/PPme.8684
20. https://machinelearningmastery.com/support-vector-machines-for-machine-learning/
21. Abreu, P.H., Santos, M.S., Abreu, M.H., Andrade, B., Silva, D.C.: Predicting breast cancer recurrence using machine learning techniques: a systematic review. ACM Comput. Surv. (CSUR) **49**(3), 52 (2016)
22. https://uberpython.wordpress.com/2012/01/01/precision-recall-sensitivity-and-specificity/

Rating Prediction of Tourist Destinations Based on Supervised Machine Learning Algorithms

Anupam Jamatia, Udayan Baidya, Sonalika Paul, Swaraj DebBarma and Subham Dey

Abstract The paper highlights the process of predicting how popular a particular tourist destination would be for a given set of features in an English Wikipedia corpus based on different places around the world. Intelligent predictions about the possible popularity of a tourist location will be very helpful for personal and commercial purposes. To predict the demand for the site, rating score on a range of 1–5 is a proper measure of the popularity of a particular location which is quantifiable and can use in mathematical algorithms for appropriate prediction. We compare the performance of different machine learning algorithms such as Decision Tree Regression, Linear Regression, Random Forest and Support Vector Machine and maximum accuracy (74.58%) obtained in both the case of Random Forest and Support Vector Machine.

Keywords Prediction · Machine learning · Decision tree regression · Linear regression · Random forest · Support vector machine

1 Introduction

Tourism has become one of the most popular industry in the world today. To get a share of this increasing tourism market, government and private agencies alike are always keen on investing in the right places that have the possibility of becoming

A. Jamatia (✉) · U. Baidya · S. Paul · S. DebBarma · S. Dey
Department of Computer Science and Engineering,
National Institute of Technology Agartala, Jirania 799046, Tripura, India
e-mail: anupamjamatia@gmail.com

U. Baidya
e-mail: udayanbaidya@gmail.com

S. Paul
e-mail: sonalikapaul@gmail.com

S. DebBarma
e-mail: swarajdebbarma175@gmail.com

S. Dey
e-mail: subham6666@gmail.com

H. S. Behera et al. (eds.), *Computational Intelligence in Data Mining*, Advances in Intelligent Systems and Computing 990,
https://doi.org/10.1007/978-981-13-8676-3_11

tourist attractions. A system developed to predict the rating of a site based on its attributes can be used by government agencies and private agencies alike while planning to invest in a tourist destination. This project attempts to solve this crucial problem by using machine learning models and data from websites and already present user ratings to predict new scores which can be used for proper planning and execution on the part of the agencies. Machine learning systems are being used in all fields in our modern world. The use of this technology in the field of tourism is needed for proper prediction of decisions.

The prime motivation for this research comes from the need to help the government and private agencies to plan and execute a tourism attraction project properly. For a proper tourist attraction, the location chosen is of prime importance. Such a system that can aid agencies in selecting the correct area by giving a predicted rating of a place would be of utmost importance and a boon for the planners.

The primary goal of our research is to create a system that can accurately predict the rating of a potential tourist location on a scale from 1–5. Such a system should be able to quickly give an accurate prediction about the possible score of a tourist location based on given attributes of a place. High accuracy is the primary target and so proper data collection, and adequate model training is of utmost importance. Decision Tree Regression, Linear Regression, Random Forest and Support Vector Machine are the machine learning algorithms that are used.

Previously, hand coding rules were involved in many language-processing tasks which did not give any powerful solution to process many natural language processing operations. The statistical machine-learning paradigm automatically learns such rules through the analysis of large corpora available in typical real-world examples. Different types of supervised and un-supervised machine learning algorithms have been applied to Natural Language Processing tasks. These algorithms work by taking a broad set of features as input. In our research, statistical approach using Decision Tree Regression, Linear Regression, Random Forest and Support Vector Machine are being used to solve the problem of rating prediction.

2 Related Works

Prediction of tourist locations and its rating is a very new field, and some very interesting algorithms have been applied to predict their popularity. One such algorithm is the One Slope Algorithm, which is a recommender, as used by Hu and Zhou [1], which describes how using the one slope algorithm, recommendations can be produced with very high efficiency. A good point about one slope algorithm is that it is effortless to implement. One slope algorithm is a unique form of item-based collaborative filtering (item-based CF) proposed by Lemire and Maclachlan [2]. But the way attributes are selected, and their values assigned in Hu and Zhou's paper is quite similar to the one we used in this project.

There are a lot of methods to predict ratings based on attributes and community provided scores, as discussed by Marović et al. [3], where they discussed various ways. One such method is through decision trees proposed by Li and Yamada [4] for movie rating prediction. Another technique uses neural networks to provide web page rating [5]. Algorithms such as k-Nearest neighbors heuristic methods can be used to cluster the ratings based on attributes and then give predictions based on these clusters. Resnick et al.[6] used such a system for news filtering based on predicted ratings. Accurate prediction of tourism data is utmost essential for the success of tourism industry. Time Series prediction has been used to significant effect in tourism industry predictions. Koutras et al.[7] suggested various linear and non-linear models using systems like multi-layer perceptron models, support vector regressor with polynomial kernel models, Support Vector Regressor with Radial Basis Functions Kernels (SVR-RBF). Item-based collaborative filtering technique has been used in the past for recommending tourist locations as seen in this paper by Chen et al. [8]. Collaborative filtering works by collecting preferences from many users, it is based on the fact that users with similar tastes in a particular paper might have the same feeling on another topic and that can be used for giving the recommendation. A few websites that use collaborative filtering system are DieToRecs, TripAdvisor, Heracles, TripSay.

Automatic feature extraction methods are increasingly being used for extracting relevant features from text for many classification and regression tasks. Automatic feature extraction has shown to produce better results than using manual features for a variety of tasks. Metrics such as word frequency, tf-idf [9] and even modified metrics using tf-idf like delta tf-idf have been used to score features in text as show by Martineau et al. [10].

3 Methodology

The objective of this paper is to present a methodology to predict the success of a city as a tourist destination. The machine learning approaches used in this paper are Decision tree regression, Linear Regression, Random Forest and Support Vector Machine.

Decision Tree: Decision Tree Learning [11] uses a decision tree data structure where decisions are made on each node of the tree to arrive at conclusions about the item's target value which are represented as the leaf nodes. It is one of the predictive modeling techniques being used in many fields of mathematics and computer science like statistics, data mining and machine learning. Tree models can be of two types—classification trees and regression trees. A tree model where the target variable can take a discrete set of values are called classification trees, in this case leaves represent class labels and branches represent conjunctions of features that lead to those class labels. Decision trees where the target variable is of continuous values are called regression trees. A decision tree, in other words, is an acyclic graph with a fixed root. A node describes an attribute in the data, and the edges define a decision based on this attribute. Even though most examples use binary decisions, it is important to note that a node can have as many edges as they want. In operations research, these trees are understood as decisions and consequences. In general, the tree is traversed from the root, using the attribute in the node to choose a new node. Depending on the task, the leaf has a different meaning. Decision trees used in data mining are two main types:

- Classification tree analysis is when the predicted outcome is the class to which the data belongs.
- Regression tree analysis is when the predicted outcome can be considered a real number (e.g., the price of a house, or a length of day at a hotel).

Decision tree regression is being used for our research. Regression differs from classification algorithms as in regression we predict a continuous output on a given range and not from a discrete set of values. Regression with decision trees is very similar to classification since the trees are taking data and classifying it to a regression model or even previously computed value.

Linear Regression is used in statistics as a linear approach for modelling the relationship between a scalar dependent variable y and one or more independent variables denoted X. Regression which consists of only one independent variable is called Simple Linear Regression. For more than one independent variable, the process is called multiple Linear Regression. Multiple Linear Regression is distinct from multivariate Linear Regression, where multiple correlated dependent variables are predicted, rather than a single dependent variable [12]. In linear models, linear predictor functions are used to model relationships in linear regression whose unknown model parameters are estimated from the data. Linear Regression focuses on the conditional probability distribution of the output given a set of input, rather than on the joint probability distribution of the all the variables inputs and outputs, which is the domain of multivariate analysis. The practical applications of Linear Regression may fall into one of the following two categories:

- Linear Regression may be used to fit a predictive model to an observed data set of y and X values if the goal is prediction or forecasting or error reduction. After developing such a model, if an additional value of X is then provided without its accompanying value of y then the fitted model can be used to make a prediction of the value of y.
- If we are provided a variable y and a number of variables X_1, ..., X_p that may be related to y, Linear Regression analysis can be applied to estimate the strength of the relationship between y and the X_j, to evaluate which X_j may have no relationship with y at all, and to identify which subsets of the X_j contain inessential information about y.

Random Forests, also defined as random decision forests are a type of ensemble learning method which can be used for classification, regression, etc; they operate by constructing a plenitude of decision trees during the training time and outputting the class that is the mode of the classes (as used in random forests for classification) or mean prediction (as used in random forests for regression) of the individual trees. Random decision forests are used to overcome decision trees' habit of over fitting to their training set. Random Forests use the popular method of bagging to train the model. It works by taking a majority vote on each of the decision tree values and outputting the majority voted prediction as the prediction of the model [13].

In machine learning, Support Vector Machines (SVMs, also Support Vector Networks) are supervised learning models used for classification and regression analysis. SVM uses the training data and optimizes support vector points to create a boundary separating various classes in a 2D space. Each new point is mapped to an existing class in this 2D space and that is how classification works in a Support Vector Machine [14]. SVMs can be used to resolve various real world problems such as:

- SVMs can be used for text and hypertext categorization as their usage can significantly reduce the need for labeled training specimens in both the standard inductive and transductive settings.
- Image classification can also be done using SVMs. Experimental results show that SVMs are capable of achieving significantly higher search accuracy than traditional query refinement schemes after just three to four rounds of relevant feedback.
- SVM can be used to recognize hand-written characters.
- SVM have been used in Biology and other sciences where they've been used to classify proteins with an higher accuracy of the compounds.

4 Datasets, Experiments and Results

This section provides data description, experimental implementation and results. The experiments were conducted on a dataset whose data points were collected from www.mouthshut.com.

4.1 Data sets

For the experiment, we used Beautiful Soup which is python package for parsing to collect the details of a total of 590 cities from www.wikipedia.org. The percentage of cities chosen from various regions are shown in Table 1.

Table 1 Percentage of cities from each continent

North America (%)	South America (%)	Africa (%)	Europe (%)	Asia (excluding India) (%)	Australia (%)	Antarctica (%)	India (%)
2.71	0.68	3.39	10	9.83	2.20	0	71.19

Table 2 Distribution of ratings of the tourist destinations when floored to integer values

Classification	1	2	3	4
% of Distribution	1.36	3.79	25.08	69.83

Table 3 Sample of the final dataset

City	Attributes					
	Mountain	Desert	Waterfall	...	Concert	Rating
Cape Town	1	0	0	...	1	4.83
Cairo	0	1	0	...	1	4.71
Lanzarote	1	1	0	...	0	4
Mauritius	1	0	0	...	1	3.84
...	...	...	...	...	...	...
Manchester	1	0	0	...	1	4.33
London	1	0	0	...	1	4.4

Ratings from 1 to 5 for the corresponding cities were collected from www.mouthshut.com. We have chosen a total of 20 attributes such as Mountains, River, Waterfall etc, that can be used to form our dataset which would be fed into the machine learning algorithm. Table 2 shows the distribution of ratings of the tourist destinations when floored to integer values.

To prepare the dataset for the experiment, we checked that if the cities have those pre-selected attributes present. A true and false value in the dataset indicates the presence or absence of the attributes for a particular city. The ratings are also inserted in the dataset which looks like the table shown in Table 3:

In total 20 attributes were chosen for the experiment. These attributes are mountain, desert, waterfall, beach, river, worship place, climate, zoo, park, travel, archaeological site, festival, pollution, tourist, cuisine, safety, museum, stadium, market, concert. These particular attributes have been chosen among the various other features of a location keeping in mind the factors which mostly affect tourism patterns. Natural Beauty such as mountain, beach, waterfall, desert etc are favourable. Hill stations generally have scenic beauty and cool climate which attract many tourists. Hence cool climate is added as one feature. A place with good connectivity will also have more tourists visiting that place. Amenities like Zoo, Park, Museum, Stadium will boost tourism. Further pollution and safety of people also determine number of tourist visiting. Concerts and grand festivals also contribute to tourism growth. In addition, tourists are quite likely to visit archaeological sites. A place with superb local cuisine is more likely to attract tourists. Keeping these factors in mind, we have chosen the aforementioned attributes.

The data set is used to train the machine learning model. The data set is split into two parts of training and test data set in an 80:20 ratio. Splitting of the data set is done on a random basis. Initially, the training set is selected as 80% of the corpus,

Table 4 Test set prediction instances for manual features

City	Golden value	Predicted value
Cape Town	4.83	4.01
Mauritius	3.84	4.01
Johannesburg	3.6	4.09
Nairobi	4.57	4.1
…	…	…
Oxford	4.0	4.21
London	4.4	4.21

and then the remaining 20% part is used as the test set. The training data set is used to train the various machine learning model. For each of the test set instances, the rating is predicted based on how the machine learning model is trained. A sample of the test set prediction instances for manual features using Decision Tree Regression is shown in Table 4.

4.2 Experiments

Using automatic feature extraction would make our algorithm compatible with a variety of machine learning tasks and variety of different corpus. Thus our algorithm would be a general fit for many tasks and in the process would usually yield better accuracy than manual selection of features. Keeping that in mind we decided to go for automatic feature extraction using tf-idf.

TFIDF or tf-idf stands for term frequency-inverse document frequency. In information retrieval, it indicates how important a word is to a document in a collection or corpus. It is most often used as a weighting factor in searches of information retrieval, text mining. The value of tf-idf is proportional to the number of times a word appears in the document and is offset by the frequency of the word in the corpus, this helps to decrease the score for the words that appear in many documents in the corpus.

The tf-idf is the product of two statistics, term frequency and inverse document frequency.

- Term frequency The term frequency indicates raw count of a term in a document, i.e. the number of times that term t occurs in document d.
  Therefore, Term Frequency

$$tf_{t,d} = \frac{\text{Number of times the given token } t \text{ appears in the document } d}{\text{Total number of tokens in the document } d}$$

Table 5 20 attributes selected by automatic feature extraction

Serial No.	Feature	Score
1	Area	0.131
2	Population	0.131
3	State	0.121
4	District	0.114
5	World	0.112
6	Century	0.097
7	Town	0.090
8	Government	0.089
9	South	0.085
10	Temple	0.079
11	Capital	0.073
12	North	0.073
13	University	0.067
14	Region	0.065
15	River	0.064
16	Year	0.063
17	East	0.062
18	Island	0.614
19	School	0.059
20	Airport	0.058

– Inverse document frequency The inverse document frequency is a measure of how much information the word provides, that is, whether the term is common or rare across all documents. It can be calculated as the logarithm of the fraction of the number of documents in the corpus to the number of documents containing the given term.
Therefore, inverse domain frequency

$$idf_t = \log_{10} \frac{\text{Total number of documents in the corpus}}{\text{Number of documents in which the term } t \text{ appears}}$$

The 20 attributes that were automatically selected by the tf-idf feature extractor from our corpus are listed in Table 5.

The next step was the dataset creation which was done the same way as the manual features were used for creating the dataset. After the dataset was created the next step was the machine learning model creation using the new dataset.

4.3 Results

The algorithm is run through different metrics and the accuracy score, mean squared error, F-measure score, precision score and recall score is found out.

In real life regression problems, one significant challenge always occurs in calculating the accuracy score. A regressor model working on floating point data can give varied results which have a very low probability of exactly matching with the golden value. The predicted values were rounded off to the nearest integer and then compared them with rounded off golden values that we had acquired from www.mouthshut.com. The different metrics scores using manual features and automatically generated features are shown in Table 6.

Finally, we see that the decision tree algorithm produces an accuracy of 54.24% using manual features and 70.34% using automatic features. The accuracy is calculated by seeing the number of instances that our algorithm could predict correctly to the total number of instances given to the algorithm for testing. This, in turn, was represented as a percentage. This accuracy implies that the algorithm can be used in real life scenarios where it is meant to be used to guide and give a brief idea to decision-makers about the probability for a tourist destination being successful. The predictions give a good idea about the mood of a reviewer about a particular place based on the attributes being provided. We did achieve a mean squared error score of 38.98% which should be reduced further. We converted our ratings into floored discrete integer value for use by algorithms such as SVM and random forests and we get an accuracy of 74.58%. Such accuracy is totally valid in cases where floating points are not a concern and the user just wants a brief idea of the possible popularity of a tourist destination. The F-measure value, recall, and precision are other metrics that can also be used to measure the effectiveness of the machine learning model thus created. The score achieved for F-measure was quite low, and this should

Table 6 Different metrics scores for manual features and automatically generated features

Feature extraction style	Metrics	Decision tree algorithm (%)	Linear regression algorithm (%)	Random forest algorithm (%)	SVM algorithm (%)
Manual	Accuracy	54.24	55.08	74.58	74.58
	MSE	48.31	51.69	34.75	34.75
	Precision	27.34	27.76	18.64	18.64
	Recall	28.53	29.42	25.00	25.00
	F-measure	27.90	28.51	21.36	21.36
Automatic	Accuracy	70.34	62.71	74.58	74.58
	MSE	38.98	39.83	34.75	34.75
	Precision	29.26	26.92	18.64	18.64
	Recall	29.06	27.72	25.00	25.00
	F-measure	29.12	27.29	21.36	21.36

be increased. A low precision and recall and comparatively higher accuracy show that we have less number of true positives and comparatively more number of true negatives for some classes. A proper statistical measure of the effectiveness of a model is always helpful to find out how good an algorithm is and how much more improvement can be made over it.

5 Conclusion

Predicting user ratings for tourist places presents an interesting and well-formed problem. The report has shown that machine learning and natural language processing technologies can be used for predicting the rating of a particular location and hence gather an idea about how popular the place would be if the place would be turned into a tourism hub. Decision tree regression has given a accuracy of 54.24% with manual features and 70.34% with automatic features extraction method when using decision trees. Using algorithms such as SVM we get an accuracy of 74.58%. The experiment result shows that this simple method is effective. Our work is just beginning, we will continue to improve the system, and improve its accuracy. This study can be of great significance for the practical application of assisting Government of India in deciding about investment in locations. This research can act as a benchmark to compare and produce a more accurate system in the future.

In the future, we will focus our work on using universal dependencies and using them to not include negated words. When the description of a location contains a sentence like- "The area is not mountainous", it would not be picked up by algorithm as a mountainous area as it is negated in the current context. Universal dependencies or Stanford dependencies would work well for this purpose. The Stanford typed dependencies representation was designed to provide an easy to understand system to extract relationships between different phrases and words for people not having linguistic expertise [15]. Also better data and attributes can be acquired for the dataset to further reduce the mean squared error, increase the accuracy, precision, and F-measure.

Acknowledgements Thanks to all the anonymous reviewer for extensive and helpful comments.

References

1. Hu, H., Zhou, X.: Recommendation of tourist attractions based on slope one algorithm. In: 2017 9th International Conference on Intelligent Human-Machine Systems and Cybernetics (IHMSC), vol. 1, pp. 418–421. IEEE (2017)
2. Lemire, D., Maclachlan, A.: Slope one predictors for online rating-based collaborative filtering. In: Proceedings of the 2005 SIAM International Conference on Data Mining, pp. 471–475. SIAM (2005)

3. Marović, M., Mihoković, M., Mikša, M., Pribil, S., Tus, A.: Automatic movie ratings prediction using machine learning. In: MIPRO, 2011 Proceedings of the 34th International Convention, pp. 1640–1645. IEEE (2011)
4. Li, P., Yamada, S.: A movie recommender system based on inductive learning. In: 2004 IEEE Conference on Cybernetics and Intelligent Systems, vol. 1, pp. 318–323. IEEE (2004)
5. Pazzani, M., Billsus, D.: Learning and revising user profiles: the identification of interesting web sites. Mach. Learn. **27**(3), 313–331 (1997)
6. Resnick, P., Iacovou, N., Suchak, M., Bergstrom, P., Riedl, J.: Grouplens: an open architecture for collaborative filtering of netnews. In: Proceedings of the 1994 ACM Conference on Computer Supported Cooperative Work, pp. 175–186. ACM (1994)
7. Koutras, A., Panagopoulos, A., Nikas, I.A.: Forecasting tourism demand using linear and nonlinear prediction models. Acad. Tur.-Tour. Innov. J. **9**(1) (2017)
8. Chen, J.H., Chao, K.M., Shah, N.: Hybrid recommendation system for tourism. In: 2013 IEEE 10th International Conference on e-Business Engineering (ICEBE), pp. 156–161. IEEE (2013)
9. Ramos, J., et al.: Using tf-idf to determine word relevance in document queries. In: Proceedings of the First Instructional Conference on Machine Learning, vol. 242, pp. 133–142 (2003)
10. Martineau, J., Finin, T., Joshi, A., Patel, S.: Improving binary classification on text problems using differential word features. In: Proceedings of the 18th ACM Conference on Information and Knowledge Management, pp. 2019–2024. ACM (2009)
11. Freund, Y., Mason, L.: The alternating decision tree learning algorithm. In: icml, vol. 99, pp. 124–133 (1999)
12. Freedman, D.A.: Statistical Models: Theory and Practice. Cambridge University Press, Cambridge (2009)
13. Ho, T.K.: Random decision forests. In: Proceedings of the Third International Conference on Document analysis and Recognition, vol. 1, pp. 278–282. IEEE (1995)
14. Cortes, C., Vapnik, V.: Support-vector networks. Mach. Learn. **20**(3), 273–297 (1995)
15. De Marneffe, M.C., Manning, C.D.: The stanford typed dependencies representation. In: Coling 2008: Proceedings of the Workshop on Cross-Framework and Cross-Domain Parser Evaluation, pp. 1–8. Association for Computational Linguistics (2008)

An Automatic Intent Modeling Algorithm for Interactive Data Exploration

Vaibhav Kumar and Vikram Singh

Abstract Information plethora in recent years raises vital challenges to data science related computing communities, huge data is being generated by and for domain user/applications and thus managed in domain-specific context. Information search in these settings is a daunting task for naive user, due to lack of database/data semantics awareness, and own inability. Capturing user's search intents and utilizing them into future searches is key a research issue, which enhances cognitive satisfaction as well and thus becomes a *point-of-interest* among researchers and communities of information retrieval also. In this paper, we proposed an automatic intent modeling (AIM) algorithm, which models user's search interest (via relevance, preferences, etc.) to guide an interactive data exploration. The user applies reviews/relevance/preferences on the retrieved result for the initial data request, in order to gain more data objects. We validated the need for intent modeling and impact of user's intent in data exploration, through a prototype ESS, referred to as '*AimIIDE*'. The assessment of the system validates the fact that user's exploratory process significantly as user's confidence/knowledge enhances over the multiple interactions, and few users also claimed that interactive UI reduces their cognitive effort.

Keywords Big data analytics · Data exploration · Intent modeling · Information visualization · Human information behavior

V. Kumar (✉) · V. Singh
Computer Engineering Department, National Institute of Technology, Kurukshetra, Haryana 136119, India
e-mail: vaibhav1.kumar@nokia.com

V. Singh
e-mail: viks@nitkkr.ac.in

H. S. Behera et al. (eds.), *Computational Intelligence in Data Mining*, Advances in Intelligent Systems and Computing 990,
https://doi.org/10.1007/978-981-13-8676-3_12

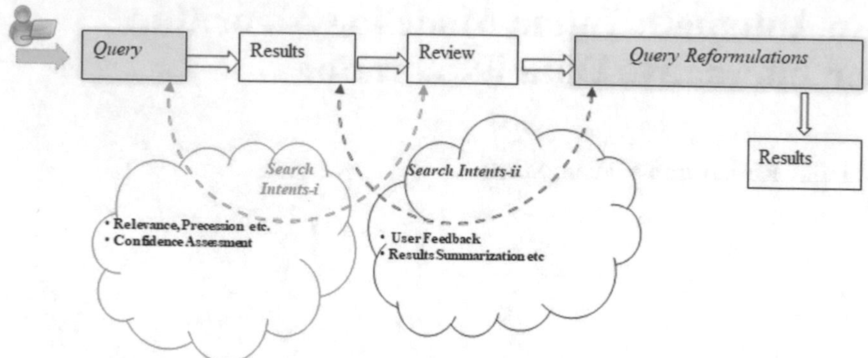

Fig. 1 Role of search intents in data exploration

1 Introduction

Nowadays, *Internet* has become a huge medium of connectivity, which results in excessive use of web search by all age groups. Information search is the core activity among most of the task that the user performs, and user information searches can be categorized: *known-item* and *exploratory* search. *Known-item* search is to find data or snippet that can satisfy users search need, thus often consists of short-queries. This search is well supported by modern search systems. While, sometimes a user poses stream of queries, and review multiple retrieved documents, that allows the user to explore additional information and lead towards his interest, referred as an *exploratory search*. Therefore, the user search task becomes complex and challenging to emulate by a search system. Formulating data request queries in these settings is cognitive challenging, as the user is unaware of data content and semantics or uncertain of real information needs [1]. In order to meet user's information needs, information system required to capable in capturing two key aspects of search: '*What user is Searching*' and '*How user's search is evolving*'. In user search, the first aspect can be captured by the simple evaluation of '*Query-Results*' and thus referred to as *search intent-i*. While to capture *how* aspect, information system requires to monitor user's interactions with results and future query reformulations, as shown in Fig. 1.

Here, various result summarization techniques are used to visualize the retrieved result set to the user, such as ontology-based summarization, clustering result set, etc. This reduces the user's effort on reviewing results and assists in the identification of precise keyword for future queries. This mainly guides the user in the formulation of data request, and identification of appropriate query variants. As next in the process, the reviews/feedbacks are submitted to be categorized, each categorized user review can mapped into a data retrieval strategy '*Query transformation/variants*'.

Effective utilization of identified users search intents in search is a key element in search of relevant data objects. We consider a key characteristic of exploratory search system to design the *AimIDE* (*Automatic Intent Modelling for Interactive*

Data Exploration) system, the sequence of queries that are submitted to find user's information. Large numbers of queries are required for several reasons: *to understand the user's search intent*, *to incorporate new evolving search goals*, etc. Earlier, in traditional IR systems, it is considered that user can at least able to formulate queries partially, as the user already have information related to his interest [2]. But studies revealed that in most cases this is not the actual real-life scenario. We focused on the users, who have less knowledge of their search domain as well as preferences, so that study on additional aspects of user's interest from the initial query posed by the user. This initial query is considered as *exemplary queries* [3].

The *exemplary queries* are having important applications in information search, and are best suited for the naïve user-driven search. *Exemplary query* helps to extract initial data samples, as a starting point to initiate search and provide feedback [4]. The user applies reviews/relevance/preferences on the retrieved result for the initial data request, in order to gain more relevant data objects. This intent-driven approach of data exploration is different from traditional search systems due to user's engagement. The proposed '*AimIDE*' is an exploratory search system (ESS), which extracts the user's intent of accuracy as close to the precision of 82% compared to human prediction. *AimIDE* is a GUI-based prototype designed to validate *the need* of intent modeling and *the impact* in data exploration. In order to achieve it, we focused more on "*Find What You Need*" and "*Understand What You Find*". An automatic intent modeling (AIM) algorithm is designed to assist the intent modeling and object extraction. The assessment of designed system establishes that user's exploratory process significantly as user's confidence increased over the multiple interactions.

Contribution: The key contribution of the paper is an *automatic intent modeling* (AIM) algorithm. We believe that algorithm assists ESS '*AimIDE*'; guides search session irrespective of domain and volume of the database, develop an exploratory search system based on this model. This model covers two aspects of data exploration process, i.e., Extraction of user's search intent and query formulation. A conceptual model has been designed which highlighted the two key aspects of data exploration: (1) role of User's feedback in data retrieval and (2) how to mine the user's intents for the relevance-based query formulation for exploratory search.

2 Related Work

The issue of search intent mining is often overlapped with other topics, such as *search topic/sub-topic mining*, *search goal mining, search query classification/categorization,* etc. The intent behind a search query can be categorized into different semantic categories as represented in Broder et al. [5] taxonomy. The taxonomy categorizes as follows:

- *Navigational*: Intent of the user is to reach a specific source.
- *Informational*: Intent of the user is to acquire certain information that may be available in one or more sources.

- *Transactional*: Intent of the user is to perform some activity.

Similarly, Rose and Levinson [6] developed a framework that can manually classify the intent of queries based on search intent classes.

2.1 Understanding User's Search Intents

Query-log Mining: In line with *Borders taxonomy*, several authors contributed to the automatic classification of the user intents. In [7, 8], works on large datasets (approx. 6,000 queries) with manually annotated, evaluated, and classified is divided into two aspects: intention (Informational, Not Informational, and Ambiguous) and topics. These tasks are done on two manners: supervised and unsupervised intent learning. Supervised learning is used to identify user's query intent based on some predefined intents and topics whereas, unsupervised learning technique is used to validate, refine, and select the most appropriate intent based on the user's requirement. Qian et al. [9] worked on mining of dynamic intents from the query logs maintained in search sessions. Hu et al. [10] developed a clustering algorithm that can automatically extract the subtopics of the query. Dang et al. [11] clustered reformulated queries generated from publicly available resources. Sadikov et al. [12] proposed a methodology of clustering the refinements of the queries submitted by the user to the web search engine.

Li et al. [13] first time proposed the concept of increasing the amount of training data for intent mining of queries. In recent advancements, systems have been incorporating commercial [14] as well as geographical [15] information.

User's Interactions: Click-through data is a collection of user's query and URLs of the web pages clicked by the user for those queries. For web search engines, click-through data can be considered as a triplet <q, r, c>, where q is the user's query, r is the ranking that is presented to the user, and c is the links that user clicks in order to get the result [16]. Lee et al. [17] concentrated his work on automatic mining of search intents of user based on the clicks that the user makes on the result offered by search engine on queries put by the user. Two major features discussed in his work consist of users past click behavior and anchor-link distribution. Moreover, queries that posses same clicked links can be considered as a similar query [18–21]. Radlinski et al. [22] suggested to initially use search session data to obtain all similar queries and then use of click-through bipartite graph to find the queries that are similar to each other and at the end grouping all the similar queries into a single cluster. A cluster containing all similar queries can be considered as a topic that refers to intent. Celikyilmaz et al. [23] proposed an approach of graph summarization that involves categorization of given speech utterance into several classes, where each class represents particular intent.

2.2 Understanding Data Intents

Results Visualization: For each query, a huge amount of dataset is retrieved. In order to provide a better understanding of retrieved results, it should be visualized in a user-friendly and interactive manner. Flamenco interface [24] provides many facets based on metadata for an exploratory search session. *mSpace* interface [25] displays "slices" of information with multidisciplinary facets to narrow down objects of interests. Relation Browser navigation system offers the associations among different data objects as facets to filter out interesting facts. *Phlatsearch* interface [26] offers flexible keyword-based search with the support of metadata browsing to narrow down information. *Stuff I've Seen* search system [27] promotes information use and searches on the basis of personal user profiles. Here, the key issue is to define facets, filters; categories and hierarchies which are very costly in terms of computational resources, hence they do not support unrestricted data exploration. In recent advancement for visual support on graphs, charts are preferred. For example, *Tile-Bars* system uses tile-based graphic aid to display frequency of query terms and their relative distribution in documents, whereas *HotMaps* uses the effective color coding scheme to visualize if the query term is present in the document or not.

Relevance Feedback: The efficiency of a search engine depends upon how well it matched the user's query terms. Additionally, user's query should be precise enough so that web search engine can map those query terms with its database and return relevant results, and then ask the user to give its feedback on the retrieved result set, this procedure is often called as *relevance feedback* (RF). In ESS, two retrieval models are used: *Boolean Model and Vector-Space Model*. Boolean model is associated with the exact match whereas in Vector-Space model is based on the best match. Based on these models, different RF techniques are employed. For the Boolean model, the key aspect is found the terms that appear more frequent in accepted results instead of discarded results [28]. In this line of work, Khoo and Poo [29] suggested to modify query terms and their connectives based on the results marked by the user. In the case of Vector-Space model, Rocchio [30] first time defined the formal definition of relevance feedback. He gave the definition of RF to find the optimal query vector, i.e., the one that maximizes similarity to relevant documents while minimizing similarity to nonrelevant documents.

3 Proposed System: *AimIDE*

We are observing '*Query-Result-Review-Query*' paradigm for the development of the system, where user's search intents are at the center. In the first step, strategically extracted data samples are used to initiate the data search. Automatic data exploration is beneficial here as the naive user is unaware and uncertain of database semantics and information needs, respectively. Next, naive user review and submit the relevance feedback. The relevance feedback is the key representative of the user's search intent, which steers towards an interesting pattern or relationship out of the data set.

3.1 Search Intents

According to Oxford online dictionary, intent is defined as '*a task that we need to achieve; a goal or purpose*'. Similarly, Jansen and Booth [3] defined search intent '*as an effective way to attain a cognitive goal while interacting with the web search engine*'. A data exploration process, the intents are captured and incorporated at two points: first, as user's query and its reformulations and next in the form of relevance feedback. *Search Intent-I*, involves that *how relevant search results have been generated based on the user's first query*. Certain measures are involved to judge the efficiency of search results that are produced; these measures may involve relevance, precession, etc. While *Search intent-ii* tries to extract user's search intent by applying different techniques such as analyzing user's feedback and click-through data. Access patterns generated from the user's search click is treated as search intent. Datasets generated from each query variants can be analyzed in a better way by applying result summarization techniques in order to decrease the cognitive effort of the user.

3.2 Proposed System

An ESS '*AimIDE*' is developed to assist the user in pursuit of his information needs. The architecture of *AimIDE* is based on client–server model, where client-side manages the user interface that captures user queries as well as visualizes results and server-side handles all the data exploitation related tasks. The workflow of the system is defined below:

- User starts with typing the query in the *Query Box*, followed by clicking on the Search button. Results are fetched from the database and visualized on the *Result Section* along with various operations.
- User can provide feedback on the retrieved results, by simply clicking on the desired buttons, i.e., user can make result permanent by clicking on *keepIncheck-box*.
- In keyword sections, most frequent keywords will be displayed, where user can directly drag and drop into the query box for sake of query refinement.
- A bar chart is associated in the *Keyword section* to provide a better understanding to the user for the appropriate keyword selection. Additionally, user can also go through the history section to get all the queries made in a particular search session.

User Interface: ESS developed consists of four key sections as shown in Fig. 2: *Query Section*, *Result Section*, *Suggested Keyword Section,* and *Recent Searches Section*. Each section has designated functionality to assist the user in exploratory search sessions and to reduce the cognitive effort.

Query Section: This section requires the user to type his query in the query box and hitting, this mechanism is common as the user can put his query either in form

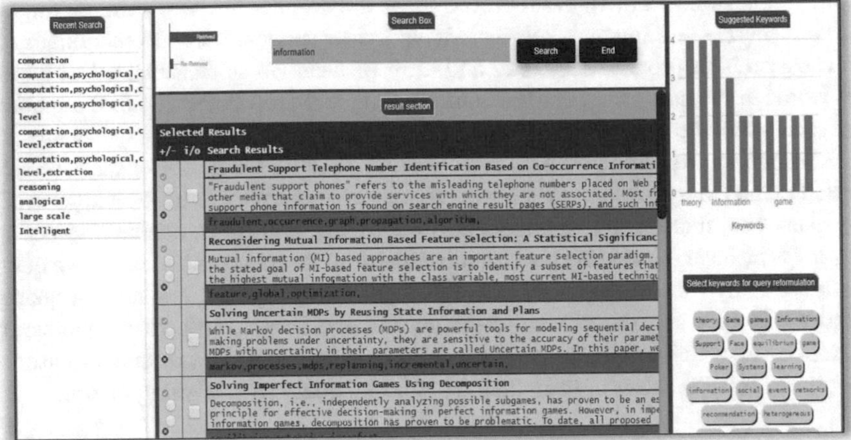

Fig. 2 *AimIDE* system's user interface

keywords or text. This section also helps in query refinement feature by appending dragged keywords. A relevance graph is provided beside the query box, which represents the frequency of new as well as re-retrieved documents and shows the overlap between the current and the previous query. This helps the user in understanding the context of the current query and of the previous query, also whether both queries are variants.

Result Section: This section displays the retrieved results for the user query. This section has two parts: *Selected Results section* and *Search Result Section. Search Result section* has three columns, i.e., "±", "i/0" and Search Result. The first column (±) is used for relevance feedback; here user can click on the appropriate radio button for positive feedback (tick button) and negative feedback (cross button). These feedbacks will be compiled and extracted to help in query reformulation. Second column (i/o) is used to move that result in Selected Result section so that the result do not get lost in the exploratory search process. Last Column, i.e., Search Result displays the results obtained from the user query. This column has three rows based on the backend database that displays title, abstract and keywords, respectively. The entire result section helps the user in visualizing results as well as submit their feedback on the retrieved dataset.

Suggested Keyword Section: This section provides an overview of the keywords that are extracted based on the user's relevance feedback, This section is divided into two parts. The first part represents the relative distribution of the most frequent keywords on a bar chart. Each bar in the chart represents a keyword. The major goal of this graph is to provide an interactive way to choose which keyword is more appropriate to be appended in the search query. The second part consists of a set of keywords which are most frequent in the search result and provides two major functionalities; first, a list of keywords is presented to the user in decreasing order of their frequency in the retrieved dataset. Frequent keywords are colored in darker, whereas less frequent keywords have lighter shades. Second functionality of this section is to provide easy drag and drop mechanism so that instead of typing keywords in the query box, user can simply drag and drop keywords into the query box, this saves the user from the hustle of typing the keywords in the query box.

Recent Searches Section: This section consists of all the queries user have submitted in the *Query Section*. Whenever the user types a query or drops a keyword in the query box, this section highlights the matching queries from the history section, this helps the user in avoiding the repetitive queries. Whenever the user types a query or drops a keyword in the query box, this section highlights the matching queries from the history section, this helps the user in avoiding the repetitive queries. In certain cases, previous queries can be treated as sample queries presented to user so that it can assist the user in formulating his initial query by analyzing queries.

3.3 AIM Algorithm

We presume a binary, non-noisy, feedback system 'for relevance feedback', in which the user simply indicates whether a data object is relevant or not. This categorization will be important in the following iterations. The feedback starts an iterative steering process [8, 31] and labeled data samples are used to characterize the initial data model. The data model may use any subset of attribute or predicate to define the user interest. However, domain experts could restrict the attribute set on which the exploration is performed [9]. Each iteration of the proposed model will add samples and presented to the user. The sample acquisition is purely intent driven, as data items labeled as relevant are primarily used to extract a new sample and train the data model. The user feedback and trained data model acts as a representative of the user search intents.

Automatic intent modeling (AIM) algorithm as shown in Fig. 3 leverage the current data model to create and identify the possible variants or reformulations, and further, retrieve the data samples. New labeled objects are incorporated with the already labeled sample set and a new classification model is built. The data objects meet his search interest or sufficient numbers of samples are observed, and thus the user can explicitly terminate it. Finally, *AIM* algorithm transforms the classification model that results in query expansion.

This query will extract the data object that is marked relevant by the data model. Hence, the main aspect of AIM is to converge to a model that captures the user interest. The more effort invested in this iterative process, the better the accuracy achieved by the classification model. The algorithmic explanation of the proposed 'Automatic Intent Modeling (AIM)' algorithm is presented. Overall, the user decides on the effort he/she is willing to invest, while the AIM algorithm leverages his intentions (feedback and data model) to capitalize on the correctness of the data model.

```
Automatic Intent Mining (AIM) Algorithm
Input: Initial User Query Qi and Initialized data space
       Ds
Output: Data Retrieval Query (Qi+1)
Step1:User submits initial Query Qi on database D and
       retrievesinitial data samples Ds
Step 2:User reviews these data samples Ds and label as
       relevant/ non-relevant, like DSrel/DSnrel
Step 3:Evaluate degree of search intent incorporated,
       and  For each Dsrel object in relevant sample,
       {Initialize, each data point on d-dimension data
       space as Cluster, further Neighborhood is ex-
       plored,
       {if the density of neighboring Cell is greater
       than thresholdτ,
       Then Merge that Cell and Form a Cluster at lower
       level } }
Step 4: For each higher dimension k ← 1 to d
       {If, overlapping cluster is dense at k-1 dimen-
       sion and density is greater than thresholdτ,
       Then Intersection again saved/merge as cluster}
Merge_cluster (c1, c2,..cn),
       { If intersection of clusters has greater then
       threshold τ,
       Then Form_new_cluster(c1,2,..n),
Remove lower dimension clusters from cluster set C }
Step 5: Define New Data retrieval model, based on In-
       tents {labelled Samples, Old Data model variants)

Step 6:Extract new samples Dsnew, for each reformulated
       query Qi+1
```

Fig. 3 AIM algorithm

3.4 Working Example

Considering the following schema as shown in Fig. 4 that is based on movies, User tries to extract certain data by writing SQL queries. The entire search session consists of four iterations as shown below:

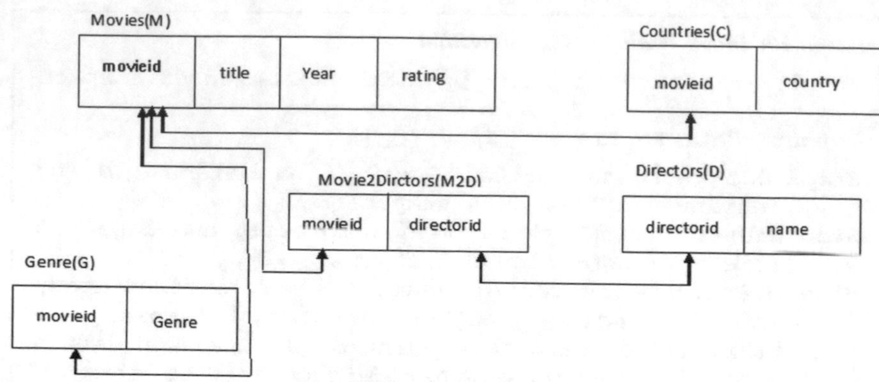

Fig. 4 Logical schema for *movie* database

Query1: Initial user queries, where the user tries to extract *title*, *year* and *genre* of the movies that were directed by director "M. Scorsese".

Initial Query
SELECT *M.title, M.year, G.genre*
FROM *D,M2D, M,G*
WHERE *D.name = "M. Scorsese" **AND** D.directorid = M2D.directorid **AND** M2D.movieid=M.movieid **AND** M.movieid = G.movieid*

Result

M.Title	M.Year	G.Genre
The Aviator	2004	Biography
Gangs of New York	2002	Drama
Goodfellas	1990	Biography
Casino	1995	Drama
Shutter Island	2004	Thriller
M.Jackson	1995	Drama
The last waltz	1978	Biography
Raging Bull	1980	Documentary

After reviewing the result set, the user decides to add one more constraint, i.e., genre = "Drama".

Query Variant 1:
SELECT *M.title, M.year, G.genre*
FROM *D, M2D, M,G*
WHERE*G.genre = "Drama" **AND** D.name = "M. Scorse-se" **AND** D.directorid = M2D.directorid**AND** M2D.movieid=M.movieid**AND** M.movieid = G.movieid*

Result

M.Title	M.Year	G.Genre
Gangs of New York	2002	Drama
Casino	1995	Drama
M.Jackson	1995	Drama

Next, the intention of the user gets changed in the exploratory search, and now he tries to retrieve *title*, year and *genre* of all movies whose *genre* is drama.

Query Variant 2
SELECT *M.title, M.year, G.genre*
FROM *M,G*
WHERE *G.genre = "Drama"*
AND*M.movieid = G.movieid*

Result

M.Title	M.Year	G.Genre
Gangs of New York	2002	Drama
Casino	1995	Drama
M.Jackson	1995	Drama
The Godfather	1972	Drama
Titanic	1997	Drama
Fight Club	1999	Drama
Taxi Driver	1976	Drama

Now, the user appends one more constraint to the query, i.e., releasing year of movies >1960.

Query Var.3:
SELECT *M.title, M.year, G.genre*
FROM *M,G*
WHERE *G.genre = "Drama"* ***AND*** *M.year>1960*
AND *M.movieid = G.movieid*

Result

M.Title	M.Year	G.Genre
Gangs of New York	2002	Drama
The Godfather	1972	Drama
Titanic	1997	Drama
Fight Club	1999	Drama
Taxi Driver	1976	Drama

In the above example, we observed that the intent of the user gets evolved in an exploratory search session and enhances the knowledge domain of the user as well. The goal of the *AIM* algorithm is to successfully incorporate the evolving intents and reduce cognitive efforts of the user.

4 Experimental Analysis

For the assessment, 10 users of variable knowledge of domain are considered, each user poses search tasks, i.e., T1 (Machine Learning), T2 (Computational Intelligence), and T3 (Information Retrieval). Ademo of the *AimIDE* is conducted for a better understanding of the system, and then each user is allowed to perform a search on T1 and T2 on until their information needs to be fulfilled. Whereas, task T3 is performed on a system which does not have any visual aids.

Search Time: In order to evaluate search time, completion time and number of iterations of each user are noted for all the search tasks. The result is shown in Table 1. We observe that task T1 and T2 took less search time than task T3 which was performed on a system with no visual aids. Hence we can conclude that T3 puts more cognitive effort on user.

System Performance: To measure the performance of the system, recall and precision are calculated. Traditionally, Precision can be defined as the number of correct results divided by the number of all returned results. Recall is the number of correct results divided by the number of results that should have been returned. Mathematically, it can be represented as F-score came out to be good. Precision and recall graph is plotted in Fig. 5 for all the three search tasks.

User Engagement: To ensure the engagement of the user for the system, different actions performed by the users are tracked and recorded. We observed that most users enjoyed the interactions. Table 2 shows the max and min count of the activities performed by users.

Table 1 Measuring search time

Search tasks	Search time (minutes)	
	Max	Min
T1	5.20	3.56
T2	6	4.25
T3	13.15	8

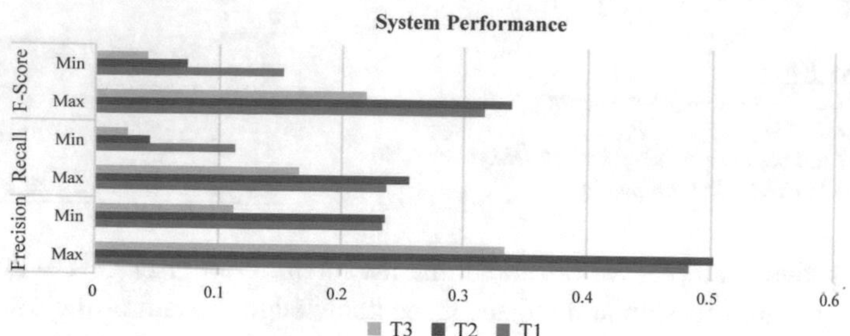

Fig. 5 System performance graph

Table 2 Measuring User Engagement

User engagement	Task T1		Task T2		Task T3	
	Max	Min	Max	Min	Max	Min
Positive feedback	10	2	19	9	NA	
Negative feedback	4	0	6	3	NA	
Queries entered	8	2	16	7	27	15
Dragged keywords	18	7	20	13	NA	

In this, it is observed that most of the users are involved in different activities in task T1 and T2 whereas, T3 have been performed on a system with no visual aid hence the number of queries entered has be listed down.

Cognitive Effort: Users are provided with the set of questions to extract feedback, containing categories like task complexity, mental effort, user satisfaction, system guidance. Evaluation of feedback obtained from all user's shows that search task T1 and T2 put less cognitive effort than search task T3.

5 Conclusion

Effectiveness of a search system largely relies on the user's ability of query formulation, and often a user fails to do so. In these situations, it becomes crucial for a search system to extract user's search intent for the retrieval of relevant data. We designed '*AimIDE*', an ESS and perform an assessment to validate its ineffectiveness for the assisting user's search. The performance evaluation is focused to understand the usability of the system and also provides insight on *how and when* the user interacts with the system. The assessment validates the difference in user behaviors for the same set of features, might be due to a different level of knowledge. Most of the users were able to reach their information goal in less search time, as systems interactive UI reduces their cognitive effort. The developed *AimIDE* can be further improved in terms of UI layout by including more user-interaction mechanism. Feedback Mechanism employed in the system has a scope of improvement.

References

1. Dubois, D., Prade, H.: Using fuzzy sets in flexible querying: why and how. In: Proceedings of the FQAS'1997. Kluwer Academic Publishers (1997)
2. Jain, K., et al.: Data clustering: a review. ACM Comput. Surv. **31**(3) (1999)
3. Jansen, B.J., Booth, D.L., Spink, A.: Determining the informational, navigational, and transactional intent of web queries. Inf. Process. Manag. **44**(3), 1251–1266 (2008)
4. Broder, A.: Taxonomy of web search. In: ACM SIGIR Forum, vol. 36, pp 3–10. ACM (2002)
5. Broder, A.Z., Fontoura, M., Gabrilovich, E., Joshi, A., Josifovski, V., Zhang, T.: Robust classification of rare queries using web knowledge. In: Proceedings of the 30th ACM SIGIR RDIR, pp. 231–238 (2007)
6. Rose, D.E., Levinson, D.: Understanding user goals in web search. In: Proceedings of the 13th International Conference on WWW, pp. 13–19. ACM (2004)
7. Jansen, B.J., Booth, D.: Classifying web queries by topic and user intent. In: SIGCHI'10 Extended Abstracts on Human Factors in Computing Systems, pp. 4285–4290. ACM (2010)
8. Liu, Y., Zhang, M., Ru, L., Ma, S.: Automatic query type identification based on click through information. In: Information Retrieval Technology, pp. 593–600. Springer (2006)
9. Qian, Y., Sakai, T., Ye, J., Zheng, Q., Li, C.: Dynamic query intent mining from a search log stream. In: Proceedings of the 22nd ACM CIKM, pp. 1205–1208 (2013)
10. Hu, Y., Qian, Y., Li, H., Jiang, D., Pei, J., Zheng, Q.: Mining query subtopics from search log data. In: Proceedings of the 35th international ACM SIGIR RDIR, pp. 305–314 (2002)
11. Dang, V., Xue, X., Croft, W.B.: Inferring query aspects from reformulations using clustering. In: Proceedings of the 20th ACM CIKM, pp. 2117–2120 (2011)
12. Sadikov, E., Madhavan, J., Wang, L., Halevy, A.: Clustering query refinements by user intent. In: Proceedings of the 19th International Conference on WWW, pp. 841–850 ACM (2010)
13. Li, X., Wang, Y.-Y., Acero, A.: Learning query intent from regularized click graphs. In: Proceedings of the 31st Annual International ACM SIGIR RDIR, pp. 339–346 (2008)
14. Ashkan, A., Clarke, C.L., Agichtein, E., Guo, Q.: Classifying and characterizing query intent. In: Advances in Information Retrieval, pp. 578–586 (2009)
15. Gan, Q., Attenberg, J., Markowetz, A., Suel, T.: Analysis of geographic queries in a search engine log. In: Proceedings of the 1st International Workshop on Location and the Web, pp. 49–56. ACM (2008)
16. Joachims, T.: Optimizing search engines using click through data. In: Proceedings of the 8th ACM SIGKDDKDDM, pp. 133–142 (2002)
17. Lee, U., Liu, Z., Cho, J.: Automatic identification of user goals in web search. In: Proceedings of the WWW ACM, pp. 391–400 (2005)
18. Beeferman, D., Berger, A.: Agglomerative clustering of a search engine query log. In: Proceedings of the 6th ACM SIGKDD KDDM, pp. 407–416 (2000)
19. Cao, H., Jiang, D., Pei, J., He, Q., Liao, Z., Chen, E., Li, H.: Context-aware query suggestion by mining click-through and session data. In: Proceedings of the 14th ACM SIGKDD KDDM, pp. 875–883. ACM (2008)
20. Craswell, N., Szummer, M.: Random walks on the click graph. In: Proceedings of the 30th Annual International ACM SIGIR RDIR, pp. 239–246 (2007)
21. Fujita, S., Machinaga, K., and Dupret, G. Click-graph modeling for facet attributes estimation of web search queries. Information Document, 2010, pp. 190–197
22. Radlinski, F., Dumais, S.: Improving personalized web search using result diversification. In: Proceedings of the 29th Annual ACM SIGIR RDIR, pp. 691–692 (2006)
23. Celikyilmaz, A., Hakkani-Tür, D., Tür, G.: Leveraging web query logs to learn user intent via bayesian discrete latent variable model. In: Proceedings of the ICML (2011)
24. Yee, K.P., et al.: Faceted metadata for image search and browsing. In: Proceedings of the SIGCHI Conference on Human Factors in Computing Systems, pp. 401–408. ACM (2003)
25. Wilson, M., et al.: mSpace: improving information access to multimedia domains with multimodal exploratory search. Commun. ACM **49**(4), 47–49 (2006)

26. Cutrell, E., et al.: Fast, flexible filtering with phlat. In: Proceedings of the SIGCHI Conference on Human Factors in Computing Systems. ACM, 261–270 Apr 2006
27. Dumais, S., et al.: Stuff I've seen: a system for personal information retrieval and re-use. In: ACM SIGIR Forum, vol. 49, No. 2. ACM, 28–35 Jan 2016
28. Harman, D.: Relevance feedback and other query modification techniques (1992)
29. Khoo, C.S., Poo, D.C.: An expert system approach to online catalog subject searching. Inf. Process. Manag. **30**(2), 223–238 (1994)
30. Rocchio, J.J.: Relevance feedback in information retrieval (1971)
31. Clarke, C.L., Craswell, N., Soboroff, I.: Overview of the TREC 2009 Web Track. Technical report, DTIC Document (2009)
32. Capra, R.G., Marchionini, G.: The relation browser tool for faceted exploratory search. In: Proceedings of the 8th ACM/IEEE-CS Joint Conference on Digital Libraries, pp. 420–420. ACM (2008)

Prediction of Arteriovenous Nicking for Hypertensive Retinopathy Using Deep Learning

Himanshu Kriplani, Mahima Patel and Sudipta Roy

Abstract Retinal arteriovenous nicking is a phenomenon where artery crosses a vein resulting in the compression of the vein or bulging of vein on one side. Being one of the causes of hypertensive blood pressure, it is needed to be diagnosed at an early stage. This paper explains a method devised using deep learning to classify arteriovenous nicking using the retinal images of the patient. The dataset provided by the Structured Analysis of the Retina project (STARE) has been used as training data and then classified in the presence and absence of arteriovenous nicking by using the proposed model. Proposed model, i.e., aggregate residual neural network (ResNeXt) which has been evaluated with performance metric and achieved more than 94% accuracy on the test dataset.

Keywords Arteriovenous nicking · Convolutional neural network · Residual neural network · Deep learning

---

H. Kriplani (✉) · M. Patel
Department of Computer Science and Engineering, UV Patel College of Engineering, Ganpat University, Mehsana 384012, Gujarat, India
e-mail: kriplani.himanshu96@gmail.com

M. Patel
e-mail: patelmahima1998@gmail.com

S. Roy
Department of Computer Science and Engineering, Institute of Computer Technology, Ganpat University, Mehsana 384012, Gujarat, India
e-mail: sudiptaroy01@yahoo.com

Department of Computer Science & Engineering, Calcutta University Technology Campus, Kolkata 700098, India

H. S. Behera et al. (eds.), *Computational Intelligence in Data Mining*, Advances in Intelligent Systems and Computing 990,
https://doi.org/10.1007/978-981-13-8676-3_13

1 Introduction

Damage to the retina and retinal circulation is seen due to high blood pressure (hypertension), which is also called hypertensive retinopathy. It is difficult to diagnose hypertensive retinopathy as there are no explicit symptoms other than blurred vision and headache. Currently, arteriovenous (AV) nicking is assessed by comparing standard photographs for detecting abnormalities which are classified as absent, mild, moderate, or severe [1] done by humans. An automated method based on Colored Fundus Images was proposed to detect AV nicking [2]. More details on the severity of AV nicking which are not even available on manual grading by a human can be revealed using complex measurements. Such information may help to study the relationship between AV nicking and correlated diseases such as stroke and hypertension. Another recent method to improve the accuracy proposed enhancement of non-uniformly lightened fundus image of retina across thin layer of cataract [3]. Gamut less image enhancement by intensity histogram equalization in adapted Hue Saturation Intensity (HSI) color image space was the part of the restoration. The resulting image has problem with appropriate brightness and saturation, enhance contrast, reinstate the color space information and maintain complete information. Another computerized method to detect the presence of AV nicking was developed [4]. The algorithm uses a multi-scale line detector to segment the blood vessels. The presence or absence of AV nicking is detected by AV cross points and classified with a support vector machine model. That approach was used in distinguishing evident or severe cases from the normal AV nicking cases. A Haar-based discrete wavelet transformation was used to identify the AV nicking [5]. Then, with the given dimension for finding circular regions to localize the point, Hausdorff distance based pattern matching was used. Another method [6] captured brighter circular structures using a line operator. For each pixel, evaluation of variation of image brightness around 20 line segments was done in the specific direction using the proposed line operator. Current automatic detection method suffers from high accuracy with different types of images. The proposed method has solved this accuracy problem.

The rest of the paper is organized as follows: Proposed method has been described in Sect. 2 with a brief explanation and data set. Results of the proposed method with a brief explanation have been described in Sect. 3. Finally, the conclusion is in Sect. 4.

2 Methodology

Two image classifiers have been used. One of which is the convolution neural networks used for image classification in many deep learning models was implemented to build a deep learning model. The other is the PyTorch implementation of aggregated residual transformations for a deep neural network. The different stages with a brief explanation of training data have been described below.

2.1 The STARE Data

The Project STARE by Goldbaum M. M.D., was established in 1975 at the UC, San Diego. Over 30 people belonging to medicine background as well as science and engineering background contributed to the project. The Shiley Eye Center at the UC, San Diego and the Veterans Administration Medical Center in San Diego are the major sources of images and clinical data [7]. An ophthalmologist detects the object that is observable in the eyes of the subject in a clinical test. These explorations are then utilized by the ophthalmologist to justify the subject's health. For instance, a patient demonstrates narrowing of the blood vessels in the retina or discoloration of the optic nerve. With the above information, an ophthalmologist can diagnose the patient suffering from various retinal diseases. The main concern of this research project is to automate the detection of the human eye diseases.

2.2 Image Dataset Used for the Deep Learning Model

The image dataset used for the Classification of a Retinal Images is available as "A-V Change" on the website of the STARE project [7]. A case of the presence of AV nicking in the retinal image is shown in Fig. 1a, b shows the absence of AV nicking. It had 68 images which was classified into presence of AV nicking and 294 retinal images as absence of AV nicking.

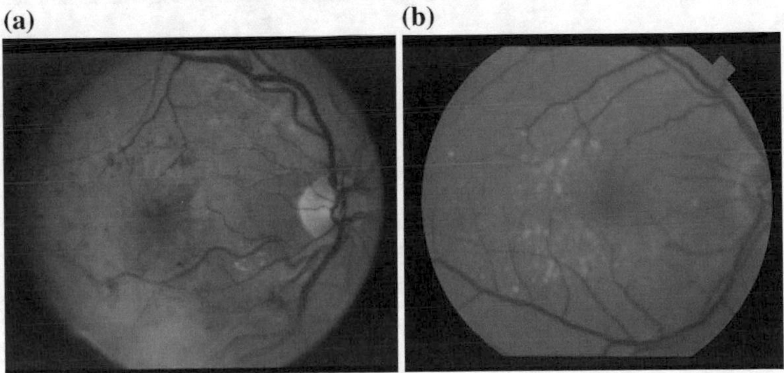

Fig. 1 **a** AV nicking present, **b** AV nicking absent

2.3 Preprocessing

The retinal image dataset of 402 images was not fully classified. 35 images were discarded due to lack of classification as deep learning uses classified images. The images were in portable pixmap image which were converted to standard JPG format for better usage. Raw images were used for CNN implementation.

2.4 Convolutional Neural Network

The convolution network consists of two layers and a ReLU function for the internal network and then for final output layer has a sigmoid function. Each of the two layers contains a convolutional layer which generates a Feature matrix. The Max pool layers does the max pooling, i.e., extract maximum weighted features from the feature matrix. With such two layers, the model gets an accuracy of 99.4% with a 90–10 test ratio of training versus testing dataset. Both the layers have ReLU as their activation function.

Every raw image is taken as input to the convolutional neural network. In each convolutional layer, there are neurons which take input from the previous layer of neurons. In convolutional layer, weights specify convolution filter. There are several grids which take input from all previous layers using different filters. The pooling layer takes the input from the convolutional layer and records it to produce a pooled featured map. In the end, after flattening its output is given to the artificial neural network which does the job of classifying the features from images into the resultant classes.

2.5 Residual Network

Deep neural networks those of with more number of layers are complex and difficult to train. ResNet is a residual learning architecture which eases the process of training deeper networks than those used previously. Instead of learning unreferenced functions, with reference to the layer inputs it specifically reformulates the layers as learning residual functions. The comprehensive empirical evidence provided by this architecture shows that the optimization of these residual networks are easier, and can help gain accuracy from greatly increased depth. The results of ResNet won the first place on the ILSVRC 2015 classification task. For many visual recognition jobs the depth of layers is vital. Merely due to immensely deep layers, COCO object detection is improved to 28% relatively as stated on the website [8].

ResNet architecture has basic building block which is made up of residual modules. It mainly does two jobs: performs a set of functions on input or it skips all the steps together. Unlike other architectures, there is no reduction in the performance

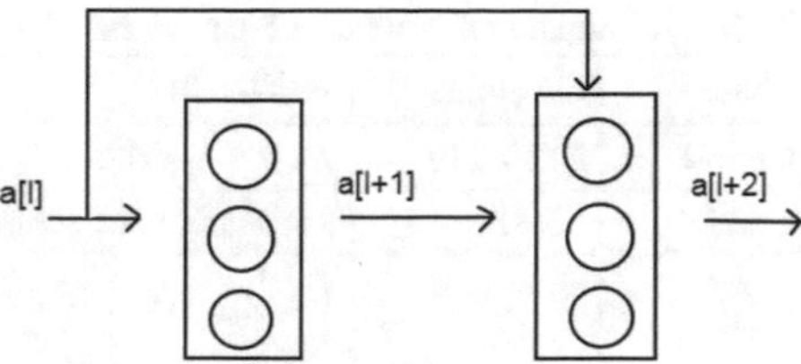

Fig. 2 Residual Network

upgrades as numbers of layers are increased. In contrast to other architectures, a new technique that is introduced in the ResNet architecture changes in preprocessing the input, i.e., before feeding input into the network, it is divided into the patches. CNN architecture is widely used in object detection, image classification but ResNet takes CNN one level up by little but significant change in the architecture. ResNet architecture requires less extra parameters by increasing the depth of the network instead of widening the network. Also training speed of deep networks increases along with the reduction in vanishing gradient problem. Higher Accuracy is achieved in the performance of the network in image classification.

As shown in figure, there are two layers where activation in layers starts from a[l] then it goes to a[l + 1] and the deactivation after two layers is a[l + 2]. The computation starts by applying linear operator which is handled by Eq. 1, where z[l] is obtained (linear model) is obtained by multiplying a[l] with weight matrix (W[l + 1]) and adding the bias vector (b[l + 1]). Then after to get a[l + 1]'s value ReLU nonlinearity is applied. Thus in the next step, again linear operation is performed and handled by ReLU nonlinearity function followed by a ReLU operation which is handled by Eq. 1 which gives z[l + 2]. Hence, the information goes through all these steps to flow from a[l] to a[l + 2] which is actually the main path for this set of layers. But the Residual network just takes a[l] and fast forward it and then copy much further into the second layer as shown in Fig. 2. Hence, a[l] is fed after the linear operation but before the ReLU operation, i.e., the step of z[l + 2] is eliminated and the output obtained for a[l + 2] is a ReLU nonlinearity applied to the summation of z[l + 2] and a[l]. a[l] is added as a residual, therefore, it is called a residual block. Hence, to process the information deep into the neural network, a[l] skips over the layer allowing to train much deep into the neural network.

$$z[l+1] = W[l+1] * a[l] + b[l+1] \tag{1}$$

2.6 *Aggregate Residual Transformations for Deep Neural Network (ResNeXt)*

It is an image classifier that has high modularized network architecture. It is built by repeating the building blocks that aggregate a set of transformation with the same topology. The design is homogeneous and multi-branch architecture with few hyper-

Table 1 Comparison of ResNet 50 and ResNeXt 50 architectures

Phase	Outcome	ResNet 50	ResNeXt 50 (32 • 4d)
Conv-1	112 • 112	7 • 7, 64, stride 2	7 • 7, 64, stride 2
Conv-2	56 • 56	3 • 3 max pool, stride 2	3 • 3 max pool, stride 2
		$\begin{bmatrix} 1 \times 1 & 64 \\ 3 \times 3 & 64 \\ 1 \times 1 & 256 \end{bmatrix} \times 3$	$\begin{bmatrix} 1 \times 1 & 128 & \\ 3 \times 3 & 128 & C = 32 \\ 1 \times 1 & 256 & \end{bmatrix} \times 3$
Conv-3	28 • 28	$\begin{bmatrix} 1 \times 1 & 128 \\ 3 \times 3 & 128 \\ 1 \times 1 & 512 \end{bmatrix} \times 4$	$\begin{bmatrix} 1 \times 1 & 256 & \\ 3 \times 3 & 256 & C = 32 \\ 1 \times 1 & 512 & \end{bmatrix} \times 4$
Conv-4	14 • 14	$\begin{bmatrix} 1 \times 1 & 256 \\ 3 \times 3 & 256 \\ 1 \times 1 & 1024 \end{bmatrix} \times 6$	$\begin{bmatrix} 1 \times 1 & 512 & \\ 3 \times 3 & 512 & C = 32 \\ 1 \times 1 & 1024 & \end{bmatrix} \times 6$
Conv-5	7 • 7	$\begin{bmatrix} 1 \times 1 & 512 \\ 3 \times 3 & 512 \\ 1 \times 1 & 2048 \end{bmatrix} \times 3$	$\begin{bmatrix} 1 \times 1 & 1024 & \\ 3 \times 3 & 1024 & C = 32 \\ 1 \times 1 & 2048 & \end{bmatrix} \times 3$
	1 • 1	Global average pooling 1000 d FC, softmax	Global average pooling 1000 d FC, softmax
Parameter		**25.0** • 106	**25.5** • 106
FLOPs		**4.1** • 109	**4.2** • 10

parameters which exposes to the new parameter which is termed as cardinality (size of the set of transformations). The dimensions of a residual block are inside the brackets and the number of layered blocks in a phase are outside the brackets. "C = 32" suggests 32 convolutions grouped. The numbers of parameters and also the FLOPs of these two models are approximately equal as shown in Table 1. Experiments have shown that increasing cardinality results in improved classification accuracy rather than increasing the depth and the width of the network when the capacity needs to be incremented.

Table 1 is the comparison between regular ResNeT and the Aggregate version, i.e., ResNeXt. Hence, there is no performance degradation in comparison to regular ResNeXt.

The matrix in Table 1 convolutional layer 2, 3, 4 and 5 shows the shape of the residual block. The number multiplied by the matrix in Table 1 shows the number of residual blocks stacked together at each layer. You can see them with the similar fashion of inputs and layers in both different architecture doesn't affect the flops much but decreases the error rate in training as well as the validation phase. Global average pooling refers to taking an average of value instead of taking the max pool as in max pooling layer of the CNN. The global average pool takes the global average

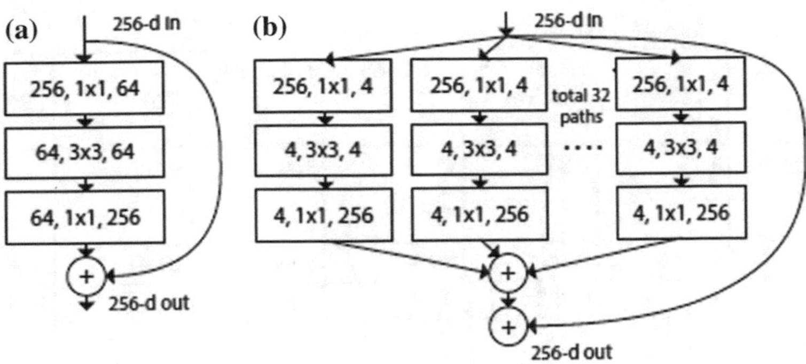

Fig. 3 **a** a simple ResNet, **b** A ResNeXt with cardinality = 32

pooling of all the output found after the fifth convolutional layers. The number of flops in ResNeT is 4.1×10^9 and that of ResNeXt is 4.2×10^9, which doesn't affect the usage of the resources.

2.6.1 Working of Aggregate Residual Transformations for Deep Neural Networks

Figure 3a is the structure of a single ResNet. Figure 3b is a combination of a number of such single ResNets with different transformations. That number is defined by the cardinality. The changes in time and space complexity are negligible compared to the improvement in accuracy. The new parameter introduced here shows more significance which is proved by the experiment. The grouped layers don't work individually as in the ensemble. So it is different from the ensemble methods.

The inputs and output shape is similar in both ResNet and ResNeXt. According to the ResNeXt50 gives an error rate of 1.7% less than that of ResNet50 with no significant increase in the number of flops. ResNeXt also has lesser training errors than that of ResNet.

For image classification of the dataset, ResNeXt architecture is implemented. It showed a bit of better accuracy than that of the ResNet with similar hyper-parameters and a cardinality of 32.

3 Results and Discussion

All the three models are implemented on the dataset and an accuracy of 85.7, 91.4, and 94.28% for CNN, ResNet and ResNeXt, respectively, as in Table 2. The evaluation matrices and confusion matrix found using the convolutional neural network are as in Table 2 and Table 3. The true positive rate, false positive rate, precision, recall, and f-measure are shown as in Table 2. The values of all the measures as in Table 2

Table 2 Metrics for Evaluating Classifier Performance

	True positive rate	False positive rate	Precision (%)	Recall (%)	F-measure (%)	Accuracy (%)
Convolutional neural network	1	0.833	85.29	100	92.00	85.71
ResNet	1	0.500	90.62	100	95.07	91.42
ResNeXt	1	0.333	93.54	100	96.66	94.28

Table 3 Confusion matrix with 2 epochs

	AV Nicking absent	AV Nicking present
Convolutional neural network	29(TP)	0(FN)
	5(FP)	1(TN)
ResNet	29(TP)	0(FN)
	3(FP)	3(TN)
ResNeXt	29(TP)	0(FN)
	2(FP)	4(TN)

are calculated using the Eq. 2, 3, and 4. The true positive rate is an average of all true positive values in each epoch. Similarly, the value of false positive rate is calculated for each architecture.

Following equations have been used to study the given three architectures for image classification:

$$\text{Accuracy} = \frac{tp + tn}{tp + fp + fn + tn} \tag{2}$$

$$\text{Precision} = \frac{tp}{tp + fp} \tag{3}$$

$$\text{Recall} = \frac{tp}{tp + fn} \tag{4}$$

With 100 epochs, the following confusion matrix is obtained.

Table 3 shows the confusion matrix of all the three architectures. With 29 True predictions in all architectures, i.e., all images without AV nicking were classified correctly for the absence of AV nicking. The False positive rate differs in all algorithm. False negative counts are 5, 3, 2 for respective architectures which shows ResNeXt have been performing better in the testing environment. The test set used for the prediction was distinct from that of the training dataset.

4 Conclusions

This paper proposes a method to automate AV nicking prediction using deep learning with high accuracy. With the use of pre-trained model, the screening of a large number of patients can be carried out in a more efficient and faster way. The proposed solution can enhance the study of other diseases such as hypertension in which AV nicking plays a role. This scalable and rapid method can be deployed at any medical system which holds the potential for substantial impact on clinical treatment including fastening clinical decision for doctors.

References

1. Hubbard, L.D., Brothers, R.J., King, W.N., Clegg, L.X., Klein, L.X., Cooper, L.S., Sharrett, A.R., Davis, M.D., Cai, J.: Methods for evaluation of retinal microvascular abnormalities associated with hypertension/sclerosis in the atherosclerosis risk in communities study. Ophthalmology **106**(12), 2269–2280 (1999)
2. Uyen, T.V.N., Bhuiyan, A., Laurence, A.F.P., Kawasaki, R., Wong, T., Jie, J. Wang, J.J., Mitchell, P., Kotagiri, R.: An Automated Method for Retinal Arteriovenous Nicking Quantification from Colour Fundus Images (2013)
3. Mitra, A., Roy, S., Roy, S., Setua. S.: Enhancement and restoration of non-uniform illuminated fundus image of retina obtained through thin layer of cataract. Computer Methods and Programs in Biomedicine, vol. 156, Issue March, pp. 169–178, ELSEVIER (2018)
4. Pereira, C., Veiga, D., Gonçalves, L., Ferreira, M.: Automatic arteriovenous nicking identification by color fundus images analysis. In: Campilho, A., Kamel, M. (eds.) Image Analysis and Recognition. ICIAR 2014. Lecture Notes in Computer Science, vol. 8815. Springer, Cham (2014)
5. Lalonde, M., Beaulieu, M., Gagnon, L.: Fast and robust optic disc detection using pyramidal decomposition and Hausdorff-based template matching. IEEE Trans. Med. Imaging **2001**, 1193–200 (2011)
6. Lu, S., Lim, J.H.: Automatic optic disc detection from retinal images by a line operator. IEEE Trans. Biomed. Eng. **58**(1), 88–94 (2011)
7. Structured Analysis of the Retina: http://cecas.clemson.edu/~ahoover/stare/. Accessed on Jan 2018
8. Deep Residual Learning for Image Recognition: https://arxiv.org/abs/1512.03385. Accessed May 2018

ANFIS-Based Modeling for Prediction of Surface Roughness in Powder Mixed Electric Discharge Machining

Pragyan Paramita Mohanty, Deepak Mohapatra, Asit Mohanty and Swagat Nayak

Abstract In the present work, an artificial neural network approach is involved to model the machining parameters within experimental investigation in order to improve the output performance. Adaptive neuro fuzzy inference system (ANFIS) is a powerful network that has been involved in order to reduce the process variability. In addition, ANFIS was used to predict the output parameters such as surface roughness (SR) through trial-and-error basis. The architecture of ANFIS through its competitive algorithm results a minimum prediction error in order to evaluate the performance of SR values. Results indicated that the PMEDM has superiority in prediction of output as an alternative to conventional EDM. The average prediction error for SR was $5.2569e^{-06}$ for conventional EDM and $4.5549e^{-006}$ PMEDM, respectively.

Keywords EDM · PMEDM · ANFIS · Surface roughness

1 Introduction

In present days, different areas such as chemical, aerospace, and marine industries utilize nickel-based superalloys for retaining mechanical properties at higher temperature. Further properties like strain hardening with the presence of hard and abrasive phrases avoid these materials to machine through conventional machining processes. Therefore, the aim of the current work is set to improve the surface integrity or quality of machined surface of a nickel-based superalloy such as Inconel 625, etc. PMEDM enhances the capabilities of EDM with the addition of suitable material with a fine powder shape such as graphite, aluminum, and silicon to dielectric during electric

P. P. Mohanty (✉) · D. Mohapatra · S. Nayak
Department of Mechanical Engineering, VSSUT, Burla 768018, Odisha, India
e-mail: pragyanmohanty.design@gmail.com

A. Mohanty
Department of Electrical Engineering, CET Bhubaneswar, Bhubaneswar 751003, Odisha, India
e-mail: asithimansu@gmail.com

H. S. Behera et al. (eds.), *Computational Intelligence in Data Mining*, Advances in Intelligent Systems and Computing 990,
https://doi.org/10.1007/978-981-13-8676-3_14

discharge machining (EDM). The electrically conductive powder affects the EDM process and its performance.

In present days, artificial intelligence techniques are getting good attention for modeling experimental behavior and for optimization of performance in terms of quality and productivity. Further, due to advanced computing capability, artificial intelligence techniques are getting importance in modeling of machining processes. Nonlinear relationship of machining parameters with performance parameter is not explained by empirical or any experimental models developed [1–3]. Artificial intelligence as a replacement to conventional approach has been implemented nowadays by many researchers [4–11].

Surface roughness is nothing but surface texture, which is due to the interaction between the work material and surrounding environment. It is always difficult to have a direct and online measurement of surface roughness. Therefore, a prediction model based on artificial neural network is used to investigate the influence of cutting parameters on surface texture. In this context, an intelligent technique such as ANFIS (Adaptive neuro fuzzy system) can be suggested for predicting modeling of surface roughness during machining. Further, a dataset based on experimentation can be prepared for taking the machine parameters.

2 Experimental Procedure

An experimental EDM set up (Electronica, India: Model: EMS-5535) as depicted in Fig. 1 having positive polarity with servo head has been used.

Dielectric fluid like commercial grade EDM oil having freezing temperature 94 °C and with specific gravity value 0.763 is considered for experimentation. A "Digital Storage Oscilloscope" is used to measure the current and voltage. Further, the recir-

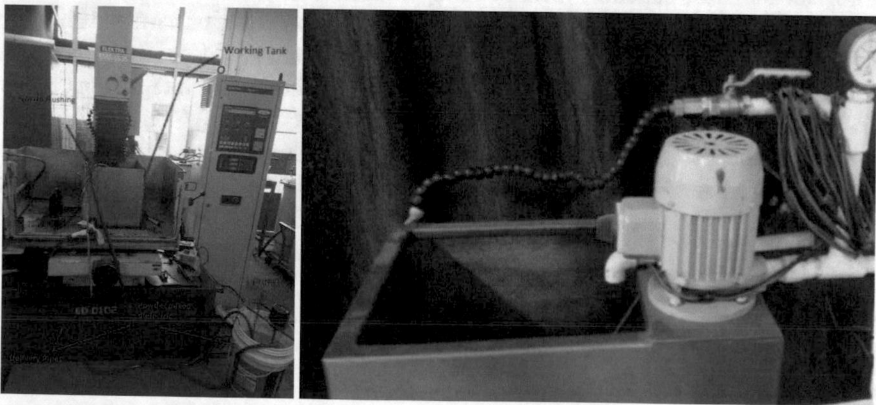

Fig. 1 Experimental setup

Table 1 Composition of received Inconel 625

Element	Weight (%)
Ni	62.13
Cr	22.18
Nb	3.59
Mo	8.47
Al	0.32
Si	0.35

culation system is made of a cylindrical working tank with 15 L capacity. Due to continuous circulation, the powder particles never settle inside the flushing system. Inconel 625 (nickel-based superalloy, possessing excellent corrosion resistance) is considered as workpiece material in shape of thin plates with a dimension of 50 × 50 × 7 mm with the following composition as shown in Table 1. An electrolyte copper tool having a dimension of 13*70*70 has been used to perform experiments.

Here graphite (Gr) powder was used as an additive in EDM oil dielectric. The average particle size of this powder claimed by the manufacturer is ~45 μm.

Effective method like Taguchi is suggested for designing the responses, which are influenced by many input variables. The technique reduces the frequency of experiments to save overall time and cost. In this work, Table 2 discusses the domain of experiment. A L9 design Table 3 is chosen to conduct the experiment having four PMEDM machining parameters. The experiment has been conducted to study the effect of cutting parameters on surface roughness. Figure 2 shows the machined workpiece sample.

Surface roughness expressed in terms of center line average (Ra), which is nothing but the average departure of roughness profile from the center line. The expression of Ra is given below in Eq. 1.

$$R_a = \frac{1}{l} \int |y(x)| dx \quad (1)$$

Table 2 Domain of experiment

	Symbol	Unit	Level 1	Level 2	Level 3
Pulse on time (A)	Ip	mAmp	5	10	20
Peak current (B)	Ton	μsec	200	400	800
Duty factor (C)	τ	——	2	4	8
Voltage (D)	v	volt	10	20	30

Table 3 Layout of L9 orthogonal array for experiment

	Ip (mA)	Ton (μsec)	DF (τ)	V (volt)
1	1	1	1	1
2	1	2	2	2
3	1	3	3	3
4	2	1	2	3
5	2	2	3	1
6	2	3	1	2
7	3	1	3	2
8	3	2	1	3
9	3	3	2	1

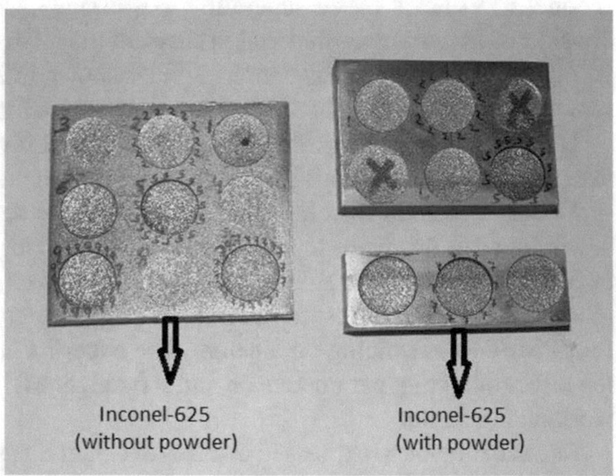

Fig. 2 Samples of EDM and PMEDM machined workpiece

In this case, "*l*" represents the sampling length, "y" denotes the heights of peaks and valleys of the roughness profile. Further, "x" stands for the profile direction and "Ra" denotes the measurement of the center line average, which was captured through a small portable type (stylus) of profilometer [Taylor Hobson, Model-Surtronic].

Initially a set of experiments were conducted by using pure dielectric and a set of experiments was conducted. After that in the next segment, graphite powder is added into the dielectric as additive with a constant concentration of about 4 g/l and stirred continuously, until it became well mixed. Then, the mixture solution is allowed to pump into the working tank and nuzzled onto the job. With this arrangement, a number of experiments were conducted. Each experiment was carried out for 5 min with constant flushing pressure of 0.5 kg/cm^2. Table 4 represents the calculated performance parameter (surface roughness) after experimentation.

Table 4 Experimental results

Sl. No	Input parameters				Response (without powder)	Response (with powder)
	IP(A)	TON	DF	V	SR (μm)	SR (μm)
1	1	1	1	1	5.93	0.05
2	1	2	2	2	1.02	0.07
3	1	3	3	3	0.06	0.06
4	2	1	2	3	9.04	0.07
5	2	2	3	1	7.83	6.85
6	2	3	1	2	0.206	0.08
7	3	1	3	2	7.28	5.78
8	3	2	1	3	0.07	8.94
9	3	3	2	1	4.20	3.28

3 Adaptive Neuro Fuzzy Inference System (ANFIS)

ANFIS combines the working patterns of both Fuzzy and Neural network with the learning abilities with nonlinear relationship [12, 13]. Fuzzy inference system depends upon components such as fuzzy rule, membership function, and reasoning mechanism.

ANFIS happens to be a FIS in framework of the Neural Network, which improves the system performance [as shown in Fig. 3]. Based on the first-order fuzzy inference system, a fuzzy model can be designed with a couple of rules [14, 15] as shown below:

According to rule no 1: If $X = A_1$, $y = B_1$ then $f_1 = p_1x + q_1y + r_1$

According to rule no 2: If $X = A_2$, $y = B_2$ then $f_2 = p_2x + q_2y + r_2$

When f_1 and f_2 become constants, zero-order TSK fuzzy models are generated in place of linear equations. Generally, node functions of the same type of layer belong to the same functional family. It is seen that O_i^j stands for the "*i*th" node in the layer of "j".

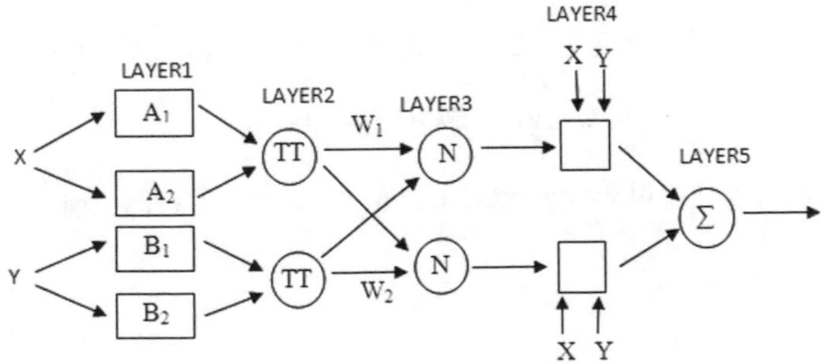

Fig. 3 ANFIS structure

Layer number 1: In this case, every node produces a membership grade in the layer of linguistic label as is used for *i*th node in Eq. 2.

$$O_i^j = \mu A_i(x) = \frac{1}{1 + \left[\left(\frac{x - c_i}{a_i}\right)^2\right]^{b_i}} \tag{2}$$

In this case, x stands for the input going toward the node 1 and A_i stands for linguistic label (small and large) attached to this node, $\{a_i, b_i, c_i\}$ denotes parameter set.

Layer number 2: Every node in the particular layer calculates the firing power by using Eq. 3.

$$O_i^2 = w_i = \mu A_i(x) \times B_i(y), i = 1, 2 \tag{3}$$

Layer number 3: In this layer, as shown in Eq. 4, the *i*th node finds out the ratio of *i*th rule's firing power to the sum of all firing power.

$$O_i^3 = \overline{w_i} = \frac{w_i}{w_1 + w_2}, i = 1, 2 \tag{4}$$

Output of layer is known as the normalized firing which circulates.

Layer number 4: In this case, every node i is a squared node with a node function that obtained from Eq. 5.

$$O_i^4 = \overline{w_i} f_i = \overline{w_i}(p_i + q_i y + r_i) \tag{5}$$

where W_i stands for the output of the layer number 3. All parameters in the layer are known as "Consequent Parameters".

Layer number 5: Using Eqs. 6 and 7, a single circle-type node calculated the output.

$$O_i^5 = \sum_i^n \overline{w_i} f_i = \frac{\sum_i w_i f_i}{\sum_i w_i} \tag{6}$$

$$\begin{aligned} f = {} & \frac{w_1}{w_1 + w_2} f_1 + \frac{w_1}{w_1 + w_2} f_2 = \overline{w} f_1 + \overline{w} f_2 (\overline{w} x) p_1 + (\overline{w} y) q_1 \\ & + (\overline{w}_1) r_1 + (\overline{w}_2 x) p_2 + (\overline{w}_2 y) q_2 + (w_2) r_2 \end{aligned} \tag{7}$$

is a linear one because of the consequent parameters $p_1, q_1, r_1, p_2, q_2, r_2$.

4 Model Evaluation

Several membership functions have been used for training ANFIS. Out of these, Gaussian membership function produces the best result. In this case, both artificial neural network (ANN) and fuzzy logic (FL) have been used to construct the model. Further after 30 epochs in the training process, the error gets minimized, and then it is followed by the testing process. Further, the fuzzy rule architecture consists of nine fuzzy rules generated from input–output data set and is based on Sugeno model. Furthermore, 9 performance measures have been used to conduct 30 cycles of learning. The ANFIS network, as shown in Fig. 4 composes of many nodes connected by the help of directional links. The average value of error is $5.2569e^{0.006}$ for the conventional EDM and is $4.5549e^{0.006}$ for the PMEDM.

The actual values in blue dots and predicted values in red dots for the two models are shown in Fig. 5. The deviation analysis is done for the predicted value of the model. Also, the deviation plots for the two models have been shown in Fig. 6.

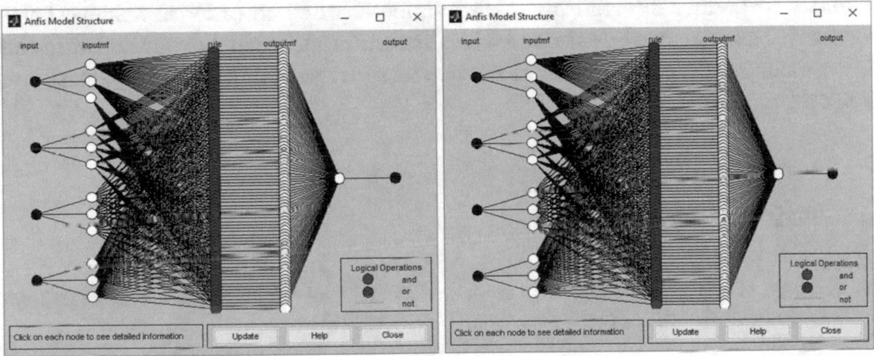

Fig. 4 ANFIS model structure

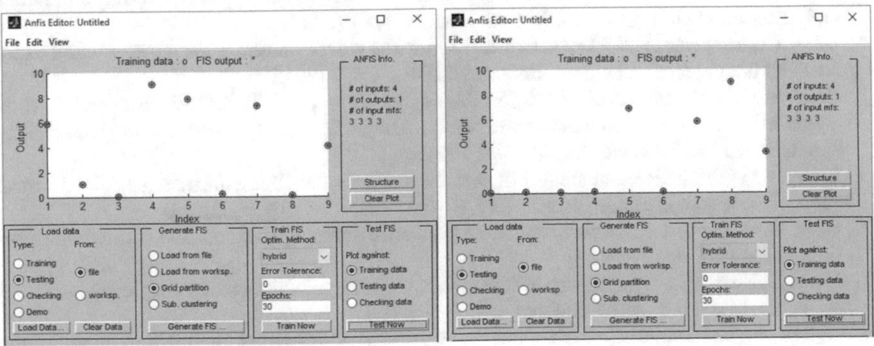

Fig. 5 Different response training

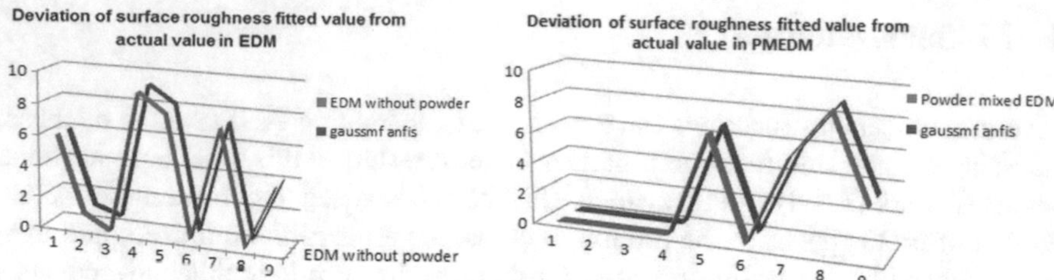

Fig. 6 Deviation of surface roughness in case of EDM and PMEDM

5 Conclusion

This paper proposed a suitable predictive process for calculation of surface roughness for simple EDM as well as advanced PMEDM. Further, this study discusses the comparison between EDM and PMEDM and establishes the superiority of PMEDM in comparison to EDM with ANFIS. ANFIS remains as one of the best soft computing methods to predict the load by an experimental study in order to achieve better accuracy. The work can be extended with simulation and prediction through different AI methods for different aspects of surface integrity such as surface roughness, and crack formation.

References

1. Kant, G., Rao, V.V., Sangwan, K.S.: Predictive modeling of turning operations using response surface methodology. Appl. Mech. Mater. **307**, 170 (2013)
2. Sangwan, K.S., Kant, G., Deshpande, A., Sharma, P.: Modeling of stresses and temperature in turning using finite element method. Appl. Mech. Mater. **307**, 174 (2013)
3. Wang, X., Da, Z.J., Balaji, A.K., Jawahir, I.S.: Performance-based predictive models an optimization methods for turning operations and applications: part 3—optimum cutting conditions and selection of cutting tools. J. Manuf. Process. **9**, 61 (2007)
4. Kant, G., Sangwan, K.S.: Predictive modeling for power consumption in machining using artificial intelligence techniques. Procedia CIRP **26**, 403 (2015)
5. Pusavec, F., Deshpande, A., Yang, S., M'Saoubi, R., Kopac, J., Dillon, O.W., Jawahir, I.S.: Sustainable machining of high temperature Nickel Alloy – Inconel 718: Part 2 – chip breakability and optimization. J. Clean. Prod. **87**, 941 (2015)
6. Karayel, D.: Prediction and control of surface roughness in CNC lathe using artificial neural network. J. Mater. Process. Technol. **209**, 3125 (2009)
7. Feng, C.X., Wang, X.F.: Surface roughness predictive modeling: neural networks versus regression. IIE Trans. **35**, 11 (2003)
8. Cus, F., Zuperl, U.: Approach to optimization of cutting conditions by using artificial neural networks. J. Mater. Process. Technol. **173**, 281 (2006)
9. Oktem, H., Erzurumlu, T., Erzincanli, F.: Prediction of minimum surface roughness in end milling mold parts using neural network and genetic algorithm. Mater. Des. **27**, 735 (2006)
10. Davim, J.P., Gaitonde, V.N., Karnik, S.R.: Investigations into the effect of cutting conditions on surface roughness in turning of free machining steel by ANN models. J. Mater. Process. Technol. **205**, 16 (2008)

11. Zain, A.M., Haron, H., Sharif, S.: Prediction of surface roughness in the end milling machining using artificial neural network. Expert Syst. Appl. **37**, 1755 (2010)
12. Jang, J.R.: Fuzzy modeling using generalized neural networks and Kalmman filter algorithm. In: Proceedings of Ninth National Conference on Artificial Intelligence, pp. 762–767 (1991)
13. Jang, J.R.S.: ANFIS: adaptive-network-based fuzzy inference system. IEEE Trans. Syst. Man Cybern. **23**(3), 665–685 (1993)
14. Sugeno, M., Kang, G.T.: Structure identification of fuzzy model. Fuzzy Sets Syst. **28**(1), 15–33 (1988)
15. Takagi, T., Sugeno, M.: Fuzzy identification of systems and its applications to modeling and control. Syst. Man Cybern. **15**(1), 116–132 (1985)

Trendingtags—Classification & Prediction of Hashtag Popularity Using Twitter Features in Machine Learning Approach Proceedings

P. Suthanthira Devi, R. Geetha and S. Karthika

Abstract Most of the tweets are associated with unique hash tags. Some hash tags become trending in a very short time. Predicting which hash tags will become trending in the near future is of significant importance for taking proper decisions in news media, marketing and social media advertising. This research work is aimed at predicting the popularity and tagging the hash tags using machine learning algorithms. It categorizes the popularity under five classes namely not popular, marginally popular, popular, very popular and extremely popular using content and contextual features. The content features can be extracted from both hashtag string and content of the tweets, whereas the contextual features can be extracted from the social network graph of the users. The features are evaluated based on the metrics such as micro-F1 score and macro-F1 score. The result shows that contextual features are more effective than content features as it has the highest prediction accuracy of 94.1%.

Keywords Twitter · Hashtag · Content features · Contextual features · Popularity · Machine learning algorithms

1 Introduction

Twitter is an online news and social networking service where users post and interact with messages, known as '*tweets*'. Each tweet is associated with an identifier (id), person who posted the tweet, date and time of posting, followers count, followings count, favorite count, retweet count, etc. Anyone sharing content on a relevant topic

P. S. Devi (✉) · R. Geetha · S. Karthika
Department of Information Technology, SSN College of Engineering,
Kalavakkam 603110, Tamil Nadu, India
e-mail: devisaran2004@gmail.com

R. Geetha
e-mail: geethakaajal@gmail.com

S. Karthika
e-mail: skarthika@ssn.edu.in

H. S. Behera et al. (eds.), *Computational Intelligence in Data Mining*, Advances in Intelligent Systems and Computing 990,
https://doi.org/10.1007/978-981-13-8676-3_15

can add a hashtag to their tweet [1]. Users create and use hashtags by placing the number/pound sign '#' in front of a string [2]. Hashtags make it possible for others to easily find messages with a specific topic [3].

Most of the tweets are associated with unique hashtags. Because of Twitter's popularity and viral nature of information sharing, some hashtags become trending in a very short time. Predicting which hashtags will become popular in the near future is of significant importance for taking proper decisions. It will also help news media and advertisers to improve the publicity in advance [4, 5].

The objective of the research work is to predict the popularity and tag the hashtag based on content and contextual features as proposed by [6, 7]. The content features are extracted from both the hashtag string and the content of the tweets whereas the contextual features are associated to twitter features and not extracted from the content of the tweets. The proposed work focuses on data collection, pre-processing and feature value computation in the extracted tweets [8, 9]. The state-of-the-art machine learning algorithms are used to analyze the popularity of the hashtag [10]. The features are evaluated based on the metrics such as accuracy, precision, recall, micro-precision, micro-recall, micro-F1 score, macro-precision, macro-recall and macro-F1 score. The Sect. 2 presents the previous works related to hashtag popularity prediction. Section 3 describes about the system design and focuses on the methods used. Section 4 highlights the results obtained. Section 5 concludes the research findings.

2 Related Works

In the paper [11], the prediction of the trend of new hashtags on Twitter was analyzed based on content and contextual features. The authors of [11], proposed models for 1-day and 2-day prediction of trending hashtags. They identify and evaluate the tweets based on the effectiveness of content and contextual features. Tweets were annotated with candidate hashtags. In the research work [12], early adoption properties including profile of tweet authors and adoption time series are used to predict a tag's later popularity level. Fourier transform (FT) spectrum and wavelet transform (WT) spectrum are used to augment the simple mean and standard deviation characteristics of the adoption time series.

The authors of [13], presented an efficient hybrid approach based on a linear regression for predicting the spread of an idea in a given time frame. They minimize the prediction error, by combining content features with temporal and topological features. The authors of [14], studied the variation in the ways that widely-used hashtags on different topics spread. They also discussed the persistence of diffusion of tweets based on topics such as idioms, political topics, etc.

3 Methodology

The proposed hashtag popularity prediction and tagging system has 4 major modules namely classification, prediction, analysis and identification of popular hashtags. The user logs-into his/her Twitter account from Twitter Archiver Tool. Tweets are collected by specifying the hashtag needed. The total dataset is divided into 3 sub-datasets for Training, Testing-1 and Testing-2. All the tweets are pre-processed using the pre-processing rules. The content and contextual features are applied for all the hashtags [9]. The popularity of training dataset is classified as Unpopular, Slightly Popular, Popular, Highly Popular or Extremely Popular based on threshold user count [15, 16].

The popularity of Testing-1 dataset is predicted by using Training dataset, while the popularity of Testing-2 dataset is predicted by using both Training and Testing-1 dataset. The classification models (Naïve Bayes, SVM, Decision Tree, and KNN) and regression model (Logistic Regression) are used for prediction. The prediction accuracy is evaluated for all the algorithms for both content and contextual features based on micro-F1 score and macro-F1 score. The sequential flow of the proposed system is presented in Fig. 1.

3.1 Twitter Data Collection

Collected twitter data can be used for data mining and sentiment analysis [17]. The data contains tweet text, tweet id, person who posted the tweet, date and time of posting the tweet, number of followers, number of followings, number of favorites, number of retweets, etc. The dataset is described in Table 1. Twitter data may be incomplete, inconsistent and noisy which can produce misleading mining results. Data pre-processing is used to resolve them by transforming unstructured tweet data from word to vector format. The pre-processing rules applied are shown in Table 2.

3.2 Data Pre-processing

Twitter data may be incomplete, inconsistent and noisy which can produce misleading mining results. Data pre-processing is used to resolve them by transforming unstructured tweet data from word to vector format. The pre-processing rules applied are shown in Table 2.

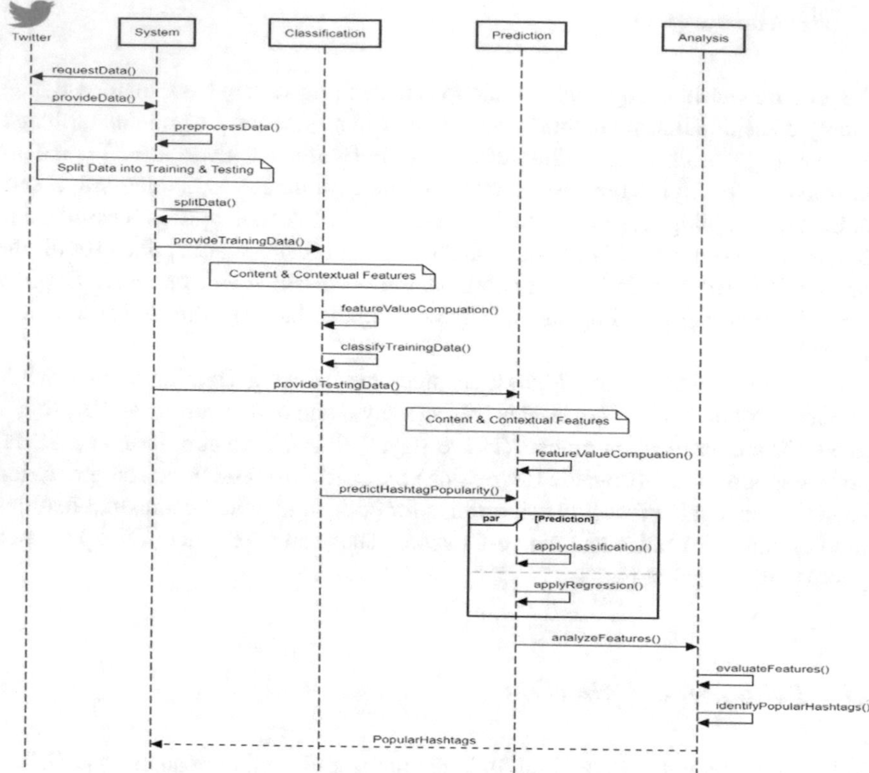

Fig. 1 Process flow for hashtag popularity

Table 1 Dataset description

S. No.	Dataset description	
1.	Data collection tool	Twitter archiver
2.	Number of hashtags chosen	18
3.	Time of data collection	09/10/2017 to 17/10/2017
4.	Number of tweets collected	2,34,088
5.	Training period (T_r)	9/10/2017 to 11/10/2017
	No. of Tweets	66,399
6.	Testing 1 period (T_{S1})	12/10/2017 to 14/10/2017
	No. of tweets	81,650
7.	Testing 2 period (T_{S2})	15/10/2017 to 17/10/2017
	No. of tweets	86,039

Table 2 Pre-processing rules

Rule No	Pre-processing rules	Rule description
1	Convert to lowercase	Transform tweet into small letters
2	RT removal	Remove RT
3	Replacement of user-mentions	Replace @-mentions with 'NAME'
4	URL replacement	Replace URLlinks with 'URL'
5	Person name removal	Stanford named entity recognizer is used to remove person names
6	Hash character removal	Remove '#' in hashtags
7	Lemmatization	Stanford core NLP is used to convert words to base form
8	Stopword removal	NLTK's English Stopword list is used to remove stop words
9	Uncommon word removal	NLTK's English word dictionary is used to remove uncommon English words

Table 3 Content features

Feature	Feature description	Short forms
N_{f1}	Digit usage	DU
N_{f2}	Segment words count	SWC
N_{f3}	Related news ratio	RNR
N_{f4}	URL ratio	UR
N_{f5}	3D-sentiment fraction	3D-SF
N_{f6}	Topic modelling	TM
N_{f7}	Hashtag clarity value	HCV
N_{f8}	Segment word clarity value	SCV

3.3 Content Features

These features are extracted from the hashtag string and content of the tweets. The content features are represented in Table 3. Figure 2 shows the content features taxonomy for feature analysis.

Digit Usage **(N_{f1})** is a binary feature that considers the presence or absence of digits in the hashtag string. The use of digits in hashtag increases its trend [6].

Segment Words Count **(N_{f2})** identifies the number of distinct words present in a hashtag [6]. When the number of segment words in a hashtag is high, its meaning will be clearer. As a result, many users will adopt it in their tweets.

Related News Ratio **(N_{f3})** identifies the number of related news happened in 't'. This feature is introduced since news has influence on the hashtag utility. RNR is computed by

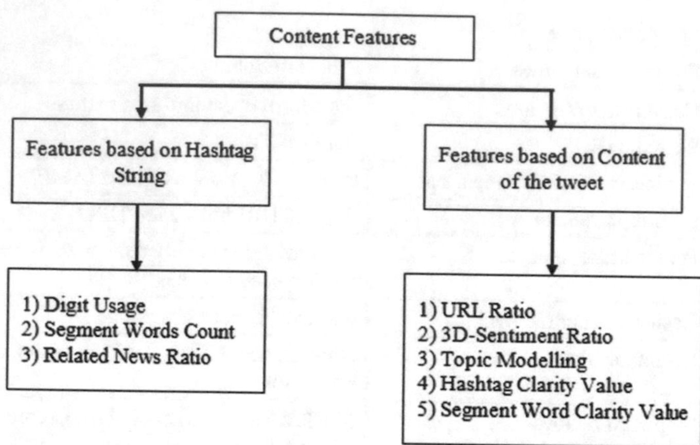

Fig. 2 Taxonomy of content

$$RNR = \frac{Number\ of\ Related\ News}{Total\ Number\ of\ News} \tag{1}$$

***URL Ratio* (N_{f4})** is the fraction of tweets containing URLs. This feature is considered since URLs linked to the tweets increase the number of reads and shares by the users [12]. UR is computed by

$$UR = \frac{Number\ of\ Tweets\ Containing\ URL}{TC} \tag{2}$$

***3D-Sentiment Fraction* <positive, negative, neutral> (N_{f5})** denotes the fraction of positive, negative, and neutral tweets. This feature is considered since popular events are typically associated with increase in sentiment strength [14]. Vader, python library for social media sentiment analysis is used for this purpose. 3D-SF can be computed as follows:

$$SF_{(POS)} = \frac{No.of\ Tweets\ Containing\ Positive\ Sentiment}{TC} \tag{3}$$

$$SF_{(NEG)} = \frac{No.of\ Tweets\ Containing\ Negative\ Sentiment}{TC} \tag{4}$$

$$SF_{(NEU)} = \frac{No.of\ Tweets\ Containing\ Neutral\ Sentiment}{TC} \tag{5}$$

***Topic Modeling* (N_{f6})** identifies the topic distribution of a hashtag. This feature is useful since hashtags on similar topics may have similar trend [11]. LDA (Latent Dirichlet Allocation) is used for topic modeling; Gensim, python library for topic modeling is used to generate bag of words.

Table 4 Contextual features

Feature	Feature description	Short forms
X_{f1}	Tweet count	TC
X_{f2}	User count	UC
X_{f3}	Mention fraction	MF
X_{f4}	Retweet fraction	RF
X_{f5}	Average authority score	AAS
X_{f6}	Graph density	GD
X_{f7}	Average edge strength	AES
X_{f8}	Triangle fraction	TF
X_{f9}	Component ratio	CR
X_{f10}	Border user count	BUC
X_{f11}	Exposure vector	EV

***Hashtag Clarity Value* (N_{f7})** denotes the clarity of hashtags. Hashtags having high clarity provide clear meaning [6]. As a result, many users will adopt them in their tweets. As defined in Eq. 6, HCV is computed using Kullback–Leibler Divergence between the unigram language models inferred from T'th and the background language model from T.

$$Clarity(h) = \sum_{w \in bag\, of\, words(h)} P\left(\frac{w}{T_t^h}\right) \log_2 \frac{P\left(\frac{w}{T_t^h}\right)}{P\left(\frac{w}{T}\right)} \tag{6}$$

***Segment Word Clarity Value* (N_{f8})** is applicable only for those hashtags which can be segmented into multiple words. It is used to evaluate the contribution of segment words in improving the bag of words, as well as the clarity [6].

3.4 *Contextual Features*

The contextual features are not extracted from the content of the hashtag or tweets. The contextual features are shown in Table 4 and its taxonomy is shown in Fig. 3.

The features are evaluated based on metrics such as accuracy, precision, and recall. Since two predictions (Testing-1 and Testing-2) have been done, there are two values for all the above metrics. In order to have a single value for the metrics as it will be easier for evaluation, the features are evaluated based on micro-F1 score and macro-F1 score. Micro-F1 Score is the harmonic mean of micro-precision and micro-recall, whereas Macro-F1 score is the harmonic mean of macro-precision and macro-recall. From the context and contextual features, specific feature can be chosen for further

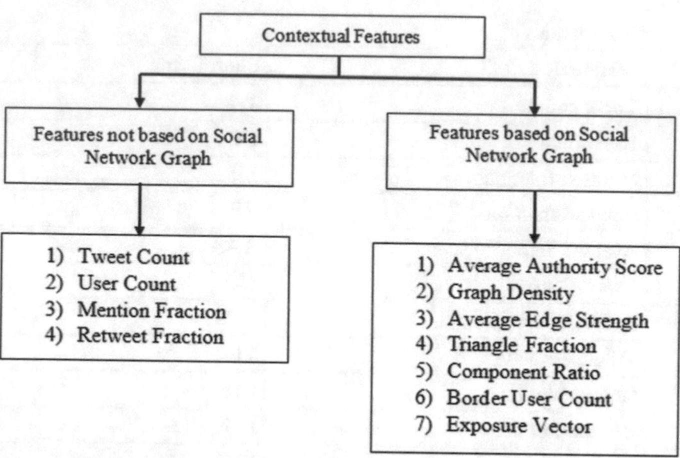

Fig. 3 Taxonomy of contextual

analysis using chi-Square test. But, in this present work all the features have been considered for analysis.

Tweet Count **(X_{f1})** is the total number of tweets for a hashtag in time period ‘t’. T_t^h is the collection of tweets posted for a hashtag ‘h’ in ‘t’. Time period (t) can be training, testing-1 or testing-2. TC determines the adoption level of a hashtag and is computed by

$$TC = \left|T_t^h\right| \tag{7}$$

User Count **(X_{f2})** is the number of users who have used the hashtag. U_t^h is the collection of users who have used a particular hashtag in ‘t’. Hashtag adoption will be more when UC is high. UC is computed by

$$UC = \emptyset_t^h = \left|U_t^h\right| \tag{8}$$

Mention Fraction **(X_{f3})** is the fraction of tweets containing @-mentions [6]. User @-mentions denote the user interaction in Twitter. If user1 mentions user2 in his/her tweet associated with hashtag ‘h’, user2 is more likely to reply. As a result, hashtag usage increases. MF is computed by

$$MF = \frac{No.of\ Tweets\ Containing\ @ - mentions}{TC} \tag{9}$$

Retweet Fraction **(X_{f4})** is the fraction of tweets containing RT [6]. It also denotes the user interaction in Twitter. If a user retweets another user’s tweet associated with a hashtag, his/her followers are more likely to retweet. As a result, hashtag usage increases. RF is computed by

$$RF = \frac{No.of\ Tweets\ Containing\ RT}{TC} \tag{10}$$

Average Authority Score (X_{f5}) measures the influential level of the community [6]. The users having high followers or followings count, or those who mentions many users or those who are mentioned by many users are influential users. These users increase the hashtag usage. Hits, python package are used for this purpose.
Graph Density (X_{f6}) measures the sparsity of graph [6]. It is the ratio between actual connections and potential connections in the graph. If the graph is denser, the social relationship among the users will be more which leads to increase in hashtag utility. GD is computed by

$$GD = \frac{EC}{UC * (UC - 1)} \tag{11}$$

Average Edge Strength (X_{f7}) measures the overall degree of user interaction in the graph. It is defined as the ratio between the total edge weights to that of total edge count (EC) in the graph [6]. The larger the edge weights, the more interaction among the users. AES is computed by

$$AES = \frac{Total\ Edge\ Weights\ in\ the\ Graph}{EC} \tag{12}$$

Triangle Fraction (X_{f8}) measures the overall degree of user interaction in the graph. It is defined as the fraction of users forming triangles in the graph in't' [6]. The larger the triangle fraction, the more interaction among the users. TF is computed by

$$TF = \frac{Number\ of\ Users\ forming\ Triangles}{UC} \tag{13}$$

Component Ratio (X_{f9}) measures the overall degree of user interaction in the graph. It is defined as the ratio between the number of connected components and UC in't' [6]. The smaller its value, the more interaction among the users. CR is computed by

$$CR = \frac{No.of\ Connected\ Components\ in\ the\ Graph}{UC} \tag{14}$$

Border User Count (X_{f10}) denotes the number of users, who have not adopted the hashtag, but are mentioned by hashtag users in 't' [6]. Hashtag users induce the border users to use the hashtag in their tweets. BUC is computed by

$$BUC = \left|BU_t^h\right| \tag{15}$$

Exposure Vector (X_{f11}) is a 6-dimensional probability vector that denotes the fraction of border users having k-edges from the hashtag users [6]. The number of exposures is the number of edges a border user has from the hashtag users. The border users with more exposures to a hashtag are expected to be more likely to adopt the hashtag. The exposure probability can be computed by

$$P[k] = \frac{Number\ of\ Border\ Users\ having\ k - edges}{TC} \quad (16)$$

4 Results and Discussion

A dataset of 234,088 tweets related to 18 unique hashtags has been collected by using Twitter Archiver. The tweets are divided into single phase training and two phases of testing datasets. The collected tweets are further pre-processed based on the rules discussed in the previous section and the Table 5 presents the rule-wise as well as the overall pre-processed sample tweet captured based on #rallyforrivers.

The prediction of the popular hashtags has been analyzed upon 18 unique hashtags namely #rallyforrivers, #window10, #diwali, #yourbestdiwaligift, #apple, #iphone, #iphone8, #neet, #indian, #innovatorsrace50, #makeinindia, #happydiwali, #film, #air, #times, #people, #modi, #mersal. The collected tweets are further pre-processed based on the rules discussed in the previous section and the Table 6 presents the rule-wise as well as the overall pre-processed sample tweet captured based on #rallyforrivers.

4.1 Popularity Prediction

The collected dataset used for predicting the popularity of Hashtag. So the data collection time frame is considered in 3 phases, where phase 1 for training and phase 2 and phase 3 for testing of the hashtag popularity. The training phase classifies the popular hashtag.

The testing-1 phase predicts the near-time popularity of the already identified hashtags. The testing-2 phase predicts the long-run popularity of the same set of hashtag. The hashtag popularity over different phases using Logistic Regression algorithm is shown in Table 6. The result column denotes whether the popularity increases, decreases or maintains from Testing-1 to Testing-2 or not. The training data for each hashtag is classified as one of the popularity classes (UnPopular, Slightly Popular, Popular, Highly Popular, and Extremely Popular) based on threshold user count (Φ) and is shown in Table 7. Based on the dataset, Φ is set to 100.

Table 5 Pre-processing of sample tweet

Sample tweet	RT @SirKenRobinson: Huge congratulations to Sadhguru & all involved in historic momentum of #RallyForRivers. https://t.co/N2WwJ81VKV	
Rule No.	Rule name	Pre-processed tweet
1	Convert to lowercase	rt @sirkenrobinson: huge congratulations to sadhguru& all involved in historic momentum of #rallyforrivers. https://t.co/n2wwj81vkv
2	RT removal	@sirkenrobinson: huge congratulations to sadhguru& all involved in historic momentum of #rallyforrivers. https://t.co/n2wwj81vkv
3	Replacement of user mentions	NAME: huge congratulations to sadhguru& all involved in historic momentum of #rallyforrivers. https://t.co/n2wwj81vkv
4	URL Replacement	NAME: huge congratulations to sadhguru& all involved in historic momentum of #rallyforrivers. URL
5	Person name removal	NAME: huge congratulations to & all involved in historic momentum of #rallyforrivers. URL
6	Hash character removal	NAME: huge congratulations to & all involved in historic momentum of rallyforrivers. URL
7	Removal of punctuations and symbols	NAME huge congratulations to all involved in historic momentum of rallyforrivers URL
8	Lemmatization	NAME huge congratulation to all involve in history momentum of rallyforrivers URL
9	Stopword removal	NAME huge congratulation involve history momentum rallyforrivers URL
10	Uncommon word removal	NAME huge congratulation involve history momentum URL
Pre-processed tweet	NAME huge congratulation involve history momentum URL	

Table 6 Hashtag popularity over different phases

Hashtag	Training	Testing-1	Testing-2	Result
#rallyforrivers	Popular	Popular	Very	Increase
#windows10	Popular	Extremely	Very	Decrease
#diwali	Extremely	Extremely	Extremely	Maintain
#yourbestdiwaligift	Popular	Very	Popular	Decrease
#apple	Extremely	Extremely	Extremely	Maintain
#iphone	Extremely	Extremely	Extremely	Maintain
#iphone8	Very	Very	Very	Maintain
#neet	Marginally	Marginally	Marginally	Maintain
#indian	Very	Extremely	Extremely	Maintain
#innovatorsrace50	Not	Marginally	Not	Decrease
#makeinindia	Very	Very	Very	Maintain
#happydiwali	Marginally	Extremely	Extremely	Maintain
#film	Extremely	Extremely	Extremely	Maintain
#air	Very	Very	Very	Maintain
#times	Marginally	Popular	Marginally	Decrease
#people	Very	Extremely	Extremely	Maintain
#modi	Very	Very	Very	Maintain

Table 7 Hashtag distribution

Popularity classes	UnPopular	Slightly popular	Popular	Highly popular	Extremely popular
Range	$[0, \Phi)$	$[\Phi, 5\Phi)$	$[5\Phi, 10\Phi)$	$[10\Phi, 30\Phi)$	$[30\Phi, \infty)$
No. of hashtags	1	3	3	6	5

4.2 Evaluation of Features

The Content and Contextual features are evaluated using Weka Tool. The classification algorithms Support Vector Machine, KNN, Naive Bayes and Decision Tree along with the logistic regression algorithms have been applied. For KNN, the value of k is set to 4. Because in general the social reach ability from one node at time 't' is 4 [3]. Based on the training dataset, the popularity classes for the testing dataset are predicted. The evaluation metrics for Content, Contextual and both content and contextual features are shown in Tables 8, 9 and 10 respectively. Micro-F1 score and Macro-F1 score are high for contextual features, and comparatively low for content features. It is because the social network graph improves information diffusion, but the content features do not. The content features are represented as count values. As a result, these features alone do not contribute much to the prediction of hashtag

Table 8 Evaluation metrics for content features

Classification Algorithm	Average accuracy	Precision 1	Recall 1	Precision 2	Recall 2	Micro-precision	Micro-recall	Micro-F1 score	Macro-precision	Macro-recall	Macro-F1 score
Naïve Bayes	0.855	0.611	0.611	0.538	0.538	0.58	0.58	0.58	0.574	0.574	0.574
Decision tree	0.822	0.388	0.388	0.615	0.615	0.483	0.483	0.483	0.502	0.502	0.502
SVM	0.866	0.5	0.5	0.769	0.769	0.612	0.612	0.612	0.634	0.634	0.634
KNN (k = 4)	0.866	0.5	0.5	0.769	0.769	0.612	0.612	0.612	0.634	0.634	0.634
Logistic regression	0.877	0.666	0.666	0.615	0.615	0.645	0.645	0.645	0.641	0.641	0.641

Table 9 Evaluation metrics for contextual features

Classification algorithm	Average accuracy	Precision 1	Recall 1	Precision 2	Recall 2	Micro-precision	Micro-recall	Micro-F1 score	Macro-precision	Macro-recall	Macro-F1 score
Naïve Bayes	0.911	0.666	0.666	0.888	0.888	0.777	0.777	0.777	0.777	0.777	0.777
Decision tree	0.944	0.833	0.833	0.888	0.888	0.861	0.861	0.861	0.861	0.861	0.861
SVM	0.922	0.833	0.833	0.777	0.777	0.805	0.805	0.805	0.805	0.805	0.805
KNN (k = 4)	0.866	0.722	0.722	0.611	0.611	0.666	0.666	0.666	0.666	0.666	0.666
Logistic regression	0.944	0.888	0.888	0.833	0.833	0.861	0.861	0.861	0.861	0.861	0.861

Table 10 Evaluation metrics for both content & contextual features

Classification algorithm	Average accuracy	Precision 1	Recall 1	Precision 2	Recall 2	Micro-precision	Micro-recall	Micro-F1 score	Macro-precision	Macro-recall	Macro-F1 score
Naïve Bayes	0.888	0.666	0.666	0.777	0.777	0.722	0.722	0.722	0.722	0.722	0.722
Decision tree	0.944	0.833	0.833	0.888	0.888	0.861	0.861	0.861	0.861	0.861	0.861
SVM	0.922	0.833	0.833	0.777	0.777	0.805	0.805	0.805	0.805	0.805	0.805
KNN (k = 4)	0.9	0.777	0.777	0.722	0.722	0.75	0.75	0.75	0.75	0.75	0.75
Logistic regression	0.933	0.944	0.944	0.722	0.722	0.833	0.833	0.833	0.833	0.833	0.833

popularity. Using Contextual features, Decision Tree and Logistic Regression algorithms perform well among others as they have highest accuracy of 94.4%. Using both content and contextual features, the F1 scores are optimal. It is because the poor F1 score values of content features had an influence on the F1 score values of both content & contextual features. Decision Tree algorithm performs well among the five algorithms. The ranked precision values are used for evaluation. It is because decision tree algorithm depends much on the features. From the result, it is concluded that either contextual features alone or both content & contextual features can be used for prediction of hashtag popularity.

5 Conclusion

This research work predicts the popularity of Twitter hashtags using content and contextual features. The predicted popular hashtags for our dataset are 18 unique hashtags. From the obtained results, it is clear that the system performs better than the previous works. Decision tree and Logistic regression are found to be the best classification and regression algorithm respectively. Among all features, contextual features perform better prediction. It out-performs content features and has an accuracy of 94.4%. The major contribution of this research work is the improvement of the existing content features like hashtag segmentation and the introduction of a new content feature—Related News Ratio. The authors have made effort to automatically segment the hashtags, which showed better results than earlier manual segmentation. The new content feature also clearly exhibits the popularity of hashtags. The authors have also analyzed the popularity of hashtags based on digit usage, border users and relationship between hashtags. These analyses are not done in the previous works.

References

1. Huberman, B.A., Romero, D.M., Wu, F.: Social networks that matter: twitter under the microscope. arXiv preprint arXiv:0812.1045 (2008)
2. Newman, T.P.: Tracking the release of IPCC AR5 on Twitter: users, comments, and sources following the release of the working group I summary for policymakers. Public Underst. Sci. **26**(7), 815–825 (2017)
3. Romero, D.M., Meeder, B., Kleinberg, J.: Differences in the mechanics of information diffusion across topics: idioms, political hashtags, and complex contagion on twitter. In: Proceedings of the 20th International Conference on World Wide Web. ACM (2011)
4. Kwak, H., et al.: What is Twitter, a social network or a news media?. In: Proceedings of the 19th International Conference on World Wide Web. ACM (2010)
5. Kreiss, Daniel: Seizing the moment: the presidential campaigns' use of Twitter during the 2012 electoral cycle. New Media Soc. **18**(8), 1473–1490 (2016)
6. Ma, Z., Sun, A., Cong, G.: On predicting the popularity of newly emerging hashtags in twitter. J. Assoc. Inf. Sci. Technol. 1399–1410 (2013)

7. Li, C., Sun, A., Datta, A.: Twevent: segment-based event detection from tweets. In: Proceedings of the 21st ACM International Conference on Information and Knowledge Management. ACM (2012)
8. Tsur, O., Rappoport, A.: What's in a hashtag? content based prediction of the spread of ideas in microblogging communities. In: Proceedings of the Fifth ACM International Conference on Web Search and Data Mining. ACM (2011)
9. Jeon, J., et al.: A framework to predict the quality of answers with non-textual features. In: Proceedings of the 29th Annual International ACM SIGIR Conference on Research and Development in Information Retrieval. ACM (2006)
10. Doong, S.H.: Predicting Twitter hashtags popularity level. In: 49th Hawaii International Conference on System Sciences (HICSS). IEEE (2011)
11. Blei, D.M., Ng, A.Y., Jordan, M.I.: Latent Dirichlet allocation. J. Mach. Learn. Res. **3**, 993–1022 (2003)
12. Dong, A., et al.: Time is of the essence: improving recency ranking using twitter data. In: Proceedings of the 19th International Conference on World wide web. ACM (2010)
13. Kasiviswanathan, S.P., et al.: Emerging topic detection using dictionary learning. In: Proceedings of the 20th ACM International Conference on Information and Knowledge Management. ACM (2014)
14. Thelwall, M., Buckley, K., Paltoglou, G.: Sentiment in Twitter events. J. Assoc. Inf. Sci. Technol. **62**(2), 406–418 (2014)
15. Ma, Z., Sun, A, Cong, G.: Will this #hashtag be popular tomorrow?. In: Proceedings of the 35th International ACM SIGIR Conference on Research and Development in Information Retrieval. ACM (2012)
16. Yang, Xiao, Macdonald, Craig, Ounis, Iadh: Using word embeddings in Twitter election classification. Inf. Retr. J. **21**(2-3), 183–207 (2018)
17. Calais Guerra, P.H., et al.: From bias to opinion: a transfer-learning approach to real-time sentiment analysis. In: Proceedings of the 17th ACM SIGKDD International Conference on Knowledge Discovery and Data Mining. ACM (2011)

Enhancement of Identification Accuracy by Handling Outlier Feature Values Within a Signature Case Base

Uttam Kumar Das, Shisna Sanyal, Anindita De Sarkar and Chitrita Chaudhuri

Abstract This is a study of applying Case Based Reasoning techniques to identify a person using offline signature images. Classification related to proper identification is achieved by comparing distance measures of new test cases with existing cases within the base. The patterns pertaining to the images are captured by fusing some standard global features with some indigenously developed local feature sets. Outlier values in both these sets are handled to maintain statistical tolerable limits. The effect of outlier handling within feature values is found to enhance identification accuracy for two standards and one indigenously collected offline signature sets utilized in the experimental phase.

Keywords Case based reasoning (CBR) · Classification · Distance measures · Global and local feature values · Outliers

1 Introduction

The handwritten signature of any person is an important biometric characteristic which is used prevalently for identification in financial and business transactions. Invigilation in examination halls are also conducted via signature identification and authentication. System fraudulence and unauthorised intrusion are crimes which are on the rise these days. One time-tested method of protection against these is by the

U. K. Das · S. Sanyal (✉) · A. De Sarkar · C. Chaudhuri
Department of Computer Science and Engineering, Jadavpur University, Kolkata 700032, India
e-mail: sanyal.shisna2@gmail.com

U. K. Das
e-mail: uttamkrdas95@gmail.com

A. De Sarkar
e-mail: aninditadesarkar@gmail.com

C. Chaudhuri
e-mail: cchitrita@gmail.com

H. S. Behera et al. (eds.), *Computational Intelligence in Data Mining*, Advances in Intelligent Systems and Computing 990,
https://doi.org/10.1007/978-981-13-8676-3_16

use of an identification system based on handwritten signatures [1, 2]. This method is cost-efficient and simple as compared to other biometric methods.

Handwritten signature, strictly defined, is full or part of a name written in ones' own handwriting. But in reality signatures are composed of special characters and flourishes and therefore most of the time they can be unreadable. Hence the main objective of this research is to introduce machine learning techniques to improve readability and subsequently the identification of the signatory. This in itself is a challenging job with little benchmarking history available for large datasets [3] and provided the primary motivation of the present research work.

Signature identification systems are broadly classified in two ways: online and offline. As the name signifies, online signatures are directly signed onto the electronic tablets or the screen using some appliance like a digital pen. Thus the technique involves sophisticated and costly tools.

Here we are restricting ourselves only to offline techniques for person identification. In this technique, a signature is made on an external media such as a piece of paper and scanned to produce a digitized copy to be stored within a computer system. Due to the relative ease of use of an offline system, a number of applications worldwide prefer to use this system for person verification.

In the proposed system, the preserved signatures of an individual are used to train the machine to recognize the person when an authentic signature of the same person is presented. Signatures are preserved as attributes or features extracted from the individual's signature images. These features are considered as the input, which are then processed based on some predefined standardized methods to produce a class value. The class value gives the identity of the signatory.

So the present work is oriented towards building a classifier that helps to detect the identity of a person based on some offline signatures of that person preserved for training. Ideally, the training set needs to be incorporated in a manner so that it enhances the knowledge base of the system, keeping it in a constantly updated state. The techniques associated with CBR have been employed here to utilize incremental learning procedure.

In a machine learning environment, features play a vital role as they are the determining factor on which recognition depends. Here the effect of detecting and handling outlier values within the features, using some prevailing statistical techniques, have substantially contributed towards improving identification accuracy. Utilization of some indigenously developed attributes such as the interior to outline pixel ratio, number of significant components and local angular distance features also affected the success rate of the system positively.

The following section is utilized to present literature review as well as some basic concepts associated with the proposed system. Section 3 describes the different methodologies adopted to implement classification and enhance accuracy of identification. Section 4 contains the descriptions of datasets and system configurations. It also contains tables and bar-charts to depict the outcome of the experiments. The final section concludes with inferences drawn from the results, as well as some future directions of research which may further augment the accuracy and performance of the system. The references of all works are provided at the end.

2 Related Work and Concept Building

In this section we present some existing research work in the current domain. We also elaborate on some machine learning and statistical techniques adopted in our system.

2.1 Literature Review

Muhammad Reza Pourshahabi et al., in their paper [1], suggest a new offline handwritten signature identification and verification system based on Contourlet transform (CT). CT is used as feature extractor in their system. Signature image is enhanced by removing noise and then it is normalized by size. They have obtained reliable results for both Persian and English signature sets. However, the dataset sizes are too small to utilize them fruitfully for bench-marking purpose in spite of their claimed high identification accuracy.

Ghazali Sulong et al., in their paper [2] utilize techniques that address the challenge of focusing on the meaning of the handwritten signature and also on the individuality of the writer, by using Adaptive Window Positioning. The process has produced efficient and reliable method for accurate signature feature extraction for the identification of offline handwritten signatures with a GPDS data set containing 4870 signature samples from 90 different writers. They further claim contribution towards detecting signatures signed under emotional duress. They have primarily presented only verification results for comparison and no documentation could be found to benchmark the identification process under consideration.

Meenakshi K. Kalera et al., in their paper [3] described a novel approach for signature verification and identification in an offline environment based on a quasi-multi-resolution technique using GSC (Gradient, Structural and Concavity) features for feature extraction, reporting results with accuracies of 78% and 93% for verification and identification, respectively. Yet, their method, based on obtaining statistical distance distributions within and between writers, were tested on a small dataset. So, it does not provide a proper challenge to the present identification system.

2.2 Overview of CBR Techniques

In the present work, a CBR classifier is utilized to identify a person by comparing distances between vectors comprising of feature values to represent offline handwritten signature images of that person. There are many available classification techniques amongst which CBR is employed here.

CBR classifiers treat every problem-solution pair as a case and each such case is stored in a base. An unsolved problem is supplemented with its correct solution which represents its class value. Often, a case-base, besides a detailed statement of the problem and its solution, also houses the necessary meta-data required for the problem.

As mentioned in the work by Farhan et al. [4], CBR brings some important advantages to the problem-solving strategy. It can reduce the processing time significantly and also be very useful when domain knowledge is not completely available or not easy to obtain, although extensive knowledge and expertise in the field always helps while modifying the similar solutions to produce a new solution. It also accommodates incremental learning techniques by allowing new cases to upgrade the system knowledge overall.

Most importantly, potential errors can be avoided and past mistakes rectified in similar cases, while attending to the problem in hand. Search time may be reduced by a fool-proof indexing technique.

Thus CBR is an AI technique that considers old cases to take decision for new situations. The old cases constitute past experiences on which one can rely, rather than on rules, during the decision making process. CBR works by recalling similar cases to find solution to new problems.

If an exact match is found for the present problem, there is no need for insertion. Otherwise, cases constituting the nearest matches are considered, and their class information is collected to provide the new case with a suitable class value. The new case with its new solution is now ready for insertion into the case-base.

The CBR processes include four main steps described below [5]:

(1) **Retrieve**: Given a target problem, retrieve from memory cases relevant to solving it. For fast retrieval the pre-requisite is an efficient indexing technique.
(2) **Reuse**: The solution(s) from the retrieved case(s) need to be mapped to the target problem. This may involve adaptation or reuse of the solution(s) as needed to fit the new situation.
(3) **Revise**: Having mapped the retrieved solution(s) to the target situation, the new solution generated has to be tested in the real world (or a simulation) and, if necessary, need to be revised.
(4) **Retain**: After the solution has been successfully adapted and revised for the target problem, retain the resulting experience as a new case in memory.

2.3 *Distance Measures*

The distance measures give us the proximities between objects [6]. In our identification problem, we treat each signature as a vector in an n-dimensional space, where every element of the vector is represented by one of the n feature values. So to calculate the similarity (or proximity) between two signature images i.e., two points in the n-dimensional space, we utilize the Manhattan distance measure. As most of the features exist originally as real values in different ranges, we have further normalized these values to discrete levels within a specific range.

3 Proposed System Methodology

Here CBR is being utilized to establish a person's identity with the help of offline digitized signatures. A fresh signature of the person is taken and matched against the authentic signatures already preserved within a case-base to seek the accurate identity of the person. A set of authentic signatures of a person need to be scanned, digitized and pre-processed prior to extraction of feature values in the form of discrete numbers. The normalized feature values can now be stored in a case-base with other general relevant information about the person, as depicted in Fig. 1, to introduce a starting expertise level within the knowledge-base.

The detail of each person consists of an index value, authentic signature list and other general information. Following Fig. 2 shows how for every signature the normalized feature values and the collection time stamp (if available) are stored.

Data is often represented by the first quartile (Q1) or 25th percentile, the second quartile (Q2) or the median or 50th percentile, and the third quartile (Q3) or 75th percentile [7]. The Inter Quartile Range (IQR) is obtained by the expression Q3 − Q1. The minimum and the maximum values are measured as values falling 1.5 × IQR below Q1 or above Q3. These are marked as the minimum and maximum of the tolerable range [6]. By a common rule of thumb, all values beyond the minimum and maximum may be singled out as suspected outliers. All these are stored in a feature-base shown in Fig. 3.

A digital image of a signature is pre-processed for image binarization, noise reduction, image cropping, skeletonization and resizing. The final description of

Person 1	Authentic Signature List	General Information
Person 2	Authentic Signature List	General Information
:	:	:
Person P	Authentic Signature List	General Information

Fig. 1 Structure of the case base

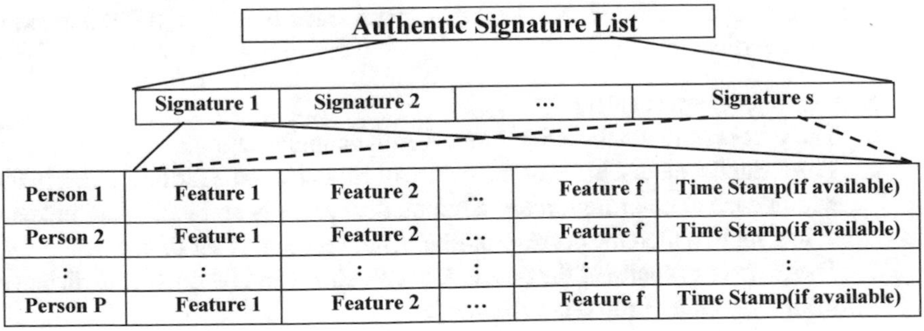

Authentic Signature List

Signature 1	Signature 2	...	Signature s

Person 1	Feature 1	Feature 2	...	Feature f	Time Stamp(if available)
Person 2	Feature 1	Feature 2	...	Feature f	Time Stamp(if available)
:	:	:	:	:	:
Person P	Feature 1	Feature 2	...	Feature f	Time Stamp(if available)

Fig. 2 Authentic signature list details

Feature 1	Q1	Q2	Q3	IQR	Tolerable Range
Feature 2	Q1	Q2	Q3	IQR	Tolerable Range
:	:	:	:	:	:
Feature f	Q1	Q2	Q3	IQR	Tolerable Range

Fig. 3 Structure of the feature base

the image is obtained by the process of feature extraction [8]. Machine learning techniques utilize this description as attribute sets, which are presented below in the following Sect. 3.1.

3.1 Attribute Sets

An attribute is an aspect of an instance or a specification that defines a property of an object. Attributes are often called features in Machine Learning. A special attribute is the class label that defines the class the instance belongs to [9]. The total set of attributes comprises of two types of features extracted from the digital image:

Global Attributes: A set of 23 attributes obtained from the whole of the image comprises of the following particulars of the digitized signature: [10–12].

Signature height, Pure width, Image area, Vertical centre, Horizontal centre, Maximum vertical projection, Maximum horizontal projection, Number of vertical projection peaks, Number of horizontal projection peaks, Baseline shift, Number of edge points, Number of cross points, Number of closed loops, Top heaviness, Horizontal dispersion, Interior to outline pixel ratio, Mean ascending height, Mean descending height, Reduced no of components, Number of significant components, Global slant, Local slant, Mean slant.

Local attributes: 40 angular distance features, indigenously developed as follows:

I. Binarized, cropped and thinned signature image re-sized to 400 × 100 pixels.
II. Next the image is divided into 4 vertical parts, each of 100 × 100 dimension. For each vertical part:

   a. Centre of mass is calculated.
   b. The part is further sub-divided into 8 equal angular regions.
   c. The 8 Euclidean distances of the centre of mass from the farthest black pixel within each angular region are computed. If there is no black pixel present in any region, then the corresponding distance value is set to 0.
   d. The geometric centre of the part is also calculated and its Euclidean distance from the centre of mass is computed.

e. The tenth attribute of the part is obtained by calculating the angle with the horizon subtended by the line joining the centre of mass and the geometric centre of the part.

Thus we have $10 \times 4 = 40$ local attributes for each signature image.

Combined Set of Attributes = Global (23) + Local (40) = Overall (63)

All these attribute values are next normalized. Normalization, in the present context, is the process of converting continuous attributes into discrete ones for a machine learning algorithm [13].

3.2 *Algorithm: Identification Accuracy*

```
Input:
P – Total number of persons in the dataset
F – Total number of features
S – Total number of training signatures per person
ST – Total number of test signatures per person
FV[P,S,F] – Corresponding feature vector
MAX[F] – Vector storing maximum value for all features
MIN[F] – Vector storing minimum value for all features
D – Discretization level
LARGE – System defined largest integer

Output: Percentage of Identification accuracy

Method – 1 // Main method with capping of features
BEGIN
1. Call Max_Min_IQR ( )
2. Call Feature_Capping ( )
3. Call Discretization ( )
4. Call Identification_Accuracy ( )
END

Method – 2 // Main method without capping of features
BEGIN
1. Call Max_Min ( )
2. Call Discretization ( )
3. Call Identification_Accuracy ( )
END
```

```
Procedure Max_Min ( ) // Highest & lowest feature values
BEGIN
For each f = 1 to F
    Set min = max = FV [1,1,f ]
    For each p = 1 to P
        For each s = 1 to S
            If (FV[p,s,f] < min) min = FV[p,s,f]
            If (FV[p,s,f] > max) max = FV[p,s,f]
        EndFor
    EndFor
    MIN[f]= min
    MAX[f]= max
EndFor
END
```

```
Procedure Max_Min_IQR ( ) // IQR-based Highest & lowest
BEGIN
For each f =1 to F
    Set i = – 1
    For each p = 1 to P
        For each s = 1 to S
            t = t + 1
            M[t]=FV[p,s,f]
        EndFor
    EndFor
    Vector M is sorted in ascending order.
    Q1[f] = 1st Quartile or 25th percentile value of M
    Q2[f] = 2nd Quartile or 50th percentile value of M
    Q3[f] = 3rd Quartile or 75th percentile value of M
    Inter-Quartile Range IQR[f]= Q3[f]– Q1[f]
    Lower-Limit MIN[f]= Q1– 1.5*IQR[f]
    Upper-Limit MAX[f]=Q3+1.5*IQR[f]
EndFor
END
```

```
Procedure Feature_Capping ( ) // Values limited to range
BEGIN
For each  p = 1 to P
        For each  s = 1 to S
                For each  f = 1 to F
                        If FV[p,s,f]<MIN[f] then
                                FV[p,s,f]=MIN[f]
                        EndIF
                        If FV[p,s,f]>MAX[f] then
                                FV[p,s,f]=MAX[f]
                        EndIF
                EndFor
        EndFor
EndFor
END
```

```
Procedure Discretization ( ) // Convert to Discrete values
BEGIN
For each p = 1 to P
 For each s = 1 to S+ST
        For each f = 1 to F
          If  MAX[f] == MIN[f] then
                FV[p,s,f]=0
          Else
                FV[p,s,f]
                = int(((FV[p,s,f]– MIN[f]) / (MAX[f]– MIN[f]))*D)
          EndIF
        EndFor
 EndFor
EndFor
END
```

```
Procedure Identification_Accuracy ( ) // Calculate Accuracy
BEGIN
Set count=0
For each p = 1 to P
 For each s = 1 to ST
        U– Test signature s of person p
        Set dist = LARGE
        For each p1 = 1 to P
         For each s1 = 1 to S
                V – Authentic signature s1 of person p1
                d = distance between signatures U and V
                If d<dist then
                        dist = d
                        id = p1
                EndIF
         EndFor
        EndFor
        If id==p then
                count = count + 1
        EndIF
 EndFor
EndFor
Evaluate accuracy = (count / (P * ST )) * 100
END
```

4 Experimental Setup and Results

We provide below the description of the datasets and results obtained through the experiments carried out using MATLAB R2017a for processing images.

4.1 Datasets

Following is a description of the 3 datasets used to prove the viability of the system and sample signatures from each set are depicted in Fig. 4.

Dataset 1 (Designated as OUR Dataset): This signature set was collected from 121 volunteers within University students with 20 authentic signatures per person. All signatures were scanned at a resolution of 200 dpi to obtain grey scale images.

Dataset 2 (Designated as ATVS Dataset): This one consists of 2250 signatures belonging to the standard MCYT Bimodal Biometric Database [14] scanned at a resolution of 300 dpi with 15 genuine and 15 skilled forgeries for each of 75 persons.

Dataset 3 (Designated as Persian Dataset): A Persian Dataset [15] containing handwritten signatures of 115 persons, with 27 genuine signatures per person collected from graduate students of University of Tehran and Sharif University of Technology and reported to have been scanned at 600 dpi resolutions.

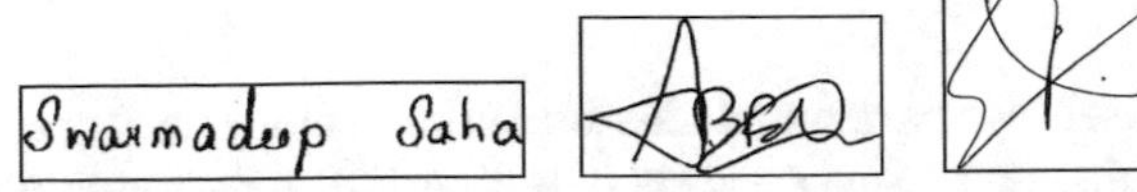

Fig. 4 Sample signature of dataset 1, 2 and 3

4.2 Results

See Tables 1, 2.

Table 1 Accuracy percentage obtained with the proposed system

Dataset	No. of persons	Training set size	Test set size	Method	Discrete level	Manhattan accuracy (%)	Euclidian accuracy (%)
1 (OUR)	121	121 × 15	121 × 5	With capping	12	**97.1901**	94.7107
				W/O capping	15	**97.1901**	94.8760
2 (ATVS)	75	75 × 10	75 × 5	With capping	22	**82.6667**	72.5333
				W/O capping	31	81.3333	72.5333
3 (PERSIAN)	115	115 × 20	115 × 7	With capping	17	**87.5776**	73.2919
				W/O capping	22	83.4783	71.6770

Table 2 Identification accuracy % of proposed system compared with other works

Reference	Dataset name	No. of persons	Training signatures per person	Test signatures per person	Best accuracy level (%)
Proposed	OUR	121	15	5	97.19
[1]	Stellenbosch	22	10	20	93.20
[3]	Database A	55	16	8	93.18

5 Conclusion and Future Scope

For all three datasets, we have found identification accuracy level to be the best using normalized training and tested datasets based on IQR calculation with capping of attribute values within the training set. But this technique does not seem to affect the accuracy level in any way for dataset 1 (indigenous OUR dataset). If outlier attributes are not capped, the performance of the system has been found to deteriorate for both the standard datasets 2 and 3 (Fig. 5). The performance of the proposed system on OUR indigenous dataset 1 has been found to be better than those reported by the referred related works on their own datasets.

So the usage of statistical techniques calculated over the training set with capping applied on the same to effectively handle outlier attributes, is clearly indicated for best performance overall. As experimentally proved, the Manhattan distance measure was selected for comparison based on the findings of research work done on the same sets of data. For the indigenous dataset, we achieve 97.19% of identification accuracy utilizing these techniques. The Persian dataset establishes that signatures are treated as patterns by the proposed system irrespective of language.

A future scope in this domain may involve updation of the case-base by inserting newly identified signatures either by appending (if there is scope) or in lieu of existing ones otherwise. Old signatures can be replaced in two ways—first by calculating difference in time stamps of the two signatures and replacing with the new only if this difference exceeds some predefined threshold value. The second method would compare proximity of the median signature of a person with all the existing authentic signatures of that person, and replace the most distant one with the new, if it is nearer.

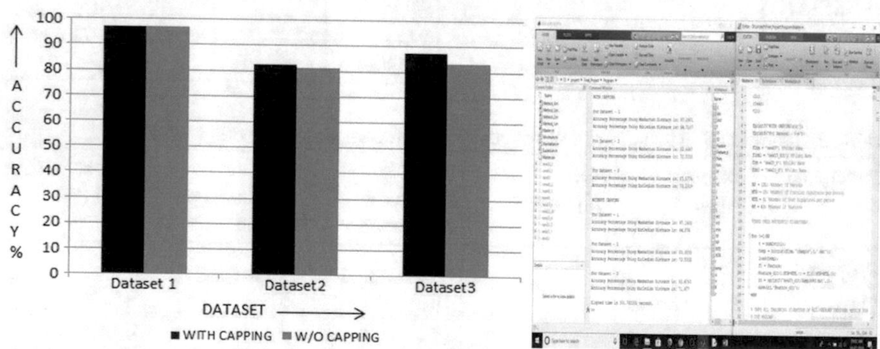

Fig. 5 Accuracy bar-charts and command window snapshots for dataset 1, 2 and 3

Acknowledgements The project is supported by 'Mobile Computing and Innovation Applications' funded by UGC UPE II of Jadavpur University, India.

References

1. Pourshahabi, M.R., Sigari, M.H., Pourreza, H.R.: Offline handwritten signature identification and verification using contourlet transform. In: International Conference of Soft Computing and Pattern Recognition, pp. 670–673 (2009)
2. Sulong, G., Ebrahim, A.Y., Jehanzeb, M.: Offline handwritten signature identification using adaptive window positioning techniques. Signal Image Process. Int. J. (SIPIJ) **5** (2014)
3. Kalera, M.K., Srihari, S., Xu, A.: Offline signature verification and identification using distance statistics. Int. J. Pattern Recognit. Artif. Intell. **18**, 1339–1360 (2004)
4. Farhan, U., Tolouei-Rad, M., Osseiran, A.: Indexing and retrieval using case-based reasoning in special purpose machine designs. Int. J. Adv. Manuf. Technol. 1–15 (2017)
5. Watson, I.: Applying Case-based Reasoning—Techniques for Enterprise Systems. Morgan Kaufmann Publishers Inc., San Francisco, CA, USA (1998)
6. Han, J., Kamber, M.: Data Mining—Concepts and Techniques. 2nd edn. Morgan Kaufmann Publishers (2006)
7. Freedman, D., Pisani, R., Purves, R.: Statistics. 3rd edn. W. W. Norton & Co. (1997)
8. Gonzalez, R.C., Woods, R.E.: Digital Image Processing, 3rd edn. Pearson Education, Inc., Pearson Prentice Hall (2008)
9. Mitchell, T.M.: Machine Learning. International edn. McGraw-Hill (1997)
10. Huang, K., Yan, H.: Off-line signature verification based on geometric feature extraction and neural network classification. Pattern Recognit. **30**, 9–17 (1997)
11. Baltzakis, H., Papamarkos, N.: A new signature verification technique based on a two-stage neural network classifier. Eng. Appl. Artif. Intell. **14**, 95–103 (2001)
12. McCabe, A., Trevathan, J., Read, W.: Neural network-based handwritten signature verification. J. Comput. **3**, 9–22 (2008)
13. Liu, X., Wang, H.: A discretization algorithm based on a heterogeneity criterion. IEEE Trans. Knowl. Data Eng. **17** (2005)
14. http://atvs.ii.uam.es/mcyt75so.html (ATVS—Biometric Recognition Group >>Databases >>MCYT—SignatureOff—75)
15. Soleimani, A., Fouladi, K., Araabi, B.N.: UtSig—A persian offline signature dataset. Inst. Eng. Technol. Biom. **6**, 1–8 (2017)

A Comparison Study of Different Privacy Preserving Techniques in Collaborative Filtering Based Recommender System

Sitikantha Mallik and Abhaya Kumar Sahoo

Abstract Recommender System is a productive tool which has been used in various fields for personalizing the applications so that we get recommended items such as e-commerce sites, shopping sites like Amazon, Netflix, research journals, mobile-based platforms and TV-based platforms. Collaborative Filtering (CF) is the most popular widely used tool on the Internet for recommender systems to search for new items that fit users' interest by finding similar users' opinion expressed on other similar items. However, there are many disadvantages for CF algorithm, which are data privacy risk and data sparsity. In this paper, we compare and analyze different privacy preserving collaborating filtering techniques that have greatly captured the attention of researchers i.e. Matrix Factorization (MF) and Singular Value Decomposition (SVD) with finding accuracy. Compared with SVD and MF, WBSVD achieves better recommendation quality and is more accurate from experimental results.

Keywords Collaborative filtering · Matrix factorization · Privacy preserving · Prediction · Recommender system · Singular value decomposition

1 Introduction

World Wide Web has changed the whole internet content and its use. With the rapid development of the cyberspace, large quantities of information are produced daily. Recommender systems are becoming popular due to the need for filtering large datasets and huge quantities of data on the internet. To solve information overload, recommendation systems have become essential. In daily life, we depend upon the recommendations from others by reviews, surveys, opinion polls. Recommender

S. Mallik · A. K. Sahoo (✉)
School of Computer Engineering, Kalinga Institute of Industrial Technology, Deemed to be University, Bhubaneswar 751024, India
e-mail: abhayakumarsahoo2012@gmail.com

S. Mallik
e-mail: sitikanthamallik@gmail.com

H. S. Behera et al. (eds.), *Computational Intelligence in Data Mining*, Advances in Intelligent Systems and Computing 990,
https://doi.org/10.1007/978-981-13-8676-3_17

systems assist the users by studying their preferences in order to make recommendations to them. Collaborative filtering is one of the widely used techniques used for filtering recommendations. With the enormous growth in the bulk of user data and people visiting shopping sites in recent years, we are facing some challenges and problems. These systems are generating preferences and recommendations for millions of users at a time and achieving high accuracy and results [1]. In primitive recommendation systems, information increases with the number of users in the web. As a result, information overload is one of many security concerns because of the difficulty of handling huge amount of data. There is a need for new innovations and cutting-edge technologies which can generate high-quality recommendations for high scalable problems. One approach to cope with the scalability, privacy and sparsity issues created by CF is to leverage a latent factor model to compare the preferences between different users. We conduct experiments to gain better accuracy in prediction of user ratings. One of the widely used metrics is Root Mean Square Error (RMSE). The lesser is the RMSE, the better is the accuracy. Since there are some unrated items, we will neglect them. Our purpose is to minimize RMSE to get better accuracy. In this paper, we discuss various filtering techniques such as Matrix Factorization (MF), Singular Value Decomposition (SVD) and Basic Singular Value Decomposition (BSVD) etc.

The rest of the paper is organized as follows: Sect. 2 describes about the overview of recommendation system and its basic concepts and applications. Section 3 presents privacy preserving collaborative filtering techniques. Section 4 tells about comparison study and experimental result analysis among different privacy preserving methods. Section 5 contains the conclusion and future work.

2 Recommendation System

It refers to a filtering system which identifies and predicts recommended content and preferences for users. It helps us to predict whether the active user would choose an item or not based on its previous purchase history or preferences of similar users.

2.1 Different Phases of Recommender System

Figure 1 shows different phases of recommendation system. There are three phases required to build a recommendation model.

Information collection phase

This phase gathers vital knowledge about users and prepares user profile based on the user's attribute, behaviors or resources accessed by users. Without constructing a well-defined user profile, the recommendation engine cannot work properly. Recommender system is based on inputs which are collected in different ways, such as explicit feedback, implicit feedback and hybrid feedback [2].

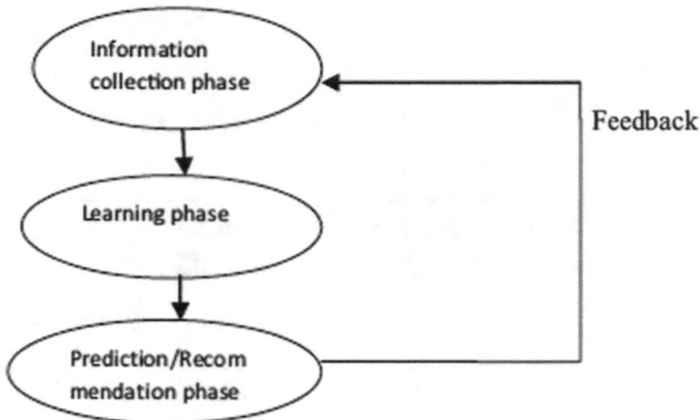

Fig. 1 Different phases of recommendation engine [2]

Learning phase
This phase takes feedback gathered in the information collection phase as input and processes this feedback by using learning algorithm and exploits user's features as output [2].
Prediction/recommendation phase
Preferable items are recommended for users in this phase. By analyzing feedback gathered in the first phase, prediction can be made which is happening through model or memory-based or observed activities and preferences [2].

2.2 *Collaborative Filtering*

Collaborative filtering is an approach commonly used in predictive recommendation systems for building personalized recommendations on the web. It is an amalgamation of two words filtering and collaborating. The filtering process is the process of collecting preferences of active users to predict right items based on item-item or user-user similarity where as collaborating is the participation of many users buying similar kinds of items. It is based on the principle that user who choose or like a particular item in the past, will choose a similar item in the future, and that they will like items liked by a similar group of users. This filtering technique works by creating a database (user-item matrix) containing user's ratings. After that, it compares users with similar interest and preferences by computing the similarities between items they purchased and finally recommendations are generated. Such a group of users is known as neighborhood [2], which is displayed in Fig. 3.

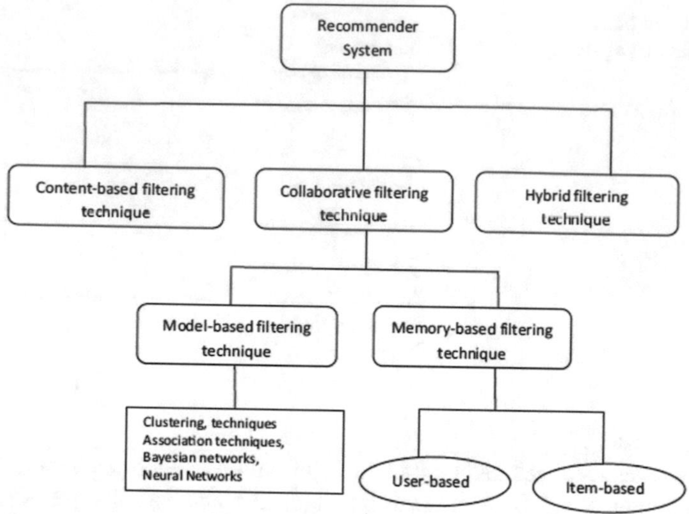

Fig. 2 Different types of filtering based recommender system [2]

2.3 *Types of Collaborative Filtering Based Recommender System*

These are different types of filtering techniques used in recommendation system which are shown in Fig. 2. Collaborative filtering is classified into memory based and model based filtering techniques.

Memory-based collaborative filtering

This type of filtering makes use of user rating data to compute the similarity between users and predicts top rated items. It's popularly used among other techniques and also simpler to use and implement. These are dependent on human ratings. Item-based CF, nearest neighbor and user-based CF are some of the common techniques [3].

Model-based collaborative filtering

In this technique, models are built based on users ratings data to predict user's interests and preferences. Algorithms like data mining, Bayesian networks are used. Examples of this type of filtering are dimensionality reduction models such as PCA, matrix factorization models, SVD.

Hybrid collaborative filtering

It combines both model and memory-based collaborative filtering. It uses the features of both types of collaborative filtering. It overcomes limitations of most of earlier CF algorithms. Hybrid recommenders can combine the expertise of different recommenders that exploit different properties of input space, to increase the overall prediction performance. Different types of recommendation techniques like knowledge-based algorithm are less affected by the cold start problem (Fig. 2).

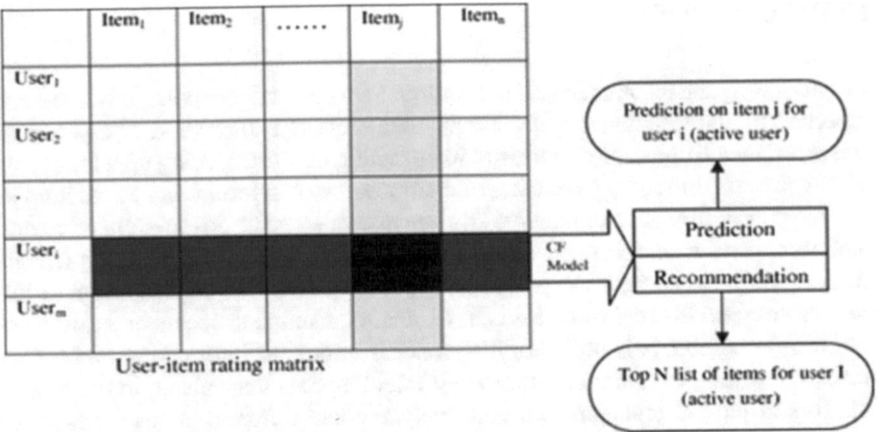

Fig. 3 Collaborative Filtering process [2]

2.4 Applications

Recommender systems decrease transaction costs of filtering and choosing items in an e-commerce site. Recommendation systems also speed up decision-making system and quality. Many websites provide recommendations (e.g. Amazon, Netflix, and Pandora).

These have become very popular these days, and are used in various fields including movies sites, shopping sites like Amazon, research journals, mobile-based platforms and TV-based platforms. Its applications are based on academic purposes, finding good restaurants around the area, business services, marketing, and sentiment analysis in twitter.

2.5 Collaborative Filtering Algorithm

This algorithm consists of following steps.

i. for every item m that a has no rating for yet
ii. for every other user b that has a rating for m
iii. calculate similarity (sim) between a and b
iv. add b's preference for m, weighted by sim, to a running average
v. return top-ranked items, ranked by weighted average (Fig. 3).

3 Related Work

As data mining and recommendation techniques are becoming popular, it has become important to find techniques that preserve data security and privacy. The personalized recommendations, may be fruitful in attracting new users, bring up a lot of data privacy matters. Based on researches and surveys, we can infer that new customers may look away from an e-commerce site because of some of privacy-related issues. Therefore, there are in escapable security matters which can heighten through illegal and unauthorized access while providing the recommendation services from different data sources. In order to shield against unauthorized data access and misuse of data & information, it is obligatory to conceal important information while concurrently creating the recommendation by effecting data unavailable to the system [4]. To sort out the problems, we will analyze about different privacy preserving based recommender system like matrix factorization model, SVD, etc. by which the recommendations can be generated maintaining data privacy.

3.1 Privacy Preserving Collaborative Filtering

Matrix factorization model

This is the most successful latent model which assists in solving high sparsity problems. In its primary form, it contains both items and users by an array of factors derived from user ratings patterns. This technique has become favorable due to better scalability and predictive accuracy. It is an influential technique to uncover the hidden structure behind data. It is used for processing large databases and providing scalability solutions. This is used in the field of information retrieval. This technique maps both users and items to a joint latent factor space of dimensionality [5, 6]. Each item can be represented by vector q_i. Similarly, each user can be resented by vector p_u. The elements of q and p measure the extent to which item will be chosen by the user. The dot product q_i^T p_u represents the interaction between user and item which is shown in Eq. (1).

$$u_i = q_i^T p_u \tag{1}$$

Singular Value Decomposition (SVD)

SVD is a well-known method for establishing latent factors in the scope of recommendation systems to work on problems faced by collaborating filtering technique. These techniques have become popular due to good scalability and better predictive accuracy [7]. It is a popular recommendation filtering technique that decomposes m× n matrix A into three matrices as $A = U\,SV^T$ where U and V are two orthogonal matrices of size $m \times r$ and $m \times y$, respectively; y is the rank of the matrix A. S is a diagonal matrix of size $r \times r$. Diagonal elements contain all singular values of matrix

$$\underset{m \times n}{\overset{\hat{X}}{\begin{pmatrix} x_{11} & x_{12} & \cdots & x_{1n} \\ x_{21} & x_{22} & \cdots & \\ \vdots & \vdots & \ddots & \\ x_{m1} & & & x_{mn} \end{pmatrix}}} \approx \underset{m \times r}{\overset{U}{\begin{pmatrix} u_{11} & \cdots & u_{1r} \\ \vdots & \ddots & \\ u_{m1} & & u_{mr} \end{pmatrix}}} \underset{r \times r}{\overset{S}{\begin{pmatrix} s_{11} & 0 & \cdots \\ 0 & \ddots & \\ \vdots & & s_{rr} \end{pmatrix}}} \underset{r \times n}{\overset{V^{\mathsf{T}}}{\begin{pmatrix} v_{11} & \cdots & v_{1n} \\ \vdots & \ddots & \\ v_{r1} & & v_{rn} \end{pmatrix}}}$$

Fig. 4 Singular value decomposition

A [8]. It then decomposes *S* matrix to get a matrix S_k, $k < r$ showing top k rated items (largest diagonal values) which is shown in Fig. 4.

The average value of user ratings is utilized for filling up sparse locations in ratings matrix (*A*) to secure a fruitful latent relationship. The normalization of the matrix is done by changing user ratings to z-scores. Using SVD, the normalized matrix (A_{norm}) is divided into *U*, *S*, and *V*. Then the matrix S_k is acquired by taking only *k* largest singular values resulting in the reduction of dimensions of matrices *U* and *V*. After that, $U_{k,}$ $S_{k,}$ and $S_k V^T$ are determined. The final matrices can be used to find the prediction for the active user. It can uncover the hidden matrix structure. This filtering is used mainly in CF recommender systems to detect the feature of user and item [9, 10].

Variable Weighted BSVD (WBSD)

This type of collaborative filtering is introduced for solving problems in SVD based collaborative filtering and improve its accuracy and privacy. Here, we are using a variable weight so that we can change the weights as per our requirements. If active users are concerned about data privacy, they will disturb the data as per requirements. By disturbing data so that unknown user cannot access data, privacy can be preserved. The users receive many disturbing information and fail to classify items on basis of disturbing nature of data, and they cannot determine that which items are rated by particular individuals, and which are not [9]. Change in weight, is computed using Eq. (2).

$$\omega_i = e^{\delta_{max} - \delta_i} \tag{2}$$

where σ_i disturb weight of user i, σ_{max} represents the maximum disturb weight, and ω_i denotes the variable weight of user i. The value of variable weight lies between 0 and 1. The new A_k can be calculated as per the Eq. (3).

$$A_k = AV_k V_k^T \tag{3}$$

where the A_k represents the disturbed matrix, the row represents the data whereas column represents the item. This column behaves as the feature vectors according to the largest k eigenvalue of the matrix C_ω. V_k is an n * k matrix and V_k^T is the transposition of matrix V_k. The C_w is calculated as per Eq. (4).

$$C_{\omega} = \sum_{I=1}^{m} \omega_i x_i^T x_i \tag{4}$$

4 Comparison Study Among Different Privacy Preserving Methods

While collaborative filtering techniques calculate similarities between users and items only and these are simple and intuitive, unlike matrix factorization models which deal with mapping features of user and item to latent factor space and then calculating similarity, therefore, they are higher productive because these techniques permit us to uncover the hidden characteristics underlying the relationship between users and items. Among all these models, SVD is considered best in terms of accuracy and solving high-level sparsity and cold start problems found predominantly in recommender systems [11].

4.1 Dataset

We conducted experiments on the data set of MovieLens. This dataset is provided by GroupLens at Minnesota University. The MovieLens dataset is consists of 10,000 ratings containing 943 users and 1682 movies [12]. Movie.dat file stores the movie names and their genres in the dataset. The range of ratings varies from 1 to 5. The highest value is the most preferred value. The main reason we chose datasets from MovieLens is that, they are widely used in multiple reports and preferred when predicting user ratings. The sparsity density is 6.305%, that is, only 6.305% items have ratings [9]. During testing, 80% dataset is taken as training data and the rest 20% is selected as test data.

4.2 Parameters

We need to choose parameter k carefully because it is very critical in SVD. If k is very less, the data would lose the vital information, and if k is too big, it will lose its data privacy property. Therefore, parameter k should be selected properly. The approach of Root Mean Absolute Error (RMSE) is used here because it can be easily identified and measured so that we can compute the quality aspect of recommendations easily. It is utilized in this paper to exhibit the performance of the different techniques and their accuracy.

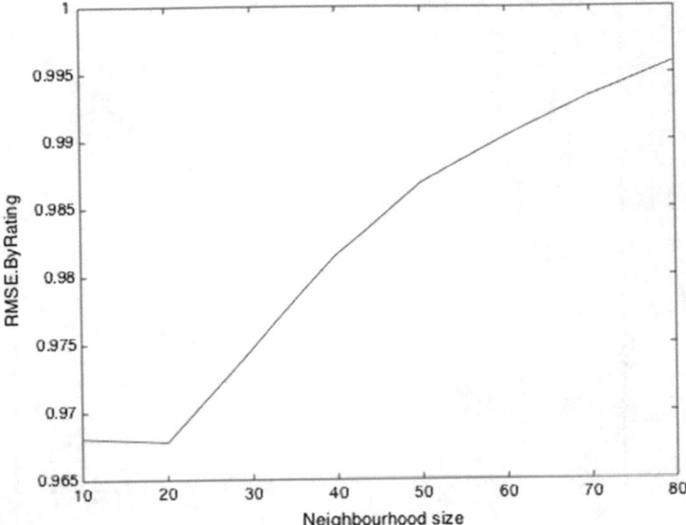

Fig. 5 Cosine-based similarity with accuracy

Table 1 RMSE values based on cosine coefficient & Pearson correlation similarity of CF based recommender system

Sl. No.	K	RMSE	
		Cosine coefficient	Pearson correlation
1.	10	0.968	1.095
2.	15	0.967	1.085
3.	20	0.966	1.079
4.	25	0.972	1.081
5.	30	0.975	1.09

4.3 Experimental Results

Different similarity measures such as Cosine coefficient and Pearson correlation are used to calculate the similarity by using number of nearest neighbors. Table 1 shows the results of RMSE by using above two similarity methods. Evaluation values for each of these three methods are tested with different neighborhood sizes. Each method is evaluated with RMSE (Fig. 5) (Table 1).

The minimum value is computed for both RMSE and MAE when neighborhood size is 20 shown in Fig. 6.

From Fig. 7, it can be seen that the RMSE of WSVD-Collaborative filtering method varies with variable k, and gets perfect value when k is equal to 10. The lesser the value is, accuracy is higher. This leads to better recommendation quality.

Table 2 depicts for different k values, specifically, the results indicate that WSVD CF is best in terms of RMSE for $k = 10$. The experimental results indicate that WSVD technique is best in terms of accuracy. As a result, WSVD should be the

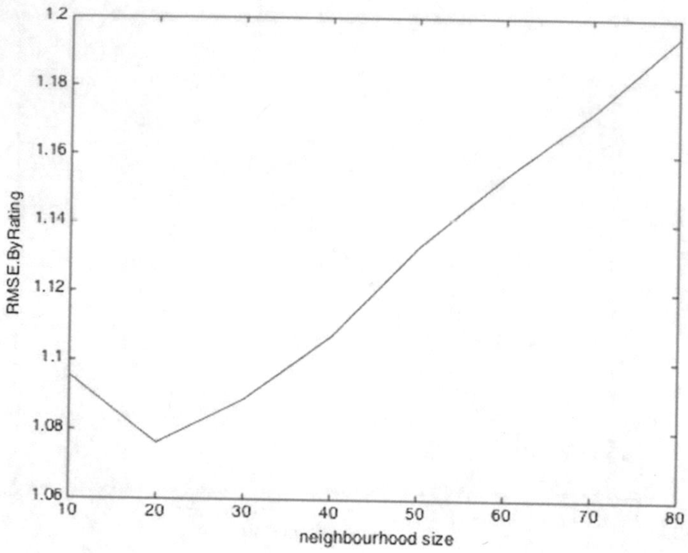

Fig. 6 Pearson correlation similarity with accuracy

Fig. 7 Comparison graph between RMSE and K

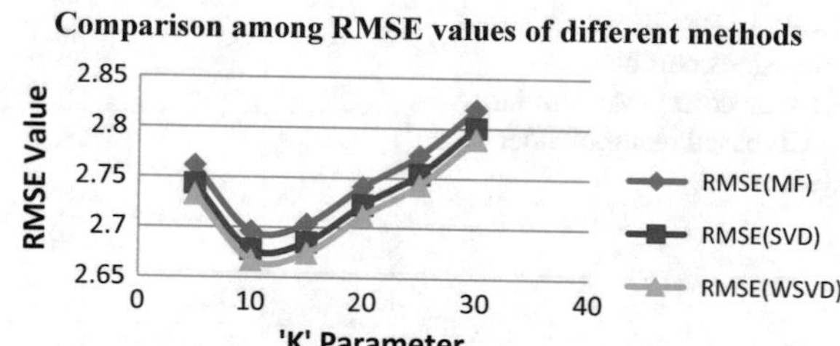

Table 2 RMSE measurements of different methods

SL NO.	K	RMSE(MF)	RMSE(SVD)	RMSE(WSVD)
1	5	2.76137	2.74313	2.73219
2	10	2.69592	2.67776	2.66688
3	15	2.70339	2.68520	2.67413
4	20	2.74089	2.72274	2.71169
5	25	2.77257	2.75391	2.74288
6	30	2.81879	2.80047	2.78935

preferred method of collaborative filtering. There are other metrics which can also measure recommendation quality such as Mean Absolute Error (MAE), precision, Recall and ROC etc.

5 Conclusion

Recommendation systems are one of the new prevailing technologies for deriving supplementary information for a business from customer data. These systems find recommended items by calculating the similarity between similar items for different users. Therefore, they play an important role in business by increasing sales and revenues. Collaborative filtering recommender systems are swiftly becoming an important tool on World Wide Web. These systems are being put under pressure by the growing quantities of information overload in institutions and organizations. There is a growing worldwide concern for data privacy so that data can't be misused by any third party organization without knowledge of the user. Modern state of art technologies are needed that can definitely resolve the data security and privacy problem found in C.F. recommender systems. In this paper, different privacy preserving collaborative filtering methods like Singular Value Decomposition (SVD) based collaborative filtering, variable weight based SVD (WBSVD), matrix factorization model are presented. Compared with SVD and MF, WBSVD works better on privacy preserving than others.

References

Shrivastav, L.K.: An intelligent & enhanced hydra: a web recommender system. Int. J. Eng. Appl. Sci. Technol. **1**, 165–169(2016)

Isinkaye, O., Folajimi, Y.O., Ojokoh, B.A.: Recommendation systems: principles, methods, and evaluation. Egypt. Inform. J. **16**(3), 261–273 (2015)

Yao, G., Cai, L.: User-Based & Item-Based Collaborative Filtering Recommendation Algorithms Design. University of California, San Diego (2015)

Badsha, S., Khalil, Y., Xun, A.: Practical privacy-preserving recommender system. Data Sci. Eng. **1**, 161–177 (2016)

Bokdea, D., Giraseb, S., Mukhopadhyay, D.: MatrixFactorization model in collaborative filtering algorithms. Procedia Comput. Sci. **49**, 136–146 (2015)

Canny, J.: Collaborative filtering with privacy via factor analysis. In: Proceedings of the 25th Annual International ACM SIGIR Conference on Research and Development in Information Retrieval, pp. 238–245. ACM (2002)

Xian, Z., Li, Q., Li, G.: New collaborative filtering algorithms based on SVD ++ and differential privacy. Math. Probl. Eng. 1–14 (2017)

Sarwar, B., Karypis, G., Konstan, J.: Application of dimensionality reduction in recommender system-a case study. In: ACM Web KDD 2000 Workshop (2000)

Wu, J., Yang, L., Li, Z.: Variable weighted BSVD-based privacy-preserving collaborative filtering. In: 10th International Conference on Intelligent Systems and Knowledge Engineering (ISKE) (2015)

Polat, H., Du, W.: SVD-based collaborative filtering with privacy. In: ACM Symposium on Applied Computing SAC, USA, 13–17 Mar 2005

Gope, J. Jain, S.K.: A survey on solving cold start problem in recommender systems. In: 2017 International Conference on Computing, Communication, and Automation (ICCCA), pp. 133–138, Greater Noida (2017)

GroupLens Research project: MovieLens DataSet. http://www.grouplens.org/

Artificial Intelligence Based Multi-object Inspection System

Santosh Kumar Sahoo

Abstract The proposed research focuses on multi object inspection system in an efficient manner. The proposed model is suitable for multi-object investigation considering the various features like linearity, Circularity and dimensionality etc. This model is an intelligent one by employing various nature-inspired algorithms and classifiers like ANN (Artificial Neural Network), KNN (Kth nearest neighbor), SVM (Support-Vector-Machine) and LSSVM (Least-Square-Support-Vector-Machine). Overall performance of the proposed model is analyzed with the conventional procedures. Hence the proposed model is best choice for inspecting multi objects with high accuracy.

Keywords Multi-objects · Linearity · Dimensionality · Accuracy · Inspection system

1 Introduction

Inspection task is a very difficult task and plays a crucial role in all manufacturing unit. Without this, the product or production unit can't be sustained and its automation process may not be achieved. Hence, the proposed model emphasized on a smart or an intelligent inspection system which is suitable for multi-object's inspection. This model will be more smoothing and suitable for inspection of an object it trained well [1]. Wang [2] has implemented a concept for multi instance classifier for mines recognition on sea bed by considering multiple side scan sonar views. Similarly, Chena [3] reviewed the development of wavelet transform and its application in rotating machinery fault diagnosis. In the meantime, a dual wavelet structure has developed by Li for edge detection [4]. Likewise, Sharma [5] designed a complex wavelet transform for the image processing in a different manner. Compression of wavelet transforms and thresholding schemes has been addressed by Alarcon [6]

S. K. Sahoo (✉)
Department of Electronics & Instrumentation Engineering, CVR College of Engineering, Hyderabad 501510, Telengana, India
e-mail: santosh.kr.sahoo@gmail.com

H. S. Behera et al. (eds.), *Computational Intelligence in Data Mining*, Advances in Intelligent Systems and Computing 990,
https://doi.org/10.1007/978-981-13-8676-3_18

for proper processing of images in a smooth manner. Wu [7] explained the various arrhythmia-electro-cardio-graphic signal applied for quantization scheme optimization through wavelet-techniques for E-C-G data. Many researchers have been used a lifting-based wavelet for compressing images in both side where each and every part of the pixels corresponding to the images has analyzed for better response [8–10]. By considering the above facts the proposed work is planned for inspection of multiple objects. In simplicity point of view, three different objects like battery, washer and PCB boards are considered here and proper analysis related to its performance has been studied. In view of designing the structure, a database has been developed related to various features of the three different types of objects as stated above. This data set is validated with the proposed model developed with an artificial neural network and LSSVM (Least-Square-Support-Vector-Machine) classifier.

2 Proposed Inspection Structure

Image of the objects to be inspected (here battery, washer and PCB are used as object) are captured by NI-Smart camera-1762 with high resolution. The model is designed for inspecting multi-object to satisfying the objective criteria of proposed work. The captured image is preprocessed by application of Gaussian and median filters. After this various feature (Energy, variance and skewness) are extracted using wavelet transforms and also for proper information related to desire object, the dimension is optimized with the help of PCA. The proposed plan is represented in Fig. 1 and this scheme response is represented as an artificial manner which uses various schemes for its response.

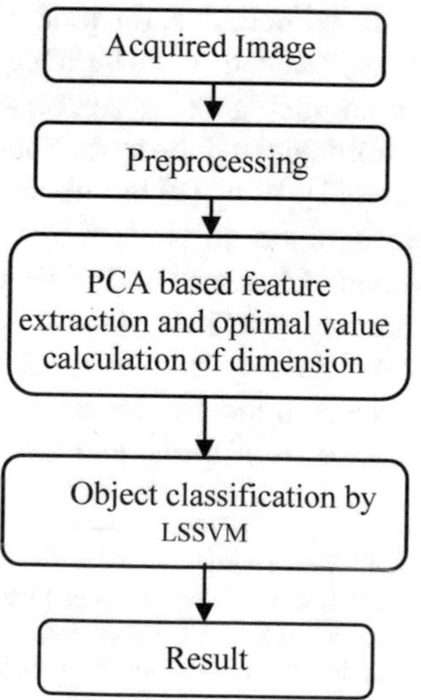

Fig. 1 Proposed scheme

(a) (b) (c)

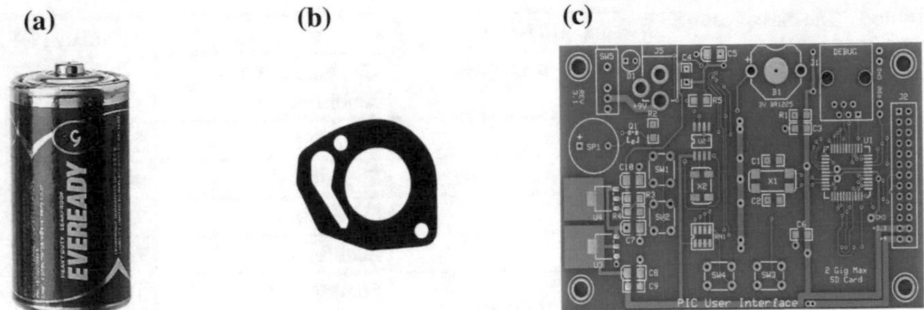

Fig. 2 **a** Battery. **b** Washer. **c** PCB

Fig. 3 Representation diagram displays of three levels wavelet transforms

<table>
<tr><td>ll3</td><td>hl3</td><td rowspan="2">hl2</td><td rowspan="3">hl1</td></tr>
<tr><td>lh3</td><td>hh3</td></tr>
<tr><td colspan="2">lh2</td><td>hh2</td></tr>
<tr><td colspan="3">lh1</td><td>hh1</td></tr>
</table>

Here around 100 samples of each object in different condition like good and defective is considered and trained accordingly. Mat-Lab is used for analysis of afore said segmentation and feature extraction data. Then after a standard data set is formed for further processing because a standard data set for the proposed object is not available. After formation of training and testing data the model is taken for proper inspection of different objects. Every classifier response in-terms-of Classification Rate (CR) is presented in Eq. (1).

$$CR = TP + TN/TP + TN + FP + FN \quad (1)$$

In above expression TP = True-Positive, TN = True-Negative, FP = False-Positive, FN = False-Negative. Different object considered here for sample inspection is shown in Fig. 2a–c respectively.

After acquiring the image of these objects, a wavelet-based feature extraction is followed which gives the accurate vale to corresponding objects.

A three-level two-dimensional wavelet structure is used for evaluating various feature coefficients in different angle like vertical, horizontal and diagonal ways as per Fig. 3. Constants of wavelet decomposition are presented [11] in Eqs. (2) and (3).

$$w_{\varphi}(J_0, M, N) = \frac{1}{\sqrt{m \times n}} \sum_{x=0}^{m-1} \sum_{y=0}^{n-1} F(x, y)\varphi_{J_0 M,N}(X, Y) \quad (2)$$

$$w_{\psi}^{I}(J, M, N) = \frac{1}{\sqrt{m \times n}} \sum_{x=0}^{m-1} \sum_{y=0}^{n-1} F(x, y) f_{J,M,N}^{I}(X, Y)$$

$$I = \{h, v, d\} \quad (3)$$

Table 1 The classification response for Kth nearest-neighbor

For Kth nearest-neighbor		Accuracy (%)
Nearest-neighbor value	Gap (smallest) condition	
1	Euclidian	81.22
2	Euclidian	84.43
3	Euclidian	79.89
4	Euclidian	79.78
5	Euclidian	81.53
6	Euclidian	81.43
7	Euclidian	82.78

where $\phi^{I}_{J,M,N}(X, Y) = 2^{J/2}\{\varphi\}(\rho^{J}\chi - m, \rho^{J}\gamma - n)$ and $\varphi_{J_{o}mn}(x, y) = 2^{J/2}\varphi(P^{J}\chi - m, \rho^{J}\gamma - n)$ are scrambled and converted base functions. Now, energy is intended laterally in all the ways like, vertical, horizontal and diagonal as 3 level of decomposition is cast-off so the energy for individual wavelet coefficient is according to [12] Eq. (4)

$$e_{cl} = \frac{1}{l \times l}\sum_{I=0}^{l}\sum_{J=0}^{ll}(w_{c,l,I,J})^{2} \tag{4}$$

In single wavelet coefficient 9 features can be attained similarly for 3 level around eighty-one features are extracted. The projected scheme is repetitive together for defective and defect free objects. After collecting all the features, the trained and test data sets are formed for proper inspection as per desires.

3 Result and Discussion

The reduced features obtained through PCA are applied to classifier for distinguishing the desired class like defective and defect free class. For better response by using KNN (Kth nearest neighbor) classifier the parameters like nearest neighbors (K) and smallest gap from nearest neighbors are followed. Then the KNN response is observed for conclusions. During this experimentation the classification rate is summarized as per Table 1, in which the best accuracy is achieved at K = 2. Therefore, the developed model is designed using KNN with K = 2.

Figure 4 signifies the mean square error with respect to different sample objects taken during the course of inspection. At the time of investigation, the performance of SVM classifier is observed in order to validate the effort of the proposed model.

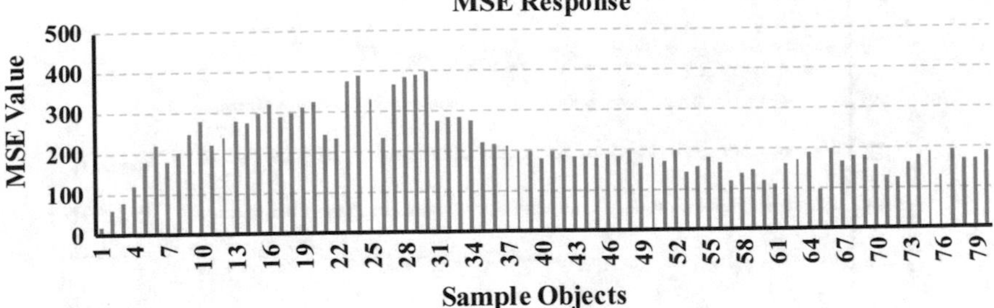

Fig. 4 Error response

Table 2 Classification performance table

Classifiers	TP	TN	FP	FN	Classification rate (%)
Multi-feed-neural-network	27	1	10	4	90.88
Linear-Kernel-support-vector-machine	30	9	3	1	87.56
RBF with SVM	29	1	11	2	92.65
LSSVM with Kernel	30	0	11	2	93.12
LSSVM with RBF Kernel	32	0	12	1	96.75

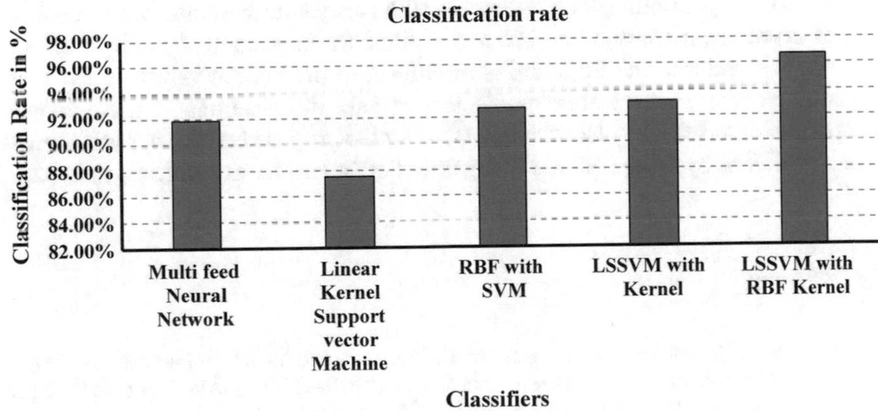

Fig. 5 Classification rate analysis

Here the projected SVM scheme is capable with linear and kernel Radial Basis Functions for better information. The investigation data are articulated as classification rate expressed in Table 2 and Fig. 5. At last the classification rate analysis considering TP, TN, FP and FN parameters are presented in Fig. 6.

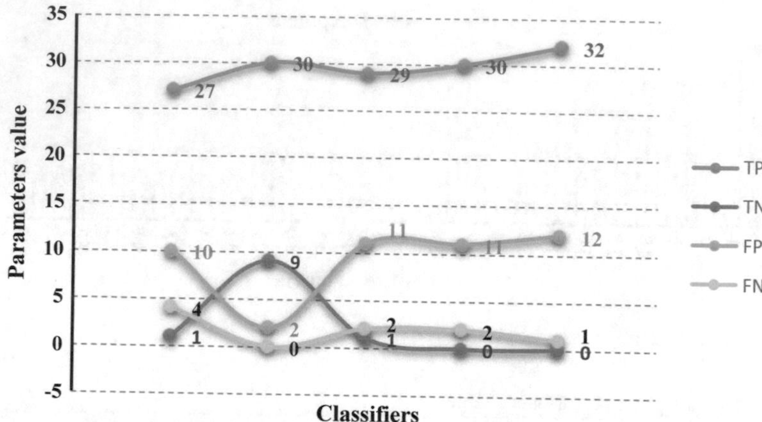

Fig. 6 Classification rate analysis considering TP, TN, FP and FN parameters

4 Conclusion

Multi-objects are classified or inspected perfectly and categorized in different classes accordingly by using the proposed scheme with a smooth manner. Nowadays many researchers also continuing their research on this area for further improvement but the proposed scheme presented here is a new method for inspection of multiple objects in a single structure. The Structure is intelligent to predict the various noise optimization procedures and feature parameters to make the system more accurate with an optimum interval. The classification label for the proposed model has achieved as a rate of 96.75% by using LSSVM along with RBF classifier as presented in Table 2.

References

1. Sahoo, S.K., Choudhury, B.B.: Dynamic bottle inspection system. In: 4th International Conference on Computational Intelligence in Data Mining (ICCIDM-2017), AISC series of Springer, Veer Surendra Sai University of Technology (VSSUT) (2017)
2. Wang, X., Liu, X., Japkowicz, N., Matwin, S.: Automated approach to classification of mine-like objects using multiple-aspect sonar images. J. Artif. Intell. Soft Comput. Res. **4**(2), 133–148 (2015)
3. Chena, J., Lia, Z., Pana, J., Chena, G., Zia, Y., Yuanb, J., Chenc, B., Hea, Z.: Wavelet transform based on inner product in fault diagnosis of rotating machinery: A review, vol. 70–71, pp. 1–35. Elsevier (2015)
4. Li, Y.: A construction of dual wavelet and its applications in image edge detection. Image Process. Pattern Recognit. **8**(7), 151–160 (2015)
5. Sharma, N., Agarwal, A., Khyalia, P.K.: Squeezing of color image using dual tree complex wavelet transform based preliminary plan. Signal & Image Process., Int. J. (SIPIJ) **5**(3) (June 2014)

6. Alarcon-Aquino, V., Ramirez-Cortes, J.M., Gomez-Gil, P., Starostenko, O., Lobato-Morales, H.: Lossy image compression using discrete wavelet transform and thresholding techniques. The Open Cybern. Syst. J. **7**, 32–38 (2013)
7. Wu T.-C., Hung K.-C., Liu J.-H., Liu, T.-K.: Wavelet-based ECG data compression optimization with genetic algorithm, pp. 746–753. Scientific Research (2013)
8. Cai1, B., Liu, Z., Wang, J., Zhu, Y.: Int. J. Signal Process. **8**(7), 51–62 (2015)
9. Sahoo, S.K., Choudhury, B.B.: Artificial intelligence (AI) based classification of an object using principal images. In: International conference on Computational Intelligence in Data mining, Ronald, Odisha, India, pp. 143–151 (2015).
10. Renjini, L., Jyothi, R.L.: Int. J. Comput. Trends Technol. (IJCTT) **V-21**(3), 134–140 (2015)
11. Panat, A., Patil, A., Deshmukh, G.: IRAJ Conference (2014)
12. Samanta, S., Mandal, A., Jackson Singh, T.: Procedia Mater. Sci. **6**, 926–930 (2014)

PSO Based Path Planning of a Six-Axis Industrial Robot

Supriya Sahu and B. B. Choudhury

Abstract Aristo robot is a 6-axis articulated robot widely used in industries to lift small parts with greater accuracy. The use of robot in production sector depends upon the efficiency with which it performs a given task with shortest possible time. This research presents a proposal, to optimize the path for a particular task by adopting Particle Swarm Optimization (PSO) algorithm. For the analysis of optimal path, total ten numbers of objects have been defined and the robot has to cover up all objects with all possible paths with minimum cycle time. The optimal solution is obtained using robomaster simulation software and the global best solution is obtained using PSO. The program for optimization of path length is written in MATLAB software. The global best result obtained using PSO is a simple method for solving path planning problems for any industrial applications.

Keywords Aristo robot · PSO · Path planning · Path length · Robomaster simulation

1 Introduction

The requirement of high efficiency and performance with greater accuracy and precision with minimum machining time has introduced the use of robots in different production industries. In order to decrease the machining time while using the robot in industries the path through which it performs the given task with minimum time is required to be defined. The selection of path with minimum path length can be achieved by use of different optimization techniques and PSO is one of the best techniques that has been adopted to solve different path planning problems. Multi

---

S. Sahu · B. B. Choudhury (✉)
Department of Mechanical Engineering, I.G.I.T Sarang, Dhenkanal, Odisha, India
e-mail: bbcigit@gmail.com

S. Sahu
e-mail: supriyaigit24@gmail.com

H. S. Behera et al. (eds.), *Computational Intelligence in Data Mining*, Advances in Intelligent Systems and Computing 990,
https://doi.org/10.1007/978-981-13-8676-3_19

objective algorithms may also be used for path planning but PSO is a simple but powerful optimization technique to solve multi dimensional optimization problems.

Kennedy et al. [1] discussed PSO in terms of social and cognitive behavior as a problem-solving optimization technique for engineering applications and presented a commonly used version operated in a space of real numbers. Deepak et al. [2] presented a novel method for PSO for the path planning task of mobile robot. The model of a new fitness function satisfied the navigation system and was also able to find out the optimal path. Panda et al. [3] used a new search algorithm (PSO) where optimization of the workspace is important, especially for highly constrained environments. Ahmadzadeh et al. [4] proposed an intelligent method for robot path finding in an unknown environment. Particle Swarm Optimization (PSO) method was used to find out proper minimum value of path. For every particle in PSO an evaluation function was calculated based on the position of the goal. It is a very flexible method where any parameter can be changed to reach the goal. Masehian et al. [5] proposed an improved novel method of PSO which not only worked for the shortness but also worked for the smoothness of the path. The comparison of standard PSO method and the standard roadmap method was able to give a good computational time and path quality. Tang et al. [6] addressed the extremum seeking of mechanical particle swarm optimization in robot cooperative search. To aid mechanical PSO, extremum seeking which is a new scheme was designed to avoid robot localization and noise and reaches the searched target cooperatively. Estevez et al. [7] presented the performance of an areal robot for a hose transportation system and demonstrated Particle swarm optimization (PSO) tuned method with a robust control system with more than one solution and was a fast method. Nedjah et al. [8] proposed a PSO based distributed control algorithm in a swarm robotic environment for dynamic allocation of task. The PSO inspired algorithm developed was a best solution for task allocation. Ulricha et al. [9] generated an automated path planning algorithm to reduce cycle time and improved the flexibility of inspection system. Das et al. [10] proposed a swarm optimization with dynamic distribution for multiple robots. The path planning was done to find out the optimal path which is free of collision for each robot with the use of two optima detectors. Dewang et al. [11] proposed an adoptive particle swarm optimization for obtaining a solution for path planning problems of mobile robot and compared with that of conventional PSO.

2 Robomaster Simulation

The setup of robot tasks in dynamic environment is carried out by robomaster simulation software by the use of robot master's extensive library. By the help of robot simulator it validates and optimizes the programs and translate the tool path data into optimized 6-axis robot programs.

The cycle time is taken as the parameter for optimization. For the analysis of optimal cycle time, ten numbers of similar objects have been defined which are

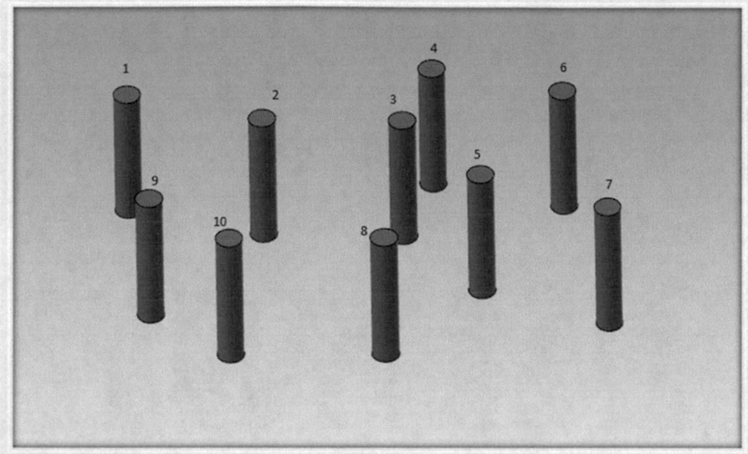

Fig. 1 Arrangement of task

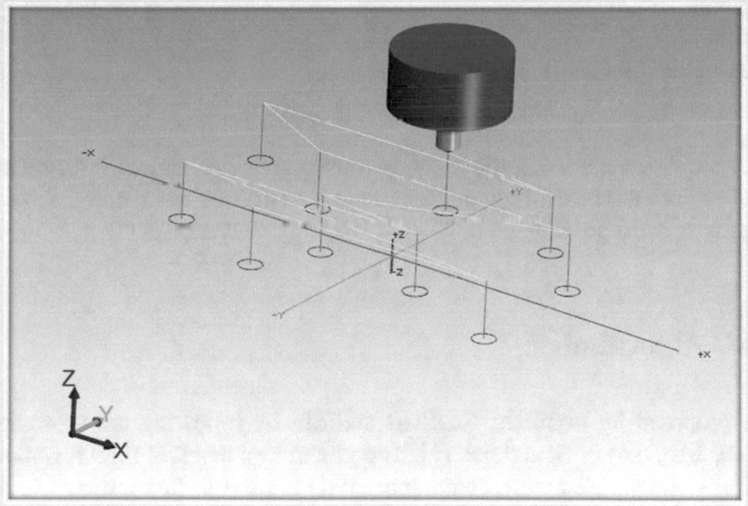

Fig. 2 Movement of robot endeffector in X, Y and Y coordinates

numbered from 1 to 10, as shown in Fig. 1 and the robot has to cover up all objects with all possible paths with minimum cycle time.

Figure 2 describes the movement of robot end effector. As an example, drilling tool is used in the simulation process which travels in X, Y and Z coordinates.

The tool cover ups all the defined objects by choosing different sort options available for different paths. The yellow line in Fig. 2 shows the path followed by the tool covering all the ten objects.

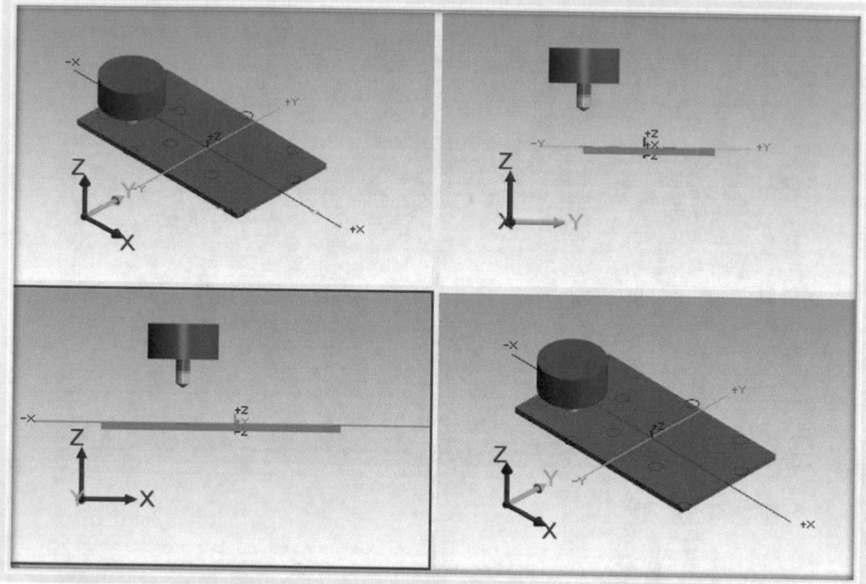

Fig. 3 Isometric view of movement of robot endeffector

Figure 3 shows the isometric view of movement of robot end effector in X, Y and Z coordinates. The drilling tool used here is to drill holes in all ten objects after finding the shortest possible path.

3 PSO Algorithm

PSO is proposed by Kennedy, Eberhart and Shi by following the movements of organisms like, bird flock and fish [1]. The global best result in PSO is obtained by following some programming commands. PSO is used in this research to solve a path planning problem to find out the optimal path for a task defined to a industrial robot The coordinate points of the ten objects that are used in the task to find out the global solution are shown in Fig. 4. The global optimum path is estimated using Eqs. 1 and 2.

$$\mathrm{PrPo_{ij}} = \mathrm{PrPo_{ij}} - 1 + \mathrm{PrV_{ij}} \tag{1}$$

$$\mathrm{PrV_{ij}} = \left[\mathrm{w.PrPo_{ij}} - 1 + c_1 r_1 \left(\mathrm{pbest_{ij}} - 1 - \mathrm{PrPo_{ij}} - 1\right) + c_2 r_2 \left(\mathrm{gbest_i} - 1 - \mathrm{PrPo_{ij}} - 1\right)\right] \tag{2}$$

where

$\mathrm{PrPo_{ij}}$ = Position of jth particle for ith iteration

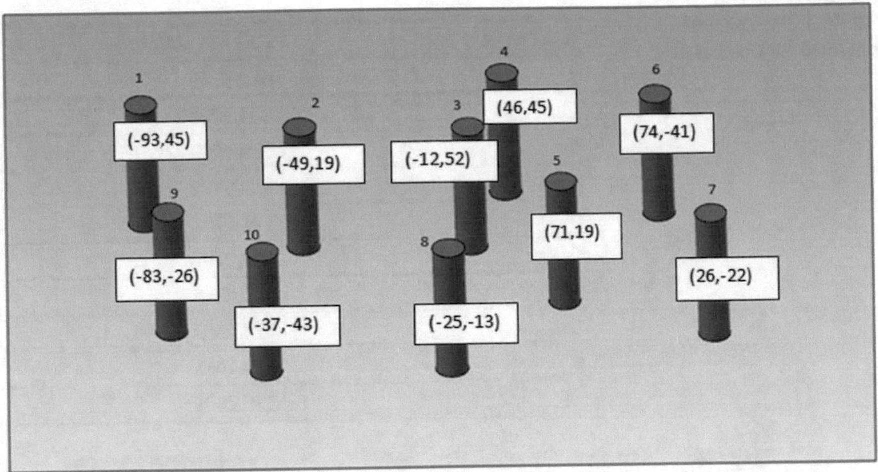

Fig. 4 Co-ordinate points of the objects

Table 1 Parameters of PSO

Sl. No	Parameter	Value
1	Number of particles (i)	10
2	Number of iterations (it)	50
3	c_1	1.5
4	c_2	1.5
5	w	1

PrV_{ij} = Velocity of jth particle for ith iteration
$pbest_{ij}$ = The best position of jth particle till the ith iteration
$gbest_i$ = The best position in the swarm
c_1, c_2 = Acceleration coefficient
r_1, r_2 = Random numbers between 0 and 1
w = Constant (Inertia Weight).

The parameters of PSO algorithm are presented in Table 1.

4 Results and Discussions

The best value of cycle time is obtained by running the simulation several times and the performance of robot is analyzed. After analyzing different paths followed by the robot, the path with minimum path length is obtained. The minimum value of path length obtained using Robomaster simulation software is 513 mm.

Several test runs are carried out to obtain the best solution of shortest path followed by the robot. Total 50 number of iterations are taken in the program for optimization.

Table 2 Path lengths obtained for one iteration

Number of test	Path length (mm)
1	612.3818
2	824.2008
3	780.2608
4	855.9877
5	571.6783
6	508.079
7	564.4242
8	578.0857
9	571.7063
10	539.2400

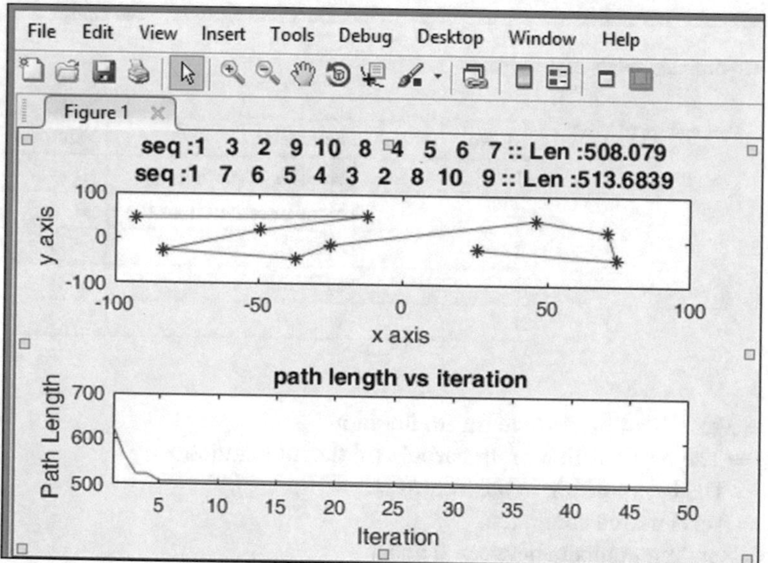

Fig. 5 Shortest path obtained using PSO

The path length in every iteration for the best results are observed which is shown in Table 2 and the minimum value of path length obtained is for run 6 with a value of 508.079 mm.

Figure 5 shows the path along with its sequence for global best solution which is of the value of 508.079. The sequence of global optimum path is 1-3-2-9-10-8-4-5-6-7. It also shows the optimal sequence that has been obtained using Robomaster simulation software with a value of 513.6839 mm. Path length is updated for every iteration. In the graph the path length is taken as the y-axis and number of iteration as the x-axis. It is observed from the graph that the path length decreases up to iteration 5 and then after it remains constant with a value of 508 mm.

Table 3 Workspace of PSO

Name	Value
No. of particles	10
Max_iter	50
Optimal length	513.6839
Optimal sequence	[1, 7, 6, 5, 4, 3, 2, 8, 10, 9]
gbest	508.0790
gsol	[1, 3, 2, 9, 10, 8, 4, 5, 6, 7]

Table 3 shows the parameters in workspace of PSO. It shows the optimal value and global best value along their sequences.

5 Conclusions

The proposed path planning algorithm of PSO can effectively used for computing a shortest path by obtaining the global best solution which can not be obtained by other algorithms like GA. PSO is more effective compared to other algorithms because it is easy to implement and few parameters have to be adjusted to achieve the global best solution. The proposed method can also be improved in future by adopting multi-objective algorithms like cuckoo search etc and a comparative study can be done to increase the effectiveness of industrial robot.

From this analysis it is observed that the path planning of robot in any industrial application can be enhanced by the use of PSO. Therefore, this analysis can effectively be used for solving any path planning problems and to increase the efficiency of robot for industrial purposes.

References

1. Kennedy, J., Eberhart, R.C., Shi, Y.: Particle Swarm. The Morgan Kaufmann Series in Artificial Intelligence, Swarm Intelligence, pp. 287–325 (2001)
2. Deepak, B.B.V.L., Parhi, D.R.: PSO based path planner of an autonomous mobile robot. Cent. Eur. J. Comput. Sci. **2**(2), 152–168 (2012)
3. Panda, S., Mishra, D., Biswal, B.B.: A Multi-Objective Workspace Optimization of 3R Manipulator Using Modified PSO, SEMCCO, LNCS 7677, pp. 90–97 (2012)
4. Ahmadzadeh, S., Ghanavati, M.: Navigation of mobile robot using the PSO particle swarm optimization. J. Acad. Appl. Stud. (JAAS) **2**(1), 32–38 (2012)
5. Masehian, E., Sedighizadeh, D.: An improved particle swarm optimization method for motion planning of multiple robots. In: Martinoli, A. (eds.) Distributed autonomous robotic system, STAR vol. 83, pp. 175–188 (2013)
6. Tang, Q., Eberhard, P.: Mechanical PSO aided by extremum seeking for Swarm robotics cooperative search. In: Tan, Y., Shi, Y., Mo, H. (eds.) International Conference in Swarm Intelligence. ICSI, Lecture Notes in Computer Science, vol. 7928, pp. 64–71 (2013)

7. Estevez, J., Grana, M.: Robust control turning by PSO of aerial robots hose transportation. In: Ferrández Vicente, J.M. (eds.) IWINAC 2015: Bioinspired Computation in Artificial systems, Part II, LNCS 9108, pp. 291–300. Springer (2015)
8. Nedjah, N., Mendonc, R.M., Mourelle, M.: PSO based distributed algorithm for dynamic task allocation in a robotic swarm. Procedia Comput. Sci. **51**, 326–335 (2015)
9. Ulricha, M., Lux, G., Reinhart, G.: Automated and cycle time optimized path planning for robot based inspection system. Procedia CIRP. **44**, 377–382(2016)
10. Das, P.K., Behera, H.S., Jena, P.K., Panigrahi, B.K.: Multi-robot path planning in a dynamic environment using improved gravitational search algorithm. J. Electr. Syst. Inf. Technol. **3**(2), 295–313 (2016)
11. Dewang, H.S., Mohanty, P.K., Kundu, S.: A robust path planning for mobile robot using smart particle swarm optimization. Procedia Comput. Sci. **133**, 290–297 (2018)

Selection of Industrial Robot Using Fuzzy Logic Approach

Swagat Nayak, S. Pattanayak, B. B. Choudhury and N. Kumar

Abstract This paper introduces a modified fuzzy technique (FUZZY TOPSIS) for the selection of best Industrial robot according to the assigned performance rating. Both conflicting quantitative and qualitative evaluation criteria are considered during the selection process. A collective index is prepared using weighted average method for preparing the ranking of rule base. Triangular and Gaussian membership function is used to describe the weight of each criterion (input parameters) and rating of each alternatives (ranking of robots). From comparison study, it is found that the Gaussian membership function is most effective for closeness measurement as its surface plot shows a good agreement with the output result. This approach confirms that the fuzzy membership function is a suitable decision making tool for the Manufacturing decisions with an object lesson in the robot selection process.

Keywords Industrial robots · Attributes · Selection · Fuzzy membership functions

1 Introduction

Industrial requirement of precision in repetitive works, reduction of man power and their associated cost, reduction of lead and idle time and working in hazardous situation, demands automation. Industrial robotics is the most flexible and fully automated technique used to fulfill such demands. Selection of industrial robot is a significant problem to the designer/manufacturer. Selection of robot is always task specific. Controlling a robot in varying operational environment like production work is always laborious. The environment "where subjective is not accurate and uncertain information available about the source", then Fuzzy logic controller will be

S. Nayak · N. Kumar
Suresh Gyan Vihar University, Jaipur, Rajasthan, India
e-mail: sasmitacet@rediffmail.com

S. Pattanayak · B. B. Choudhury (✉)
Indira Gandhi Institute of Technology, Sarang, Dhenkanal, India
e-mail: bbcigit@gmail.com

H. S. Behera et al. (eds.), *Computational Intelligence in Data Mining*, Advances in Intelligent Systems and Computing 990,
https://doi.org/10.1007/978-981-13-8676-3_20

preferred over conventional controller. Because fuzzy logic executes some numerical computation using linguistic labels based on membership functions. The fuzzy logic translates qualitative and imprecise linguistic statements into some clear cut computer statements.

Chu et al. [1] proposed a fuzzy TOPSIS technique for picking the suitable industrial robot based on specified operation, using the ratings of different objective and subjective criteria. Triangular membership function was used to train each criterion in linguistic terms. The values of objective criteria is first converted into some unit less indices and then compared with the ratings of subjective criteria. Purchase cost, load carrying capacity and positional accuracy of robot were considered as objective criterion. Man machine interface, programming flexibility and service contract of vendor were taken as subjective criterion. A reliable and exhaustive data base was prepared by Bhangale et al. [2], by considering various admissible robot attributes. There are 83 attributes from various aspects of robot such as 'physical appearance, performance, structure, application, control and availability' are considered during this evaluation work. This data base is found convenient in determining the robot selection procedure for a distinct manufacturing operation. This approach not only saves time but also hand over the right robot, for right work in right place and time. Digraph and Matrix approach is adopted, to select the handiest robot among wide range of alternative industrial robots by Rao et al. [3]. Digraph method prepares a selection index for the attributes like purchase cost, velocity, repeatability, degree of freedom and man-machine-interface of robot. That evaluates and ranks the industrial robot for a specified application. Kumar et al. [4] developed a deterministic quantitative model based on Distance Based Approach (DBA) to evaluate, select and ranks the robots. The criticality of attributes was resolved by conducting sensitive analysis test. Rao et al. [5] developed an integrated MADM (Multiple Attribute Decision Making) approach for the choosing the appropriate application specific robot. The attributes recognized in this study are load carrying capacity, repeatability, speed; memory capacity and manipulator reach of robot. Tansel et al. [6] proposed a ROBSEL system that assists the manufacturer regarding the robot selection. It was reported that, the ROBSEL approach was found to be appropriate, practicable and an easy tool for robot selection and it also improvised the decision related to robot election. Khandekar et al. [7] explained the Fuzzy axiomatic design (FAD) principle to clarify the problem associated with robot selection in real time. In this study seven robots were selected and ranked, which are employed for the lighter assembly application. PROMETHEE-II technique is implemented by Sen et al. [8] for industrial robot selection. It was reported that PROMETHEE-II not only avoids the errors arises during decision making, but also presents the entire ranking order of all available alternatives deliberately. Parameshwaran et al. [9] adopted the FDM, FAHP, Fuzzy VIKOR or fuzzy modified TOPSIS, and Brown-Gibson model for selecting the robots based on objective and subjective criterions. Important criterions were listed using Fuzzy Delphi Method (FDM). FAHP assigns weight to each criterion. Then the robots were ranked using Fuzzy VIKOR approach by considering all factors. Brown-Gibson model was implemented for the estimation of robot selection index. Ghorabaee [10] figure out the disputes associated with robot selection using VIKOR methodology that combines with interval type- 2 fuzzy member. The rank presented by this approach was nearly consistent with others and this approach also holds good stability during handling of different weight criterions. Analytic Hierarchy Process (AHP) was demonstrated by Breaz et al. [11] to select the suitable industrial robot for milling operation. A simulation exercise was carried out to estimate milling efficiency of the robotic structure.

This work presents an advanced fuzzy modified TOPSIS approach to overcome the obstacles arises during robot selection. The performance rating values of each alternative are expressed by triangular and Gaussian membership function under the selected weight criteria. Then a comparison is established among the membership functions to clinch the leading technique that contributes towards nearly closeness result.

2 Robot Selection Attributes

Convenient recognition of robot selection attribute is perilously essential, when a comparison is established among different available alternative robots for a specific industrial operation. This identification always creates conflict for the manufacturer. If the attributes are selected in proper manner, then it will reflect accuracy in the operation that was carried out by the selected robot. That ultimately enhances the industrial productivity. There are 32 manipulator parameters (attributes) are available for the selection of industrial robot. Out of which only seven attributes are taken up during this investigation work, which affects the performance of robot. Because it is not possible to estimate the final score for all parameters due to uncertainty in measurement of these factors, lacuna in interpretation of acceptable range for each parameter and method used to integrate these dissimilar factors. The predicted manipulator attributes are both quantitative as well as qualitative in nature. The minimum essential attributes (parameter) needed for a material handling robot is summarized in Table 1.

Table 1 Standard parameter requirement of a material handling robot

Sl. No.	Parameters (manipulator attributes)	Values
1	Load capacity	Minimum 4 kg
2	Repeatability	0.5 mm
3	Speed	At least 800 mm/s
4	Memory capacity	–
5	Reach	500 mm
6	Degree of freedom	At least 4
7	Swept area	2700

3 Methodology

Fuzzy logic is a computation based reasoning approach that holds good for decision making when the information about the input data's is incomplete/uncertain [1]. Fuzzy logic first converts the input, output variables into fuzzy variables, and then establishes a rule base among them to produce the fuzzy result. Here fuzzy logic is introduced to characterize the industrial robot selection for material handling operation. After predicting the manipulator attributes, a feed forward fuzzy rule is established using various membership functions (Triangular and Gaussian) in linguistic terms. The weightage of each input parameter and the ranking range for robot are illustrated in Tables 2 and 3 respectively.

The capacity of industrial Robots is expressed in terms of two membership functions like triangular and Gaussian as input to the fuzzy controller. Three different linguistic terms are used for each parameter of different membership function. Table 4 represents the linguistics terms used for membership functions.

3.1 Triangular Membership Function

Triangular membership function is expressed by combination of three points that constitute a triangle as shown in Fig. 1. Triangular membership function has got popularity due its simplicity, ease of arithmetic computation and ease of specifying the input range. Triangular-shaped membership function is mathematically expressed as:

Table 2 Weightage value of parameters

Sl. No.	Parameter/attribute name	Weightage range
Input-1	Load carrying capacity	[−7 7]
Input-2	Repeatability	[−1 1]
Input-3	Tip speed	[−5 5]
Input-4	Memory capacity	[−6 6]
Input-5	Manipulator reach	[−4 4]
Input-6	Degree of freedom	[−3 3]
Input-7	Swept area	[−1 1]

Table 3 Factors required for ranking range of robot

Sl. No.	Factors	Range assigned	Ranking range
1	Low	−1	−1 to −0.333
2	Medium	0	−0.333 to +0.333
3	High	+1	+0.333 to +1

Table 4 Linguistic terms used for membership function

Linguistic terms of membership functions	Linguistic term descriptions
LCL, LCM, LCH	Load capacity value is low, medium and high
REL, REM, REH	Repeatability value is low, medium and high
MTSL, MTSM, MTSH	Max tip speed value is low, medium and high
MCL, MCM, MCH	Memory capacity value is low, medium and high
MRL, MRM, MRH	Manipulator reach value is low, medium and high
DOFL, DOFM, DOFH	Degree of freedom load value is low, medium and high
SAL, SAM, SAH	Swept area load value is low, medium and high

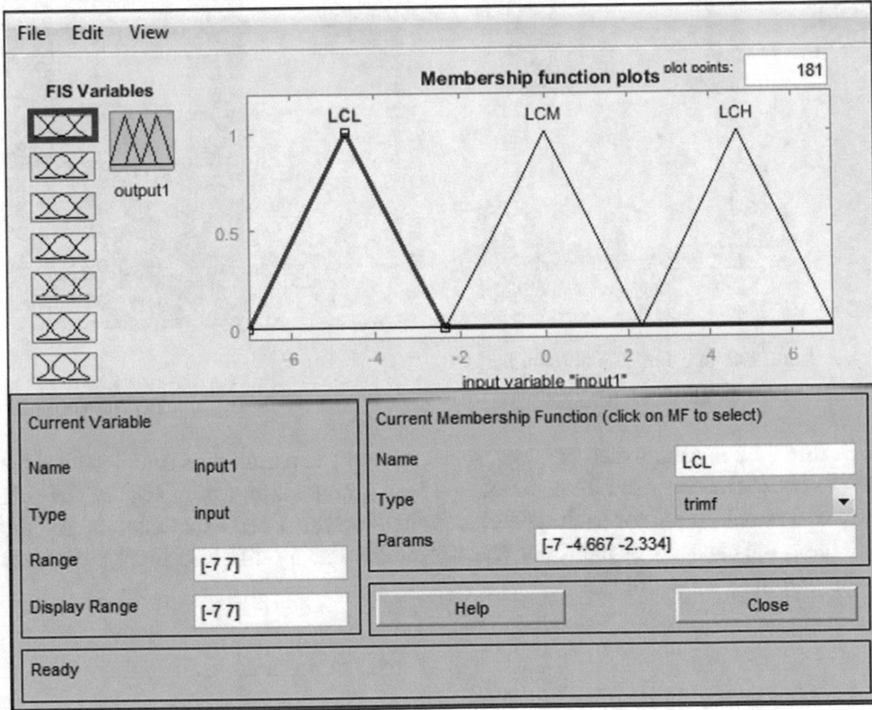

Fig. 1 Input parameters with triangular membership function

$$y = trimf(x[a, b, c]) \tag{1}$$

where, ‘a’ and ‘c’ is the staring and end point of the range and ‘b’ represents the mid value of the range.

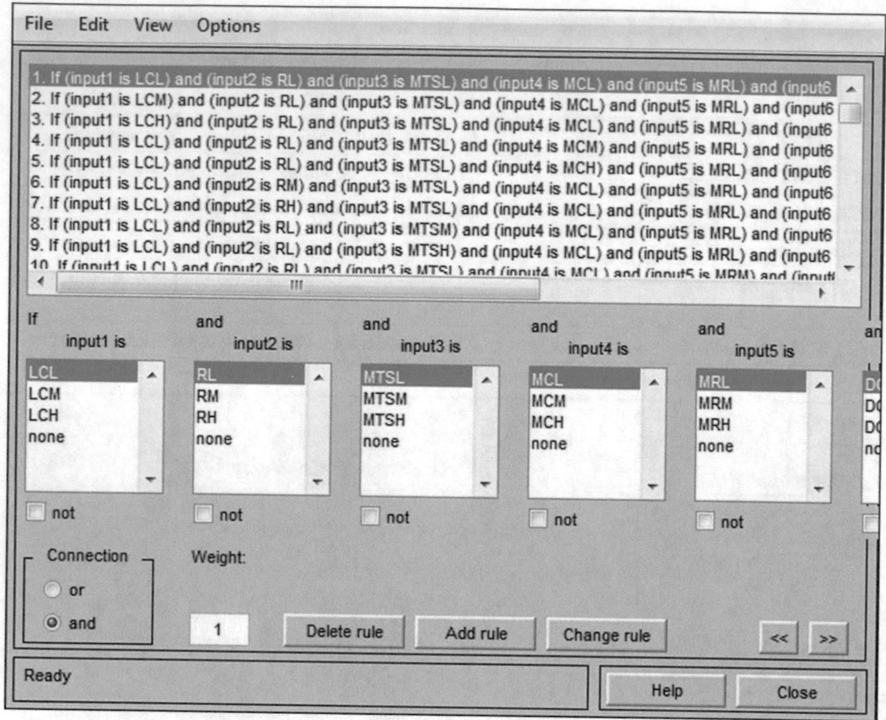

Fig. 2 Rule base for estimation of robot ranks

A rule base is established between input and output parameters to determine the fuzzy result. The rule used for estimation of robot ranks is shown in Fig. 2. The rule base will remain same for each type of membership function. After establishing the rule base, the capacity of the robot was determined using different input parameter combination as shown in Fig. 3.

3.2 *Gaussian Membership Function*

Two input value is needed to represent a Gaussian membership function. Where first value is responsible for the shape of the curve and second value specifies the midpoint of the range as presented in Fig. 4. Most of the research papers are available with Gaussian membership function, because of their rigorous result. Mathematically the Gaussian membership function can express as follows:

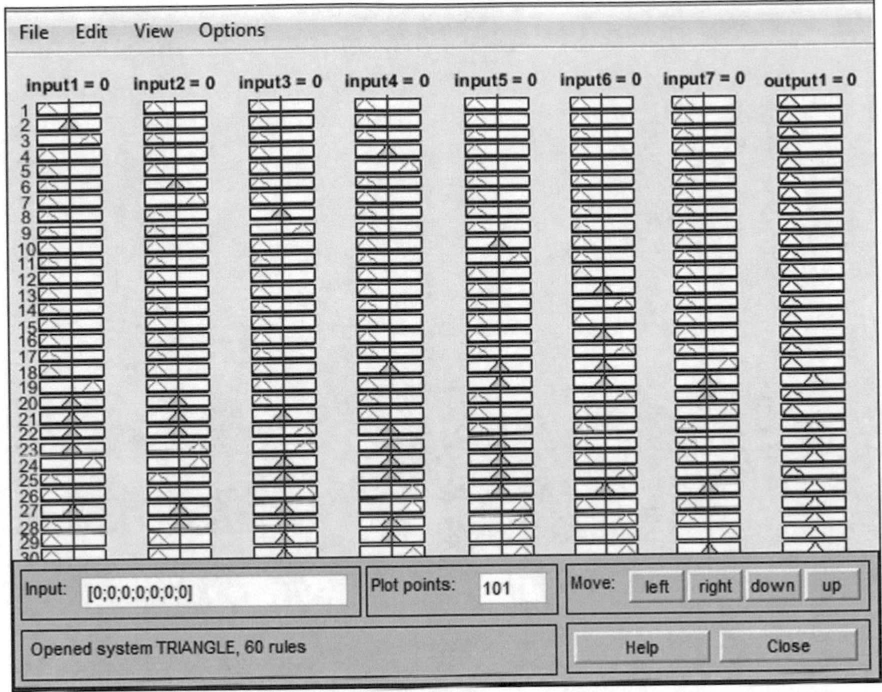

Fig. 3 Graphical representation of input variables and output with triangular membership function

$$y = gaussmf(x, [\sigma \ c]) \tag{2}$$

After defining the rule base (same as in Triangular membership function), different input factors are arranged to evaluate the capacity of robot. Figure 5 graphically illustrates the rapport between input variables and output using Gaussian membership function for estimating the robot capacity.

4 Result and Discussion

Different input attribute combinations are used to figure out the ranking range of robots for each membership function. The input parameter combinations will remain same for all membership function (Triangular and Gaussian). Since this work aims at finding, the most effective approach among these techniques. The input and output data set combination for each approach is presented in Table 5.

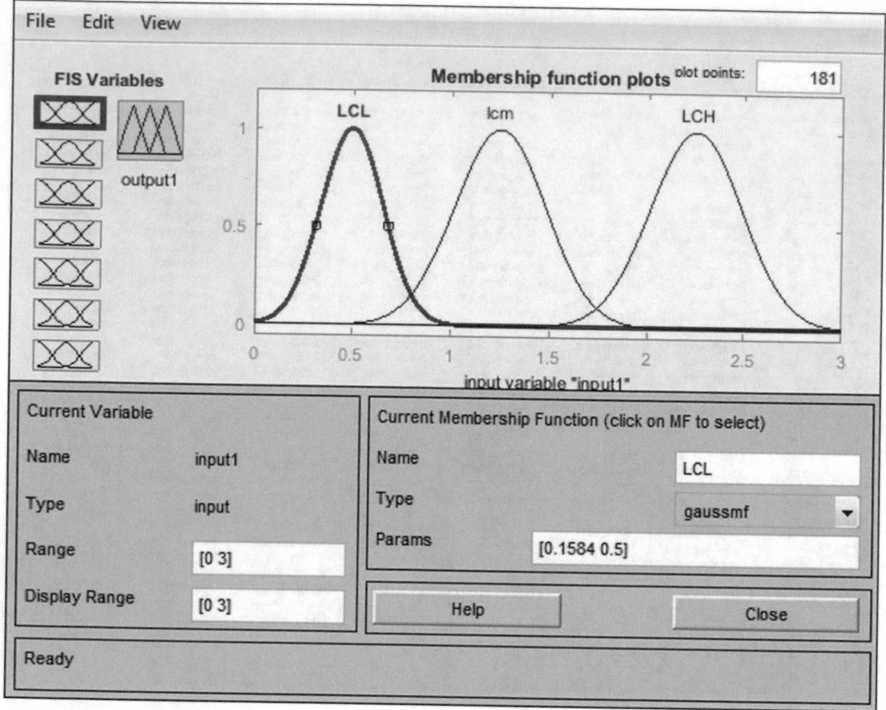

Fig. 4 Input parameters with Gaussian membership function

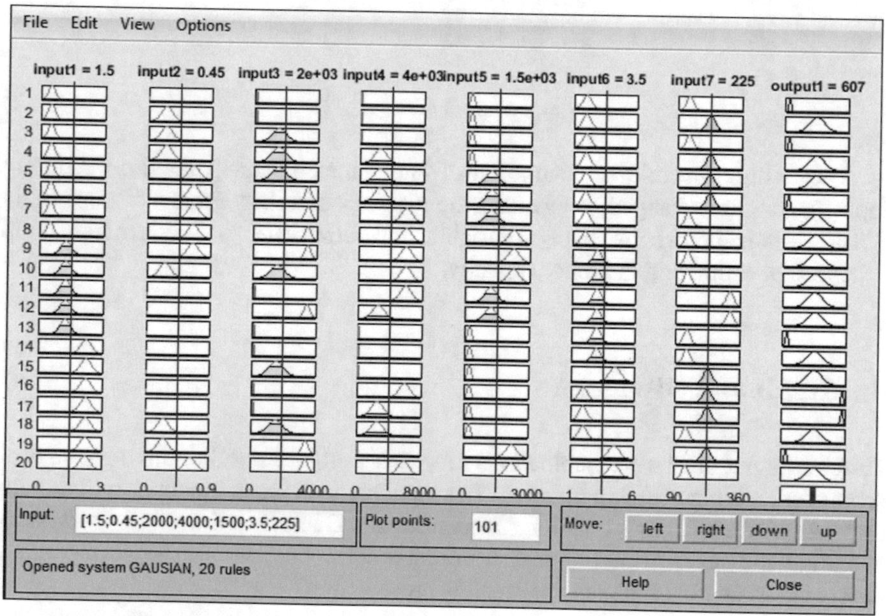

Fig. 5 Graphical representation of input variables and output with Gaussian membership function

Table 5 Comparison among different membership functions

Sl. No.	LC	RE	MTS	MC	MR	DOF	SA	Triangular	Gaussian
1	−7	−1	−5	−6	−4	−3	−2	0	−0.654
2	−6.07	−0.866	−4.333	−5.2	−3.466	−2.6	−1.733	−0.666	−0.666
3	−5.14	−0.732	−3.666	−4.4	−2.932	−2.2	−1.466	−0.666	−0.666
4	−4.21	−0.598	−2.999	−3.6	−2.398	−1.8	−1.199	−0.666	−0.666
5	−3.28	−0.464	−2.332	−2.8	−1.864	−1.4	−0.932	0	−0.666
6	−2.35	−0.33	−1.665	−2	−1.33	−1	−0.665	0	−0.655
7	−1.42	−0.196	−0.998	−1.2	−0.796	−0.6	−0.398	0	−0.594
8	−0.49	−0.062	−0.331	−0.4	−0.262	−0.2	−0.131	0	−0.53
9	0.44	0.072	0.336	0.4	0.272	0.2	0.136	0	0.524
10	1.37	0.206	1.003	1.2	0.806	0.6	0.403	0	0.591
11	2.3	0.34	1.67	2	1.34	1.0	0.67	0	0.654
12	3.23	0.474	2.337	2.8	1.874	1.4	0.937	−3.36 e −0.18	0.666
13	4.16	0.608	3.004	3.6	2.408	1.8	1.204	−2.49 e −0.17	0.666
14	5.09	0.742	3.671	4.4	2.942	2.2	1.471	−1.46 e −0.17	0.666
15	6.02	0.876	4.338	5.2	3.476	2.6	1.738	−1.44 e −0.17	0.666
16	7	1	5	6	4	3	2	0	0.654

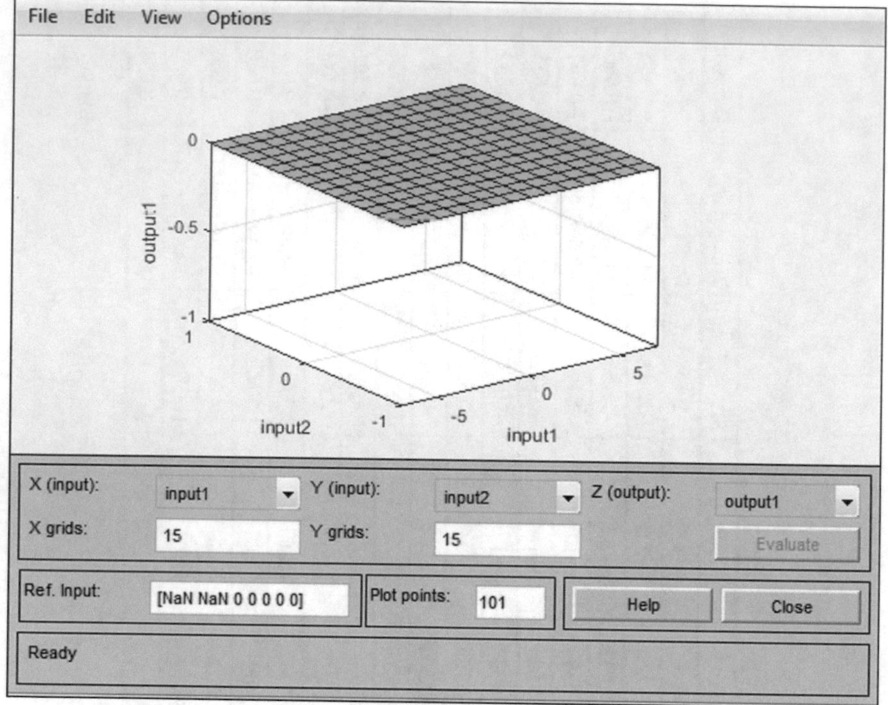

Fig. 6 Surface plot for triangular membership function

From the comparison study, it is concluded that, the Gaussian approach provides much consistent and produces qualitative results in comparison to triangular method. In addition to the above facts the proposed method also offers a more objective, easy to use, and simple robot selection process. To make a clear understanding among these approaches, some 3D surface graphs are plotted for the membership functions. Figures 6 and 7 represents the surface plot for triangular and Gaussian membership function respectively. The 3D surface plot also provides conclusive evidence regarding the effectiveness of Gaussian approach than triangular approach.

5 Conclusion

The ranking of industrial robot is effortlessly assessed by recognizing numerous manipulator attributes. The robots ranks are computed by establishing the fuzzy optimization approach with some fuzzy rules. Then these rankings are considered for predicting the closeness of robots. A very good agreement is found between fuzzy result and the result obtained from theoretical analysis. From comparison study (Table 5) it is concluded that, the Gaussian membership function shows more

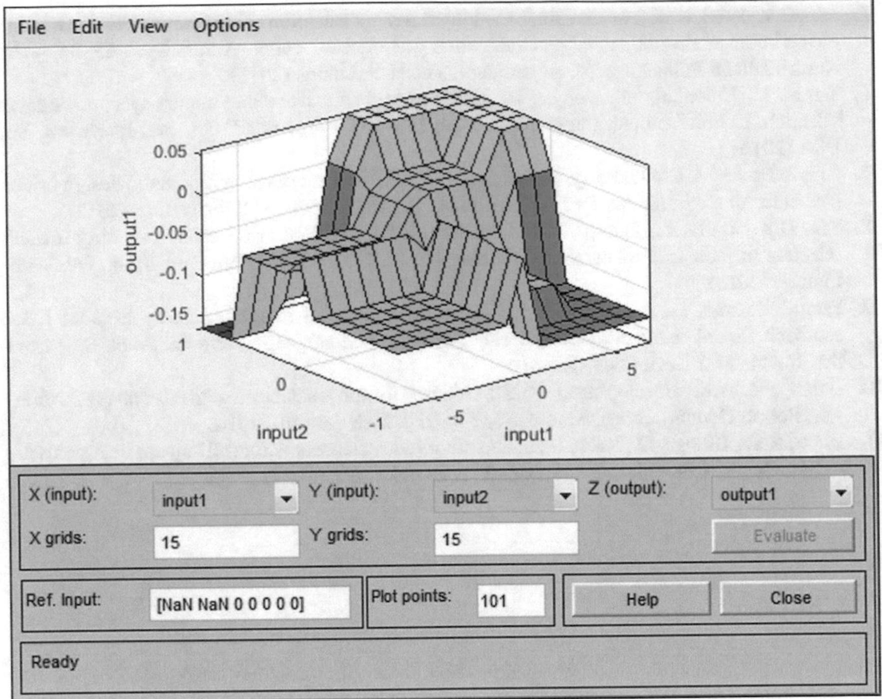

Fig. 7 Surface plot for Gaussian membership function

precise outcomes as compared to Triangular membership function. This shows the robustness of the proposed fuzzy controller, while making decisions related to robot selection. The application of this proposed Gaussian approach not only limited as a decision making tool in robot selection but also can be implemented into other optimization works like optimization of machining parameters.

References

1. Chu, T.C., Lin, Y.C.: A fuzzy TOPSIS method for robot selection. Int. J. Adv. Manuf. Technol. **21**(4), 284–290. Springer-Verlag London Limited (2003)
2. Bhangale, P.P., Agrawal, V.P., Saha, S.K.: Attribute based specification, comparison and selection of a robot. Mech. Mach. Theory **39**, 1345–1366. Elsevier Ltd. (2004)
3. Rao, R.V., Padmanabhan, K.K.: Selection, identification and comparison of industrial robots using digraph and matrix methods. Robot. Comput. Integr. Manuf. **22**(4), 373–383. Elsevier Ltd. (2006)
4. Kumar, R., Garg, R.K.: Optimal selection of robots by using distance based approach method. Robot. Comput. Integr. Manuf. **26**(5), 500–506. Pergamon Press, Inc. Tarrytown, NY, USA (2010)

5. Rao, R.V., Patel, B.K., Parnichkun, M.: Industrial robot selection using a novel decision making method considering objective and subjective preferences. Robot. Auton. Syst. **59**, 367–375. North-Holland Publishing Co. Amsterdam, The Netherlands (2011)
6. Tansel, Y., Yurdakul, M., Dengiz, B.: Development of a decision support system for robot selection. Robot. Comput. Integr. Manuf. **29**, 142–157. Pergamon Press, Inc. Tarrytown, NY, USA (2013)
7. Khandekar, A.V., Chakraborty, S.: Selection of industrial robot using axiomatic design principles in fuzzy environment. Decis. Sci. Lett. **4**, 181–192. Growing Science Ltd. (2015)
8. Sen, D.K., Datta, S., Patel, S.K., Mahapatra, S.S.: Multi-criteria decision making towards selection of industrial robot. Benchmarking Int. J. **22**(3), 65–487. Emerald Group Publishing Limited (2015)
9. Parameshwaran, R., Kumar, S.P., Saravanakumar, K.: An integrated fuzzy MCDM based approach for robot selection considering objective and subjective criteria. Appl. Soft Comput. **26**, 31–41. Elsevier B.V (2015)
10. Ghorabaee, M.K.: Developing an MCDM method for robot selection with interval type-2 fuzzy sets. Robot. Comput. Integr. Manuf. **37**, 221–232. Elsevier Ltd. (2016)
11. Breaz, R.U., Bologa, O., Racz, S.G.: Selecting industrial robots for milling applications using AHP. Procedia Comput. Sci. **122**, 346–353. Elsevier B.V. (2017)

Efficient Analysis of Clinical Genomic Database

Sarbani Dasgupta and Banani Saha

Abstract Clinical genomic database is a database consisting of human genomic data which can be analysed to get information about diagnosis of a particular disease. Data mining is the process of extracting meaningful information from data stored in huge database. Among several data mining algorithms Association rule mining algorithm is used for frequent pattern analysis. In case of Clinical genomic database, Association rule mining algorithm can be used for determination of genes which are responsible for a particular disease. However sequential implementation will consume a lot of processing time. In order to increase the speed of processing parallel implementation Association Rule mining algorithm can be used. In this paper parallel implementation of Association Rule mining algorithm using analytical big data technology have been proposed for efficient analysis of Clinical genomic database.

Keywords Clinical genomic database · Association rule mining algorithm · Big data technologies

1 Introduction

The main aim of clinical genomic database is to increase the knowledge of the physicians about various diseases. With the advancement in technology a lot of human genomic sequencing is available. In order to analyse the genomic data, so that it can help in the process clinical of intervention by the physician, a Clinical Genomic database (CGD) is created. A CGD is a freely available, searchable as well as Web

S. Dasgupta (✉)
Techno International New Town, Block-DG, Action Area 1 New Town, Kolkata 700156, India
e-mail: only4sarbani@gmail.com

B. Saha
Computer Science and Engineering Department, University of Calcutta, 2, JD Block, Sector III, Salt Lake City, Kolkata 700098, West Bengal, India
e-mail: bsaha_29@yahoo.com

H. S. Behera et al. (eds.), *Computational Intelligence in Data Mining*, Advances in Intelligent Systems and Computing 990,
https://doi.org/10.1007/978-981-13-8676-3_21

accessible database [1]. This database may be used for efficient analysis of individual human genome sequencing to obtain clinically significant health information. The application of data mining techniques to extract useful knowledge from datasets has been widely applied in various fields. By mining frequent patterns from huge amount of data, association among various data items in a database can be obtained in form of rules. In case of vast integrated biological or clinical dataset, association rule mining algorithm can help in the process of analysing individual human genome for obtaining information about various diseases. Among the various Association rule mining algorithm, Apriori algorithm one of the basic association rule mining algorithm can is used. However the original algorithm is modified for efficient analysis of CGD. Another problem with this algorithm is that its sequential implementation will consume a lot of processing time. In order to overcome the problem of sequential algorithm, in this paper parallel implementation of association rule mining algorithm i.e. parallel implementation of modified Apriori algorithm [2] for efficient analysis of CGD have been proposed.

2 Overview of Clinical Genomics

Clinical genomics or Clinicogenomic refers to the study of genomic data to gather knowledge about its Clinical outcome. In case of Clinical genomics, the entire genome of an individual can be used for detection of several diseases. Whole genome testing can be used for detection of all the diseases which can help physicians to take decision about the process of treatment of a particular patient. By analysing the whole genome, a physician is able to construct medical plans based on an individual patient's genome rather than generic plans for all patients with the same diagnosis [3]. Furthermore, whole genomic testing can help physician for deciding on the medications and treatments that work best on particular disease. It is particularly helpful for detection of cancer-causing mutations, which can be used by the physician to treat future patients with same genomic sequencing [4].

Due to advancement of technology, a lot of human genomic data is obtained. In order to obtain valuable information from this genomic data, a database of genetic data consisting of information about clinical utility of genetic diagnosis and specific medical intervention is constructed [1]. Actually it is a collection of datasets of genes mainly categorised into two types, namely manifestation and interventions. Each category consists information about genes of twenty different types of clinical conditions.

3 Association Rule Mining and Its Applications in CGD

Data mining is defined as the process of extracting valuable information from a huge amount of data stored in a database [5]. Among several data mining techniques, association mining is used for discovering valuable relations between variables in large dataset. Association rule mining is also known as frequent pattern analysis which is used for finding correlations among frequent patterns in a huge dataset. Like several applications of Association rule mining, it is also used for mining medical data records [6, 7].

Association rule is generally represented in the form $\alpha \Rightarrow \beta$ where α, β subsets of item set γ and $\alpha \cap \beta = \emptyset$. The rules $\alpha \Rightarrow \beta$ holds in transaction set denoted as λ with support s, if s% of the transactions in λ contain both α and β. Similarly the rule $\alpha \Rightarrow \beta$ holds in λ with confidence c, if c% of the transactions in λ supporting α also support β [8].

There are several Association rule mining algorithm namely Apriori algorithm, FP-growth algorithm, Éclat algorithm. Among them Apriori algorithm is the traditional Association rule mining algorithm which is originally proposed by [8] for market basket analysis can be used for the purpose of analysing clinical genomic database. But the original Association rule is modified and is represented in the form of [IF-THEN] structure. The algorithm uses a level-wise search, where n-item sets (an item set that contains n items) is used to explore (n + 1) item set. Simply, a first-item set generated, will be used to generate the second-item set, in turn generate the third item set until no more n-item set can be found [2].

4 Proposed Approach

In this section the proposed approach has been discussed in detail. This section gives an elaborate description of parallel implementation of Association Rule Mining and its application in analysis of Clinical Genomic Database (CGD).

4.1 Basics

Previously sequential association analysis of genomic data have been proposed in various literature [1]. The association rule mining algorithm that is used for this purpose is a modified version of Apriori algorithm. The rules are represented as IF-THEN-ELSE structure. The main problem with this approach is that a lot of processing time is required as the amount of data increases. In order to overcome this problem of sequential implementation of modified Apriori algorithm for the purpose of analysing CGD, parallel implementation of modified Apriori algorithm have been proposed in this paper. The proposed algorithm has parallelised using big data technologies [9, 10].

There are two types of analytical big data technologies. One of them performs processing in batch mode. They use the concept of Google Map Reduce programming framework [11]. But the main problem with this framework is that it does not support data reuse for multiple computation [4]. Several machine learning and data mining algorithms require interactive processing. For this purpose a new data structure Resilient Distributed Datasets (RDD) has been proposed, which is fault tolerant and can provide in-memory storage. It is very essential for interactive distributed computing.

4.2 Dataset Used for the Experimental Purpose

The dataset that is used is the Clinical genomic database [1] which is a condition-based database consisting of information about 2616 genes organised according to the disease caused. This database is categorised into two different types namely manifestation and intervention. As stated in [12] the manifestation category consist of genes of organ systems that are affected by gene mutation. The intervention category consist of genes of organ systems for which information about clinical intervention is available [12]. The database consists of information about the genes of infected conditions into different datasets. These are Allergy/Immunology/Infectious, Audiology/Otolaryngology, Biochemical, Cardiovascular, Craniofacial, Dental, Dermatologic, Endocrine, Gastrointestinal, General, Genitourinary Hematologic, Musculoskeletal, Neurologic, Obstetric, Oncologic, Ophthalmologic, Pharmacogenomic, Pulmonary and Renal [12].

Each row of single dataset in Clinical genomic database in both manifestation and intervention category, consist attributes namely, "gene", "entrez gene id", "age group", "allelic conditions", "Manifestations categories", "intervention categories", "interventions/rationale" and "references".

4.3 Proposed Methodology

The association rule used in CGD is the detection gene of responsible for "cancer". The support count is specified by the user. The user can detect the gene and the condition which is responsible for cancer condition. As stated earlier, the proposed algorithm is parallelized using analytical big data technology. For this purpose a lighting fast technology, Apache Spark is used. Its main feature is in memory cluster computing that increases the speed of processing. Apache Spark supports interactive query and stream processing which is not supported by Hadoop. It helps in faster execution of Map Reduce operation. The architecture of Apache Spark is shown in Fig. 1.

Spark SQL is a programming module provided by Apache Spark for structured data processing. It follows certain architecture. For our proposed algorithm we have used Spark SQL. It follows an interactive approach.

In this case the support count of the' genes' responsible for condition "cancer" are calculated using Sparksql.

The proposed approach follows two steps:

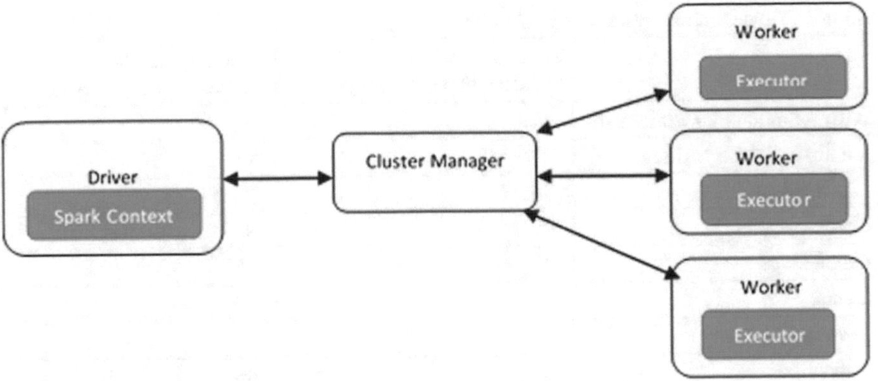

Fig. 1 Architecture of apache spark

i. Frequent item set generation.
ii. Modified Association rule is generated from frequent item sets is obtained.

In the first step frequent item set is generated from the candidate set. In this case two attributes of clinical genomic database for every datasets are considered for the generation of frequent item sets. They are gene symbol and condition data which is obtained for a particular gene. In the process of frequent item set generation the user defined minimum support count is required. The number of "genes" responsible for the condition "cancer/malignant/Leukoencephalopathy/Beginin" is calculated. In order to increase the speed of processing this step is calculated with the help of Spark sql. After obtaining the support count of dataset the result it is checked with the minimum support count. If the support count is greater than user specified minimum support count then that gene is consider for next step. This process will continue until no more item set can be generated.

Next step is the generation of the Association rule. In this case the rule is modified and is represented as IF gene "FH" THEN "Hereditary leiomyomatosis and renal cell cancer". The rule obtained satisfies the minimum confidence.

4.4 Experimental Results

At first the sequential implementation of the proposed algorithm is performed. It is done on Windows 10 operating system using Oracle 10 g and JAVA software version 1.8.0_91. The time required by the algorithm is given in Table 1.

For parallel algorithm, one driver node and four worker node for Apache Spark platform with Intel Corei3 processor 2.00 GHz CPU(X4) and 8 GB Ram is used. The Operating system used for this purpose is same as that of the sequential one i.e. Ubuntu 16.04 LTS. Since for the purpose of parallel processing the datasets have to be load into Hadoop Distributed File System (HDFS), so Hadoop version 2.6.0

Table 1 Time required by sequential algorithm

Area	Time required for manifestation category (s)	Time required for intervention category (s)
Allergy/immunology/infectious	64	70
Audiology/otolaryngology	82	71
Biochemical	90	70
Cardiovascular	95	83
Craniofacial	98	0
Dental	61	0
Dermatologic	88	23
Endocrine	85	66
Gastrointestinal	89	65
General	30	180
Genitourinary	70	20
Hematologic	90	83
Musculoskeletal	120	30
Neurologic	180	30
Obstetric	32	33
Oncologic	63	59
Ophthalmologic	102	72
Pharmacogenomics	0	67
Pulmonary	65	60
Renal	90	59

along with Spark version 1.5 and eclipse version 4.6.0 (neon) is used as software. The time required for parallel algorithm is given in Table 2.

5 Evaluation of Experimental Result and Discussion

It can be seen from the above tables that the processing time of the parallel algorithm is much less than that of the sequential algorithm.

In order to evaluate the performance of any parallel algorithm three factors are considered. They are speed up [13], scale up [14] and size up [13].

The speedup factor is defined as the ratio of runtime of the sequential algorithm for solving a problem to the time taken by the parallel implementation of the same algorithm to solve the same problem on k number of processors.

$$\text{Speedup} = \frac{\text{RT}_1}{\text{RT}_\text{k}}. \tag{1}$$

Table 2 Time required by parallel algorithm

Area	Time required for manifestation category (s)	Time required for intervention category (s)
Allergy/immunology/infectious	0.4	0.4
Audiology/otolaryngology	0.4	0.2
Biochemical	0.527	0.15
Cardiovascular	0.6	0.4
Craniofacial	0.4	0
Dental	0.1	0
Dermatologic	0.5	0.05
Endocrine	0.5	0.157
Gastrointestinal	0.52	0.2
General	0.003	0.9
Genitourinary	0.2	0.001
Hematologic	0.518	0.3
Musculoskeletal	0.6	0.0001
Neurologic	0.8	0.003
Obstetric	0.05	0.05
Oncologic	0.3	0.3
Ophthalmologic	0.5	0.02
Pharmacogenomics	0	0.001
Pulmonary	0.1	0.005
Renal	0.5	0.2

RT_1 is time required by a single processor and RT_k is the processing time of the parallel implementation of the algorithm with k number of processors. The proposed approach has been performed with 8 processors.

Next, the parallel implementation of the proposed algorithm is measured in terms of scale up factor. The idea of the scale up factor is that, with increase of the size of the dataset the time required for processing will remain same as the runtime for processing smaller dataset, if the number of processor is increased. Actually it is the measurement of capacity of a larger system to perform larger jobs at the same run time as that of a single system. Therefore,

$$\text{Scaleup} = \frac{T_1}{T_{nn}}. \tag{2}$$

T_1 is the runtime of algorithm obtained for processing smaller size dataset in this case 128 MB on a single core machine. T_{nn} is the time required for processing n times larger dataset in this case 2 GB with a machine with n cores.

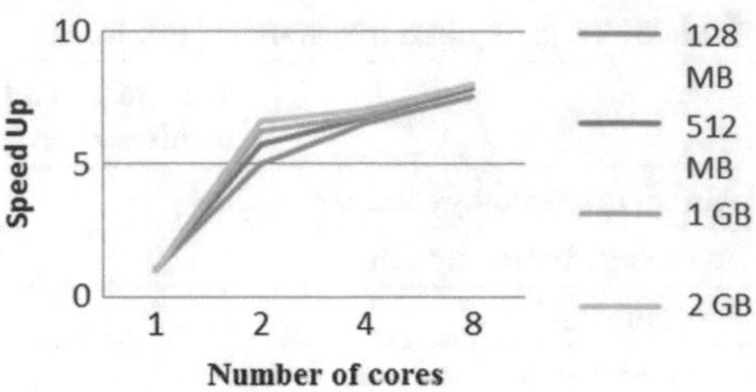

Fig. 2 Speed up factor

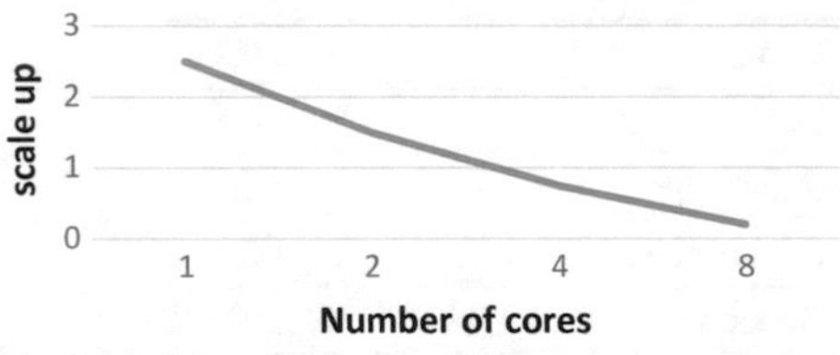

Fig. 3 Scale up factor

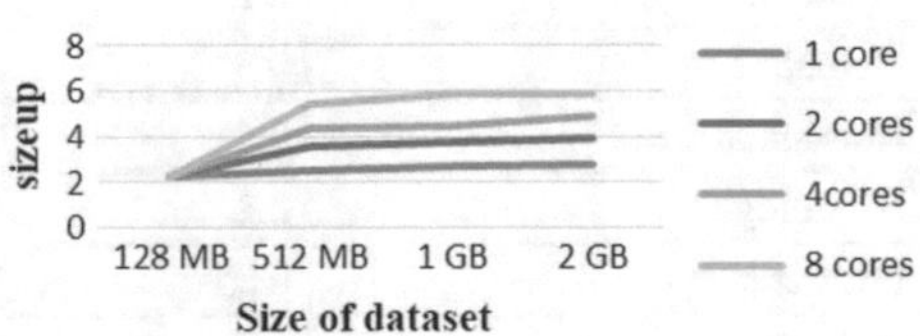

Fig. 4 Size up factor

Finally, in case of size up measurement, it is shown that the time required for execution of n times larger dataset times will not change if the number of processor is increased.

Size up indicates the ratio of the work increase.

Size up factor is defined as

$$\text{Sizeup} = \frac{T_n}{T_1}. \tag{3}$$

T_n is the time required for execution of the algorithm with larger datasets and T_1 is the time required for execution of smaller dataset.

In this case the size of the dataset varies from 128 MB to 2 GB.

As stated earlier to evaluate the speedup as given in Eq. 1, scale up in Eq. 2 and size up factor in Eq. 3, the experimental analysis is performed on 1–8 cores. The results obtained by calculating the three factors are shown graphically in Figs. 2, 3, and 4 respectively.

6 Conclusion

Clinical genomic database allows to analyse genomic data in a clinically meaningful way. Association analysis of Clinical genomic database helps to identify genes responsible for various diseases. In this paper, association rule mining algorithm is used for the purpose of identification of genes responsible for 'cancer'. A parallel approach of association rule mining algorithm has been proposed for faster analysis of Clinical genomic database. For this purpose, Apache Spark which is based on cluster computing techniques is used. The original Association rule mining algorithm is modified and the rules obtained are represented in different form. For experimental purpose CGD is downloaded. It has been shown parallel implementation of association rule mining algorithm for faster analysis of CGD shows more efficient result.

References

1. Clinical Genomic Database: https://rsearch.nhgri.nih.gov/CGD/download/. Accessed on 07 July 2018
2. Tsai, F.S.: Network intrusion detection using association rules. Int. J. Recent Trends Eng. **2**(2) (2009)
3. Wikipedia: https://en.wikipedia.org/wiki/Clinicogenomics. Accessed on 7 July 2018
4. Zaharia, M., Chowdhury, M., Das, T., Dave, A., Ma, J., McCauley, M., Franklin, M.J., Shenker, S., Stoica, I.: Resilient Distributed Datasets: A Fault-Tolerant Abstraction for In-Memory Cluster Computing. In: Proceedings of the 9th USENIX conference on Networked Systems Design and Implementation, pp. 2–2. USENIX Association (2012, April)
5. Tan, P.-N., Steinbach, M., Kumar, V.: Introduction to Data Mining. Addison Wesley, New York (2006)
6. McCormick, T.H., Rudin, C., Madigan, D.: A hierarchical model for association rule mining of sequential events: an approach to automated medical symptom prediction [Submitted to the Annals of Applied Statistics] (2000)
7. Xu, X., Qiuhong, S., Hongtao, Z., Lei, W., Liu, Y.: Study and application on the method of association rules mining based on genetic algorithm. In: The 2nd International Conference on Computer Application and System Modeling (2012)
8. Aggarwal, R., Imielinski, T., Swami, A.: Mining association rules between sets of items in very large database. In: Proceedings of ACM SIGMOD Conference (1993)
9. Dasgupta, S., Saha, B.: A framework for discovering important patterns through parallel mining of protein–protein interaction network. In: Proceedings of the Second International Conference on Computer and Communication Technologies IC3T 2015, vol. 3, pp. 397–406, ISSN 2194-5357, ISSN 2194-5365 (electronic)
10. Li, N., Zeng, L., He, Q., Shi, Z.: Parallel implementation of Apriori algorithm based on MapReduce. In: Proceedings of 13th ACIS International Conference on Software Engineering, Artificial Intelligence, Networking and Parallel/Distributed Computing, @IEEE (2012)
11. Dean, J., Ghemawat, S.: MapReduce: simplified data processing on large clusters. In: OSDI'04, 6th Symposium on Operating Systems Design and Implementation, Sponsored by USENIX, in cooperation with ACM SIGOPS, pp. 137–150 (2004)
12. Solomon, B.D., Nguyen, A.-D., Bear, K.A., Wolfsberg, T.G.: Clinical Genomic Database. www.pnas.org/cgi/doi/10.1073/pnas.1302575110
13. Sun, X.-H., Gustafson, J.L.: Toward a better parallel performance metric. Parallel Comput. **17**, 1093–1109 (1991)
14. Alexandrov, V.: Parallel Scalable Algorithms-Performance Parameters. www.bsc.es

Database Intrusion Detection Using Adaptive Resonance Network Theory Model

Anitarani Brahma and Suvasini Panigrahi

Abstract In the current automation world, every organizations starting from education to industry and also in research organization are maintaining database and several security breaches are found in these databases. Traditional database security mechanism cannot handle the malicious access toward database. Although, various researches have been done in database intrusion detection, but most of the researches are limited in efficiency and accuracy. Inspired by human cognitive system, we present a database intrusion detection system using Adaptive Resonance Theory which is accompanied with some data mining techniques for preprocessing data. The proposed model can learn easily and cope up with the dynamic environment which entitles the system to detect both known and unknown patterns accurately with low false positive cost. The calculated simulation result shows that the database intrusion detection based on Adaptive Resonance Theory can accelerate the detection process with higher accuracy as compared to Self Organizing Map and Radial basis functional neural network.

Keywords Database intrusion detection · Adaptive resonance theory · Self organizing map · Radial basis functional neural network · Back propagation neural network

1 Introduction

There are two learning process in artificial neural network to train any model namely supervised and unsupervised. Supervised learning process requires target output or label for preparing the model to learn; but unsupervised neural network model learns

A. Brahma (✉) · S. Panigrahi
Department of Computer Science & Engineering, Veer Surendra Sai University of Technology, Burla, Odisha, India
e-mail: brahmaanita00@gmail.com

S. Panigrahi
e-mail: suvasini26@gmail.com

H. S. Behera et al. (eds.), *Computational Intelligence in Data Mining*, Advances in Intelligent Systems and Computing 990,
https://doi.org/10.1007/978-981-13-8676-3_22

by itself with the help of its surrounding or compete with itself [1]. Supervised Neural Network Model like Adaptive Linear Neural Network (ADALINE), Back Propagation Neural Network (BPNN) has been used in the field of classification and pattern recognition or in the field of intrusion detection. But there are many situations where supervised learning process fails to handle the received data viz.

- Where target output is not available.
- Very difficult to label the available data.
- Where the external environment for the model is dynamic.

In such type of situations, unsupervised learning plays very important role to model the problem. Database Intrusion detection can be assumed to become under the above problem and serious issues to be solved as the database plays a key role in any information system. When an intruder attempts to break and get access into the database system is called Intrusion. The intruder having authorization to access the database is more vulnerable than the intruder having no authorization [2]; the intrusion detection system detects the intrusion and reports it to the proper authority. Anomaly Detection and Misuse detection are the two categories of intrusion detection. Anomaly detection identifies the intrusion based on the behavioral pattern of the user where as misuse detection identifies by comparing it with the stored intrusive patterns.

In this investigation, we tried to implement one unsupervised learning neural network, Adaptive Resonance Theory1 (ART1) to categorically classify the malicious transactions from the normal one in a database which has already shown its magic in pattern recognition field [3]. We also made a comparison with Self Organizing Map (SOM) and Radial Basis Functional Neural Network (RBFN) based database intrusion detection model.

The rest of the paper is organized as follows; related work of this research is discussed in Sect. 2. Section 3 briefly describes about the proposed approach where as Sect. 4 include implementation details. Finally we conclude at Sect. 5 with highlighting some future work.

2 Related Work

In these years, research toward Database Intrusion Detection System (DIDS) techniques has received little attention. Basically, most of the researches are based on anomaly intrusion detection. Lee et al. proposed a database intrusion detection method that utilized the frequent access of temporal data [4]. Rakesh et al. analyzed the recorded data of accessing habit of database users and compared them with the real time to detect illegitimate database access [5]. By using Hidden Markov Model and Time series, a database behavioral model was developed by Barbara et al. that can encounter the dynamic behavior of users [6]. Zhong et al. mined the user profiles by using query template [7]. Bertino et al. used c, m, and f-tuple to represent any query and proposed a Role Base Access Control (RBAC) based DIDS [8]. Panigrahi et al.

used both inter and intra transactional features of transactions to detect insider threat in a database system [9]. All the above research is based on statistical methods. Very limited research has been done with computational intelligence techniques in this research circumference. But these computational intelligence techniques redeemed its potential in network-based intrusion detection [10]. S. Cho proposed a probabilistic database intrusion detection technique by using Self Organizing map and fuzzy logic [11]. Panigrahi et al. used fuzzy logic in database intrusion detection behavioral model to reduce false alarm [12].

Hence, this investigation focused on developing a DIDS model in pattern recognition point of view by using ensemble computational intelligence techniques such as Adaptive Resonance Theory (ART) and Back Propagation Learning Algorithm.

3 Proposed Approach

The objective of our DIDS model is to identify the fault cases based on two parameters: fault location and fault space. Two different soft computing techniques have been used in the investigation. First, ART1 uses the biometric training and test dataset for cluster classification. Second, Back Propagation neural network verifies the classification process by using the biometric cluster data. ART1 contributed a mathematical model of cognitive and neural theory and are widely used in medical database prediction, remote sensing, and control of autonomous robot. Amini et al. [13] exploited the concept of ART in case of intrusion detection in network. In this experiment, we used biometric dataset of an educational institution, which consist of in-time, out-time, average working hour of 1600 students and staff. The threat in Biometric database system is to modification of in-time and out-time; which in turn affects the average working hour of individual and made fraud to the institution while processing the qualifying attendance and salary. In addition to this, there is a possibility that intruder can steal the biometric of users for criminal misuse. By considering the novel property of controlled discovery of clusters and also the property of accommodate new cluster without affecting the storage and recall capabilities for cluster already learned, we exploit the concept of ART in database intrusion detection field. In this case, we viewed the database intrusion detection problem as pattern recognition point of view which recognizes the pattern of normal usage of users toward the biometric machine. It analyzes the normal access pattern to the system by taking a model developed through ART1 with Back Propagation classification network and identified the malicious access; the performance of the system is analyzed through false positive rate, accuracy, elapsed time matrices and compared against its counterpart Self Organizing Map and Radial Basis Functional Network. The following paragraph describes the proposed ART1 based DIDS.

ART1 based DIDS: Generally, ART1 can classify binary input into number of categories based on the similarity with the previous input. If the similarity condition is not satisfied, a new cluster is created. A suitable match is to be found between

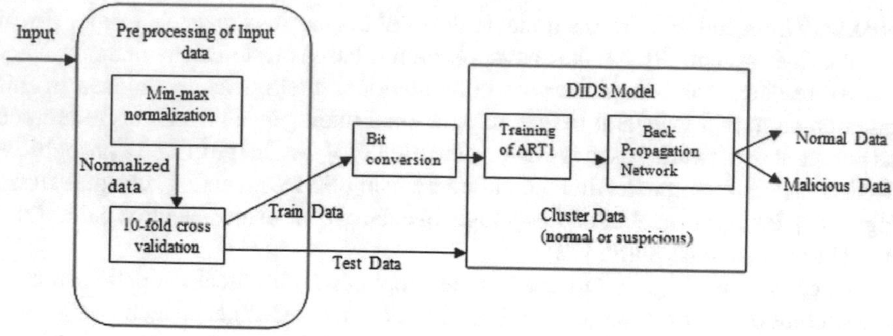

Fig. 1 Block diagram of ARTI DIDS model

binary input patterns with existing prototype vector; else a new category is to be created [14]. Function of ART1 DIDS classifier is reflected in the Fig. 1.

Initially, binary input vector is presented to the input layer and the total input signal is fed to the interface layer. The interface layer sends the activation signal to the competitive layer and calculates the net input. The unit with largest input is called Winner and this winning neuron will learn the pattern and send the signal multiplied by top down weight (as equated in Eq. (1)) to the interface layer and norm of input is calculated (equated in Eq. (2)).

$$x_i = s_i t_{ij} \tag{1}$$

$$\|x\| = \sum\nolimits_i x_i,\ \|s\| = \sum\nolimits_i s_i \tag{2}$$

If the ratio of norm of vector ||x|| and norm of input ||s|| is 1, then it is called match. And

If ||x||/||s|| >= ρ, then weight updation can be done and is called reset condition that means the learning is possible.

If ||x||/||s|| < ρ, then current winning unit is rejected and another unit should be chosen.

The learning is done through weight updation process as equated in (3) and (4).

$$\text{Bottom up weight} = \frac{\alpha\, x_i}{\alpha - 1 + \|x\|} \tag{3}$$

$$\text{Top down weight} = x_i \tag{4}$$

This process is repeated until a satisfactory condition match is found or until all the units are inhibited.

In our proposed model, the input biometric details are first pre-processed through min-max normalization and 10-fold cross-validation process to generate generalized training and test data set and before feeding them to the ART1 classifier, they first converted to bits. The weighted inter connections of the network are stored as prototype patterns represent the normal behavior of user of the biometric of the Long Term Memory traces. In the first stage, the degree of similarity is determined between the training input patterns from input layer with the output node or prototype present in the upper cognition layer. This is a rough classification of the input in one of the active cluster. The input vector belongs to a cluster indicates active state of that cluster or output unit. Vigilance parameter (value ranges from 0 to 1) control the similarity check between feedback pattern from the cluster layer with the input pattern at interface layer and is done at the second stage. In this investigation, by varying vigilance parameter, we can see the performance of ART1. The above comparison is called as vigilance test; if all categories fail at vigilance test, then a new category is created. In this way, the proposed DIDS system models normal behavior of biometric user. The output of this ART1 network is again classified into normal and malicious category by applying to back propagation learning algorithm in the third phase. In testing phase, the intrusive pattern is classified by the ART1 classifier denoted as small clusters as similarity criteria is not satisfied and is treated as suspicious and finally it declared to intrusive by applying it to Back Propagation Network. Hence, we can view this database intrusion detection model as the combination of misuse and anomaly intrusion detection system. While testing, the intrusive patterns are classified as one of the existing cluster.

4 Implementation and Experimental Results

In our experiment, we performed two-class classification. The applied biometric data set contains 80,000 rows describes about the accessing details of different biometric users. The applied biometric data set is initially fetched for pre processing, where after pre processing the pre processed data are faced to the ART1 for normal behavioral modeling of biometric user as clusters. These trained clusters are classified into intrusive and normal by Back propagation learning algorithm. In this section, we have compared the ART1 based DIDS with SOM based and RBFN based DIDS with different performance measures.

The performance metrics of any Intrusion Detection System can be categorized based on a threshold value, ranking order, and probability metrics. By varying a threshold value, this metrics explains the performance of IDS. False Positive Rate (FPR), Detection Rate (DR), Precision metrics are under ranking order category which measure how well the attack instances are ordered before normal instances and can be viewed as summary of performance of the model across all possible threshold. Following section describes how these performance measures can be calculated from the value of confusion matrix. Confusion matrix describes the result of classification in IDS.

TP (True Positive): Number of malicious transactions rightly classified as malicious which needs to be maintained for every IDS.

Table 1 Confusion matrix

Actual	Predicted Attack	Predicted Normal
Attack	TP	FN
Normal	FP	TN

FP (False Positive): Number of normal transactions misclassified as attack which is very crucial for any IDS and should be 0 for an ideal IDS.

FN (False Negative): Number of malicious transactions misclassified as normal transactions which should be reduced.

TN (True Negative): Number of Normal transactions accurately classified as Normal transactions.

From the above confusion matrix Table 1, the following performance metrics are determined. Equations (5), (6) and (7) describe the calculation of the listed performance measures.

Detection Rate (DR): It is the ratio between correctly detected attack and total numbers of attacks.

$$DR = \frac{TP}{(TP + FN)} \tag{5}$$

False Positive Rate (FPR): Ratio of total number of normal instances detected as attack and the total number of normal instances.

$$FPR = \frac{FP}{(TN + FP)} \tag{6}$$

Precision (PR): Fraction of instances predicted as positive that are actually positive

$$PR = \frac{TP}{(TP + FP)} \tag{7}$$

Recall: Equivalent to DR which is the missing part of precision. And it should be high in any IDS.

Area under ROC: Describes the trade off between TPR and FPR.

All of the above experiments are implemented in MATLAB 13 tool in a dual core processor machine. ART1 based DIDS has more than 98% accuracy in all of the testing sets. Time consumption for classification is 100 ms. The summary of the total results is described in Tables 2 and 3.

Three unsupervised neural network based DIDS are tested against biometric dataset and from the results, we can see ART1 produces impressive results compared to other two methods in the specified performance measures. We can see from the Table 2 that FPR complement the ART1 based DIDS with minimal elapsed time.

Table 2 Performance analysis of proposed model

Method	FPR (in %)	Accuracy (in %)	Precision (in %)	Elapsed Time (in ms)
ART1 based DIDS	4.53	98.06	99.09	45.23
SOM based DIDS	10.22	91.23	90.17	40.11
RBFN based DIDS	6.76	94.42	95.79	41.23

Table 3 Performance analysis of proposed model based on vigilance parameter

Method	$\rho = 0.3$		$\rho = 0.5$		$\rho = 0.7$		$\rho = 0.9$	
	TPR	FPR	TPR	FPR	TPR	FPR	TPR	FPR
ART1 based DIDS	91	9.05	93	7.43	98	4.53	95	5.87

It is also noteworthy that the performance of the proposed of DIDS is high when its vigilance parameter is set to 0.7 while referring to Table 3.

On the negative side, the FPR value is needed to be more negligible while maintaining the accuracy of the proposed DIDS and we have least focused on the stability issue of the model, which we plan to address in future work.

5 Conclusion

In this research, we have investigated a new technique for database intrusion detection and evaluated its performance using a biometric dataset to identify the insider intrusive access to the biometric database where biometric information of all the individuals present. The proposed approach combines the complementary features of ART1 and back propagation network. From the empirical results, we found that, the proposed approach ART1 based DIDS outperforms compared to SOM and RBFN based DIDS method. Our research work directed toward developing more accurate classifiers.

References

1. Zadeh, L.A.: Soft computing and fuzzy logic. IEEE Softw. **11**(6), 48–56 (1994)
2. Mathew, S., Petropoulos, M., Ngo, H.Q., Upadhyaya, S.: A Data-Centric Approach to Insider Attack Detection in Database Systems. LNCS, vol. 6307, pp. 382–401 (2010)
3. Carpenter, G.A., Grossberg, S.: ART 2: self-organization of stable category recognition codes for analog input patterns. Appl. Opt. **26**(23), 4919–4930 (http://cnsweb.bu.edu/Profiles/Grossberg/CarGro1987AppliedOptics.pdf) (1987)
4. Lee, V., Stankovic, J., Son, S.: Intrusion detection in real time databases via time signatures. In: Proceedings of the 6th IEEE Real-Time Technology and Applications Symposium, RTAS, pp. 124–133 (2000)

5. Rakesh, A., Ramakrishnan, S.: Fast algorithms for mining association rules. In: Proceedings of 20th International Conference on Very Large Data Bases. Morgan Kaufmann, Berlin, pp. 487–499 (1994)
6. Barbara, D., Goel, R., Jajodia, S.: Mining malicious data corruption with hidden Markov models. In: Proceedings of the 16th Annual IFIP WG 11.3 Working Conference on Data and Application Security, July 2002, pp. 175–189 (2002)
7. Zhong, Y., Qin, X.: Database intrusion detection based on user query frequent itemsets mining with item constraints. In: Proceeding of the 3rd International Conference on Information Security, pp. 224–225 (2004)
8. Bertino, E., Terzi, E., Kamra, A., Vakali, A.: Intrusion detection in RBAC-administered databases. In: Proceedings of the 21st Annual Computer Security Applications Conference (ACSAC), pp. 170–182 (2005)
9. Panigrahi, S., Sural, S., Majumdar, A.K.: Two-stage database intrusion detection by combining multiple evidence and belief update. Inf. Syst. Front. (2010)
10. Wu, S.X., Banzhaf, W.: The use of computational intelligence in intrusion detection systems: a review. Appl. Soft Comput. **10**, 1–35 (2010)
11. Cho, S.: Incorporating soft computing techniques into a probabilistic intrusion detection system. IEEE Trans. Syst. Man Cybern. Part C Appl. Rev. **32**, 154–160 (2002)
12. Panigrahi, S., Sural, S.: Detection of Database Intrusion Using Two-Stage Fuzzy System. LNCS, vol. 5735, pp. 107–120 (2009)
13. Amini, M., Jalili, R., Shahriari, H.R.: RT-UNNID: a practical solution to real-time network-based intrusion detection using unsupervised neural networks. Comput. Secur. **25**(6), 459–468 (2006)
14. Jang, J.S.R.: Neuro-Fuzzy and Soft Computing, 1st edn. PHI Publication. Pearson Education, New Delhi (2015)

Medical Image Classification Through Deep Learning

Kaushik Raghupathruni and Madhavi Dabbiru

Abstract The authors investigate the problem of image classification. Earlier, the task of image classification is accomplished by traditional machine learning techniques and other shallower neural network models. Later with the evolution of deeper networks, convolutional neural networks have gained without importance due to its outstanding accuracy in various domains. Unlike in real-world datasets for performing the classification of various images under different categories, the job of biomedical image classification of chest X-rays is quite tedious due to overlapping characteristics of X-ray images. The objective of this paper is to classify the images of chest X-rays and predict the pneumonia traces in lungs. Inception V3 model, with transfer learning is applied on this medical dataset. The model is implemented in Keras as front-end library with tensor flow framework. The training on this dataset to generate a custom model file on GTX 1070 video card consumed 30 min yielding 98% training accuracy and 99% validation accuracy.

Keywords Image classification · Deep learning · Convolution neural networks · Transfer learning

1 Introduction

This Image classification techniques play a vital role in the arena of artificial intelligence as the features of every image is unique. Deep learning is subset of machine learning methods in contrast to task-specific algorithms. Deep learning uses a layered structure of algorithms called an artificial neural network (ANN). Deep learning is

K. Raghupathruni (✉)
Department of CSSE, Andhra University, Visakhapatnam, Andhra Pradesh, India
e-mail: kaushikraghupathruni@gmail.com

M. Dabbiru
Department of CSE, Dr. L. Bullayya College of Engineering for Women, Visakhapatnam, Andhra Pradesh, India
e-mail: drlbcse@gmail.com

H. S. Behera et al. (eds.), *Computational Intelligence in Data Mining*, Advances in Intelligent Systems and Computing 990,
https://doi.org/10.1007/978-981-13-8676-3_23

popular today due to the Convolution neural networks [1]. The crucial parts of a CNN are convolutional layers. They will identify and capture important features in the pixels of the image. Layers that are deep will learn to detect simple features whereas higher layers will combine basic features into more high-level features and showcase classification predictions [2].

The chest X-ray is one of the important tests performed by radiologists whose report is helpful in diagnosing many lung diseases. It is very tiresome process to make clinical diagnoses with chest X-rays than with CT scan images. With computerized diagnosis, radiologists can make chest X-ray diagnoses more quickly and accurately. Thus the problem is a binary classification where the input images are chest X-ray images and the output belongs to either of the two class labels: pneumonia or non-pneumonia. The authors investigate how to train a CNN that can perform binary classification of pneumonia X-rays with high accuracy. The authors used the Chest X-ray dataset [3]. The input data set is then trained and then it is tested and validated in order to find the prediction accuracy results. The paper is organized as follows: Sect. 2 discusses the related work, Sect. 3 narrates the terminology and methodology, Sect. 4 analyzes the experimentation results and Sect. 5 gives the results and then conclusions followed by references.

2 Related Work

The intimate relationship between the layers and spatial information in CNNs renders them well suited for image processing and understanding, and they generally perform well at autonomously extracting salient features from images. Samet Akcay et al. [4], implemented transfer learning on deep convolutional neural network for object classification in X-ray baggage imagery data for identifying hand guns. Aniruddha [5] explored the process of transfer learning on the computer vision problems in plant phenotyping database. Benjamin Antin et al. [6], worked on supervised learning techniques to perform binary classification on pneumonia detection using chest X-rays.

3 Procedure

Functionalities-

1. Data pre-processing: In this we collect the datasets of biomedical images and then we pre-process them for testing.
2. Train the model: The given datasets should be trained properly to test the datasets for validation.
3. Validate the model: Now the tested models should be validated through its prediction result.

InceptionV3
The power of transfer learning by adapting existing image classification architecture (Inception V3) to a custom task: categorizing biomedical images to help doctors, radiologists, and other hospital staff in organizing various biomedical images generated at medical labs is the proposed system.

Transfer learning
Retraining deep learning models for customized image classification [7]. When classifying images, model is built from initial stage for the best fit. Transfer learning may not be as efficient as a model built from blueprint which undergoes full training, but is surprisingly effective for many applications. It allows model creation with significantly reduced training data and time by modifying any of the standard deep learning models. Here Inception V3 is retrained with chest X-ray images (Fig. 1) [8].

The inception V3 architecture is made up of modules called inception modules, as described in Fig. 2, which contain a set of convolutional operations. Convolution layer performs this by using multiple filters where each filter moves around the input

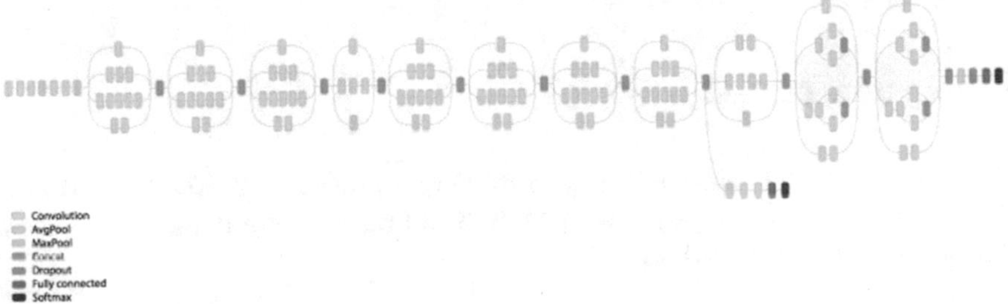

Fig. 1 Architecture of inception v3

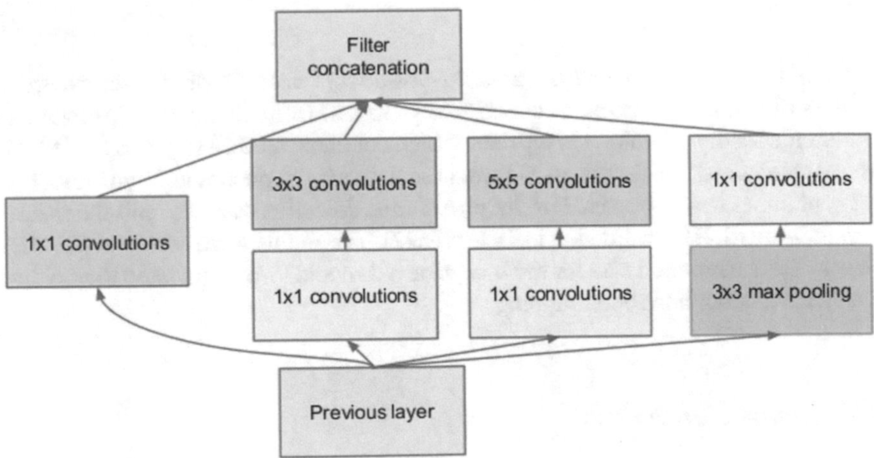

Fig. 2 Inception module

image and prepares activation map. Feature maps are built from the activation maps generated by different filters. Inception module consists of 1×1, 3×3 and 5×5 convolutions along with a 3×3 max pooling. Pooling layers performs subsampling of the input image and contributes to the reduction of input image size. The 1×1 convolution layer reduces the dimensionality of its feature map, by passing the feature map through the Rectified Linear Unit (ReLU). The aim of ReLU layer is to induce non linearity to the given system. ReLUs work with great accuracy and speed. With this dimensionality reduction happens which is less by a factor of 10, saving a lot of computations and time.

In this approach based on inception architecture, the last layer of Inception has been retrained using Softmax Regression where we generate probabilities based on our evidences extracted from the images. The evidences are computed based on a sum of weights detected by the intensity of pixels, with added bias.

The pneumonia identification from chest X-rays is a binary classification problem, where the input is a chest Xray image X and the output is a binary label $z \in \{0,1\}$indicating the absence or presence of pneumonia. We optimize the weighted binary cross entropy loss.

4 Experimentation

The experimentation is carried using i5 4500 rpm processor, 8 GB RAM, Nvidia GTX 1070 video card on windows 10 OS. The Programs are implemented using Tensorflow, Anaconda, Python.

4.1 Training

We used the Chest X ray images (anterior-posterior) selected from the retrospective cohorts of pediatric patients of one to five years old from Guangzhou Women and Children's Medical Center, Guangzhou which contains 5865 X-Ray images (JPEG) of two categories. The first set of X-Rays is contained with pneumonia and the second set containing healthy cases. For the pneumonia detection task we split the dataset into training (5,219), validation (19), test (627). The inputs were scaled to 229*229 size as it is the required size for the Inception V3 model. We augmented the training data with random horizontal flipping.

4.2 Hyper Parameters

The Incpetion V3 model pre-trained on Imagenet dataset is retrained on the dataset with total parameters of 21,802,784, out of which 21,768,352 are trainable parameters. A batch size of 32 is selected and iterated for 100 epochs. We choose the learning rate as 0.001 and used Adam Optimizer. The loss calculation was categorical cross entropy.

5 Results

We investigate how to train a CNN that can classify medical images with high accuracy. We considered the biomedical database which consists of 2 classes. The dataset is split into training data set and test data set. The input dataset is then trained and is validated in order to find the prediction accuracy results. We implemented this medical image classifier by training inception v3 with a custom biomedical image dataset.

The dataset [3, 9] is organized into 3 folders (train, test, val) and contains subfolders for each image category (Pneumonia/Normal). There are 5,863 X-Ray images (JPEG) and 2 categories (Pneumonia/Normal) (Fig. 3) [9].

We have trained our networks with a batch size 32 for 100 epochs. The training time for 100 epochs took around 30 min on the specified hardware. The values shown in Table 1 can be inferred from Fig. 4. In Fig. 4, the X-axis shows the number of epochs, ranging from 0 to 100. The Y-axis shows the training accuracy ranging from 0.550 to 1.00 when multiplied by 100 gives training accuracy at the epoch. This graph is obtained by tensor board visualization during the training of the model through 100 epochs.

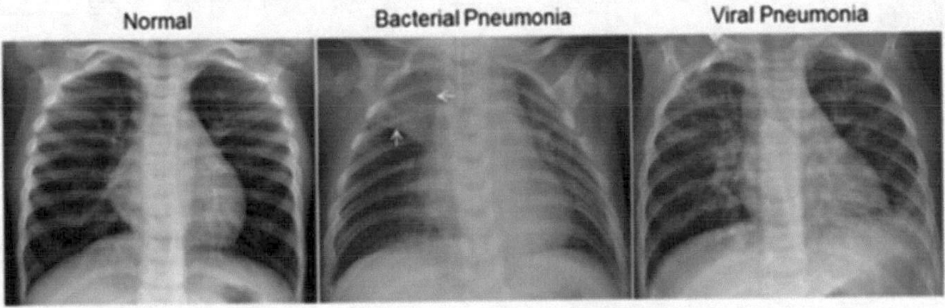

Fig. 3 Illustrative examples of chest X-Rays in patients with pneumonia

Table 1 Accuracy of image classifier

No. of epochs	Training accuracy (%)
30	75
50	86
100	99

Evaluation Metrics

The training accuracy gives us the percentage of images used in the current training batch that were labeled correctly. The validation accuracy is the precision on a randomly selected group of images from a different set during training. The training accuracy is less reflective of the performance of the classifier during because it is based on images that have already been learned from and hence, the network is at risk of over-fitting. A better measure is the validation accuracy. If there is significant mismatch between the training accuracy and validation accuracy, then that is indicative that the network is memorizing potentially unhelpful features that don't generalize well. Inception splits the training data into 3 parts where 80% is used as the training set, 10% is used as validation set, and 10% is used as a testing set during the training. In that way over-fitting is avoided and bottlenecks are fine-tuned. Then the validation set has been passed through for tuning and the test set has been sent through for classification. Figures 4, 5, 6, 7 and 8 depict the evaluation metrics of Inception v3 model.

In Fig. 5, the X-axis shows the number of epochs ranging from 0 to 100. The Y-axis shows the training loss which ranges from 0.0 to 1.0, when multiplied by100 gives training loss at the current epoch. This graph is obtained by tensor board visualization during the training of the model through 100 epochs.

In Fig. 6, the X-axis shows the number of epochs from 0 to 100. The Y-axis shows the validation accuracy from 0.820 to 100. When multiplied by 100 gives validation accuracy at the current epoch. This graph is obtained by tensor board visualization during the training of the model through 100 epochs. In Fig. 7, the X-axis shows the number of epochs from 0 to 100. The Y-axis shows the validation accuracy from 0.820 to 100. When multiplied by 100 gives validation loss at the current epoch. This graph is obtained by tensor board visualization during the training of the model through 100 epochs. Figure 8 depicts line plot for the categorical accuracy for 100 epochs.

Fig. 4 Training accuracy graph from tensor board visualization

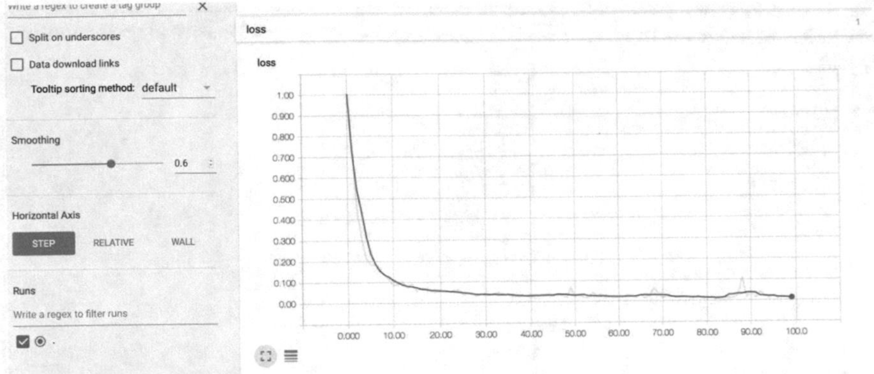

Fig. 5 Training validation graph from tensor board visualization

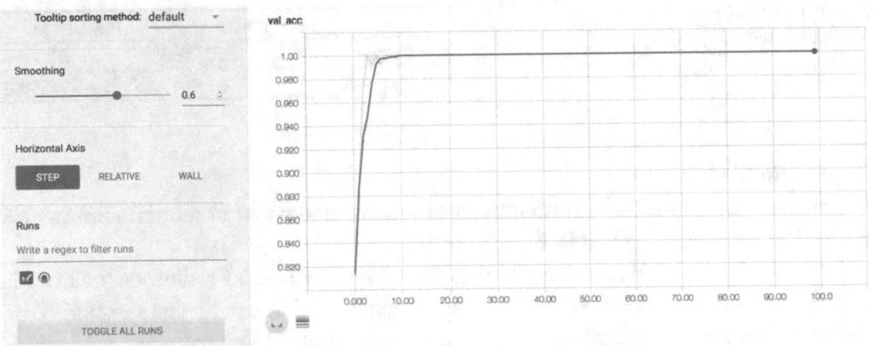

Fig. 6 Validation accuracy graph from tensor board visualization

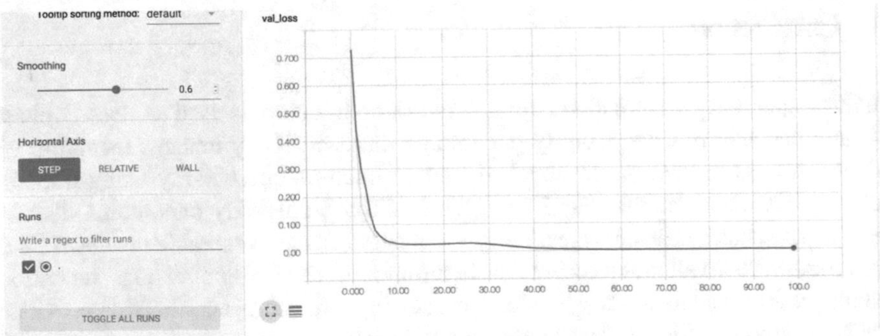

Fig. 7 Validation loss graph from tensor board visualization

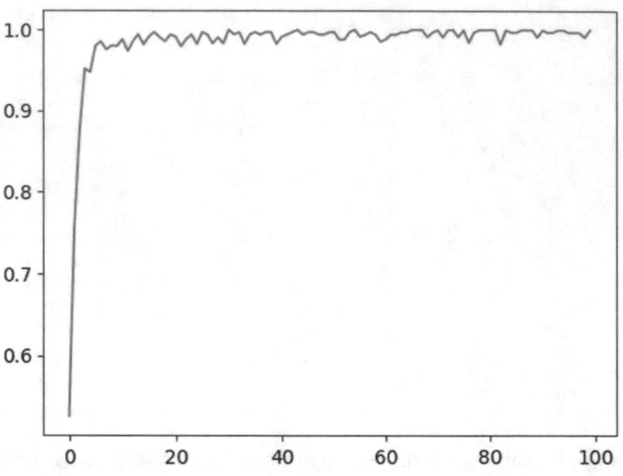

Fig. 8 Line Plot for the categorical accuracy for 100 epochs

Fig. 9 Confusion matrix for prediction accuracy

	Pred.Nor	Pred. Pneu
GT Nor	67.70%	32.20%
GT Pneu	21.40%	78.50%

Confusion Matrix
To evaluate the quality of our image classifier, we make use of a confusion matrix. Prediction accuracy is given in Fig. 9.

Pred.Nor indicates Predictive Normal, Pred. Pneu indicates Predictive Pneumonia, GT Nor indicates Ground Truth Normal image, GT Pneu indicates Ground Truth Pneumonia image. The tabulated values are average of 20 images.

6 Conclusion

In this paper, deep convolution neural network architecture is used to classify chest X-ray images in order to identify pneumonia diseased X-ray images. Inception v3 architecture is trained on the medical images such as chest X-ray using transfer learning for classification into 2 categories of images namely pneumonia disease images and healthy chest images. This model has delivered venerable results of 98% training and 99% validation accuracies. In the arena of clinical applications, Inception v3 model is proved to produce greater accuracy in performing classification provided decent training, sufficient dataset and adequate training time.

Very often it is quite difficult for the radiologists to understand images of tests due to their overlapping behavior and consumes a lot of time in reading them and end up with precise reports. With the advent of deep learning technology, it became easy for the radiologists to analyze the images and automatically generates reports with accuracy within a short span of time.

Acknowledgements The data that supports the findings of this study are available in National Institutes of Health [9]. These data are approved from the following resources available in the public domain at https://www.nih.gov/news-events/news-releases/nih-clinicalcenter-provides-one-largest-publicly-available-chest-x-ray-datasets-scientific-community. A standard ethical committee has approved this data set and the dataset has no conflict of interest/ethical issues with any other public source or domain.

References

1. Szegedy, C., Liu, W., Jia, Y., Sermanet, P., Reed, S.: Anguelov, D., Erhan, D., et.al.: Going deeper with convolutions (2014)
2. He, K., Zhang, X., Ren, S., Sun, J.: Deep residual learning for image recognition (2015)
3. http://www.cell.com/cell/fulltext/S0092–8674(18)30154-5
4. Akcay, S., Kundegorski, M.E, Devereux, M., Breckon, T.P.: Transfer learning using convolutional neural networks for object classification within X-Ray baggage security imagery. IEEE (2016)
5. Tapas, A.: Transfer learning for image classification and plant phenotyping. Int. J. Adv. Res. Comput. Eng. Technol. **5**(11) (2016)
6. Antin, B., Kravitz, J., Martayan, E.: Detecting pneumonia in chest X-Rays with supervised learning, http://cs229.stanford.edu/proj2017/final-reports/5231221.pdf (2017)
7. Nguyen, L.D.: Deep CNNs for microscopic image classification by exploiting transfer learning and feature concatenation. In: IEEE International Symposium on Circuits and Systems (ISCAS) (2018)
8. Szegedy, C., Vanhoucke, V., Ioffe, S., Shlens, J., et.al.: Rethinking the inception architecture for computer vision (2015)
9. Data set https://www.nih.gov/news-events/news-releases/nih-clinical-center-provides-one-largest-publicly-available-chest-x-ray-datasets-scientific-community

Novel Wavelet Domain Based Adaptive Thresholding Using Bat Algorithm for Image Compression

V. Manohar, G. Laxminarayana and T. Satya Savithri

Abstract Image compression is the significant study in the arena of image processing owing to its enormous usages and its ability to reduce the storage prerequisite and communication bandwidth. Thresholding is a kind of image compression in which computational time increases for multilevel thresholding and hence optimization techniques are applied. The quality of reconstructed image is superior when discrete wavelet transform based thresholding is used as compared to when it is not applied. Both particle swarm optimization and fire fly algorithm becomes unstable when the velocity of the particle becomes maximum and when there is no bright firefly in the search space respectively. To overcome the above-mentioned drawbacks bat algorithm based thresholding in frequency domain is proposed. Echolocation is the sort of sonar used by micro-bats. The way they throng their prey, overcoming the hurdles they come across, pinpointing nestling gaps have become the main motivation research in artificial intelligence. With the feature of frequency tuning and having the benefit of automatic zooming, bat algorithm produces superior PSNR values and quality in reconstructed image and also results in fast convergence rate as compared to state of art of optimization techniques.

Keywords Image compression · Thresholding · DWT · Bat algorithm · PSO · Firefly algorithm

V. Manohar (✉) · T. Satya Savithri
Department of Electronics & Communication Engineering, JNTUH, Hyderabad, Telangana, India
e-mail: manoharvu@gmail.com

T. Satya Savithri
e-mail: tirumalasatya@jntuh.ac.in

G. Laxminarayana
Department of Electronics & Communication Engineering, Anurag College of Engineering, Hyderabad, Telangana, India
e-mail: gln9855@gmail.com

H. S. Behera et al. (eds.), *Computational Intelligence in Data Mining*, Advances in Intelligent Systems and Computing 990,
https://doi.org/10.1007/978-981-13-8676-3_24

1 Introduction

The aim of image compression is the transmission of images over communication channels with limited bandwidth. It is essential and important in multimedia applications such as Mobile, Bluetooth, Internet browsing, computer to computer communication etc. The image compression applications also include bio-medical, satellite, aerial surveillance, reconnaissance, multimedia communication and ground water survey etc. The most commonly used image compression techniques are Joint Photographic Experts Group (JPEG), JPEG-2000, TIF and PNG, the first two techniques use Discrete Cosine Transform (DCT) and Discrete Wavelet Transform (DWT) respectively. Among all the image compression techniques, DWT based image compression has shown better compression ratio and image quality. Bing Fei Wu et al. [1] proposed fast convolution algorithm (FCA) for DWT calculation by observing the symmetric properties of filters and hence the computational complexity is reduced. To reduce computational complexity further, wavelets were progressed on resized, cropped, resized-average and cropped-average images [2]. However, symmetric and orthogonal filters design is critical, so multi-wavelets were introduced which offer supplementary filters with desired properties [3]. These filter coefficients are further partitioned into blocks of unique size and based on the coefficient variance a bit assignment table was computed and blocks of individual class were coded using bit assignment table [4]. In addition, fast orientation prediction-based DWT is also used to improve coding performance and to reduce computational complexity by designing a new orientation map and orientation prediction model for high-spatial-resolution remote sensing images [5]. Farid et al. [6] have modified JPEG by Baseline Sequential Coding which is based on near optimum transfer and entropy coding and trade-off between reconstructed image quality and compression ratio is controlled by quantizers. So there is a need of hybrid approach which would offer higher compression ratio than the wavelet transform (WT) alone keeping the quality of reproduced image equal in both cases. The hybrid combinations for medical image compression such as DWT and Artificial Neural Network (ANN) [7], DWT and DCT [8], hybridization of DWT, log-polar mapping (LPM) and phase correlation [9], hybridization of empirical wavelet transform (EWT) along with DWT has been used for compression of the ECG signals [10]. The hybrid wavelet combines the properties of existing orthogonal transforms and these wavelets have unique properties that they can be generated for various sizes and types by using different component transforms and varying the number of components at each level of resolution [11], Hybridization of wavelet transforms and vector quantization (VQ) for medical ultrasound (US) images. In this hybridization, the sub-band DWT coefficients are grouped into clusters with the help of vector quantization [12].

On the other side, image compression can also be performed with the non-transformed techniques such as vector quantization and image thresholding. Kaur et al. [13] proposed image compression that models the wavelet coefficients by generalized Gaussian distribution (GGD) and suitable sub-bands are selected with suitable quantizer. In order to increase the performance of quantizer, threshold is

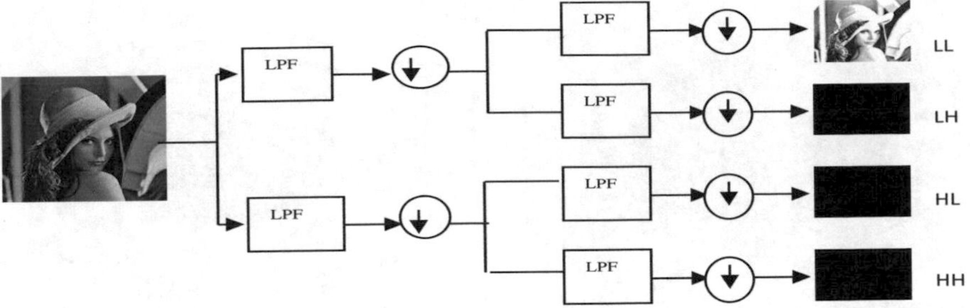

Fig. 1 First level wavelet decomposition

chosen adaptively to zero-out the unimportant wavelet coefficients in the detail sub-bands before quantization. Kaveh et al. [14] proposed a novel image compression technique which is based on adaptive thresholding in wavelet domain using particle swarm optimization (PSO). In multi-level thresholding, thresholds are optimized without transforming the image and thresholds are optimized with computational intelligence techniques.

In this paper, for the first time the application of optimization techniques for selection of the optimal thresholds is explored which reduces the distortion between the input image and reconstructed image. The aim of this work is the selections of optimal thresholds which zero-out the irrelevant DWT coefficients in all sub-bands. The performance of different optimization techniques and their optimal variable parameters are compared.

The paper consists of five sections with the inclusion of introduction. In Sect. 2 adaptive thresholding and its proposed frame work for Image compression is discussed. In Sect. 3 the proposed method of thresholding using Bat algorithm is discussed and Sect. 4 presented the results. Furthermore in Sect. 5 the conclusion is presented.

2 Proposed Framework of Adaptive Thresholding for Image Compression

2.1 2-D Discrete Wavelet Transform

DWT positioned the image compression at a subsequent stage. The advantage of DWT compared to that of DCT is that it provides not only spatial but also frequency content of the image. Image is decomposed into four coefficients by DWT in Fig. 1 and is explained in [4].

The 3-D view of approximation—horizontal, vertical and diagonal coefficients of a Lena image is explained in Fig. 2. Approximation coefficients carry much information about the input image when compared to the rest of the coefficients, which leads

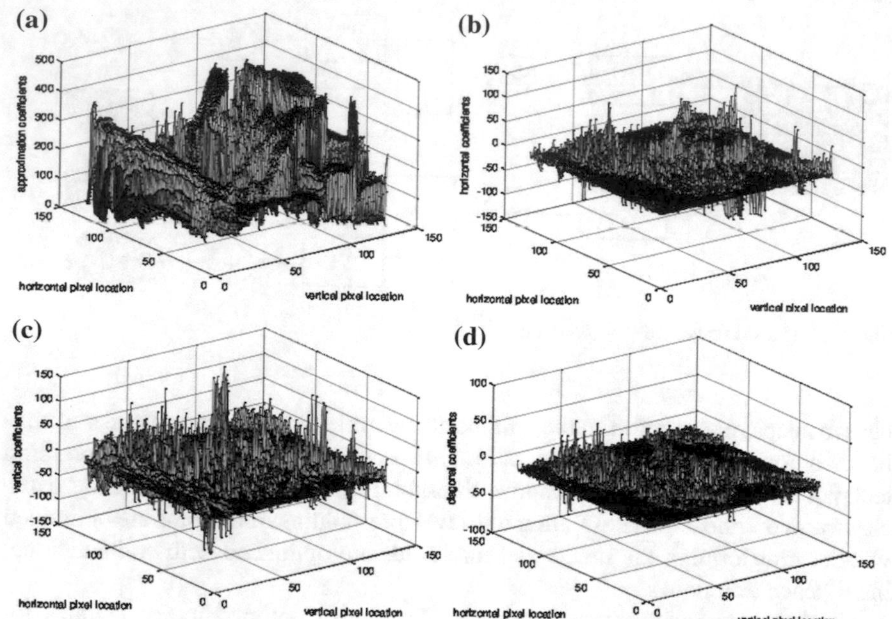

Fig. 2 3-D view of approximation, horizontal, vertical and diagonal coefficients of a Lena image

to better image compression as explained in Fig. 2d. The next level (2nd level) of decomposition the LL band is decomposed into four coefficients as like in first level of decomposition so achieved a LL2, LH2, HL2 and HH2. This process is repeated for next level (3rd level) of decomposition so achieved a ten sub-images in total. Figure 4 shows a 3rd level decomposition of Lena image. As that of JPEG-2000, bi-orthogonal wavelet is chosen for our work since it has simple design and symmetric functions. Three and five level decomposition is used for the fidelity of reconstructed image quality and comparison with the published work. In this proposed method, optimization technique spent much time on thresholding of approximation coefficients as the reconstructed image quality highly depends on it.

2.2 Thresholding

Image thresholding can be defined as extracting the objects in a scene from the background which is helpful to analyze and understand the image. Thresholding can be classified into two types: Global Thresholding and Level Dependent/Local Thresholding [12]. In global thresholding-image compression is obtained using a single threshold value for all the decomposition levels, whilst in level dependent thresholding-image compression is achieved using different threshold values for different decomposition level. The energy levels being different in decomposition

(a)

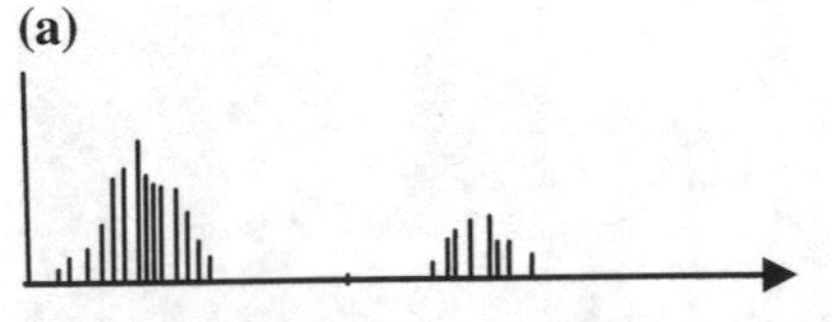

(b)

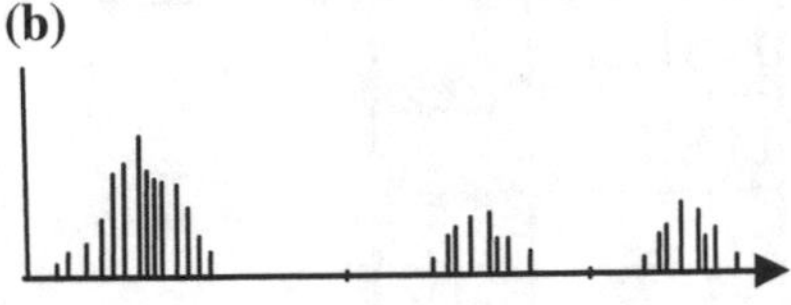

Fig. 3 Image histogram of an image f(m, n)

levels of the image are the main criterion for the application of the level thresholding method in this paper.

The histogram of an image $f(m, n)$ that is composed of several light objects on a dark background may represent two dominate modes as shown in Fig. 3. The two modes can be separated by selecting an appropriate threshold T and hence the object information can be extracted from the background. If the image is comprised of several objects then different thresholds are required to separate the object classes as shown in Fig. 3. If $f(m, n)$ lies between T_1 and T_2 the object may be classified as belonging to one object class. If $f(m, n)$ is greater than T_2 the object belongs to a second class. If T_1 is greater than or equal to $f(m, n)$, then object belongs to the background. As compared to single level thresholding, this process of threshold selection to obtain the object class is usually less reliable and it may be viewed as an operation that tests against a given function of the form as given in Eq. 1 [4]:

$$T = T[m, n, \mathrm{p}(m, n), f(m, n)] \tag{1}$$

where, $f(m, n)$ is the gray level of point (m, n); $\mathrm{p}(m, n)$ is some local property of the input i.e. the average gray level of a neighbourhood around (m, n). The thresholded image is given by

$$c(m, n) = \begin{cases} c(m, n) & \textit{if } c(m, n) > T \\ T & \textit{if } c(m, n) \leq T \end{cases} \tag{2}$$

The objects and background are represented by pixels labelled 1 (or any other convenient threshold value T) and pixels labelled 0 respectively. The threshold is termed as global threshold if T depends only on $f(m, n)$ and local threshold if T depends on $f(m, n)$ and $p(m, n)$. The threshold is called dynamic, if T depends on the spatial coordinates (m, n), in addition to the above cases. For instance, priory if certain information regarding the nature of the object is known a local threshold may be utilized, whereas a dynamic thresholding may be used if object illumination is non-uniform.

In this proposed method of thresholding in wavelet domain, different thresholds are assigned for different sub-bands. In 3 level decomposition, there are 10 sub-bands i.e. LL3, LH3, HL3, HH3, LH2, HL2, HH2, LH1, HL1 and HH1as shown in Fig. 4. Among all ten sub-bands LL3, LH3, HL3 and HH3 possess very high energy and hence these are assigned to four individual thresholds (i.e. *T1*, *T2*, *T3* and *T4*)

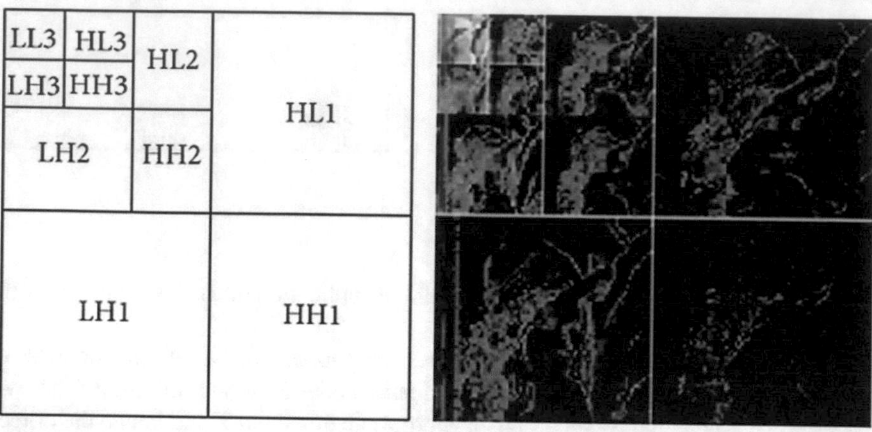

Fig. 4 3-level DWT of Lena image

and play a very significant role in the reconstructed image quality at the decoder section. As more number of thresholds are made use for the image reconstruction it increases the computational time of the optimization techniques, but the quality of the image is emphasized more than the computational time. The remaining sub-bands (LH2, HL2 and HH2) and (LH1, HL1 and HH1) have the less energy level as compared to sub-bands LL3, LH3, HL3 and HH3. Therefore the sub-bands LH2, HL2, HH2 are assigned to single threshold *T5* and the LH1, HL1 and HH1 are assigned to another single threshold *T6*. Also the same procedure is adapted for five level decomposition of the image. It consists of sixteen sub-bands in total, out of which four low frequency sub-bands are assigned to four individual thresholds (*T1* to *T4*) and the remaining twelve sub-bands are partitioned into four groups which consists of three sub-bands each and these four groups are assigned to four thresholds (*T5* to *T8*). In this work much prominence is given to low frequency sub-bands as mostly reconstructed image quality depends on the low frequency sub-bands and moreover these sub-bands carry very high energy of the input image. After the initialization of the thresholds is completed, these are optimized with various optimization techniques by maximizing/minimizing the objective function or fitness function as defined in Eq. (3) The main aim of optimization techniques is to find a better threshold that reduces the distortion between original image and reconstructed image. The optimization technique that produces thresholds with less distortion is treated as a superior optimization technique. In this work, objective function/fitness function that is used for the selection of optimal thresholds is a combination of the entropy and PSNR values in order to obtain high compression ratio and better reconstructed image quality. Here entropy is assumed as compression ratio. Hence, the fitness function is as follows [14]:

$$Fitness = k \times entropy + l/PSNR \tag{3}$$

where k and l are adjustable arbitrary and are varied as needed by user for obtaining required level of compression ratio (CR) and distortion respectively. In general, maximum and minimum values of the population in a optimization technique is a constant value whereas in this context maximum and minimum values lie between maximum and minimum value of the respective sub-band image coefficients because of the selection of different thresholds for different sub-bands, during the computational procedure of the algorithm. The optimal thresholds are obtained successfully with the application of the proposed Bat algorithm. If the coefficients of the corresponding sub-band is less than the corresponding threshold then replace the coefficients with the respective threshold else coefficients remains the same.

Threshold value for a particular sub-band is represented by T then its corresponding coefficients follow the Eq. (2) to generate thresholded image. Huffman coding follows run-length coding (RLE) to code thresholding image. The aim of RLE is to decrease the data required to store and transmit. It reduces thresholded image to just two pixel values when all the pixel values in the thresholded image are unique and if the pixel values of threshold image are not unique then it doubles the size of the original image.

Huffman coding is used to compress the outcome of RLE. As the RLE produces repetitive outcomes, the probability of a repeated outcome is defined as the desired outcome, which can be obtained by integrating RLE and Huffman coding techniques. Repeated outcomes are represented by fewer bits and infrequent outcomes are represented by higher bits.

3 Thresholding Using Bat Algorithm

Bat algorithm (BA) is developed to represent a global threshold with less iterations [15, 16]. Bat algorithm works with the three assumptions: first of the three is that bats use echolocation to measure the area. Bats know difference between food/prey and background barriers. Second of the three is that in search for prey Bats fly randomly with velocity v_i at position x_i with a fixed frequency Q_{min}, varying wavelength and loudness A_0 and wavelength is auto adjusted. Lastly, even though the loudness can differ in various ways, we assume that the loudness differs from A_0 to A_{min} which are maximum and minimum respectively. Thresholds are assumed to be Bats. The complete Bat algorithm is as follows:

Step 1: (Initialize the bat and parameters): Initialize the number of thresholds (n) and arbitrarily pick the threshold values i.e., X_i, ($i = 1, 2, 3, \ldots, n$), loudness (A), velocity (V), pulse rate (R), minimum frequency (Q_{min}) and maximum frequency (Q_{max}).

Step 2: (Figure out the best threshold): Eq. 3 is used to calculate the fitness of all thresholds, and the obtained finest fitness threshold is the X_{best}.

Step 3: (Auto zooming of threshold towards X_{best}): Individual threshold is zoomed as per Eq. (6) by means of adjusted frequency in Eq. (4) and also velocities in Eq. (5).

Frequency update:

$$Q_i(t+1) = Q_{max}(t) + (Q_{min}(t) - Q_{max}(t)) * \beta \quad (4)$$

where β is random number [0 to 1].

Velocity update:

$$v_i(t+1) = v_i(t) + (X_i - X_{best}) * Q_i(t+1) \quad (5)$$

$$X_i(t+1) = X_i(t) + v_i(t+1) \quad (6)$$

Step 4: (Random walks step size selection): In case of randomly produced number higher than pulse rate 'R' then shift the threshold towards preferred finest threshold depending on Eq. (7).

$$X_i(t+1) = X_{best}(t) + w * R_i \quad (7)$$

where R_i = random numbers, w = step size for random walk.

Step 5: (Generate a new threshold): If the generated random numbers is less than loudness and accept the new threshold if its fitness is better than the old one.

Step 6: Ranks are given to the bats to search the current best X_{best}.

Step 7: Repeat step (2)–(6) until maximum number of iterations.

4 Results and Discussions

The proposed method of image compression in discrete wavelet domain using bat algorithm based optimal thresholding has been executed in MATLAB environment with a laptop of HP Probook 4430s, Intel core i5 processor, 2 GB random access memory (RAM). The performance of the algorithm is tested for the readily available standard images obtained from the link www.imageprocessingplace.com. Five popular images which are extensively used for compression test images like Barbara, Gold Hill, Cameraman, Lena, and Peppers are selected to test the performance of the proposed algorithm. The experimental results of published methods for the above said images are available in the literature. For the simulation of PSO algorithm, the Trelea type-2 PSO toolbox is used. The same parameters as in paper [14] are selected for PSO as well as to QPSO that are number of particles, maximum number of epochs, distortion and CR parameter (a, b). Similarly for firefly algorithm a particular value to tuning parameters is selected where its performance is better in compression ratio and reconstructed image quality. The parameter values are finalized after the implementation of firefly algorithm to the problem stated. The parameters which are set for tuning of firefly algorithm are alpha (α) = 0.01, beta minimum (β_0) = 1, gamma (γ) = 1. The maxima of maximum of average PSNR obtained from the repeated

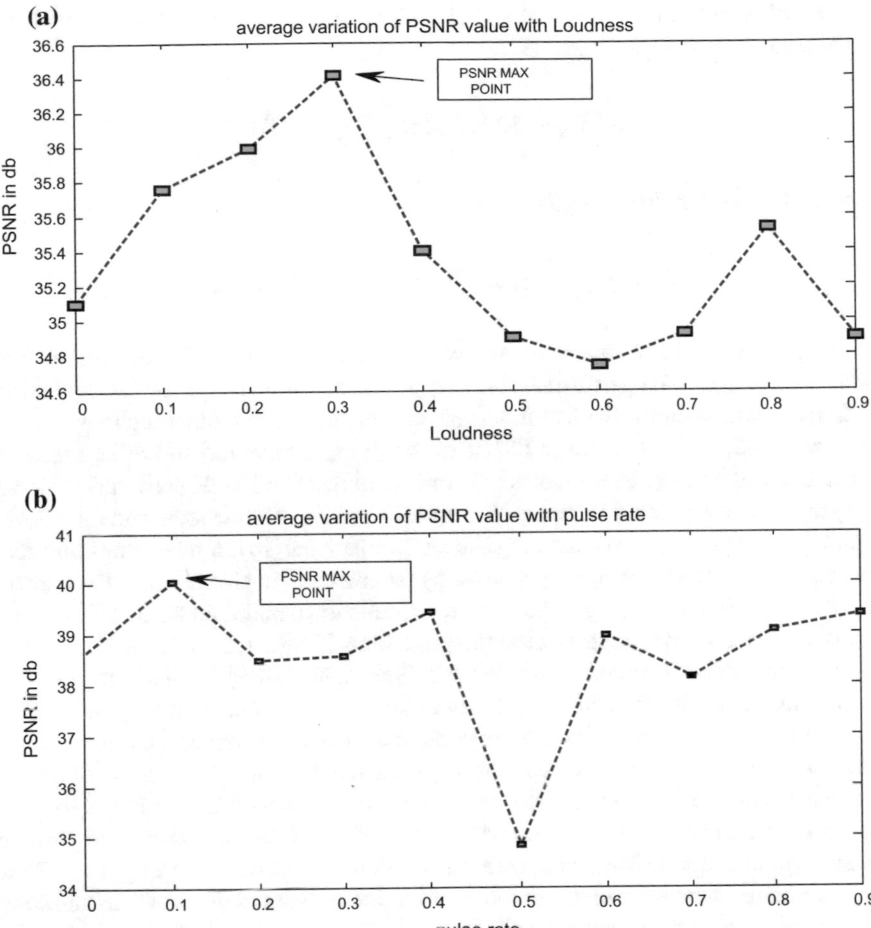

Fig. 5 Average PSNR of Lena image performed for 5 times to select parameter **a** loudness and **b** pulse rate

experimental results for loudness and pulse rate with respect to PSNR as shown in Fig. 5, is chosen as the Bat algorithm parameter value for simulating the Lena image. From experimental shown in Fig. 5, it can be observed that Loudness = 0.3 and Pulse rate = 0.1, and the same parameters values are used for later experiments. The remaining parameters: frequency $Q_{min} = 0.3$, $Q_{max} = 0.9$ and random walks step size = 0.36 is preferred randomly.

The performance measuring parameter for evaluation of effectiveness and efficiency of the proposed method is Peak Signal to Noise Ratio (PSNR) and objective function/fitness function. The equation for fitness function is as given in Eq (3) and

PSNR value and MSE of original signal f(M, N) and reconstructed image $g(M, N)$ is calculated using Eqs. (8) and (9).

$$PSNR = 10 \times 10\log\left(\frac{255^2}{MSE}\right)(\text{dB}) \tag{8}$$

where Mean Square Error is given as

$$MSE = \frac{1}{X \times Y}\sum_{M}^{X}\sum_{N}^{Y}\{f(M, N) - g(M, N)\}^2 \tag{9}$$

where X × Y is size of image, M and N represents the pixel value of original and processed images. The performance of the proposed method of image compression in frequency domain for the five images is compared with the other techniques.

Tables 1, 2, 3, 4 and 5 shows PSNR of the proposed method and other methods for five tested images. From tables it is observed that the PSNR achieved with the proposed method is better than the PSO, QPSO and FA. As five level and three level decomposition are used so, for the sake of comparison [14], a three level and five level decomposition with appropriate (a, b) parameters are also chosen. Parameters (a, b) are adjusted according to the user requirements to maintain trade-off between compression ratio and reconstructed image feature. Tables 1, 2, 3, 4 and 5 show the (a, b) values to achieve corresponding Bits Per Pixels (BPP) for five images. The major drawback with the PSO and firefly is that large number of tuning parameters and improper parameter tuning causes performance degradation of the PSO and FA, whereas in bat algorithm only two tuning parameters are enough, so effect of tuning parameters on BA is much smaller compared to PSO and FA. The bat algorithm optimization process combines the advantages of simulated annealing and particle swarm optimization process and with the concept of frequency tuning, the PSNR values which indicates the image quality is superior to PSO and FA as well as harmony search. With suitable simplifications PSO and harmony search becomes the special cases of Bat algorithm. In addition the bat algorithm possesses the benefit of automatic zooming that is accompanied by the automatic switch from explorative moves to local intensive exploitation, which results in fast convergence rate as compared to other techniques.

5 Conclusions

This paper presents a novel approach to obtain image compression in discrete wavelet domain by optimizing threshold values using bat algorithm for the first time. The coefficients obtained by applying discrete wavelet transform for the image to be compressed are classified into different groups by making use of the optimal thresholds, these optimal thresholds are obtained using bat algorithm, keeping a balance between the quality of the reconstructed image and compression ratio. The proposed

Table 1 Comparison of PSNR for Barbara image at different BPP

BPP	(a, b)		PSO		QPSO		FA		BA	
	3-level	5-level	3-level	5-level	3-level	5-level	3-level	5-level	3-level	5-level
0.125	(11, 1)	(11, 1)	25.60	25.72	25.63	25.82	25.71	25.92	26.05	26.32
0.250	(10, 1)	(9, 1)	28.32	28.42	28.39	28.47	28.43	28.64	28.83	28.93
0.500	(8, 1)	(8, 2)	32.15	32.34	32.55	32.71	32.78	32.98	32.96	33.23
1.000	(4, 1)	(4, 1)	37.10	37.12	37.15	37.31	37.39	37.64	37.52	37.83
1.250	(2, 1)	(2, 1)	39.98	40.04	40.08	40.24	40.34	40.53	40.44	40.87

Table 2 Comparison of PSNR for Gold Hill image at different BPP

BPP	(a, b)		PSO		QPSO		FA		BA	
	3-level	5-level	3-level	5-level	3-level	5-level	3-level	5-level	3-level	5-level
0.125	(7, 1)	(4, 1)	29.12	29.19	29.28	29.42	29.41	29.62	29.44	29.67
0.250	(5, 1)	(3, 2)	31.28	31.43	31.39	31.51	31.58	31.64	31.61	31.77
0.500	(3, 1)	(2, 3)	33.81	33.93	33.92	34.13	33.99	34.41	34.05	34.51
1.000	(2, 2)	(1, 5)	37.27	37.40	37.31	37.58	37.55	37.75	37.65	37.90
1.250	(1, 5)	(1, 7)	41.75	41.90	41.82	41.95	41.96	42.11	42.01	42.32

Table 3 Comparison of PSNR for Cameraman image at different BPP

BPP	(a,b)		PSO		QPSO		FA		BA	
	3-level	5-level	3-level	5-level	3-level	5-level	3-level	5-level	3-level	5-level
0.125	(18, 1)	(16, 1)	26.17	26.53	26.21	26.60	26.37	26.71	26.43	26.74
0.250	(12, 1)	(11, 1)	30.05	30.10	30.09	30.15	30.17	30.41	30.27	30.56
0.500	(10, 1)	(10, 2)	33.46	33.52	33.51	33.63	33.55	33.69	33.58	33.73
1.000	(3, 1)	(3, 2)	38.02	38.13	38.09	38.19	38.21	38.37	38.33	38.39
1.250	(1, 1)	(1, 1)	40.11	40.23	40.13	40.28	40.33	40.32	40.43	40.51

Table 4 Comparison of PSNR for Lena image at different BPP

BPP	(a, b)		PSO		QPSO		FA		BA	
	3-level	5-level	3-level	5-level	3-level	5-level	3-level	5-level	3-level	5-level
0.125	(15, 1)	(15, 1)	30.82	30.98	30.98	31.08	31.11	31.15	31.23	31.43
0.250	(11, 1)	(10, 1)	35.17	35.26	35.26	35.32	35.34	35.46	35.45	35.68
0.500	(7, 1)	(7, 1)	37.91	38.23	38.23	38.28	38.18	38.55	38.35	38.70
1.000	(4, 1)	(4, 2)	40.94	41.42	41.42	41.53	41.27	41.67	41.33	41.71
1.250	(2, 1)	(2, 3)	42.15	42.31	42.31	42.37	42.30	42.46	42.47	42.62

Table 5 Comparison of PSNR for Pepper image at various BPP

BPP	(a, b)		PSO		QPSO		FA		BA	
	3-level	5-level	3-level	5-level	3-level	5-level	3-level	5-level	3-level	5-level
0.125	(10, 1)	(9, 1)	34.81	34.98	34.82	35.01	34.89	35.16	34.90	35.21
0.250	(5, 1)	(4, 1)	37.58	37.82	37.63	37.89	37.69	37.98	37.73	38.18
0.500	(1, 1)	(1, 1)	39.75	39.92	39.84	40.12	39.92	40.27	40.12	40.44
1.000	(3, 1)	(1, 4)	42.92	43.21	42.98	43.33	43.04	43.46	43.26	43.61
1.250	(1, 5)	(1, 5)	43.10	43.42	43.19	43.43	43.29	43.52	43.38	43.74

technique is simple and adaptive in nature as individual thresholds are assigned to high energy sub-bands individually and for rest of the sub-bands a common threshold are assigned. The successful thresholded image is additionally coded with RLE followed by Huffman coding. It is observed that for multilevel thresholding bat algorithm produces superior PSNR values, good quality of the reconstructed image with less convergence time as compared to PSO and firefly as the later techniques are unstable if the particle velocity is maximum and no brighter firefly in the search space respectively. The algorithm convergence time is further improved by the fine adjustment of the pulse emission and loudness parameters and time delay between pulse emission and the echo.

References

1. Bing-Fei Wu, Chorng-Yann Su.: A fast convolution algorithm for biorthogonal wavelet image compression. J. Chin. Inst. Eng. **22**(2), 179–192 (1999)
2. Shanavaz, K.T., Mythili, P.: Faster techniques to evolve wavelet coefficients for better fingerprint image compression. Int. J. Electron. **100**(5), 655–668 (2013)
3. Deshmukh, P.R., Ghatol, A.: Multiwavelet and image compression. IETE J. Res. **48**(3–4), 217–220 (2002)
4. Singh, V.K.: Discrete wavelet transform based image compression. Int. J. Remote Sens. **20**(17), 3399–3405 (1999)
5. Libao, Z., Bingchang, Q.: Fast orientation prediction-based discrete wavelet transform for remote sensing image compression. Remote Sens. Lett. **4**(12), 1156–1165 (2013)
6. Farid, G., Ekram, K., Sadiqa, H.: A modified JPEG image compression technique. IETE J. Res. **46**(5), 331–337 (2000)
7. Panda, G., Meher, S.K.: An efficient hybrid image compression scheme using DWT and ANN techniques. IETE J. Res. **52**(1), 17–26 (2006)
8. Singha, S., Kumar, K., Verma, H.K.: DWT–DCT hybrid scheme for medical image compression. J. Med. Eng. Technol. **31**(2), 109–122 (2007)
9. Ahmed, L., Ammar, D.: Watermarking method resilient to RST and compression based on DWT, LPM and phase correlation. Int. J. Comput. Appl. **35**(1), 36–43 (2013)
10. Rakesh, K., Indu, S.: Empirical wavelet transform based ECG signal compression. IETE J. Res. **60**(6), 423–431 (2014)
11. Hemant, B., Tanuja, K.S., Rekha, V.: A new multi-resolution hybrid wavelet for analysis and image compression. Int. J. Electron. **102**(12), 2108–2126 (2015)

12. Chiranjeevi, K., Umaranjan, J.: SAR image compression using adaptive differential evolution and pattern search based K-means vector quantization. Image Anal. Ster. **37**(1) (2018)
13. Kaur, L., Chauhan, R.S., Saxena, S.C.: Joint thresholding and quantizer selection for compression of medical ultrasound images in the wavelet domain. J. Med. Eng. Technol. **30**(1), 17–24 (2006)
14. Kaveh, A., Ahmad, Y.J., Ezzatollah, S.: An efficient compression scheme based on adaptive thresholding in wavelet domain using particle swarm optimization. Sig. Process. Image Commun. **32**(1), 33–39 (2015)
15. Chiranjeevi, K., Umaranjan, J.: Fast vector quantization using a Bat algorithm for image compression. Eng. Sci. Technol. Int. J. **19**(1), 769–781 (2016)
16. Chiranjeevi, K., Umaranjan, J.: Hybrid gravitational search and pattern search based image thresholding by optimizing shannon and fuzzy entropy for image compression. Int. J. Image Data Fusion. **3**(3), 236–269. Taylor & Francis (2017)

Prototype Selection of Online Handwritten Telugu Characters Recognition Using Voronoi Tessellation

Srilakshmi Inuganti and Rajeshwara Rao Ramisetty

Abstract This paper illustrated various types of prototype selection for Online Handwritten Telugu Characters. The prototypes can be regarded as diverse styles of writing or expressing a specific character. Subsequently, a novel algorithm is proposed for prototype selection, in which the number of prototypes will be independently governed by the Voronoi Tessellation. Later, the K-Nearest Neighbor algorithm is applied on these prototypes, using parameters like angle/direction and coordinates of the pen. The HP labs dataset hpl-Telugu-ISO-char-online-1.0 is utilized for evaluation of proposed algorithm, which is found to achieve a recognition accuracy of over 93%. The average throughput is found to be 0.18 characters per second. The HP labs dataset contains 166 character classes which are aggregated from numerous writers on ACECAD Digimemo, with the help of Acecad Digi memo DCT application. Each of 166 Telugu characters contains 270 samples, which are written by native Telugu writers.

Keywords Online handwritten character · Voronoi tessellation · Clustering · Prototypes · Davies-Bouldin index · Silhouette index

1 Introduction

By each day, the handheld devices are becoming more and more popular. They have become a predominant part of our everyday life. In recent days, digital pen have replaced the conventional button keyboard, more specifically for mobile devices such as tablet, PDA, etc. The digital pen captures the handwriting of a user. This handwritten information will be converted to digital data, which enables the translated data to

S. Inuganti (✉)
JNTU, Kakinada, Andhra Pradesh, India
e-mail: inuganti.srilakshmi@gmail.com

R. R. Ramisetty
JNTUK, UCEV, Vizianagaram, Andhra Pradesh, India
e-mail: raob4u@yahoo.com

H. S. Behera et al. (eds.), *Computational Intelligence in Data Mining*, Advances in Intelligent Systems and Computing 990,
https://doi.org/10.1007/978-981-13-8676-3_25

be used for numerous other applications. Thus, Handwritten Character Recognition (HCR) becomes a challenging task. There are two categories of HCR, Online HCR and Offline HCR. In online HCR, the characters/handwriting will be recognized and translated to digital data as the user is writing. It requires some sort of transducer within the capturing device, to dynamically convert the handwriting into digital file. The translated information is dynamic in nature, and it is composed of alphabets, numbers, length, order, direction of writing, and speed of stroke. Some devices even take note of applied pressure while writing. There are numerous research and survey in the field of handwriting recognition [1, 2], but still, not much progress is made in the area of regional languages of India. As the Indian languages are quite old, they have large character set, hence OHCR is more suitable option. Details of some recent literature has given in Table 1. Here each researcher has used his own database for their experimentation that leads to result comparison difficult.

As far as mobile devices are concerned, the prototype selection plays a significant role in Online Handwritten Character Recognition techniques. This technique is helpful for devices with limited resources. The recognition accuracy is influenced by the quality of prototype, and the recognition time is affected by the total number of the selected prototypes affects recognition time. To achieve good classification of characters, a trade-off has to be made amongst class variations, with a facility for large within-class variations. Within-class variation can be made quite larger, if the recognition technique has to be applied on large set of handwriting samples. The variance in a specific character class can be considered to be composed of two modules: (1) the variations between different writing styles, (2) the variations within a given writing style. Figures 1 and 2 demonstrates this concept. The main objective of our algorithm is to study these writing styles from a given data set. This aides in building sufficient character models. In this paper, the prototypes are manually fixed by using dynamic time warping (DTW) as distance measure. Later, a novel technique is proposed which can minimize the required training set, by working on prototypes which lies closer to class boundaries. The following section provides details about these techniques. A nearest neighbor classifier is used for recognition of a test character using features of pen coordinate and pen direction angle.

2 Data Base

Telugu is an old Dravidian language. It is the official language for the Indian states of Telangana and Andhra Pradesh. The government of India declared the Telugu as a classical language, which is one among six other classical languages in India. According to 2011 census, Telugu is spoken and used by over 75 million citizens. Being an old and popular language, there are numerous vowels, consonants in the syllables of Telugu script. The usual forms of syllables are V, CV, CCV and CCCV. Hence, they are generally represented in the form of C * V. Therefore, the basic units of character of script are $O(10^2)$, these units forms $O(10^4)$ number of composite characters. The data set for Telugu language can be obtained from Hp Labs data in UNIPEN

Table 1 Review of different techniques of online handwriting

Author	Type	Features	Classifier	Dataset (size)	Accuracy (%)
Das et al. [3]	Assamese Aksharas	(x, y) coordinates and their first and second derivatives	HMM SVM	147 Aksharas (Each 1000 samples)	94 96
Ramakrishnan and Bhargava Urala [4]	Numerals and tamil characters	Local + global features	SVM	HP Labs IWFHR dataset and Alpaydin database	95
Raghavendra et al. [5]	Tamil symbols	Preprocessed (x, y) coordinates	DTW prototype selection used	1.5 char/s 7774 templates selected from 62,400 samples	94
Hariharan et al. [6]	Telugu characters	Pre-processed (x, y) coordinates	SVM	37,817 samples collected from 92 users using the Superpen, a prod	83
Prasanth et al. [7]	Telugu characters	(x, y) features + tangent angle + shape context feature + generalized shape context (GSC) feature	DTW	38,393 data samples from 143 writers buy using ACECAD Digimemo	90

format. This dataset is composed of over 270 samples of Telugu characters, with 166 characters, which are written by native Telugu writers [8]. The AcecadDigimemo electronic clipboard devices are used for collecting the samples, with the help of Digimemo-DCT application. The 166 symbols in Telugu are given in Fig. 3. These symbols/characters are collected from over 146 writers in two attempts, which aggregates to around 45,219 samples. Among this vast dataset of samples, only 33,897 are selected for training and the rest are used for testing and evaluation purposes.

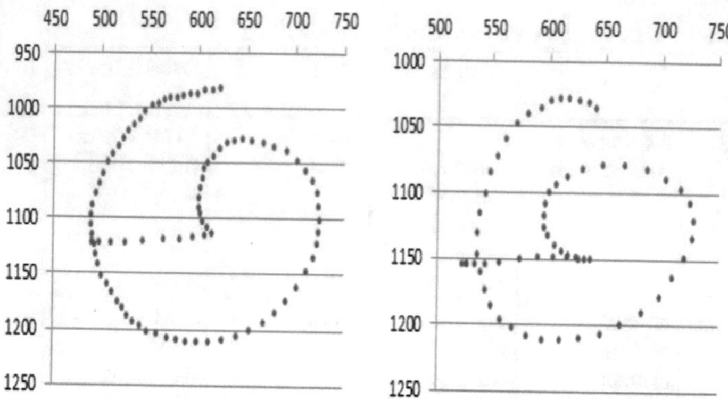

Fig. 1 Variants of "Id_0" written by User1

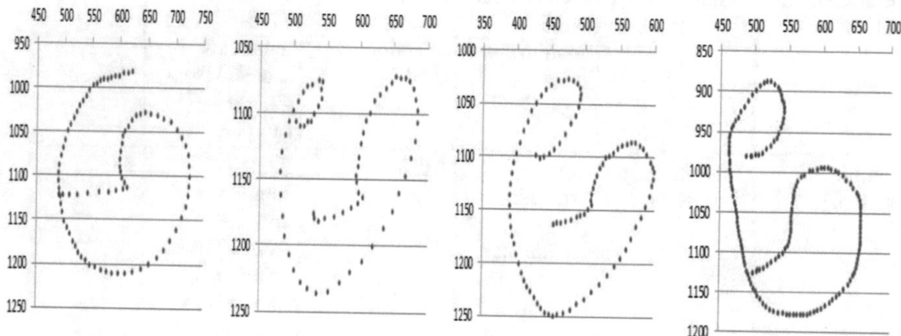

Fig. 2 "Id_0" written by 4 different users

3 Preprocessing and Feature Extraction

In a recognition technique, it is essential to translate the stroke into a standard window as the size of writing strokes varies. In this particular context, our database has a standardised window size of 200 × 250. As the speed of handwriting increases, the strokes will be affected by missing points. Nevertheless, the missing points can be completed with the help of linear interpolation. The data is later resampled to a constant number of points with uniform spacing. For the given samples in our database, the sampling results in 64 points for all samples [9].

In any Online Handwriting Recognition, Feature selection is the fundamental segment. This segment has a profound effect on the overall performance of subsequent phases. Our features consists of position of pen-tip and tangent angle at the pen-tip. It helps in preservation of direction of the path. In our database, the feature vector for any stroke is given in Eq. 1 as,

$$F = \left[(p_1, \theta_{12}), (p_2, \theta_{23}), \ldots\ldots\ldots.(p_l, \theta_{l-1l})\right] \tag{1}$$

Fig. 3 Distinct symbols in Telugu

$$where\ \theta_{l-1l} = arctan(\frac{y_l - y_{l-1}}{x_l - x_{l-1}})$$

4 Prototype Selection

4.1 Prototype Selection Manually

In this method predefined value of clusters formed and retains the prototypes closest to each cluster center. A pattern matrix is constructed using the dynamic time warping distance for each class of characters. The pattern which has cumulative smallest distance with all other pattern of the same class is selected as the prototype of the class [10]. The predefined number of cumulative smallest distance patterns is selected from each class as prototypes and this method is referred as prototype learning manually.

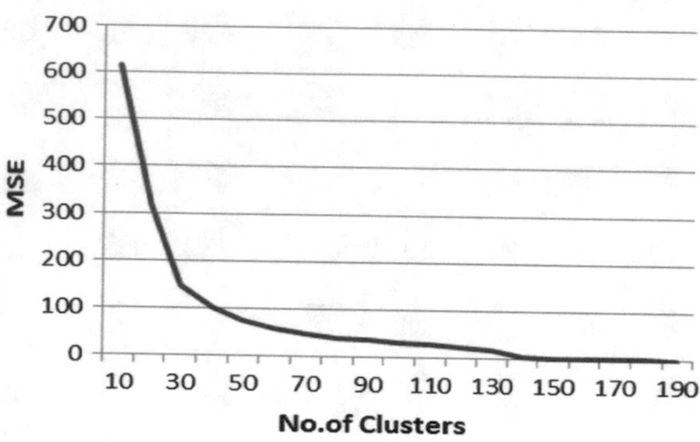

Fig. 4 MSE for different clusters for character class "Id_0"

4.2 *Prototype Selection Automatically*

The fore mentioned method produces the best results for the given value of K, but identifying the best value of K is a very difficult task. Figure 4 Shows the Mean Square Error (MSE) for different values of K for character class "Id_0". The value of MSE decreases as the value of K increases. Automatically determining the number of clusters is also a difficult task, which is addressed in this section. Here we have experimented with two cluster indices to determine the number of clusters automatically [11]. On the basis of experimental results we state that a good set of prototypes can be selected by the combination of two cluster indices. However, with some human intervention, the best results can be obtained from Prototype Learning Automatically, but automatically determining no. of prototypes present in a datasets is very tough problem [12].

4.2.1 Clustering Indices

Two clustering indices are used, for determining the number of clusters automatically: Davies-Bouldin index (DBI) and Silhouette index (SI) [12]. Both indices are implemented using DTW-based distance between two samples or sample and prototype.

Davies-Bouldin Index: Davies-Bouldin Index is one of the most commonly used clustering indices [13]. It minimizes intra class variance and maximizes inter class distance. The index, DBI is defined in Eq. 2 as:

$$DBI = \frac{1}{n}\sum\nolimits_{i=1,i\neq j}^{n} \max\left(\frac{\mathrm{MAD_i} + \mathrm{MAD_j}}{\mathrm{d(C_i, C_j)}}\right) \tag{2}$$

where, d (C_i, C_j) suggests the distance between cluster centers C_i and C_j, MAD_i is the mean distance of all patterns in cluster I from their cluster center C_i, n is the number of clusters, MAD_j is the mean distance of all patterns in cluster J from their

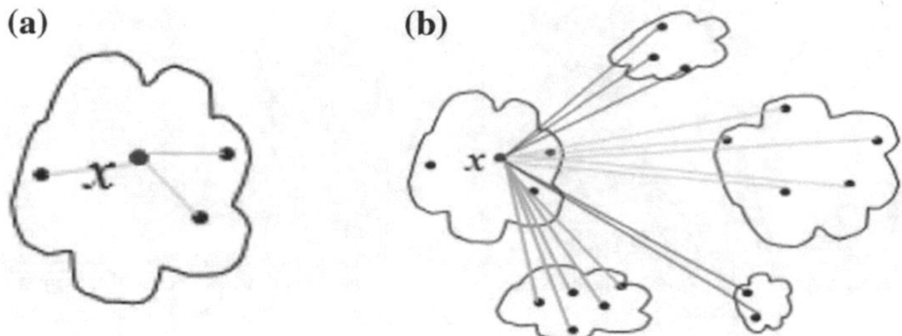

Fig. 5 **a** a(x): mean distance in the cluster, **b** b(x): minimum of mean distance to other clusters

cluster center C_j. The least value of DBI signifies the optimal number of clusters, which are closely packed and the cluster centers are located very far off from each other.

Silhouette index: Silhouette Index is selected as it out performed several other cluster indices, when in combination with the other cluster index [14]. SI measures closeness of patterns in a cluster and distinctness of one cluster from other clusters. The maximized value of SI gives the optimal number of clusters. SI index is calculated using Eq. 3.

SI is defined as:

$$s(x) = \frac{b(x) - a(x)}{\max(b(x), a(x))} \tag{3}$$

where b(x) is the mean distance of x to the patterns of each of other clusters, Minimum among these values is considered, a(x) is the mean distance of x to all other pattern in the same cluster, s(x) is known as silhouette width at point x. Figure 5 Depict a(x) and b(x). s(x) lies between −1 and 1: a value near 1 indicates that the point is appropriately clustered, near −1 indicates, it would be appropriate if it is clustered in its neighbor cluster.

The mean of the silhouette widths for a given cluster C_k is called mean silhouette and is denoted as s_k in Eq. 4:

$$s_k = \frac{1}{n_k} \sum\nolimits_{i \in I_k}^{n} s(i) \tag{4}$$

Finally the Silhouette Index is defined as in Eq. 5:

$$SI = \frac{1}{K} \sum\nolimits_{k=1}^{K} s_k \tag{5}$$

where K number of clusters. Figure 6a, b shows the graph of DBI and SI for varying number of clusters of four sample classes Id_0, Id_1, Id_2 and Id_3. Table 2 shows the number of clusters selected for each of vowel classes using manual and automatic

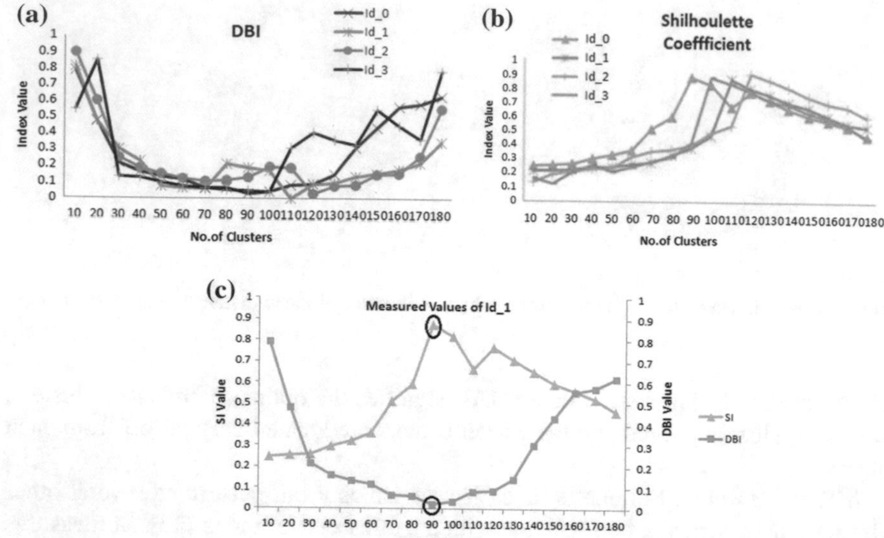

Fig. 6 **a** DBI for different clusters for character classes "Id_0" to "Id_3", **b** SI for different clusters for character classes "Id_0" to "Id_3", **c** maximum SI and minimum DBI for "Id_1"

selection based on minimum DBI and maximum SI. Figure 6c depicts the number of prototypes selected for character class "Id_1" as 90 as illustrated in Table 2.

4.3 Prototype Selection Using Voronoi Tessellation

Voronoi Tessellation (VT) can be described as the Voronoi diagram of a set of points. VT involves partitioning a plane into several regions on the basis of the distance to points in a particular subset of the plane. These set of points will be predefined. Each seed contains a matching region which is composed of all points closer to that seed among all other seeds. The association of Vornoi region R_k (calculated using Eq. 6) is with the site P_k signifies the set of all points in A, whose distance to P_k is lower than their distance to other sites P_j. Here, j denotes any other index apart from k. i.e.,

$$R_K = \{a \in A | dis(a, P_k) \leq dis(a, P_j) \quad \text{for all} \quad j \neq k\} \tag{6}$$

where, dis(a, X) signifies the distance between the subset A and point x.

In this section we address reducing training set using Voronoi Tessellation, in which patterns that lie along the class boundaries are selected as prototypes and remaining patterns are removed [10]. These chosen prototypes then give the reduced training set. We have applied the following algorithm for retaining the border patterns.

Table 2 Number of prototypes selected by different methods for 14 vowels

Class ID	Vowel	Prototype selection method		
	Telugu symbol	K manual	K automatic	Voronoi tessellation
Id_0	అ	69	110	120
Id_1	ఆ	70	90	160
Id_2	ఇ	70	120	144
Id_3	ఈ	69	110	130
Id_4	ఉ	65	90	120
Id_5	ఊ	65	80	100
Id_6	ఋ	66	110	80
Id_7	ౠ	68	140	70
Id_8	ఎ	62	100	110
Id_9	ఏ	68	120	110
Id_10	ఐ	68	130	90
Id_11	ఒ	63	110	110
Id_12	ఓ	69	130	130
Id_13	ఔ	67	150	130

An algorithm for finding Reduced Training Set Using Voronoi Tessellation

$T_o \equiv$ Set of Training Samples
$T_R \equiv$ Reduced Training set, initially empty
$S \equiv$ Set of class labels
$S(i) \equiv$ The class of pattern i
$C_c(i) \equiv$ The closest neighbor to pattern i that is from class S
$\forall x_k \in T_o$
$\forall s \in S$ where $c \neq S(x_k)$
    If $C_c(x_k)$ does not belong to T_R
        add $C_c(x_k)$ to T_R

The output of the above algorithm gives prototypes which reside close to the boundary; however, this reduced set retains more samples than the previous two approaches. Table 2 shows the number of prototypes obtained for each of vowel classes using Voronoi Tessellation.

5 Recognition

The identified prototypes have been utilized to recognize test samples. Here we illustrate K-Nearest Neighbor classifier using DTW as a distance measure. This distance measure is useful for comparing two patterns of equal or unequal length. In our approach the test sample is compared with all the prototypes in the database. Here each event in the pattern is represented with three measurements: the x and y coordinates, and the tangent angle at that point. The distance between two stroke d(i, j) between pair of events, $e_i^A = \left(x_i^A, y_i^A, \theta_i^A\right)$ of stroke A and, $e_j^B = \left(x_j^B, y_j^B, \theta_j^B\right)$ of stroke B is defined using Eq. 7 as:

$$d(i, j) = \sqrt{\left(x_i^A - x_j^B\right)^2 + \left(y_i^A - y_j^B\right)^2 + (\theta_i^A - \theta_j^B)^2} \quad (7)$$

The distance between two strokes can be estimated by considering all probable alignments between events and obtaining the alignment using dynamic programming which results in minimum distance. The calculation of distance between two strokes A and B are given in Eq. 8 as:

$$\mathrm{D(i, j) = d(i, j) + min(D(i - 1, j - 1) + D(i - 1, j) + D(i, j - 1))} \quad (8)$$

where d(i, j) is the local distance at (i, j) and D(i, j) is the global distance up to (i, j).

$$DTW(A, B) = D(N_A, N_B)$$

where $\mathrm{N_A}$ is the length of stroke A and $\mathrm{N_B}$ is the length of stroke B. In our process the class of test character is determined to be the class of its nearest neighbor in the train database.

6 Experimental Results

This section presents the results of various experiments conducted with different approaches of prototype learning. The experiments are performed on Intel i5 processor with a clock speed of 3.10 GHz and 4 GB RAM. The experiments are carried out on the Telugu Database obtained from HP Labs. The database contains samples

Table 3 Results of different prototype learning approaches

Method	No. of templates	Avg No. of templates	Throughput	Recognition accuracy (%)
Full training set	33,897	204	0.09 char/s	93.88
K-means (K manually selected)	11,248	68	0.22 char/s	81.45
K-means (K automatically selected)	17,700	106	0.16 char/s	86.85
Reduced training set using Voronoi tessellation	18,600	112	0.19 char/s	93.54

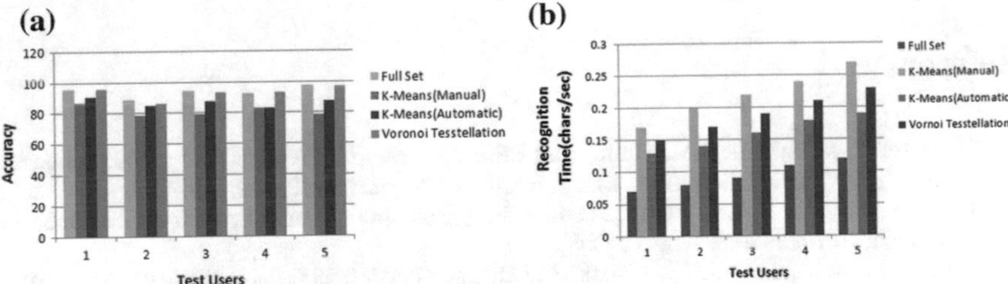

Fig. 7 Performance of Voronoi Tessellation with different prototype learning approaches. **a** Accuracy. **b** Recognition time per character

of 166 characters from 146 users. 131 samples are used for training, and the rest 35 samples are used for testing purposes. For each approach, few parameters are evaluated, namely, number of prototypes, recognition time, and recognition accuracy. The average recognition time is the ratio of total number of recognized characters and total time taken to recognize those test samples. The obtained results are presented in Table 3. The results are extracted from testing a 1-nearest neighbor classifier with the help of four different training sets. The complete training set contains 33,897 samples and prototype learning exploits 11,248 prototypes. Over 17,700 prototypes are acquired by choosing clusters with the assistance of Silhouette Index and Davies-Bouldin cluster indices. Over 18,600 prototypes are exploited for minimized training set with the help of Vornoi Tessellation. From the experimental results, it is apparent that the training set for Vornoi Tessellation is greatly minimized by over 54%. The computational time is also improved by over 48%. However, other important such as, recognition accuracy is unaffected. The performance of Vornoi Tessellation with various prototype learning approaches is depicted in Fig. 7.

7 Conclusion

In this paper, data reduction approach proposed and validated its efficacy for online handwritten Telugu character recognition. In this method Voronoi Tessellation is used to select prototypes near to class boundary, which gives reduced training set and computational time, while maintaining at most same recognition accuracy with the full training set. We have also showed that proposed approach is superior to selecting prototypes manually and automatically in recognition accuracy. Overall, the success rate is about 93.54% in less than 5 min per character on average. Further potential areas of research are to use other classification models such as SVM, HMM and Neural Networks for online character recognition problem, to possibly improve the overall classification accuracy.

References

1. Plamondon, R., Srihari, S.N.: Online and offline handwriting recognition: a comprehensive survey. IEEE Trans. Pattern Anal. Mach. Intell. (PAMI) **22**(1), 105–115 (2000)
2. Bharath, A., Madhvanath, s.: Online Handwriting Recognition for Indic Scripts, HP Laboratories, India, HPL-2008-45, May 5 (2008)
3. Das, D., Devi, R., Prasanna, S., Ghosh, S., Naik, K.: Performance comparison of online handwriting recognition system for assamese language based on HMM and SVM modelling. Int. J. Comput. Sci. Inf. Technol. **6**(5), 87–95 (2014)
4. Ramakrishnan, A.G., Urala, B.: Global and local features for recognition of online handwritten numerals and Tamil characters. In: MOCR2013: Proceedings of International Workshop on Multilingual OCR, Washington DC, USA, pp. 84–90 (2013)
5. Raghavendra, B.S., Narayanan, C.K., Sita, G., Ramakrishnan, A.G., Sriganesh, M.: Prototype learning methods for online handwriting recognition. In: ICDAR'05: Proceedings of International Conference on Document Analysis and Recognition, Seoul, Korea, pp. 287–291 (2005)
6. Swethalakshmi, H., Jayaraman, A., Srinivasa Chakravarthy, V., Chandra Sekhar, C.: Online handwritten character recognition of Devanagiri and telugu characters using support vector machines. In: IWFHR: Proceedings of 10th International Workshop on Frontiers in Handwriting Recognition, La Baule, France, pp. 83–89 (2006)
7. Prasanth, L., Babu, V., Sharma, R., Rao, G.V., Dinesh, M.: Elastic matching of online handwritten tamil and telugu scripts using local features. In: ICDAR '07: Proceedings of the Ninth International Conference on Document Analysis and Recognition, Curitiba, Brazil, vol. 2, pp. 1028–1032 (2007)
8. Aparna, K.H., Subramanian, V., Kasirajan, M., Vijay Prakash, G., Chakravarthy, V.S., Madhvanath, S.: Online handwriting recognition for Tamil. In: IWFHR' 04: Proceedings of the Ninth International Workshop on Frontiers in Handwriting Recognition, Tokyo, Japan, pp. 438–443 (2004)
9. Srilakshmi, I., Rajeshwara Rao, R.: Preprocessing of HP data set telugu strokes in online handwritten telugu character recognition. Int. J. Control Theory Appl. **8**(5), 1939–1945 (2015)
10. Connell, S.D., Jain, A.K.: Template-based on-line character recognition. Pattern Recogn. **34**, 1–14 (2001)
11. Vuori, V., Laaksonen, J.: A comparison for automatic clustering of handwritten characters'. In: ICPR: Proceedings of the International Conference on Pattern Recognition, vol. 3, pp. 168–171 (2002)
12. Jain, A.K., Dubes, R.C.: Algorithms for Clustering Data, pp. 110–115. Prentice-Hall, Englewood Cliffs, NJ (1988)

13. Davies, D.L., Bouldin, D.W.: A cluster separation measure. In: IEEE Trans. Pattern Anal. Mach. Intell. **1**(2), 224–227 (1979)
14. Rousseeuw, P.J.: Silhouettes: a graphical aid to the interpretation and validation of cluster analysis. J. Comput. Appl. Math. **20**, 53–65 (1987)

Assessment of Cosmetic Product Awareness Among Female Students Using Data Mining Technique

Aniket Muley and Atish Tangawade

Abstract This paper explores a special application to determine cosmetic awareness among hostel resident girls using data mining technique. Various significant factors, viz., parent's occupation, monthly income, education, family members, brand of the products, pocket money of girls, self-experience, scheme and offers, etc., affecting their selection of cosmetics were considered. The unsupervised and supervised learning techniques were carried out to identify the significant parameters and to discover the hidden measure of the selection of products among girl's hostel students. Also, the test of independence and test of significance is performed and further, factor analysis is carried out. The data reveals that brand, advertisement, scheme, and offer are found to be significant in the selection of cosmetic brands.

Keywords Data mining · Cosmetics · Supervised learning · Awareness · Nanded

1 Introduction

In real life, human being applies makeup for good looks or altering one's manifestation. Cosmetics are defined as substances which are applied externally to improve the appearances. They enhance one's beauty and self-confidence by looking, smelling, and feeling good. Food and Drug Administration (FDA), which regulates cosmetics, defines cosmetics as deliberate to be applied to the human body for cleansing, beautifying, encouraging attractiveness, or altering the impression without affecting the body's structure or functions. Due to awareness, cosmetics are nowadays essential part of a consumer's daily life [2]. Some researchers studied the usefulness of the famous person's approval on a meta-analytic grade athwart a selection of trial. More-

A. Muley (✉) · A. Tangawade
School of Mathematical Sciences, Swami Ramanand Teerth Marathwada University, Nanded 431606, Maharashtra, India
e-mail: aniket.muley@gmail.com

A. Tangawade
e-mail: tangawadeatish@rediffmail.com

H. S. Behera et al. (eds.), *Computational Intelligence in Data Mining*, Advances in Intelligent Systems and Computing 990,
https://doi.org/10.1007/978-981-13-8676-3_26

over, the outcome shows effects on approach and behavior of male actor choosing the endorsed object and articulate his support implicitly [13, 16, 20]. Cosmetology is the science of beautifying skin [1, 10, 11, 20]. In India, most of the individuals focused on the natural phenomenon for beautifications. Also, due to modernization folks are moving toward cosmetic use [12].

The income level is statistically significant with social, cultural aspects and insignificant with personal and psychological parameters [14]. Yuvaraj [23] studied brand choice decisions by consumers with reference to cosmetics [7, 10, 11, 18, 19]. The performed survey based on the study of consciousness of cosmetic contagion among teenager females and spread awareness regarding the different ways to prevent the harmful effects that these contaminated cosmetics on the health of the users [7, 10, 11, 18, 19]. Unsupervised erudition problems additionally assemblage into clustering and association problems. Supervised learning is used to gain knowledge of the mapping function to the get the output, i.e., classification, factor analysis [8, 22].

The acquiring cosmetic product by consumers was investigated in Kerala [15]. The brand loyalty of skin care products' influencing factors evaluated. The result showed that, in rural area brand name, brand price, and brand quality are the main determinants of brand loyalty of skin care products. Also, it is exposed that favorable name with the suitable price of the product and expected quality presence in the product enhances the brand loyalty in the product and the customers in rural areas will repeat their purchase of the same skin care product. However, promotion, distribution, and package of the products have an insignificant relationship with the brand loyalty [4, 5, 12]. According to them, the number of cosmetic products and its frequency of use were important predictors for contrary events [3]. The influence of the attitude toward buying cosmetic manufactured goods, and is also identified the important factors that determine the buying behavior. Their study result highlights a positive impact due to epoch, profession, matrimonial status in the direction of cosmetic products, but income does not have any consequence [9]. Some researchers studied about probing the theory of exploitation value and satisfaction by cosmetic retail products containing meat. Their results explore that, there is a need to investigate the functional, social, emotional, conditional, and epistemic dimensions to a greater extent. Also, to make operational of these dimensions, so that it can be better to predict and understand consumer's choice and reactions [21].

In this study, unsupervised and supervised learning data mining approach is used to identify cosmetic awareness among hostel resident girls. The aim of this study is to explore the use of cosmetic products by hostel resident girls; and to find out the attentiveness of cosmetic among hostel residents girls using data mining. To explore the information to identify the diversity among the data based on various parameters. To test the significance of use of cosmetics by girls with their father's and mother's educational qualification. To test the significance of use of cosmetics by hostel resident girls with their father's as well as the mother's occupation. To test significance is monthly income and use of cosmetic products by hostels living girls. To test the significance of girl's pocket money from their parents is associated with the use of makeup, type of makeup, use of applicator, daily time spending on makeup, borrowing/sharing makeup products, and use of particular product duration in months. Further, to perform factor analysis for identifying the significant parameters affecting purchasing a particular product. In the next consequent sections: technique, result discussion, and conclusion are discussed in detail.

1.1 Study Area

Our investigation focused on the population choice are from age 18 and above in the SRTM University campus girls. The nature of the study involved a large population (350) for having a good generalization. In this study, census survey methodology is used with a structured questionnaire for the data collection. Here, the purposive sampling method is used at the initial level to perform the data collection. This study is limited only on girl's hostel resident students in SRTM University campus only. The population of the study was categorized into four main age groups: 18–21, 22–25, 26–29, and >29.

2 Methodology

In this study, a structured questionnaire was prepared for primary data collection containing 19 questions administrated to SRTM University Campus, Nanded. The survey was performed during the months of January 2017. In this survey, to avoid redundancy, the questionnaire was numbered according to the number of hostels as well as room numbers available in the corresponding hostel. Kaiser–Meyer–Olkin (KMO) is used to determine the suitability of the data and is computed by Eq. (1):

$$KMO = \frac{\sum_{i \neq j}^{n} r_{ij}^{2}}{\sum_{i \neq j}^{n} r_{ij}^{2} + \sum_{i \neq j}^{n} u} \quad (1)$$

where $\mathbf{R} = [r_{ij}]$—Correlation matrix and, $\mathbf{U} = [u_{ij}]$—Partial covariance matrix.

The KMO value close to 1 indicates that it is useful with our data. If the value is less than 0.50, the results of the factor analysis probably won't be very useful. The Bartlett test's small value (<0.05) of the significance level specifies that factor useful with our data. A factor solution exhibits the simple structure with one or a few common factors and single common factor is highly correlated with only a few variables (Tables 5, 6 and 7). The factor solution is reoriented through a process called rotation (Table 7). The orthogonal rotation preserves perpendicularity of an axis that is rotated factor remains uncorrelated. The varimax rotation method is used, which strives and achieves simple structure by focusing on columns of the factor loading matrix.

Table 1 Summary of techniques used for similar studies

Authors	Year	Techniques
Bearden and Etzel [1]	1982	Convenience sampling, multivariate ANOVA
Bhatt and Sankhla [2]	2016	One sample t-test, frequency analysis, multiple response analysis
Bilal et al. [3]	2017	Cross-sectional study, multivariate logistic regression
Biswas et al. [4]	2009	Demographic and cross-cultural study
Dunham [6]	2006	Data mining techniques
Ingavale [8]	2016	Z-test, chi-square test, reliability test, Cronbach's alpha
Jawahar and Tamizhjyothi [9]	2013	Analysis of variance
Junaid et al. [10]	2013	Correlation, factor analysis, test of significance
Khan and Khan [11]	2013	Chi-square test
Khraim [12]	2011	Descriptive analysis, ANOVA, Pearson's correlation
Knoll and Matthes [13]	2017	Meta-analytic procedure
Kumar et al. [14]	2014	Descriptive statistics, mean standard deviation, ANOVA
Nair [15]	2007	Diagrammatic representation of data
Naz et al. [16]	2012	ANOVA
Parmar [17]	2014	Convenient sampling method, testing of hypotheses, Chi-square test, Garrett ranking method, and descriptive statistics
Raghu [18]	2013	Chi-square test, hypothesis testing
Rahim et al. [19]	2015	Descriptive and inferential analysis, correlation and regression analysis
Ilankizhai et al. [7]	2016	Survey-based cross-sectional study, means of data collection, Pie chart
Szagun and Pavlov [20]	1995	MANOVA, test of significance
Yeo et al. [21]	2016	Theory of consumption value
Yousaf et al. [22]	2012	Cronbach's alpha statistic, regression, correlation analysis
Yuvaraj [23]	2014	Frequency analysis, sampling technique

Supervised knowledge is one of the methods connected with machine learning, which involves allocating labeled data so that a certain pattern or function can be deduced from that data [6]. Unsupervised erudition is the method of machine learning algorithm used for exploration of data analysis. The summary of the review of previous studies is represented in Table 1.

In this study, MS Excel, SPSS 22.0v software was used to tabulation, coding and analyzing the collected dataset.

3 Result and Discussion

Initially, unsupervised learning is performed and ethe given data set and represented in Figs. 1, 2, 3, 4, and 5. Figure 1 explores that, out of 350 girls most of them, i.e., 207 hostel girls are of 22–25 age groups and so the maximum 59% response of the 22–25 age groups. Figure 2 represents the diversity of the respondent from postgraduate hostel girls about 81.1% out of 350 cases.

Figure 3 reveals that 76% respondent's pocket money is between Rs. 1000–2000. 17% response get from whose pocket money is between Rs. 2100–3000. Figure 4 expresses the number of females share or borrows makeup, and it is observed that, out of 350 of the students, 51.1% girls share or borrows their makeup and 49.9% girls do not share or borrow makeup.

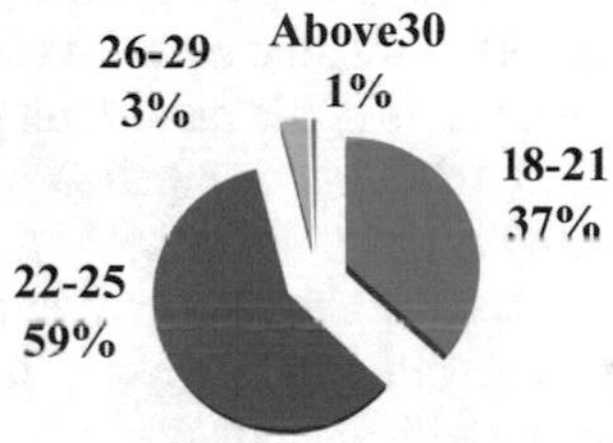

Fig. 1 Pie chart for age wise distribution

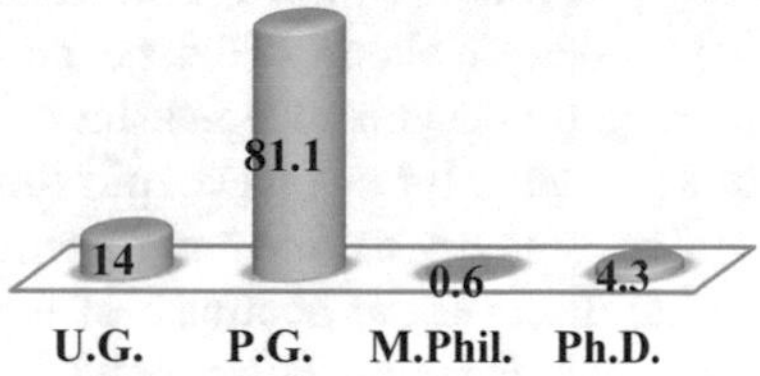

Fig. 2 Diversity of educational qualification

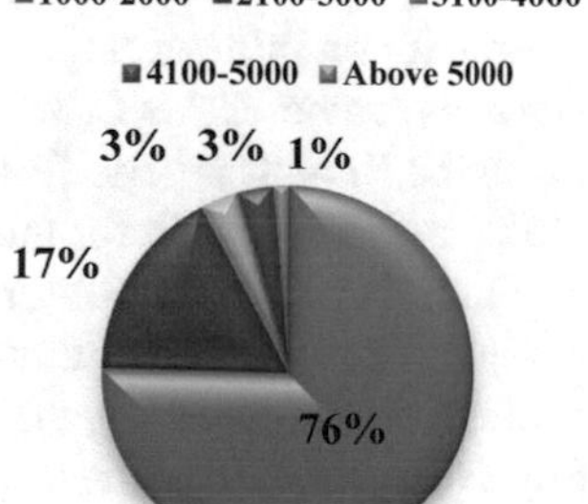

Fig. 3 Monthly pocket money

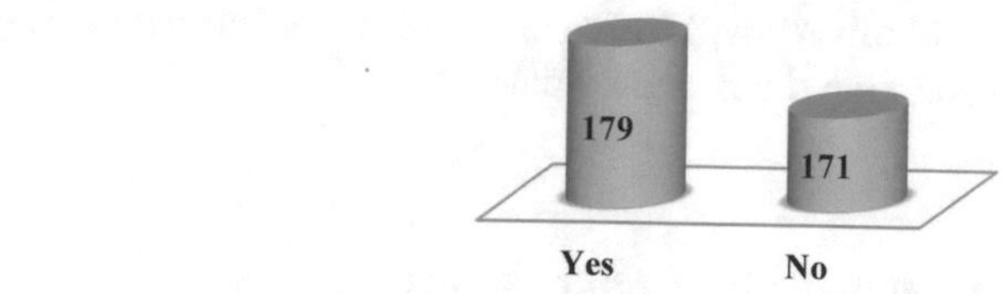

Fig. 4 Diversity of sharing or borrowing makeup

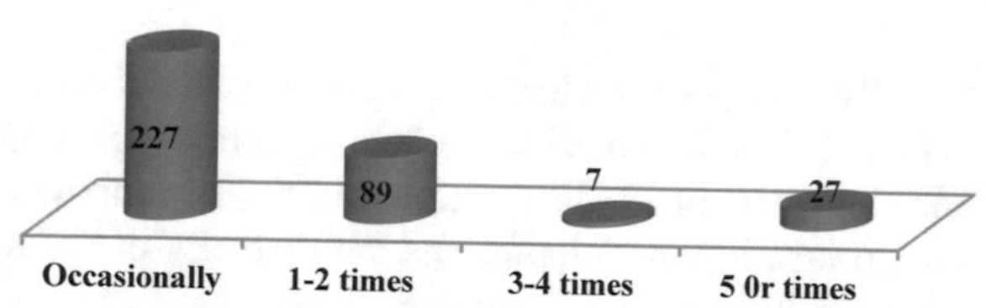

Fig. 5 Diversity of usc of makeup in a week

Figure 5 reveals that out of 350 hostel living girls, 227 occasionally in a week uses the makeup. Also, it is observed that 58% of hostel girls use the applicator finger for makeup.

To test the significant difference between various factors parents educational qualification, parents occupation and cosmetic product.

Table 2 reveals that father's educational qualification and use of cosmetic products by girls is significant but it is insignificant with a mother's educational qualification. There is a significant difference between father's occupation and cosmetic products. Also, mother's occupation and cosmetic products use by girls. Further, a survey result explores that family monthly income and cosmetic products is insignificant. Further, Chi-square test of independence is performed and is observed that for pocket money with use of makeup is independent ($p = 0.581$); the makeup variety is associated with the money given by parents other than fees for monthly expenditure ($p = 0.000$); applicator's use is dependent on pocket money ($p = 0.001$); the time spending on daily makeup is independent of pocket money (0.957); sharing/borrowing makeup products is independent of pocket money (0.142); and the duration of use of a particular type of makeup is independent of pocket money (0.339).

This study is to find the parameter which may affect for cosmetics selection, and to discover the approach of the respondent buying the products and the does their advertisement affects on it? To discover these problems here factor analysis is performed.

Cosmetics on Influences:
Here, the value of KMO test is 0.552 indicates the size of the sample is sufficient. From the Bartlett criteria (P-value = 0.00), variables are correlated with each other (Table 3), so data is fit for the factor analysis.

From Table 4, it is observed that the entire variable has a correlation with a component advertisement, brand name, self-experience, scheme, or offers.

Table 5 confirms the actual number of extracted factors. If we look at the Rotation Sums of Squared Loadings, the present study was two factors whose Eigenvalue is >1 and it is represented in Fig. 6 by Scree plot. The 30.09% of variance enlighten us the

Table 2 Variation study

Variables	S. V.	S. S.	DF	MSE	F	Sig.
Father's edu. qualification	Between groups	17.696	4	4.424	3.385	0.010
	Within groups	450.958	345	1.307		
	Total	468.654	349			
Mother's edu. qualification	Between groups	2.901	4	0.725	1.121	0.346
	Within groups	223.099	345	0.647		
	Total	226.000	349			
Father's profession	Between groups	43.172	4	10.793	3.434	0.009
	Within groups	1084.188	345	3.143		
	Total	1127.360	349			
Mother's profession	Between groups	1.592	4	0.398	1.687	0.152
	Within groups	81.368	345	0.236		
	Total	82.960	349			
Monthly earnings	Between groups	0.923	4	0.231	0.362	0.836
	Within groups	220.291	345	0.639		
	Total	221.214	349			

Table 3 KMO and Bartlett's test

Sampling capability		0.552
Bartlett's test	Approx. Chi-square	116.896
	Df	10
	Sig.	0.000

Table 4 Correlation

	Initial	Extraction
Advertisement	1.000	0.627
Family member	1.000	0.322
Brand	1.000	0.644
Self-experience	1.000	0.628
Scheme/offer	1.000	0.614

total variability. Factor 1 reported 30.09% of the inconsistency among 5 parameters, and so on.

From Table 6, we observed that Advertisement, Brand, Family member, and self-experience are most affecting factors observed in the first component. While advertisement, scheme/offers are related with the second component.

After revolution loading of the variable remains the same on both components. Rotated component matrix (Table 7) represent that self-experience and brand are related with the first component. While advertisement and schemes/offers are related with the second component.

Table 5 Total variance

Element	Initial Eigenvalues			Extraction sum of squared loadings			Rotation sum of squared loadings		
	Sum	% of var.	Cum. %	Sum	% of var.	Cum. %	Sum	% of var.	Cum. %
1	1.601	32.014	32.014	1.601	32.014	32.014	1.505	30.097	30.097
2	1.234	24.677	56.692	1.234	24.677	56.692	1.33	26.595	56.692
3	0.87	17.39	74.082						
4	0.713	14.268	88.35						
5	0.582	11.65	100						

Fig. 6 Scree plot

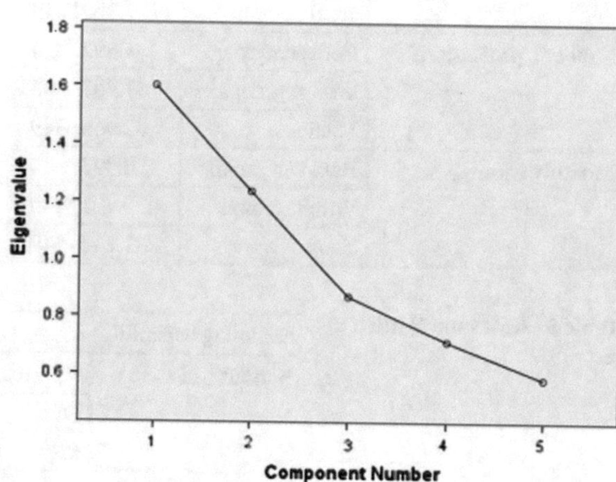

Table 6 Component matrix

Influence to buy	Component	
	1	2
Advertisement	**0.546**	**0.573**
Family member	**0.562**	0.081
Brand	**0.72**	−0.355
Self-experience	**0.628**	−0.484
Scheme/offer	0.274	**0.734**

Table 7 Rotated component matrix

	Component	
	1	2
Advertisement	0.177	0.772
Family member	0.441	0.357
Brand	0.8	0.063
Self-experience	0.787	−0.095
Scheme/offer	−0.14	0.771

4 Conclusion

This study is a special application of combination of unsupervised and supervised learning. The data is collected through the questionnaire from SRTM University hostel resident girls. It is observed that the proportion of urban users and rural users are same. The survey reveals that the cosmetic user students are independent of their faculty, viz., Arts, Commerce, Science, and other. Father's educational qualification and occupation affect the use of cosmetic products. The correlation between educational performance and the use of cosmetics is not significant. The most common type of makeup used is powder and in monsoon season it is most commonly used. In summer and winter season, the girls used lotion and creams. The hammering of an advertisement affects on buying the cosmetic product. Most of the hostel living girls having dry skin so, they used cosmetics. While purchasing cosmetic products, initially girls referred to scheme or discount offers on it or not. The data reveals that the use of the type of makeup and applicator brand depends on their pocket money. It shows the cost-saving tendency among girls. These results may be helpful to cosmetic manufacturing businessmen in terms of developing branded products with schemes and offers which should be affordable to all the girls and so that their sharing/borrowing tendency should be minimized. The limitation of this study is sample size restricted with university hostel resident girls. In the future, this work can be expanded at a large extent to get more precise results in terms of identification of cosmetic product buying behavior among girls according to a business development perspective.

Acknowledgements The authors would like to acknowledge Ms. K. B. Chaugule, Ms. S. H. Ghatage, Ms. A. S. Shinde, and Ms. S. S. Yelhekar of M.Sc. Statistics students, who helped during the collection of dataset.

References

1. Bearden, W.O., Etzel, M.J.: Reference group influence on product and brand purchase decisions. J. Consum. Res. **9**(2), 183–194 (1982)
2. Bhatt, K., Sankhla, P.: A study on consumer buying behavior towards cosmetic products. Int. J. Eng. Technol. Sci. Res. **4**(12), 1244–1249 (2017)
3. Bilal, A.I., Tilahun, Z., Osman, E.D., Mulugeta, A., Shekabdulahi, M., Berhe, D.F.: Cosmetics use-related adverse events and determinants among Jigjiga town residents, Eastern Ethiopia. Dermatol. Ther. **7**(1), 143–153 (2017)
4. Biswas, S., Hussain, M., O'Donnell, K.: Celebrity endorsements in advertisements and consumer perceptions: a cross-cultural study. J. Glob. Mark. **22**(2), 121–137 (2009)
5. Devi, S.R.: https://www.scribd.com/document/335759691/Determinants-of-Brand-Loyalty-of-Skin-Care-Products-in-Rural-Areas
6. Dunham, M.H.: Data Mining: Introductory and Advanced Topics. Pearson Education, India (2006)
7. Ilankizhai, R.J., Gayathri, R., Vishnupriya, V.: Cosmetic contamination awareness among adolescent females. Asian J. Pharm. Clin. Res. **9**(5), 117–120 (2016)
8. Ingavale, D.: Celebrity endorsement of cosmetics: a study of consumer's perception and buying preferences. Indian J. Appl. Res. **6**(2), 213–215 (2016)

9. Jawahar, J.V., Tamizhjyothi, K.: Consumer attitude towards cosmetic products. Int. J. Exclus. Manag. Res. **3**(6), 608–625 (2013)
10. Junaid, A.B., Nasreen, R., Ahmed, F.: A study on the purchase behavior and cosmetic consumption pattern among young females in Delhi and NCR. J. Soc. Dev. Sci. **4**(5), 205 (2013)
11. Khan, A.F., Khan, M.F.: A study on the awareness of product ingredients among women skincare users in state of Madhya Pradesh. IOSR J. Bus. Manag. **14**(4), 65–72 (2013)
12. Khraim, H.S.: The influence of brand loyalty on cosmetics buying behavior of UAE female consumers. Int. J. Mark. Stud. **3**(2), 123 (2011)
13. Knoll, J., Matthes, J.: The effectiveness of celebrity endorsements: a meta-analysis. J. Acad. Mark. Sci. **45**(1), 55–75 (2017)
14. Kumar, H.A.H., John, S.F., Senith, S.: A study on factors influencing consumer buying behavior in cosmetic products. Int. J. Sci. Res. Publ. **4**(9), 1–6 (2014)
15. Nair, V.K.: A study on purchase pattern of cosmetics among consumers in Kerala (2007)
16. Naz, S., Iqtedar, M., Ul Ain, Q., Aftab, K.: Incidence of human skin pathogens from cosmetic tools used in beauty saloons in different areas of Lahore, Pakistan. J. Sci. Res. **4**(2), 523 (2012)
17. Parmar, S.M.: A study of brand loyalty for cosmetic products among youth. Int. J. Res. Manag. Pharm. **3**(6), 9–21 (2014)
18. Raghu, G.: A Study on Consumer Awareness of Karachi Skincare Wet Wipes, 1–64 (2013)
19. Rahim, N.F.b., Shafii, Z., Shahwan, S.: Awareness and perception of Muslim consumers on halal cosmetics and personal care products. Int. J. Bus. Econ. Manag. **2**(1), 1–14 (2015)
20. Szagun, G., Pavlov, V.I.: Environmental awareness: a comparative study of German and Russian adolescents. Youth Soc. **27**(1), 93–112 (1995)
21. Yeo, B.L., Mohamed, R.H.N., Muda, M.: A study of Malaysian customers purchase motivation of Halal cosmetics retail products: examining theory of consumption value and customer satisfaction. Procedia Econ. Finance **37**, 176–182 (2016)
22. Yousaf, U., Zulfiqar, R., Aslam, M., Altaf, M.: Studying brand loyalty in the cosmetics industry. Log Forum **8**(4), 327–337 (2012)
23. Yuvaraj, S.: A study on the brand choice decisions of consumers with reference to cosmetics. Indian J. Appl. Res. **4**(6), 91–95 (2014)

Evolutionary Games for Cloud, Fog and Edge Computing—A Comprehensive Study

Ganesh Neelakanta Iyer

Abstract Evolutionary games combine game theory, evolutions and system dynamics to interpret the interactions of biological agents. It has been used in various domains to study the interaction of different agents in large populations. In such modelling, agents adapt (evolve) the chosen strategy based on their fitness (payoff). Evolutionary games have been used heavily to model various situations in computer science and engineering domains. In this work, I examine the effects of evolutionary game theory in modern utility computing systems such as cloud computing, fog computing and edge computing systems. They are used in those domains to study various research issues such as resource allocation, cloud service selection, deployment, etc. This study shows that evolutionary game theoretic principles are very useful for studying those issues.

Keywords Evolutionary games · Cloud computing · Edge computing · Fog computing · Deployment · Resource allocation

1 Introduction

Cloud Computing has evolved very quickly over the past decade as a disruptive paradigm which changed the way how software development has revolutionized [1, 2]. It changed the way someone develops, tests, deploys, sells and manages software and it changed how anyone buys and uses software and it also changed where the software has been deployed and delivered to customers. Some of these changes include getting software as a service without needing to be worrying about buying it and setting it up on a server, deploying the software in a shared public environment. Cloud resource vendors have to sort out several issues including efficient resource alloca-

G. N. Iyer (✉)
Department of Computer Science and Engineering, Amrita School of Engineering, Amrita Vishwa Vidyapeetham, Coimbatore, India
e-mail: ni_ganesh@cb.amrita.edu

H. S. Behera et al. (eds.), *Computational Intelligence in Data Mining*, Advances in Intelligent Systems and Computing 990,
https://doi.org/10.1007/978-981-13-8676-3_27

tion, security issues that are caused by sharing resources among several customers, software availability as per the service level agreement (SLA), etc.

Fog [3] and Edge computing [4] paradigms are emerged recently with Internet of Things (IoT) concepts to alleviate some of the issues raised with Cloud resources being remote. When the end nodes capture the data, they send it to the cloud node which involves some latency depending upon where the Cloud nodes are located. For mission critical applications requiring urgent responses, they can leverage the concepts such as fog and edge computing in order to reduce the latency involved.

On the other side, game theory [5, 6] has been used for solving several research issues in various domains including economics, politics, sports, biology, electronics and computer science. Game theory is a great aid to solve issues wherever there are conflicting situations and/or when someone needs to make decisions for social welfare and/or when somebody wants to enforce cooperation among a set of agents for a certain purpose. Hence game theory gained much popularity in solving real world issues in various domains.

Evolutionary game [7] in particular typically examines the behaviour of a large population of agents who engage in strategic interactions in a repeated fashion. As the name suggests, evolutionary biology largely uses evolutionary game theory principles where the primary idea is that an organism's genes typically determine its observable characteristics, and hence its fitness in a given environment. Organisms that are fitter tend to produce more offspring leading fitter genes to increase their presence in the larger population.

One of the most important points to remember in evolutionary game theory is that [8], many behaviours by agents participating in an interaction involve multiple organisms in the population, and the survival of any one of these organisms exclusively depends on how its behaviour interacts with the behaviour of other agents. Thus, someone can map analogy between natural non-cooperative games and evolutionary games. An organism's genetically determined characteristics and behaviours are like strategy in a non-cooperative game and fitness is analogous to payoff.

Given the nature of Cloud computing which is very dynamic and constantly evolving, evolutionary game theory is a power tool to analyze various issues in Cloud computing domain. There are a few decision-making processes in Cloud paradigm where evolutionary games are specifically seen applicable such as cloud resource allocation and cloud deployment. In this survey, I examine various ways evolutionary game theoretic principles are applied for different issues in Cloud computing.

Rest of this paper is organized as follows. Section 2 describes the basic principles of Evolutionary games, Sect. 3 describes different research issues in Cloud computing and Fog computing where evolutionary games are used, Sect. 4 outlines state-of-the-art work on evolutionary games for Cloud and Fog computing and finally Sect. 5 concludes this paper.

2 Evolutionary Games

The idea of evolutionary game [7] theory was first articulated by John Maynard Smith and G R Price in the early 1970s [9, 10]. It combines game theory, evolutions and system dynamics to interpret the interactions of biological agents [11]. As mentioned before, this concept has been used in various domains to study the interaction of different agents in large populations. In such modelling, agents adapt (evolve) the chosen strategy based on its fitness (payoff). It has several advantages [12] over traditional non-cooperative game theory:

- Nash equilibrium can be refined using Evolutionary Stable Strategies (ESS)
- Regular games assume strong rationality which may not be required in evolutionary games since they model the behaviour of biological agents
- Over time, the dynamics in nature can be modelled using evolutionary games which in turn can capture the adaption of players to change their strategies

Evolutionary games are games in which the game is repeatedly played by the players who are picked from a huge population. Evolutionary game has two components—mutation and selection. Mutation is a mechanism of modifying the characteristics of an agent (e.g. genes of the individual or strategy of player), and agents with new characteristics are introduced into the population. The selection mechanism is then applied to retain the agents with high fitness while eliminating agents with low fitness. In evolutionary game, from a static system point of view, evolutionary stable strategies (ESS)are used to describe mutation [13] and from a dynamic system point of view, the replicator dynamics describes the selection mechanism.

ESS is one of the most important concepts in the evolutionary process [13] in which when mutation has been applied, a set of players choosing a specific strategy will not be replaced by other players choosing a dissimilar strategy. The initial group of agents in a population choose incumbent strategy *s*. A small group of agents whose population share is ε choose a different mutant strategy *s'*. Strategy s is called evolutionary stable if Eq. (1) below is satisfied.

$$u\big(s, \epsilon s' + (1-\epsilon)s\big) > u(s', \epsilon s' + (1-\epsilon s), \quad (1)$$

where $u(s, s')$ denotes the payoff of strategy s given that the opponent chooses strategy s'.

Someone can divide population into multiple groups, and each group espouses a different pure strategy. Replicator dynamics model how the group size evolves over time [14] (unlike ESS, players will play only pure strategies in replicator dynamics). The proportion or fraction of agents using pure strategy s (i.e. population share) is denoted by xs(t) whose vector is x(t). Let payoff of an agent using strategy s given the population state x be denoted by u(s, x). Average payoff of the population, which is the payoff of an agent selected randomly from a population, is given by in Eq. (2):

$$\overline{u}(x) = \sum\nolimits_{s \in S} x_s u(s, x) \quad (2)$$

The reproduction rate of each agent (i.e. the rate at which the agent switches from one strategy to another) depends on the payoff (agents will switch to strategy that leads to higher payoff). Group size of agents ensuring higher payoff will grow over time because the agents having low payoff will switch their strategies. Dynamics (time derivative) of the population share can be described as in Eq. (3):

$$\dot{x}_s = x_s(u(s, x) - \bar{u}(x)) \tag{3}$$

Evolutionary equilibrium can be determined at $\dot{x}_s = 0$ where actions of the population choosing different strategies cease to change. It is important to analyze the stability of the replicator dynamics to determine the evolutionary equilibrium. Evolutionary equilibrium can be stable (i.e. equilibrium is robust to the local perturbation) in the following two cases:

(1) Given the initial point of replicator dynamics sufficiently close to the evolutionary equilibrium, the solution path of replicator dynamics will remain arbitrarily close to the equilibrium (Lyapunov stability [15])
(2) Given the initial point of replicator dynamics close to the evolutionary equilibrium, the solution path of replicator dynamics converges to the equilibrium (asymptotic stability)

Two main approaches to prove the stability of evolutionary equilibrium are based on the Lyapunov function and the eigenvalue of the corresponding matrix.

3 Cloud, Edge and Fog Computing

This section introduces some of the basic concepts of Cloud and Fog computing and then focuses on some of the relevant research issues in those domains where evolutionary games can be potentially used.

Cloud Computing has emerged as a disruptive technology which revolutionized the way someone develop, deploy and manage software. Essentially, it is based on the concepts of utility computing where resources are provided like utilities and anybody use the resources according to their need and finally pay according to the usage. This is very similar to how someone uses utilities such as electricity and water. Services offered through cloud can be of multiple types such as software, computing units, databases, etc.

One important issue for a cloud provider is allocating resources in the right way. It is important to efficiently allocate resources so as to minimize the running costs, maximize the resource utilization, etc. Another important and related issue to be considered is application deployment in cloud. Deployment is related to resource allocation in a way that ultimately deployment involves choosing the right set of resources on which the application can be deployed in the cloud.

From a cloud user perspective, an important issue to be addressed is in selecting the right cloud provider in order to meet certain objectives such as cost, deadline

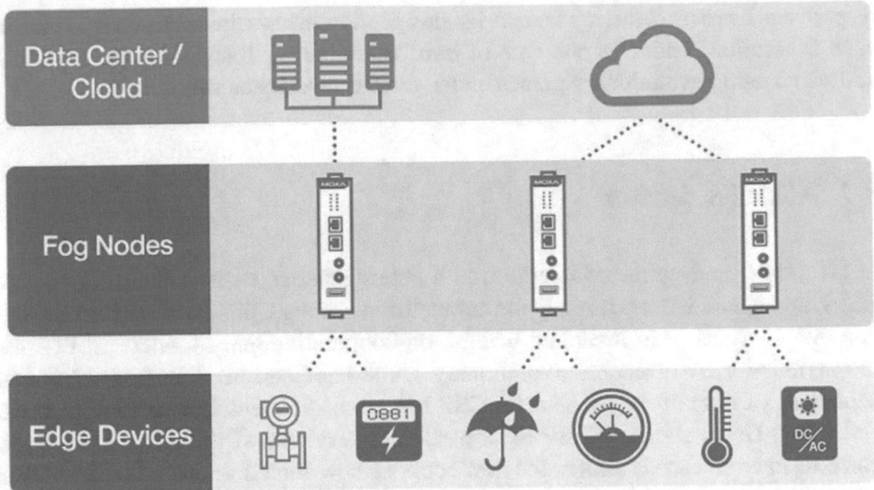

Fig. 1 Differences between cloud, edge and fog computing models [14]

requirements, etc. This also means that cloud providers need to come up with innovate and dynamic pricing strategies which can be adapted based on market conditions to maximize their revenue and resource utilization.

Other extreme is when the computation happens at the edge nodes itself where the data is being collected. This is called edge computing. In such systems, the latency has been minimized in calculations since data transfer latency is kept minimum by doing the computations at the edge itself. This is at a trade-off with the energy consumption at the edge nodes. When a user can afford to wait, or when the energy efficiency is very important, the computation needs to be offloaded to cloud systems or some other systems as such.

In the recent days, Fog computing is emerging to complement cloud in a way to offload the computation from edge devices to intermediate routers instead of sending them all the way to cloud systems. It is formulated as a mechanism to reduce the latency in data transmission from the edge devices to cloud systems and then back to the edge systems. These make me think of possible security attacks to those fog nodes. The differences between cloud, edge and fog computing models are depicted in Fig. 1.

4 Evolutionary Games for Cloud, Edge and Fog Computing Systems

In this section, I do a comprehensive survey of various research works that are done for cloud and fog computing using evolutionary games. The purpose is to explore and understand the usefulness of evolutionary games for various issues and how researchers have used them. I categorize this section into various subsections and in each subsection, I address one type of issue in cloud/fog computing that has been addressed using evolutionary games under various circumstances.

4.1 Cloud Selection

In [16], they study price competition in a heterogeneous market cloud formed by CSPs, brokers and users. Initially the competition among CSPs in selling the service opportunities has been modelled using Non-cooperative games where CSPs tries to maximize their revenues. Evolutionary game has been used to study dynamic behaviour of cloud users to select a CSP based on different factors such as price and delay. Users choose a CSP for a particular service and this service selection game repeats. In each iteration, the user adopts a new selection that minimizes their cost. This replicator dynamics helps players to learn about their environment and converges to equilibrium.

In [17], the same authors extend the same work to consider the presence of multiple cloud providers at the same time. They have used Wardrop equilibrium [18] concepts and replicator dynamics in order to calculate the equilibrium and for choosing the cloud service, they characterized its convergence properties. Some of the practical considerations they need to make in order to ensure their studies are really applicable include considering multiple cloud service delivery models as against a single service they have considered, SLA agreements possibly exist between users and providers and finally, operational costs in addition to pricing strategies which leads to considering optimized resource allocation to reduce operating costs.

4.2 Deployment in Cloud Environments

In [19] for VM deployment under objectives such as energy efficiency, budget and deadline, evolutionary games are used. The initial optimal solutions are evolved over time to minimize the loss in performance and existence of Nash equilibrium under certain conditions are specified. They design an evolutionary game mechanism which can change the multiplexed stratagems of the original optimal elucidations of different participants with minimizing their performance losses.

The works in [20–22] use evolutionary game theory to deploy a set of applications in a set of hosts based on certain performance objectives such as CPU and bandwidth availability, response time, power consumption, etc. For each application, they maintain a population of deployment strategies and over time, they reach a steady state which is evolutionarily stable. The most important objectives these works look at are the adaptability and stability. Adaptability means that based on given performance objectives, the applications should be deployed maximizing the resource usage. Stability means that the adaptation decision should have minimum oscillations.

In [23], authors use evolutionary game principles for application deployment in cloud environments. Their mechanism lets the applications to adapt resource allocation and their locations based on workload and resource availability in the cloud. They show that applications perform an evolutionary stable deployment strategy considering certain objectives such as response time. Two important application deployment properties they considered are adaptability and stability.

4.3 *Security Issues in Fog Computing*

As described earlier, in fog computing, the computation offloading happens from the edge devices to intermediate devices such as routers in order to find a trade-off between latency and energy efficiency. When computation is performed at the fog nodes, they are more susceptible to security vulnerabilities due to their diverse and distributed nature. In [24], the authors analyze the security issues in such an environment using evolutionary games. Replicator dynamics are used to understand the behavioural strategy selection. They show that when normal nodes show cooperative strategy, the malicious nodes are forced to show 'non-attack' strategy.

4.4 *Optimal Sensor Configuration in Edge Computing*

The work in [25] uses game theory for configuring Body Sensor Networks (BSNs) to be used with Cloud based on operational conditions which depends on different constraints such as resource consumption and data yield. Their concept is based on a layered architecture where cloud provider has virtual sensors and physical sensors are operated through their cloud-based virtual counterparts. They primarily use evolutionary games to study fine-tuning sensing intervals and selection rates for sensors. Similarly, to study data broadcast intervals for sensors and sink nodes. This helps them adjust BSN configurations according to operational conditions and also based on performance objectives such as energy and bandwidth consumptions. Further to adaptability, they study stability in minimizing oscillation decisions in adaption decisions.

4.5 *Resource Allocation in Cloud Systems*

While performing computation offloading for mobile edge computing, two important aspects to be considered are the cloud resource allocation and the wireless resource scheduling. This has been addressed in [26] using evolutionary games. They consider different factors when calculating the offloading time delay. Similarly, they consider several factors such as energy consumption, monetary cost and time delay when calculating the overhead for mobile edge computation. Joint cloud and wireless resource allocation problem has been modelled using evolutionary games.

A Quality of Service (QoS) constrained Cloud resource allocation has been modelled in [27, 28] using evolutionary games. They consider different parameters such as computation time, budget, etc. in their evolutionary optimization and finds Nash Equilibrium. They initially solved the using binary integer programming method and then use evolutionary game principles for getting a socially optimal solution.

A part of the work in [29] uses evolutionary game. They present a framework for resource allocation among the cloud, mobile social users and broker. They use evolutionary games for studying the mobile social user's behaviour. The utility of the mobile user depends on the price of the user and the processing rate of the allocated resource.

5 Conclusions

Evolutionary games are used to model various research issues in multi-disciplinary areas including engineering, biology, etc. In this paper, I studied various research issues in cloud, fog and edge computing which are addressed using the powerful techniques of evolutionary games. I understood that evolutionary game theoretic principles are very useful in tackling issues such as resource allocation, security issues, cloud resource selection, application deployment and more. A detailed comparison has been performed (Refer Table 1) on all the existing research works.

Table 1 Evolutionary games for cloud, edge and fog computing—table of comparison

Work	Type of issue	Basic concept	System	Objectives	Limitations
[16]	Cloud selection	Price selection in heterogeneous market cloud	Cloud	Price, delay	Study is mostly on a duopoly setup
[17]	Cloud selection	Price selection in the presence of multiple cloud providers	Cloud	Price, delay	Multiple service delivery models, SLA agreements and operations costs need to be considered

(continued)

Table 1 (continued)

Work	Type of issue	Basic concept	System	Objectives	Limitations
[19]	VM deployment	Optimal VM deployment based on several performance objectives	Cloud	Energy efficiency, budget, deadline	
[20–22]	Application deployment	Deploy a set of applications in a set of hosts based on certain performance objectives. They study adaptability and stability	Cloud	CPU and bandwidth availability, response time, power consumption	
[23]	Application deployment	Help applications to choose their locations and in allocating resources based on different characteristics	Cloud	Response time	Several other objectives are important such as energy efficiency and price
[24]	Security	Security issues in fog computing environments.	Fog	Consumption cost, profit from attacks	Lack real performance studies
[25]	Sensor configuration	Configuring BSNs to be used with Cloud based on operational conditions with respect to different constraints	Edge, Cloud	Resource consumption, data yield	
[26]	Computational offloading	Computational offloading for mobile edge computing	Edge, Cloud	Time, energy consumption, monetary cost	

(continued)

Table 1 (continued)

Work	Type of issue	Basic concept	System	Objectives	Limitations
[27, 28]	Resource allocation	QoS constrained cloud resource allocation	Cloud	Budget, computation time	
[29]	Resource allocation	Allocating resources within the cloud mobile social users and broker	Cloud	Price, Processing rate	

References

1. Prakash, P., Darshaun, K.G., Yaazhlene, P., Ganesh, M.V., Vasudha, B.: Fog computing: issues, challenges and future directions. Int. J. Electr. Comput. Eng. **7**(6), 3669–3673 (2017)
2. Niyato, D., Hossain, E., Issariyakul, T.: An adaptive WiFi/WiMAX networking platform for cognitive vehicular networks. In: Yu, F. (ed.) Cognitive Radio Mobile Ad Hoc Networks. Springer, New York, NY (2011)
3. Shi, W., Cao, J., Zhang, Q., Li, Y., Xu, L.: Edge computing: vision and challenges. IEEE Internet Things J. **3**(5), 637–646 (2016). https://doi.org/10.1109/JIOT.2016.2579198
4. Aliprantis, C.D., Chakrabarti, S.K.: Games and Decision Making. Oxford Press (2010)
5. Nissan, N., et al.: Algorithmic Game Theory. Cambridge University Press (2007)
6. Samuelson, L.: Evolutionary Games and Equilibrium Selection, vol. 1, 1st edn., number 0262692198. MIT Press Books, The MIT Press (1998)
7. Shevitz, D., Paden, B.: Lyapunov stability theory of nonsmooth systems. IEEE Trans. Autom. Control **39**(9), 1910–1914 (1994). https://doi.org/10.1109/9.317122
8. Easley, D., Kleinberg, J.: Networks, Crowds, and Markets: Reasoning About a Highly Connected World. Cambridge University Press, New York (2010)
9. Smith, J.M.: On Evolution. Edinburgh University Press, California (1972). ISBN: 0852242298
10. Price, G.R., Smith, J.M.: The logic of animal conflict. Nature **246**, 15–18 (1973). https://doi.org/10.1038/246015a0 (5427) (USA)
11. Rees, T.: An Introduction to Evolutionary Game Theory. Department of Computer Science, UBC. Technical Report, UBC, Canada (2004)
12. Farzaneh, N., Yaghmaee, M.H.: An adaptive competitive resource control protocol for alleviating congestion in wireless sensor networks: an evolutionary game theory approach. Wirel. Pers. Commun. (2015)
13. Anastasopoulos, N.P., Anastasopoulos, M.P.: The evolutionary dynamics of audit. Eur. J. Oper. Res. (2012)
14. Should you consider Fog Computing for your IIOT?, whitepaper, Nov 2017. https://www.moxa.com/newsletter/connection/2017/11/feat_02.htm
15. Ozdaglar, A., Menache, I.: Network games: theory, models, and dynamics. In: Network Games: Theory, Models, and Dynamics, vol. 1, pp. 159–. Morgan & Claypool (2011)
16. Do, C.T., et al.: Toward service selection game in a heterogeneous market cloud computing. In: IFIP/IEEE International Symposium on Integrated Network Management (IM) (2015)
17. Do, C.T., Tran, N.H., Huh, E.-N., Hong, C.S., Niyato, D., Han, Z.: Dynamics of service selection and provider pricing game in heterogeneous cloud market. J. Netw. Comput. Appl. **69** (2016) (Elsevier)
18. Sangeetha, K.S., Prakash, P.: Big data and cloud: a survey. Adv. Intell. Syst. Comput. **325**, 773–778 (2015)

19. Han, K., Cai, X., Rong, H.: An evolutionary game theoretic approach for efficient virtual machine deployment in green cloud. In: 2015 International Conference on Computer Science and Mechanical Automation (CSMA), pp. 1–4. IEEE, Hangzhou (2015)
20. Ren, Y., Suzuki, J., Vasilakos, A., Omura, S., Oba, K.: Cielo: an evolutionary game theoretic framework for virtual machine placement in clouds. In: 2014 International Conference on Future Internet of Things and Cloud, Barcelona, pp. 1–8. IEEE (2014). https://doi.org/10.1109/ficloud.2014.11
21. Ren, Y., Suzuki, J., Lee, C., Vasilakos, A.V., Omura, S., Oba, K.: Balancing performance, resource efficiency and energy efficiency for virtual machine deployment in DVFS-enabled clouds: an evolutionary game theoretic approach. In: Proceedings of Conference on Genetic and Evolutionary Computation (GECCO Comp '14), pp. 1205–1212
22. Cheng, Y.R.: Evolutionary Game theoretic multi-objective optimizaion algorithms and their applications. Computer Science, University of Massachusetts Boston (2017)
23. Lee, C., et al.: An evolutionary game theoretic approach to adaptive and stable application deployment in clouds. In: Proceedings of the 2nd Workshop on Bio-inspired Algorithms for Distributed Systems (BADS '10), pp. 29–38
24. Sun, Y., Lin, F., Zhang, N.: A security mechanism based on evolutionary game in fog computing. Saudi J. Biol. Sci. **25**(2), 237–241 (2018). ISSN 1319-562X (ScienceDirect, Saudi Arabia)
25. Ren, Y.C., et al.: An evolutionary game theoretic approach for configuring cloud-integrated body sensor networks. In: 2014 IEEE 13th International Symposium on Network Computing and Applications, pp. 277–281, Cambridge, MA
26. Wei, G., Vasilakos, A.V., Zheng, Y., Xiong, N.: A game-theoretic method of fair resource allocation for cloud computing services. J. Supercomput. **54**(2), 252–269 (2010) (Springer)
27. Wei, G., Vasilakos, A.V., Xiong, N.: Scheduling parallel cloud computing services: an evolutional game. In: First International Conference on Information Science and Engineering (2009)
28. Su, Z., Xu, Q., Fei, M., Dong, M.: Game theoretic resource allocation in media cloud with mobile social users. IEEE Trans. Multimed. **18**(8), 1650–1660 (2016). https://doi.org/10.1109/tmm.2016.2566584 (IEEE, USA)
29. Hwang, K., Dongarra, J., Fox, G.C.: Distributed and Cloud Computing: From Parallel Processing to the Internet of Things, 1st edn. Morgan Kaufmann Publishers Inc., San Francisco, CA, USA (2011)

ACRRFLN: Artificial Chemical Reaction of Recurrent Functional Link Networks for Improved Stock Market Prediction

Sarat Chandra Nayak, Koppula Vijaya Kumar and Karthik Jilla

Abstract The intrinsic capability of functional link artificial neural network (FLANN) to recognize the complex nonlinear relationship present in the historical stock data made it popular and got wide applications for stock market prediction. Contrasting to multilayer neural networks, FLANN uses functional expansion units to expand the input space into higher dimensions, thus generating hyperplanes which offer better discrimination capability in the input space. The feedback properties of recurrent neural networks make them more proficient and dynamic to model nonlinear systems accurately. Artificial chemical reaction optimization (ACRO) requires less number of tuning parameters with faster convergence speed. This article develops an ACRO-based recurrent functional link neural network (RFLN) termed as ACRRFLN, in which optimal structure and parameters of a RFLN are efficiently searched by ACRO. Also two evolutionary optimization techniques, i.e., particle swarm optimization (PSO) and genetic algorithm (GA) are employed to train RFLN separately. All the models are experimented and validated on forecasting stock closing prices of five stock markets. Results from extensive simulation studies clearly reveal the outperformance of ACRRFLN over other models similarly trained. Further, the Deibold-Mariano test justifies the statistical significance of the proposed model.

Keywords Functional link artificial neural network · Recurrent neural network · Artificial chemical reaction optimization · Stock market prediction · Recurrent functional link neural network · Particle swarm optimization · Genetic algorithm · Differential evolution

---

S. C. Nayak (✉) · K. V. Kumar · K. Jilla
Department of Computer Science & Engineering, CMR College of Engineering & Technology (Autonomous), Hyderabad 501401, India
e-mail: saratnayak234@gmail.com

K. V. Kumar
e-mail: vijaykoppula@gmail.com

K. Jilla
e-mail: jilla.karthik@gmail.com

H. S. Behera et al. (eds.), *Computational Intelligence in Data Mining*, Advances in Intelligent Systems and Computing 990,
https://doi.org/10.1007/978-981-13-8676-3_28

1 Introduction

Stock market behavior is highly unpredictable due to the influence of uncertainties, nonlinearities, high volatility, and discontinuities associated with it. Also, it is highly affected by the movement of other stock markets, political scenarios, general economic situation of a country, bank interest rate, exchange rates, macroeconomical factors as well as individual psychology, even sometimes rumor. Prediction of such irregular and highly fluctuating stock data is very much complex and typically subject to large error. Hence, more accurate and efficient forecasting methods are intensely desirable for financial judgment building process of speculators, financial managers, naive investors, etc. A realistic precise forecast of stock market may possibly capitulate sky-scraping financial benefits [1, 2]. Several statistical models have been discussed in the literature and used to study the pattern of stock movement. These were simple and widely used by mathematicians and statisticians. However they suffered from several shortcomings while dealing with stock data which is highly nonlinear by nature.

Artificial neural networks (ANN) are the computational intelligence systems which mimic the human brain's learning process. It can emulate the process of human's actions to resolve complex real-life problems. They are considered to be an effective modeling practice where input–output mapping contains regularities as well as exceptions. Their superior learning and approximation ability makes them appropriate for successful application toward a wide range of financial forecasting problems which includes volatility prediction, index and exchange rate prediction, credit scoring, bankruptcy, interest rate and return, portfolio management, etc.

Among the ANN-based methods, multilayer perceptrons (MLP) have been used frequently for stock market prediction. They have inherent capacity to approximate any nonlinear function with high degree of accuracy [3, 4]. ANN-based forecasting models necessitate many number of layers and abundance of neurons in each layer of the network hence suffer from computational complexity [5]. Also, it suffers from the limitations like black box nature, overfitting and gets trapped in local minima [6]. To overcome these limitations, Pao [7] proposed a single layer ANN with less complex architecture called FLANN. It overcomes the need of hidden layers by incorporating functional expansion units. These expansion units help to promote the low dimensional input space into a high dimension, thus generating hyperplanes which then provide superior discrimination capability in the input pattern. FLANN-based models have been well experimented in the literature. Few of these applications include pattern classification and recognition [8–10], system identification and control [11, 12], and stock market forecasting [13–15]. A variety of hybrid FLANNs using orthogonal trigonometric functions for data classification and stock market prediction have been discussed in the literature [15–18].

To prevail over the shortcomings of gradient descent-based learning, a good number of nature-inspired learning algorithms such as GA, memetic algorithm (MA), ant colony optimization (ACO), bee colony optimization (BCO), PSO, DE, and harmony search (HS), etc. have been proposed in the literature and their applications have grown incredibly fast. However, their efficiency mainly depends on fine-tuning different learning parameters. In order to land at the global optimum, an algorithm requires suitable selection of parameters which ultimately makes the use difficult. Therefore, selecting a suitable optimization technique for solving a problem involves numerous trial and error methods. This paves a path toward choosing an optimization technique requiring fewer learning parameters without compromising the approximation capability. This is the motivation behind choosing ACRO for obtaining optimal RFLN-based forecasting model.

The objective of this work is to build a robust forecasting model with an improved accuracy. Firstly, a RFLN is developed as the feedback properties make it more proficient and dynamic to model nonlinear systems accurately. Secondly, for choosing the optimal RFLN, ACRO has been employed hence forming a hybrid model called ACRRFLN. The proposed model is validated on predicting next day's closing indices of five real stock markets such as NASDAQ, DJIA, BSE, TAIEX, and FTSE.

The rest of the article is organized as follows: an introduction about the problem area and literature are presented in Sect. 1. Section 2 talks about the methods and methodologies. Proposed ACRRFLN method is explained in Sect. 3. Section 4 presents the experimental results and analysis. The concluding remarks are presented in Sect. 5 followed by a list of references.

2 Methods and Methodologies

This section describes shortly about the base model, i.e., FLANN and evolutionary search techniques such as PSO, GA, and ACRO.

2.1 FLANN

FLANN is a higher order neural network which avoids the use of hidden layers by introducing higher order effects through nonlinear functional transforms via links. The architecture consists of a single layer feed forward network. The functional expansion of input signals increases the dimensionality of the input space hence generating hyperplanes. These hyperplanes provide larger discrimination potential in the input pattern space. Attributes of each input sample arepassed through the expansion unit consisting of several nonlinear basis functions. The weighted sum of the output signals of these units is then passed on to the sigmoid activation at the output neuron to produce an estimate. The error signal is calculated by comparing this estimate with the actual output and used to train the model. The optimal parameters of the model can be obtained iteratively based on the training samples and using the principle of gradient descent learning algorithm. The model estimation is calculated as follows:

We denote a set of basis function f and weight set W to represent an approximating function $\varphi_w(X)$. The problem is to find the optimal weight set which can be obtained through updating it iteratively that can give the best possible approximation of f on the set of input–output sample. Let there be k training pattern denoted by $\langle X_i : Y_i \rangle, 1 \leq i \leq k$ to be applied. Let at ith instant $(1 \leq i \leq k)$, the N-dimensional input- output pattern be given by $X_i = \langle x_{i1}, x_{i2}, \ldots, x_{iD} \rangle, \widehat{Y_i} = [\widehat{y_i}], 1 \leq i \leq k$. On applying a set of basis function f, the input space dimension increased from N to N' and represented as:

$$\begin{aligned} f(X_i) &= [f_1(x_{i1}), f_2(x_{i1}), \ldots, f_1(x_{i2}), f_2(x_{i2}), \ldots, f_1(x_{iN}), f_2(x_{iN}), \ldots,] \\ &= [f_1(x_{i1}), f_2(x_{i2}), \ldots, f_N(x_{iN})] \end{aligned} \tag{1}$$

$W = [W_1, W_2, \ldots, W_k]^T$ is a $k \times N$ dimensional matrix where W_i is the weight vector associated with the ith output and given by $W_1 = [w_{i1}, w_{i2}, \ldots, w_{iN}]$. The ith output of the model is given by

$$\widehat{y_i}(t) = \varphi\left(\sum_{j=1}^{N'} f_j(x_{ij}) * w_{ij}\right) = \varphi\big(W_i * f^{\mathrm{T}}(X_i)\big), \forall i \tag{2}$$

The error signal associated with the output is calculated by Eq. 3.

$$e_i(t) = y_i(t) - \widehat{y_i}(t) \tag{3}$$

The weight is then updated using adaptive learning rule as in Eq. 4.

$$\begin{aligned} &w_{ij}(t+1) = w_{ij}(t) + \mu * \Delta(t), \\ &\Delta(t) = \delta(t) * [(x_i)] \end{aligned} \tag{4}$$

where $\delta(t) = [\delta_1(t), \delta_2(t), \ldots, \delta_k(t)]$, $\delta_i(t) = \left(1 - \widehat{y_i}^2(t) * e_i(t)\right)$ and μ is the learning parameter.

2.2 ACRO

ACRO is a meta-heuristic proposed by Lam and Li [19] inspired by natural phenomena of a chemical reaction. The concept mimics properties of a natural chemical reaction and slackly couples mathematical optimization techniques with it. A chemical reaction is a natural phenomenon of transforming unstable chemical substances to stable ones through intermediate reactions. A reaction starts with unstable molecules with excessive energy. The molecules interact with each other through a sequence of elementary reactions and producing some products with lower energy. During a chemical reaction the energy associated with a molecule changes with the change in intra-molecular structure and becomes stable at one point, i.e., equilibrium point. Termination condition is checked by performing chemical equilibrium (inertness) test. If the newly generated reactant has better function value, it is included and the worse reactant is excluded else a reversible reaction is applied. Several applications of ACRO for classification and financial time series prediction are found in the literature [13, 19, 20].

2.3 PSO

PSO is another nature-inspired meta-heuristic technique designed by mimicking the simulated social behavior of swarms [21] which is capable to find the global best solution. PSO starts with a set of randomly generated initial swarms or particles each representing a candidate solution in the search space. Each particle moves in the search space according to an adaptable velocity associated with it. It memorizes the best position of the search space visited so far. At each generation, the best solution can be found by adjusting the route of each particle (i) toward its best location and (ii) toward the best particle of the population [21]. Individuals of a swarm communicate their information and adjusting position and velocity using their cluster information according to the best information appeared in the recent movement of the swarm. In this way, the initial solution propagates through the search space and gradually moves toward the global optimum over a number of generations. The standard PSO algorithm consists of computational steps as follows: (1) initialization of particles' positions and velocities, (2) updating the position and velocity of each particle.

2.4 GA

Genetic algorithms are popular global search technique works on a population of potential solutions. The individual of GA is represented in the form of chromosomes. The optimal solution can be obtained through the process of artificial evolution. GA is based on mimicking genetic evolutionary theory to solve optimization problems with encoding parameter rather than the parameter itself. It consists of repeated genetic operations such as *evaluation*, *selection*, *crossover*, and *mutation*. For application of GA in financial prediction prospective readers may follow articles in literature [22–26].

3 Proposed ACRRFLN

Recurrent FLANN (RFLN) developed here is a supervised machine learning model made of FLANN and one feedback loop. The feedback loop is a recurrent cycle over sequence. The feedback enables the FLANN to remember past signals for a long time. The recurrent connection learns the dependencies among the time series data. The RFLN model maps an input–output sequence at the current time step followed by predicting the sequence in the next time step. The lagged output samples are used as input therefore helping to sustain a correlation connecting the training samples. The RFLN model is shown in Fig. 1. With the hope of providing enhanced forecasting accuracy, one step delayed output sample is fed to the input layer. To expand the original input signals into higher dimensions, trigonometric basis functions are used here. An input value x_i expanded using seven expansion functions such as:

$$\begin{aligned} c_1(x_i) &= (x_i), c_2(x_i) = sin(x_i), c_3(x_i) = cos(x_i), \\ c_4(x_i) &= sin(\pi x_i), c_5(x_i) = cos(\pi x_i), \\ c_6(x_i) &= sin(2\pi x_i), c_7(x_i) = cos(2\pi x_i). \end{aligned} \tag{5}$$

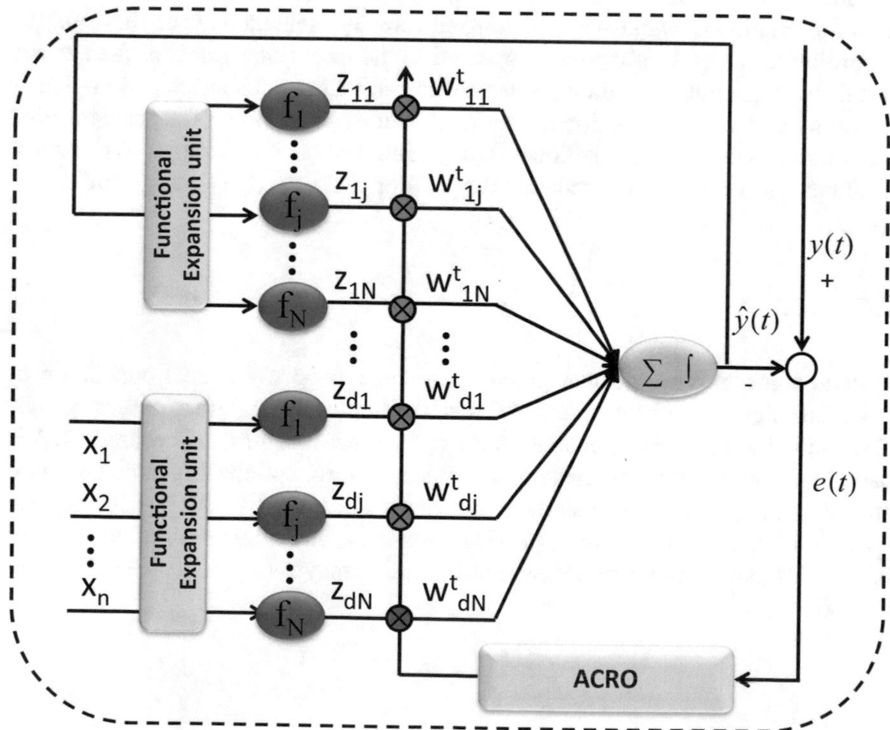

Fig. 1 ACRRFLN-based forecasting

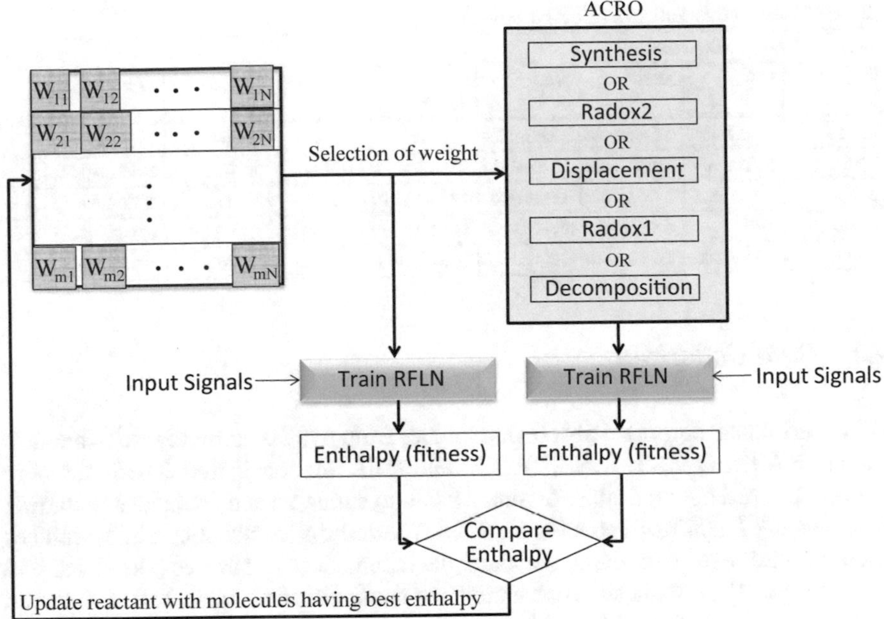

Fig. 2 ACRRFLN training process

The model combines the power of ACRO for selecting optimal RFLN to diminish the local optimality as well as accelerate the convergence rate. Here, the weights are encoded into individuals for stochastic search as the performances of the RFLN mostly rely on weight. A potential weight vector for the RFLN model is encoded as a binary string which represents a molecule. As the model is intended to forecast one value at a time, there is a single neuron at the output layer. In this way, a reactant pool represents a set of potential solutions in the search space. Input signals (closing prices) with the molecule value are fed to a set of RFLN models to produce an estimate. The absolute difference between the desired and the estimated output is considered as the fitness of an individual. The less the error signal or enthalpy of an individual, better is its fitness. The process of ACRRFLN is shown in Fig. 2.

4 Experimental Study and Result Analysis

This section describes data collection, performance metrics, input selection, preprocessing of data particularly normalization, results and finally conducts the significance test.

Table 1 Descriptive statistics from five stock indices

Stock index	Descriptive statistics						
	Min.	Max.	Mean	Std. deviation	Skewness	Kurtosis	JB test statistics
NASDAQ	1.1152e+003	4.5982e+003	2.3855e+003	709.7888	1.0392	4.0027	764.3863(h=1)
DJIA	5.9470e+003	1.7148e+004	1.1400e+004	2.1801e+003	0.6644	3.0512	254.8134(h=1)
BSE	792.1802	1.1024e+004	4.6234e+003	2.6944e+003	0.1155	1.7908	237.0430(h=1)
FTSE	3286.80	6.8765e+003	5.4155e+003	835.2382	−0.2837	2.1378	157.4568(h=1)
TAIEX	3.4463e+003	1.0202e+004	6.9835e+003	1.4846e+003	−0.1775	2.0435	159.9786(h=1)

4.1 Data Collection

For experimental purpose daily closing indices from five fast-growing stock markets such as NASDAQ, BSE, FTSE, TAIEX, and DJIA are considered. The indices are collected from the source https://in.finance.yahoo.com/ for each financial day starting from January 1, 2003 to September 12, 2016. The descriptive statistics of these indices such as minimum, maximum, standard deviation, mean, skewness, kurtosis, and Jarque-Bera (JB) statistics are calculated and summarized in Table 1. It is observed that DJIA, NASDAQ, and BSE data sets have shown positive skewness values, i.e., spread out more toward the right which suggests investment opportunities in these markets.

To select the training and test patterns for the forecasting models a sliding window technique is used. A window of fixed size (five in this experiment) is moved on the series by one step each time. In each move a new pattern has been formed which can be used as an input vector. The size of the window can be decided experimentally. The original input measures are normalized by sigmoid method [26] as follows:

$$x_{norm} = \frac{1}{1 + e^{-\left(\frac{x_i - x_{min}}{x_{max} - x_{min}}\right)}} \tag{6}$$

where x_{norm} and x_i are the normalized and original price respectively. x_{max} and x_{min} are the maximum and minimum price of the window respectively.

4.2 Performance Metrics

The Mean Absolute Percentage Errors (MAPE), Average Relative Variance (ARV), and coefficient of determination (R^2) are used as the performance metrics whose formulas are as follows:

$$\mathrm{MAPE} = \frac{1}{N}\sum_{i=1}^{N}\frac{|x_i - \hat{x}_i|}{x_i} \times 100\% \tag{7}$$

$$\mathrm{ARV} = \frac{\sum_{i=1}^{N} \left(\hat{x}_i - x_i\right)^2}{\sum_{i=1}^{N} \left(\hat{x}_i - \overline{X}\right)^2} \tag{8}$$

$$R^2 = 1 - \frac{SS_{res}}{SS_{tot}} \tag{9}$$

where $\mathrm{SS_{res}}$ is the sum of squares of residuals = $SS_{res} = \sum_{i=1}^{N} \left(x_i - \hat{x}_i\right)^2$ and $\mathrm{SS_{tot}}$ is the total sum of squares and calculated as $SS_{tot} = \sum_{i=1}^{N} \left(x_i - \overline{X}\right)^2$. In these measures x_i and $\hat{x}_i$ are the actual and estimated prices respectively. N is the total number of test data.

4.3 Result and Analysis

The input data after normalized are used to form a training bed for the network. To ascertain the performance of the ACRRFLN, five other forecasting models such as back propagation neural network (BPNN), RBFNN, PSO-based RFLN (PSO-RFLN), GA-based RFLN (GA-RFLN), and gradient descent RFLN (RFLN-GD) are developed and considered for comparative analysis. The same training and test sets are fed to all six models and respective error signals are recorded. A model that generates lower error signal is considered as a better forecasting model. Since the neural network-based models are stochastic in nature, to avoid that each model is simulated 20 times considering each training dataset. The average of 20 simulations is considered as the performance of that model. The results obtained from the next day's closing price prediction are summarized in Table 2. The best value for error statistics is in shown in bold face. It can be noticed that except few cases, the proposed ACRRFLN generates better error statistic value for all data sets. Figures 3, 4, 5, 6, 7 show the plots for the actual verses predicted closing prices by ACRRFLN for all data sets. For the sake of visibility these prices are plotted for the first fifty financial days only.

4.4 Deibold-Mariano (DM) Test

We then conducted a pairwise comparison, i.e., DM test for statistical significance of the proposed forecasting model. We represent the actual time series as $\{y_t; t = 1, \ldots, T\}$ and the two forecasts are $\{\hat{y}_{1t}; t = 1, \ldots, T\}$ and $\{\hat{y}_{2t}; t = 1, \ldots, T\}$, respectively. The objective is to test how far the forecasts are different. The DM statistic is calculated using the formula as in Eq. 10.

$$\mathrm{DM} = \frac{\overline{d}}{\sqrt{\frac{\hat{\gamma}_d(0) + 2\sum_{k=1}^{h-1} \hat{\gamma}_d(k)}{T}}} \tag{10}$$

Table 2 Performance of all models from next day's closing price prediction

Stock index	Error statistic	Forecasting models					
		ACRRFLN	PSO-RFLN	RFLN-GD	GA-RFLN	RBFNN	BPNN
NASDAQ	MAPE	0.003692	0.007403	0.025260	0.006855	0.151035	0.361844
	ARV	0.015162	0.015527	0.017283	0.041226	0.017285	0.127507
	R^2	0.998821	0.999323	0.998325	0.998450	0.998141	0.994480
DJIA	MAPE	0.004233	0.006014	0.052176	0.021172	0.055028	0.223566
	ARV	0.015125	0.015520	0.017281	0.017281	0.041226	0.045088
	R^2	0.996755	0.994532	0.994092	0.995604	0.994150	0.985415
BSE	MAPE	0.005233	0.005972	0.012106	0.006739	0.013962	0.056577
	ARV	0.031708	0.031705	0.064166	0.021809	0.055795	0.072848
	R^2	0.995604	0.994532	0.994538	0.996707	0.993580	0.985411
TAIEX	MAPE	0.009431	0.011225	0.011263	0.012109	0.009661	0.052175
	ARV	0.028300	0.034841	0.077374	0.071336	0.500322	0.532609
	R^2	0.995066	0.994625	0.990590	0.992382	0.989771	0.989126
FTSE	MAPE	0.050733	0.054152	0.053155	0.064590	0.053158	0.380919
	ARV	0.068826	0.062908	0.210245	0.073685	0.076126	0.270600
	R^2	0.995945	0.992873	0.992555	0.993240	0.992552	0.992320

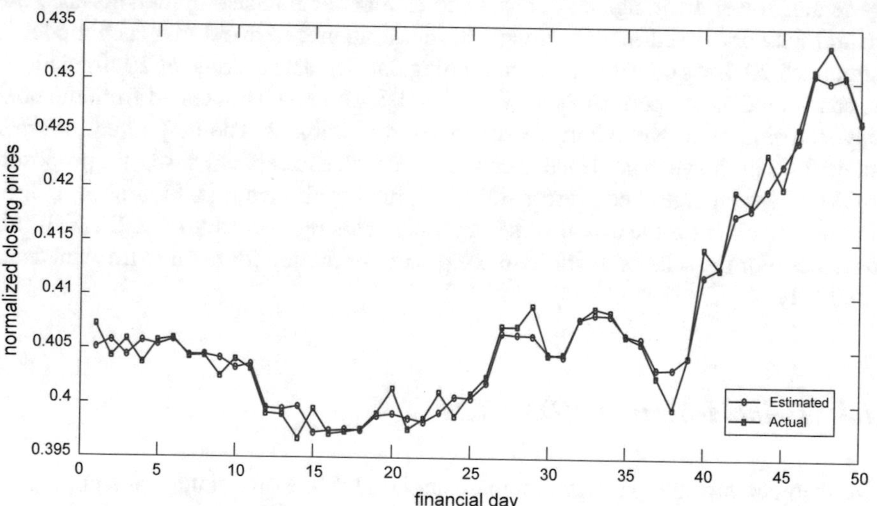

Fig. 3 Predicted versus actual prices from DJIA

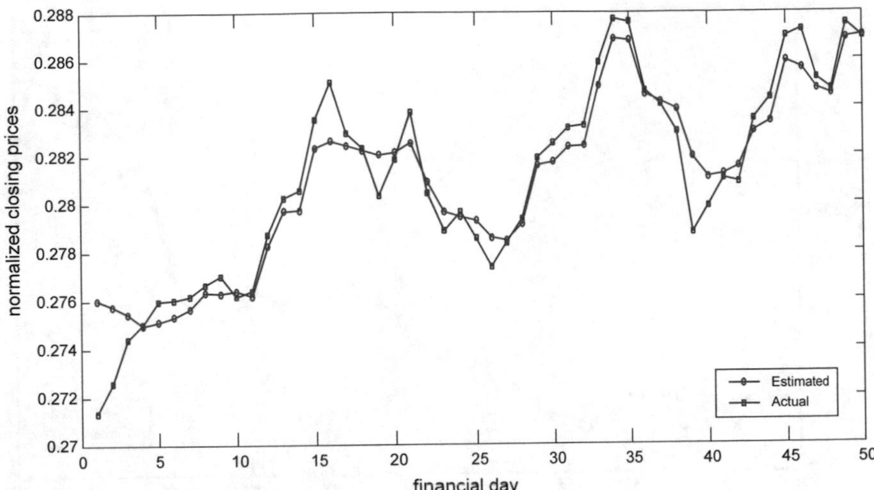

Fig. 4 Predicted versus actual closing prices from BSE

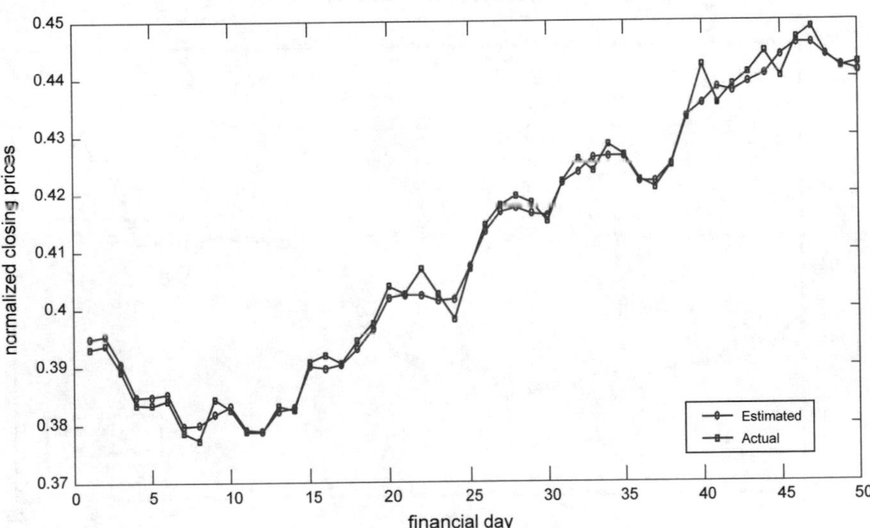

Fig. 5 Predicted versus actual closing prices from NASDAQ

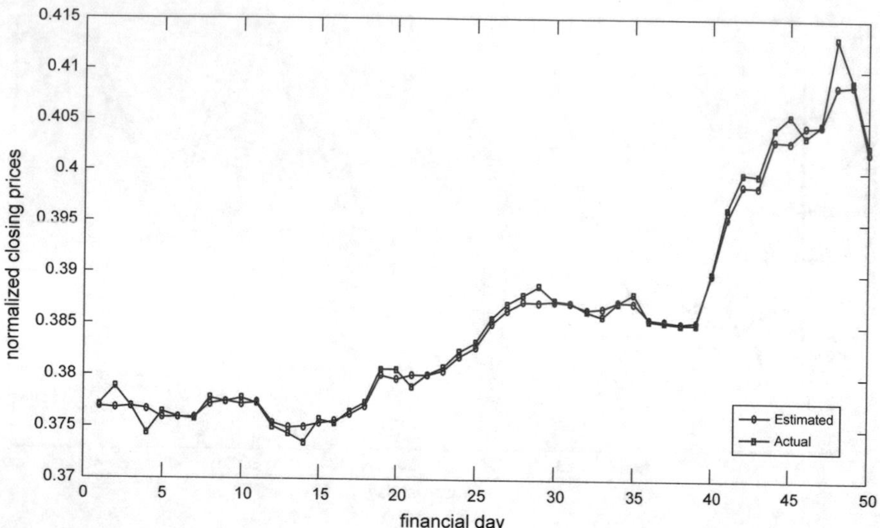

Fig. 6 Predicted versus actual closing prices from TAIEX

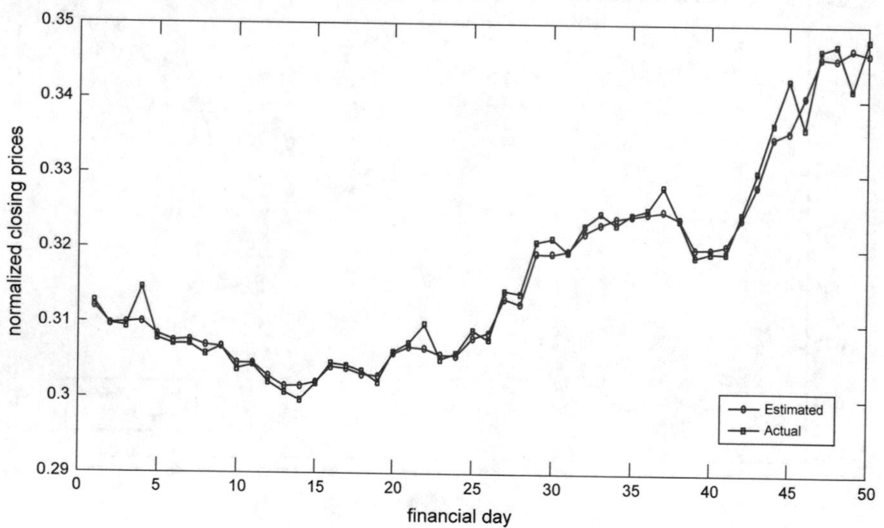

Fig. 7 Predicted versus actual closing prices from FTSE

Table 3 Computed DM statistics from five financial time series

Indices	ACRRFLN	PSO-RFLN	RFLN-GD	GA-RFLN	RBFNN	BPNN
NASDAQ		2.0152	1.9945	−2.3345	2.0003	−2.7543
DJIA		1.9087	2.5533	3.1517	−4.1286	−3.1265
BSE		2.5066	2.8164	−2.4545	3.1090	3.6316
TAIEX		−3.4212	−2.2659	1.9975	−2.5533	−3.4137
FTSE		−2.6473	2.2575	−2.4552	2.7677	−3.6808

where $\overline{d}$ is the sample mean of the loss differential, *h* is the forecast horizon, $\widehat{\gamma}_d(k)$ is an estimate to the auto-covariance of the loss differential $\gamma_d(k)$ at lag k. If $|DM| > z_{\alpha/2}$, the null hypothesis of no difference will be rejected. Considering the significance level of the test $\alpha = 0.05$, the lower critical z-value corresponds to −0.0255 is −1.965 and upper critical z-value corresponds to 0.9755 is 1.965. Therefore if the computed statistic is either smaller than −1.965 or greater than +1.965, the null hypothesis of no difference will be rejected. The DM statistics considering six models and five datasets are computed and summarized in the Table 3. It can be noted that the computed DM statistic values for all cases are lying outside the critical range. These observations are in support of rejection of null hypothesis.

5 Conclusions

This article proposes a hybrid forecasting model termed as ACRRFLN, incorporating the proficiency and dynamism of an RFLN with faster convergence and lesser tunable parameter of ACRO. The model exhibits less computationally complex architecture and achieves better forecasting accuracy. The efficiency of the proposed model is tested on predicting one-day-ahead closing price of five real stock market indices. It is found competitive to five other neural network as well as hybrid models and quite capable to capture the uncertainties and dynamism coupled with the stock markets. The overall prediction accuracy of ACRRFLN is found much better than others. Also, it is found statistically significant through conducting the DM test. The work can be extended by using other neural network-based models and application to other data mining problems.

References

1. Barak, S., Modarres, M.: Developing an approach to evaluate stocks by forecasting effective features with data mining methods. Expert Syst. Appl. **42**(3), 1325–1339 (2015)
2. Ballings, M., Van den Poel, D., Hespeels, N., Gryp, R.: Evaluating multiple classifiers for stock price direction prediction. Expert Syst. Appl. **42**(20), 7046–7056 (2015)

3. Yu, L.Q., Rong, F.S.: Stock market forecasting research based on neural network and pattern matching. In: 2010 International Conference on E-Business and E-Government (ICEE), pp. 1940–1943. IEEE (2010, May)
4. Tahersima, H., Tahersima, M., Fesharaki, M., Hamedi, N.: Forecasting stock exchange movements using neural networks: a case study. In: 2011 International Conference on Future Computer Sciences and Application (ICFCSA), pp. 123–126. IEEE (2011, June)
5. Dutta, G., Jha, P., Laha, A.K., Mohan, N.: Artificial neural network models for forecasting stock price index in the Bombay stock exchange. J. Emerg. Market Financ. **5**(3), 283–295 (2006)
6. Huang, F.Y.: Integration of an improved particle swarm algorithm and fuzzy neural network for Shanghai stock market prediction. In: Workshop on Power Electronics and Intelligent Transportation System, 2008. PEITS'08, pp. 242–247. IEEE (2008, August)
7. Pao, Y.H., Takefuji, Y.: Functional-link net computing: theory, system architecture, and functionalities. Computer **25**(5), 76–79 (1992). Mishra, B.B., Dehuri, S.: Functional link artificial neural network for classification task in data mining. J. Comput. Sci. **3**, 948–955 (2007)
8. Mishra, B.B., Dehuri, S., Panda, G., Dash, P.K.: Fuzzy swarm net (FSN) for classification in data mining. CSI J. Comput. Sci. Eng. **5**(2&4 (b)), 1–8 (2008)
9. Dehuri, S., Cho, S.B.: A hybrid genetic based functional link artificial neural network with a statistical comparison of classifiers over multiple datasets. Neural Comput. Appl. **19**(2), 317–328 (2010)
10. Majhi, R., Majhi, B., Panda, G.: Development and performance evaluation of neural network classifiers for Indian internet shoppers. Expert Syst. Appl. **39**(2), 2112–2118 (2012)
11. Purwar, S., Kar, I.N., Jha, A.N.: On-line system identification of complex systems using Chebyshev neural networks. Appl. Soft Comput. **7**(1), 364–372 (2007)
12. Majhi, R., Panda, G., Sahoo, G.: Development and performance evaluation of FLANN based model for forecasting of stock markets. Expert Syst. Appl. **36**(3), 6800–6808 (2009)
13. Nayak, S.C., Misra, B.B., Behera, H.S.: ACFLN: artificial chemical functional link network for prediction of stock market index. Evolving Systems, pp. 1–26 (2018)
14. Nayak, S.C., Misra, B.B., Behera, H.S.: Comparison of performance of different functions in functional link artificial neural network: a case study on stock index forecasting. In: Computational Intelligence in Data Mining-Volume 1, pp. 479–487. Springer, New Delhi (2015)
15. Patra, J.C., Thanh, N.C., Meher, P.K.: Computationally efficient FLANN-based intelligent stock price prediction system. In: International Joint Conference on Neural Networks, 2009. IJCNN 2009, pp. 2431–2438. IEEE (2009, June)
16. Dehuri, S., Roy, R., Cho, S.B., Ghosh, A.: An improved swarm optimized functional link artificial neural network (ISO-FLANN) for classification. J. Syst. Softw. **85**(6), 1333–1345 (2012)
17. Mili, F., Hamdi, M.: A hybrid evolutionary functional link artificial neural network for data mining and classification. In: 2012 6th International Conference on Sciences of Electronics, Technologies of Information and Telecommunications (SETIT), pp. 917–924. IEEE (2012, March)
18. Lam, A.Y., Li, V.O.: Chemical-reaction-inspired metaheuristic for optimization. IEEE Trans. Evol. Comput. **14**(3), 381–399 (2010)
19. Alatas, B.: A novel chemistry based metaheuristic optimization method for mining of classification rules. Expert Syst. Appl. **39**(12), 11080–11088 (2012)
20. Nayak, S.C., Misra, B.B., Behera, H.S.: Artificial chemical reaction optimization of neural networks for efficient prediction of stock market indices. Ain Shams Eng. J. (2015)
21. Kennedy, J.: Particle swarm optimization. In: Encyclopedia of Machine Learning, pp. 760–766. Springer, US (2011)
22. Nayak, S.C., Misra, B.B., Behera, H.S.: Index prediction with neuro-genetic hybrid network: a comparative analysis of performance. In: 2012 International Conference on Computing, Communication and Applications (ICCCA), pp. 1–6. IEEE (2012, February)
23. Nayak, S.C., Misra, B.B., Behera, H.S.: An adaptive second order neural network with genetic-algorithm-based training (ASONN-GA) to forecast the closing prices of the stock market. Int. J. Appl. Metaheuristic Comput. (IJAMC) **7**(2), 39–57 (2016)

24. Nayak, S.C., Misra, B.B., Behera, H.S.: Efficient financial time series prediction with evolutionary virtual data position exploration. Neural Comput. Appl., 1–22 (2017)
25. Nayak, S.C., Misra, B.B., Behera, H.S.: On developing and performance evaluation of adaptive second order neural network with GA-based training (ASONN-GA) for financial time series prediction. In: Advancements in Applied Metaheuristic Computing, pp. 231–263. IGI Global (2018)
26. Nayak, S.C., Misra, B.B., Behera, H.S.: Impact of data normalization on stock index forecasting. Int. J. Comp. Inf. Syst. Ind. Manag. Appl. **6**, 357–369 (2014)

Graph-Based Crime Reports Clustering Using Relations Extracted from Named Entities

Priyanka Das and Asit Kumar Das

Abstract The proposed work demonstrates a graph-based clustering approach for partitioning crime reports using extracted relations from named entities present in a crime corpus of India. The proposed relation extraction scheme initiates with choosing different types of entity pairs followed by computing similarity among the named entity pairs depending on the intermediate context words. A graph (weighted and undirected) is formed with entity pairs representing the nodes and similarity between entities as the weight of the edges linked to the nodes. The clustering coefficient is calculated for all the nodes and the average clustering coefficient has been set as the threshold to partition the graph. Iterative application of the proposed clustering algorithm generates multiple clusters comprising named entity pairs. Finally, the newly formed clusters got labelled according to the most frequent context word of the cluster which results in analysing few aspects of crime performed against women in India. The result with high cluster validation indices illustrates the effectiveness of the proposed method.

Keywords Entity recognition · Relation extraction · Graph-based clustering · Clustering coefficient

1 Introduction

Crime pattern analysis via relation extraction is one of the applications of information extraction task. Relation extraction considers a sentence of two named entities with some intervening context words providing an insight on the relationship between the entities. Say, a sentence '**Kamal** has been accused of killing **Sima**', the

P. Das (✉) · A. K. Das
Computer Science and Technology, Indian Institute of Engineering Science and Technology, Shibpur, Howrah 711103, West Bengal, India
e-mail: priyankadas700@gmail.com

A. K. Das
e-mail: akdas@cs.iiests.ac.in

H. S. Behera et al. (eds.), *Computational Intelligence in Data Mining*, Advances in Intelligent Systems and Computing 990,
https://doi.org/10.1007/978-981-13-8676-3_29

relational tuple is considered as $< U, x, V >$, where 'x' being the intervening context words, express the relation of U (Kamal) with V (Sima). Domain specific knowledge helps in improvement of relation recognition task. The 6th Message Understanding Conference (MUC-6) introduced seven basic named entities like, PER (person), ORG (organisation), LOC (location), TIME, DATE, MONEY and GPE (geo-political entity). Sekine et al. [1] observed that heterogeneous data contain many subtypes of the existing named entities and acknowledging them proved to be quite beneficial for relation detection task. Weir et al. [2] explored newspapers for obtaining named entities. Brin et al. [3] introduced a scheme called 'DIPRE' (Dual Iterative Pattern Relation Expansion), which involved bootstrapping for extracting relations from the World Wide Web. The associated demerits of this method were solved in 'Snowball' [4] involving similar elementary concepts of DIPRE and discovered novel methods for pattern extraction. An unsupervised approach presented in [5] considered The New York Times (1995) newspaper corpus for identifying the named entities and used a hierarchical clustering for discovering the relations. But this approach performed the experiment with newspaper articles for one year, which failed to extract less frequent relations between the entity pairs and consecutively could not find paraphrases from them. Basili et al. [6] introduced 'REVEAL', a system that incorporated the use of support vector machine (SVM) for extracting relations for crime inspection. Arulanandam et al. [7] considered three different newspapers from three different countries and executed the task of crime location extraction by using the conditional random field model.

The present work proposes a graph-based crime reports clustering by determining relations of the named entities from a crime corpus containing crime information of Indian states and union territories. The crime dataset for the present experiment has been crawled from electronic version of some classified newspapers. Several named entities like persons, locations (states, districts, cities, geo-political entities), organisations have been recognised from the dataset by using an available named entity tagger. Then relations among the named entities have been discovered by a top-down hierarchical graph-based clustering technique. PER-PER (person-person), PER-LOC (person-location), ORG-PER (organisation-person) and PER-ORG (person-organisation) domain of entity pairs are chosen for envisioning the crime trend. Now, cosine similarity factor (based on the context words) provides a similarity score for each entity pair. Then the similarity score has been used to generate a weighted undirected graph. The newly generated graph contains the entity pairs as the nodes and the computed cosine similarity score is allotted as the weights of connected edges. This primarily generated graph has been considered as a single partition and clustering coefficient has been calculated for all the nodes present in the single partitioned graph. The average value of the clustering coefficient is set as the threshold score. Two different subgraphs have been created based on the threshold. The first subgraph contains the nodes having equal and more clustering coefficient than the threshold, whereas the second subgraph comprises the nodes with clustering coefficient below the threshold. The resultant disconnected graph may be the collection of several components which has been applied separately as input to the next iteration of the clustering algorithm. The threshold has been updated individually for each component and they are further partitioned into more compact subgraphs. At each level of iteration, several cluster validation indices have been measured and the process continues only if the cluster quality improves. Finally, the graph clustering algorithm helps in the formation of multiple clusters comprising the entity pairs. Then each cluster has been labelled with the most frequent context word present in it. The clusters of PER-PER domain have been labelled with crime terms, such as, for example, 'kidnap', 'dowry harassment', 'murder', 'rape', etc. Similarly, the clusters of PER-LOC domain have been labelled with the context words describing the social status of the victim/offender. The clusters from ORG-PER domain are labelled with the terms relating to the actions taken by the court or police against a criminal involved in crime and the clusters of PER-ORG domain shows the steps taken by a person victimised in crime. The clusters have been evaluated by using several external and internal cluster validation indices.

The remaining part of paper is organised as follows: Sect. 2 discusses the proposed method elaborately. The result is being described in Sect. 3 and finally, Sect. 4 provides conclusion and future work.

2 Present Methodology

The primary goal of the proposed work involves recognition of the relations existing between the named entities from the crime corpus using a graph-based clustering technique. The full information about the proposed methodology is given in the following subsections.

2.1 Experimental Data Collection and Data Preprocessing

A couple of classified e-papers like 'The Times of India', 'The Hindu' and 'The Indian Express' have been chosen for collecting the news reports on crime against women in Indian states and in its union territories. A customised Python-based site crawler has been used to crawl the news reports from the above-mentioned websites. For crawling, terms related to crime like 'rape', 'kidnap', 'abuse', etc. have been used as the keywords and reports containing any of the keywords have been crawled from the e-papers. The extracted data is based on different crimes committed against women in India and comprises a total of 140,110 reports for 29 states and 7 union territories of India for over a time period of 2008–2016. It is observed that the crawled reports provide information based on cities, districts, etc. The collected reports are based on the crimes that took place all over India and hence, it is quiet necessary to sort the reports according to the corresponding states where the crimes occurred. Google Map's Geocoding API has been used to arrange all the crime reports according to their corresponding states.

All the stop words present in the crime corpus have been removed followed by stemming that obtains the root words discarding the prefixes and suffixes. All the preprocessing steps have been achieved using the Natural Language Tool Kit (NLTK) module in Python [8].

2.2 *Entity Recognition and Measuring Similarity Among Entities*

The proposed work has recognised the named entities for further relation extraction scheme. The preprocessed data is split into multiple sentences. Each word in those sentences are considered as tokens and parts-of-speech tagging is done for each word. All of these steps have been performed using the NLTK's default sentence segmenter, word tokenizer and parts-of-speech tagger respectively. Next, noun phrase chunking has been done that searches for the noun phrases present in a sentence.

The named entities identified are paired with each other that facilitate the crime analysis task. The common words shared between the entities (context words) reflect their interrelationship and hence, entity pairs with at least four intermediate context words have been chosen for the present work because less number of context words will fail to recognise the relation properly. The four different domain of entity pairs chosen for the present work are PER-PER, PER-LOC, ORG-PER and PER-ORG. For each entity pair, relation of the first one has been determined with the latter. A context vector has been created for each entity pair and values of the vector have been filled up by computing the term frequency and inverse document frequency (TF-IDF). Say, for example, C be the collection of crawled crime reports, w be a shared word between the entities and $c \in C$ be an individual crime report, then the TF-IDF score of the shared word w in the report c is computed using (1).

$$w_c = f_{w,c} * \log\left(\frac{|C|}{|\{c \in C : w \in c\}|}\right) \tag{1}$$

where $f_{w,c}$ is the frequency of w in report c and $|\{c \in C : w \in c\}|$ is the number of reports in which the word w appears.

Depending on the context words, the *Cosine Similarity Factor* has been computed for entities p and q using (2). It basically compares the context of say, one ORG-PER pair with another ORG-PER pair and same for other entity pairs of different domains.

$$Sim_{pq} = \frac{p \cdot q}{||p||||q||} \tag{2}$$

Based on this cosine similarity score, a weighted undirected graph $G = (N, E, W)$ is constructed, where N defines the set of nodes (entity pairs) present in G, E is the set of edges and W defines the weights of the edges. More the similarity score between p and q, more the weight assigned to their corresponding edges. Cosine similarity factor ranges in [0, 1] where, 1 expresses the entity pairs having the most similar context words and 0 reflects least similar words.

2.3 Graph-Based Clustering Algorithm for Relation Extraction

Here, a graph-based clustering algorithm has been developed by using the clustering coefficient of the nodes or entity pairs present in the graph. Clustering coefficient of a node in any graph defines how well the node can form a group. For unweighted graphs, the clustering coefficient of a particular node namely ‘A’, is the fraction of possible triangles formed in the surroundings of that node. But for weighted graphs, the clustering coefficient is defined as the geometric average of the subgraph edge weights which is being measured using (3).

$$c_A = \frac{1}{deg(A)(deg(A)-1))} \sum_{AB} \left(\hat{w}_{AB}\hat{w}_{AC}\hat{w}_{BC}\right)^{1/3} \tag{3}$$

where, the edge weight w_{AB} for edge (A, B) is normalised to $\hat{w}_{AB}$ by the maximum weight in the network as $\hat{w}_{AB} = w_{AB}/\max(w)$. The value of c_A is assigned 0 if $deg(A) < 2$.

Upon constructing the complete graph $G = (N, E, W)$, the clustering coefficient has been calculated for all the nodes present in G using (3) and initially, this graph G is considered as a single partition. Again, the average clustering coefficient is calculated using (4) and it has been considered as a threshold t.

$$c_{avg} = \frac{1}{m}\sum_{i=1}^{m} c_i \tag{4}$$

where m is the number of nodes present in the graph G.

Based on the threshold t, the initial complete graph has been partitioned into two subgraphs:

1. $G_1 \rightarrow$ subgraph with nodes having clustering coefficient at least threshold t
2. $G_2 \rightarrow$ subgraph with nodes having clustering coefficient below the threshold t

Here, G_1 and G_2 may be disconnected graphs and thus after the application of this clustering algorithm, a disconnected graph with many connected components has been obtained and there may also exist many isolated nodes. Thus, after this step, each connected component including isolated vertices is considered as a cluster. Then the components have been provided as the input to the next iteration of the clustering algorithm. Again, the threshold of average clustering coefficient has been calculated for every component of the graph and accordingly they have been partitioned into more compact subgraphs.

Algorithm 1: Graph based clustering for relation extraction from entity pairs

```
Input: Undirected graph G = (N, E, W),
where N = Node set (i.e. set of entity pairs),
E = Edge set and
W = Weight set of the edges in E;
Output: Clusters C_new of Entity Pairs;
begin
    for all nodes n ∈ N do
        Calculate the clustering coefficient of the weighted graph G using (3)
    end
    C_old ← G, //The whole graph is a single partition//;
    Calculate cluster validation indices Dunn_old, Dav_old, Xie_old and Sil_old for cluster C_old
    using (5)–(8);
    repeat
        C_new ← ϕ;
        while C_old is not empty do
            Remove next graph from C_old and name it as G;
            for all nodes n ∈ N(G) do
                Calculate the average clustering coefficient c_avg using (4);
            end
            //Assign average clustering coefficient to a threshold t//
            t ← c_avg;
            //Partition graph G into two subgraphs G1 and G2//
            G1 ← ϕ, //empty graph without any node and edge//
            G2 ← ϕ;
            for every node n ∈ N do
                if c_n ≥ t then
                    G1 = G1 ∪ {n}
                end
                else
                    G2 = G2 ∪ {n}
                end
            end
            Consider each component of G1 and G2 as a new subgraph;
            C_new = C_new ∪ {allsubgraphs};
        end
        Calculate cluster validation indices Dunn_new, Dav_new, Xie_new and Sil_new for
        cluster C_new;
        if C_new is better than C_old with respect to all four indices then
            C_old ← C_new;
            C_new ← ϕ;
            Dunn_old ← Dunn_new; Dav_old ← Dav_new;
            Xie_old ← Xie_new; Sil_old ← Sil_new ;
        end
        else
            return C_new
        end
    until C_new is a null graph;
    //Null graph is the graph with all isolated vertices//
end
```

The method is mainly a top-down hierarchical approach, where after each iteration new group of clusters are formed from the previous group of clusters. After every level of clustering, various cluster validation indices are measured and next level of iteration is continued only if the cluster quality improves.

In the paper, we have terminated the process when the maximum values for Dunn and Silhouette indices and minimum value for Davis–Bouldin and Xie-Beni indices are achieved for a group of clusters. The definitions of used cluster validation indices are given in Sect. 3.1. The index values show that the resultant clusters contain pairs of named entities having contextual similarities. The methodology of entity pair clustering is described in Algorithm 1.

2.4 *Cluster Labelling*

The frequency of each context word present in the entity pairs of each cluster has been calculated and as a result the clusters have been labelled with the most frequent context word. The clusters of PER-PER domain are labelled with different crime types like 'rape', 'abduction', etc. Clusters from PER-LOC domain are labelled as the social status of the victim/offender. Likewise, the clusters from ORG-PER domain are labelled by the terms relating to the actions taken by the court or police against a criminal involved in crime and PER-ORG for steps taken by a person victimised in crime.

3 Experimental Results

The proposed work has performed the experiment on Indian crime reports existing for a time period of 2008–2016. There were 3,244, 3,135, 2,246 and 2,018 number of entity pairs for PER-PER, PER-LOC, ORG-PER and PER-ORG domain, respectively. The proposed graph -based clustering technique has generated several clusters of named entity pairs. Though many isolated clusters have been formed, but we have considered the clusters with at least 15 entity pairs in them. Table 1 shows number of newly generated clusters and number of clusters considered for further processing. Figure 1 shows the original graph and some of the resulting subgraphs formed by the proposed graph-based clustering technique. This figure has been generated by considering 100 entity pairs from PER-PER domain as total of 3,244 pairs are difficult to visualise by a plot.

The newly generated clusters are composed of the entity pairs and the words shared between the entity pairs establish their interrelationship. Now, a distinct context word with the maximum count is selected for labelling the cluster. This method has been repeated for relational labelling of all clusters. Tables 2 and 3 shows the distribution of the pairs and their corresponding relational labelling for the four different domains. The relational labelling task caters various aspects of crime analysis. It not only

Table 1 Number of clusters formed and considered by proposed graph-based clustering technique

Domain	Clusters formed	Clusters considered
PER-PER	30	12
PER-LOC	25	10
ORG-PER	21	10
PER-ORG	23	6

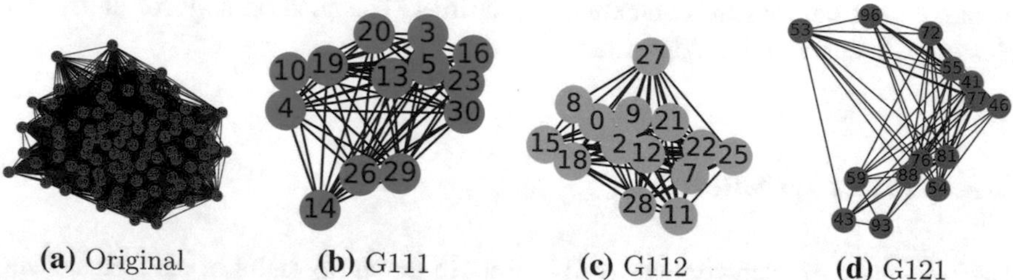

(a) Original **(b)** G111 **(c)** G112 **(d)** G121

Fig. 1 Original graph and some subgraphs generated by the proposed graph based clustering for PER-PER named entity pairs

Table 2 Number of entity pairs for relation labelling in PER-PER and PER-LOC domain

Type	NE Pairs	Type	NE Pairs
Murder	270	Domestic violence	276
Rape	348	Acid attack	152
Molestation	285	Abuse	238
Kidnap	189	Human trafficking	125
Sexual harassment	219	Street harassment	137
Dowry death	332	Foeticide	128
Teacher	282	Bishop	247
Driver	326	Labourer	318
Student	343	Techie	176
Housewife	274	Party leader	215
Doctor	211	Businessman	253

focuses on the crime types but also emphasises on the social status of the victims or actions taken by both the victims and governmental organisations for prevention of the crime. This analysis part holds the main significance for the proposed relation detection scheme.

Table 3 Number of entity pairs for relation labelling in ORG-PER and PER-ORG domain

Type	NE Pairs	Type	NE Pairs
Arrested	315	Investigation	321
Convicted	287	Order probe	153
Penalty	244	Seize property	161
Death sentence lifetime	185	Order DNA test produce	119
Imprisonment	236	Chargesheet	107
Lodge complaint	347	Counselling	152
Bail	278	Medical care	269
Seek justice	261	Financial help	255

Table 4 Results for internal cluster evaluation indices in (%)

Domain	Dunn	Dav	XB	Sil
PER-PER	82	36	45	77
PER-LOC	77	52	49	74
ORG-PER	73	45	48	78
PER-ORG	62	64	45	75

3.1 Evaluation

The clusters representing various crime aspects have been assessed by the ground truth clusters obtained from domain experts. The internal cluster evaluation techniques like Dunn Index (Dunn), Davis-Bouldin Index (Dav), Xie-Beni Index (XB) and Silhouette Index (Sil) have been used for evaluating the effectiveness of the proposed framework. These indices are defined in (5)–(8) and Table 4 provides the result for internal cluster evaluation indices used in the present framework. This result provides the insight on how good the clusters have been formed. It is observed that Dunn index provides the best result for PER-PER domain, whereas, Silhouette index provides good result almost in all cases.

$$Dunn = \min_{1 \le a \le u} \left\{ \min \left\{ \frac{d(x_a, x_b)}{\max_{1 \le k \le u}(d(x_k))} \right\} \right\} \tag{5}$$

where, $d(x_a, x_b)$ is the inter-cluster distance between the clusters x_a and x_b, $d(x_k)$ is the intra-cluster distance of cluster x_k and u is the number of clusters. Higher values of Dunn index indicate good clustering.

$$Dav = \frac{1}{u} \sum_{a=1}^{u} \max_{a \ne b} \left\{ \frac{d(x_a) + d(x_b)}{d(x_a, x_b)} \right\} \tag{6}$$

where, a and b are cluster labels, u is the number of clusters. $d(x_a)$ and $d(x_b)$ are intra-cluster distance of clusters X_a and X_b respectively and inter-cluster distance $d(x_a, x_b)$ between clusters x_a and x_b is measured as the distance between the cluster centroids. The minimum value for DB index denotes good clustering.

$$XB = \frac{1}{u} \frac{W}{\min_{a<b} d(x_a, x_b)} \tag{7}$$

where, W is the mean of the squared distances of all the points. The minimum value of Xie-Beni index provides better cluster evaluation.

$$Sil = \frac{(b-a)}{\max(a, b)} \tag{8}$$

where, a is the mean intra-cluster distance and b is the mean nearest-cluster distance for each sample. The maximum value of Silhouette index is a requirement for well-formed clusters.

4 Conclusion and Future Work

The proposed methodology develops a graph based clustering technique for recognising interrelationship of named entities from crime reports. To the best of the authors' knowledge this is for the first time a graph based method has been devised for extracting relations from a crime corpora. Significant relations among the named entity pairs have been identified that helps in crime analysis. Four different aspects of crime performed against women in India are brought into light by this experimental research work. Apart from the chosen domain of entity pairs, other different domains can be considered as future research work. Improvisations in the methodology will further provide a vast description of crime related activities by exploring other aspects of crime pattern analysis and eventually it will help the law enforcement agencies analyse the crime at a faster pace.

References

1. Sekine, S., Sudo K., Nobata, C.: Extended named entity hierarchy. In: Third International Conference on Language Resources and Evaluation (LREC-2002), pp. 1818–1824 (2002)
2. Weir, G., Anagnostou, N.: Exploring newspapers: A case study in corpus analysis. In: ICTATLL Workshop (2007)
3. Brin, S.: Extracting patterns and relations from the world wide web. In: Selected Papers from the International Workshop on The World Wide Web and Databases, pp. 172–183 (1999)
4. Agichtein, E., Gravano, L.: Snowball: extracting relations from large plain-text collections. In: Proceedings of the Fifth ACM Conference on Digital Libraries, pp. 85–94 (2000)

5. Hasegawa, T., Sekine, S., Grishman, R.: Discovering relations among named entities from large corpora. In: Proceedings of the 42nd Annual Meeting on Association for Computational Linguistics, ACL '04, p. 415. Association for Computational Linguistics, Stroudsburg, PA, USA (2004)
6. Basili, R., Giannone, C., Del Vescovo, C., Moschitti, A., Naggar, P.: Kernel-based relation extraction for crime investigation. In: AI*IA, pp. 161–171. Citeseer (2009)
7. Arulanandam, R., Savarimuthu, B.T.R., Purvis, M.A.: Extracting crime information from online newspaper articles. In: Second Australasian Web Conference (AWC 2014), vol. 155, pp. 31–38 (2014)
8. Loper, E., Bird, S.: NLTK: the natural language toolkit (2002). arXiv:cs.CL/0205028

Spectral–Spatial Active Learning Techniques for Hyperspectral Image Classification

Sushmita Rajbanshi, Kaushal Bhardwaj and Swarnajyoti Patra

Abstract Designing active learning (AL) techniques to determine informative training samples for hyperspectral image (HSI) classification is an open research issue. In this chapter, several spectral–spatial AL techniques are designed by exploiting both, an interesting approach to fuse spectral and spatial information and different combinations of existing uncertainty and diversity criteria. In order to fuse spectral and spatial information, the dimensionality of HSI is reduced and mean filtering (for incorporation of spatial information) is applied to each component image in the reduced domain of HSI considering multiple windows of different sizes. The filtered images are concatenated with original component images to form an extended spatial profile for the HSI. These spectral–spatial features are used with different combinations of uncertainty and diversity criteria to design several spectral–spatial AL techniques for determining most informative pixels. The experiments carried out on two real HSI data sets show that the spectral–spatial AL methods are more robust than the AL methods based on spectral measurements only.

Keywords Hyperspectral image · Mean filtering · Active learning · Support vector machine · Spectral-spatial classification

1 Introduction

Hyperspectral Images are represented by hundreds of spectral channels, which provide rich information for accurate recognition of different materials on the earth's surface. The analysis of HSI is useful for several applications like land cover/uses,

S. Rajbanshi · K. Bhardwaj · S. Patra (✉)
Tezpur University, Tezpur 784028, Assam, India
e-mail: swpatra@tezu.ernet.in

S. Rajbanshi
e-mail: rajbanshisushmita16@gmail.com

K. Bhardwaj
e-mail: kauscsp@tezu.ernet.in

H. S. Behera et al. (eds.), *Computational Intelligence in Data Mining*, Advances in Intelligent Systems and Computing 990,
https://doi.org/10.1007/978-981-13-8676-3_30

agriculture, inland water, mineralogy, surveillance, etc. [7]. Classification of HSI is a challenging task since it has limited availability of labeled samples. Labeling a pixel requires manual inspection or photo interpretation, which makes it costly in terms of time and effort. To address this problem, two possible approaches exist in the literature: one is semi-supervised learning and another is active learning (AL). Semi-supervised approach uses both labeled and unlabeled patterns for training to obtain a better discriminating function [4, 6]. On the other hand, AL techniques queried most informative samples from unlabeled pool to train the classifier. It is an iterative process, where in each iteration the most informative unlabeled samples (single most or batch of informative samples) are selected for manual labeling using a query function and the classifier is retrained with the additional labeled samples. This process is continued until a stable classification result is obtained [12, 13].

The main component of active learning is to design a query function by integrating multiple criteria for selecting informative samples from a set of unlabeled samples [13–17]. In the literature, several query functions are presented based on different criteria. Uncertainty criterion aims at selecting a batch of unlabeled samples that have the lowest classification confidence among the samples in the unlabeled pool. Several uncertainty criteria such as margin sampling (MS), multiclass level uncertainty (MCLU), and breaking ties are widely used in the literature [5, 9]. Diversity criteria select the unlabeled samples having dissimilar properties in concerned feature space and thus reduces redundancy among the selected samples. Angle, closest support vector, clustering are well-known diversity criterion widely used in the AL literature [3, 17, 18]. All the aforementioned AL methods are based on spectral measurements only. However, the neighboring pixels in an image are correlated due to the homogeneous nature of land cover. Two neighboring pixels are most probably of the same class. Thus, the spatial information collected from the neighboring location together with the spectral feature can adequately decrease the uncertainty of class assignment and help in finding most informative patterns. In the literature, few AL methods are present that exploits spectral and spatial information. The spatial information can be incorporated using several techniques including Markov random field, logistic regression, and construction of spectral–spatial profiles [2, 10, 11].

In this chapter, several spectral–spatial AL techniques are designed by using an interesting yet simple approach to fuse spatial information with the spectral content of an HSI and different combinations of existing uncertainty and diversity criteria. In order to fuse spectral and spatial information of the HSI, an extended spatial profile (ESP) is constructed for it. To this end, the dimension of HSI is reduced using principal component analysis (PCA) and the component images in reduced domain are filtered by applying the mean filtering operation (for incorporation of spatial information) considering multiple windows of different sizes. The filtered images are concatenated with the original component images to form an ESP. The ESP constructed for an HSI represents more class-discriminative features than the spectral features alone. Based on these spectral–spatial features and different combinations of existing uncertainty and diversity criteria, pixels are ranked for selection and labeling in each iteration of AL. Experiments are carried out on two real hyperspectral data sets considering different combination of uncertainty and diversity criteria of AL using only spectral profile and the proposed spectral–spatial profile. The spectral–spatial information based AL techniques have outperformed the AL techniques based on spectral measurement only.

The remainder of this chapter is organized as follows. Section 2 describes the proposed spectral–spatial AL techniques. Experimental results are demonstrated in Sect. 3. Finally, Sect. 4 concludes the chapter.

2 Proposed Spectral–Spatial AL Techniques

In this chapter, we present spectral–spatial active learning techniques for the classification of HSI using small number of labeled patterns. The overall architecture of spectral–spatial AL technique is presented in Fig. 1a. The dimension of HSI is decreased using PCA and considering first few PCs, the proposed extended spatial profile (ESP) is constructed by applying mean filtering operations. Then, each pixel of HSI is represented using features of ESP and are divided into two groups. Few samples from each class are initially labeled and stored in training set and rest of the samples are stored in an unlabeled pool. Next, the AL procedure is started where informative samples are selected using different combinations of uncertainty and diversity criteria iteratively until the stopping criteria is met (i.e., obtained accuracy is stable). The construction of ESP and selection of informative samples by combination of uncertainty and diversity criteria are described below.

Construction of spectral–spatial profile: The proposed spectral–spatial profile is constructed by applying mean filtering operation on component images. The mean filter is a sliding-window spatial filter that replaces the center value with the mean (Average) of all the pixel values in the window. The window defines the shape and size of the neighborhood to be considered while filtering operation. The mean filter is a simple, intuitive, and easy to implement method for smoothing images, i.e., reducing the amount of intensity variation among the neighboring pixels. For a pixel p the spatial mean (SM) considering window of size $b \times b$ is computed as shown in Eq. (1),

$$SM_w(p) = \frac{1}{n}\sum_{i=1}^{n} x_i \tag{1}$$

where x_i , $i = 1, 2, \ldots, n$ are the pixels around the pixel p including itself and n is equal to $b \times b$.

For a given image I, Mean Filtering (MF) considering window w is computed as shown in Eq. (2)

$$MF_w(I) = SM_w(p_j), j = 1, 2, \ldots R \tag{2}$$

where R is the number of pixels in the image. The mean filtering results obtained for first PC of the Indian Pines data set (Fig. 1b) considering a 3×3 window is

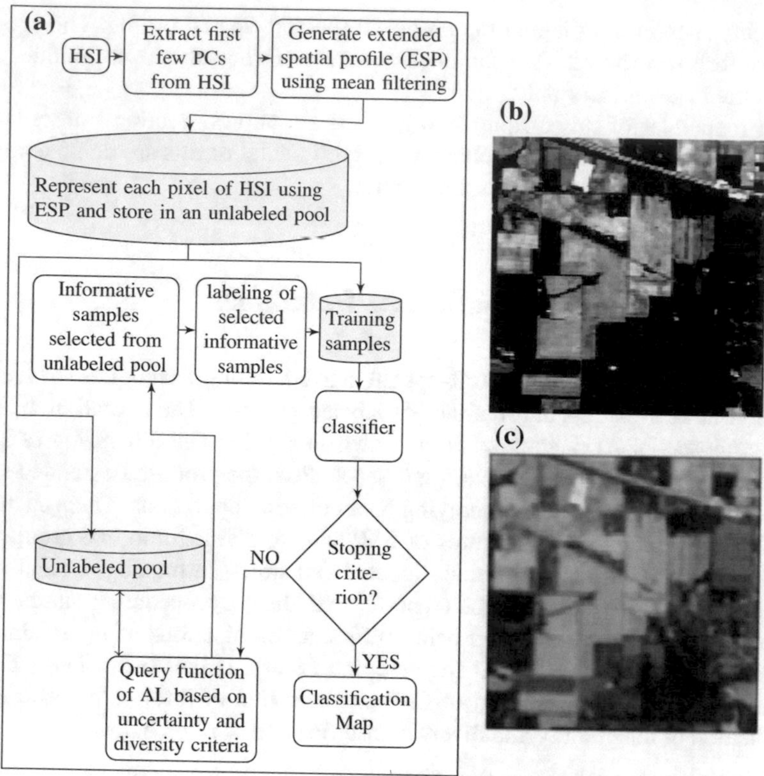

Fig. 1 **a** Proposed framework, **b** 1st PC of the Indian Pines data set and **c** corresponding mean filtering result using window size 3×3

shown in Fig. 1c. By varying the window size multiple mean filtering results can be obtained for same image, which can reduce the intensity variation at different levels. For a given image I, a spatial profile (SP) is constructed as shown in Eq. (3) by concatenating image I with filtering results obtained using k different window sizes.

$$SP(I) = \{I, MF_{w_1}(I), MF_{w_2}(I), \ldots, MF_{w_k}(I)\} \tag{3}$$

In case of an HSI, one can combine spectral and spatial information by constructing SP for all channels, but this increases the dimension of SP tremendously. To counter this situation, dimensionality reduction approach is applied before constructing profile. In the literature, PCA is the widely accepted method to reduce the dimension of an HSI for constructing spectral–spatial profile [1, 12]. In the proposed method, the dimensionality of HSI is reduced using PCA and first q PCs having largest eigenvalues are selected. An extended spatial profile (ESP) is generated for an HSI H by concatenating the SP constructed corresponding to each of the q selected PCs as shown in Eq. (4).

$$ESP(H) = \{SP(PC_1), SP(PC_2), \ldots, SP(PC_q)\} \quad (4)$$

The constructed ESP has small dimension and is rich in spectral and spatial information, which is helpful in class discrimination and selecting informative samples. After construction of the ESP, all the pixels of HSI are represented by their spectral–spatial features and stored in a pool from where few pixels are labeled and kept in a training set L. Rest of the pixels are stored in an unlabeled pool U. Then the procedure for selecting the informative samples from U is started.

Selection of Informative Samples: An AL technique selects a batch of most informative samples iteratively by using a combination of uncertainty and diversity criteria until a stopping criterion is met. In our work, a batch of h samples are selected from U at each iteration which are labeled and appended to the training set. In case where only uncertainty criterion is used the h most uncertain samples are selected in each iteration. When a combination of uncertainty and diversity criteria is used m most uncertain samples are selected and out of m samples h $(h < m)$ samples are finally selected using diversity criterion. In this chapter, two uncertainty criteria, two diversity criteria and there possible combinations are explored. Uncertainty criterion selects a batch of samples those have lowest classification confidence among U. First uncertainty criteria is margin sampling (MS) that select samples nearest to the separating hyperplane. Considering multiclass scenario, a one against all SVM architecture is used to create s decision hyperplanes (one for each class) and for each unlabeled sample x, functional distances $f_i(x), i = 1, 2 \ldots s$ from s decision hyperplanes are obtained. The minimum distance associates to the classification confidence (CC) of sample x which is computed using Eq. (5).

$$CC(x) = \min_{i=1,2\ldots s} \{|f_i(x)|\} \quad (5)$$

Second uncertainty criterion MCLU is estimated using Eq. (6) where two most largest distance value to the hyperplane is measured and then the difference between these two values is used to compute the classification confidence $CC(x)$ of x, $(x \in U)$.

$$\begin{aligned} r_{mx1} &= \arg \max_{i=1,2\ldots s} f_i(x) \\ r_{mx2} &= \arg \max_{\substack{j=1,2\ldots s \\ j \neq r_{mx1}}} f_j(x) \\ CC(x) &= f_{r_{mx1}}(x) - f_{r_{mx2}}(x) \end{aligned} \quad (6)$$

By applying uncertainty criterion m samples with lowest value of CC are selected as most uncertain. Out of these, h samples are selected by applying diversity criterion. Two diversity criteria are explored in this chapter. One is clustering based diversity (CBD) and another is angle based diversity (ABD) [3]. In CBD the m selected uncertain unlabeled samples are clustered into h groups and one representative sample is selected from each cluster [18]. In ABD angle between m selected uncertain samples is computed and h samples with maximum angle are selected. The angle A^{ABD}

between two samples x_i and x_j is computed considering a kernel function $K(x_i, x_j)$ using Eq. (7).

$$A^{ABD}(x_i, x_j) = \frac{K(x_i, x_j)}{\sqrt{K(x_i, x_i)K(x_j, x_j)}} \tag{7}$$

The possible combination of uncertainty and diversity criteria are MS-CBD, MS-ABD, MCLU-CBD, MCLU-ABD. When the proposed ESP is considered for integration of spectral and spatial information, the methods are referred to as SP-MS-CBD, SP-MS-ABD, SP-MCLU-CBD, and SP-MCLU-ABD, respectively.

3 Experimental Results

Data set Description: Experiments are carried out on two hyperspectral image datasets.[1] The first data set shown in Fig. 2a is a hyperspectral image acquired by the Airborne Visible/Infrared Imaging Spectrometer (AVIRIS) sensor over the Indian Pines area of Indiana in 1992. This image has a spectral range of 0.4–2.5 μm, spectral resolution of 10 nm, size of 145 by 145 pixels, 220 spectral band (200 after removal of noise and water absorption bands), has spatial resolution of 20 m per pixel and contains 16 classes.

The second data set is an HSI acquired by the AVIRIS sensor over the Salinas Valley, California shown in Fig. 2b. This image is characterized by high spatial resolution of 3.7 m, size of 512 by 217 and 224 spectral bands (204 after removal of noise and water absorption bands) and contains 16 classes. The class name and corresponding available reference samples of both the data sets are available online.[1]

Design of Experiments: Experiments are carried out on the abovementioned hyperspectral data sets. For fusing spectral and spatial information the proposed ESP is constructed. To this end, the dimension of considered HSI is reduced by applying PCA and first ten PCs having highest eigenvalues are selected. For each component image proposed SP is computed considering six different window sizes namely, $3 \times 3, 5 \times 5, 7 \times 7, 9 \times 9, 11 \times 11$ and 20×20. The constructed SP for each component image is concatenated together to form an ESP that has 70 features (7 for each component image). Each pixel of the HSI is represented with the proposed ESP features for further processing of AL technique. Six batch mode AL query functions are designed using two uncertainty (MS and MCLU) and two diversity (CBD and ABD) criteria namely MS, MCLU, MS-CBD, MS-ABD, MCLU-CBD, MCLU-ABD. In case of first two query functions, only h most uncertain samples are selected. In case of latter four query functions, initially m $(= 3h)$ samples are selected using uncertainty criterion and among the m selected samples h samples are selected using diversity criterion. In the experiments the batch of selected samples h is set to 20.

---

[1] Available online: http://www.ehu.eus/ccwintco/index.php?title=Hyperspectral_Remote_Sensing_Scenes.

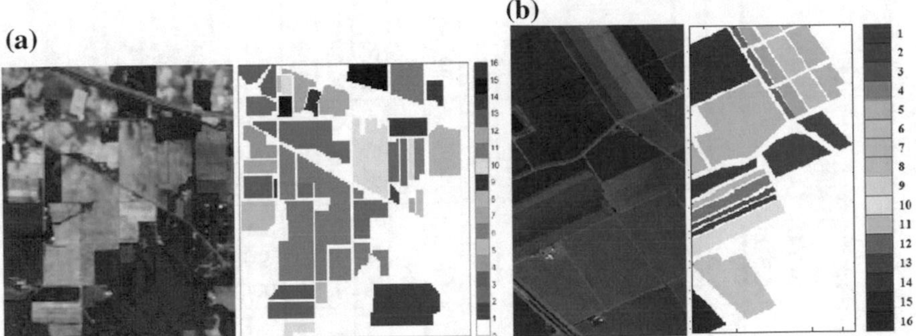

Fig. 2 Hyperspectral images and map of available reference samples for **a** the Indian Pines area, **b** the Salinas area

The proposed spectral–spatial AL techniques are referred in this paper as SS-MS, SS-MCLU, SS-MS-CBD, SS-MS-ABD, SS-MCLU-CBD, SS-MCLU-ABD.

All the AL technique presented in this chapter have been implemented in MATLAB (R2015a). One against all support vector machine classifier with radial basis function (RBF) kernel has been implemented by using LIBSVM library [8]. The SVM parameters are derived by applying a grid search with five-fold cross-validation.

Results: Indian Pines Dataset– First experiment is conducted on the Indian Pines data set in order to study and compare the performance of different AL techniques with the proposed ESP features and with spectral features only. The test set (T) consists of 10249 labeled samples from which initially 48 patterns are randomly chosen (3 patterns from each class) to construct the training set L and rest 10201 patterns are considered in the unlabeled pool U. At each iteration, 20 patterns are chosen from U for labeling. The AL process is repeated 49 times to assign the label of 1028 samples. The experiment is conducted for 10 times by selecting different initial labeled patterns.

Table 1 reports the average class-wise accuracies, average overall classification accuracy ($\overline{OA}$), its standard deviation (std) and the average kappa accuracies (kappa) obtained for different AL techniques. From the table, one can see that the spectral–spatial AL methods SS-MS, SS-MS-CBD, SS-MS-ABD, SS-MCLU, SS-MCLU-CBD, and SS-MCLU-ABD produced significantly higher accuracies (at least 7 %) than the AL methods based on spectral features only. This shows the potential of the proposed ESP for incorporation of spatial information. From the results, one can also observe that the MCLU has performed better than the MS as an uncertainty criterion. The best accuracy is obtained by the SS-MCLU-ABD. Moreover, the spectral–spatial AL methods were able to achieve higher accuracy with less number of training samples which is visible from Fig. 3a that shows the $\overline{OA}$ provided by the different methods versus the number of samples included in the training sets. This shows the usefulness of spectral–spatial AL methods.

Table 1 Class-wise classification accuracies (%), average overall classification accuracies ($\overline{OA}$), its standard deviation (std), and kappa accuracies (kappa) provided for ten runs (Indian Pines Data set). Best results are reported in bold

Methods / Class	MS	SS-MS	MS-CBD	SS-MS-CBD	MS-ABD	SS-MS-ABD	MCLU	SS-MCLU	MCLU-CBD	SS MCLU-CBD	MCLU-ABD	SS MCLU-ABD
	$\|L\| = 1028$											
Alfalfa	35.44	97.83	40	64.56	84.35	99.56	70.87	99.13	73.04	68.48	82.39	**99.78**
Corn-notill	77.85	90.99	78.98	87.39	88.82	93.20	85.79	95.27	89.88	94.26	92.36	**95.57**
Corn-min	74.36	90.22	79.66	96.01	82.97	97.71	79.08	99.81	86.08	**99.85**	89.05	99.79
Corn	79.87	86.58	82.78	88.14	88.14	91.60	76.20	**100**	85.82	**100**	92.11	**100**
Grass/pasture	91.82	95.46	93.02	97.18	95.80	99.15	92.23	99.87	93.89	99.73	97.06	**99.92**
Grass/trees	97.51	99.92	97.52	99.85	98.48	99.96	97.12	**100**	98.19	99.98	98.67	99.99
Grass/pasture-mowed	83.93	98.57	79.64	99.28	88.21	97.14	81.07	99.64	90.36	99.64	89.64	**100**
Way-windrowed	98.79	99.89	98.26	**100**	98.97	**100**	97.15	99.92	98.72	**100**	99.39	**100**
Oats	76.50	81.50	86	84	82.50	82	84	**100**	82.50	**100**	89	**100**
Soybeans-notill	83.29	91.65	82.55	90.80	87.34	96.59	83.74	**99.92**	89.43	99.71	92.26	99.28
Soybeans-min	86.47	96.86	89.39	98.01	90.08	98.82	88.14	**99.87**	90.64	**99.87**	93.09	99.84
Soybean-clean	85.87	89.78	89.07	87.66	91.30	96.86	89.11	99.76	92.24	**99.83**	95.50	99.59
Wheat	97.66	99.95	97.80	99.95	99.12	**100**	98.44	**100**	99.07	**100**	99.27	**100**
Woods	96.45	99.84	95.97	99.97	97.75	99.98	96.35	99.88	97.15	99.98	97.67	**100**
Bldg-grass-tree-drives	66.73	98.23	71.40	98.86	77.87	99.30	67.85	98.47	74.64	**99.53**	78.96	98.94
Stone-steel towers	90.54	97.74	92.15	99.68	91.72	**100**	87.10	**100**	90.75	**100**	92.69	99.89
$\overline{OA}$	85.91	95.13	87.57	95.19	91.01	97.77	88.04	99.18	91.37	99.26	93.67	**99.57**
std	1.23	0.10	0.72	.09	1.59	0.04	1.59	**0.005**	0.01	**0.005**	0.01	0.007
kappa	0.84	0.94	0.86	.94	0.89	0.97	0.86	0.99	0.90	0.99	0.93	**0.99**

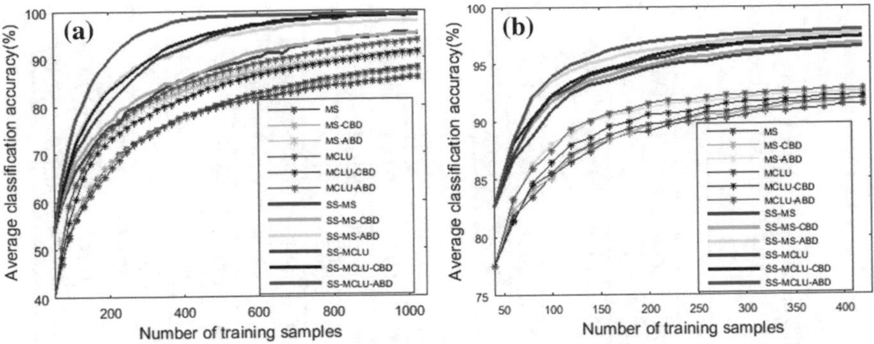

Fig. 3 Average classification accuracy ($\overline{OA}$) versus the number of labeled samples obtained by different methods considering **a** Indian Pines data set, and **b** Salinas data set

Results: Salinas Data set- The second experiment is conducted on the Salinas data set. The size of T is 54129 out of which 48 patterns (3 patterns from each class) are randomly chosen to be kept in L initially and the rest 54081 are stored in the U. The AL process has 19 iterations and 20 patterns are chosen per iteration to result in 428 patterns in the final training set. The experiments are run for ten times with different initial labeled samples and the results are reported in Table 2. Analogous to the previous results again the spectral–spatial AL methods provided significantly higher $\overline{OA}$ (at least 6 %) than the AL methods based on spectral measurements only. Again the best results are obtained by SS-MCLU-ABD. Figure 3b shows the obtained $\overline{OA}$ with respect to the number of labeled samples in L where the spectral–spatial AL methods could achieve above 94% in 150 samples whereas the AL methods based on spectral features only could not achieve 94% even in 428 samples. This confirms the superiority of spectral–spatial AL methods over the AL methods based on spectral features only.

4 Conclusion

Supervised classification of an HSI is greatly improved by integration of spectral and spatial information. This chapter presents a method for integrating spectral and spatial information of HSI based on mean filtering operations. In this chapter, several spectral–spatial AL techniques are designed for classification of HSI with small number of labeled samples and their performances are studied. The query functions of AL are designed based on different combinations of existing uncertainty and diversity criteria and the proposed spectral–spatial features. The spectral and spatial information are fused by concatenating the filtering results obtained for the component images in the reduced domain of HSI employing mean filtering operation considering multiple windows of different sizes. The experiments are conducted on two real hyperspec-

Table 2 Class-wise classification accuracies (%), average overall classification accuracies (%), its standard deviation (s), and kappa accuracies (k) provided for ten runs (Salinas Dataset). Best results are reported in bold

Class \ Methods	MS	SS-MS	MS-CBD	SS-MS-CBD	MS-ABD	SS-MS-ABD	MCLU	SS-MCLU	MCLU-CBD	SS MCLU-CBD	MCLU-ABD	SS MCLU-ABD
	$\|L\| = 428$											
Brocoli green weeds1	97.55	99.83	98.72	99.90	99.13	99.79	97.79	99.57	98.83	99.66	98.52	**99.99**
Brocoli green weeds2	99.23	99.82	99.54	99.87	99.47	99.89	99.29	**99.99**	99.67	99.98	99.72	**99.99**
Fallow	97.85	99.68	98.53	99.63	99.27	99.75	98.48	99.73	99.32	99.84	98.64	**99.89**
Fallow rough plow	98.97	98.93	99.39	99.15	99.25	99.20	99.17	99.34	99.05	98.94	99.31	**99.48**
Fallow smooth	97.87	99.05	97.61	99.02	98.42	99.26	97.62	99.12	97.87	99.26	98.32	**99.41**
Stubble	98.90	99.32	99.52	99.70	99.60	99.88	99.64	99.78	99.62	99.87	99.01	**99.97**
Celery	99.38	99.64	99.30	99.61	99.37	99.79	99.18	99.71	99.57	99.75	99.60	**99.90**
Grapes untrained	85.98	89.74	86.94	89.57	86.92	91.44	85.90	90.41	87.43	90.44	85.99	**91.51**
Soil vinyard develop	99.51	99.63	99.68	99.78	99.77	99.92	99.91	99.94	99.80	99.88	99.84	**99.98**
Corn senesced green weeds	91.41	98.11	92.18	98.70	94.18	99.56	92.74	99.46	93.63	99.65	96.22	**99.92**
Lettuce romaine 4wk	91.97	97.94	92.75	99.01	94.67	99.14	96.59	99.15	96.41	99.22	98.40	**99.70**
Lettuce romaine 5wk	99.68	98.60	99.51	99.76	99.78	99.80	99.67	99.80	99.34	99.73	99.90	**99.99**
Lettuce romaine 6wk	95.58	97.83	97.13	98.47	97.79	99.42	97.02	99.54	97.40	99.33	97.62	**99.97**
Lettuce romaine 7wk	92.94	96.34	95.22	98.41	95.40	99.26	95.06	99.31	94.87	99.78	96.26	**99.83**
Vinyard untrained	69.67	94.60	69.50	94.98	72.33	97.79	70.78	97.12	69.15	96.82	73.92	**98.44**
Vinyard vertical trellis	95.20	96.14	96.60	97.20	97.27	98.12	96.91	97.85	98.34	98.92	98.95	**99.91**
$\overline{OA}$	91.41	96.48	91.96	96.73	92.59	97.67	91.95	97.33	92.26	97.63	93.05	**97.95**
std	0.640	0.370	0.460	0.320	0.410	0.143	0.528	0.157	0.436	0.244	0.496	**0.079**
kappa	0.900	0.960	0.910	0.964	0.920	0.974	0.910	0.970	0.914	0.971	0.929	**0.977**

tral data sets where the AL methods exploiting proposed ESP features have shown significant improvement in classification accuracies. This shows the effectiveness of spectral–spatial information in determining the class-discriminative pixels for labeling. Among the different spectral–spatial AL methods, SS-MCLU-ABD achieved best results.

Acknowledgements This work is supported in part by Science and Engineering Research Board, Government of India.

References

1. Benediktsson, J.A., Palmason, J.A., Sveinsson, J.R.: Classification of hyperspectral data from urban areas based on extended morphological profiles. IEEE Trans. Geosci. Remote. Sens. **43**(3), 480–491 (2005)
2. Bhardwaj, K., Patra, S.: An unsupervised technique for optimal feature selection in attribute profiles for spectral-spatial classification of hyperspectral images. ISPRS J. Photogramm. Remote. Sens. **138**, 139–150 (2018)
3. Brinker, K.: Incorporating diversity in active learning with support vector machines. In: Proceedings of the 20th International Conference on Machine Learning (ICML-03), pp. 59–66 (2003)
4. Bruzzone, L., Chi, M., Marconcini, M.: A novel transductive SVM for semisupervised classification of remote-sensing images. IEEE Trans. Geosci. Remote. Sens. **44**(11), 3363–3373 (2006)
5. Campbell, C., Cristianini, N., Smola, A. et al.: Query learning with large margin classifiers. In: ICML, pp. 111–118 (2000)
6. Camps-Valls, G., Marsheva, T.V.B., Zhou, D.: Semi-supervised graph-based hyperspectral image classification. IEEE Trans. Geosci. Remote. Sens. **45**(10), 3044–3054 (2007)
7. Camps-Valls, G., Tuia, D., Bruzzone, L., Benediktsson, J.A.: Advances in hyperspectral image classification: Earth monitoring with statistical learning methods. IEEE Signal Process. Mag. **31**(1), 45–54 (2014)
8. Chang, C.C., Lin, C.J.: LIBSVM: A library for support vector machines. ACM Trans. Intell. Syst. Technol. (TIST) **2**(3), 27 (2011)
9. Demir, B., Persello, C., Bruzzone, L.: Batch-mode active-learning methods for the interactive classification of remote sensing images. IEEE Trans. Geosci. Remote. Sens. **49**(3), 1014–1031 (2011)
10. Li, J., Bioucas-Dias, J.M., Plaza, A.: Spectral-spatial classification of hyperspectral data using loopy belief propagation and active learning. IEEE Trans. Geosci. Remote. Sens. **51**(2), 844–856 (2013)
11. Pasolli, E., Melgani, F., Tuia, D., Pacifici, F., Emery, W.J.: SVM active learning approach for image classification using spatial information. IEEE Trans. Geosci. Remote. Sens. **52**(4), 2217–2233 (2014)
12. Patra, S., Bhardwaj, K., Bruzzone, L.: A spectral-spatial multicriteria active learning technique for hyperspectral image classification. IEEE J. Sel. Top. Appl. Earth Obs. Remote. Sens. **10**(12), 5213–5227 (2017)
13. Patra, S., Bruzzone, L.: A fast cluster-assumption based active-learning technique for classification of remote sensing images. IEEE Trans. Geosci. Remote. Sens. **49**(5), 1617–1626 (2011)
14. Patra, S., Bruzzone, L.: A batch-mode active learning technique based on multiple uncertainty for SVM classifier. IEEE Geosci. Remote. Sens. Lett. **9**(3), 497–501 (2012)

15. Patra, S., Bruzzone, L.: A novel som-svm-based active learning technique for remote sensing image classification. IEEE Trans. Geosci. Remote. Sens. **52**(11), 6899–6910 (2014)
16. Singla, A., Patra, S.: A fast partition-based batch-mode active learning technique using svm classifier. Soft Comput. **22**(14), 4627–4637 (2018)
17. Tuia, D., Ratle, F., Pacifici, F., Kanevski, M.F., Emery, W.J.: Active learning methods for remote sensing image classification. IEEE Trans. Geosci. Remote. Sens. **47**(7), 2218–2232 (2009)
18. Xu, Z., Yu, K., Tresp, V., Xu, X., Wang, J.: Representative sampling for text classification using support vector machines. In: European Conference on Information Retrieval, pp. 393–407. Springer, Berlin (2003)

A Comprehensive Analysis of Kernelized Hybrid Clustering Algorithms with Firefly and Fuzzy Firefly Algorithms

B. K. Tripathy and Anmol Agrawal

Abstract In order to handle the problem of linear separability in the early data clustering algorithms, Euclidean distance is being replaced with Kernel functions as measures of similarity. Another problem with the clustering algorithms is the selection of initial centroids randomly, which affects not only the final result but also decreases the convergence rate. Optimal selection of initial centroids through optimization algorithms like Firefly or Fuzzy Firefly algorithms provide partial solution to this problem. In this paper, we focus on two kernels; Gaussian and Hyper-tangent and use both Firefly and Fuzzy Firefly algorithms separately along with algorithms like FCM, IFCM and RFCM and analyse their efficiency using two measures DB and D. Our analysis concludes that RFCM with Hyper-tangent kernel and fuzzy firefly produce the best results with fastest convergence rate. We use the two images; MRI scan of a human brain and blood cancer cells for our analysis.

Keywords Data clustering · Image segmentation · Kernel function · Firefly · Fuzzy firefly · DB index · Dunn index

FCM	Fuzzy C-Means
RFCM	Rough Fuzzy C-Means
IFCM	Intuitionistic Fuzzy C-Means
FCMFA	Fuzzy C-Means with Firefly algorithm
FCMFFA	Fuzzy C-Means with Fuzzy Firefly algorithm
GKFCM	Gaussian Kernelized Fuzzy C-Means
HKFCM	Hyper-tangent Fuzzy C-Means
IFCMFA	Intuitionistic Fuzzy C-Means with Firefly algorithm
IFCMFFA	Intuitionistic Fuzzy C-Means with Fuzzy Firefly algorithm
RFCMFA	Rough Fuzzy C-Means with Firefly algorithm
RFCMFFA	Rough Fuzzy C-Means with Fuzzy Firefly algorithm

B. K. Tripathy · A. Agrawal (✉)
School of Information Technology and Engineering, VIT University, Vellore 632014, Tamil Nadu, India
e-mail: anmol98agr@gmail.com

B. K. Tripathy
e-mail: tripathybk@vit.ac.in

H. S. Behera et al. (eds.), *Computational Intelligence in Data Mining*, Advances in Intelligent Systems and Computing 990,
https://doi.org/10.1007/978-981-13-8676-3_31

1 Introduction

Data clustering techniques are used extensively in image segmentation over the past decade. Image segmentation involves the splitting and grouping of similar pixels of an image. There are two types of clustering methods with respect to position of elements in various clusters; namely hard-clustering and soft-clustering. In hard-clustering, one data point cannot belong to multiple clusters. In soft-clustering or fuzzy clustering, the data points may belong to multiple clusters based on certain membership values. They use uncertainty models like Fuzzy Set [17], intuitionistic fuzzy sets [2], rough sets [12] and their hybrid models. Fuzzy C-Means (FCM) [3], Intuitionistic Fuzzy C-Means (IFCM) [4] and Rough Fuzzy C-Means (RFCM) [11, 12] are some such algorithms. In all of the above-mentioned algorithms, the Euclidean distance was used as a similarity measure. The Euclidean distance-based clustering algorithms have the problem of linearly separable datasets. However, this issue was corrected by using the kernel function in which the feature space is projected into a higher dimension using an appropriate non-linear mapping function that ensures the linear separability of the clusters. Thus in an attempt to avail this generality, several kernel function-based algorithms have been developed over the years, some of which are the Kernel-based K-means clustering, kernel-based FCM, RCM [13] and RFCM [14]. A comparative analysis of kernel functions and uncertainty-based c-means algorithms was made by Tripathy and Mittal [15]. All these algorithms involved random initialization of cluster centroids. This resulted in slow convergence and hence more computational cost. In 2009, a metaheuristic based on the flashing behaviour of fireflies was proposed by Yang [16]. An improved version of this algorithm, namely, the Fuzzy Firefly algorithm was proposed by Hassanzadeh [8]. Stabilization of Rough Sets Based clustering algorithm using the Firefly algorithm was proposed by Abhay et al. in 2017 [1]. In this paper, we combine FCM, IFCM and RFCM as well as their kernelized versions with firefly and fuzzy firefly and make a comparative analysis of the performance and convergence rate of these algorithms. Two performance indices, Davis–Bouldin index (DB) [6] and Dunn index (D) [7] has been used to analyse and make a comparison of these algorithms. We have also established the relative relations between these algorithms. In Sect. 2, we describe the various algorithms used in this analysis. In Sects. 3 and 4, we discuss the methodology and analyse the results. The conclusion of our analysis is discussed in Sect. 5.

2 Definitions and Notations

Here, we define some of the terms and algorithms that has been used in the paper.

2.1 Clustering Algorithms

Clustering can be considered as the most important unsupervised learning problems. Clustering can be defined as the unsupervised classification of patterns (data items or observations) into different groups called as clusters [9]. For the comprehensive analysis in this paper, we use the Fuzzy C-Means (FCM) algorithm [3], its extension the Intuitionistic Fuzzy C-Means algorithm [4] and the Rough Fuzzy C-Means [10, 11]. We avoid reproducing these algorithms due to scarcity of space.

We use the two genetic algorithms, Firefly algorithm and Fuzzy Firefly algorithm in order to get optimal initial centroids for the algorithms to be compared.

2.1.1 Firefly algorithm

Firefly algorithm was proposed in 2009 [16]. It is a bio-inspired metaheuristic which mimics the behaviour of fireflies. Biologically, fireflies are attracted to luminous objects. Certain brightness is associated with each firefly in this algorithm and hence attracts all the other fireflies. It makes a firefly to being attracted towards another firefly that is brighter than itself (directly proportional). The movement of the brightest firefly is random. Inverse of the distance between them measures the degree of attraction between any two fireflies. Each problem has specific objective function, from which the brightness of a firefly is computed. β(attractiveness function) is determined by the following formula in Eq. (1).

$$\beta(r_{i,j}) = \beta_0 e^{-\gamma r_{i,j}^2} \tag{1}$$

Here, β_0 is the initial value of attractiveness, γ is the light absorption coefficient and the Euclidean distance between two fireflies i and j is denoted by $r_{i,j}$. In the implementation of this algorithm, we take $\beta_0 = 1$. The value of γ generally varies from 0.01 to 100.

$$x_i = x_i + \beta_0 e^{-\gamma r_{ij}^2}(x_i - x_j) + \alpha\left(rand - \frac{1}{2}\right) \tag{2}$$

The above equation (Eq. 2) is for the movement of firefly (*i*) to the brighter firefly (*j*). In the third term of the equation, $\alpha \in [0, 1]$ is the randomization parameter and rand is a random number generator function that is uniformly distributed in [0, 1]. This ensures that the fireflies are not stuck at a local optima.

The most important characteristic of Firefly algorithm is its ability to avoid the local optima. This is because random fireflies covering the whole solution space are initialized randomly. This ensures that at least one firefly has a high intensity. All the other fireflies start moving towards this brighter firefly. Moreover, with the movement of each firefly, there is an associated degree of randomness. This ensures that the whole solution space is thoroughly covered by the population of fireflies.

2.1.2 Fuzzy Firefly Algorithm

The area explored by a firefly is increased and the number of iterations was decreased in the extended version introduced by Hassanzadeh [8] and called Fuzzy Firefly algorithm. In order to achieve this, the number of brighter fireflies was increased to k so that each one of these can influence the fireflies having lesser brightness than them. The value of 'k' is decided by the user and is dependent upon the population size and the problem complexity. Taking h as a brighter firefly with fitness value '$f(p_h)$' and the local optimum firefly has its fitness value $f(p_g)$ then the degree of attractiveness of the firefly 'h' is defined as in Eq. (3).

$$\psi(h) = 1/\left(\frac{f(p_h) - f(p_g)}{\beta}\right) \tag{3}$$

Here, β is given by as

$$\beta = f\left(\frac{p_g}{l}\right) \tag{4}$$

Here (Eq. 4), 'l' is a user-defined parameter. The movement of a firefly 'i' towards another firefly 'h', is formulated as in Eq. (5).

$$x_i = x_i + \left(\beta_0 e^{-\gamma r_{ij}^2}(x_j - x_i) + \sum\nolimits_{h=1}^{k} \psi(h)\beta_0 e^{-\gamma r_{ih}^2}(x_h - x_i)\right) \cdot \alpha(rand - 1/2) \tag{5}$$

2.2 *Similarity Measures*

Measuring similarity between data points is the most important step in data clustering. The commonly used measure, Euclidean distance has the drawbacks of being sensitive to initial assignment of centroids and being stable for linearly separable data points. The second drawback can be solved by the use of Kernel functions as similarity measures. Here, we use two such kernel functions for study. These are:

2.2.1 Kernel Distance

Let a data point be denoted by 'a'. $\phi(a)$ denotes the transformation of 'a' to a feature plane of higher dimensionality. Description of inner product space is given by $K(a, b) = \phi(a)\phi(b)$

Let us assume a = $\{a_1, a_2, \ldots a_n\}$ and b = $\{b_1, b_2, \ldots b_n\}$ be two points in the n-dimensional space. There are several kernel functions available in the literature. Some of these functions (Eqs. 6 and 7) that has been used in this paper are:

i. Gaussian Kernel:

$$G(a, b) = \exp\left(-\frac{\sum_{i=1}^{n} (a_i - b_i)^2}{2\sigma^2}\right) \tag{6}$$

ii. Hyper-tangent Kernel:

$$H(a, b) = 1 - \tan h\left(-\frac{\sum_{i=1}^{n} (a_i - b_i)^2}{2\sigma^2}\right) \tag{7}$$

where,

$$\sigma^2 = \frac{1}{N}\sum_{i=1}^{N} \left\|a_i - a'\right\|^2 \text{ and } a' = \frac{1}{N}\sum_{i=1}^{N} a_i \tag{8}$$

According to Chen [5], the generalized kernel distance formula is $D(x, y) = K(x, x) + K(y, y) - 2K(x, y)$. Using the fact that K(x, x) = 1, the kernel distance can be written as D(x, y) = 2(1 – K(x, y)).

2.3 *Performance Indices*

Performances indices are used to evaluate the performance of clustering algorithms. There are several performance indices available in the literature. The Davis–Bouldin (DB) index [6] and Dunn index (D) [7] are some of the most common efficiency analysis indices. The results are dependent on the required number of clusters.

2.3.1 Davis–Bouldin (DB) Index

The DB index is the ratio of sum of distance within cluster to distance between cluster. It is given by the formula (Eq. 9),

$$DB = \frac{1}{c}\sum_{i=1}^{c} \max_{k \neq i}\left\{\frac{S(v_i) + S(v_k)}{d(v_i, v_k)}\right\}, for\ 1 < k, i < c \tag{9}$$

The aim of this index is to minimize the separation within cluster and maximize the distance between clusters. Hence, a low DB value indicates good clustering.

2.3.2 Dunn (D) Index

The D index is used to identify the clusters that are compact and separated. It is calculated using (Eq. 10):

$$Dunn = min_i \left\{ min_{k \neq i} \left\{ \frac{d(v_i, v_k)}{max_l S(v_l)} \right\} \right\} for\ 1 < k, i, l < c \tag{10}$$

The aim of Dunn index is to maximize the distance between cluster and minimizing the distance within cluster. Therefore, a high D index indicates better clustering.

3 Methodology

The swarm of fireflies is initialised to random values and the metaheuristic is allowed to calculate the intensity of each firefly. These fireflies are allowed to move around following Eq. (8) and their intensities are recalculated. At the end of this cycle, the best firefly (centroid) values are passed as the initial values of clustering algorithms. We have used this technique for the Kernelised (Gaussian and Hyper-tangent) versions of the algorithms FCM, IFCM and RFCM and made a comparative analysis among themselves and the existing clustering algorithms in this direction. It is observed that the algorithms obtained through our approach not only show significant improvement (verified through the computation of the measuring indices DUNN and DB and results obtained) but also their rates of convergences are high. In this paper, we have used two different types of images for our experimental purpose.

4 Result and Analysis

Implementation of the algorithms has been done in Python 3.6 using Spyder 3.1.4 IDE. Numpy library has been used in the implementation of algorithms and mathlotlib library has been used to plot the output figures. In the experimental analysis, we have used two different kinds of images (Fig. 1) to make our study extensive.

ORIGINAL DATASET:

Figure 1a (221 × 228) represents MRI scan of a human brain. The lighter region on the forehead region indicates presence of a tumour. Figure 2b (250 × 250) represents the proliferation of abnormal White Blood Cells (WBCs) among Red Blood Cells (RBCs) and normal WBCs.

Fig. 1 Original dataset

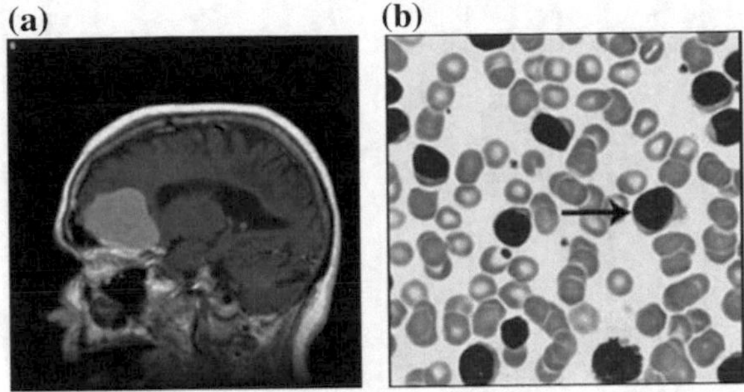

4.1 Segmentation of Tumour in Brain MRI Scan

4.1.1 Using Euclidean Distance

It can be inferred from the above output (Fig. 2) that FCM and IFCM produce roughly similar result while RFCM produces much better results. The output produced by FCM and IFCM are slightly blurred and noisy especially at the edges. Output produced by FCMFA, FCMFFA, IFCMFA, IFCMFFA, RFCMFA and RFCMFFA are marginally better than their original counterparts as is clear from the values of Dunn and DB indices and have a better convergence rate (Table 1). It can be observed that IFCM and IFCMFA works slightly better than IFCMFFA but have much higher convergence rate.

Overall, by looking at the performance indices, the following relation can be established:

$$\text{FCM} < \text{FCMFA} < \text{FCMFFA},\ \text{IFCMFFA} < \text{IFCM} < \text{IFCMFA},$$
$$\text{RFCM} \approx \text{RFCMFA} \approx \text{RFCMFFA},\ \text{FCM} < \text{IFCM} < \text{RFCM}$$

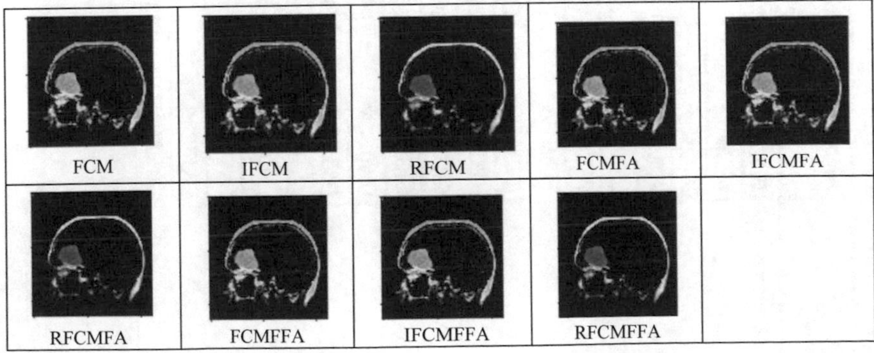

Fig. 2 Output of segmentation of tumours in MRI scan using Euclidean distance

Table 1 Performance analysis indices for brain tumour segmentation

Dist. Fn.	Algorithm	Number of clusters = 3			Number of clusters = 4		
		#Iterations	DB	Dunn	#Iterations	DB	Dunn
Euclidean distance	FCM	28	17.0474	0.0412	16	8.0970	0.0935
	IFCM	28	16.5964	0.0420	14	7.8390	0.0970
	RFCM	27	3.2600	0.2190	19	1.7927	0.5413
	FCMFA	23	17.0471	0.0412	14	8.0966	0.0935
	IFCMFA	28	16.5955	0.0419	12	7.8396	0.0969
	RFCMFA	13	3.2600	0.2190	15	1.7927	0.5412
	FCMFFA	19	17.0463	0.0412	10	8.0914	0.0935
	IFCMFFA	15	16.6106	0.0420	8	7.8447	0.0966
	RFCMFFA	21	3.2599	0.2190	16	1.7927	0.5412
Gaussian Kernel	GKFCM	20	0.0021	6.1558	23	0.0006	15.4141
	GKIFCM	23	0.1354	6.2586	18	0.0441	15.4779
	GKRFCM	48	0.0530	18.1913	21	0.0096	118.2917
	GKFCMFA	14	0.0021	6.1560	12	0.0006	15.4147
	GKIFCMFA	11	0.1354	6.2586	9	0.0440	15.4762
	GKRFCMFA	14	0.0530	18.1914	10	0.0096	118.2918
	GKFCMFFA	14	0.0021	6.1561	14	0.0006	15.4480
	GKIFCMFFA	12	0.1354	6.2586	11	0.0438	15.5861
	GKRFCMFFA	13	0.0530	18.1915	9	0.0096	118.3178

(continued)

Table 1 (continued)

Dist. Fn.	Algorithm	Number of clusters = 3			Number of clusters = 4		
		#Iterations	DB	Dunn	#Iterations	DB	Dunn
Hyper-tangent Kernel	HKFCM	18	0.0025	6.9640	31	0.0024	6.9641
	HKIFCM	18	0.1325	7.0495	13	0.0473	15.7339
	HKRFCM	33	0.0546	18.7862	15	0.0100	112.5565
	HKFCMFA	13	0.0024	6.9648	15	0.0007	15.4989
	HKIFCMFA	16	0.1324	7.0495	11	0.0439	15.5437
	HKRFCMFA	19	0.0546	18.7862	12	0.0099	115.6368
	HKFCMFFA	7	0.0024	6.9652	11	0.0007	15.5262
	HKIFCMFFA	11	3.4744	7.0494	9	3.4652	15.7382
	HKRFCMFFA	12	0.0546	18.7862	8	0.0099	115.6067

4.1.2 Using Gaussian Kernel

It is easy to infer from Table 1 that the results produced using Gaussian Kernel are much better than that produced using Euclidean distance. The differences between the outputs are difficult to notice but the indices show that firefly and fuzzy firefly versions of the algorithm works better than their conventional counterpart in all the three cases and show much better convergence rate. Thus, the following relation can be established:

GKFCM<GKFCMFA<GKFCMFFA, GKIFCM≈GKIFCMFA<GKIFCMFFA,
GKRFCM<GKRFCMFA<GKRFCMFFA, GKIFCM<GKFCM<GKRFCM

4.1.3 Using Hyper-tangent Kernel

It can be observed from Table 1 that the results produced by Hyper-tangent kernel are almost similar to the results produced by the Gaussian Kernel. Even here, the differences in the output are not noticeable and we have to rely on the performance indices to compare them. Data shows that output produced by HKFCMFA, HKFCMFAA, HKRFCMFA, HKRFCMFAA are better than the standard versions.

However, it can be observed that HKIFCMFAA does not perform well when compared to HKIFCM and HKIFCMFA. Convergence rate is better in optimized versions of the algorithm. The performance of the algorithms can be related as follows:

HKFCM<HKFCMFA<HKFCMFFA, HKIFCMFFA<HKIFCM<HKIFCMFA,
HKRFCM<HKRFCMFA<HKRFCMFFA, HKIFCM<HKFCM<HKRFCM

4.2 Segmentation of Blood Cancer Cells

4.2.1 Using Euclidean Distance

It can be easily observed from the above output (Fig. 3) that performance of FCM and IFCM are comparable. RFCM produces much better result than FCM and IFCM. The firefly and fuzzy firefly versions of all the three algorithms outperform their standard versions both in quality and the convergence rate (Table 2). RFCMFFA produces the best result amongst all the nine cases. The following relation can thus be established:

FCM<FCMFA<FCMFFA, IFCM<IFCMFA<IFCMFFA,
RFCM<RFCMFA<RFCMFFA, FCM<IFCM<RFCM

Table 2 Performance analysis indices for blood cancer cells segmentation

Distance function	Algorithm	Number of clusters = 3			Number of cluster = 4		
		#Iterations	DB	Dunn	#Iterations	DB	Dunn
Euclidean distance	FCM	11	7.8496	0.1446	34	7.2173	0.0877
	IFCM	11	7.7675	0.1480	29	7.0405	0.0901
	RFCM	23	2.8246	0.2305	20	1.8026	0.4871
	FCMFA	8	7.8494	0.1446	23	7.2094	0.0879
	IFCMFA	6	7.7673	0.1481	24	7.0409	0.0901
	RFCMFA	10	1.7249	0.9082	12	1.3202	0.5407
	FCMFFA	8	7.8486	0.1446	20	7.2090	0.0879
	IFCMFFA	5	7.7666	0.1481	20	7.0244	0.0909
	RFCMFFA	5	1.7248	0.9083	13	1.2880	0.5542
Gaussian Kernel	GKFCM	16	0.0010	19.2985	10	0.0006	8.8262
	GKIFCM	10	0.0549	19.3398	19	0.0421	8.9770
	GKRFCM	22	0.0090	167.3301	13	0.0043	81.197
	GKFCMFA	6	0.0008	19.2991	21	0.0006	8.8265
	GKIFCMFA	8	0.0549	19.3398	15	0.0420	8.8361
	GKRFCMFA	5	0.0089	167.3261	12	0.0042	84.9657
	GKFCMFFA	6	0.0008	19.2985	21	0.0006	8.8241
	GKIFCMFFA	4	0.0549	19.3408	6	0.0225	68.3863

(continued)

Table 2 (continued)

Distance function	Algorithm	Number of clusters = 3			Number of cluster = 4		
		#Iterations	DB	Dunn	#Iterations	DB	Dunn
	GKRFCMFFA	6	0.0089	167.3011	7	0.0042	84.5302
Hyper-tangent Kernel	HKFCM	9	0.0010	21.1514	20	0.0007	9.2740
	HKIFCM	9	0.0541	21.2074	18	0.0462	9.2792
	HKRFCM	20	0.0092	145.1430	12	0.0045	76.9258
	HKFCMFA	6	0.0010	21.1520	20	0.0006	9.2752
	HKIFCMFA	5	0.0541	21.2133	16	0.0462	9.2797
	HKRFCMFA	8	0.0092	166.5903	11	0.0045	76.9258
	HKFCMFFA	5	0.0009	21.1529	6	0.0004	65.2348
	HKIFCMFFA	9	2.8761	21.2130	21	6.9347	9.2775
	HKRFCMFFA	5	0.0089	168.5143	6	0.0102	137.5797

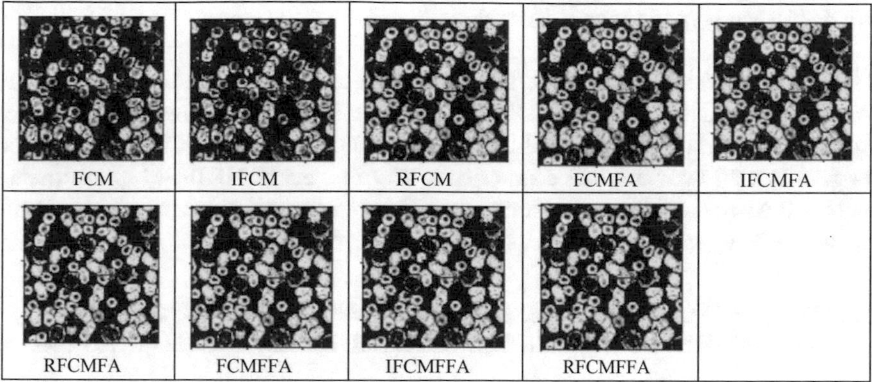

Fig. 3 Output of segmentation of blood cancer cells using Euclidean distance

4.2.2 Using Gaussian Kernel

It can be observed from Fig. 4 that the outputs are significantly improved than those using the Euclidean measures. It is also evident that FCM and RFCM produce better results than IFCM. It can be established by referring to the performance indices (Table 2) that in FCM and RFCM, both firefly and fuzzy firefly gives slightly better results. In case of IFCM, it is difficult to establish any concrete relation.

The optimized versions show better convergence rate in all the cases. We may conclude the following relations:

$$\text{GKFCM} < \text{GKFCMFA} < \text{GKFCMFFA},\ \text{GKIFCM} \approx \text{GKIFCMFA} < \text{GKIFCMFFA},$$
$$\text{GKRFCM} < \text{GKRFCMFA} < \text{GKRFCMFFA},\ \text{GKIFCM} < \text{GKFCM} < \text{GKRFCM}$$

Fig. 4 Output of segmentation of blood cancer cells using Gaussian Kernel

4.2.3 Using Hyper-tangent Kernel

These results are quite similar to those obtained using Gaussian Kernel (Table 2). Results obtained by HKFCM and HKRFCM are evidently better than those obtained from HKIFCM. HKFCMFA, HKFCMFFA, HKRFCMFA, HKRFCMFFA gives better results than their original counterparts. But in case of HKIFCM, performance increases when combined with firefly algorithm but decreases when combined with fuzzy firefly algorithm. Thus, the following relations can be established:

HKFCM<HKFCMFA<HKFCMFFA, HKIFCMFFA<HKIFCM<HKIFCMFA,
HKRFCM<HKRFCMFA<HKRFCMFFA, HKIFCM<HKFCM<HKRFCM

5 Conclusions

Summarizing the results obtained above we conclude the following:

(i) The use of Firefly and Fuzzy Firefly algorithms show improvements in both performance indices as well as the convergence rate of all the clustering algorithms handled here.
(ii) Using Fuzzy Firefly algorithm shows better results than using Firefly algorithm.
(iii) The kernelized versions provide better results than their corresponding Euclidean versions.
(iv) RFCM outperforms both IFCM and FCM.
(v) Hyper-tangent Kernel provides slightly better results than Gaussian Kernel.
(vi) The kernelized versions of both FCM and RFCM outperform IFCM.
(vii) RFCM with Hyper-tangent kernel as similarity measure and Fuzzy Firefly algorithm for initial centroid selection gives the best result amongst all the combinations considered here.
(viii) The performance of IFCM combined with firefly or fuzzy firefly does not show any stable relation.

References

1. Abhay, J., Srujan, C., Tripathy, B.K.: Stabilizing rough sets based clustering algorithms using firefly algorithm over image datasets. Informat. Commun. Technol. Intell. Syst. (ICTIS) **2**, 325–332 (2017)
2. Atanassov, K.T.: Intuitionistic fuzzy sets. Fuzzy Sets Syst. **20**(1), 87–96 (1986)
3. Bezdek, J.C., Ehrlich, R., Full, W.: FCM: the fuzzy c-means clustering algorithm. Comput. Geosci. **10**(2–3), 191–203 (1984)
4. Chaira, T.: A novel intuitionistic fuzzy C means clustering algorithm and its application to medical images. Appl. Soft Comput. **11**(2), 1711–1717 (2011)

5. Chen, D.Z.S.: Fuzzy clustering using kernel method. IEEE, Nanjing, China (2002)
6. Davis, D.L., Bouldin, D.W.: A cluster separation measure. IEEE Trans. Pattern Anal. Mach. Intell. PAMI-1(2), 224–227
7. Dunn, J.C.: A fuzzy relative of the ISODATA process and its use in detecting compact well-separated clusters. J. Cybernet. **3**(3), 32–57 (1973)
8. Hassanzadeh, T., Kanan, H.R.: Fuzzy FA: A modified firefly algorithm. Appl. Artificial Intelligence. **28**(1), 47–65 (2014)
9. Jain, A.K., Murty, M., Flynn, P.: Data clustering: a review. ACM Comput. Survey. **31**, 264–323 (1999)
10. Maji, P., Pal, S.K.: RFCM: a hybrid clustering algorithm using rough and fuzzy set. Fundamenta Informaticae **8**(4), 475–496 (2007)
11. Mitra, S., Banka, H., Pedrycz, W.: Rough-Fuzzy collaborative clustering. IEEE Trans. Syst Man Cybernet. Part B Cybernet. **36**(4), 795–805 (2006)
12. Pawlak, Z.: Rough sets. Int. J. Parallel Prog. **11**(5), 341–356 (1982)
13. Tripathy, B.K., Ghosh, A., Panda, G.K.: Kernel based k-means clustering using rough set. In: Proceedings of 2012 International Conference on Computer Communication and Informatics (ICCCI-2012), Jan. 10–12, pp. 1–5. Coimbatore, India (2012)
14. Tripathy, B.K., & Bhargav, R. (2013). Kernel based rough-fuzzy C-means. In: International Conference on Pattern Recognition and Machine Intelligence (PReMI), ISI Calcutta, December, LNCS 8251, pp. 148–157
15. Tripathy, B.K., & Mittal, D. (2015). Efficiency analysis of kernel functions in uncertainty based C-means algorithms. In: International Conference on Advances in Computing, Communications and Informatics, ICACCI 2015, Article number 7275709, pp. 807–813
16. Yang, X.S.: Firefly algorithms for multimodal optimization. In: International Symposium on Stochastic Algorithms. Springer, Berlin, Heidelberg (2009)
17. Zadeh, L.A.: Fuzzy Sets. Inf. Control **8**(3), 338–353 (1965)

Stock Market Prediction and Portfolio Optimization Using Data Analytics

Sulochana Roy

Abstract Analyzing the past records and precisely predicting the future trends is an important facet in a financially volatile market like the stock market! This paper employs Auto-ARIMA and Holt-Winters models for predicting the future value of stocks and Linear programming for portfolio optimization. Continuously training the model using the latest market data, helps the model get trained on the latest market behavior, so that it can be deployed further for more no. of portfolios in future; The trend obtained can be analyzed for the same to predict better investment plans in the stock market. Thus, helping people optimally invest in the stocks, depending upon the present market analysis, would help them fetch maximum return and suffer comparatively lesser loss.

Keywords Holt-Winters · Auto-ARIMA · Linear programming · Stock market · Portfolio optimization · Market analysis

1 Introduction

Stock market prediction involves determining the future value of a company stock being traded on an exchange. While there is risk associated with every stock, optimizing a portfolio of stocks might help in reduced risk and increased returns. Application of different algorithms to analyze the price patterns and predict stock prices and index changes, has caught the attention of many in the recent years. Intelligent trading systems are now being widely used by stock traders for predicting prices based on various situations and conditions, thus assisting them in making major investment decisions instantaneously. A constant monitoring of the stock prices is thus of utter importance as they are very dynamic and possess a susceptible nature to quick changes by virtue of the underlying trait of the financial domain. So an intelligent trader is one who

S. Roy (✉)
Department of Computer Science and Engineering, Ramaiah Institute of Technology, Bangalore 560054, India
e-mail: roy.sulochana@gmail.com

H. S. Behera et al. (eds.), *Computational Intelligence in Data Mining*, Advances in Intelligent Systems and Computing 990,
https://doi.org/10.1007/978-981-13-8676-3_32

would invest in a stock after predicting its price, so as to avoid a hike in its price before purchase or sell a stock before its price falls. However, since it isn't an easy task replacing the expertise of an experienced stock trader, designing an accurate prediction algorithm can help one gain to a higher extent in their investments. Thus, there exists a direct relationship between the accuracy of the prediction algorithm and the profit made out of it.

Also, portfolio optimization focuses on minimizing the risks involved in investing in one stock and choosing various stocks and optimizing the investments to get maximum returns. This project focuses on how individual stock values in the market can be predicted and finding optimum solutions which increases the accuracy of the predictions. It also involves optimization of a combination of stocks to get maximum returns over a fixed period of time.

The two attributes chosen here for analysis are date and open values of the stocks. We analyzed the trend of these portfolios over a span of more than 13 years, till the current date. After preprocessing the dataset fetched, a range of plots were obtained by applying the Holt-Winters algorithm and Auto-ARIMA model; Then with the help of linear programming the return values of the selected stocks were computed, which helped us predicting, how to optimally invest the money in certain stocks that are chosen here, to get the best returns and suffer less loss comparatively; This can later be implemented even on a larger scale, involving more number of stocks to be analyzed.

2 Literature Survey

Yetis et al. [1] stated that stock market prediction can be done based on the neural networks model, called artificial neural network (ANN). Using a number of input parameters of the share market, with the help of ANN approach, the NASDAQ's stock values are predicted. Also the real exchange rate values of NASDAQ's stock market index were utilized for this proposition. Stock market data for the year 2012 and 2013 were fed as the input to the generalized feedforward network to train the network and thus a high-performance prediction method was explored. Multilayer Perceptron (MLP) networks was the prime area of focus here as they are layered feedforward networks, typically trained during back propagation. However, its main disadvantage lies in the fact, that they train very slowly and require lots of training data. Further regression was used to validate the network performance which is depicted using the Figs. 1 and 2.

The model designed was not only beneficial for individuals, but also corporate investors and financial analysts as it provided them with an insight into the future behavior and movement of stock prices and thus take appropriate steps immediately, in order to gain more profit in their trading and thus reduce loss as much as possible.

Vui et al. [2] cited that investment in the stock market was a great mode of financial income, though one needs to carefully invest in it because if its volatile nature. Many researchers till date have tried out various methods in computer science and

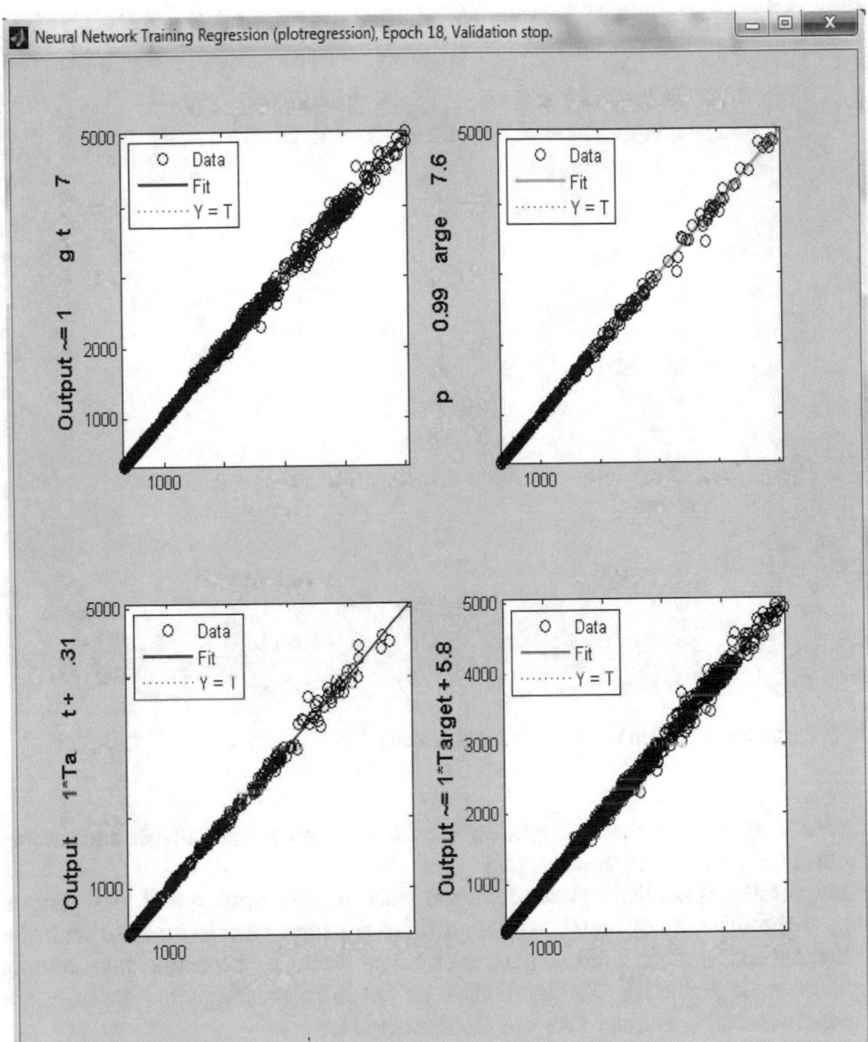

Fig. 1 Regression plot for training

economics, in order to forecast the stock market pricing and average movement but different techniques involving artificial neural networks (ANN) were investigated for the stock market prediction in this case. Thus it involved a review of the applications of artificial neural networks, which can be used for stock market prediction in future. It further led to the conclusion that the application of artificial neural networks for stock market prediction provided better accuracy than the application of general neural networks, by using a hybrid method and taking into account more number of

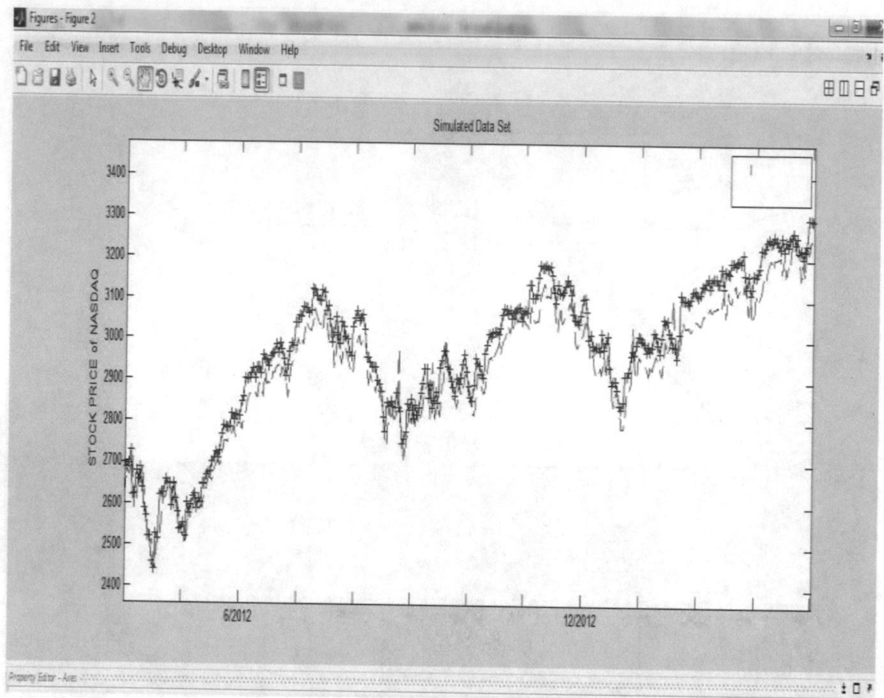

Fig. 2 Target and estimated values of annual NASDAQ data

external factors, as the dynamic stock market data is non-linear, volatile and subject to influences by many external factors.

Xing et al. [3] studied the trends in the stock market pricing that have always grabbed the eyeballs of many; Although there are high risks associated with the investment in the stock market, people continue doing so expecting high returns based on their prediction. The methods involved in forecasting stock prices often emerge in an endless stream like non-linear regression.

The method implemented here to forecast the trends in the stock market prices is based on the Hidden Markov Model, which differs from the ones already in use and fetches quite a precise result, particularly effective while predicting in a short duration of time. However, the traditional method used regression analysis for the prediction of stock prices which was based on analyzing the relationship between the market variables and dependent variables and establishing a regression equation among those variables; The same has been depicted in the Fig. 3.

The Hidden Markov Model used later is a technique used for estimating parameters and pattern recognition, which thereby helps in providing a probabilistic framework for multiple observations. The principle of HMM has been shown in Fig. 4.

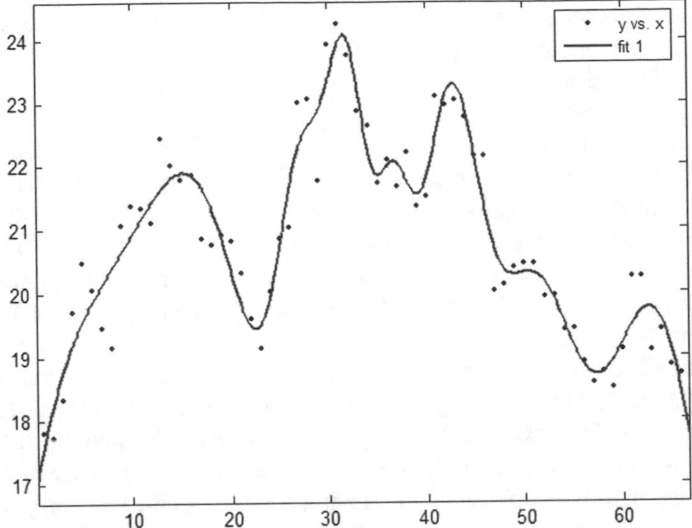

Fig. 3 Regression analysis prediction method of stock price

Fig. 4 Principle of HMM

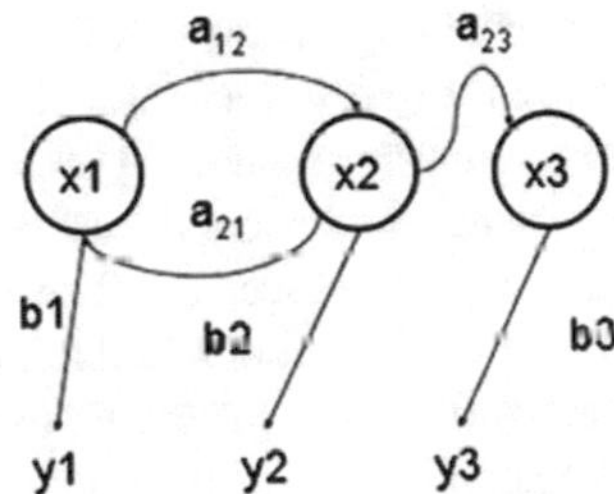

An empirical analysis was made on the Regression Analysis Prediction Method and Hidden Markov Model Prediction Method, respectively and the results were further compared to derive a conclusion. So based on the Hidden Markov Model, they selected the proper observed vector, the number of hidden state, the emission probability density functions, and identify the target time series change pattern.

Ahuja et al. [4] stated that public opinion and stock market sentiment analysis were the two major areas to be explored, in order to establish a relation between public moods and the trends in the stock market. The two main datasets used for this analysis, were namely Bombay Stock Exchange (BSE) values from June 2013 to December 2013 extracted from Yahoo Finance to obtain the open, close low and high values for various days and the publicly available Twitter data of the same period which included the timestamp, username and tweet text of each text during that period was used in this case in order to predict public emotions and past stock market data in order to help predict the future trends.

The model learning framework was fed with the mood values and BSE values which used Self Organizing Fuzzy Neural Networks (SOFNN), for learning a model to predict a future stock. Finally, BSE together with mood values are used for the portfolio management system for predicting the future stock values to make appropriate buying or selling decisions. This has not only helped in obtaining higher accuracy but also capture large scale trends in the stock market.

Khatri and Srivastava [5] stated that sentimental analysis can be used wisely in the prediction of stock market investment. "Sentiment analysis" being a term that is widely used in almost every industry these days, basically refers to the extraction of sentiments from the user's comments, that can help in detecting the view of a user for a particular company. In their research work, data for the top five companies namely Apple, Microsoft, Oracle, Google and Facebook were collected as comments and tweets from Stock Twits.

The data after extraction, was then analyzed to decide the mood of the comments made by the user and the comments were further classified under four main categories called happy, up, down and rejected. Finally, the polarity index and the market data, was made available to an artificial neural network that helped in predicting the results for all the top five companies. Also the closing prices of all the five companies were compared in order to suggest in which company an investor should invest.

Wang and Lin [6] focused on analyzing the data derived from the stock market that has considerably risen over the years; The need to understand the market trends better through analysis of the past records, for wiser financial investment in stocks, makes it really important for the investors. The main goal of their work contained two functions namely (1) Reducing stock information dimensionally and increasing efficiency and accuracy of the prediction. (2) Identifying Web robot system in the data capture procedure. The Web robot was employed here to capture the data from the stock market and henceforth the system analyzes this information to predict the stock prices in the seesaw process. So they not only employed the web robot, but also the genetic algorithm and the support vector machine here, to provide a framework for data analysis and stock market prediction.

Majumder and Hussian [7] established a computational approach for predicting the S&P CNX Nifty 50 Index. The direction in which the closing value of the index moved, could be predicted using the neural network-based model. This model could also predict the price index value of the stock market quite efficiently. An optimum model was thus proposed for forecasting purposes after studying the various characteristics of the network model. The data set chosen here for analysis comprised of data from 1st January 2000 to 31st December 2009. The model here was validated across 4 years of the trading days.

Therefore, the performance levels of the neural network models, in a highly volatile market like Indian Stock Market, reported in this paper was found to be quite beneficial. Along with this, predicting the direction of growth of the market with moderately high accuracy, also helped in guiding the investors and regulators.

Fiol-Roig et al. [8] stressed on the application of data mining techniques for stock market analysis. The problem considered here dealt with the design and validation of a data mining decision system which generates buying and selling orders in the stock market. The decision system's design mainly had three phases, namely, the data processing, data mining, and evaluation phase. The data processing phase selected a data set from the raw database. The step further following this involves the generation of the Object Attribute Table. In the data mining phase, the data contained in the OAT was converted into useful patterns, which helped yielding a decision tree. Eventually, the evaluation phase proved the pattern consistency with the help of a testing set.

Schumaker and Chen [9] emphasized on using Textual Analysis of Stock Market prediction using financial news articles. The role of financial news articles on three different textual representations: Bag of Words, Noun Phrases, and Named Entities and their ability to predict discrete stock prices after an article release was the prime aspect that they focused on. Using a support vector machine (SVM) derivative, it was proved that the model had a statistically significant impact on predicting the future stock prices, as compared to linear regression. It was also further stated that Named Entities representation scheme performed better than the de facto standard of Bag of Words.

Thus in order to answer their research questions on the effectiveness of discrete stock prediction and the best textual representation, they tested the model against a regression-based predictor using the three dimensions of analysis as mentioned earlier and finally concluded that not only their ***model predicted the stock prices significantly better than regression, but also stated that the Named Entities was better than the textual representation***.

Rajput and Bobde [10] primarily focuses on the stock market forecasting techniques and tries to figure out the best method among them. A study of different techniques was made to predict the stock price movement in the near future based on sentiment analysis from social media and data mining methodologies and out of the many, the one with maximum efficiency was chosen to predict the stock movement more accurately.

3 Forecasting Models

3.1 Autoregressive Integrated Moving Average (ARIMA)

Predicting values for a variable using its historical data points or predicting the change in one variable with respect to the value of another given variable is commonly termed as "forecasting". Forecasting can thus be categorized into qualitative and quantitative forecasting. Now noteworthy is the fact that time series forecasting which falls under the category of quantitative forecasting, has one of the most popular forecasting techniques under this category known as the ARIMA (Autoregressive Integrated Moving Average) model. This model comprises of three basic methods:

- Auto Regression: In this method, regression of the values of a given time series data occurs on their own lagged values, indicated by the "p" value.
- Differencing: In this method, the time series data is differenced to remove the trend and convert the non-stationary time series into a stationary one.

- Moving Average: In this method, the number of lagged values of the error term is measured, which is indicated by the "q" value.

The auto.arima() function used in R employs a variation of the Hyndman and Khandakar algorithm that includes the following steps:

- Finding the number of differences "d" using repeated KPSS tests.
- This is followed by minimization of the AICc and MLE, in order to fetch the values for "p" and "q", which finally yields the ARIMA model.

The degree of association between the current and past series values are measured using the autocorrelation and partial autocorrelation functions. This helps us to assess which past series values helps in prediction of the future values. Thereby using these Autocorrelation function (ACF) and Partial Autocorrelation function (PACF) values, the order of the processes in an ARIMA model can be easily determined. To be more precise,

- At lag *k*, Autocorrelation function is the correlation between series values that are *k* intervals apart.
- At lag *k*, Partial autocorrelation function is the correlation between series values that are *k* intervals apart, accounting for the values of the intervals between.

3.2 Holt-Winters

Capturing seasonality led Holt (1957) and Winters (1960) to extend the Holt's method. It is a seasonal method comprising of a forecast equation and three smoothing equations. They are namely for the level "$\ell t \ell t$", trend "bt", seasonal component denoted by "st", and smoothing parameters "α", "$\beta *$" and "γ". Also existing are two variations to this method which differ in the nature of the seasonal component. The additive method holds an upper hand when the seasonal variations are roughly constant through the series, over the multiplicative method, where the seasonal variations change proportionally to the level of the series.

4 Proposed Method

Stock market prediction involves determining the future value of a company stock (or other financial instruments) being traded on an exchange. While there is risk associated with every stock, optimizing a portfolio of stocks might help in reduced risk and increased returns. I have selected 10 stocks here from NYSE/NSE/BSE stock markets and collected their historical prices for a span of the last 13 years approximately. The stocks selected are diverse enough either by their sector or performance. Prediction of the stock prices would be for either next one week or month. For fund investors, it is essential to know how much to allocate/invest in each asset in a portfolio to maximize the returns and/or minimize the risk. Thus a portfolio, created with the top 5 performing stocks has been optimized.

Selecting an appropriate forecaster for modeling stock market trend is a critical and risky job. The ensemble framework thus suggested here attempts to overcome the risk of model selection as well as enhancing the overall prediction accuracy of the method. The framework combines two models namely, ARIMA and Holt-Wintersand the technique used for portfolio optimization is linear programming.

Algorithm 1: Prediction and Portfolio Optimization

//The two attributes chosen here for analysis are date and open values of the selected stocks.

Input: The TrainData, TestData, and n component forecasting models M_i where i = 1, 2, ..., n.

Output: The combined forecast through (1)

1. Normalization of TrainData and TestData.
2. Fit the model M_i to TrainData and update knowledge through learning.
3. Supply TestData to M_i and estimate the forecasts.
4. Record the MAPE values in the data table, calculated using (1), for all the selected stocks with respect to the chosen models.
5. Function to plot the ACF and PACF curves and forecasts using the chosen models.
6. Compute mean yearly return for each of the selected stocks .
7. Finally use Linear Programming for portfolio optimization, in order to get a wise idea of how much should be invested in each stock, in order to maximize the returns or minimize the risk, based on the return values computed in the previous step.

---

5 Experimental Results and Analysis

5.1 *Analysis of the ACF and PACF Plots*

5.1.1 Asian Paints

Figure 5 depicts the ACF and PACF plots for the stock values of Asian Paints. The ACF plot indicates that the data is highly correlated with one lag, but definitely decreasing at a very slow pace which indicates that data is non-stationary. In PACF plot, statistically significant is the first lag value, whereas not that statistically significant are the partial autocorrelations for all other lags. Thus a possible AR(1) model can be suggested for this dataset.

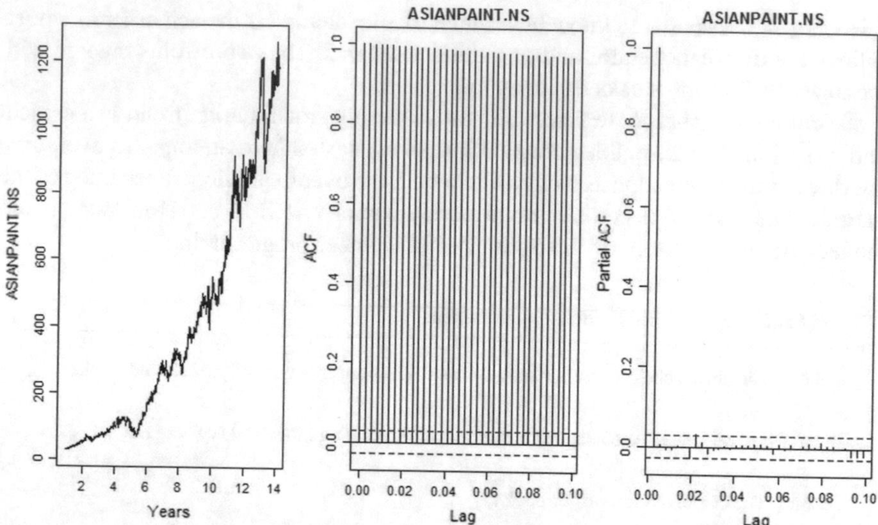

Fig. 5 ACF and PACF plots for the stock values of Asian Paints

5.1.2 HDFC Bank

The ACF and PACF plots for the stock values of Asian Paints are also obtained. The ACF plot indicates that the data is highly correlated with one lag, but definitely decreasing at a very slow pace which indicates that data is non-stationary. In PACF plot, statistically significant are some of the lag values, whereas not that statistically significant are the partial autocorrelations for all other lags. Thus, a possible AR(0) model can be suggested for this dataset as there is a single negative and positive correlation present here.

5.1.3 Hindustan Unilever

The ACF and PACF plots for the stock values of Hindustan Unilever are obtained. The ACF plot indicates that the data is highly correlated with one lag, but definitely decreasing at a very slow pace which indicates that data is non-stationary. In PACF plot, statistically significant some of the lag values, whereas not that statistically significant are the partial autocorrelations for all other lags. Thus, a possible AR(0) model can be suggested for this dataset as there is only one negative and positive correlation present here.

5.1.4 Infosys

The ACF and PACF plots for the stock values of Infosys is obtained. The ACF plot indicates that the data is highly correlated with one lag, but definitely decreasing at a very slow pace which indicates that data is non-stationary. In PACF plot, statistically significant are the first and second lag values, whereas not that statistically significant are the partial autocorrelations for all other lags. Thus, a possible AR(0) model can be suggested for this dataset as there is only one negative and positive correlation.

5.1.5 ITC

The ACF and PACF plots for the stock values of ITC are obtained. The ACF plot indicates that the data is highly correlated with one lag, but definitely decreasing at a very slow pace which indicates that data is non-stationary. In PACF plot, statistically significant are many lag values, whereas not that statistically significant are the partial autocorrelations for all other lags. Thus, a possible AR(4) model can be suggested for this dataset as there is only one negative correlation for one lag here.

5.1.6 L&T

The ACF and PACF plots for the stock values of L&T are also obtained. The ACF plot indicates that the data is highly correlated with one lag, but definitely decreasing at a very slow pace which indicates that data is non-stationary. In PACF plot, it is noteworthy that the many lag values are statistically significant, whereas partial autocorrelations for all other lags are not statistically significant. There are negative correlation as well for some of the lags. This suggests a possible AR (4) model for these data.

5.1.7 Maruti Suzuki

The ACF and PACF plots for the stock values of Maruti Suzuki are obtained. The ACF plot indicates that the data is highly correlated with one lag, but definitely decreasing at a very slow pace which indicates that data is non-stationary. In PACF plot, it is noteworthy that the many lag values are almost statistically significant, whereas partial autocorrelations for all other lags are not statistically significant. There is negative correlation for one lag. This suggests a possible AR (2) model for these data.

5.1.8 Reliance

The ACF and PACF plots for the stock values of Reliance are also obtained. The ACF plot indicates that the data is highly correlated with one lag, but definitely decreasing at a very slow pace which indicates that data is non-stationary. In PACF plot, it is noteworthy that the many lag values are statistically significant, whereas partial autocorrelations for all other lags are not statistically significant. There are negative correlations as well for some of the lags. This suggests a possible AR (4) model for these data.

5.1.9 TCS

The ACF and PACF plots for the stock values of TCS are also obtained. The ACF plot indicates that the data is highly correlated with one lag, but definitely decreasing at a very slow pace which indicates that data is non-stationary. In PACF plot, statistically significant are some of the values in the center, whereas not that statistically significant are the partial autocorrelations for all other lags. Thus, a possible AR(2) model can be suggested for this dataset.

5.1.10 Kotak Mahindra Bank

The ACF and PACF plots for the stock values of Kotak Mahindra Bank are also obtained. The ACF plot indicates that the data is highly correlated with one lag, but definitely decreasing at a very slow pace which indicates that data is non-stationary. In PACF plot, statistically significant are the first and second lag values, whereas not that statistically significant are the partial autocorrelations for all other lags. Thus, a possible AR(0) model can be suggested for this dataset as there is one negative and positive correlation present here.

5.2 *Experimental Results*

5.2.1 Stock Market Prediction

The error metric used here to calculate the error in forecasting is MAPE (Mean Absolute Percentage Error), which is defined as

$$MAPE = \frac{1}{n}\sum_{t=1}^{n}\left|\frac{\mathrm{A}_t - \mathrm{E}_t}{\mathrm{A}_t}\right| \times 100\% \quad (1)$$

Table 1 MAPE values using both the models

	Stockname	HoltWinters	Auto-ARIMA
1	ASIANPAINT.NS	1.9620919	1.247149
2	HDFCBANK.NS	3.1435497	3.423577
3	HINDUNILVR.NS	1.7659287	1.228250
4	INFY.NS	1.8697016	1.301106
5	ITC.NS	0.6487883	0.472609
6	LT.NS	5.5143472	5.576714
7	MARUTI.NS	2.9315057	2.616841
8	RELIANCE.NS	0.5645602	2.872139
9	TCS.NS	1.8829762	2.160074
10	KOTAKBANK.NS	3.9934912	4.262594

where "A_t" is the actual value and "E_t" is the forecast value and "n" is the number of test data.

MAPE values for all the 10 selected stocks have been computed using (1) after fetching the forecasted values using Holt-Winters and ARIMA models. The MAPE obtained from individual models from all the ten stock data sets are summarized in Table 1, which displays the error percentage of each of the stock's forecasted values. The data thus fetched helps us infer which among the chosen two models perform comparatively better, considering the one which has minimal error percentage. It can be observed that the performance of ARIMA is found to be better in most cases than the other model chosen here which is Holt-Winters. However, there is no single model outperforming the other over all data sets.

5.2.2 Portfolio Optimization

The portfolio optimization was achieved through Linear Programming here. Say 6000 INR was the amount initially available for investment, this technique after forecasting MAPE values, further helped us in predicting the best performing stocks out of the selected ones. For example here, based on the return percentage summarized in Table 2, it could be inferred that it would be wise to invest 2000 INR in Kotak Mahindra Bank and the rest 4000 INR could be invested equally in the remaining 4 best performing stocks chosen here.

6 Conclusion

For creation of the prediction model, various steps like data preprocessing, classification, and model evaluation were incorporated as a part of the implementation process. The Auto-ARIMA and Holt-Winters models were employed as an efficient approach

Table 2 Yearly returns of the selected stocks

	Stockname	Returns
1	ASIANPAINT.NS	0.3516342
2	HDFCBANK.NS	0.2854772
3	HINDUNILVR.NS	0.1873871
4	INFY.NS	0.1644498
5	ITC.NS	0.2045008
6	LT.NS	0.3444382
7	MARUTI.NS	0.3563149
8	RELIANCE.NS	0.3037858
9	TCS.NS	0.2723172
10	KOTAKBANK.NS	0.5154164

for stock prediction, which helped us analyze the trends and seasonality from the decomposition multiplicative time series plots, ACF and PACF plots to understand the data correlation parameters and finally the forecast plots from Holt-Winters and Auto-ARIMA model, helped us tally the forecasted and actual stock values. Eventually, linear programming was employed for portfolio optimization, to calculate the yearly returns for each of the stocks after obtaining the error percentage of each of the stock's forecasting values from the previous step; Hence this model can be very beneficial for individuals and corporate investors, financial analysts, and users of financial news. It can also provide the future behavior and movement of stock prices which would help one take the correct actions immediately and act accordingly in their trading practices to incur more profit and reduce loss as much as possible. In future work, we can further reduce the minimal errors obtained here if we can train our model on a more recent data set than very old data, as the unnecessary seasons, trends or randomness which have no significance today can then be easily avoided. The models were thus found to perform well, considering the fact that all the error percentages were very minimal in nature.

References

1. Yetis, Y., Kaplan, H., Jamshidi, M.: Stock market prediction by using artificial neural network. In: World Automation Congress (WAC), 2014, pp. 718–722. IEEE (2014)
2. Vui, C.S., Soon, G.K., On, C.K., Alfred, R., Anthony, P.: A review of stock market prediction with artificial neural network (ANN). In: 2013 IEEE International Conference on Control System, Computing and Engineering (ICCSCE), pp. 477–482. IEEE (2013)
3. Xing, T., Sun, Y., Wang, Q., Yu, G.: The analysis and prediction of stock price. In: 2013 IEEE International Conference on Granular Computing (GrC), pp. 368–373. IEEE (2013)
4. Ahuja, R., Rastogi, H., Choudhuri, A., Garg, B.: Stock market forecast using sentiment analysis. In: 2015 2nd International Conference on Computing for Sustainable Global Development (INDIACom), pp. 1008–1010. IEEE (2015)

5. Khatri, S.K., Srivastava, A.: Using sentimental analysis in prediction of stock market investment. In: 2016 5th International Conference on Reliability, Infocom Technologies and Optimization (Trends and Future Directions) (ICRITO), pp. 566–569. IEEE (2016)
6. Wang, C.-T., Lin, Y.-Y.: The prediction system for data analysis of stock market by using genetic algorithm. In: 2015 12th International Conference on Fuzzy Systems and Knowledge Discovery (FSKD), pp. 1721–1725. IEEE (2015)
7. Majumder, M., Hussian, M.A.: Forecasting of Indian stock market index using artificial neural network. Inf. Sci., 98–105 (2007)
8. Fiol-Roig, G., Miró-Julià, M., Isern-Deyà, A.P.: Applying data mining techniques to stock market analysis. In: Trends in Practical Applications of Agents and Multiagent Systems, pp. 519–527. Springer, Berlin, Heidelberg (2010)
9. Schumaker, R., Chen, H.: Textual analysis of stock market prediction using financial news articles. In: AMCIS 2006 Proceedings, p. 185 (2006)
10. Rajput, V., Bobde, S.: Stock market forecasting techniques: literature survey. Int. J. Comput. Sci. Mob. Comput. **5**(6), 500–506 (2016)

A Novel Position Concealment Audio Steganography in Insensible Frequency

Nabanita Mukherjee, Goutam Paul and Sanjoy Kumar Saha

Abstract For years, people have devised different techniques for encrypting data while others have attempted to break these encrypted codes. Steganography is a technique that allows one to hide data within an image, audio, or video. Imperceptibility of noise is an important issue for any steganographic technique. For this reason, we have proposed a new audio steganographic approach without considering the sample value less than 50 Hz and above 5000 Hz, because this range is highly sensible by ear as per biological phenomenon. The novelty of our work is that we have embedded the matching position value (determined from audio sample value itself) in 4 least significant bits rather than direct message bits flipping. Rather than performing sequential embedding, samples are selected nonsequentially based on audio feature itself that helps to withstand statistical attacks. Blind audio steganalysis techniques are hardly found—most of them are either algorithm specific or useful for bit-flipping data hiding scheme. In this context, unlike most common message bit flipping techniques, a novel bit matching and position concealment procedure has been proposed. Satisfactory theoretical and experimental outcomes have been achieved.

Keywords Audio steganography · Information security · Bit matching · Position concealment

N. Mukherjee (✉) · S. K. Saha
Department of Computer Science and Engineering, Jadavpur University,
Kolkata 700 032, India
e-mail: nabanitaganguly0@gmail.com

S. K. Saha
e-mail: sanjoy@jdvu.ac.in

G. Paul
Cryptology and Security Research Unit, R. C. Bose Centre for Cryptology and Security,
Indian Statistical Institute, Kolkata 700 108, India
e-mail: goutam.paul@isical.ac.in

H. S. Behera et al. (eds.), *Computational Intelligence in Data Mining*, Advances in Intelligent Systems and Computing 990,
https://doi.org/10.1007/978-981-13-8676-3_33

1 Introduction

In the wake of technology in the twenty-first century, especially in the field of multimedia, different types of processes have been developed in the recent past for information hiding and data manipulation. There are different types of information hiding techniques as cryptography, watermarking, steganography [5, 6]. In cryptography, secret message is encrypted itself using some keys whereas the idea of steganography is that the secret message is concealed in some open secret manners [4].

Unlike watermarking, optimizing capacity with minimum and imperceptible distortion is the goal and challenge for a good steganographic algorithm.

Carrier signal selection is an important issue in steganography with respect to robustness against attacks. For that audio is one of the suitable carriers. Binny et al. [12] have proposed least significant bit (LSB) based audio steganographic technique that limits its payload less than or equal to 1. In other words, it confers a poor payload. Bandyopadhyay et al. [11] have described a 4-bit compensation method that minimizes the distortion levels in single-bit audio steganography. Asad et al. [16] have proposed a two-way audio steganography—first, the bit number of host message is randomized and then the next secret message in the sample is randomized. Provos has shown that if embedding is done in nonsequential way, then it can withstand the statistical attacks. Most of the researchers use key for nonsequential selection of pixels in image or samples from the audio space. Paul et al. [7] have described an innovative approach of nonsequential selection based on the cover media feature itself in spatial domain, but this concept can be extended in frequency domain as well. Nugraha et al. [17] have described direct sequence spread spectrum audio steganography based on some practical use. But this is a very costly technique as compared with other LSB methods. An audio-to-audio synchronized steganographic method, as described in [2], is based on sharing a secret key, but sharing of keys is extra overhead for both sender and receiver. Gambhir et al. [9] have proposed a multilayer crypto-steganography in order to enhance the security level. A multibit grouping algorithm [10] is provided with an advanced deviation of LSB encoding method that ensures the high payload with low distortion. But in all these methods, message bits are directly embedded in the sample value in spatial domain by flipping of bits which may easily be detected by steganalysis tools.

Here, we put forward a novel audio steganography method where we avoid sample value less than 50 Hz and above 5000 Hz frequencies, because this range is highly sensible by Human Auditory System. We select samples in a nonsequential manner, which helps to withstand statistical attacks. We present a bit matching and position concealment embedding technique.

Table 1 shows the list of Abbreviations and Symbols.

Table 1 Abbreviations and Symbols

SNR:	Signal-to-noise ratio	MOS:	Mean opinion score
BER:	Bit error rate	$SSIM$:	Structural similarity index metric
NCC:	Normalized cross-correlation	FFT:	Fast Fourier transformation
M'_k	kth message bit	$\gamma_b^{(i)}$	bth bit of ith sample of cover audio file
$\gamma^{(i)}$	ith sample of the cover audio	$\gamma'^{(i)}$	ith sample of the stego audio
$\gamma_y'^{(i)}$	yth bit of ith sample of stego audio file		

2 Proposed Method

The embedding and extraction procedures are described as follows.

2.1 *Message Embedding*

Human Auditory System is very sensible in range 50 Hz–5000 Hz, so we avoid this range for audio sample selection. First, we take an 8-bit color image as a secret message. Consider a 16 bit cover audio sample as follows:

$$\underbrace{s_{15}\ s_{14}\ s_{13}\ s_{12}\ s_{11}\ s_{10}\ s_9\ s_8\ s_7\ s_6\ s_5\ s_4}_{g-h}\ \underbrace{s_3\ s_2\ s_1\ s_0}_{h}$$

Here $s_{15}s_{14}\ldots s_0$ is denoted by g and $s_3s_2s_1s_0$ is represented by h. So, $s_{15}s_{14}...s_5s_4$ is $g-h$.

Now select samples if $F = 1$ according to Eq. (1).

$$F = \begin{cases} 0\,, \text{if } 50\text{ Hz} < f(s) <= 5000\text{ Hz;} \\ 1\,, \text{otherwise,} \end{cases} \tag{1}$$

where $f(s)$ represents the frequency of the respective sample after performing FFT operation.

Then sort the samples in ascending order considering $g-h$ MSBs, of each sample (break tie based on sample number in cover audio).

Let $\gamma_b^{(i)}$ be the bth bit of ith sample of the audio file. Compare the raw data bit of secret image (say, $M'_{(i-1)}$) to be concealed in the audio sample with its $g-h$ MSBs (say, M'_i) till the first match is found. If match is found then conceal the h LSBs of the audio sample with their binary equivalent of the particular matched Most Significant Bit position. Otherwise, conceal the h LSBs of the audio sample with all 0s (indicates that the samples have no message bit). Then proceed to next audio sample for further

Input: S samples of audio file, an image with known height and width.
Output: The stego audio file.

Select samples if $F = 1$ according to Eq. (1).
Consider $g - h$ MSBs for bit matching where h is the predefined counter of bits to be embedded each audio sample
Sort the samples in increasing order of $g - h$ MSBs values (break the tie based on the sample number in cover audio).
for $i = 1$ *to* S **do**
    for $b = g$ *down to* $h + 1$ **do**
        if $M'_{(i-1)} = \gamma_b^{(i)}$ **then**
            $p = binary(b)$;
            break;
        end
        else
            $p = 0$;
        end
    end
    $\gamma'^{(i)} = (\gamma^{(i)} \;\&\; 0xFFF0) \mid p$;
end
Transform the stego audio

Algorithm 1: Embbeding algorithm

embedding. Repeat this till all the image bits are embedded. The whole embedding technique has been summarized in Algorithm 1.

2.2 Message Extraction

Consider a 16-bit stego audio sample as follows:

$$\underbrace{s_{15}\ s_{14}\ s_{13}\ s_{12}\ s_{11}\ s_{10}\ s_9\ s_8\ s_7\ s_6\ s_5\ s_4}_{g-h}\ \underbrace{s_3\ s_2\ s_1\ s_0}_{h}$$

Here $s_{15}s_{14}\ldots s_0$ is denoted by g and $s_3s_2s_1s_0$ is represented by h. So, $s_{15}s_{14}\ldots s_5s_4$ is $g - h$.

Now select samples if $F = 1$ according to Eq. (1). Here, $f(s)$ represents the frequency of the respective sample after performing FFT operation. Then sort the samples in ascending order considering $g - h$ MSBs, of each sample (break tie based on sample number in cover audio). Now from the selected ith audio sample, consider the last four bits of the audio sample (say, $\beta'^{(i)}$). If the h LSBs positions show all zeros, i.e., if $\beta'^{(i)} = 0$ then ignore it, otherwise determine its decimal equivalent (say, y). Find the bit (say, $\gamma_y'^{(i)}$) located at yth position of the stego sample value and that bit refers to the secret message bit (say, $l^{(i)}$) stored in ith sample. The extraction procedure has been summarized in Algorithm 2.

Input: The S samples stego audio file.
Output: The secret image.

Select samples if $F = 1$ according to Eq. (1);
Consider $g - h$ MSBs for bit matching where h is the predefined counts of bits to be embedded each audio sample;
Sort the samples in increasing order of $g - h$ MSBs values (break the tie based on the sample number in stego audio);
for $i = 1\ to\ S$ **do**
    $\beta'^{(i)} = (\gamma'^{(i)}\ \&\ 0x000F)$;
    if $\beta'^{(i)} \neq 0$ **then**
        $y = decimal(\beta'^{(i)})$;
        $l^{(i)} = \gamma_y'^{(i)}$;
    end
end
Extract the secret bits;

Algorithm 2: Extraction algorithm

3 Results and Analysis

First, we report a theoretical result followed by several experimental analyses.

3.1 MSB Invariance Theorem

Most significant bits remain unchanged before and after embedding.

Theorem 1 *Both of Algorithm 1 and 2 does not change the sequence of MSBs of the samples chosen in increasing order.*

Proof The $g - h$ MSBs remain unchanged before as well as after embedding. Because, here message bit and $g - h$ audio sample bit is compared. Only the h least significant bits are altered for storing the position number of the matching bit. □

3.2 Graphical Analysis

Figure 1 shows alike cover and stego audio, respectively, which are hardly detectable.

3.3 Payload Analysis

Our embedding capacity is 4 out of 16 bits audio sample, so theoretically we get a payload of 25% per sample. In the least significant bit based method [15], embedding

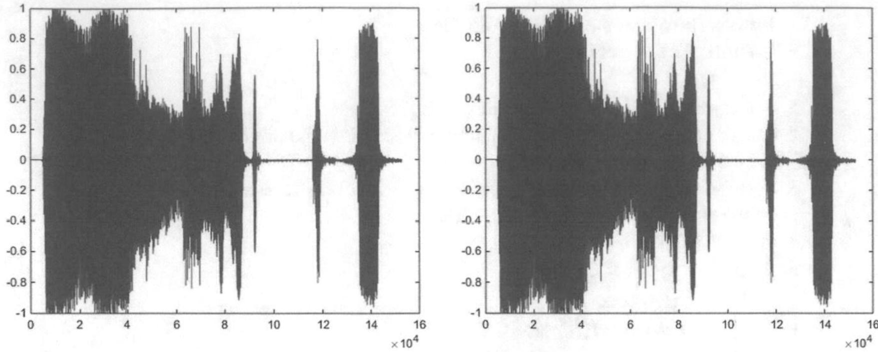

Fig. 1 Cover audio waveform (at left) and stego audio wavform (at right)

capacity is 1 out of 16 bits which give payload 6.25% of per sample. Therefore, our embedding capacity is four times better than compared to different LSB embedding methods.

3.4 Signal-to-Noise Ratio (SNR)

Definition 1 SNR is the ratio of a signal power (meaningful information) and unwanted signal power (i.e., noise) as shown in Eq. (2). Here, it is expressed in decibel.

$$\text{SNR} \triangleq 10 \log_{10} \frac{Power\ of\ signal}{Power\ of\ noise} \tag{2}$$

The tested average result of SNR are summarized in Table 3.
Fig. 2 shows the comparison of SNR of various Algorithms (Algo 1 [3], Algo 2 [13], Algo 3 [18], Algo 4 [1], Algo 5 [10] and proposed Algo).

3.5 Mean Opinion Score (MOS)

Definition 2 MOS is a statistical sample based measurement to differentiate cover and stego audio by common people based on an impairment scale as shown in Table 2.

For *M*OS analysis, a large number of audio sample files are tested by various people. Here 44.1 kHz sampled stereo audio files have been considered (i.e, represented by 16 bits per sample). The cover and the stego audio files are listened by every people in pairs. According to the people's opinion, MOS analysis is done based on 5-point impairment scale:

The tested result of MOS are shown in Table 3.

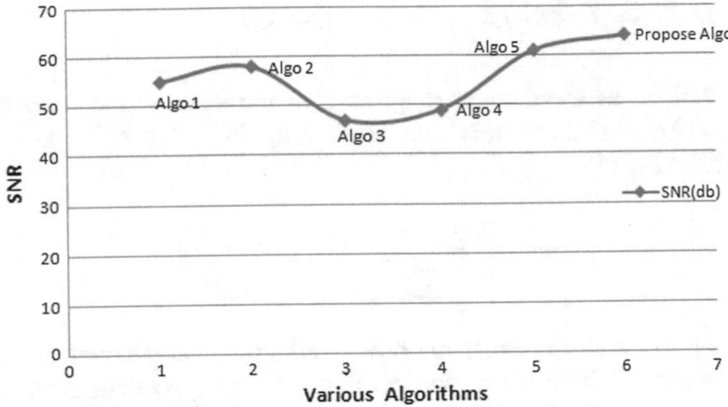

Fig. 2 Comparison of SNR for various Algorithms

Table 2 Score chart for calculating MOS

Imperceptible: 5
Perceptible but not annoying: 4
Very annoying: 3
Slightly annoying: 2
Annoying: 1

Table 3 Summarize experimental outcomes of SNR, MOS, BER, NCC, and SSIM tested on large audio of four categories, viz., 'Animal sounds', 'Melody', 'Punk' and 'Pop' music

Sound category	SNR	MOS	BER	NCC	SSIM
Animal sound	62.92	4.35	0.0411	0.9999	0.9998
Melody sound	61.58	4.50	0.0637	1.0000	0.9972
Punk sound	63.45	4.55	0.0845	0.9999	0.9999
Pop sound	64.50	4.45	0.0956	0.9999	0.9998

3.6 Bit Error Rate (BER)

Definition 3 *B*ER is the ratio of the counts of changed bits as after embedding into analyzed audio file and total count of bits in the cover version as described in Eq. (3).

$$BER \triangleq \frac{Counts\ of\ changed\ bits}{Total\ counter\ of\ bits} \tag{3}$$

The tested average BER values, as shown in Table 3, are satisfactory.

3.7 Normalized Cross-Correlation (NCC)

NCC is described as the deflection amount in the stego audio regarding cover audio [14]. A pair of same audio files NCC value is 1. The NCC value is determined from the Eq. (4)

$$NCC \triangleq \frac{\sum_{m=1}^{R}\sum_{n=1}^{W}[\theta(i,j).\theta^*(i,j)]}{\sum_{m=1}^{R}\sum_{n=1}^{W}\theta(i,j)^2} \tag{4}$$

where $\theta(i, j)$ and $\theta^*(i, j)$ denote the cover and stego audio sample, respectively. The tested average NCC value of our technique is summarized in Table 3. We find the value is very near to 1 which implies a negligible degradation caused by our embedding algorithm.

3.8 Structural Similarity Index Metric (SSIM)

Definition 4 *S*SIM [8] is a structural quality estimator measured by Eq. (5).

$$SSIM \triangleq \frac{(2 \times \overline{u} \times \overline{v} + W_1)(2 \times \lambda_{uv} + W_2)}{(\lambda_u^2 + \lambda_v^2 + W_2) \times (\overline{u^2} + \overline{v^2} + W_1)} \tag{5}$$

where $W_1 = (m_1' \cdot E)^2$ and $W_2 = (m_2' \cdot E)^2$ are two constants. E is $2^{\text{counter of bits per sample}}$ – 1. $m_1' = 0.01$ and $m_2' = 0.03$ by default.

The SSIM value varies from −1 to 1. The ideal case SSIM value for same audio files is 1. Table 3 shows the average SSIM value which is very close to 1 that evaluates the structural similarities between the cover and stego version which is very marginal deflection.

Table 3 shows the summarized outcomes of SSIM, BER, MOS, NCC, and SNR.

3.9 Payload Curve

The different payload curves (embedding capacity Versus distortion) are shown in Fig. 3. The distortion parameter includes SNR, BER, NCC, and SSIM. The payload is varied from 20% to 100%. Figure 3a shows capacity versus SNR where SNR decreases as capacity increases. Figure 3b shows capacity versus BER where BER increases as capacity increases. Figure 3c shows capacity versus NCC where NCC increases when capacity. Figure 3d shows capacity versus SSIM where SSIM decreases as capacity decreases.

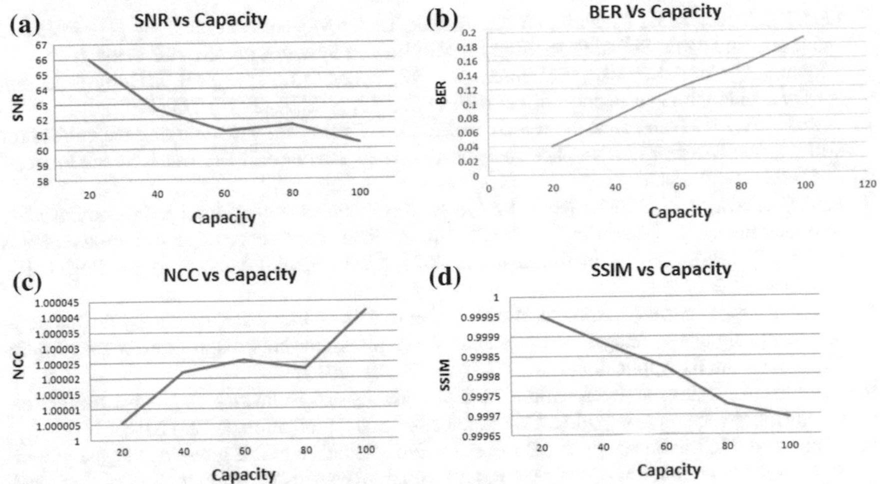

Fig. 3 Distortion parameter versus embedding capacity

4 Conclusion

Direct flipping in the LSB may be easier for the attacker to attack. In this work, we describe a multi-bit audio steganographic algorithm where matching bit position from MSB has been embedded in the last four bits from LSB of the sample value of the audio instead of flipping bits with secret message bits. Multibit blind audio steganalysis techniques hardly exist. As described, the last four bits of the cover audio is used for hiding secret bit, hence our technique can withstand against steganalysis attacks. Moreover, the samples have been chosen in nonsequential way from the sample space without any data loss at the receiver end which corresponds to the strength against first-order statistical attacks.

References

1. Sun, W., Shen, R., Yu, F., Lu, Z.: Data hiding in audio based on audio-to-image wavelet transform and vector quantization. In: Eighth International Conference on Intelligent Information Hiding and Multimedia Signal Processing (IIHMSP), , pp. 313-316. IEEE Xplore, Piraeus, China, 18–20th July 2002
2. Huang, X., Kawashima, R., Segawa, N., Abe, Y.: Design and implementation of synchronized audio to audio steganography scheme. In: International Conference on Intelligent Information Hiding and Multimedia Signal Processing (IIHMSP), pp. 331-334. IEEE Xplore. 15th–17th Aug 2008
3. Cvejic, N., Seppnen, T.: Increasing robustness of LSB audio steganography using a novel embedding method. In: Proceedings of the International Conference on Information Technology: Coding and Computing (ITCC 04), vol. 2, p. 533 (2004)

4. Cox, I.J., Miller, M.L., Linnartz, J.M.G., Kalker, T.: A review of watermarking principles and practices. In: Digital Signal Processing for Multimedia Systems, pp 461–482 (1999)
5. Bloom, J.A., Cox, I.J., Kalker, T., Linnartz, J.M.G., Miller, M.L., Traw, C.B.S.: Copy protection for DVD video. In: Proceedings of the IEEE, vol. 87, pp 1267, 1272–1275 (1999)
6. Hernandez, J.R., Amado, M., Perez-Gonzalez, F.: DCT-domain watermarking techniques for still images: Detector performance analysis and a new structure. IEEE Trans. Image Process. **9**, 55–68 (2000). Jan.
7. Paul, G., Davidson, I., Mukherjee, I., Ravi, S.S.: Keyless steganography in spatial domain using energetic pixels. In: Proceedings of the 8th International Conference on Information Systems Security (ICISS), IIT-Guwahati, India, vol. 7671, pp. 134-148. LNCS, Springer, Berlin, 15–19th Dec 2012
8. Hemlatha, S., Dinesh, A., Renuka, A., : Wavelet transform based steganography technique to hide audio signals in image. In: Elsevier B.V. Graph Algorithms, High Performance Implementations and Its Applications, vol: 47, pp. 272–281 (2015)
9. Gambhir, A., Khara, S.: Integrating RSA cryptography and audio steganography. In: International Conference on Computing, Communication and Automation, IEEE (2016)
10. Mukherjee, N., Bhattacharya, A., Bose, S.S.: Evolutionary multibit grouping steganographic algorithm. In: International Conference on Information Systems Security (ICISS), pp. 285–296. Springer, Berlin (2013)
11. Bandyopadhyay, S., Datta, B.: Higher LSB layer based audio steganography technique. Int. J. Electron. E-commun. Technol. **2**(4) (2011)
12. Binny, A., Koilkuntla, M.: Hiding secret information using LSB based audio steganography. In: International Conference on Soft computing and Machine Intelligence, IEEE (2015)
13. Divya, S., Reddy, M.: Hiding text in audio using multiple LSB steganography and provide security using cryptography. Int. J. Sci. Technol. Res. **1**(6) (2012)
14. Singh, S., Siddiqui, T.J.: A security enhanced robust steganography algorithm for data hiding. Int. J. Comput. Sci. Issues **9**(3),1694–0814 (2012). ISSN (Online)
15. Jayaram, P., Ranganatha, H., Anupama, H.: Information hiding using audio steganography "A Survey". Int. J. Multimed. Appl. (IJMA) **3**(3) (2011)
16. Asad, M., Gilani, J., Khalid, A.: An enhanced least significant bit modification technique for audio steganography. In: International Conference on Computer Networks and Information Technology (ICCNIT), pp. 143–147. 11–13th July 2011
17. Nugraha, R.M.: Implementation of direct sequence spread spectrum steganography on audio data. In: 2011 International Conference on Electrical Engineering and Informatics, Bandung, Indonesia. 17–19th July 2011
18. Zhu, J., Wang, R., Yan, D.: The sign bits of huffman codeword-based steganography for AAC audio. In: International Conference on Multimedia Technology (ICMT), pp. 1–4. IEEE Xplore, Ningbo, China, 29–31st Oct, 2010

Automated Evaluation of Outlier Performers in Programming Sessions Using Case-Based Reasoning Techniques

Anindita Desarkar, Ajanta Das and Chitrita Chaudhuri

Abstract Manual evaluation of software program is always a daunting task as it is subjective, time-consuming and requires significant amount of effort. It may also involve human bias. Automated analysis and evaluation can address the above problems adequately. In our paper, we have proposed an automated performance evaluation framework for programming assignment tasks, maintained within a Case-based corpus. The corpus should mandatorily consist of different programming techniques utilized to solve the problem set in an assignment, each one saved as a case instance. Candidate programs are maintained in a separate corpus and then statistically measured to detect outlier performance with respect to the degree of novelty of the approach, quality based on time complexity and amount of plagiarism detected. All new approaches would automatically update the existing base.

Keywords Case-Based Reasoning (CBR) · Automated evaluation · Interquartile Range (IQR) · Time complexity · Outliers

1 Introduction

Outlier detection is a technique to identify the presence of unusual patterns within a system, which doesn't conform to the general expected behavior. In the domain of software development, outlier programs can be detected based on a few selected parameters like whether they are syntactically correct, producing the desired output,

A. Desarkar (✉) · C. Chaudhuri
Department of Computer Science and Engineering, Jadavpur University, Kolkata, India
e-mail: aninditadesarkar@gmail.com

C. Chaudhuri
e-mail: cchitrita@gmail.com

A. Das
Department of Computer Science and Engineering, University of Engineering and Management, Kolkata, India
e-mail: cse.dr.ajantadas@gmail.com

H. S. Behera et al. (eds.), *Computational Intelligence in Data Mining*, Advances in Intelligent Systems and Computing 990,
https://doi.org/10.1007/978-981-13-8676-3_34

computationally efficient, and avoiding plagiarism. Scaled on a single dimension, outliers can be classified into two types: positive, if they lie beyond the maximum permissible level, and negative if they lie below the minimum permissible level. In our system, positive outliers are the outstanding programs whereas programs with very low scores have a high chance of being categorized as negative outliers.

Outlier program detection plays a decisive role in grading performers in educational institutes as well as in Industry. Outperformers get high ranks irrespective of background. But it is crucial to identify the backbenchers also in order to provide support and scope of improvement. An automated system for such gradation rules away the possibilities of human bias and proficiency. The candidates with poor performance and going through such checks, may have a better chance of upgrading themselves without causing hard feelings between team workers. Moreover, in the industry, it may also help to increase the robustness of a system if negative outliers are detected in advance.

Manual detection of flaws in programs is a herculean task, demanding huge amount of time and effort. The system proposed here is an Automated Assessment Evaluation Framework for measuring both outstanding and below standard performance of software programs and grade them accordingly, by outlier detection techniques. Two corpuses, one containing the problem assigned supported by some solution approaches for the same, and the other containing the candidate programs along with identification for each programmer, are maintained and updated by the system. CBR technology [1] is utilized to preserve both corpuses—the first is designated the PRB as it houses the problems, and the second one is designated PRG as it contains the candidate program details.

In the next Sect. 2, some previous researches in related domain are surveyed and discussed. The architecture of the proposed framework is depicted in the Sect. 3, followed by some detailing on the constituent modules. Actual results in the form of performance tables are presented in Sect. 4. Section 5 concludes by discussing several beneficial aspects of the proposed system. Some future scopes of improvements on the system are also indicated here. The reference section at the end records the foundation works based on which our research is performed.

2 Literature Review

Mentioned below are a few recent researches published in the domain of automated evaluation of software programs. Patel et al. [2] have proposed a tool to minimize manual intervention in java program evaluation. The tool makes the grading system easier by analyzing various stylometric features as well as structural behaviors of the program. Alhindawi et al. [3] have proposed a hybrid technique for java code complexity analysis. The module consists of two concepts—Halstead and McCabe. With the help of these two concepts, the tool analyzes the difference in complexity level present in the code. It is used to evaluate the proficiency level of programmers and their improvements over time. Research work has been performed on comparative analysis of software complexity of searching algorithms using code based metrics by Olabiyisi et al. [4]. The metrics include line of codes, McCabe Cyclomatic complexity metrics, and Halstead complexity metrics. They have used the following metrics to compare software complexity of linear and binary search algorithms, evaluate, rank competitive object-oriented applications like Visual Basic, C#, C++, and Java languages for these two algorithms. An automated evaluation system is presented by Saraf et al. [5] to make the grading process hassle-free. A web-based online automated scoring system is suggested by Kitaya and Inoue [6] which accepts submissions of programming assignments, assess automatically the submitted programs, and returns feedbacks to the students immediately.

3 Proposed Evaluation Methodology

The framework is primarily developed to create, maintain and upgrade two case bases, PRB and PRG, which houses the problem-solution pair details and candidate program details, respectively. Ultimately the candidate program case base PRG produces the final scores from which program gradation is evolved. Statistical evaluation of the final scores yields the outlier information, based on which some of the candidate programs are marked as outstanding, and some are noted to be below standard ones. The following subsections describe the details of the system.

3.1 Structure of Proposed Case Bases

CBR technique uses prior cases as instances to handle new situations. Cases are similar to experiences of problems preserved along with their solutions. CBR ***retrieves*** the past cases housing problems bearing maximum similarity with the present one. No new case is generated for an exact match—only the solution part is taken as the outcome. Otherwise, the solution parts of the nearest matches are ***reused*** and ***revised*** to generate the solution to the new problem. The new problem-solution pair is then ***retained*** as a new case in the Case Base [7].

A. **Problem Case Base: PRB**

This case base is utilized to establish various approaches of solving a specific problem along with the complexity and weight factors associated with each of these approaches. Time complexity of the underlying algorithmic approach for tackling the set problem is obtained either manually or by standard existing techniques [8, 9]. The descriptive parts are intentionally omitted in the present discussion. Table 1 depicts the details of this case base.

The first column identifies the ***Approach*** adopted by each solution. The next one records the ***Complexity*** analysis. The ***Initial_Complexity*** column assigns a preliminary weight to the approach within a prefixed percentage scale of the ***Score_Limit***.

Table 1 Structure of problem case base

Approach	Complexity	Initial_ Complexity	Approach Count	Final_Complexity
1	$O(2^n)$	1	2	3
2	O(n)	3	1	4
3	O(n)	3	1	4
4	O(n)	3	4	1
5	O(log n)	4	6	0.5

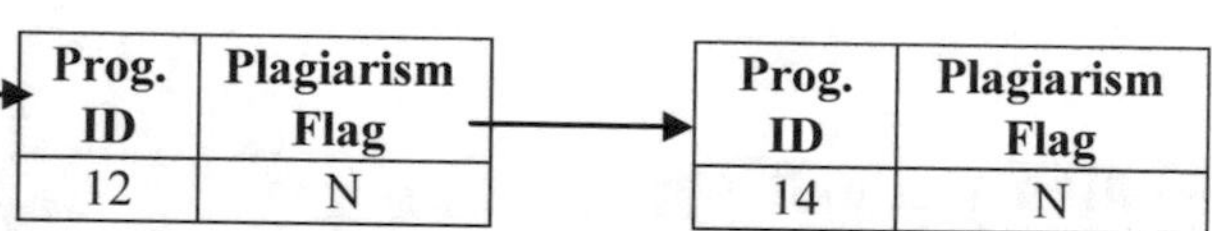

Prog. ID	Plagiarism Flag
3	Y

Prog. ID	Plagiarism Flag
4	Y

Prog. ID	Plagiarism Flag
8	Y

In this scale, the least weight is associated with the most complex approach, and the maximum with the least complex approach, the interim ones being arranged in between. The ***Approach Count*** column preserves the count of candidate programs adopting that approach. A ***Final_Complexity*** is calculated as per Eq. (1) by distributing the surplus weight equally amongst the candidates embracing that approach.

A linked table keeping the plagiarism check between such candidates is also partially shown in Table 1 to indicate how the plagiarism based ***Penalty***, as calculated in Eq. (2), is associated with the approach, if any candidate is found guilty of such a charge [10].

The complexity and penalty for each approach are calculated as follows:

$$\textit{Final\_Complexity} = (\textit{Score\_Limit} - \textit{Initial\_Complexity})/(\textit{Count\_per\_Approach}) \tag{1}$$

$$\textit{Penalty} = \textit{Initial\_Complexity}/\textit{Plagiarism\_Count\_per\_Approach} \tag{2}$$

B. Program Case Base: PRG

The Program Case Base is used to preserve all program-related details for each candidate. *Final_Score* for each candidate is obtained by applying the following expression:

$$\textit{Compile\_Score} + \textit{Run\_Score} + \textit{Initial\_Complexity} + \textit{Final\_Complexity} - \textit{Penalty} \tag{3}$$

Table 2 depicts the same.

Table 2 Structure of program case base

Prog. ID	Approach	Compile score	Run score	Initial complexity	Final complexity	Plag. penalty	Final score
1	5	2	1	4	0.5	2	5.5
2	5	2	1	4	0.5	2	5.5
3	4	2	1	3	1	1.5	5.5
4	4	2	1	3	1	1.5	5.5
5	5	2	1	4	0.5	2	5.5
6	4	2	1	3	1	1.5	5.5
7	0	0	0	0	0	0	0
8	4	2	1	3	1	1.5	5.5
9	2	2	1	3	4	0	10
10	5	2	1	4	0.5	2	5.5
11	5	2	1	4	0.5	2	5.5
12	1	2	1	1	3	0	7
13	3	2	1	3	4	0	10
14	1	2	1	1	3	0	7
15	5	2	1	4	0.5	2	5.5

3.2 *Outlier Detection Based on Statistical Techniques*

An outlier may be defined to be a data point in a system which is significantly different from all other data points belonging to that system. Statistically, such incongruous data points may be deciphered by setting up quartiles within the sorted data and generating the Interquartile Range (IQR) value of the system. All data points, lying at distances more than 1.5 times the IQR value beyond the 1st and 3rd quartile of the dataset, will be marked as outliers in this system [11].

The ones thus far beyond the 3rd quartile are truly outstanding performers—to be selected for nurturing their talents both in the academic as well as industrial environment. The poor performers—marked at an equal distance below the 1st quartile—would be provided with special attention in the academic world in the form of extra tutorials and guidance. In industry, they would need to undergo motivational counseling and changed responsibilities until they improve their skill level.

3.3 Proposed Framework

The proposed evaluation framework is presented in Fig. 1.

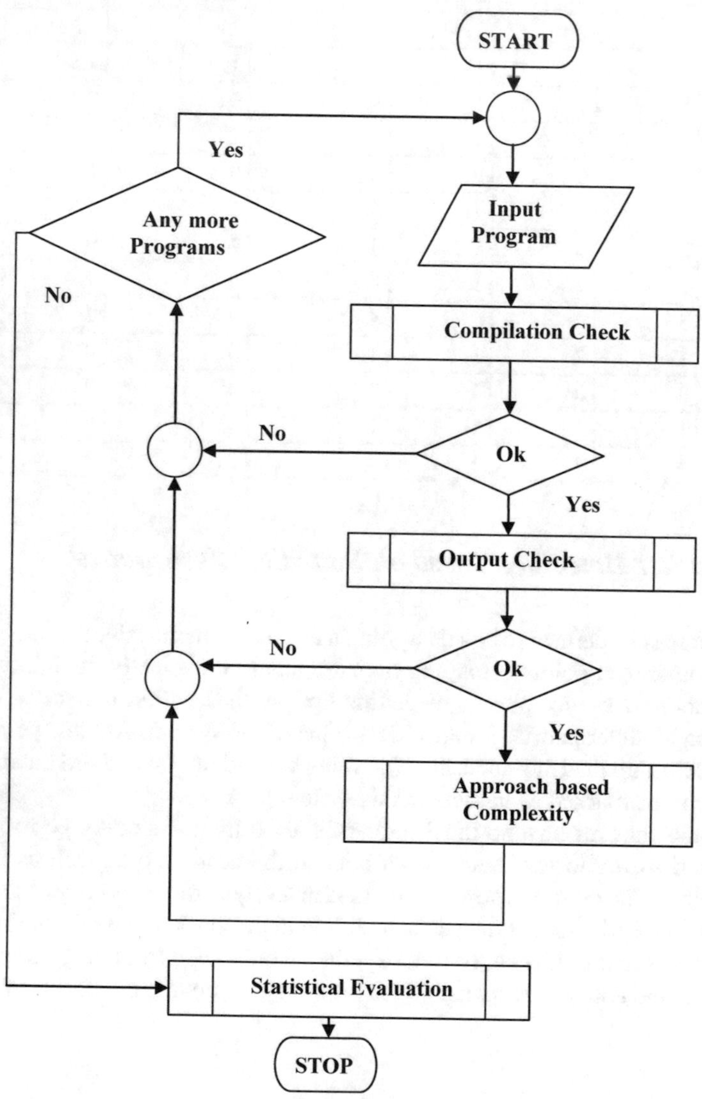

Fig. 1 Outline of assessment evaluation framework

Following is a detailed description of the algorithm adopted in the system.

Input : Problem Case Base : ***PRB*** and Program Case Base : ***PRG***
Output : (1) Problem Case Base for new approaches, if any
(2) Program Case Base with evaluation and outliers marked, if any

Procedure :

1. For each ***approach*** in ***PRB***
2. Initialize ***count[approach]*** to ***0***.
3. Setup initial and final_complexity[approach] attributes.
4. Setup linked table for candidate list.
5. EndFor approach
6. For each program in ***PRG***
7. Set ***compilation_value***.
8. If ***compilation_value*** = ***0*** goto ***19***.
9. Set ***output_value***.
10. If ***output_value*** = ***0*** goto ***19***.
11. Switch on ***approach_based_complexity*** value.
12. If approach is new, update ***PRB*** accordingly.
13. Increment count[approach] in ***PRB*** appropriately.
14. Set ***initial_complexity*** appropriately.
15. Fill up ***PRB*** linked table for each matching candidate.
16. Set ***plagiarism flag*** of program, if applicable.
17. Increment corresponding plagiarism count.
18. EndSwitch
19. EndFor program
20. For each approach in ***PRB***
21. Divide ***final_complexity*** by count[approach].
22. If applicable
23. Final ***plagiarism penalty*** obtained with division by plagiarism count.
24. Final_complexity[approach] obtained with division by count[approach].
25. EndFor approach
26. For each program in ***PRG***
27. Add ***compilation, output, initial and final complexity*** values to ***Score***.
28. Deduct ***plagiarism_penalty*** from Score.
29. EndFor program
30. //Program Case Base copy arranged in order to generate Outliers
31. Calculate Second Quartile (***Q2***) or median.
32. Calculate First Quartile (***Q1***) = median of lower half
33. Calculate Third Quartile (***Q3***) = median of upper half.
34. Calculate Inter Quartile (***IQR***) = ***Q3 – Q1***.
35. Calculate lower limit of tolerable range = ***Q1 − 1.5 x IQR***.
36. Calculate Weak performers who lie below the lower limit.
37. Calculate upper limit of the tolerable range = ***Q3 + 1.5 x IQR***
38. Calculate outstanding performers above the upper limit.

EndProcedure

4 Experimental Results

A prototype has been created based on the above methodology where "Fibonacci Number Generation" is taken as the problem statement and six approaches are detected from amongst the candidate programs. Three student groups have been tested containing 15, 25, and 35 candidates, respectively. For each of these groups, the outstanding performers as well as the below standard ones are detected and their results have been displayed in Figs. 2, 3 and 4.

```
For 15 students ---->   Q2 : [ Roll = 8 ] ---> 5.5
    Q1 : [ Roll = 4 ] ---> 5.5      Q3 : [ Roll = 14 ] ---> 7
IQR : 1.5       L = Q1 - 1.5 x IQR = 3.25       H = Q3 + 1.5 x IQR = 9.25

Srl# Roll  Approach  Compilation  Output  Initial_Complexity  Final_Complexity  Plgiarism_Penalty  Final_score
(1)   1     5          2           1         4                 0.5               2                5.5
(2)   2     5          2           1         4                 0.5               2                5.5
(3)   3     4          2           1         3                 1               1.5                5.5
(4)   4     4          2           1         3                 1               1.5                5.5
(5)   5     5          2           1         4                 0.5               2                5.5
(6)   6     4          2           1         3                 1               1.5                5.5
(7)   7     0          0           0         0                 0                 0                0 <=== Below Standard
(8)   8     4          2           1         3                 1               1.5                5.5
(9)   9     2          2           1         3                 4                 0                10 <=== Outstanding
(10) 10     5          2           1         4                  0.5              2                5.5
(11) 11     5          2           1         4                  0.5              2                5.5
(12) 12     1          2           1         1                 3                 0                 7
(13) 13     3          2           1         3                 4                 0                10 <=== Outstanding
(14) 14     1          2           1         1                 3                 0                 7
(15) 15     5          2           1         4                 0.5               2                5.5
```

Fig. 2 Results for 15 candidates

```
For 25 students ---->   Q2 : [ Roll = 2 ] ---> 4.83333
    Q1 : [ Rolls = 4 and 6 ] ---> 4.41667       Q3 : [ Rolls = 17 and 24 ] ---> 5.41667
IQR : 1     L = Q1 - 1.5 x IQR = 2.91667        H = Q3 + 1.5 x IQR = 6.91667

Srl# Roll  Approach  Compilation  Output  Initial_Complexity  Final_Complexity  Plgiarism_Penalty  Final_score
(1)   1     5          2           1         4                 0.5               2.66             4.83
(2)   2     5          2           1         4                 0.5               2.66             4.83
(3)   3     4          2           1         3                 0.66              2.25             4.42
(4)   4     4          2           1         3                 0.66              2.25             4.42
(5)   5     5          2           1         4                 0.5               2.66             4.83
(6)   6     4          2           1         3                 0.66              2.25             4.42
(7)   7     0          0           0         0                 0                 0                0 <=== Below Standard
(8)   8     4          2           1         3                 0.66              2.25             4.42
(9)   9     2          2           1         3                 1.33              1.5              5.83
(10)  10    5          2           1         4                 0.5               2.66             4.83
(11)  11    5          2           1         4                 0.5               2.66             4.83
(12)  12    1          2           1         1                 2                 0                6
(13)  13    3          2           1         3                 1                 2                5
(14)  14    1          2           1         1                 2                 0                6
(15)  15    5          2           1         4                 0.5               2.66             4.83
(16)  16    1          2           1         1                 2                 0                6
(17)  17    2          2           1         3                 1.33              1.5              5.83
(18)  18    3          2           1         3                 1                 2                 5
(19)  19    0          0           0         0                 0                 0                 0 <=== Below Standard
(20)  20    0          0           0         0                 0                 0                 0 <=== Below Standard
(21)  21    3          2           1         3                 1                 2                 5
(22)  22    4          2           1         3                 0.66              2.25              4.42
(23)  23    4          2           1         3                 0.66              2.25              4.42
(24)  24    3          2           1         3                 1                 2                 5
(25)  25    2          2           1         3                 1.33              1.5               5.83
```

Fig. 3 Results for 25 candidates

Srl#	Roll	Approach	Compilation	Output	Initial_Complexity	Final_Complexity	Plgiarism_Penalty	Final_score
(1)	1	5	2	1	4	0.33	2	5.33
(2)	2	5	2	1	4	0.33	2	5.33
(3)	3	4	2	1	3	0.57	0	6.57 <=== Outstanding
(4)	4	4	2	1	3	0.57	0	6.57 <=== Outstanding
(5)	5	5	2	1	4	0.33	2	5.33
(6)	6	4	2	1	3	0.57	0	6.57 <=== Outstanding
(7)	7	0	0	0	0	0	0	0 <=== Below Standard
(8)	8	4	2	1	3	0.57	0	6.57 <=== Outstanding
(9)	9	2	2	1	3	1	2	5
(10)	10	5	2	1	4	0.33	2	5.33
(11)	11	5	2	1	4	0.33	2	5.33
(12)	12	1	2	1	1	1.2	0.66	4.53
(13)	13	3	2	1	3	1	2	5
(14)	14	1	2	1	1	1.2	0.66	4.53
(15)	15	5	2	1	4	0.33	2	5.33
(16)	16	1	2	1	1	1.2	0.66	4.53
(17)	17	2	2	1	3	1	2	5
(18)	18	3	2	1	3	1	2	5
(19)	19	0	0	0	0	0	0	0 <=== Below Standard
(20)	20	0	0	0	0	0	0	0 <=== Below Standard
(21)	21	3	2	1	3	1	2	5
(22)	22	4	2	1	3	0.57	0	6.57 <=== Outstanding
(23)	23	4	2	1	3	0.57	0	6.57 <=== Outstanding
(24)	24	3	2	1	3	1	2	5
(25)	25	2	2	1	3	1	2	5
(26)	26	2	2	1	3	1	2	5
(27)	27	5	2	1	4	0.33	2	5.33
(28)	28	5	2	1	4	0.33	2	5.33
(29)	29	0	0	0	0	0	0	0 <=== Below Standard
(30)	30	0	0	0	0	0	0	0 <=== Below Standard
(31)	31	4	2	1	3	0.57	0	6.57 <=== Outstanding
(32)	32	5	2	1	4	0.33	2	5.33
(33)	33	1	2	1	1	1.2	0.66	4.53
(34)	34	1	2	1	1	1.2	0.66	4.53
(35)	35	0	0	0	0	0	0	0 <=== Below Standard

Fig. 4 Results for 35 candidates

5 Conclusion and Future Work

By implementing the above methodology, we have been able to grade a group of students with minimum manual intervention and bias. The system further detects extraordinarily merited performers as well as poor ones in a group. Another plus point of the evaluation system is its capability of measuring both the programming talent and the integrity of a candidate performer. The complexity-based weights measure the student's intelligence quotient, whereas plagiarism check finds out the person's ethical standards. Automatic upgrade of problem case bases results in knowledge enhancement with new approaches and statistics.

At present, it has not been possible to compare the results obtained with the present system as the survey did not yield any published work in this domain. A future scope may involve the incorporation of best practice rule set to determine the programmer's efficiency, which can then be benchmarked with existing systems.

References

1. Aamodt, A., Plaza, E.: Case-based reasoning: foundational issues, methodological variations, and system approaches. AI Commun. **7**(1), 39–59 (1994)
2. Patel, A., Panchal, D., Shah, M.: Towards improving automated evaluation of Java program. In: Emerging ICT for Bridging the Future-Proceedings of the 49th Annual Convention of the Computer Society of India (CSI), vol. 1, pp. 489–496. Springer, Cham (2015)

3. Alhindawi, N., Malkawi, R., Al-Batah, M.S., Al-Zuraiqi, A.: Hybrid technique for java code complexity analysis. Int. J. Adv. Comput. Sci. Appl. **1**;**8**(8), 379–385 (2017)
4. Olabiyisi, S.O., Omidiora, E.O., Sotonwa, K.A.: Comparative analysis of software complexity of searching algorithms using code based metrics. Int. J. Sci. Eng. Res. **4**(6) (2013)
5. Saraf, P., Ramesh, S., Patel, S., Pathak, S.: Automatic evaluation system for student Code. Int. J. Comput. Sci. Inf. Technol. (IJCSIT) **6**(2), 1869–1871 (2015)
6. Kitaya, H., Inoue, U.: An online automated scoring system for java programming assignments. Int. J. Inf. Edu. Technol. **6**(4), 275–279 (2016)
7. Farhan, U., Tolouei-Rad, M., Osseiran, A.: Indexing and retrieval using case-based reasoning in special purpose machine designs. Int. J. Adv. Manuf. Technol. **1**;**92**(5–8), 2689–2703 (2017)
8. Time Complexity Determination Tool: https://www.sonarqube.org/. Accessed 28 June 2018
9. Time Complexity Determination Tool: http://trend-prof.tigris.org/. Accessed 28 June 2018
10. Plagiarism Detection Tool: https://jplag.ipd.kit.edu/. Accessed 28 June 2018
11. Han, J., Kamber, M.: Data Mining Concepts and Techniques, 2nd edn. Mogran Kaufmann Publishers (2006)

Black Hole and k-Means Hybrid Clustering Algorithm

Shankho Subhra Pal and Somnath Pal

Abstract In recent advancements, to tackle with NP problems, many heuristic approaches have been developed. A lot of meta-heuristic methods are inspired from nature. In this paper, Black Hole optimization approach for data clustering has been hybridized with k-Means to get improved results. As compared to Black Hole optimization approach for data clustering, where the initial population is initialized randomly, in the proposed method, a part of the population is initialized with some of the better results from k-Means clustering algorithm and rest of the population is initialized randomly.

Keywords Nature-inspired clustering · k-Means Clustering · Black Hole optimization

1 Introduction

From around 1960, many nature-inspired meta-heuristic algorithms have been developed. The nature behaves in a very efficient way. These ways have been mathematically modeled to develop different nature-inspired meta-heuristic algorithms such as PSO [1] (Particle Swam Optimization), GSA [2] (Gravitational Search Algorithm), BH [3] (Black Hole), GA [4] (Genetic Algorithm), GWO [5] (Grey Wolf Optimization), ALO [6] (Ant Lion Optimization), MVO [7] (Multi-Verse Optimization) and many more.

Particle Swam Optimization has been inspired from movement of birds. Gravitational Search Algorithm has been inspired from movement of celestial bodies in the Universe due to gravity. Black Hole has been inspired by the effect of black hole

S. S. Pal (✉) · S. Pal
Department of Computer Science and Technology, Indian Institute of Engineering Science and Technology, Shibpur, Howrah 711103, India
e-mail: shankho.subhra.pal@gmail.com

S. Pal
e-mail: sp@cs.iiests.ac.in

H. S. Behera et al. (eds.), *Computational Intelligence in Data Mining*, Advances in Intelligent Systems and Computing 990,
https://doi.org/10.1007/978-981-13-8676-3_35

on other bodies in the Universe. Genetic Algorithm is inspired from the evolution of species and theory of "survival of the fittest". Grey Wolf Optimization models the hunting pattern of Grey Wolves. Ant Lion Optimization models the way insects called ant lion, traps their food such as ants. Multi-Verse Optimization models the theory of multiverse. All these meta-heuristic optimization algorithms are inspired from the nature and it can be noticed that natural processes are inherently optimized.

All the above mentioned meta-heuristic optimization algorithms are populations based. But this does not mean that all nature-inspired meta-heuristic optimization algorithms need to be population based. For example, IDA [8] (Intelligent Water Droplet Optimization) is a graph-based meta-heuristic optimization algorithm. Some other meta-heuristic optimization algorithms can be found in [9–14]. Apart from these, new meta-heuristic optimizations are being developed in the research domain.

These types of meta-heuristic optimization algorithms have been used to solve different problems in engineering domain. Some of the problems have been mentioned in [15–20].

In this work, a modified Black Hole optimization approach for data clustering has been proposed. This is a nature-inspired clustering technique. In this technique, Black Hole optimization approach for data clustering has been hybridized with k-Means.

1.1 Nature-Inspired Clustering

Different nature-inspired algorithms have been used for clustering [21–26]. Clustering is a method of grouping similar data samples inside a dataset. The stochastic nature of "meta-heuristic nature-inspired algorithms" has helped to find good clusters.

The primary criterion of a good cluster is, it should have small intra-cluster distance, i.e., the sum of distance between each sample of a dataset to its corresponding cluster center should be as small as possible.

1.2 Black Hole Clustering

Black Hole optimization approach for data clustering [27] is a nature-inspired clustering technique. In it, the effect of black hole on other celestial bodies is modeled to develop a meta-heuristic algorithm. This meta-heuristic algorithm is used to cluster different datasets.

1.3 Black Hole k-Means Clustering

In the existing method, the initial population is initialized completely randomly. In the proposed method, i.e., Black Hole k-Means Clustering, some of the objects of the population are initialized as the objects having good result obtained from k-Means clustering algorithm and rest of the objects are initialized randomly. Then the same step as Black Hole optimization is followed.

In Sect. 2, related works have been discussed. In Sect. 3, the proposed method has been discussed. In Sect. 4, experimental setup and results enumerated. Section 5 contains conclusion and ongoing future work.

2 Related Work

As mentioned in Sect. 1.1, different nature-inspired algorithms have been used for clustering such as Gravitational Search Algorithm [28], Particle Swam Optimization [29], Genetic Algorithm [30–33], Chemical Reaction Optimization (CRO) [34, 35] and so on. Most of the used meta-heuristic optimization algorithms are population based.

2.1 Black Hole Clustering

Black bole optimization approach for data clustering has been proposed in [27]. In this approach, each object of the population describes all the cluster centroids. For clustering a dataset into k clusters, k cluster centroids will be needed. So each object will have k cluster centroids. Each cluster centroid will have number of features equal to the number of features of each sample of the dataset. The initial population is generated randomly.

After initializing the population, fitness values of all the objects are found. The fittest object is the black hole for that iteration. All other objects move towards the black hole stochastically. If any object enters the event horizon of the black hole, it disintegrates and a new object is born, which is initialized randomly. These steps are repeated till stopping criteria is satisfied. The event horizon [27] is defined by Eq. (1).

$$R = f_{BH} / \sum f. \tag{1}$$

Where R is the radius of event horizon and f is the fitness function.

Here fitness value is the sum of the distance of each sample from its nearest cluster centroid. The objective is to find k cluster centroids such that the fitness value as mentioned above is minimized. Thus, this problem becomes a minimization problem which has been optimized by black hole optimization. The fitness function [27] is defined by Eq. (2).

$$f = \sum \sum (c - o)^2. \quad (2)$$

Where f is the fitness value, c is cluster centroid and o is a sample.

2.2 Gravitational Search Algorithm-Based Clustering

Gravitational Search Algorithm has been discussed in [2]. It is used for data clustering in [28]. In this existing work, each object of the population describes all the cluster centroids. For clustering a dataset into k clusters, k cluster centroids will be needed. So each object will have k cluster centroids. Each cluster centroid will have similar number of features as in black hole optimization.

To get good solution, a good initial population is used. In the good initial population, the first object is the result from k-Means, second object is minimum of the dataset, third object is maximum of the dataset, fourth object is mean of the dataset and rest objects initialized randomly.

Here fitness value is the sum of distance of each sample from its nearest cluster centroid. The objective is to find k cluster centroids such that the fitness value as define above is minimized. Thus, this problem becomes a minimization problem which has been optimized by Gravitational Search Algorithm.

In Gravitational Search Algorithm, after the initial population is initialized, fitness values of all the objects are calculated and stored. The fitness values are used to calculate the mass of each object. The mass of each object is found by dividing the difference between that object's fitness value and the worst fitness value by the difference between the best fitness value and the worst fitness value. Once the mass is found, the gravitational force on each object due to all other objects is found. From these forces, the acceleration of each object is found. Using the acceleration and previous velocity, the new velocity of each object is found. The velocity and current position of each object are used to find its new position. All these values are found using the formulae of laws of motion. Like this, the position of each object is updated again and again until stopping criteria is satisfied.

2.3 Genetic k-Means Clustering

The first attempt on evolutionary clustering using genetic k-Means Clustering algorithm was done by Krishna and Murti [30]. Genetic algorithm (GA), a stochastic optimization algorithm, uses evolutionary operators (selection, mutation, and crossover) on population of coded solutions to evolve over generations and search for globally optimized solution in a large space. But when GA was applied to clustering, it turned out to be very expensive due to crossover operation. In [30], GA was hybridized with k-Means algorithm and crossover operator was done away with. The performance on clustering was compared with k-Means, bisecting k-Means and fuzzy C-Means in [36]. It was found in [36] that when there is no overlap between clusters, Genetic k-Means algorithm gave best performance as it searches for global optima. Genetic k-Means algorithm proposed in [32], a method for eliminating empty cluster(s) is proposed. In [33], single point crossover has been introduced and a novel mutation operator is proposed. Their mutation depends on extreme points of clusters. They initialized the whole population with solutions from k-Means and they obtained improved results compared to k-Means algorithm. However, in our proposed method, the initialization is different.

2.4 Chemical Reaction Optimization k-Means Clustering

In [37], Chemical Reaction-based meta-heuristic optimization (CRO) was proposed for optimization problems. The first step of the optimization is to generate quasi-opposite molecular matrix. The fitness PE quantifies the energy of a molecular structure. Next depending upon the decomposition criterion (of chemical reaction) being satisfied, the decomposition is performed. Otherwise, on-wall ineffective collision is performed. Next depending upon synthesis (of chemical reaction) being satisfied, decomposition is performed. Otherwise, intermolecular ineffective collision is performed. The process is continued till termination condition is not met. In [38], CRO-based clustering was proposed. Finally, the solution is returned. CRO was used for clustering data in [34] by hybridizing it with k-Means algorithm. They have used k-Means algorithm for on-wall ineffective collision and CRO for diversification of intermolecular reactions. In [35], CRO is hybridized with k-medoid clustering algorithm. It has been found that their method achieved competitive results.

3 Proposed Method

In the proposed method, Black Hole clustering algorithm has been modified. In the existing black hole clustering algorithm, the initial population is initialized randomly. But in this method, a good initial population is used.

The good initial population is generated by first running k-Means multiple times. Then a part of the better results of those runs is used to initialize some of the objects for the Black Hole k-Means Clustering. The rest of the objects for Black Hole k-Means Clustering are initialized randomly.

Here the fitness value is the sum of distance of each sample from its nearest cluster centroid. The objective is to find k cluster centroids such that the fitness value as defined above is minimized. Thus this problem becomes a minimization problem which is optimized by Black Hole clustering.

As in Black Hole optimization, first the fitness values of all the objects in the population are found and stored. Then, the fittest object is selected as the black hole. All other objects move towards the black hole stochastically in each iteration. After each iteration, if a fitter object is found, it becomes the black hole for the next iteration. If any object goes too near to the black hole, i.e., if any object crosses the event horizon, then it collapses into the black hole and a new object is created, which is initialized randomly. The event horizon is a distance from the black hole, which calculated by dividing fitness of the black hole by the sum of finesses of all the objects in the population. The steps are repeated till stopping criteria is met. The stopping criteria can be that fitness value of the black hole has become less than a threshold value or some preset maximum number of iterations has been reached.

The algorithm is as follows:

Step 1: Run k-Means clustering algorithm m times.
Step 2: Select p'top most results of k-Means clustering algorithm and use them to initialize some of the objects of the population.
Step 3: Initialize p" of the objects of the population p (such that p = p' + p") randomly.
Step 4: Find fitness value of each object.
Step 5: Find the black hole.
Step 6: Move each object towards the black hole stochastically.
Step 7: If any object has crossed the event horizon, reinitialize it randomly.
Step 8: Repeat from Step 3 till maximum number of iterations, t, completes.
Step 9: Find fitness values of all the objects.
Step 10: Find the fittest object.
Step 11: Return the fittest object and its fitness value.

3.1 Time Complexity of Proposed Method

In Step 4, for each of the "p" objects in the population, for a dataset with "n" number of samples each having "a" features and with "k" classes, O(nak) computation is needed to find the fitness value. So for "p" objects of the population, O(pnak) computation is needed. To find the black hole, in Step 5, O(p) comparisons are needed. In Step 6, O(pa) computations are needed to move all the objects towards the black hole. This is repeated for "t" iterations. Thus total time complexity becomes O(tpnak).

Table 1 Characteristics of the datasets

Dataset	Number of features	Number of samples	Number of clusters	Number of missing values
Pima Indians Diabetes Database	8	768	2	0
Iris Dataset	4	150	3	0
Lower Back Pain Symptoms Dataset	12	310	2	0
Red Wine Quality	11	1599	6	0
Dataset Wine	13	178	3	0
Glass Classification	9	214	6	0

4 Experimental Setup and Results

The experiment was performed in Ubuntu 18.04 LTS operating system. Python 3.6 was used for coding. Six datasets were used. All the datasets were obtained from [39]. The characteristics of the datasets are shown in Table 1. In the data preprocessing step, before clustering all the datasets were normalized, i.e., in each column, the maximum value was set to one and minimum value was set to zero. Accordingly, all other values were scaled.

Our proposed hybrid method consisting of Black Hole Optimization and k-Means clustering algorithm is compared with original Black Hole Optimization clustering, k-Means clustering, and Gravitational Search clustering algorithms. k-Means clustering algorithm was run fifty times with random initialization with one hundred iterations each, and the best-obtained fitness value of the result of those fifty runs has been used for comparison. For the rest three algorithms, the population size was set to fifty for each of the datasets and the maximum number of iterations was set to one thousand.

The fitness value of the best object obtained using k-Means Clustering, Gravitational Search Algorithm-based Clustering, Black Hole-based Clustering and the proposed method, i.e., Black Hole k-Means-based hybrid Clustering has been presented in Table 2. The best result of each test case has been italicized.

As it can be seen from Table 2, the proposed method gives best result for Red Wine Quality and Glass Classification. For all other datasets, the proposed method gives at least as good result as some of the other methods. Although the results obtained in Table 2 are interesting in themselves, the results of Table 2 are summarized in Table 3.

As it can be seen from Table 3, the proposed method gives best result for two of the test cases and at least as good result as other methods for the other four

Table 2 Best intra-cluster distance

Dataset	Intra-cluster distance			
	Gravitational Search Algorithm-based Clustering	Black Hole-based Clustering	k-Means Clustering	Black Hole k-Means Clustering
Pima Indians Diabetes Database	121.258	121.271	*121.258*	*121.258*
Iris Dataset	*6.998*	*6.998*	7.035	*6.998*
Lower Back Pain Symptoms Dataset	172.163	*170.859*	170.862	*170.859*
Red Wine Quality	157.848	172.218	156.372	*156.371*
Dataset Wine	48.967	*48.954*	*48.954*	*48.954*
Glass Classification	19.229	19.366	18.256	*18.241*

Table 3 Summary of Table 2 based on intra-cluster distance

Criterion	Gravitational Search Algorithm-based Clustering	Black Hole-based Clustering	k-Means Clustering	Black Hole k-Means Clustering
Win-Draw-Loss	0-1-5	0-3-3	0-2-4	2-4-0
Arithmetic mean of intra-cluster distance	87.7438	89.9443	87.1228	87.1135

Win means that no other method gives better result, Draw means at least one of the other methods gives same result, Loss means at least one of the other methods gives better result

test cases. The best Arithmetic Mean considering all six datasets is obtained from proposed Black Hole k-Means Clustering. However, when the datasets are from different domains, the geometric mean is more meaningful compared to the arithmetic mean. The comparative geometric mean of the four algorithms is shown in Fig. 1. Both from Table 3 and Fig. 1, it can be seen that our proposed modification of the original black hole algorithm reduces both arithmetic mean and geometric mean of intra-cluster distance to a significant extent.

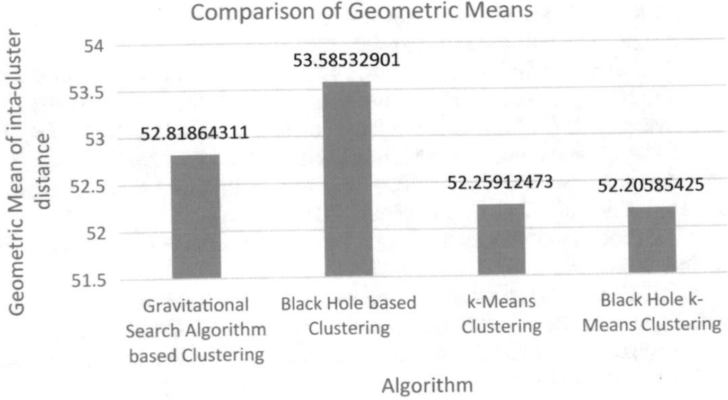

Fig. 1 Comparison of geometric means

5 Conclusion

The proposed method never gives worse result than Gravitational Search Algorithm-based Clustering, k-Means-based Clustering or Black Hole-based Clustering. This is because, in this method, the initial population is taken as some good results of k-Means. This is because, in this method, the best object found so far is never eliminated. So it gives at least as good result as k-Means. Furthermore, in this method a good initial population is selected, so it gives a better result than Black Hole-based Clustering. In Gravitational Search Algorithm-based Clustering, the next iteration may give a worse result as the best object moves. But in this method, the best object found so far does not move. Thus, this method gives better result in some cases than Gravitational Search Algorithm-based Clustering, but never gives a worse result than Gravitational Search Algorithm-based Clustering. Also our proposed modification improves results of original black hole algorithm to a significant extent. Further work is being carried out with more number of datasets and also with further hybridization possibilities for improvement.

Acknowledgements The first author acknowledges his exposure to Gravitational Search Algorithm done in his B. Tech major project with Dr. Santosh Kumar Majhi, at Veer Surendra Sai University of Technology, Burla, Odisha.

References

1. Kennedy, J.: Particle swarm optimization. In: Encyclopedia of Machine Learning, pp. 760–766. Springer US (2011)
2. Rashedi, E., Nezamabadi-Pour, H., Saryazdi, S.: GSA: a gravitational search algorithm. Inf. Sci. **179**(13), 2232–2248 (2009)

3. Piotrowski, A.P., Napiorkowski, J.J., Rowinski, P.M.: How novel is the “novel” black hole optimization approach? Inf. Sci. **267**, 191–200 (2014)
4. Holland, J.H.: Adaptation in Natural and Artificial Systems: An Introductory Analysis with Applications to Biology, Control, and Artificial Intelligence. MIT Press (1992)
5. Mirjalili, S., Mirjalili, S.M., Lewis, A.: Grey wolf optimizer. Adv. Eng. Softw. **69**, 46–61 (2014)
6. Mirjalili, S.: The ant lion optimizer. Adv. Eng. Softw. **83**, 80–98 (2015)
7. Mittal, N., Singh, U., Sohi, B.S.: Modified grey wolf optimizer for global engineering optimization. Appl. Comput. Intell. Soft Comput. **2016**, 8 (2016)
8. Hosseini, H.S.: Problem solving by intelligent water drops. In: IEEE Congress on Evolutionary Computation, 2007. CEC 2007. IEEE (2007)
9. Dorigo, M., Di Caro, G.: Ant colony optimization: a new meta-heuristic. In: Proceedings of the 1999 Congress on Evolutionary Computation, vol. 2, 1999. CEC 99. IEEE (1999)
10. Kaveh, A., Mahdavi, V.R.: Colliding bodies optimization: a novel meta-heuristic method. Comput. Struct. **139**, 18–27 (2014)
11. Mirjalili, S.: Dragonfly algorithm: a new meta-heuristic optimization technique for solving single-objective, discrete, and multi-objective problems. Neural Comput. Appl. **27**(4), 1053–1073 (2016)
12. Oftadeh, R., Mahjoob, M.J., Shariatpanahi, M.: A novel meta-heuristic optimization algorithm inspired by group hunting of animals: hunting search. Comput. Math. Appl. **60**(7), 2087–2098 (2010)
13. Sayadi, M., Ramezanian, R., Ghaffari-Nasab, N.: A discrete firefly meta-heuristic with local search for makespan minimization in permutation flow shop scheduling problems. Int. J. Indus. Eng. Comput. **1**(1), 1–10 (2010)
14. Maenhout, B., Vanhoucke, M.: An electromagnetic meta-heuristic for the nurse scheduling problem. J. Heuristics **13**(4), 359–385 (2007)
15. Pandey, S., et al.: A particle swarm optimization-based heuristic for scheduling workflow applications in cloud computing environments. In: 2010 24th IEEE International Conference on Advanced Information Networking and Applications (AINA). IEEE (2010)
16. Emary, E., Zawbaa, H.M., Hassanien, A.E.: Binary grey wolf optimization approaches for feature selection. Neurocomputing **172**, 371–381 (2016)
17. Mohanty, S., Subudhi, B., Ray, P.K.: A new MPPT design using grey wolf optimization technique for photovoltaic system under partial shading conditions. IEEE Trans. Sustain. Energy **7**(1), 181–188 (2016)
18. Sodeifian, G., et al.: Application of supercritical carbon dioxide to extract essential oil from Cleome coluteoides Boiss: experimental, response surface and grey wolf optimization methodology. J. Supercrit. Fluids **114**, 55–63 (2016)
19. Guha, D., Roy, P.K., Banerjee, S.: Load frequency control of interconnected power system using grey wolf optimization. Swarm Evol. Comput. **27**, 97–115 (2016)
20. Aljarah, I., Faris, H., Mirjalili, S.: Optimizing connection weights in neural networks using the whale optimization algorithm. Soft. Comput. **22**(1), 1–15 (2018)
21. Mahdavi, M., et al.: Novel meta-heuristic algorithms for clustering web documents. Appl. Math. Comput. **201**(1–2), 441–451 (2008)
22. Shokoufandeh, A., et al.: Spectral and meta-heuristic algorithms for software clustering. J. Syst. Softw. **77**(3), 213–223 (2005)
23. Fathian, M., Amiri, B., Maroosi, A.: Application of honey-bee mating optimization algorithm on clustering. Appl. Math. Comput. **190**(2), 1502–1513 (2007)
24. Gopal, N.N., Karnan, M.: Diagnose brain tumor through MRI using image processing clustering algorithms such as Fuzzy C Means along with intelligent optimization techniques. In: 2010 IEEE International Conference on Computational Intelligence and Computing Research (ICCIC). IEEE (2010)
25. Niknam, T., et al.: An efficient hybrid evolutionary optimization algorithm based on PSO and SA for clustering. Zhejiang Univ.-SCIENCE A **10**(4), 512–519 (2009)
26. Chu, S.-C., et al.: Constrained ant colony optimization for data clustering. In: Pacific Rim International Conference on Artificial Intelligence. Springer, Berlin, Heidelberg (2004)

27. Hatamlou, A.: Black hole: a new heuristic optimization approach for data clustering. Inf. Sci. **222**, 175–184 (2013)
28. Hatamlou, A., Abdullah, S., Nezamabadi-Pour, H.: A combined approach for clustering based on k-means and gravitational search algorithms. Swarm Evol. Comput. **6**, 47–52 (2012)
29. Van der Merwe, D.W., Engelbrecht, A.P.: Data clustering using particle swarm optimization. In: The 2003 Congress on Evolutionary Computation, 2003, vol. 1. CEC'03. IEEE (2003)
30. Krishna, K., Murty, M.N.: Genetic K-means algorithm. IEEE Trans. Syst. Man Cybern. Part B (Cybernetics) **29**(3), 433–439 (1999)
31. Maulik, U., Bandyopadhyay, S.: Genetic algorithm-based clustering technique. Pattern Recogn. **33**(9), 1455–1465 (2000)
32. Al Malki, A., et al.: Hybrid genetic algorithm with K-means for clustering problems. Open J. Optim. **5**(02), 71 (2016)
33. El-Shorbagy, M.A., et al.: A novel genetic algorithm based k-means algorithm for cluster analysis. In: International Conference on Advanced Machine Learning Technologies and Applications. Springer, Cham (2018)
34. Panigrahi, S., Balaram Rath, Kumar, P.S.: A hybrid CRO-K-means algorithm for data clustering. In: Computational Intelligence in Data Mining, vol. 3, pp. 627–639. Springer, New Delhi (2015)
35. Hudaib, A., Khanafseh, M., Surakhi, O.: An improved version of K-medoid algorithm using CRO. Mod. Appl. Sci. **12**(2), 116 (2018)
36. Banerjee, S., Choudhary, A., Pal, S.: Empirical evaluation of k-means, bisecting k-means, fuzzy c-means and genetic k-means clustering algorithms. In: 2015 IEEE International WIE Conference on Electrical and Computer Engineering (WIECON-ECE), IEEE (2015)
37. Lam, A.Y.S., Li, V.O.K.: Chemical-reaction-inspired meta-heuristic for optimization. IEEE Trans. Evolut. Comput. **14**(3), 381–399 (2010)
38. Baral, A., Behera, H.S.: A novel chemical reaction-based clustering and its performance analysis. Int. J. Bus. Intell. Data Min. **8**(2), 184–198 (2013)
39. https://www.kaggle.com/

Improving Stock Market Prediction Through Linear Combiners of Predictive Models

Sarat Chandra Nayak, Ch. Sanjeev Kumar Dash, Ajit Kumar Behera and Satchidananda Dehuri

Abstract Accurate and efficient stock market forecasting model design has been attracted attentions of researchers continuously. This leads to development of various statistical and machine learning-based models in the context. Accuracy of a method is immensely problem and domain specific. Hence, identifying a best method, in general is controversial. Combining outputs of different forecasting models to enhance overall accuracies and minimizing the risk of model selection has been suggested extensively in literature. This work presents a linear combiner of five predictive models such as ARIMA, RBFNN, MLP, SVM, and FLANN towards improving the stock market predictive accuracy. Three statistical methods such as trimmed mean, simple average, and the median, and an error-based method are used for appropriate selection of combining weights. The individual forecasts and the linear combiner are employed separately to predict the next day's closing price of five real stock markets. Extensive simulation work demonstrates the feasibility and superiority of the linear combiner vis-à-vis others.

Keywords Combining forecasts · Ensemble method · Artificial neural network · Stock market prediction · Financial time series forecasting · Multilayer perceptron

---

S. C. Nayak (✉)
Department of Computer Science and Engineering, CMR College of Engineering & Technology (Autonomous), Hyderabad, India
e-mail: saratnayak234@gmail.com; drsaratchandranayak@cmrcet.org

Ch. Sanjeev Kumar Dash · A. K. Behera
Department of Computer Science & Engineering, Silicon Institute of Technology, Bhubaneswar, India
e-mail: sanjeev_dash@yahoo.com

A. K. Behera
e-mail: ajit_behera@hotmail.com

S. Dehuri
Department of ICT, Fakir Mohan University, Balasore, India
e-mail: satchi.lapa@gmail.com

H. S. Behera et al. (eds.), *Computational Intelligence in Data Mining*, Advances in Intelligent Systems and Computing 990,
https://doi.org/10.1007/978-981-13-8676-3_36

1 Introduction

Over the last few decades, a large number of statistical as well as machine learning approaches have been formulated and suggested in the literature of stock market forecasting. However, the accuracy of a model is problem specific and varies among data sets. Though few methods claimed superior over others, achieving improved forecasting accuracy is still a question. The selection of most promising forecasting model is itself a challenging task. With the increase in the number of available forecasting techniques, the studies interrelated to their comparative assessment are also growing incessantly. In recent years, methods of combining different forecasting models and aggregating their outputs using a suitable combination technique have attracted the attention of researchers for modeling time series [1–4]. The method of combining multiple forecasts has several advantages such as benefit from the strength of constituent models, reduced error from faulty assumptions, bias in the data set [5]. Also, different individual models can offer complementary information about unknown instances and energies the quality of overall prediction [6, 7]. Several combining techniques have been proposed by researcher during last decades. However, selection of most promising combination scheme is a challenging issue [8]. The outperformance of combined methods over individual and achieving improved overall forecasting accuracies has been shown in the literature. Majority of forecast combination techniques form a weighted linear combination of constituent forecasts. The most basic ensemble methods are based on statistical averaging techniques such as simple average, trimmed mean, median, and the error-based method. Even these simple methods are found outperformed more advanced combining schemes [2, 9]. The statistical methods are simple and computationally economical. However, these methods are unable to retain the performance history of individual models. Since they ignore the relative dependence among the forecasts the combined forecast may be inefficient [2]. An alternative is the error base method which assigns weights to component models on the basis of their past performance. Other schemes such as differential weighting, outperformance method, etc., are found in the literature for time series forecasting.

Stock market behavior is highly unpredictable due to the influence of uncertainties, nonlinearities, high volatility, and discontinuities associated with it. Also, it is affected by movement of other stocks, political influences, macro-economical factors as well as psychology of individual, even sometimes rumor. Prediction of such irregular and highly fluctuating stock data is very much complex and typically subject to large error. Hence, robust forecasting methods are intensely desirable for decision making of speculators, financial managers, naive investors, etc., A small improvement in prediction may probable leads to elevated financial benefits.

Artificial neural networks (ANN) are mimicking the human brain's way of learning and emulate human's behavior for solving nonlinear complex problems. They are well known as an effective modeling procedure to solve real problems, particularly when the input–output mapping contains both regularities and exceptions. Their superior learning and approximation capabilities make them suitable for successful application to financial forecasting such as index prediction, foreign exchange rate prediction, business failure and bankruptcy prediction, scoring of credit limit, interest rate forecasting, portfolio management, and option and future prices prediction, etc. Among the ANN-based methods, multilayer perceptrons (MLP) have been used frequently for their inherent capabilities to approximate any nonlinear function to a high degree of accuracy [10, 11]. ANN-based forecasting models have necessitates many numbers of layers and abundance of neurons in each layer of the network hence suffers from computational complexity. Again it suffers from the drawbacks such as overfitting and black box technique. It may get trapped in local minima. As an alternative a single layer ANN is proposed by Pao and Takefuji [12] which is less complex in architecture and termed as functional link artificial neural network (FLANN). Basically, it is a single layer network and the need of hidden layers can be compensated with the incorporation of functional expansions of the input signal set. The functional expansions help to increase the dimensionality of the input vector and generate hyperplanes. These hyperplanes provide greater discrimination capability to FLANN in the input pattern space. The performance of FLANN is accepted by several research works in the literature. There have been several applications of FLANN to stock market forecasting [13–17].

The objective of this work is to form a homogeneous ensemble of ANNs with varying architectures in order to achieving better overall accuracy. One statistical forecasting model, i.e., autoregressive integrated moving average (ARIMA) and three neural-based models such as MLP, RBFNN, and FLANN, and SVM are used as the constituent models. Four techniques, i.e., an error-based method, simple average, trimmed mean, the median are used for assignment of combination weights. The proposed ensemble method is tested on five real stock data sets and its performances are compared with the five individual models in terms of two error statistics.

The rest of the article is organized as follows. Description about the stock market prediction, artificial neural network-based models for stock market trend analysis and the importance of ensemble schemes are discussed in Sect. 1. The working principles of component forecasting models are presented by Sect. 2. Section 3 presents the proposed ensemble framework for stock market prediction. Experimental results and analysis are presented by Sect. 4. Section 5 gives the concluding remarks followed by a list of references.

2 Forecasting Models

2.1 ARIMA

In the domain financial time series forecasting, ARIMA models are the very common statistical models used. These are proposed by Box et al. [18] and normally known as Box-Jenkins models. The hypothesis considers a linear combination of historical observations and a random white noise term in order to generate the associated time series. Mathematically this can be represented as:

$$\Phi(s)(1-s)^d(y_t) = \theta(s)\varepsilon_t \tag{1}$$

where

$$\Phi(S) = 1 - \sum_{i=1}^{p} \theta_i S^i, \ \Phi(S) = 1 + \sum_{j=1}^{q} \theta_j S^j, \tag{2}$$

Here q is moving average terms, $p =$ number of autoregressive, $d =$ degree of differencing, ε_t is the random white noise term (satisfying *i.i.d* property) and y_t is the actual observations. These classes of models are typically termed as ARIMA (p, d, q) models. We determined the suitable parameters as per the Box-Jenkins model build specifications [18].

2.2 MLP

MLPs are the most widely implemented neural networks for stock market prediction. We considered here a MLP with one hidden layer and single output unit. The neurons in the input layer use a linear transfer function. The neurons in the hidden and output layer use a sigmoid function. The first layer is the input layer and contains one node for each input signal. So the number of neurons in this layer is equal to the size of input vector. The second layer also called as the hidden layer try to capture the nonlinear relationships among data points in the financial time series. The output neuron calculates the model estimation and compared with the actual output. The difference is termed as error signal generated by the model. Less the error signal better is the model. The root mean squared error obtained from the training patterns is then used to propagate back to previous layers in order to train the MLP. The weight and other parameters are updated based on the principle of gradient descent rule. Prospective readers are suggested to refer [10, 11].

2.3 RBFNN

RBFNNs are found quite capable in terms of approximating functions and recognizing patterns [19]. In RBFNN an activation of a hidden neuron is determined based on the distance between an input and a sample vector. A RBFNN has two layers. Each unit of hidden layer implements a basis function called radial activation function. Each neuron of output layer computes a weighted sum of outputs obtained from hidden units'. The centers of clusters formed by the patterns in the input space are called as sample or prototype vector. The hidden and output layers are interconnected and each such connection is associated with some weight value. A summation unit at the output layer estimates the network output. The input signals that connect the network to the environment determine the size of input layer. A nonlinear transformation is carried out by a set of kernel units in the hidden layer from the input space to the hidden space. During the model training, the centers of prototype vectors are determined. Selection of appropriate quantity of basis functions which controls the approximation and the generalization ability of the network is an important issue in RBFNN. We used the Gaussian functions as basis function as presented in Eq. 3.

$$\Phi_i(\mathrm{x}) = \exp\left(-\frac{\|x - \mu_i\|^2}{2\sigma_i^2}\right), \tag{3}$$

where $\|\ldots\|$ is the Euclidean norm and x is the input vector. μ_i, σ_i and $\Phi_i(x)$ are the center, spread, and output of the ith hidden node, respectively. The output is calculated as in Eq. 4.

$$y = f(x) = \sum_{k=1}^{N} w_k \Phi_k(\|x - c_k\|), \tag{4}$$

where y and x represents the network output and input vector, respectively. $w = [w_1, w_2, \ldots, w_N]^T$ is the weight vector, the number of hidden neurons is N, and $\Phi_k(.)$ is the basis function. The bandwidth of the basis function is k. The vector $\mathrm{c}_k = (\mathrm{c}_{k1}, \mathrm{c}_{k2}, \ldots, \mathrm{c}_{km})^T$ is the center vector for kth node, and the input vector size is m.

2.4 FLANN

In FLANN the higher order effects at nodes are introduced through nonlinear functional transforms via links. Each attribute of the input vector is allowed to pass through an expansion unit. Trigonometric basis functions of *sine* and *cosine* are used here to expand the original input into higher dimensions. An input value x_i expanded to several terms by using the trigonometric expansion functions as

$$\begin{aligned} c_1(x_i) &= (x_i),\ c_2(x_i) = sin(x_i), \\ c_3(x_i) &= cos(x_i),\ c_4(x_i) = sin(\pi x_i),\ c_5(x_i) \\ &= cos(\pi x_i),\ c_6(x_i) = sin(2\pi x_i),\ c_7(x_i) = cos(2\pi x_i). \end{aligned}$$

The weighted sum of outputs obtained from the functional expansion units is then passed on to a sigmoid activation function at output unit. The estimated output is then compared with the target to obtain the error signal which is utilized to train the

FLANN using the principle of gradient descent technique. Details about FLANN can be found in [12].

2.5 SVM

SVMs are supervised learning methods have been used widely for various data mining problems. It constructs a set of hyperplanes in a high dimensional space which can be used for classification or regression task. The hyperplanes create a decision boundary so that most of the data points belong to one category fall on one side and data points belong to other class fall on the other side. The optimal hyperplane is one which maximizes the distance between the plane and any point termed as margin. Maximum the margin better is the splitting of data points. The selection of hyperplane depends upon the crucial elements closest to the boundary called as support vectors, hence the hyperplane is known as support vector classifier. We used the SVM implementation in Matlab with the radial kernel. The details about SVM can be found in literature [20].

3 Proposed Method

Selection of appropriate forecaster for modeling stock market trend is a critical and risky task. The ensemble framework suggested here attempts to overcome the risk of model selection as well as enhancing the overall prediction accuracy of the method. The framework combines five individual models such as ARIMA, RBFNN, SVM, MLP, and FLANN as shown in Fig. 1. The process can be explained as follows. Let $X_i = [\mathrm{x}_1, \mathrm{x}_2, \ldots \mathrm{x}_n]^T$ be the current training set generated from the original financial time series by sliding a window of fixed size through one step. The training set is then normalized using sigmoid data normalization method as ANN-based models are performing better on normalized data. Now the normalized data is fed to individual forecasting models separately. Each forecaster transforms the input signal X_i to the corresponding forecast $\hat{y}_i (i = 1, 2, \ldots, 5)$. A combination of the forecasts can be expressed as:

$$\begin{aligned} \hat{\mathrm{Y}} &= f(ARIMA(X_i), RBFNN(X_i), SVM(X_i), MLP(X_i), FLANN(X_i), W) \\ &= f(\hat{y}_1, \hat{y}_2, \ldots, \hat{y}_5, \mathrm{W}) \end{aligned} \tag{5}$$

where f is a linear combination function and $W = [w_1, w_2, \ldots, w_5]^T$ is the weight vector assigned to the linear combiner. The linear combination can be calculated as:

$$\hat{y}_k = \sum_{i=1}^{n} w_i * \hat{y}_k^i, \quad \forall k = 1, 2, \ldots, N \tag{6}$$

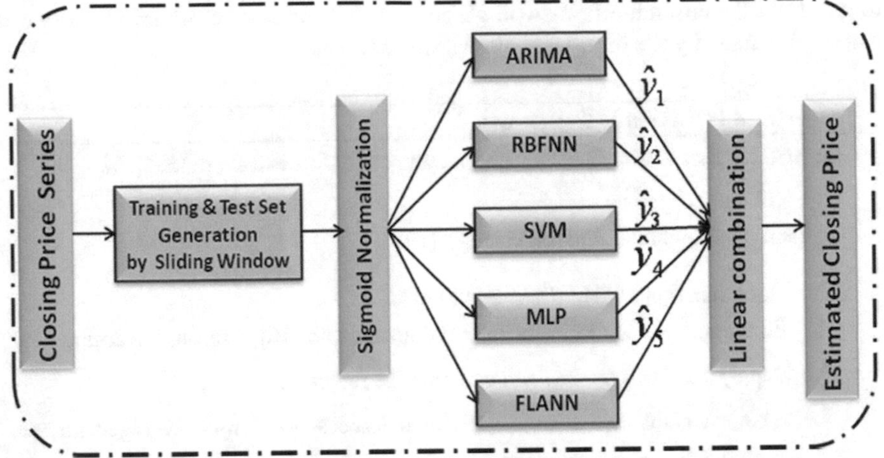

Fig. 1 Linear combination of forecasters

Three statistical methods such as simple average, trimmed mean, and the median, as well as an error-based method are used for appropriate selection of combining weights. These weights are often constrained to be nonnegative as well as unbiased. The simple average method assigns equal weight to all individual forecasting models. Though it is more sensitive to extreme values, still it found performing better [8, 9]. The alternative statistical methods are trimmed mean and the median. Trimmed mean method averages the forecasts excluding $\beta\%$ worst performing of the models. A 10–40% trimming is recommended [8, 9] and calculated as $\beta = 100 * (2\alpha/n)$. A β value zero corresponds to simple average and $n/2$ corresponds to median.

The trimmed mean is calculated as

$$t_k(\alpha) = \frac{1}{n-2\alpha} \sum_{i=\alpha+1}^{n-\alpha} \hat{y}_k^i, \ \forall 0 \leq \frac{n}{2}; \quad k = 1, 2, \ldots, N \tag{7}$$

The combining weight to individual forecast in error-based method is considered as the inversely proportional of the error generated by the corresponding model [5] and can be calculated as

$$w_i = \frac{err_i}{\sum_{i=1}^{n} err_i^{-1}} \tag{8}$$

where err_i is the error generated by ith forecasting model. This method assigns less weight to a model with more error and vice versa.

It was observed from the literature that best accuracies are often obtained if forecasts are made through independent models and the combination method consists four to five such models [5]. As per this guideline, we considered five models here to construct the ensemble framework. The overall process of ensemble framework can be explained by the high-level algorithm as follows:

Algorithm 1: Ensemble framework

Input: The *TrainData*, *TestData*, and *n* component forecasting models M_i ($i = 1, 2, ..., n$).

Output: The combined forecast through (10)

1. Normalization of *TrainData* and *TestData*
2. Fit the model M_i to *TrainData* and update knowledge through learning
3. Supply *TestData* to M_i and estimate the forecasts
4. Assign weight w_i to each individual forecast by simple average/trimmed mean/median/error base method
5. Obtain the final combined forecast as $\hat{y}_k = \sum_{i=1}^{n} w_i * \hat{y}_k^i$

---

4 Experimental Results and Discussions

For experimental purpose, daily closing indices from five fast-growing stock markets such as DJIA, BSE, TAIEX, NASDAQ and FTSE are considered. The indices are collected from the source https://in.finance.yahoo.com/ for each financial day starting from 1st January 2003 to 12th September 2016. The descriptive statistics of these indices such as standard deviation, mean, minimum, maximum, skewness, kurtosis, and Jarque-Bera (JB) statistics are calculated and summarized in Table 1. it can be observed that DJIA, NASDAQ, and BSE data sets have shown positive skewness values and spread out more toward right further suggesting investment opportunities in these markets.

Table 1 Descriptive statistics from five stocks indices

Stock index	Descriptive statistics						
	Min.	Max.	Mean	Std. deviation	Skewness	Kurtosis	JB test statistics (h = 1)
BSE	793.1801	1.1025e+004	4.6255e+003	2.6937e+003	0.12	1.79	236.14
DJIA	6.5470e+003	1.7138e+004	1.1403e+004	2.1800e+003	0.66	3.05	253.81
NASDAQ	1.1151e+003	4.5981e+003	2.3857e+003	709.7689	1.04	4.00	763.37
FTSE	3286	6.8784e+003	5.4166e+003	836.2382	−0.28	2.14	158.46
TAIEX	3.4462e+003	1.0201e+004	6.9834e+003	1.4846e+003	−0.18	2.05	159.98

The patterns for training and testing the models are generated by using a sliding window technique. A window of fixed size is move over the financial time series by one step each time. In each move, a new pattern has been formed which can be used as an input vector. The size of window can be decided experimentally. The original input data are then normalized using sigmoid normalization [21, 22] which is represented as in Eq. 9.

$$x_{norm} = \frac{1}{1 + e^{-\left(\frac{x_i - x_{\min}}{x_{\max} - x_{\min}}\right)}} \tag{9}$$

Here, x_{norm} and x_i are normalized and current day closing price, respectively, x_{max} and x_{min} are the maximum and minimum prices of the current pattern, respectively. The MAPE (Mean Absolute Percentage Errors) and ARV (Average Relative Variance) are used as the performance metrics whose formulas are as follows:

$$MAPE = \frac{1}{N}\sum_{i=1}^{N}\frac{\left|x_i - \hat{x}_i\right|}{x_i} \times 100\% \tag{10}$$

$$ARV = \frac{\sum_{i=1}^{N}(\hat{x}_i - x_i)^2}{\sum_{i=1}^{N}(\hat{x}_i - \overline{X})^2} \tag{11}$$

In these measures x_i and $\hat{x}_i$ are the actual and estimated closing prices, $\overline{X}$ is the mean of data set. N is the total number of test data.

As mentioned earlier, the experimental work is carried out with five forecasting models on five stock datasets. The performance of each model is calculated in terms of two metrics. For each financial time series, the models are fed with same training and testing data. The MAPE and ARV obtained from individual models from five stock data sets are summarized in Table 2. The best ARV and MAPE value across a column (particular data set) are in boldface letter. It can be observed that the performance of FLANN is better than other models followed by RBFNN. However, there is no single method outperforming others over all data sets.

Similarly, the performances of combination methods from all data sets are summarized in Table 3. The best metric values are highlighted with boldface letter. It can be observed that the ARV and MAPE values from all combination methods are found better than the best performing individual model which justifies improvement in prediction accuracies and overcome the risk of model selection. Even a small improvement in forecasting accuracy can create a reasonable profit, particularly the case of financial time series forecasting. Though the performance of error base combination method found better in case of three data sets, still there is lacking of a single method performing best across the data sets. Therefore, utmost care should be taken while selecting a combination method over individual.

Table 2 Performance of individual models on five data sets

Model	Error statistic	DJIA	BSE	NASDAQ	FTSE	TAIEX
ARIMA	ARV	0.5223	0.4035	0.4875	0.4099	0.4987
	MAPE	0.6275	0.3506	0.1009	0.3423	0.1096
RBFNN	ARV	**0.05393**	0.31575	0.0824	**0.07166**	0.08355
	MAPE	0.07747	**0.05228**	0.07712	0.38364	**0.07928**
MLP	ARV	0.41225	0.30685	0.47681	0.4139	0.5235
	MAPE	0.5052	0.09045	0.09273	0.3168	0.08615
SVM	ARV	0.41969	0.30633	**0.05827**	0.40681	0.40066
	MAPE	0.19974	0.09242	0.07263	0.09273	0.08856
FLANN	ARV	0.05842	**0.05668**	0.07832	0.08562	**0.08102**
	MAPE	**0.04789**	0.06828	**0.05231**	**0.06954**	**0.07923**

Table 3 Performance of ensemble methods on five data sets

Combination method	Error statistic	DJIA	BSE	NASDAQ	FTSE	TAIEX
Simple avg.	ARV	0.05349	0.05624	**0.05703**	0.07156	0.07902
	MAPE	0.04286	0.05028	0.05002	0.06344	0.07275
Trimmed mean	ARV	0.03964	**0.01856**	0.0527	0.06023	0.08087
	MAPE	0.03942	0.03987	**0.03907**	0.06595	0.07721
Median	ARV	0.05085	0.05393	0.05809	0.05945	0.07909
	MAPE	0.04727	0.03943	0.03957	0.06794	0.07886
Error base	ARV	**0.02534**	0.02562	0.05805	**0.05037**	**0.07452**
	MAPE	**0.03178**	**0.03751**	0.03952	**0.04433**	**0.06806**

In order to obtain more clarity about the relative performance of combination methods, we follow another measure called as relative worth of a model. It is the average reductions % in the prediction errors of the best performing individual model by a particular method over all datasets. Then, the relative worth RW_j of the jth combination method is defined as follows:

$$RW_j = \frac{1}{D}\sum_{i=1}^{D}\left[\left(\frac{err_i^w - err_{ij}}{err_i^w}\right) * 100\right] \quad \forall j = 1, 2, \ldots, n, \tag{12}$$

where err_{ij} is the forecasting error generated by the jth method on the ith data, err_i^w is the error of best individual model for the same dataset, and D is the total number of datasets. The relative worth values (in terms of ARV and MAPE) are summarized in Table 4.

It can be observed that the combining methods achieving 1.2–29.1% ARV reduction over the best performing individual models. Similarly, they are generating

Table 4 Relative worth values of combination methods

Combination method	Relative worth value	
	ARV	MAPE
Simple average	1.265652	7.131438
Trimmed mean (35%)	21.09834	14.88932
Median	23.61359	10.59929
Error base	29.18441	27.33848

7.1–27.3% less MAPE over the best performing individual models. These observations establish the fact that the ensemble methods are able to provide about 29.1% better ARV than a best individual forecast and about 27.3% better MAPE as compared to that of a best individual.

5 Conclusions

Analysis of stock market trend is a challenging and complex task as it is highly associated with uncertainties, nonlinearity, etc., and talks to the current economical as well as political scenarios. Achieving improved forecasting accuracy is the key point while designing a model for stock prediction. Also, accuracy of an individual method is very much problem specific and exact identification of a best method is controversial. Combining outputs of different forecasting models to enhance overall accuracies and minimizing the risk of model selection has been suggested extensively in literature. We suggest an ensemble framework of five individual ANN-based models where the combination weights are assigned by simple statistical methods such as trimmed mean, simple average, median, and an error-based method. To maintain a trade-off between computational cost and predictive accuracy, we use simple methods for aggregating the individual models. Five state-of-the-art models such as ARIMA, RBFNN, MLP, SVM, and FLANN are used as the component model for the ensemble framework. The individual as well as combination models have been evaluated on predicting next day's daily closing prices of five real stock markets. From extensive simulation work it is clearly observed that the linear combiner methods producing better accuracies over all individual models. Particularly, the error base combination method found performing best among all which justified the approach of combining multiple forecasts as an alternative to individual. The work can be extended by choosing other neural base component models and exploring other aggregating methods for weight selection.

References

1. Terui, N., Van Dijk, H.K.: Combined forecasts from linear and nonlinear time series models. Int. J. Forecast. **18**(3), 421–438 (2002)
2. De Gooijer, J.G., Hyndman, R.J.: 25 years of time series forecasting. Int. J. Forecast. **22**(3), 443–473 (2006)
3. Andrawis, R.R., Atiya, A.F., El-Shishiny, H.: Forecast combinations of computational intelligence and linear models for the NN5 time series forecasting competition. Int. J. Forecast. **27**(3), 672–688 (2011)
4. Adhikari, R., Agrawal, R.K.: Performance evaluation of weights selection schemes for linear combination of multiple forecasts. Artif. Intell. Rev. **42**(4), 529–548 (2014)
5. Armstrong, J.S. (ed.): Principles of Forecasting: A Handbook for Researchers and Practitioners, vol. 30. Springer (2001)
6. Kuncheva, L.I.: Combining Pattern Classifiers: Methods and Algorithms. Wiley (2004)
7. Tan, P.N.: Introduction to Data Mining. Pearson Education India (2006)
8. Lemke, C., Gabrys, B.: Meta-learning for time series forecasting and forecast combination. Neurocomputing **73**(10–12), 2006–2016 (2010)
9. Jose, V.R.R., Winkler, R.L.: Simple robust averages of forecasts: some empirical results. Int. J. Forecast. **24**(1), 163–169 (2008)
10. Yu, L.Q., Rong, F.S.: Stock market forecasting research based on neural network and pattern matching. In: International Conference on E-Business and E-Government (ICEE), pp. 1940–1943 (2010)
11. Tahersima, H., Tahersima, M., Fesharaki, M., Hamedi, N.: Forecasting stock exchange movements using neural networks. In: International Conference on Future Computer Sciences and Application (ICFCSA), pp. 123–126. IEEE (2011)
12. Pao, Y.H., Takefuji, Y.: Functional-link net computing: theory, system architecture, and functionalities. Computer **25**(5), 76–79 (1992)
13. Majhi, R., Panda, G., Sahoo, G.: Development and performance evaluation of FLANN based model for forecasting of stock markets. Expert Syst. Appl. **36**(3), 6800–6808 (2009)
14. Nayak, S.C., Misra, B.B., Behera, H.S.: ACFLN: artificial chemical functional link network for prediction of stock market index. Evol. Syst., 1–26 (2018)
15. Nayak, S.C., Misra, B.B., Behera, H.S.: Comparison of performance of different functions in functional link artificial neural network: a case study on stock index forecasting. In: Computational Intelligence in Data Mining, vol. 1, pp. 479–487. Springer, New Delhi (2015)
16. Sahu, K.K., Sahu, S.R., Nayak, S.C., Behera, H.S.: Forecasting foreign exchange rates using CRO based different variants of FLANN and performance analysis. Int. J. Comput. Syst. Eng. **2**(4), 190–208 (2016)
17. Nayak, S.C., Misra, B.B., Behera, H.S.: Index prediction with neuro-genetic hybrid network: a comparative analysis of performance. In: 2012 International Conference on Computing, Communication and Applications (ICCCA), pp. 1–6. IEEE (2012, February)
18. Box, G.E., Jenkins, G.M., Reinsel, G.C., Ljung, G.M.: Time Series Analysis: Forecasting and Control. Wiley (2015)
19. Dash, C.S.K., Dash, A.P., Dehuri, S., Cho, S.B., Wang, G.N.: DE+RBFNs based classification: a special attention to removal of inconsistency and irrelevant features. Eng. Appl. Artif. Intell. **26**(10), 2315–2326 (2013)
20. Halls-Moore, M.: Support Vector Machines: A Guide for Beginners (2014)
21. Nayak, S.C., Misra, B.B., Behera, H.S.: Impact of data normalization on stock index forecasting. Int. J. Comp. Inf. Syst. Ind. Manag. Appl. **6**, 357–369 (2014)
22. Nayak, S.C., Misra, B.B., Behera, H.S.: Evaluation of normalization methods on neuro-genetic models for stock index forecasting. In: 2012 World Congress on Information and Communication Technologies (WICT), pp. 602–607. IEEE (2012, October)

ML-IDS: MAC Layer Trust-Based Intrusion Detection System for Wireless Sensor Networks

U. Ghugar and J. Pradhan

Abstract Over the last few years, security concern has a very important role over the wireless sensor networks (WSNs). There are lot of attacks on WSN due to its open deployment and wireless communication between the nodes. In this paper, trust-based intrusion detection system is proposed to calculate the trust by using MAC layer protocol. The trust value of each node in WSN is evaluated by deviation of key factor at the MAC layer. The ML-IDS is impressive system to detect the malicious nodes in wireless sensor network. The malicious nodes attack the MAC layer by Back-off manipulation attack. Mainly Back-off manipulation attack is used to reduce the back-off time to get the channel priority. We have implemented the back-off manipulation attack at MAC layer for performance evaluation. Results show that ML-IDS perform higher quality as compared to Wang et al. (Sensors 17(6):1227, 2017, [1]) based on detection accuracy and false alarm rate.

Keywords Trust · Cluster head (CH) · Sensor node (SN) · Intrusion detection system · MAC protocol layer

1 Introduction

Wireless sensor network (WSNs) are universally used in diversity of area such as area observing, forest fire observing, military surveillance, health care, home affirmation, water quality management, ocean extracting and satellite communication. It is a distributive, automatic governing network and it consists of sensor nodes classify in a particular environment. A sensor node is a tiny and computing device which has a limited measurement resource [2, 3]. These nodes sense the natural conditions, such as humidity, compression, heat, sound, wave, and direction at different areas [4, 5].

U. Ghugar (✉) · J. Pradhan
Department of Computer Science, Berhampur University, Berhampur 761006, Odisha, India
e-mail: ughugar@gmail.com

J. Pradhan
e-mail: jayarampradhan@hotmail.com

H. S. Behera et al. (eds.), *Computational Intelligence in Data Mining*, Advances in Intelligent Systems and Computing 990,
https://doi.org/10.1007/978-981-13-8676-3_37

On the view of security concern, Intrusion detection plays a vital role in protecting the network. In WSN, Intrusion detection system detects the unauthorized access, therefore it specify the activity of a node in the network and it implements the system to defend against the intruder [6, 7]. Trust management system is an impressive method to identify the abnormal node in the network. In the recent year, lot of work has been done on intrusion detection system using trust theory and its application [8]. Feng et al. [9], proposed a concept, where trust calculation algorithm (NBBTE) is used to evaluate the trust (direct and indirect) value of each node and its neighboring nodes in the network using Fuzzy set theory.

We are strongly motivated from the proposed method by Wang et al. [1]. As per the authors, they build a detection model for WSN using trust concept. As per the authors trust is calculated of each sensor node at physical layer, MAC layer and network layer using trust metrics. Finally, each layer trust is combined and forms a single trust value. According to that single trust value, the monitoring node can calculate the trust of its neighboring node. That trust value is called as direct trust. In their model, where monitoring node calculates the trust of his monitored node. However, we have considered trust calculation of monitored node from the experience of his neighboring node.

In this paper, our proposed system is to detect the intruder, where intrusion is detected by using the trust concept over trust metrics of MAC protocol layer by using deviation factor. The major points are listed as follows:

1. ML-IDS are designed to detect Back-off manipulation attack at MAC layer over network.
2. Back-off time parameter and successful message transmission parameter is used to calculate the trust at the MAC layer. Here the deviation parameter of monitored node is calculated from direct and recommendation of neighbor nodes experiences.
3. Our proposed system is based on weighting method concept. In this model, the monitoring node monitors the key factor of the monitored node at the MAC layer and finally individual trust value is calculated. Then the trust value is forwarded periodically at the Cluster head (CH). Then CH declares whether the SN is genuine or abnormal node by using a threshold value (trust value).
4. The performance is evaluated by the Matlab R2015a, and as per the numerical results, ML-IDS shows the quality performance than Wang et al. [1] in the basis of DA and FAR.

The rest of paper is described in several parts, part-2 represents the system model, part-3 represents the discussion and numerical results, and finally part-4 represents the conclusion with upcoming work.

2 System Model

Basically, system model is based on network architecture and attack scenario. So we have discussed about the network topology and communication flow for network architecture and the attack scenario at MAC layer.

2.1 Network Model

Here, the network has been constructed by multiple clusters (Fig. 1). Each cluster having multiple sensor nodes (SNs) and is controlled by one cluster head (CH). The sensor node in the network can send or receive the data to their respective clustered head (CHs) over several intermediate sensor nodes (SNs). As usual Base Station (BS) received the combined data from all Clustered Head (CH). Here we have considered the Sensor Node to Clustered Head transmission scenario [10, 11].

Here, we have calculated the trust at the MAC layer and considered a time slot (Δt). Basically (Δt) is to represent the update the trust.

Let $T_{pq}(t)$ represents the trust value is calculated. Here sensor node (*p*) calculates the trust value of sensor node (*q*) at the time period (t) and represented as Eq. 1:

$$T_{pq}(t) = T_{pq}^{MAC}(t) \tag{1}$$

where $T_{pq}^{MAC}(t)$ represents the trust value of sensor node, which is evaluated by sensor node (*p*) for sensor node (*q*) for Media Access Control layer.

Equation 2 used for overall trust of monitoring node *p* on monitored node *q*.

$$T_{pq}(t) = \gamma T_{pq}(t - nt) + (1 - \gamma) T_{pq}(t) \tag{2}$$

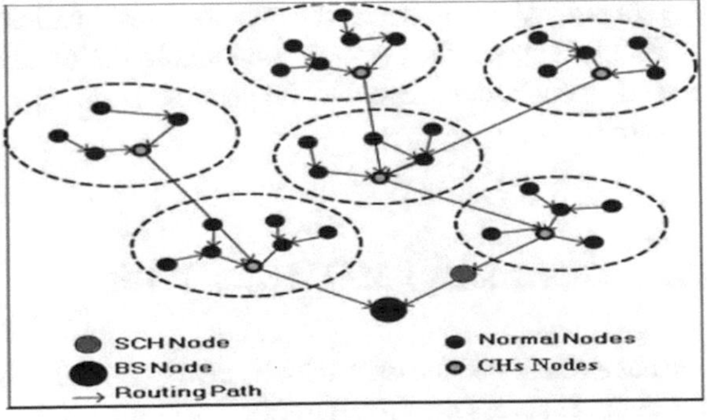

Fig. 1 Clustered architecture for WSNs

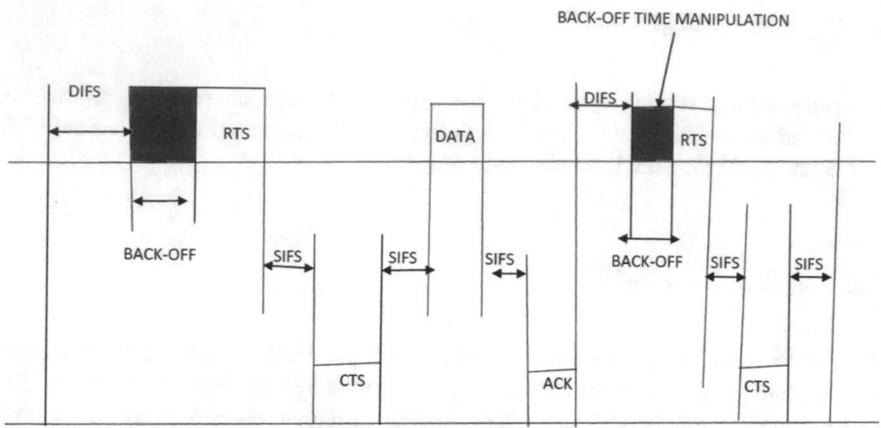

Fig. 2 Back-off manipulation attack at MAC Layer

where $T_{pq}(t - \Delta t)$ shows the history of previous trust of monitoring node *p* on monitored node *q* and $\gamma[0,1]$ indicates the value based on weighting method for earlier trust value [12].

2.2 Back-Off Manipulation Attack

In this layer, the abnormal node decreases the back-off time and manipulated for quicker access of the channel. So that it can easily send and receive the message to its nearest nodes with a maximum priority value. Here, back-off time is random in nature. It is manipulated by lowering it, so that priority of getting the channel access increases. This increases the number of successful transmissions (Smt). In this layer, back-off time (Bot) parameter and successful message transmission (Smt) parameter are considered as i to detect the unauthorized (malicious) node in the network.

Figure 2 shows the back-off time manipulation attack where the malicious node *p* sends a continuous signal to the genuine node *q* by getting the channel priority in less time.

2.3 Estimation of MAC Layer Trust

We have calculated the trust at MAC layer. Numbers of attacks are arising for getting the channel access priority. Here, Back-off manipulation attack is considered so that the time can be manipulated.

In this work, back-off time (Bot) and successful message transmission (Smt) are considered as trust parameter to detect the unauthorized activity in the network. The primary objective of this attack is channel access. Here the monitoring node p collects the recommendation of other neighboring at the time period (Δt) on monitored node q using the trust value of (Bot) and (Smt) [12].

The average recommendation is calculated by Bot and Smt as in Eq. 3:

$$\overline{\Delta \mathrm{Rec(Bot)}} = 1/n \sum_{p=1}^{n} \mathrm{Bot} \tag{3}$$

where, $\overline{\Delta \mathrm{Rec(Bot)}}$ (Eq. 4) represents the average number of the recommendation, which is collected by the neighbor nodes (n) at (Δt) time.

$$\overline{\Delta \mathrm{Rec(Smt)}} = 1/n \sum_{q=1}^{n} \mathrm{Smt} \tag{4}$$

where, $\overline{\Delta \mathrm{Rec(Smt)}}$ is the average number of the recommendation, which is collected by the neighbor nodes (n) at (Δt) time.

Now the relative deviation (Eq. 5) is of trust parameter is calculated of monitored node q as follows:

$$\mathrm{R_{Dev}(Bot)} = \frac{\overline{\Delta \mathrm{Rec(Bot)(t)}} - \Delta\mathrm{(Bot)q(t)}}{\overline{\Delta \mathrm{Rec(Bot)(t)}}} \tag{5}$$

$$\mathrm{R_{Dev}(Smt)} = \frac{\overline{\Delta \mathrm{Rec(Smt)(t)}} - \Delta\mathrm{(Smt)q(t)}}{\overline{\Delta \mathrm{Rec(Smt)(t)(t)}}} \tag{6}$$

where, $\mathrm{R_{Dev}}$(Bot) and $\mathrm{R_{Dev}}$(Smt) (Eq. 6) are the relative deviation of trust parameter. $\mathrm{(Bot)_q}$(t) reperents the back-off time at time period (t). $\Delta\mathrm{(Bot)_q}$(t) represents the back-off time at (Δt) time between monitoring and monitored node.

Finally, we have calculated the trust of each node using these two metrics as in Eqs. 7 and 8 as follows:

$$\mathrm{T_{pq}Bot(t)} = \begin{cases} 1 - R_{\mathrm{Dev(Bot)}(t), if \Delta \mathrm{(Bot)q(t)} < \overline{\Delta \mathrm{Rec(Bot)(t)}}} \\ 1, \\ Else \end{cases} \tag{7}$$

$$\mathrm{T_{pq}Smt(t)} = \begin{cases} 1 - R_{\mathrm{Dev(Smt)}(t), if\, \Delta \mathrm{(Smt)q(t)} > \overline{\Delta \mathrm{Rec(Smt)(t)}}} \\ 1, \\ Else \end{cases} \tag{8}$$

From the Eqs. (8) and (9), we know that if the back-off time at (Δt) time is decreased then recommended average trust means node is not trusted and if the successful message transmission at (Δt) time is increased then recommended average trust means node is trusted.

The overall trust of MAC layer is as follows:

$$T_{pq}^{MAC}(t) = b_1.T_{pq}^{Bot}(t) + b_2.T_{pq}^{Smt}(t) \quad (9)$$

where b1 and b2 is weighting parameter, for which summation value is 1. Here the weighting factors value is based on the system under implementation. Sensor node *(p)* calculates the trust value of sensor node (q) at (Δt) time.

3 Results and Discussion

To study the performance of our proposed model, we have used Matlab R2015a, PC of configuration 4 GB RAM, Core i3 processor, and 64 bit OS. The ML-IDS scheme is compared with Wang et al. [1], which calculates the trust layer-wise. The performance of our proposed model is analyzed by various parameters like Detection Accuracy (DA) and False Alarm Rate (FAR).

- Detection Accuracy (DA): The number of sensor node detected as abnormal from the total number of sensor node detected as malicious in the network.
- False Alarm Rate (FAR): The number of genuine sensor node detected as malicious from the total number of genuine sensor node.

This proposed scheme is simulated in Matlab R2015a. The network is randomly deployed an area of 100 × 100 m^2 in 50 nodes and communication range is set be 20 m. Here we have implemented back-off manipulation attack according to the model discussed previously. The values such as position of sensor (x,y), Bot, and Smt are randomly generated using randi(). Table 1 shows the performance analysis parameter.

Figure 3 shows the combined result comparison of ML-IDs and Wang et al. [1]. It is observed that when the malicious node uses back-off manipulation attack the

Table 1 Performance analysis parameters

Parameters	Values
Size of the network	100 × 100 m2
Number of cluster head	01
Number of SNs node	50
Number of malicious nodes	5–40% of SNs
Communication range	20 m
Message size	04 bytes
Bot for genuine node	20–30 μs
Bot for malicious node	10–25 μs
Smt for genuine node	1–3
Smt for malicious node	1–10
b1, b2	0.5, 0.5
Number of iteration	10

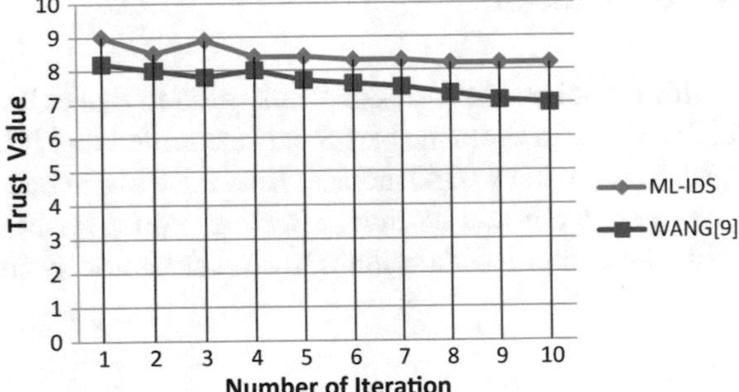

Fig. 3 Trust value comparison between ML-IDS and Wang et al. [1]

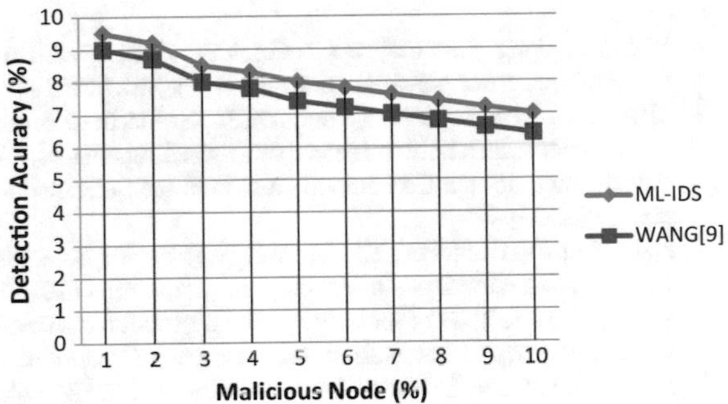

Fig. 4 Detection accuracy

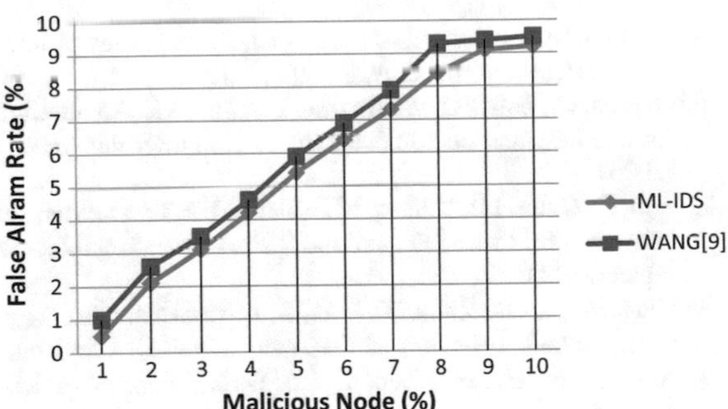

Fig. 5 False alarm rate

trust values calculated for a node decreases. It is also observed that when iteration increases then the trust value is gaining the stability as compared to Wang et al. [1].

Figure 4 shows the combined DA result comparison of ML-IDs and Wang et al. [1] under back-off manipulation attack. It observed that detection accuracy of ML-IDS is greater than Wang et al. [1]. It is also observed that when malicious nodes increases then the DA also decreases.

Figure 5 shows the combined FAR result comparison of ML-IDs and Wang et al. [1] under back-off manipulation attack. It is observed that FAR of ML-IDS is less than Wang et al. [1]. It is also observed that when number of malicious nodes increases in the network, FAR also increases.

4 Conclusion

In this model, we have designed a system to detect the back-off manipulation attack in the network. From the part-3, it is examined that if the node density increases then detection accuracy (DA) increases and false alarm rate (FAR) reduces and it indicates that our system will be more effective model for MAC layer attack. In future, we will plan to construct a better intrusion detection system other layers in the network.

References

1. Wang, J., Jiang, S., Fapojuwo, A.O.: A protocol layer trust-based intrusion detection scheme for wireless sensor networks. Sensors **17**(6), 1227 (2017)
2. Tiwari, M., Veer Arya, K., Choudhari, R., Sidharth Choudhary, K.: Designing intrusion detection to detect black hole and selective forwarding attack in WSN based on local Information. In: Fourth International Conference on Computer Sciences and Convergence Information Technology (2009)
3. Nam Huh, E., Hong Hai, T.: Lightweight intrusion detection for wireless sensor networks. In: Intrusion detection System. Intech publisher, China (2011)
4. Du, J., Li, J.: A study of security routing protocol for wireless sensor network. In: International Conference on Instrumentation, Measurement, Computer, Communication and Control (2011)
5. Bao, F., Ray Chen, I., Jeong Chang, M., Cho, J.-H.: Hierarchical trust management for wireless sensor networks and its applications to trust-based routing and intrusion detection. IEEE Trans. Netw. Serv. Manag. (2012)
6. Rassam, M.A., Maarof, M.A., Zainal, A.: A survey of intrusion detection schemes in wireless sensor networks. Am. J. Appl. Sci. (2012)
7. Depren, O., Topallar, M., Anarim, E., Ciliz, M.: An intelligent intrusion detection system (IDS) for anomaly and misuse detection in computer networks. Expert Syst. Appl. **29**(4), 713–722 (2005)
8. Bao, F., Chen, I.R., Chang, M., Chao, J.H.: Trust-based intrusion detection in wireless sensor networks. In: Proceedings of the IEEE International Conference on Communications, Kyoto, Japan (2011)
9. Feng, R., Xu, X., Zhou, X., Wan, J.: A trust evaluation algorithm for wireless sensor networks based on node behaviors and DS evidence theory. Sensors **11**(2), 1345–1360 (2011)
10. Atakli, I.M., Hu, H., Chen, Y., Ku, W.S., Su, Z.: Malicious node detection in wireless sensor networks using weighted trust evaluation. In: Proceedings of the Spring Simulation Multi Conference Society for Computer Simulation International, pp. 836–843 (2008)
11. Ghugar, U, Pradhan, J, Bhoi, S., Sahoo, R., Panda, S: PL-IDS: physical layer trust based intrusion detection system for wireless s sensor networks. IJIT, Springer **10**(4), 489–494 (2018)
12. Wu, R., Deng, X., Lu, R., Shen, X.: Trust-based anomaly detection in wireless sensor networks. In: 1st IEEE International Conference on Communications in China (ICCC), pp. 203–207 (2012)

Restoration of Mammograms by Using Deep Convolutional Denoising Auto-Encoders

Swarup Kr. Ghosh, Biswajit Biswas and Anupam Ghosh

Abstract To obtain better image visibility of noisy mammograms, a deep unsupervised learning method for degrading mammogram restoration scheme has been adopted in this paper. A deep convolutional denoising autoencoder method based on total variational multi-norm loss function minimization approach has been introduced for the restoration of mammograms. The suggested model can extract relevant features and can reduce the dimensionality of the image data while preserving the key features that have been applied to restore image data in feature space. Later, an unsupervised learning is performed based on the relation between key features and the weighted parameters of the network model. Experimental results authenticate that the suggested method outplays a number of state-of-the-art methods. The proposed scheme is not just pronc to decrease noise but also restores acceptable structural details of mammograms. After fine-tuning the network, its execution speed for target noisy images increase as compared to other algorithms.

Keywords Deep learning · Image denoising · Convolutional autoencoder · Variational method

1 Introduction

Recent trend medical image analysis plays a crucial role for the diagnosis of different diseases like breast cancer, lung cancer, liver cancer, brain tumor, etc. Image denoising is a significant preprocessing step during the processing of medical images [17].

S. Kr. Ghosh (✉)
Brainware University, Barasat, Kolkata 700125, India
e-mail: swarupg1@gmail.com

B. Biswas
University of Calcutta, Kolkata 700098, India
e-mail: biswajit.cu.08@gmail.com

A. Ghosh
Netaji Subhash Engineering College, Kolkata 700152, India
e-mail: anupam.ghosh@rediffmail.com

H. S. Behera et al. (eds.), *Computational Intelligence in Data Mining*, Advances in Intelligent Systems and Computing 990,
https://doi.org/10.1007/978-981-13-8676-3_38

Many vital structures are not perceptible properly since medical images are poorly illuminated. Lots of uncertainties are present in the image in imprecise form, which leads to inaccurate diagnosis. Thus, it is essential to remove noise from the image or to restore the image. The objective of denoising of an image is to renovate an image to another form that is more appropriate for further processing [15]. A number of image denoising algorithms have been established in the past few decades. Recently, deep learning based image denoising scheme has presented outstanding performance compared to other straight image denoising methods. In this exploration, we present an image denoising strategy in the light of a convolutional denoising autoencoder and assess clinical applications by looking at existing image denoising methods.

An autoencoder (AE) is an alternative type of feedforward neural network, where the number of input layer and output layer is indistinguishable, called an autoassociator, designed for unsupervised learning or one kind of MLP (Multilayer Perceptron) in information system. Basically, AE is prepared to encode the input data and to turn it into an abstract illustration by encoder function so that the input can be reconstructed from that representation [9, 12]. Vincent et al. suggested the denoising autoencoder (DAE), which deliberately adds noise into the training data and trains the AE with these adulterated information. DAE can recover the original data and make it noise-free through the training process leading to superior robustness [18].

The determination of image denoising is to restore maximum of the original image details by eliminating surplus noise. Numerous denoising algorithms have been suggested by several researchers such as median filtering, total variation minimization (TV), and non-local means (NLM), etc. Lovedeep Gondara [3] has proposed a denoising scheme using convolutional denoising autoencoders on medical image and the algorithm performed on both MRI and X-ray images had a low performance on high-resolution images. A novel artifact suppressed dictionary learning method presented by Chen et al. [1] is used for low-dose CT image processing. Zhang et al. [21] described an adaptive bilateral filter-based framework for image denoising at which entropy-based feature extraction is used for restoration of medical images. A deep convolutional networks for large-scale image recognition suggested by Simonyan et al. [16] at which very small (3×3) convolution filters were used as the increase of the depth of the layer leads to improvement on the prior configurations. Mao et al. [10] described an image restoration method by using Convolutional Autoencoders in which symmetric skip connection is used to extract the features from the images.

In this work, we deliberate a denoise autoencoder (DA) based on variational technique to restoring mammograms. We measure multiple-loss function to reach to fast optimization in convolutional model during training periods. We verified overall execution of the proposed methodology and compared with other three state-of-the-art methods using a popular public benchmark database in a training session. The empirical results show that the proposed method not only restores mammograms but also preserves image details more accurately than most privileged methods. Main contributions of this paper are given below:

1. We review the deep-based denoising autoencoder algorithms and propose a Deep Conventional Denoised Autoencoder (CDDA) for mammogram restoration to exploit the merits of learning fields.
2. Denoising Autoencoder (DA) exploiting the multi-level loss function using Kullback–Leibler Divergence (KLD) in the proposed scheme to address restoration for degraded mammograms.
3. We have designed DA based on total variation model to process artifacts in mammograms for efficient elimination and the perfection of the proposed scheme outperforms the state-of-the-art algorithms while achieving best restoration results.

The rest of this paper is organized as follows: In Sect. 2, we have discussed the preliminaries and mathematical model. The proposed methodology and designed mammogram restoration algorithm in details have been discussed in Sect. 3. The experimental results and discussions have been explored in Sect. 4 and finally we conclude this work in Sect. 5.

2 Preliminaries

This section concisely discusses the theories of Convolutional Autoencoder (CAE) [3] and the basic framework of mammograms denoisification scheme [10]. The proposed algorithm, denoise autoencoder method (DAM), is developed on the basis of following conception.

2.1 *Denoise Model for Mammogram Images*

According to mathematical definition [18], we consider that a given $\mathbf{x} \in \mathbb{R}^{m \times n}$ noisy mammogram image and $\mathbf{y} \in \mathbb{R}^{m \times n}$ be its corresponding noise-free image. Then, the interrelation between $\mathbf{x}$ and $\mathbf{y}$ can be written as follows [3]:

$$\mathbf{x} = \eta\left(\mathbf{y}\right)$$

where $\eta : \mathbb{R}^{m \times n} \rightarrow \mathbb{R}^{m \times n}$ denotes the extortion of the additive noise that distorts the normal mammogram. To recover the corrupted image, the noise reduction algorithm (viz., a function) mainly converted into the original once while finding a function ϕ that hold the following criterion [1]:

$$\phi = \arg\min \|\phi\left(\mathbf{x}\right) - \mathbf{y}\| \tag{1}$$

where ϕ is deploys as the best approximation of η^{-1} and $\| \cdot \|$ is l_1 norm.

2.2 Denoise Autoencoder (DA) for Noisy Mammogram

To restore a mammogram, a denoising autoencoder(DA) can be the best choice to regenerate a noise-free image from its noisy version with self-learning [10, 18]. The feedforward functions for a noised image $\mathbf{x}$ and corresponding normal image $\mathbf{y}$ of the Denoising Autoencoder is illustrated as [3]

$$h(\mathbf{x}) = \mathcal{H}(\mathcal{W} \cdot \mathbf{x} + \mathbf{b}) \tag{2}$$

$$s(\mathbf{x}) = \mathcal{H}\left(\hat{\mathcal{W}} \cdot h(\mathbf{x}) + \hat{\mathbf{b}}\right) \tag{3}$$

where the set of the weights and biases are $\Theta = \{\mathcal{W}, \mathbf{b}, \hat{\mathcal{W}}, \hat{\mathbf{b}}\}$, the 'sigmoid' function is $\mathcal{H}(x) = \left(1 + e^{(-x)}\right)^{-1}$, the hidden layer activation is $h(\mathbf{x})$, and $s(\mathbf{x})$ is a prediction of $\mathbf{y}$.

For the mammogram image in training data set $\mathbf{X} = \{(x_i, y_i)\}$, where $i = 1, 2, \cdots, N$ where N is the number of training examples. The backpropagation algorithm has been used in training phase of the denoising autoencoder (DA), which minimizes the sparsity regularization and regeneration loss function from Eq. (1) with l_1 norm [22]:

$$\mathcal{Z} = \frac{1}{N}\sum_{i=1}^{N}\frac{1}{2}\|s\left(x^i\right) - y^i\| + \beta \cdot \mathrm{KLD}\left(\hat{\kappa}\|\kappa\right) + \frac{\lambda}{2}\left(\|\mathcal{W}_F^2\| + \|\hat{\mathcal{W}}_F^2\|\right) \tag{4}$$

where λ is the weight decomposition parameter and β is the sparsity term parameter. The sparsity is the Kullback–Leibler Divergence (KLD) [10, 22]

$$\mathrm{KLD}\left(\hat{\kappa}\|\kappa\right) = \sum_{l=1}^{|\hat{\kappa}|}\kappa \log\frac{\kappa}{\hat{\kappa}_l} + (1-\kappa)\log\frac{(1-\kappa)}{\left(1-\hat{\kappa}_l\right)}$$

where $\hat{\kappa}$ is mean activation of the hidden layer and κ is the target activation.

3 Mammogram Reconstruction by Deep Denoising Autoencoder Method

The proposed image reconstruction scheme is based on a deep learning approach that determines the sequences of noisy image as concurrent input and provides the sequences of noise-free output image (normal image) on a huge number of training–testing data pairs, which consists of a train input sequence and also the desired target sequence.

Here, we have designed a regeneration scheme using a convolutional neural network with ordered encoder–decoder functions [12, 13] described in Eqs. (2) and (3).

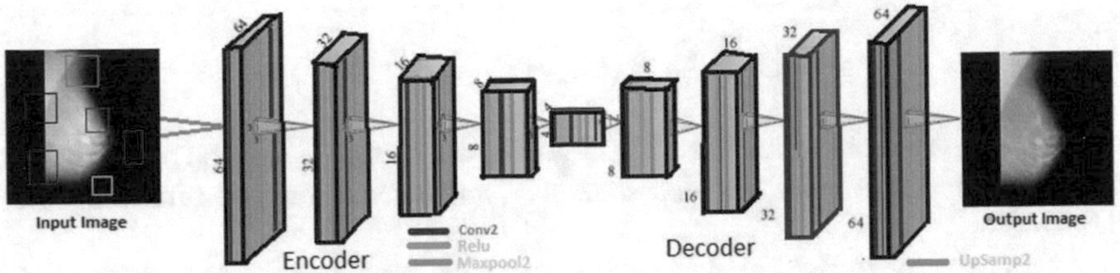

Fig. 1 Architecture of the proposed autoencoder

To achieve the best performance, the model has 30 layers with 3×3 kernel support, each in operation on identical spatial resolution. A dropout layer association is supplementary from each third layer to the corresponding layer last layer of the network (each case 0.5 is used as a dropout).

We address deficiency by updating the existing DA architecture to incorporate subsampling and up-sampling stages by batch normalization theme. Later, we have a tendency to continue by deploying the preparation of training data and also the correct loss function with optimizer throughout training. Figure 1 shown the schematic block diagram of the proposed denoise autoencoder method(DAM).

3.1 Denoising Autoencoder (DA) with Batch Normalization(BN) and Dropout

As mentioned in Fig. 1, the network architecture consists of individual encoder and decoder phases that care for decreasing and increasing abstract resolutions, respectively. It is basically the same as the U-Net [14] and Flownet [2] architectures that provides image segmentation and good result in optical flow estimation, severally, and additionally emphasizes the association to denoising autoencoders [10]. Autoencoders are networks that determine to rebuild their input from some internal organization, and the proposed denoising autoencoders learn to get rid of noise from input data. We have a tendency to use the term denoising autoencoder and as a result we would like to reconstruct from noisy input images [3].

Each convolution layer is composed of a 3×3-kernel operators with spatial support. Every single layer deals with the features as input from the preceding layer of the DA. Then, it integrates the outcomes with the features from the previous hidden layer, and pass it to rest of the convolution layers. The output generates each of the first hidden network and also the result of the successive layers. It gives a recovered accepting field along with the multi-layered cascade layers, which permit efficient tracking and hold image features fleeting at fine scales. Furthermore, convolution layer with batch normalization and dropout operate on identical input layer of same quantity of features and it passes on to the next layer in the encoding step. However, the hidden state along with the output layer can be expressed using a deep heuristic expression [16]

$$s_i = \mathcal{O}_i = \mathcal{K}_{3\times3}\left(\mathcal{K}_{3\times3}\left(\mathcal{K}_{3\times3}\right)\left(\mathbf{I}_i \odot s_{i-1}\right)\right)$$

where $\mathcal{K}_{3\times3}$ is a kernel which has a 3×3-pixel spatial support, $\mathbf{I}_i$ is the present input, O_i is the output, ith input's hidden state is s_i, and convolution operator is denoted by $\odot$.

As a precursor to the DA design, three-channel image sequences as input and target has been set up in the training network. This reduces temporal a flicker (batch normalization), and increases visibility in sequence of training sample.

3.2 Training

In this section we have explained the construction of training data called data augmentation has been illustrated. We begin with 250 slices of mammograms for each train sets during training period. For each image, we initiate a noisy image at one frame per data set by adding the Gaussian additive noise in training phase [21]. We make 512×512 images during explanation, but the training is tested with resizing as 128×128 patch which have been randomly chosen for each training sequence. A sequence of 50 consecutive frames has been used to give abundant temporal substances in the training phase.

In data normalization, we have chosen randomly small rotation of the training sequence in 2–3° in train increasing capture more temporal details of changing directions. We additionally choose a random gamma correction within the range [0, 1] one by one to every color channel, and have put on the network to the complete sequence.

3.3 Loss Function

The computed error between training targets and outputs from network during the training phase is described as a loss function. The most ordinary MSE (mean squared error) from the train image S and the target image T is defined by the $\mathcal{L}_2$ loss functions. However, another loss function, $\mathcal{L}_1$ loss instead of $\mathcal{L}_2$ can reduce the blemish in artifacts from regenerated images [5]. As mentioned above, the first loss term is a spatial $\mathcal{L}_1$ loss in the basis of KLD, denoted as $\mathcal{L}_z$ applicable for a distinct image in the training phase [1]. The second loss term is a spatial $\mathcal{L}_1$ loss, stands for $\mathcal{L}_s$ for a particular image in the training stage [3]:

$$\mathcal{L}_s = \frac{1}{N}\sum_{j=1}^{N} \|\mathcal{S}_j - \mathcal{T}_j\| \tag{5}$$

where $\mathcal{S}_j$ and $\mathcal{T}_j$ are jth pixel of the resultant and goal image correspondingly.

The $\mathcal{L}_2$ spatial loss gives a good image metric, which is used to measure the overall outliers [11]. To highlight the variances in adequate information (edge, corner, etc.), we use a Laplacian operator with square root of absolute $\mathcal{L}_1$ in the domain of gradient loss [5]

$$\mathcal{L}_l = \frac{1}{N}\sum_{j=1}^{N}\sqrt{\|\nabla\mathcal{S}_j - \nabla\mathcal{T}_j\|} \quad (6)$$

where individual gradient $\nabla(\cdot)$ is calculated based on High-Frequency Error Norm (HFEN), which is a metric used for medical image comparison [22]. The Laplacian of Gaussian kernel (3 × 3) have been used for edge-detection, but it is sensitive to noise and hence to make image pre-smoothed, first Gaussian filter is used for edge detection [1, 11]. So in the parameter setup, we suggested $\sigma = 1.25$ for Gaussian kernel size and the final training loss $\mathcal{L}$ by using the weighted combination of losses from Eqs. (5) and (6), determined as follows:

$$\mathcal{L} = w_z \cdot \mathcal{L}_l + w_s \cdot \mathcal{L}_l + w_l \cdot \mathcal{L}_l \quad (7)$$

where $w_{z,s,l}$ are the adaptable weights that restrict the contribution of each loss. We have taken $w_{z,s,l} = 0.25, 0.75, 0.12$ to approximately adjust the scales, which enhanced the convergence.

To authenticate that the combined loss ends up in an enhancement above the spatial loss $\mathcal{L}_s$, it has been calibrated the structural similarity metric (SSIM) [20] on tuning sequence of mammograms after 50 epochs with 128 batch-size of training. SSIM showed an enhancement from 0.9509 for $\mathcal{L}_s$ to 0.9838 for the fused loss.

4 Experiments

The effectiveness of our algorithm, along with comparisons is demonstrated on two standard databases which are publicly available on the Internet, viz., MedPix [6] and CBIS-DDSM [7] shows in this section.

4.1 Experimental Setup and Validation

A total of 2400 ordinary mammogram images of size 1024 × 1024 from 420 patients collected from the National Library of Medicine presents MedPix [6] and CBIS-DDSM [7] archive as normal-dose dataset which are shown in Fig. 2a–c. As stated above, the corresponding dataset of noise mammogram images have been made by introducing Gaussian noise into the routine mammogram images for the simulation of the proposed method DAM. The salt-and-pepper noise plus Gaussian noise is utilized to test the performances of the proposed method DAM. In the initial configuration,

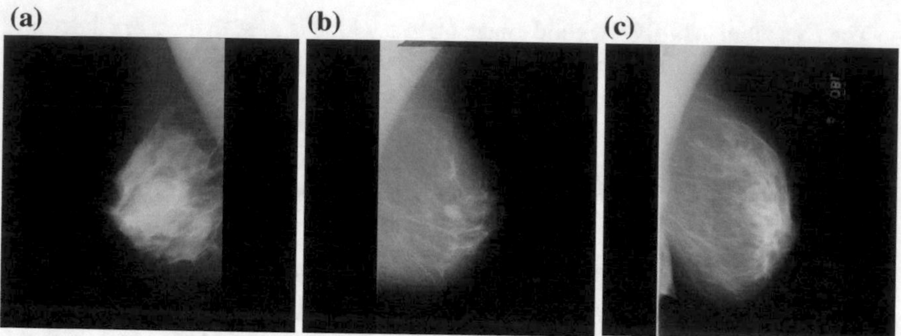

Fig. 2 Examples from the normal mammogram dataset

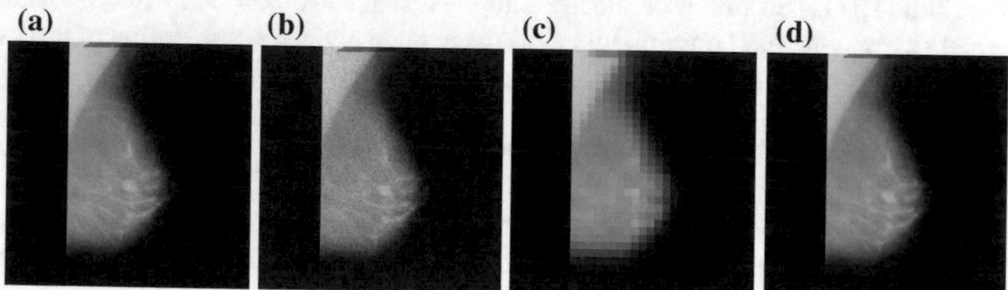

Fig. 3 Simulation with the proposed scheme. **a** the original image. **b** noise image **c** reconstructed image (epoch = 300, batch-size = 64), and **d** denoised result (epoch = 1000, batch-size = 128)

$\mu, \sigma = 0, 30$ control the noise level of images. Each mammogram image is corrupted by uniform impulse noise $p = 20\%$ and Gaussian noise $\sigma = 30$, which is shown in Fig. 3b, for example.

In the training stage, 300 normal (cleaned) and corresponding additive noise mammogram pairs have been selected randomly. Patch size p has been set to 128×128 pixels. We have a tendency to evaluate with many parameter accumulations, and therefore the following parameter settings have been finalized as κ which has set to 0.08, λ set to 10^{-5}, and β set to 5×10^{-3} in cost function Eq. (4). With increase in hidden layers and neurons, duration of training increases without noticeable performance gain, therefore the number of layers L is set to 16 for all cases. The projected loss estimation has been accustomed to optimize the loss function w_l by using Eq. (7).

In test phase, we have decomposed each mammogram into patches are managed by trained proposed DAM, and then used the resultant patches to regenerate the approximated reconstructed image. Three metrics have been applied to statistical quantification. The primary one is that the root mean square error (**MSE**) [4], and later ones are Picture quality score (PQS) [20] universal quality index (**UQI**) [19] and Structural Similarity Index Metric (**SSIM**) [4, 20], respectively. The highest value among them indicates minimum distortion and better reconstruction except MSE.

Three different baseline methods have been compared for the performance evaluation, such as "An adaptive bilateral filter based framework for image denoising" abbreviated for convention as (ABF) [21], "Artifact suppressed dictionary learning for low-dose CT image processing" (SDA) [1], and "Medical image denoising using convolutional denoising autoencoders" (CDA) [3]. ABF is most fashionable image domain-based denoising strategies that used bilateral filter recently. CDA has been projected low-dose CT restoration ways supported by neural networks. All the connected parameters within the baseline techniques have been fixed as per the proposal within the primary investigation. The networks of CDA have been trained with similar dataset as DAM.

4.2 Analysis of Different Parameters

The efficiency of the proposed DAM scheme depends on different parameters. Here, we examine the contact of two key factors: the amount of training instance, and noise level of the dataset have been discussed in this section.

The training data: We amplify the effect of the size of training data from two databases. In the first phase, 300 image pairs have been arbitrarily chosen from the database as training set. In the second phase, data augmentation has been made to magnify the training set supporting the initial 300 image pairs using Keras API. The transformations enclosed rotation (by 2–3), flipping (vertical and horizontal), and scaling (varying scale factors 0.25–0.5). Finally, similar operations for data augmentation have been executed on the new elite a hundred image pairs, and a complete of 2400 images are achieved.

Noise level: In practice, consistent to noise levels from the training and testing sets should be dissimilar; consequently, it is essential to justify the validity of the suggested scheme completely for various datasets with different noise levels. However, one testing set with a Gaussian noise level, $\mu = 0, \sigma = 30$ and salt-and-pepper noise with $p = 25\%$ area unit utilized in the analysis. The training set preserved a similar noise level CDA [3]. The quantitative results are shown in Table 1. It is determined that for a given noise level with large epochs the CDA and DAM has higher MSE, PQS, and UQI values than different baseline methods. In Fig. 3 pictured the reconstruction of mammograms (a) with low (near 300) (c) and higher (d) (1000) epochs by the DAM, where (b) is a noisy version of (a).

Table 1 Quantitative evaluation of various methods

Image	Method	MSE	PQS (dB)	UQI	SSIM
Figure 4	ABF [21]	0.0278	26.82	0.844	0.825
	SDA [1]	0.0226	27.32	0.861	0.843
	CDA [3]	0.0193	28.66	0.878	0.882
	DAM	**0.0188**	**29.10**	**0.883**	**0.891**

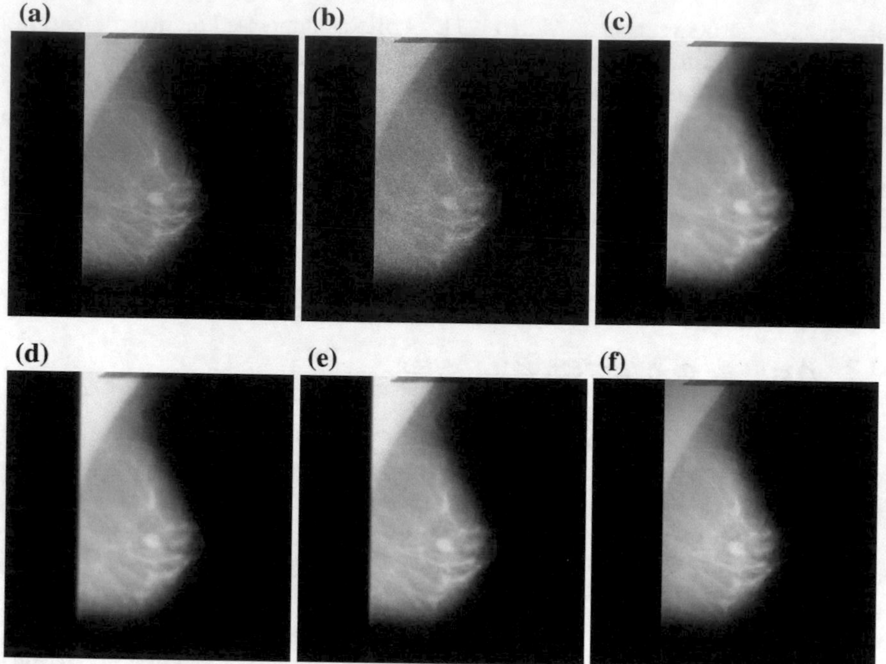

Fig. 4 Results of a mammogram (in Fig. 2b) image for comparison. **a** Cleaned image; **b** noised image; **c** ABF; **d** SDA; **e** CDA; and **f** DAM

4.3 Subjective Analysis

In test phase, one sequence frame from the test set has been picked to assess the performance of the proposed DAM. Figure 4 demonstrated the results received a mammogram image by all methods. In Fig. 4b, the noise mammogram with noise artifacts appear near the bright tissue with distorted breast regions, such as nodules. All baseline methods can efficiently suppress the noise; however, in Fig. 4c, ABF is affected by conspicuous square effect and over-smoothed a few essential details in the breast interior regions. CDA retain major details than SDA, but the artifacts close the interior of the breast still exist the noisy appearance with the low-contrast levels. The deep learning based techniques such as CDA can protect more subtle elements such as noise and artifacts. Nonetheless, CDA over-smoothed few low-contrast domains as shown in Fig. 4d.

Three baseline strategies (ABF, SDA, and CDA) have been established. It is detected in Fig. 4c–e that three strategies preserve the soft tissues abundantly distinct than different methods. DAM additionally obtained the most effective performance within the low distinction region if we have a tendency to circumstantially ascertained all resultant data. Additional methods are mainly ABF and SDA which failed in attaining to eliminate impulse noise completely. The proposed methodology of DAM can preserve additional unchanged noise-free pixels supported on multi-loss function utilization and hence we can conclude that the details of the restored image by the proposed methodology and DAM produces the best visual effect compared to other strategies.

4.4 Objective Analysis

Quantitative performance of the proposed scheme DAM has been compared with three existing methods as mentioned above and as shown in Fig. 4. Comparable statistical results for the four metrics (MSE, PQS, UQI, and SSIM) are given in Table 1 with boldfont with the highest score of DAM. For all the images, DAM achieved better scores of PSNR and PQS, UQI and SSIM and interestingly, CDA also achieved better performance with respect to MSE, PQS, UQI, and SSIM, which is the same as DAM and it indicates that both are able to preserve more detail in image reconstruction task. The quantitative outcomes for all resultant images are presented in Table 1. Table 1 shows the statistical validation of Fig. 4c–f in test phase. Table 1 shows that the suggested DAM method obtained competitive performance with CDA and excellent restoration than other baseline methods.

The performances of the proposed techniques of DAM are compared with three baseline methods such as ABF [21], SDA [1], and CDA [3] in Table 1. The UQI obtained by baseline methods such as ABF and SDA are in the range of 0.8–0.88, whereas deep learning based techniques CDA in UQI close to the proposed algorithm. SSIM for proposed technique DAM is greater than ABF, SDA and CDA and close to 0.9. So we conclude that edges are not properly preserved using ABF and SDA whereas DA and proposed technique DAM are able. So we can say that Table 1 reveals

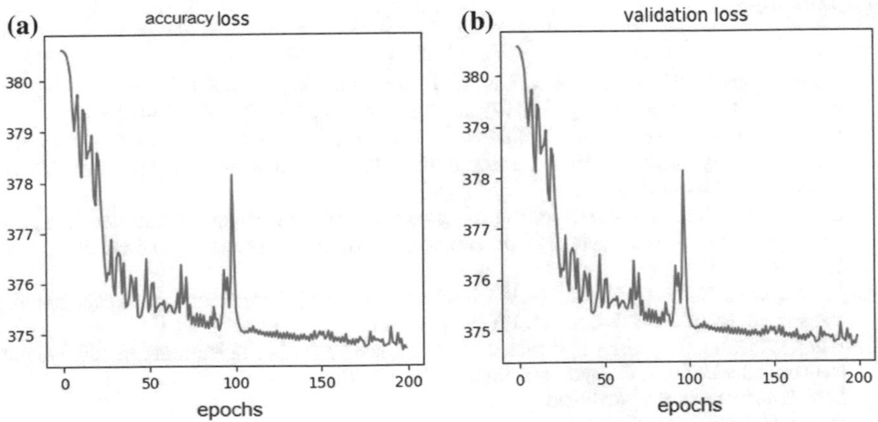

Fig. 5 Performance evaluation of the proposed method. **a** curve for accuracy, and **b** curve for validation

that all image quality metrics are improved in CDA and the proposed technique DAM than ABF and SDA methods. The metric indices of performance metric in Table 1 shows superior noise reduction capabilities of the proposed DAM technique. In Fig. 5a–b the test and validation loss of the DAM with a certain number of epochs is shown, respectively.

For the experiment, we have used publicly available two standard databases, viz., MedPix [6] and CBIS-DDSM [7], due to page limitation we have given few images in this paper. For further details of more result, refer [8].

5 Conclusion

A deep network based on denoising autoencoder has been presented for mammogram imaging in this work. Once trained offline with existing customary and mammogram image pairs, the proposed method can enhance image quality proficiently and rapidly. Both the subjective and quantitative outcomes have been exhibited in this work and we have figured out its capacity to diminish noise to preserve the structure. Simulation results demonstrate that the proposed method is significantly superior to some popular methods both visually and tentatively.

In future work, striking deep learning into other issues of mammogram image analysis area together with sparse view reconstruction, interior deformation measure in mammogram imaging, and structural superfluity reduction, will be exploited. It will focus on artifact anomaly detection such as lumps, swollen lymph nodes in mammogram, or early stage of breast cancer detection. In this context, more deep learning methods will be explored.

References

1. Chen, Y., Shi, L., Feng, O., Yang, J., Shu, H., et al.: Artifact suppressed dictionary learning for low-dose CT image processing. IEEE Trans. Med. Imaging **33**(12), 2271–2292 (2014)
2. Fischer, P., Dosovitskiy, A., Ilg, E., Häusser, P., Hazırbaş, C., Golkov, V., Smagt, P.V., Cremers, D., Dosovitskiy, T.B.: FlowNet: learning optical flow with convolutional networks (2015). arxiv.org/abs/1504.06852
3. Gondara, L.: Medical image denoising using convolutional denoising autoencoders. In: Proceedings of the IEEE International Conference Data Mining Workshops (ICDMW 2016), pp. 241–246 (2016)
4. Gu, K., Zhai, G., Yang, X., Zhang, W.: Using free energy principle for blind image quality assessment. IEEE Trans. Multimed. **17**(1), 50–63 (2015)
5. Hang, Z., Gallo, O., Frosio, I., Kautz, J.: Loss Functions for Image Restoration with Neural Networks. IEEE Trans. Comput. Imaging (2016). (ACM)
6. https://medpix.nlm.nih.gov/home
7. https://wiki.cancerimagingarchive.net/display/Public/CBIS-DDSM
8. https://github.com/biswajitcsecu
9. Liu, W., Wang, Z., Liu, X., Zeng, N., Liu, Y.: A survey of deep neural network architectures and their applications. Neurocomputing **234**, 11–26 (2017)

10. Mao, X.J., Shen, C., Yang Y.B.: Image restoration using convolutional autoencoders with symmetric skip connections. In: Proceeding of the NIPS (2016)
11. Mittal, A., Bovik, A.C.: No-reference image quality assessment in the spatial domain. IEEE Trans. Image Process. **21**(12), 4695–4708 (2012)
12. Pathak, D., Krahenbuhl, P., Donahue, J., Darrell, T., Efros A.: Context encoders: feature learning by inpainting. In: Proceeding of the CVPR (2016)
13. Patraucean, V., Handa, A., Cipolla, R.: Spatio-temporal video autoencoder with differentiable memory (2015)
14. Ronneberger, O., Fischer, P., Brox, T.: U-Net: convolutional networks for biomedical image segmentation (2015). arXiv:1505.04597
15. Sawant, H.K., Deore, M.: A comprehensive review of image enhancement techniques. Int. J. Comput. Technol. Electron. Eng. **1**(2), 39–44 (2010)
16. Simonyan, K., Zisserman, A.: Very deep convolutional networks for large-scale image recognition. In ICLR, (2015)
17. Sonka, M., Hlavac, V., Boyle, R.: Image processing analysis and machine vision, Thomson learn (2001)
18. Vincent, P.: A connection between score matching and denoising autoencoders. Neural Comput. **23**(7), 1661–1674 (2011)
19. Wang, Z., Bovik, A. C.: A universal image quality index. IEEE Signal Process. Lett. **9**(3) (2002)
20. Wang, Z., Bovik, A. C., Sheikh, H. R., Simoncelli, E. P.: Image Quality Assessment: From Error Visibility to Structural Similarity. IEEE Trans. Image Process. **13**(4) (2004)
21. Zhang, Y., Tian, X., Ren, P.: An adaptive bilateral filter based framework for image denoising. Neurocomputing (2014)
22. Zhang, Y., Zhang, W. H., Chen, H., Yang, N. L., Li, T., Y., Zhou J. L.: Few-view image reconstruction combining total variation and a high-order norm. Int. J. Imaging Syst. Technol. **23**(3), 249–255 (2013)

A Dynamic Panel Data Study on ICT and Technical Higher Education in India

Uday Chandra Mukherjee, Madhabendra Sinha and Partha Pratim Sengupta

Abstract The study explores the dynamic relationship between growing trends in information and communication technology (ICT) and performance of technical higher education in India during the last two decades. Impacts of ICT on the higher education sector, especially in the field of technical education in India, during a vibrant period of digitalization in the era of globalization has been a burning topic of research. ICT enables training providers withgreater exposure and better knowledge in technical institutions. In this context, we build up a panel of various technical education institutes over the period of 2004–2017 with the relevant variables and apply GMM estimation as suitable for the dynamic panel. Our empirical findings refer that ICT has significant impacts on technical higher education indicators.

Keywords ICT · Technical higher education · India · Panel data · GMM estimator

1 Introduction

It has been empirically validated that education not only improves the quality of life in a society but it also modernizes the way of thinking, attitudes and behaviours of the citizens of a country. Educated people are better equipped for significant economic production, which is important for rapid growth of the economy. The efficiency and effectiveness of the workers can be improved by imparting higher technical education to use sophisticated advanced technologies of production. Typically, the education system is associated with a very high degree of direct personal contacts

U. C. Mukherjee (✉) · M. Sinha · P. P. Sengupta
Department of Humanities and Social Sciences, National Institute of Technology Durgapur, Durgapur 713209, West Bengal, India
e-mail: ucmukherjee@gmail.com

M. Sinha
e-mail: madhabendras@gmail.com

P. P. Sengupta
e-mail: parthapratim.sengupta@hu.nitdgp.ac.in

H. S. Behera et al. (eds.), *Computational Intelligence in Data Mining*, Advances in Intelligent Systems and Computing 990,
https://doi.org/10.1007/978-981-13-8676-3_39

of teachers with the learners along with teacher-centered delivery of instruction to classes of students who are the receivers of information. The introduction of information and communication technology (ICT) in the system of education is lending itself to additional student-oriented settings of learning. During the last two decades, there has been a fundamental change in the use of ICT, reflected in terms of procedures and practices of almost every form of attempt within governances and business practices. Particularly in the technical higher education sector ICT has started its existence; however, its immediate impacts are not captured as anticipated, and have been quite progressive because of several factors like lack of proper policies, fundamental infrastructure, financial resources, and teachers and learners with their requisite skills.

By ICT-enabled learning we mean various technological tools and resources used for communication, creation, dissemination, storing and managing the information. It is a process of learning using ICT to support, enhance and optimize the delivery of information. ICT-enabled learners use several technological tools in the traditional classroom, in addition to various data storage devices like DVD or CD-ROM, videotape or over a television channel and even Internet. An ICT-enabled classroom may use different technological tools like radio, television, telephony (landlines/mobiles), computers, internet, etc. Method of ICT-enabled learning is different from e-Learning which refers to a learning procedure using electronic technologies in order to access the educational curricula outside of a traditional classroom. The e-Learning includes learning courses or getting degrees, diplomas or certificates that are solely delivered via Internet ranging from distance education for computerized electronic learning, online learning, etc. Contrary to the courses delivered through a DVD or CD-ROM, videotape or over a television channel e-Learning is interactive in that the participants can also communicate with the teachers or other students in the class.

Higher education institutions might persuade development in several areas, where the approach of ICT impacts. The areas include vision, mission, philosophy and pedagogy of education, curricula designing, strategies of development, various resources and amenities, professional development, employees of institutes, involvement of community, evaluation, etc. It is a fact that ICT becomes not only the backbone of the information community, but also a significant catalyst and instrument to induce the educational system reforms which transform learners into workers with creative knowledge. Different studies confirm that ICT-enabled learning structure has positive and significant impacts on both students and teachers. As the world population crossed already 750 crores in 2017, it has an intense effect on almost all the domains where ICT should be given priority [13]. Size of the population in India has also crossed 130 crores and high proportions of that are young. As a result of which the demand for higher and technical education in India has escalated in recent years as education is still regarded as the most sought after tool to bridge the socio-economic gap in developing countries. Currently, most of the developed and developing nations are the users of ICT-enabled system of education as a tool for quality education and to fill the huge gap between traditional methods and new methods of teaching and learning. It is already proved that methods of ICT-enabled learning offer better exposure and knowledge to students and teachers in technical institutes located in underdeveloped regions in India.

Different studies and analysis suggest that initial attempts to bring ICT to education follow a conventional story of automation rather than overall modification/evolution of the new concept. IT is primarily used for automating the delivery system of information in classrooms. In the absence of radical changes to the conventional teaching and learning process, such automated classrooms may do little but speed up ineffective processes and methods of teaching. Most of the countries are now emphasizing on the practical aspects of ICT and its complementary methods and mastering fundamental skills and ideas of ICT as a component of mainstream curricula of learning. The National Mission on Education in India highlights the functions of ICT for enhancing the present rate of enrolment in higher education from 13.5 to 21% during the twelve plan period (2012–17). India is a vast country, where the knowledge and the expertise are lying in a scattered way in different demographical regions where ICT presents a huge opportunity to all experts and educators to pool their collective knowledge to benefit the Indian students.

Massive Open Online Courses (MOOCs) has brought about a paradigm shift in the field of Distance Learning. Macquarie University's global classroom has offered access to the students around the world to have free access to their contents through the launch of the new 'Big History: Connecting Knowledge' MOOC on Courses. The course provides students a new framework to evaluate and approach problems in fundamentally new ways which are as critical to the challenges faced by business leaders today as it is for scientific research, politics and entrepreneurship. The Ministry of Human Resource Development has launched a portal SWAYAM (Study Webs of Active learning for Youths Aspiring Minds) where MOOCs is available to Indian students. An initiative by Indian Institute of Science (IISc) and seven old Indian Institutes of Technology (IIT Bombay, Delhi, Guwahati, Kanpur, Kharagpur, Madras and Roorkee) for creation of course contents in the field of engineering and science is very much popular now and known as National Programme on Technology Enhanced Learning (NPTEL). Total 923 nos. of courses with video contents are available at NPTEL. Based on model curriculum as suggested by All India Council for Technical Education (AICTE) most of the course content of NPTEL is developed.

ICT acts as an augmenter for capacity building efforts in higher educational institutions with continuous value addition and it becomes essential to strengthen the high rate of economic growth of the country through knowledge sharing. Mooij [10] expressed that various ICT-based systems of learning could be supposed to offer larger validity, reliability and competence of data collection and also better ease of assessment, analysis and explanation at any learning level. As India is speedily moving towards 'Digital India', the significance of ICT in learning has been more relevant. ICT has revolutionized the knowledge dissemination process in traditional classrooms teaching and as a whole the teaching pedagogy. The changes in the role of teachers and learners as the results of ICT in classroom teaching are represented in Table 1.

The student, the industry and the government can be immensely benefitted by the application of ICT in the education sector as per UNESCO [12]. The ICT-enabled

Table 1 Role of teachers and students in traditional and ICT-enabled classrooms

	Traditional classroom	ICT-enabled classroom
Teacher's role	Educator, tutor, instructor, imparting knowledge	Sharing of knowledge, capacity builder, guidance to students, advisor
	Focus on developing academic knowledge	Focus on developing professional capabilities
Student's role	Static receiver of knowledge	Contributor to the teaching-learning process
	Teaching-learning as an individual activity	Teaching-learning is a group activity requiring collaborations, consolidation etc.

Source Authors' own presentation

teaching-learning approach, not only cut down the cost of learning, but can also reach the farthest corner of the society for the benefit of the masses. In this context, the study explores the dynamic inter-linkage between the trends of ICT and progress of higher education particularly in technical field in India during the post-economic reform period. The remainder of the paper is organized as follows. The next section reviews the present evidences in brief followed by a discussion about the scenario of higher education and ICT in post-reform India. The sources of data used in the study and the methodological issues are accurately documented. The final section concludes the paper after analyzing the empirical findings.

2 Survey of Literature

In the recent past, we witnessed a complete transformation as the result of fast development of ICT. McGorry [9] expressed the exponential growth of home and office uses of Internet. The study also recommended that ICT may be utilized as an instrument to defeat the issues linked to cost, time, distance and quality barriers in education. Various studies like Amutabi and Oketch [2] viewed that skyrocket demand for education in developing economies like India is still regarded as a vital bridge of socio-economic and political mobility. UNESCO [12], Chandra and Patkar [5] and Bhattacharya and Sharma [4] document that working culture and procedure in several industries, and methods of interactions and works of people in the society are influenced by ICT. Entire education sector could be immensely benefited by the implementation of ICT techniques in teaching pedagogy, which can successfully eliminate the barriers, causing the difficulties of the low rate of education in countries.

Bhattacharya and Sharma [4] also point out the existence of socio-economic, demographic, linguistic, infrastructural and physical barriers in India to people who would like to access primary and higher education. Human resources in terms of quantity and quality in the Indian education system and proper infrastructural facilities also pose a challenge in this sector. Both union and state governments in India reserve about 3.5% of national income as compared to intended 6% for education (MHRD, Government of India, 2007). Tilak [11] argued that public expenditure on education as proportion of national income increased to 3.5% in 2004–05 from 1.8% in 1965–66, even could not be doubled. India has a high illiteracy rate (25.96) which is unevenly scattered in various states, largely on children and adult people with a staggering 47 million young citizens dropped out of school by the tenth standard. ICT-enabled education is the only remedy for such huge unevenly scattered uneducated youth population which can easily remove the infrastructural, socio-economic, linguistic and physical barriers with affordable cost.

3 Higher Education and ICT: A Brief Scenario in India

As one of the largest one in the world, the higher education system in India comprises of a total of 677 universities and 37,204 colleges [14]. The comparison below shows the tremendous increase in the number of universities and colleges in post-reform India. However, this is a fact that these quantitative growths don't reflect the satisfactory progress in terms of quality and delivery of higher education in India. The Indian education system is suffered a lot from technological obsolescence and lack of advanced facilities. However, off late the adoptions and applications of ICT in various forms has brought the much-needed diversification and affordability in the Indian higher education system for the masses. ICT is being widely used to modernize the content delivery system. DD Gyan Darshan 1 under Doordarshan, the largest public-owned television network, broadcasts various educational programmes for students of schools, colleges and university since 2000. As educational FM radio station located at different cities in India, Gyan Vani regularly broadcasts various kinds of educational programmes of IGNOU. Open and Distance Learning Institutions developed a National Digital Repository named eGyanKosh to store, preserve, index, share and distribute digital learning resources through digitalization of printed study materials, question papers, etc.

National Programme on Technology Enhanced Learning is a scheme initiated by seven oldest Indian Institutes of Technology (IITs) and Indian Institute of Science (IISc) to prepare course curricula in science and engineering disciplines using modern technology. National Mission on Education through Information and Communication Technology is a centrally funded scheme for enhancing ICT applicability in the learning and teaching process to benefit the students at higher education institutions. 'Train 10 thousand teachers' (T10KT) is a commendable, digital-based, joint initiative of IIT Kharagpur and IIT Bombay under NMEICT to improve pedagogy, techniques and skills of teachers in various engineering and science disciplines in India. T10KT has become more popular in remote areas nowadays. It is envisaged that ICT can revolutionize India towards becoming the knowledge society by strengthening Digital India movement. However, the concern also raises the question that whether technology alone can improve the Indian higher education quality? It is a fact that due to poor IT infrastructure facilities, applications of ICT in Indian institutes are negligible. A major portion of the population of India that live in rural areas don't have Internet access, so mass education programmes using advanced ICT need rural digitization.

India experienced low Internet penetration of around 19% as compared to 90% in other developing nations in 2014. Consultants at KPMG [7] reported that 'the number of mobile internet users is expected to grow to 314 million by 2017 with the compounded annual growth rate of about 28% over the period of 2013–2017'. 'This impressive growth would drive India to become one of the leading internet markets in the world with more than 50% of internet user base being mobile-only internet users'. To ensure the availability of public services, facilities and subsidies to all citizens located in remote areas, recently Government of India launched the Digital India campaign promoting Indian transformation into digital and knowledge economy. Another challenge for implementing ICT in higher education is linguistic diversities. The domiciles of a particular state in the country speak a particular language. A majority of the 68.84% rural population of India do not speak English. Therefore, the ICT course contents are to be developed in almost all the 22 languages the Constitution has approved.

4 Data Sources and Methodology

The basic objective of the study is the empirical exploration of long-run relationships among various measures of expansion of ICT and different indicators of the performance of technical higher education in India. In this regard, we test the null hypotheses as follows:

- *ICT does have any impact on technical higher education in India*

Data on the progress of ICT in India measured by aggregate expenditure on ICT and several qualitative (PTL) and quantitative (PTN) parameters of performances of various technical higher educational institutes in India collecting from various secondary sources like National Project Implementation Unit (NPIU) of MHRD, Government of India and Reserve Bank of India (RBI), etc. over the period from 2004 to 2017 for major Indian technical institutes across the states.

This study employs Arellano and Bond [3] developed approach of generalized method of moments (GMM) for the dynamic panel data to control the endogeneity in regression equations which are being estimated. Panel data models contain more information since these models are having more efficiency and more degrees of freedom. The panel data model allows us to manage the heterogeneity among the individual units, and enables us to recognize the results which cannot be recognized in the case of time series regression or cross section regression. Before estimating the aforementioned dynamic panel regression equation, for scrutinizing the stochastic properties of the variables we invoke various unit root tests for panel data as proposed by Levin, Lin and Chu (LLC) [8] and Im, Pesaran and Shin (IPS) [6]. The panel unit root tests are not identical to the unit root tests for time series data, but the purpose of their application is quite similar. For testing the panel unit root the equation of panel data regression according to the ADF unit root test is shown in Eq. (1).

$$\Delta y_{it} = \rho y_{i,t-1} + \sum_{j=1}^{p_i} \eta_{ij} \Delta y_{i,t-j} + X'_{it}\delta + \varepsilon_{it} \tag{1}$$

According to the statistical properties of the LLC test, intercepts, time trends, variances of the residual and the order of autocorrelation may be freely varied across the cross section units. However, application of LLC test requires common autocorrelation coefficient for all independently generated time series with identical sample size, and for all individual Autoregressive order one series (AR(1)). The lag order (p_i) can be varied across individual corporate companies. The suitable order of the lag is selected by permitting the maximum lag order, and then by applying the t-statistics for η_{ij}. The estimation of the autocorrelation coefficient (ρ) cannot be directly obtained by estimating the Eq. (1). Moreover, the proxies for Δy_{it} and y_{it} used in the analysis have been standardized and also made them free of autocorrelations deterministic components.

The study uses Arellano and Bond [3] proposed GMM estimator for the dynamic panel regression having two steps. At first, the fixed effects in the equations are removed through the introduction of the first difference form of the variables in the equations used for estimation instead of the level forms of variables. After this conversion of the forms of the variables, the equations comprising the difference forms of the variables are estimated by introducing instrumental variables for all difference form variables included in the equations. As there is a correlation between the lagged values of the difference forms of the endogenous variable or any other variable/s and error term in differenced term, and we use instruments for the former, all the levels of lags of the variables are used, starting from the lag two, then potentially go backside to the point of the beginning of the sample. The Sargan test of over-identifying restrictions is used to examine the overall strength of the instruments used for the variables of the analysis. The equation of the simplest form of the dynamic panel model having one period lag is presented in the following Eq. (2):

$$y_{it} = \alpha_i + \theta_t + \beta y_{i,t-1} + x'_{it}\eta + \varepsilon_{it} \tag{2}$$

The fixed effect is represented by α_i, time dummy is represented by θ_t, x_{it} is a vector having order $(k-1) \times 1$ of the exogenous regressors, where $\varepsilon_{it} \sim N(0, \sigma^2)$ is the random disturbance. It is important to state that the fixed effect model is relatively more appropriate compared to the random effect model in the panel frame, since in macro panel analysis most of the sectors are included for consideration, which reduces the likelihood of randomness of the sample. The fundamental estimators in the GMM panel estimation method, $\delta = (z'x)^{-1}z'y$, are based on the moments of the Eq. (3):

$$g(\delta) = \sum_{i=1}^{N} g_i(\delta) = \sum_{i=1}^{N} z_i' \varepsilon_i(\delta) \tag{3}$$

where, z_i is a $T_i \times p$ instruments matrix for the cross section, *i*, and, then it is represented by Eq. (4).

$$\varepsilon_i(\delta) = (y_i - f(x_{it}, \delta)) \tag{4}$$

The GMM estimation does minimize the quadratic form as shown in Eq. (5):

$$S(\delta) = \left(\sum_{i-1}^{N} z_i' \varepsilon_i(\delta)\right)' H \left(\sum_{i-1}^{N} z_i' \varepsilon_i(\delta)\right) \tag{5}$$

With respect to δ, for the weighting matrix, 'H' is suitably chosen. Therefore, the fundamental of the GMM estimator is to identify instruments, and the selection of 'H', and determination of an estimator. Based on the above-mentioned method specified for panel data, first we verify the stochastic properties of the panel variables (ICT, PTN and PTL) considered in this study and then the impacts of ICT on PTN and PTL are explored using the Arellano and Bond [3] refereed GMM estimation.

5 Empirical Results

Using a dynamic panel structure to estimate the impact of ICT on technical higher education performances, first we conduct LLC [8] and IPS [6] unit root tests and test statistics are calculated for all underlying panels, reported in Table 2. The lag lengths are decided by the minimum Akaike [1] information criterion (AIC). Both the individual and linear trends effects as exogenous regressor have been incorporated in estimated regression equations. All variables are found to be non-stationary at level; however, their first differences become unit root free as both LLC [8] and IPS [6] panel unit root tests statistics report.

Table 2 Results of panel unit root tests

Series	LLC (2002) test		IPS (2003) test	
	Level	First difference	Level	First difference
ICT	1.87	−6.65*	−1.53	−6.88*
PTN	−1.78	−7.91*	−1.41	−7.61*
PTL	−1.69	−5.96*	−1.62	−6.04*

*Represents level of significance at 5% level
Source Own estimations of authors using MHRD and RBI data

Table 3 Estimated results of GMM estimator of PTN and ICT relationship

Dependent variable: ΔPTN			
Balanced panel observations: 342			
Variables	Coefficients	t-statistic	Probability
ΔPTN	0.21	18.49	0.0000
ΔICT	0.29	13.41	0.0001
J-statistic	13.27 (0.3900)	Instrument rank	14

Source Own estimations of authors using MHRD and RBI data

Table 4 Estimated results of GMM estimator of PTL and ICT relationship

Dependent variable: ΔPTL			
Balanced panel observations: 315			
Variables	Coefficients	t-statistic	Probability
ΔPTL	0.22	17.92	0.0000
ΔICT (it)	0.11	10.33	0.0001
J-statistic	12.94 (0.4300)	Instrument rank	14

Source Own estimations of authors using MHRD and RBI data

We use first differenced GMM estimators developed by Arellano and Bond [3] for controlling the unobserved heterogeneities involved in estimated relationships. PTN and PTL are considered as the dependent variable in the separate dynamic panel equations and ICT behaves as an independent variable. The existence of lagged dependent variable in estimated regression equation captures the dynamisms. The t-statistic represents the significance of the corresponding regressor in a particular regression equation, whereas the overall validity of the GMM estimate of dynamic panel model can be checked by Sargan test represented by J-statistic. Both t-statistic and j-statistic are calculated on the basis of usual statistical formulas. Table 3 represents the estimated coefficients of PTN and ICT relationship in a dynamic panel equation containing a few relevant variables as instruments. We observe that the growth of ICT significantly impacts the quantitative measure of technical education performance (PTN). Table 4 represents the relationship of estimated coefficients PTL and ICT in the same kind of dynamic panel equation. Using the first difference GMM estimator after incorporating instruments, results imply a significant positive impact of ICT on PTL in the considered dynamic panel structure for selected institutes in India. However, it mustbe noted that the degree of impact of ICT on the quantitative measure is much higher than the qualitative measure of technical higher education performance in India.

6 Summary and Conclusion

The study explores the trends and patterns of ICT and its impacts on higher education sector with special emphasis on technical education in India during a vibrant period of digitalization in the era of globalization. Traditionally the education system has been associated with a very high degree of interpersonal linkages between teachers and learners along with teacher-centred delivery of instruction to classes of students who receive various information. ICT in education lends itself as a setting of students-centred learning. ICT also enables training providers with a better exposure and knowledge in technical institutions located in various developing countries like India, where ICT is given more importance mainly in learning and allied fields nowadays. In this context, present study investigates the relationship between growing trends of ICT and performance of technical higher education simply measured in terms of expenditure towards enhancement of ICT and also some qualitative and quantitative parameters of performances of various technical higher educational institutes in India on collecting data from various secondary sources over 2004–2017 for major Indian states. The stochastic properties are looked into by carrying out panel unit root tests on forming a balance panel by using state-wise data on variables followed by the GMM estimation. Our findings imply that ICT is significantly influencing the various performance indication of higher education mainly in the area of technological meadows and performance measured in quantitative aspects is much better than qualitative aspects in India. So, the government of India as well as state governments also should give more emphasis on ICT and ICT-based activities for a significant promotion of technical education to a greater extent. Finally it should be mentioned here that our study is limited to India only with the basic measures of ICT and its possible qualitative and quantitative impacts on top selected technical higher education in India due to time and data constraints. Researchers can extend this study to global aspects with a special focus on fastest growing developing countries and even in case of India also, with more accurate measures of ICT and its impacts' parameters with more data over a longer period of time.

References

1. Akaike, H.: Fitting autoregressive models for prediction. Ann. Inst. Stat. Math. **21**, 243–247 (1969)
2. Amutabi, M.N., Oketch, M.O.: Experimenting in distance education: the African Virtual University (AVU) and the paradox of the World Bank in Kenya'. Int. J. Educ. Dev. **23**(1), 57–73 (2003)
3. Arellano, M., Bond, S.: Some tests of specification for panel data: Monte Carlo evidence and an application to employment equations. Rev. Econ. Stud. **58**, 277–297 (1991)
4. Bhattacharya, I., Sharma, K.: India in the knowledge economy – an electronic paradigm. Int. J. Educ. Manag. **21**(6), 543–568 (2007)
5. Chandra, S., Patkar, V.: ICTS: a catalyst for enriching the learning process and library services in India. Int. Inf. Libr. Rev. **39**(1), 1–11 (2007)

6. Im, K.S., Pesaran, M.H., Shin, Y.: Testing Unit Roots in heterogeneous panels. J. Econ. **115**(1), 53–74 (2003)
7. Klynveld Peat Marwick Goerdeler. (KPMG): Telecommunications. Netherlands (2017)
8. Levin, A., Lin, C.F., Chu, C.: Unit root tests in panel data: asymptotic and finite-sample properties. J. Econ. **108**(1), 1–24 (2002)
9. McGorry, S.Y.: Online, but on target? Internet-based MBA courses: a case study. Internet High. Educ. **5**(2), 167–175 (2002)
10. Mooij, T.: Design of educational and ICT conditions to integrate differences in learning: contextual learning theory and a first transformation step in early education. Comput. Hum. Behav. **23**(3), 1499–1530 (2007)
11. Tilak, J.: On Allocating 6 Per Cent of GDP to Education. Econ. Polit. Wkly. 613–618 (2006). Accessed 18 Feb 2016
12. UNESCO: ICT in Teacher Education - A Planning Guide, Report. Paris, France (2002)
13. UNESCO: Technology, Broadband and Education: Advancing the Education for all Agenda. Paris, France (2013)
14. University Grants Commission. (UGC): Ministry of Human Resource Development, Government of India. www.ugc.ac.in (2014)

Optimization of Posture Prediction Using MOO in Brick Stacking Operation

Biswaranjan Rout, P. P. Tripathy, R. R. Dash and D. Dhupal

Abstract This article is based on the planning of human movements in different joints to reduce energy and fatigue of the working subjects. For this a simulation model is generated which complies with the constraints of the process. It requires verifying the capability of human postures and movements in different working conditions for the evaluation of effectiveness of the new workplace design. Here, a simple human performance measure is introduced which enables the mathematical model for evaluation of a cost function, i.e., discomfort function and energy expenditure rate. The basic study is to evaluate human performance in the form of cost functions. It aims to optimize the movement of the limbs with the above-mentioned cost factors. The method is tested with a case study of a fly ash brick manufacturing unit using two subjects for stacking bricks in stacking pan. MOO method is used for posture prediction. For an optimized posture prediction cost functions are minimized with less joint discomfort and less energy expenditure.

Keywords Human posture · Optimization · Discomfort function · Energy rate

---

B. Rout (✉) · D. Dhupal
Department of Production Engineering, VSSUT, Burla 768018, Odisha, India
e-mail: biswa.pp@gmail.com

D. Dhupal
e-mail: debabratadhupal@gmail.com

P. P. Tripathy
Department of Mechanical Engineering, MIET, Bhubaneswar 751015, Odisha, India
e-mail: priyam.tripathy1@gmail.com

R. R. Dash
Department of Mechanical Engineering, CET, Bhubaneswar 751003, Odisha, India
e-mail: ratiranjan.dash@gmail.com

H. S. Behera et al. (eds.), *Computational Intelligence in Data Mining*, Advances in Intelligent Systems and Computing 990,
https://doi.org/10.1007/978-981-13-8676-3_40

1 Introduction

Ergonomics generally used to study the relation between subjects and their surroundings or environment. This checks the subject's behavior while doing the job. At the same time, it also looks at physiological drawbacks and ability of the subject to get better output during work. The possibility of improvement is to get a new work surrounding depending upon the abilities and drawbacks of the subject as well as the surrounding. These can be done after serious study of the drawbacks and abilities of both subject and working surrounding. The stress level of the work should be considered to ensure the health and safety level of the subject. Hence the main purpose of ergonomics is to ensure health and safety of the subjects for the productive output.

Modeling and simulation for the ergonomic evaluation of relation between man and machine is done with the use of computer hardware and required software. Use of virtual environment cuts down the cost of producing the prototype nowadays. This virtual environment not only helps to evaluate the procedure by changing the location, posture and work level but also to simulate the position and posture of subjects. For a particular workplace, ergonomist simulate a virtual human model for its posture prediction.

The subject's ability is enhanced by considering better position changing of the limbs. The idea is obtaining some criteria for minimizing the losses in energy due to the subject's limb movement. Again the discomfort level of the subject should be minimum depending upon the motion of the limbs from the neutral position. Due to this cause a normal human model is accepted for the simulated workplace. The main aim is to analyze the performance in the shape of some factors relying on the motion of the limbs and required energy for the limb motion [1].

2 Modeling of Limbs

For this work we have taken a 20 DOF simple human model as in Fig. 1. Here the human body is modeled as a series of rigid bodies as a kinematic system and joined by revolute joints. These systems of joints are ankle joint, knee joint, hip joint, trunk joint, shoulder joint, elbow joint, and wrist joint in a human musculoskeletal system. To describe each joint and its movable capacity it is assigned to different reference coordinate system in order and the z-axis represents the rotational direction. The 20 degrees of freedom of the present model is described in Table 1 [2]. In this case ankle, knee, hip, trunk, spine, and elbow are modeled as one DOF revolute joint. The shoulder is modeled as three DOF revolute joint. The wrist is modeled as two DOF.

In this effort, the human model is having 20 DOF as shown in Fig. 1. The different movements of body parts are dorsal flexion/planter flexion, flexion/extension, abduction/adduction, and supination/pronation. The ankle is modeled as one DOF and the movement is dorsal flexion/planter flexion. Knee joint, hip joint, trunk, and spine is modeled as one DOF each with flexion/extension movement. The shoulder is modeled

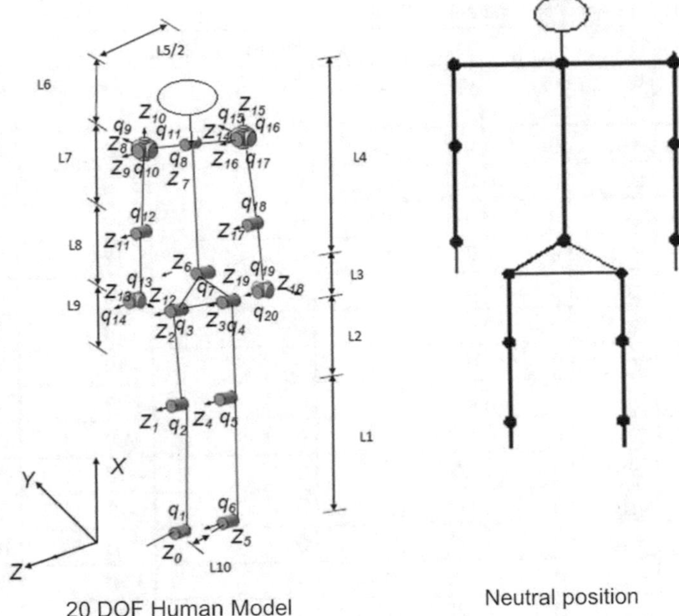

Fig. 1 The schematic representation of a human

as three DOF and the movements are flexion/extension, abduction/adduction, and supination/pronation. Elbow is modeled as one DOF with flexion/extension movements. The wrist is modeled as two DOF with movements of flexion/extension and abduction/adduction [3].

The most neutral position is basically defined as a person stands vertically and his arm hangs toward the torso as in Fig. 1. It is considered to be the most comfortable position with minimum energy requirement.

For mathematical modeling of human DH method is used for transformation of adjacent joints as in robotics. DH method is a proven method for mathematical modeling in the field of robotics and human biomechanics. In this method four factors are used, two link factors, i.e., θ_i and d_i and two joint factors a_i and α_i to explain the position and orientation [4].

For a human model of *n*-DOF, the necessary position vector is explained in terms of joint variable as shown in "Eq. (1)." [5]:

$$X = x(q). \tag{1}$$

where $\mathbf{q} \in \mathbf{R}n$ is the generalized n degrees of freedom coordinate system. The matrix arises from successive multiplication of transform matrices is x(q) which depending upon n-DOF expressed as in "Eq. (2)." [6]:

Table 1 DH parameters and Neutral position

D-H parameters					Neutral positions		
Joints DOF	θ_i	d_i	α_i	a_i	Neutral position	Lower limit	Upper limit
1	$-q_1$	0	0	L_1	0	−50	38
2	q_2	0	0	L_2	0	0	135
3	q_3	0	$-\pi$	0	0	−18	113
4	$-q_4$	0	0	L_1	0	−18	113
5	q_5	0	0	L_2	0	0	135
6	q_6	0	$-\pi$	0	0	−50	38
7	q_7	0	0	L_3	0	−19	56
8	q_8	0	0	0	0	−30	40
9	q_9	0	$\pi/2$	0	0	−60	170
10	$-q_{10}$	0	$\pi/2$	0	0	−18	80
11	q_{11}	0	$\pi/2$	−L7	0	−20	97
12	$-q_{12}$	0	$\pi/2$	$-L_8$	0	0	140
13	$-q_{13}$	0	$\pi/2$	0	0	−70	80
14	q_{14}	0	$\pi/2$	$-L_9$	0	−30	20
15	q_{15}	0	$\pi/2$	0	0	−60	170
16	$-q_{16}$	0	$\pi/2$	0	0	−18	80
17	q_{17}	0	$\pi/2$	−L7	0	−20	97
18	$-q_{18}$	0	$\pi/2$	$-L_8$	0	0	140
19	$-q_{19}$	0	$\pi/2$	0	0	−70	80
20	q_{20}	0	$\pi/2$	$-L_9$	0	−30	20

$$^0T_n = {}^0T_1{}^0T_2\cdots{}^{n-1}T_n = \begin{bmatrix} {}^0R_n(q) & x(q) \\ 0 & 1 \end{bmatrix}. \tag{2}$$

DH parameter for this model is shown in Table 1.

3 Optimization Formulations

Initially we take the path points as input data from the brick stacking operation. Using these path points, from the program we get the output as angles. For optimization of the angles of posture prediction multi-objective optimization (MOO) method is used, which is determined as in "Eq. (3).":

$$\textit{find} : q \in R^{DOF}.$$

$$\begin{aligned}&To\min imize: f(q)=[f_1(q)f_2(q)\ldots .f_k(q)]^T.\\&Subject\ to: q_i(q)\le 0i=1,2,\ldots..,m\quad h_j(q)=oj=1.2,\ldots,e\end{aligned}\tag{3}$$

In "Eq. (3)." m denotes total number of inequality constraints, k denotes total no. of objective functions, e denotes total no. of equality constrains and q ϵ E^{DOF} is the vector of design variables.

The optimal posture can be obtained by solving the optimization condition. For this case, the problem can be defined as in "Eq. (4).":

$$\begin{aligned}&find: q\in R^{10}.\\&\text{To minimize : Cost functions } \|P(hand)-P(object)\|\le\in.\\&\text{Subject to : } q_i^L\le q_i\le q_i^U, \text{i}=1,2,\ldots,20.\end{aligned}\tag{4}$$

In this work, cost functions refer to joint discomfort function and Energy function. q_i^U and q_i^L defines the upper and lower limits for q_i joint in that order. Limits are given for not assuming an unrealistic posture. Discomfort function generally measures the difference of CG to the desired position with respect to the neutral position. This model gives joint discomfort function at the rotational positions.

It is found that the joint discomfort function is minimum in its neutral position where no movement is there and is higher when going toward the upper limit of the joint and lower limit of the joint by the addition of QU and QL, where QL and QU are especially designed provisions for the lower limit and the upper limit of each joint. "Equation (5)." states the mathematical expression for joint discomfort function.

$$f_{discomfort}=\frac{1}{G}=\sum_{i=1}^{DOF}\left[\gamma_i\left(\Delta q_i^{norm}\right)^2+G*QU_i+G*QL_i\right].\tag{5}$$

where, "Eq. (6)." states the normal of the ith joint.

$$\Delta q_i^{norm}=\frac{q_i-q_i^N}{q_i^U-q_i^L}.\tag{6}$$

"Equation (7)." states the upper limit and lower limit of the ith joint.

$$QU_i=(0.5\sin\left(\frac{5.0\left(q_i^U-q_i\right)}{q_i^U-q_i^L}+1.571\right)+1)^{100}, QL_i=\left(0.5\sin\left(\frac{5.0\left(q_i-q_i^L\right)}{q_i^U-q_i^L}+1.571\right)+\right.\tag{7}$$

The total energy expenditure rate (E), which is expressed in Watts/kilogram, is given as shown in "Eq. (8).":

$$\dot{E}=E_{\dot{W}}+E_{\dot{M}}+E_{\dot{S}}+E_{\dot{B}}.\tag{8}$$

where, $E_{\dot{W}}$ is the power of muscle, $E_{\dot{M}}$ is the heat rate of muscle, $E_{\dot{S}}$ is the shortening heat rate of muscle, and $E_{\dot{B}}$ is the basal metabolic rate (BMR).

Normally, in the joint space only the force is considered for energy expenditure in the human model. As per Kim et al. [8], the definition of mechanical power is given as dot product of joint torque (τ_i) and joint velocity (q_i). The total mechanical power E_M is expressed as shown in "Eq. (9).":

$$-E_{\dot{W}} = \sum_{i=1}^{DOF} \left|\tau_i \dot{q}_i\right|. \tag{9}$$

The muscle maintenance heat rate for a muscle is evaluated as the product of the muscle activation and a constant. As per the joint space, with reference to the work of Anderson et al. [7] and Kim et al. [8], it is approximately found to be proportional to the joint torque. It is expressed as shown in "Eq. (10).":

$$E_{\dot{M}} \approx \sum_{i=1}^{DOF} \xi_i |\tau_i|. \tag{10}$$

where, ζ_i is defined as the coefficient of maintenance heat rate taken at joint i, and it is reciprocal of joint torque limits.

The basal metabolic rate, i.e., BMR is taken as the metabolic rate of a human being at rest. It means the minimum quantity of energy necessary for functioning of human, without doing any kind of external work. The BMR is expressed as shown in "Eq. (11).":

$$E_{\dot{B}} = 0.685BW + 29.8. \tag{11}$$

where, BW is the body weight (kilogram).

The muscle shortening heat rate is neglected in this analysis. The overall metabolic energy expenses rate is taken as sum of BMR and maintenance heat rate, the expression is given as shown in "Eq. (12).":

$$f_{metabolic} \approx \sum_{i=1}^{DOF} \xi_i |\tau_i| + E_{\dot{B}}. \tag{12}$$

Torque basically depends on various-link static model that is especially applicable for symmetric activities of sagittal plane.

In this case, for equilibrium condition the forces are measured to act in parallel direction vertically. The moment equilibrium condition is expressed as shown in "Eq. (13).":

$$M_j = M_{j-1} + \left[\overline{jCM_L} \cos\theta_j W_L\right]\left[\overline{jj-1} \cos\theta_j R_{j-1}\right]. \tag{13}$$

4 Case Study

The discomfort and energy function of a subject for this particular job was obtained using the previously discussed method. To check off the usefulness of the model, a subject in a fly ash brick producing unit is taken for study. Two subjects were taken, one moving in clockwise and second subject moving in counter clockwise direction to pick up the brick and to keep it in the stack. During this period many videos are taken and different movements are captured. The movements are looked into for getting posture prediction and its related position of every joint, which are ergonomically analyzed.

The stacking process is carried out by concern subjects. Three hundred and sixty bricks of 6.5 kg weight and of size 22.86 mm × 12.7 mm × 7.62 mm were arranged in 3 stacking pans of size 900 mm × 600 mm × 152.4 mm. The conventional diagram bricks stacking configuration are shown in the figure.

Here two subjects having heights of 162.5 cm, 175 cm and weights 68 kg, 75 kg are taken for our study. The first subject puts the bricks in stack A1, A2, and A3 and the second subject stack in B1, B2, and B3 [9] (Fig. 2).

In the stacking pan, there are sixty bricks in each pan. There are six numbers of rows, the first five rows contain eleven bricks each and the sixth row contains five bricks. The subjects had put the bricks continuously in the stack. We have taken a number of videos while the stacking process is carried out for two subjects for the ergonomic analysis. The cost functions taken for analysis are "discomfort function" and "energy function."

Figure 3 shows the full 3D view of the brick stacking process.

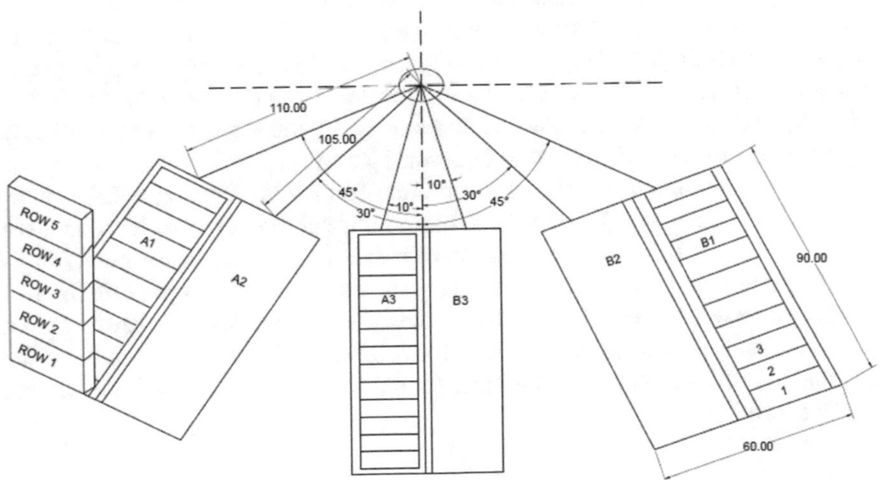

Fig. 2 Conventional diagram of bricks stacking layout

Fig. 3 3D view of fly ash brick stacking

In this case study, we have taken three different paths taken by the two subjects to carry out stacking of bricks in the specific stacking pans.

5 Results and Discussion

The two objective functions are checked for the output of simulation, i.e., discomfort function and energy expenditure rate for the particular configuration of the subjects. When the brick is kept below the subject's upper extremity reach area, a bend posture is necessary as a part of workplace layout. In order to avoid excess bend in the trunk, the lower parts hip, knee, and ankle have to bend to get the required bend posture.

Discomfort functions for two subjects are given in Figs. 4 and 5 for the three different paths followed by the subjects. Figs. 6 and 7 give energy expenditure rate of two subjects for three different paths. It is evident from the figure that the path-2 is having less discomfort function and less energy expenditure rate, hence path-2 is recommended for two subjects taken into considerations.

We have taken simultaneously discomfort function, energy function, and the multi-objective function in Figs. 8 and 9 for subject-1 and subject-2, respectively. In these figures we have taken hip displacement in y-axis and the reference points in x-axis. The optimum posture with less energy expenditure and less joint discomfort is obtained by MOO method. From the figures it is evident that there are less joint discomfort and less energy expenditure for subject-2. The three curves show the adopting of a tall subject for bend posture for reduced amount of the discomfort and energy expenditure.

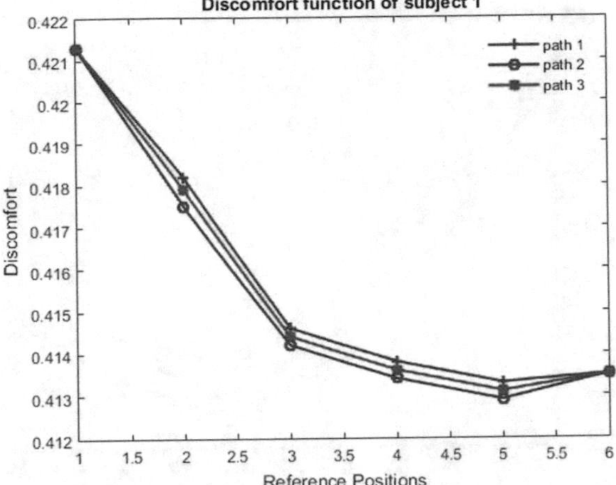

Fig. 4 Discomfort function for subject-1

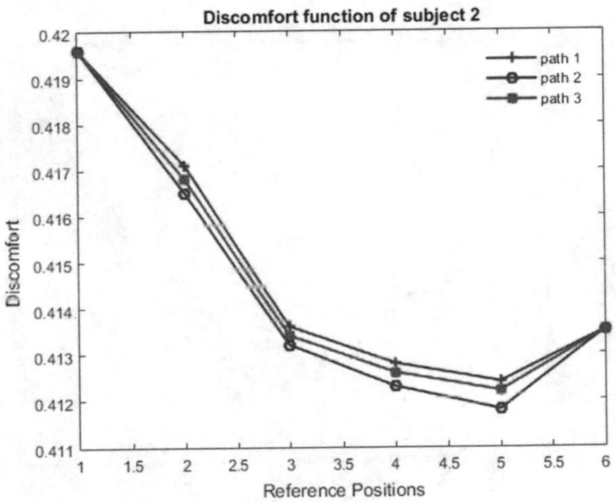

Fig. 5 Discomfort function for subject-2

6 Conclusion

From the simulated model discussed earlier the movement of the subject's body parts in bricks piling operation of fly ash bricks is looked at. The sequence for the cause is elaborated and optimized parameters are derived. For the posture optimization MOO method is utilized. The result is plotted graphically and it indicates the effectiveness in enhancement in the objective cost functions which means an improvement in discomfort and energy functions.

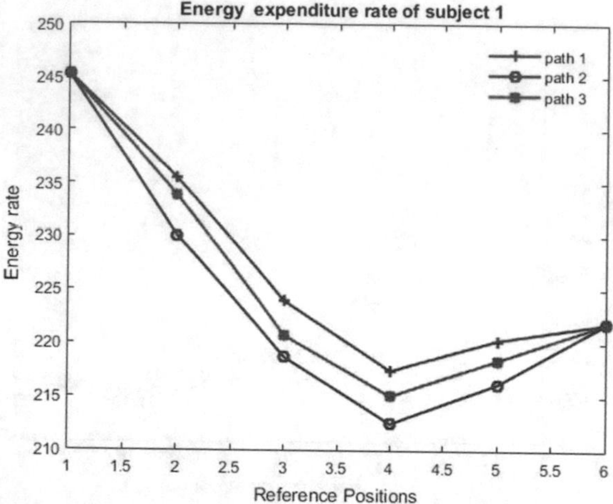

Fig. 6 Energy expenditure rate for subject-1

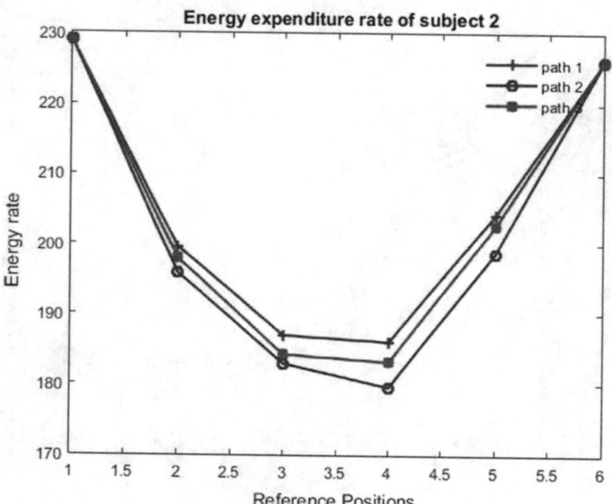

Fig. 7 Energy expenditure rate for subject-2

The same problem can be taken for analysis with more number of degrees of freedom for the object taken. This analysis can be taken for various other workplaces as future scope of application of multi-objective optimization (MOO).

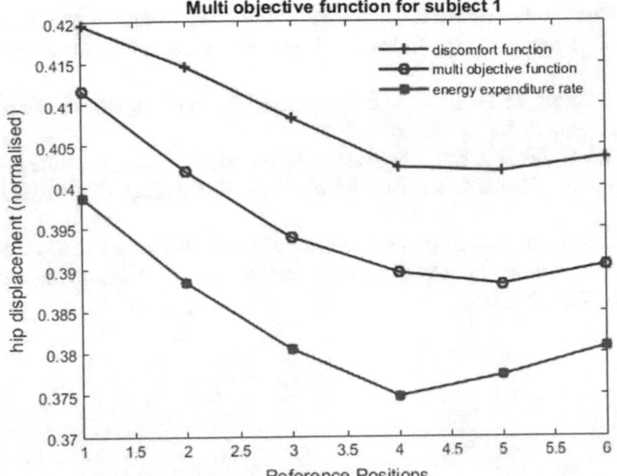

Fig. 8 Multi-objective function for subject-1

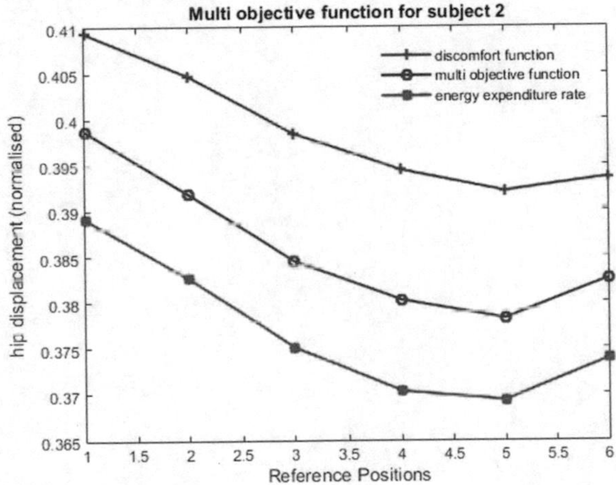

Fig. 9 Multi-objective function for subject-2

References

1. Mavrikios, D., Karabatsou, V., Alexopoulos, K., Pappas, M., Gogos, P., Chryssolouris, G.: An approach to human motion analysis and modelling. Int. J. Ind. Ergon. **36**, 979–989 (2006)
2. Michalewicz, Z.: Genetic Algorithm Data Structures Evolution Programs, 3rd edn. Springer, New York (1996)
3. Abdel-Malek, K., Yang, J., Brand, R., Tanbour, E.: Towards understanding the workspace of the upper extremities. SAE
4. Denavit, J., Hartenberg, R.S.: A kinematic rotation for inner pair mechanism based on matrices. J. Appl. Mech. **77**, 215–221 (1955)
5. Yang, J.: Human Modeling and Simulation. Human Factors and Ergonomics chapter 29 (2008)

6. Faraway, J.J., Chaffin, D., Woolley, C., Wang, Y., Park, W.: Simulating industrial reach motions for biomechanical analyses. In: Industrial Engineering Research Conference, Phoenix, AZ (1999)
7. Anderson, F.C., Pandy, M.G.: Static and dynamic optimization solutions for gait are practically equivalent. J. Biomech. **34**(2), 153–161 (2001)
8. Kim, J.H., Abdel-Malek, K., Yang, J., Marler, T., Nebel, K.: Lifting posture analysis in material handling using virtual humans. Am. Soc. Mech. Eng. Manuf. Eng. Div. MED **16**(2), 1445–1453 (2005)
9. Rout, B., et al.: Effective work procedure design using discomfort and effort factor in brick stacking operation-a case study. IOP Conference Series: Materials Science and Engineering, vol. 310, pp. 012020 (2018)

Sentiment Analysis and Extractive Summarization Based Recommendation System

Rajendra Kumar Roul and Jajati Keshari Sahoo

Abstract With the commencement of new technology and increase of online shopping companies like Amazon, E-Bay, and Flipkart, people give a wide range of reviews for the products they purchase. Some are too long, some too short, some difficult to understand while some are totally irrelevant. Thus, there is a pressing need to reduce the diversity of reviews for a particular product and to show the users only the gist of most useful and important reviews about a product. The proposed work focuses to summarize the reviews for movies purchased from Amazon using a combination of four state-of-the-arts algorithms and a feature selection technique. Sentiment analysis has been performed to categorize the reviews into positives and negatives. Also, a novel method named *hierarchical summarization* is proposed to summarize large reviews into summary of few sentences. The results of this summary are compared with the existing algorithms using the ROUGE score to determine the best summary. Experimental results show that the proposed approach is promising.

Keywords Extractive · Hierarchical · Recommendation · Sentiment · Summarization · TF-IDF

1 Introduction

With an ever-increasing trend of online shopping, people tend to rate their purchases by giving reviews of a product on the e-commerce companies websites. It also provides a medium through which people can share their experiences and feedbacks. These reviews are seldom informative along and are quite large in number. Thus,

R. K. Roul (✉)
Department of Computer Science, Thapar Institute of Engineering and Technology, Patiala 147004, Punjab, India
e-mail: raj.roul@thapar.edu

J. K. Sahoo
Department of Mathematics, BITS-Pilani, K.K. Birla Goa Campus, Zuarinagar, Mormugao 403726, Goa, India
e-mail: jksahoo@goa.bits-pilani.ac.in

H. S. Behera et al. (eds.), *Computational Intelligence in Data Mining*, Advances in Intelligent Systems and Computing 990,
https://doi.org/10.1007/978-981-13-8676-3_41

there is a need to combine these reviews and represent them in a short summary by extracting the important essence of all the reviews. Hundreds and thousands of review comments on any product can be easily found in the social website, but reading them all is a time-consuming process which we called as *information overload* and it sometimes confuses the new user to take an appropriate decision [1–3] while purchasing a product. By summarizing reviews and then capturing its gist will not only help the reader to get the detailed information about the product but also save precious time. This technique of summarizing all reviews into a short summary is known as text summarization. Many researchers have worked on text summarization [4–7]. Generally, text Summarization algorithms are classified into two categories [8]: *Extractive* and *Abstractive*. *Extractive summarization* is where the sentences in a document (i.e., review) are ranked based on their relative importance in the document and some top sentences are chosen to make the summary. *Abstractive summarization* on the other hand summarizes the document in its own words analogs to how we summarize an event to our friend in our own words. In order to achieve this, abstractive summarization requires lots of data and complex algorithms thereby making it a difficult task to achieve. However, in order to judge the accuracy of the generated summaries, we have "human-generated summaries" which are summaries generated by humans according to what they found important in a document. The predicted summary is then compared to these human-generated summaries to check how much important information is being retained by the predicted summaries. But in the real world, it is difficult to find human-generated summaries for every dataset because it is a tedious and cumbersome procedure to generate these summaries. Especially in the case of Amazon reviews, the reviews are written in an unstructured manner, so it is quite difficult to generate human summaries for such a large number of reviews. Thus, in order to overcome this problem, we have implemented a novel approach to summarize the Amazon movies review dataset using a technique that we called *hierarchical summarization* which is an extractive summarization technique. Four different extractive summarization algorithms are used to generate summaries from which important sentences are extracted using a feature selection technique and it generates the human-generated gold summaries. Prior to these, we classify the data into positive and negative reviews using sentiment analysis that improves the summarization of reviews and makes it easier for the user to get detailed information about a product. Empirical results show the effectiveness of the proposed work.

The paper is organized as follows. In Sect. 2, we talk about the experimental design where we describe each step of the pipeline used to generate the gold summaries using hierarchical summarization. In Sect. 3, we present the results and an analysis of the summaries by comparing it with benchmark algorithms using the ROUGE score. Finally, the paper is concluded with some future enhancement in Sect. 4.

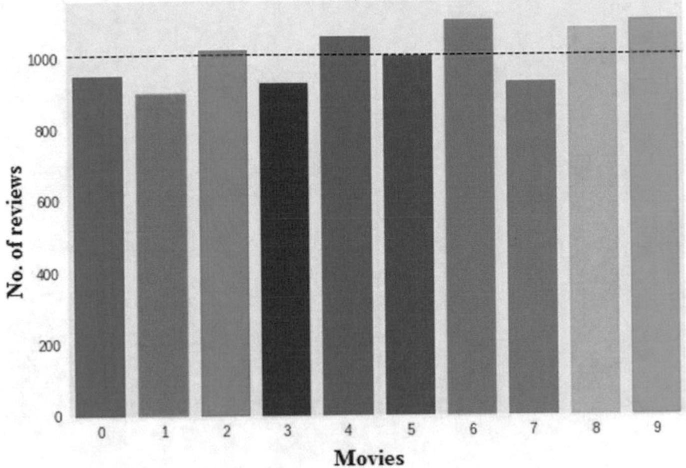

Fig. 1 Number of reviews for the top ten selected movies

2 Methodology

The following steps are used to summarize the reviews of Amazon dataset.

2.1 *Data Preprocessing*

The proposed work has been done on Amazon review dataset.[1] In the Amazon review dataset since the reviews are user-generated, the data are highly unstructured. There was no general structure to punctuations, spaces, etc. As a result, extensive pre-processing of the data is done before further use for obtaining the accurate results. Regular expressions and tokenizers are used to convert the data into a well-defined format. Keeping in mind that our aim is to perform sentiment analysis and extractive summarization, the proposed approach selected the top ten movies based on the maximum number of reviews. On an average, each selected movie had thousand reviews with average review length being two hundred fifty words and each review had an average of 15 sentences.

The plot in Fig. 1 shows the average number of reviews for the selected movie set stands at around 1000. These are the movies with the highest number of reviews in the Amazon dataset that we chose for analysis. The graph plotted in Figs. 2 and 3 elaborates the fact that the selected movie set had an overall 15 sentences per review per movie along with 250 words per review on an average. The movies have been selected in such a way that there is enough data available, comparatively, for analysis

[1] http://jmcauley.ucsd.edu/data/amazon/.

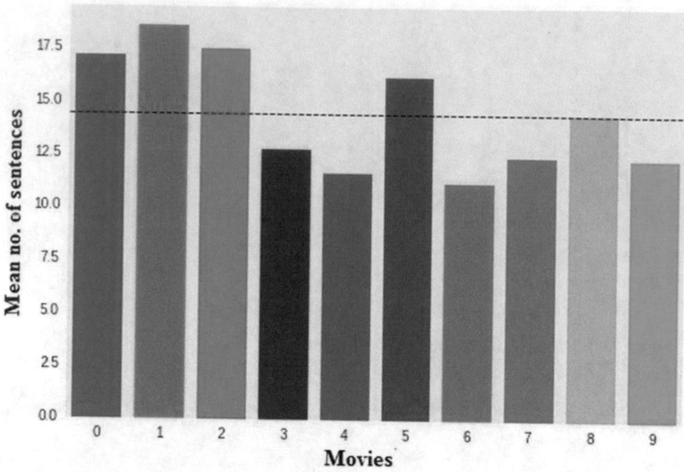

Fig. 2 Mean number of sentences per review

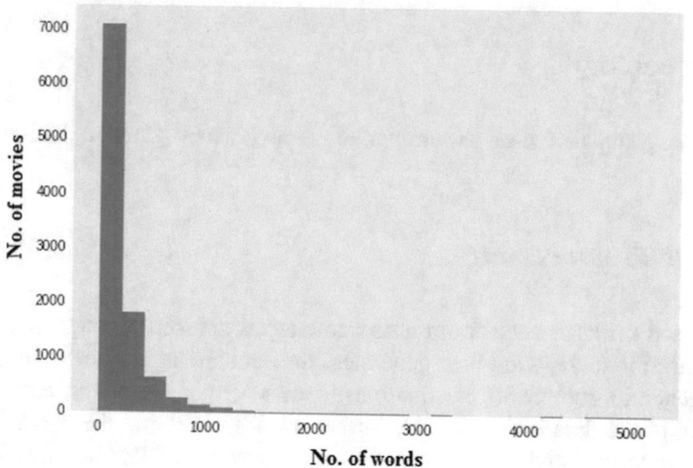

Fig. 3 Distribution of number of words in reviews

to generate out the positive and negative classes for the movies, as shown in Fig. 4. Algorithm 1 discusses the preprocessing of reviews.

2.2 Sentiment Analysis

The reviews have been categorized on the basis of the polarity into positive and negative classes by carrying out a sentiment analysis on the preprocessed dataset. To achieve this, a pretrained Naive Bayes classifier is run on movie reviews dataset by

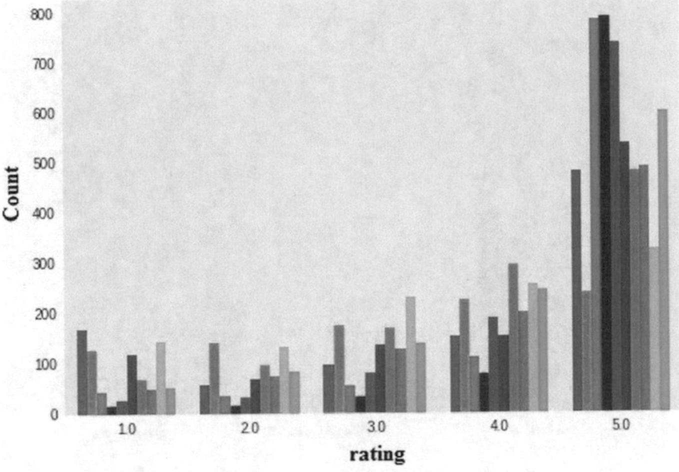

Fig. 4 Distribution of ratings of movies given on a scale of 1 to 5

Algorithm 1: Preprocessing of Reviews

```
Data: Preprocessed reviews (PR) for various movies
Result: Two lists of positive and negative pool of reviews
positive_list ← ϕ
negative_list ← ϕ
for each review ⊂ PR do
    polarity ← sentiment_analysis(review)
    if polarity > 0 then
        add review to positive_list
    else
        add review to negative_list
    end
end
return positive_list, negative_list
```

using the TextBlob library.[2] The model is trained on features like number of positive and negative words which help in determining the polarity of an input review/sentence and is shown in Eq. 1.

$$P(A|B) = \frac{P(B|A)P(A)}{P(B)} \quad (1)$$

Here, A is the class (positive/negative) that the review B will be classified into on the basis of the features that B contains and have been learned by the model. After doing the sentiment analysis on the preprocessed dataset, the distributions of reviews are shown in Fig. 5. We discarded the neutral reviews, because the main focus is to concentrate on the top positive and negative aspects of the movie and do the analysis on the same.

[2]https://textblob.readthedocs.io/en/dev/.

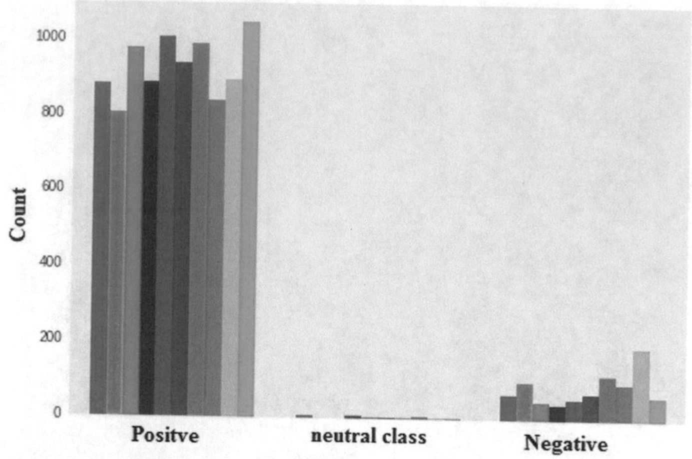

Fig. 5 Sentiment distribution per movie

2.3 Hierarchical Summarization

The following four algorithms have been used for hierarchal summarization to get four summaries of 20 sentences each which are the *system-generated summaries* that we used for comparative analysis in the later stage. The threshold for polarity was set to 0.5[3] for getting featured in one of the positive or negative lists. This was done to ensure that the summarization is biased to the most positive and negative reviews. The main motive for using hierarchical summarization is to ensure the prevention of data loss that would have occurred if we used each algorithm for summarization in one go. Also, it was found that summaries generated using this technique is rich in information and thus, gave dependable results for our specific dataset. Detail steps are shown in Algorithm 2.

- *LexRank*: This algorithm computes sentence importance based on the concept of eigenvector centrality in a graph representation of sentences [9]. In this model, a connectivity matrix based on intra-sentence cosine similarity is used as the adjacency matrix of the graph representation of sentences.
- *TextRank*: This algorithm is a graphical text ranking algorithm in which text units best defining the tasks are added as vertices in the graph [10]. The relationship between the text units is represented by the edges in the graph. This graph-based ranking algorithm is run until convergence and vertices are sorted based on their final score.
- *Latent Semantic Analysis (LSA)*: LSA [11] is an algorithm that analyze the relationships between a set of documents and the terms by producing a set of concepts related to the documents and terms. It helps in reducing the dimensions by selecting the most relevant features using singular value decomposition.

[3]Decided by experiment for which the best result is obtained.

Algorithm 2: Hierarchical summarization

```
Data: List of reviews (LOR)
Result: Summarization of list of reviews in 20 sentences
Initialization:
final_extractive_summary ← ϕ
one_sentence_summary ← ϕ
for each review ∈ LOR do
    temp ← review
    while numberOfSentences(temp) > 1 do
        length ← numberOfSentences(temp)
        temp ← SummarizationAlgorithm(temp, length/2)
    end
    add temp to one_sentence_summary
end
filtered_summary ← ϕ
for each review in one_sentence_summary do
    polarity ← sentiment_analysis(review)
    if |polarity| > threshold then
        append review to filtered_summary
    end
end
while numberOfSentences(filtered_summary) > 20 do
    length ← numberOfSentences(filtered_summary)
    filtered_summary ← SummarizationAlgorithm(filtered_summary,length/2)
end
return filtered_summary
```

- *SumBasic*: This algorithm uses the probabilistic approach in which sentence weight is calculated by taking the mean of the probabilistic distribution of the words appearing in that sentence [12]. Best scoring sentence containing the highest probability word is selected.

It can be seen from Fig. 6 that applying summarization just once gives an average sentiment score of 0.2. However, as the study is more focused upon getting the best positive and negative summaries, it is necessary to choose the best sentences at each step. Thus, after applying the hierarchical summarization, we can see that the average sentiment score for positive reviews is around 0.66 (Fig. 7) while for negative reviews it is around 0.35 (Fig. 8). This captures the essence of motive behind the study.

2.4 *Feature Selection*

In order to compare the results more effectively, a benchmark is necessary which could mimic the human written summary. To achieve this, the four summaries obtained above are concatenated and a feature selection technique is applied to obtain the 20 most important sentences for each of the positive and negative classes. The

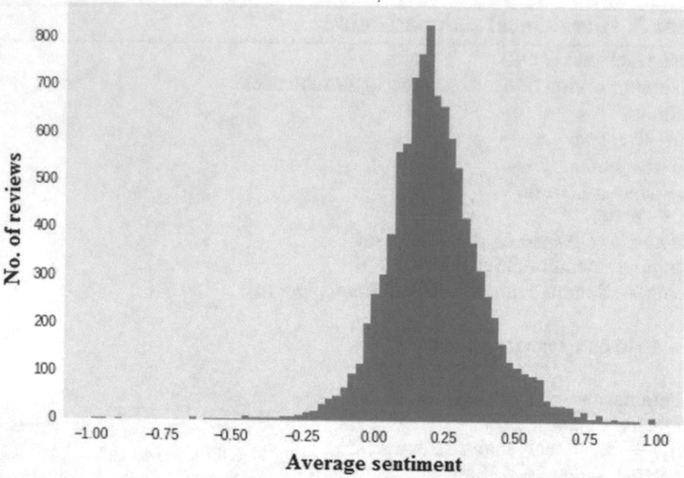

Fig. 6 Distribution of sentiment polarities over all reviews

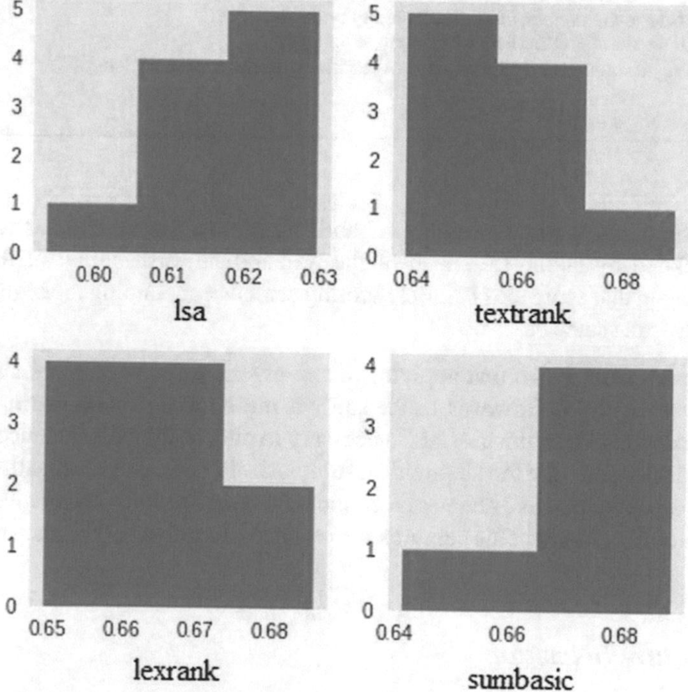

Fig. 7 Segment polarities of positive summaries

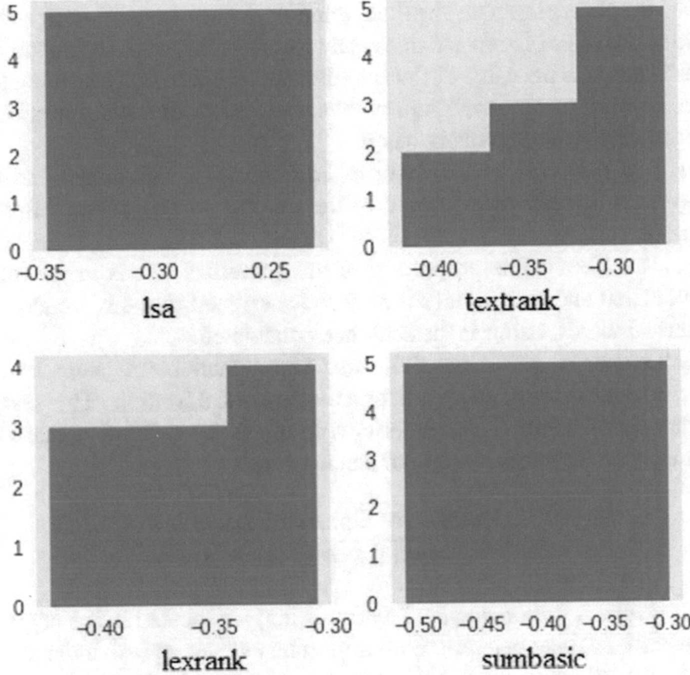

Fig. 8 Segment polarities of negative summaries

important features which are used to rank these sentences for selecting the top 20 sentences (called *human-generated gold summaries*) are as follows:

i. *Length of Sentence*: This feature counts the number of words in a sentence s. First the normalized length of s is calculated. Sentence s is awarded less score if its length is very small or very large based on a threshold.[4]
ii. *Positional Feature*: If s is at the beginning of the document d or at the end of d then it has been given more score. This has been done keeping in mind that the increased amount of importance given to the abstract or conclusion in any piece of text.
iii. *Weight of Sentence*: This feature finds the weight of s by summing up the individual TF-IDF weight of each word of s, and depending on the frequency of the word in the sentence, the weight of the sentence is calculated.
iv. *Quoted Text*: If some part of s is within the quotation marks then it asserts something specific and that information is likely to be important. If the quote exists in the sentence, then s received a score of one else zero.
v. *Upper Case Text*: Sentences containing more uppercase words have been given more weight. This has been done as the upper case words represent the sentiments of the users more strongly.

[4]Decided by experiment.

vi. *Sentiment Polarities*: The absolute value of the sentiment polarity of the sentence on the scale of zero and one is taken as the score of the sentence. Absolute value is taken as the value of the negative sentiment polarity is going to be negative. As we are summarizing reviews, the sentences having higher absolute polarity values are obviously better.
vii. *Density of sentence*: The number of key phrases is calculated and the score assigned is directly proportional to the number of key phrases found in the sentence.
viii. *Numerical Words*: The proportion of the numerical words in the total words is calculated and accordingly the score is assigned to the sentence. More the numerical words, better is the sentence considered.
ix. *Alphanumeric words in the sentence*: The alphanumeric word is useful for describing a password, keyword, or a mathematical formula. This feature takes into account whether alphanumeric word is present in the sentence s or not. If present, then it returns score f_a of s using Eq. 2.

$$f_a = \frac{\text{number of alphanumric words in } s}{\text{total length of the review}} \tag{2}$$

x. *Relative offset of the sentence*: Sentences that are located in the beginning and at the end of a document are more important as they represent the abstract and conclusion of the document. Hence, we calculate the location of a sentence called the offset to find such sentences.

3 Experimental Result

In this Section, we perform experiments on Amazon review dataset.[5] The dataset contains reviews of movies and TV shows purchased from amazon. The statistics of the dataset is as follows:

Number of reviews: 7,911,684, Number of users: 889,176, Number of products: 253,059, Users with > 50 reviews: 16,341, Timespan: Aug 1997–Oct 2012.

The attributes of the dataset are reviewerID, asin, reviewerName, helpful, reviewText, overall, summary, reviewTime. Some of the important attributes include reviewerID: unique ID of the reviewer, asin: unique ID of the movie, reviewText: The review given by the reviewer corresponding to the reviewerID for the product with unique ID as asin, summary: The heading of the reviewText.

To measure the performance of text summarization, *ROUGE* or Recall-Oriented Understudy for Gisting Evaluation score [13] is used by the proposed approach. ROUGE-N is for n-gram recall between the summary generated by the system and the human-generated gold summary. ROUGE-1 and ROUGE-2 are used to compute the recall value for unigram and bigram matching, respectively [13]. ROUGE toolkit

[5] http://jmcauley.ucsd.edu/data/amazon/.

Table 1 ROUGE-1 (Positive)

Movies	LexRank	TextRank	LSA	SumBasic
1	0.5347	0.2298	0.5281	0.4812
2	0.5519	0.6275	0.5817	0.3633
3	0.4506	0.4928	0.4898	0.4207
4	0.2933	0.5081	0.0775	0.3246
5	0.4886	0.4229	0.1289	0.4285
6	0.5534	0.3644	0.1594	0.4227
7	0.4400	0.4040	0.0898	0.3027
8	0.4347	0.4358	0.4839	0.3597
9	0.3732	0.4464	0.5420	0.3537
10	0.5360	0.6304	0.0834	0.2446

takes as input both system-generated summary "sgs" and human-generated gold summary "gs" and evaluates the different scores based on the n-gram matching. ROUGE-N is computed using Eq. 3.

$$recall = \frac{\sum_{s \in gs} \sum_{gram_n \in s} Count_{match}(gram_n)}{\sum_{s \in gs} \sum_{gram_n \in s} Count(gram_n)} \tag{3}$$

where n stands for the length of the n-gram, $gram_n$, and $Count_{match}(gram_n)$ is the maximum number of n-grams overlap between sgs and the gs. Equation 3 shows that ROUGE-N relies on n-gram recall because the denominator of the equation denotes the total sum of the number of n-grams present in the human-generated gold summary. It can also be noted that the denominator of the Eq. 3 can be increased when one adds more summaries because there might exist more human generated gold summaries. The numerator of the Eq. 3 finds the sum over the huam generated gold summaries. This eventually gives more importance to those n-grams which are common to both the system- and human-generated gold summaries. Hence, the system-generated summaries that contain words shared by more sentences in human-generated gold summary will be preferred by ROUGE-N measure. The precision is defined in Eq. 4.

$$precision = \frac{\sum_{s \in gs} \sum_{gram_n \in s} Count_{match}(gram_n)}{\sum_{s \in sgs} \sum_{gram_n \in s} Count(gram_n)} \tag{4}$$

All the symbols have the same meaning as in Eq. 3. Here the denominator denotes the total sum of the number of n-grams present in the system-generated summary, i.e., sgs.

Table 2 ROUGE-2 (Positive)

Movies	LexRank	TextRank	LSA	SumBasic
1	0.3714	0.1039	0.3670	0.3258
2	0.1334	0.4167	0.4216	0.1331
3	0.2647	0.2209	0.3798	0.2202
4	0.1307	0.3310	0.0126	0.1854
5	0.2430	0.2032	0.0223	0.2717
6	0.3741	0.2150	0.0406	0.2238
7	0.1875	0.2658	0.0178	0.2134
8	0.2648	0.1799	0.3588	0.2106
9	0.2081	0.2457	0.4490	0.1795
10	0.3636	0.4861	0.0088	0.1216

Table 3 ROUGE-1 (Negative)

Movies	LexRank	TextRank	LSA	SumBasic
1	0.3000	0.2168	0.0805	0.2122
2	0.5254	0.4724	0.5837	0.3549
3	0.4337	0.2904	0.1597	0.3750
4	0.3406	0.1970	0.1144	0.4403
5	0.1341	0.0812	0.0494	0.4500
6	0.4130	0.4804	0.1231	0.3943
7	0.2894	0.2197	0.1213	0.4567
8	0.4669	0.2490	0.1207	0.4502
9	0.3421	0.2463	0.1674	0.4529
10	0.3953	0.2886	0.1838	0.3280

Table 4 ROUGE-2 (Negative)

Movies	LexRank	TextRank	LSA	SumBasic
1	0.1666	0.0547	0.0071	0.1132
2	0.4040	0.2688	0.4754	0.1089
3	0.3013	0.1629	0.0251	0.2435
4	0.1913	0.0157	0.0123	0.3636
5	0.0277	0.0333	0.0012	0.3666
6	0.2926	0.2767	0.0103	0.2459
7	0.1923	0.0711	0.0255	0.3380
8	0.3381	0.0711	0.0152	0.3274
9	0.1538	0.0546	0.0138	0.4666
10	0.2605	0.1549	0.0234	0.1739

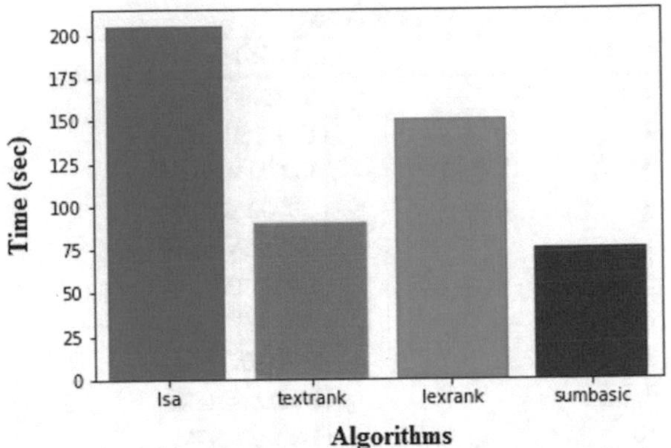

Fig. 9 Time for review summarization

3.1 Comparison of Summaries of Different Algorithms Using Rouge Score

ROUGE score is used as a metric to compare the summaries generated using the four algorithms with the human-generated gold summary. The Rouge score (F-measure) obtained after the comparison of positive and negative summaries with the human-generated gold summary are shown in Tables 1, 2, 3 and 4 , respectively. The top ten movies which are recommended by the approach are as follows: i. Star Wars Trilogy THX Digitally Mastered Edition, ii. Star Wars—Episode I, The Phantom Menace VHS, iii. The Lord of the Rings: The Fellowship of the Ring, iv. Firefly: The Complete Series, v. Marvel's: The Avengers, vi. Avatar, vii. The Hunger Games, viii. The Hobbit: An Unexpected Journey, ix. Prometheus, x. Star Trek Into Darkness. These movies are selected due to their popularities and high number of reviews are available so that there is more amount of information to summarize. After analyzing the values it can be concluded that LSA despite taking the longest time is giving the least accurate summary in most of the cases. Another observation that can be made is that the cases where the number of summaries is comparatively more (in the summarization of positive reviews of movies), lexrank and textrank are functioning better, but the cases where the number of summaries is less (in the summarization of negative reviews of the movies), the sumbasic algorithm is performed quite well. Figure 9 shows the timing profile of each of the four summarization algorithms incorporated in our model. SumBasic, one of the most widely used summarization technique takes the least time while LSA takes the longest because of the matrix calculus associated with it. Tables 5, 6 and 7 show the number of reviews per movie, average words and sentences per review of each movie, respectively.

Table 5 No. of reviews per movie

Sr.No	Movie Id	No. of reviews
1	0793906091	949
2	630575067X	901
3	B00003CWT6	1022
4	B0000AQS0F	928
5	B001KVZ6HK	1058
6	B002VPE1AW	1004
7	B003EYVXV4	1105
8	B0059XTU1S	933
9	B005LA1HXQ	1081
10	B009934S5M	1107

Table 6 Avg. words per review

Sr.No	Movie Id	Avg_words
1	0793906091	269.91
2	630575067X	264.02
3	B00003CWT6	268.89
4	B0000AQS0F	181.54
5	B001KVZ6HK	152.02
6	B002VPE1AW	251.46
7	B003EYVXV4	160.95
8	B0059XTU1S	187.56
9	B005LA1HXQ	219.28
10	B009934S5M	176.71

Table 7 Avg. sentences per review

Sr.No	Movie Id	Avg_sentences
1	0793906091	17.15
2	630575067X	18.59
3	B00003CWT6	17.49
4	B0000AQS0F	12.73
5	B001KVZ6HK	11.57
6	B002VPE1AW	16.13
7	B003EYVXV4	11.10
8	B0059XTU1S	12.37
9	B005LA1HXQ	14.41
10	B009934S5M	12.33

4 Conclusion

The proposed work aims to summarize the reviews for movies purchased from Amazon by combining four traditional machine learning algorithms with a feature selection technique. Sentiment analysis has been performed to categorize reviews into positives and negatives. A novel method called hierarchical summarization is proposed which summarizes the large reviews into summary of few sentences. The results of this summary are compared with the traditional algorithms using the ROUGE score to decide the best summary. Experimental results show the effectiveness of the proposed approach. The work can be further extended by introducing deep learning to summarize the reviews. Also, abstractive text summarization can be done on those reviews to give better results.

References

1. Liu, C.-L., Hsaio, W.-H., Lee, C.-H., Lu, G.-C., Jou, E.: Movie rating and review summarization in mobile environment. IEEE Trans. Syst., Man, Cybern. Part C (Applications and Reviews) **42**(3), 397–407 (2012)
2. Gugnani, S., Roul, R.K.: Triple indexing: an efficient technique for fast phrase query evaluation. Int. J. Comput. Appl. **87**(13), 9–13 (2014)
3. Peetz, M.-H., de Rijke, M., Kaptein, R.: Estimating reputation polarity on microblog posts. Inf. Process. Manag. **52**(2), 193–216 (2016)
4. Fang, C., Mu, D., Deng, Z., Wu, Z.: Word-sentence co-ranking for automatic extractive text summarization. Expert Syst. Appl. **72**, 189–195 (2017)
5. Roul, R.K., Sahoo, J.K., Goel, R.: Deep learning in the domain of multi-document text summarization. In: International Conference on Pattern Recognition and Machine Intelligence, pp. 575–581. Springer, Berlin (2017)
6. Liu, F., Flanigan, J., Thomson, S., Sadeh, N., Smith, N.A.: Toward abstractive summarization using semantic representations. In: Proceeding of Human Language Technologies: The 2015 Annual Conference of the North American Chapter of the ACL, pp. 1077–1086 (2015)
7. Narayan, S., Cohen, S.B., Lapata, M., Ranking sentences for extractive summarization with reinforcement learning. In: Proceedings of NAACL-HLT 2018, Association for Computational Linguistics, pp. 1747–1759 (2018)
8. Ganesan, K., Zhai, C., Han, J.: Opinosis: a graph-based approach to abstractive summarization of highly redundant opinions. In: Proceedings of the 23rd International Conference on Computational Linguistics. Association for Computational Linguistics, pp. 340–348 (2010)
9. Erkan, G., Radev, D.R.: Lexrank: Graph-based lexical centrality as salience in text summarization. J. Artif. Intell. Res. **22**, 457–479 (2004)
10. Mihalcea, R., Tarau, P.: Textrank: bringing order into text. In: Proceedings of the 2004 Conference on Empirical Methods in Natural Language Processing (2004)
11. Landauer, T.K., Foltz, P.W., Laham, D.: An introduction to latent semantic analysis. Discourse Process. **25**(2–3), 259–284 (1998)
12. Vanderwende, L., Suzuki, H., Brockett, C., Nenkova, A.: Beyond sumbasic: task-focused summarization with sentence simplification and lexical expansion. Inf. Process. Manag. **43**(6), 1606–1618 (2007)
13. Lin, C.-Y.: Rouge: a package for automatic evaluation of summaries. In: Text Summarization Branches Out: Proceedings of the ACL-04 Workshop, vol. 8, pp. 74–81 (2004)

Significance of the Background Image Texture in CBIR System

Charulata Palai, Satya Ranjan Pattanaik and Pradeep Kumar Jena

Abstract The Content Based Image Retrieval (CBIR) using local texture feature is extensively studied for decades. The local texture features such as LBP, LTP, LDP, and other derived local texture features uniquely describes the content of an image. The performances of the local texture features with different classifiers are well exposed by the researchers. The purpose of this paper is to explore the effect of the LBP texture feature due to simple and complex background in the image. The experiments are done with the Wang database using various distance measures. The results are promising and it shows the LBP feature is biased by the complex background and it affects the performance of the CBIR system.

Keywords CBIR · LBP · Complex background image · ARR · PR-curve

1 Introduction

The image retrieval is an active research area [1, 2] in the domain of image processing, machine learning, and pattern recognition. Various methods used for image retrieval are text-based, content-based, and semantic-based retrieval method. The semantic-based retrieval techniques, which focus on extracting and matching features from the images, have been extensively explored for many years. The semantic feature of the image is a vector which uniquely describes the content of the image. The feature vector is treated as the descriptor of the respective image and significantly differentiates intraclass and interclass images. The visual property of an image is highly

C. Palai (✉) · S. R. Pattanaik · P. K. Jena
National Institute of Science and Technology, Palur Hills, Golanthara, Berhampur 761008, Odisha, India
e-mail: charulatapalai@gmail.com

S. R. Pattanaik
e-mail: srp.nist@gmail.com

P. K. Jena
e-mail: pradeep1_nist@yahoo.com

H. S. Behera et al. (eds.), *Computational Intelligence in Data Mining*, Advances in Intelligent Systems and Computing 990,
https://doi.org/10.1007/978-981-13-8676-3_42

dominated by the high-level semantics of the image. However it is difficult to derive the high-level semantic features. Hence most of the works are based on low-level semantic features. The low-level semantics shows the arrangement of the neighboring pixels with respect to the center pixel rather than the actual value contained in the center pixel. The Local Binary Pattern (LBP), Local Ternary Pattern (LTP), Center-Symmetric LBP (CS-LBP), Orthogonal Combination LBP (OC-LBP), Local Derivative Pattern (LDP), etc. [3, 4] are some popular local semantic features [5]. Many times the local semantic features fail to describe the actual object contained in the image. This is referred to as semantic gap. The semantic gap restricts the performance [6, 7] of the CBIR system. It is obvious that the low-level semantics features such as LBP and others are also biased by the background image and supposed to affect the performance of the retrieval rate. The motivation of this work is to analyze the significance of the background image texture on performance of CBIR system [8–10].

Performance of CBIR systems depend upon two major steps: The feature selection and matching technique used. It is observed here that the performances of few classes are more dominating than the others irrespective of the feature selection and similarity measures used. In this work we tried to explore the significance of the texture of the image in the background.

2 Proposed Method

The proposed work is to verify significance of the background image texture in the CBIR system. The proposed model is shown in Fig. 1. We considered the Wang database which consists of 1000 images of 10 different classes. The LBP features are extracted and stored in the feature database. The LBP features vector is the histogram of LBP image. It is a unique image descriptor with dimension 1×256 that describes the low-level texture information of a gray-scale image. The images of same class consist of same kind of objects or items. The low-level textures of all the images of a class are expected to have more similarity rather than the other class images. This assumption is biased by two issues.

Firstly, there are many images in the database having more background scene than the foreground object, that results in the texture pattern beingmore dominated by the background scene. Secondly, the background scenes differ drastically for the same class images. Due to this the dimension of the feature vector is stretched and allows the objects of other class to intervene. This hinders the performance of the recognition process. To test the significance of the background image texture during the recognition process we select images from two classes, one the best performing class, i.e., the Dinosaur class and other is a below average performing class the Elephant class according to the results shown in the class-wise performance analysis. Figure 1 shows model of the CBIR system with two types of query inputs. In the first

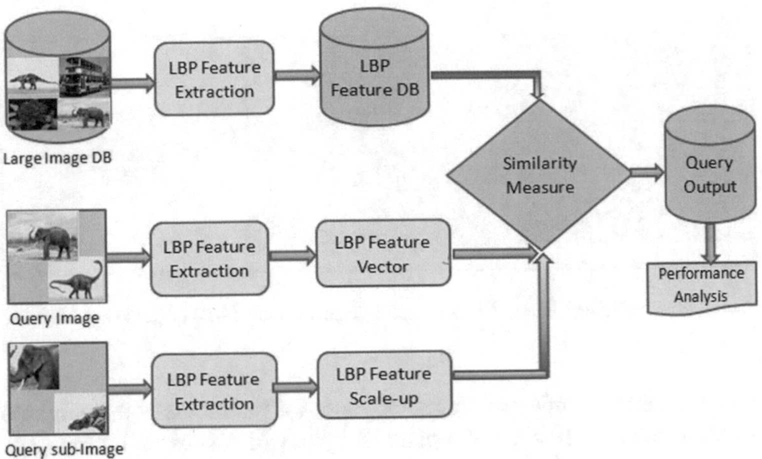

Fig. 1 Schematic diagram of the proposed model

phase, we took all the images as it is from the database and shown the performance of the query output. In the second phase, we select the sub-image parts highly occupied by the foreground object for the dinosaur class as well as the elephant class. Then we compute the LBP feature of the sub-image and scale it up as per the size of the original image. The LBP feature is derived from the histogram of the LBP image which is frequency count information. In consequence, the LBP feature is biased by the size of the image. The LBP feature of the query image (LBPQ) scale up using Eq. (1). The results of both the phases are discussed in the Sect. 4.

$$\mathrm{LBPQ} = \mathrm{LBP}_{\text{Sub-Image}} \times \mathrm{S} \tag{1}$$

where S is the scaling factor calculated as

$$\mathrm{S} = \mathrm{R}/\mathrm{R1}$$

R is the resolution of the original image
R1 is the resolution of the query sub-image.

2.1 LBP Features

The LBP pattern extracts the information of a center pixel and based on this the distribution of edges are coded using two directions (positive or negative direction).

In this approach a center pixel in the image considered as the threshold value, the LBP value is computed by comparing its gray value with its neighbors as mentioned below in Fig. 2. Here g_c is the gray value of the center pixel, g_p is the gray value of

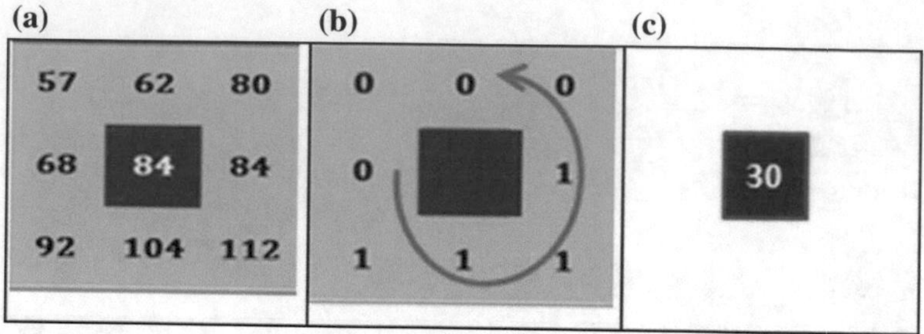

Fig. 2 Calculation of LBP value **a** 3 × 3 Image window **b** Local binary pattern **c** LBP value

its neighbors, P is the number of neighbors, and R is the radius of the neighborhood. It shows the value of LBP code of a pixel (x_c, y_c), which is given by Eq. (2):

$$LBP_{P,R} = \sum_{p=0}^{P-1} S(g_p - g_C)\, 2^P \tag{2}$$

$$where\ S(x) = \begin{cases} 1, & if\ x \geq 0; \\ 0, & otherwise; \end{cases}$$

2.2 *Distance Measures*

The various distance measures [1] plays an important role with a classifier for finding the interclass and intraclass similarity. The distance measures are broadly classified as square distance, absolute distance, weighted distance and angular distance. In this case, we have used five distance measures such as Euclidean, City block, Canberra, Chi-square, and Extended-Canberra distance. These distance measures are expressed using Eqs. (3)–(7).

Euclidean distance:

$$D_{ED} = \sqrt{\sum_{i=0}^{L-1} \left(F_i^q - F_i^t\right)^2} \tag{3}$$

City Block distance

$$D_{CT} = \sum_{i=0}^{L-1} \left|F_i^q - F_i^t\right| \tag{4}$$

Canberra distance:

$$D_{WT} = \sum_{i=0}^{L-1} \frac{\left|F_i^q - F_i^t\right|}{F_i^q + F_i^t} \tag{5}$$

Chi-square

$$D_{Chi} = \sum\nolimits_{i=0}^{L-1} \frac{\left(F_i^q - F_i^t\right)^2}{F_i^q + F_i^t} \tag{6}$$

Extended-Canberra distance:

$$D_{ECD} = \sum\nolimits_{i=0}^{L-1} \frac{\left|F_i^q - F_i^t\right|}{\left(F_i^q + \mu^q\right) + \left(F_i^t + \mu^t\right)}, \tag{7}$$

$$where\ \mu^q - \frac{1}{L}\sum_{i=0}^{L-1} F_i^q\ and\ \mu^t - \frac{1}{L}\sum_{i=0}^{L-1} F_i^t$$

The above-mentioned distance measures are used with the LBP features for performance analysis.

2.3 Performance Measures

The performance of the CBIR system is estimated using the average precision value *P(L)* [1] and the average recall value *R (L)* [1]. The performance of a recognition system is measured with the percentage of correct prediction with respect to the test data set. The average retrieval rate *(ARR)* [1] shows the overall performance of a classifier. Higher the value of ARR implies the better the retrieval performance.

The average precision $P(L) = mean\left(\frac{number\ of\ relevant\ images\ retrieved}{number\ of\ retrieved\ images}\right)$

The average recall $R(L) = mean\left(\frac{number\ of\ relevant\ images\ retrieved}{number\ of\ relevant\ images}\right)$

This is mathematically formulated [1] in Eqs. (8)–(10) as follows:

$$P(L) = \frac{1}{N_t L}\sum_{q=1}^{N_t} n_q(L) \tag{8}$$

$$R(L) = \frac{1}{N_t N_R}\sum_{q=1}^{N_t} n_q(L) \tag{9}$$

$$ARR = \frac{1}{N_t N_R}\sum_{q=1}^{N_t} n_q(N_R) \tag{10}$$

Fig. 3 Elephant images with different backgrounds **a** in the Water **b** in the desert **c** in the Forest

where

L	*denote the retrieved image set*
$n_q(L)$	*is the number of correctly retrieved images with a query image q*
N_t	*show the number of images in the database*
N_R	*denote the number of relevant images in the database.*

2.4 Dataset Description

The experiments for the significance of the background image texture is carried on the Wang database of 1000 images, which consists of 10 classes such as People, Beach, Building, Bus, Dinosaur, Elephant, Flower, Horse, Mountain, and Food. Each class consists of 100 images. The size of the images are either 384 × 256 or 256 × 384 hence the resolution of all the images are same. Hence the LBP feature vectors are consistent and comparable. As the images are captured naturally the size of the foreground object differs from image to image. The amount of artifacts in the background images are very dominating and inconsistence in many classes. Fig. 3 shows three images of the elephant class having different background (a) elephant in the water (b) elephant in the desert (c) elephant in the forest. As the image contains more background information, the LBP texture is biased by the background artifacts.

Figure 4 shows the original images with its histogram and their LBP images with LBP histogram that shows the variance in the LBP feature vector with respect to the change in background and image class.

3 Results and Discussions

The performance of a recognition system is described by its average precision and average recall values. In Fig. 5a, we have shown the performance of the Wang database LBP features with different distance measures for the top 20 retrieved images with Precision-Recall curve (PR-curve). The PR-curve performance is sig-

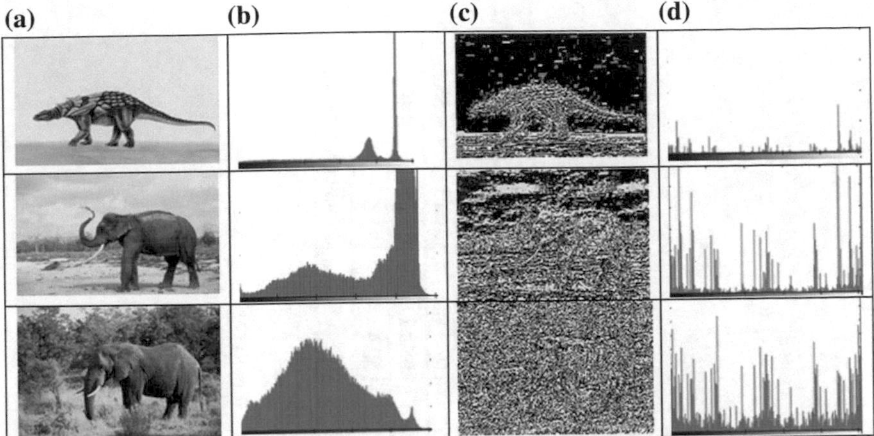

Fig. 4 **a** Original image **b** Histogram of original image **c** LBP image **d** Histogram of LBP image

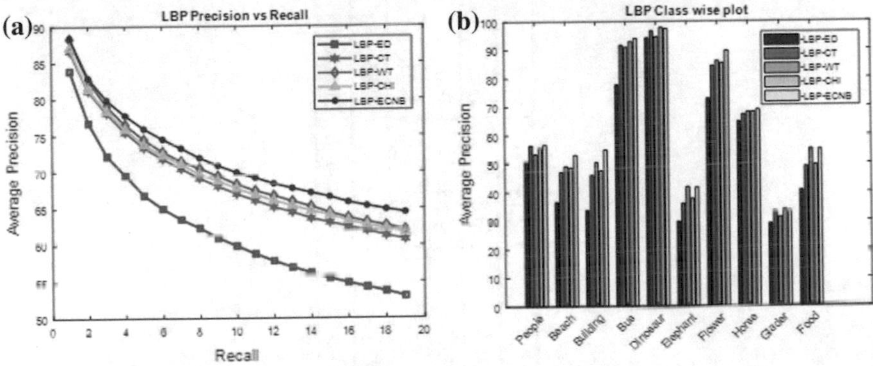

Fig. 5 **a** Precision and recall curve **b** Class-wise average precision with different distance measure

nificant using Extended-Canberra distance and the performance is poor with the Euclidean distance and all other distance measures performance are well comparable. Fig. 5b shows the class-wise average precision values using the five cited distance measures. As it is observed the retrieval performance is better Extended-Canberra distance, for all the remaining part of our analysis we have used the Extended-Canberra distance measure.

The average precision values of the various classes are shown in Table 1 according to the different distance measures. It is observed that the Extended-Canberra distance measure outperforms in comparison to all the other distances then followed by Canberra and Chi-square distance. Among the different classes the maximum average precision value of the Dinosaur class is 97.8 using Chi-square distance whereas the values of the Mountain and Elephant classes are 33.85 with Chi-square and 41.8 with Canberra distance, respectively.

Table 1 Class-wise average precision values

	People	Beach	Building	Bus	Dinosaur	Elephant	Flower	Horse	Mountain	Food
Euclidean	51	36.55	33.7	77.8	94.15	29.7	72.85	64.65	28.9	40.6
City_Block	56.45	47.1	45.95	91.6	96.6	36	84.35	67.3	33.35	48.9
Canberra	53.5	49.15	50.5	90.9	94.55	**41.8**	86.05	68.15	31.15	55.1
Chi-***squere***	55.7	48.65	47.55	92.9	**97.8**	37.75	85.3	68.15	33.85	49.5
Ext_Canberra	**56.75**	**53.1**	**54.9**	**94**	97.4	41.65	**89.55**	68.9	33.5	55

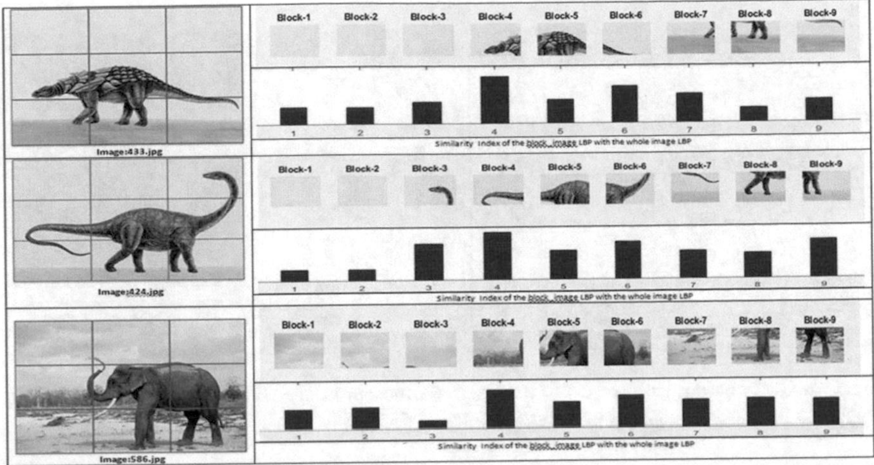

Fig. 6 Image divided into 3×3 sub-block images, comparison of the similarity of LBP of the sub-block image versus the LBP of the original image

To measure the significance of the background we divide the query image into 3×3, i.e., nine equal image sub-blocks compute the LBP, scale-up LBP by nine times. We further compare the similarity of the LBP feature vector of the original image with the feature vectors of image sub-blocks. Similarity index are computed and shown in Fig. 6. It is interesting to observe that the image sub-block with substantial foreground and background contents are more similar to the original image.

The similarity index values shown in Fig. 6 signify that the LBP feature is having a good amount of correlation with the background contents on the image. To justify the above significance the LBP feature of the different image sub-block are given for query and the results are shown in the result sections.

As the performance of the dinosaur class was excellent in overall performance, in this section we tried to explore the feature property of the dinosaur class. In Fig. 7a the output of the top 20 images retrieved when the LBP feature of original dinosaur image "433.jpg" is given as the input, all the outputs are of the same class. On the other hand in Fig. 7b the LBP feature of block-4 of the same image is given as the input with proper scale up. The output shows all the top 20 images retrieved are from the same class, despite the fact that the orders and few images are different.

As the elephant class was one of the poor performing class, in this section our attempt is to understand the feature property of the elephant class. In Fig. 8a the query image is the elephant in the water and out of 20 top retrieved images 13 are from the same class, i.e., 65% success rate. Whereas in Fig. 8b the query image is the elephant in the forest and out of 20 top retrieved images only 4 are from the same class, i.e., 20% success rate. Whereas images retrieved from the horse class is 12 out of 20, i.e., 60% of the total recognition, moreover all the horses are in the forest. This show the LBP feature is more dominated by the background image texture.

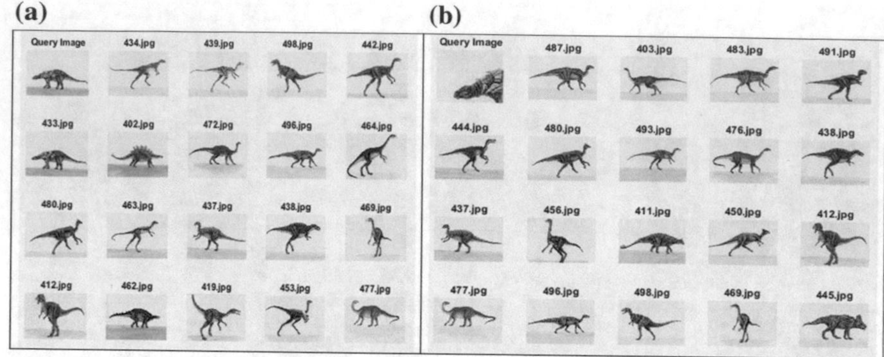

Fig. 7 **a** Query output with respect to the original dinosaur image "433.jpg" **b** Query output with respect to the block_image4 of the dinosaur image "433.jpg"

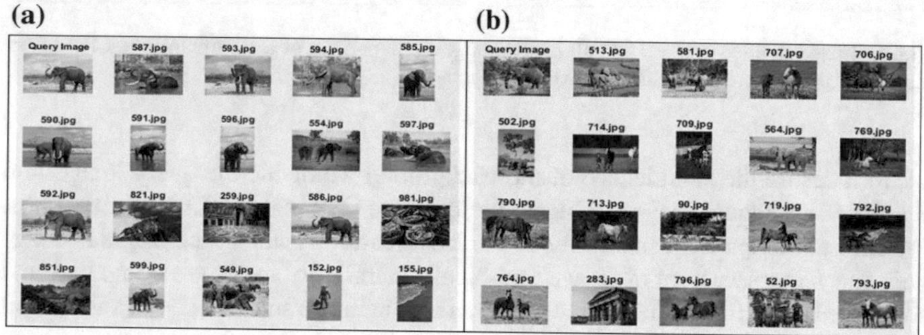

Fig. 8 **a** Query output with respect to the elephant image "586.jpg" in the water **b** Query output with respect to the elephant "512.jpg" in the forest

Figure 9 shows the output for the query sub-image in which mostly the background image part is cropped. In Fig. 9a the input image is the body part of the same dinosaur image and in Fig. 9b the input is a whole body part of the above elephant which is in water. Before using for query the LBP features are scaled up appropriately using Eq. (1) for both the images. The output result shown in Fig. 9 signify the performance of the query sub-images reduced in both the cases, although these are highly dominated by the foreground contents.

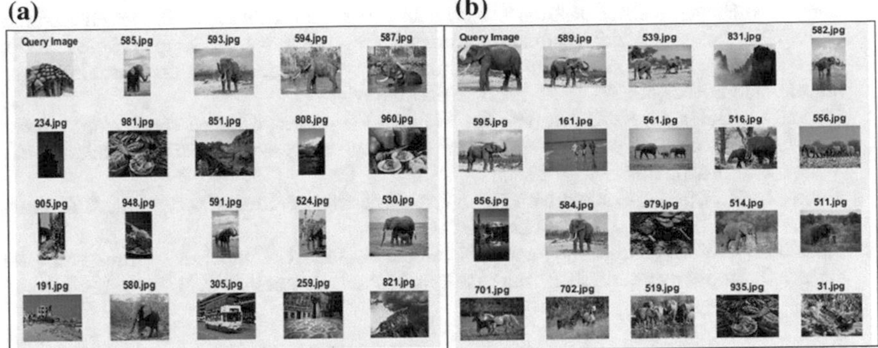

Fig. 9 Query output with respect to major foreground and minor background object **a** major body part of the dinosaur image **b** full body part of the elephant image

4 Conclusion

In this paper, we analyze the performance of CBIR system for the Wang database with LBP features using different distance measures. It is worth to mention here that the performance of CBIR system using Extended-Canberra distance measure is better than the performances using other distance measures. The LBP feature is biased by the texture of the background image as well as the size of the foreground object. This fact is observed in the results as the performance of the dinosaur class is better than the other classes as it is having a simple background. The performance of the elephant and mountain class is deprived due to their complex and variant background. This analysis shows the texture of the background image is also having adequate significance in the performance of the CBIR system. Hence to improve the performance of CBIR system further analysis is required to deal with the images having complex background and other artifacts present in the image database.

References

1. Singh, C., Walia, E., Kaur, K.P.: Color texture description with novel local binary patterns for effective image retrieval. Pattern Recognit. **76**, 50–68 (2018). Elsevier
2. Singh, C., Walia, E., Kaur, K.P.: Enhancing color image retrieval performance with feature fusion and non-linear support vector machine classifier. Opt. Int. J. Light Electron Opt. **158**,127–141 (2018). Elsevier
3. Celik, C., Bilge, H.S.: Content based image retrieval with sparse representations and local feature descriptors: a comparative study. Pattern Recognit. **68**, 1–13 (2017). Elsevier
4. Liu, P., Guo, J., Chamnongthai, K., Prasetyo, H.: Fusion of color histogram and LBP-based features for texture image retrieval and classification. Inf. Sci. **390**, 95–111 (2017). Elsevier
5. Fadaei, S., Amirfattahi, R., Ahmadzadeh, M.: Local derivative radial patterns: a new texture descriptor for content-based image retrieval. Signal Process. **137**, 274–286 (2017). Elsevier

6. Zhou, X.S., Huang, T.S.: CBIR: from low-level features to high level semantics. In: Proceedings of the SPIE, Image and Video Communication and Processing, vol. 3974, pp. 426–431 (2000)
7. Liu, G.H., Yang, J.Y., Li, Z.: Content-based image retrieval using computational visual attention model. Pattern Recognit. **48**, 2554–2566 (2015). Elsevier
8. Carson, C., Belongie, S., Greenspan, H., Malik, J.: Blobworld: image segmentation using expectation-maximization and its application for image querying. IEEE Trans. Pattern Anal. Mach. Intell. **24**(8), 1026–1038 (2002)
9. Yue, J., Li, Z., Liu, L.: Content-based image retrieval using color and texture fused features. Math. Comput. Model. **54**, 1121–1127 (2011)
10. Smeulders, A.W.M., Santini, S., Worring, M., Gupta, A., Jain, R.: Content-based image retrieval at the end of the early years. IEEE Trans. Pattern Anal. Mach. Intell. **22**(12), 1349–1380 (2000)

Supervised and Unsupervised Data Mining Techniques for Speaker Verification Using Prosodic + Spectral Features

Basanta Kumar Swain, Sanghamitra Mohanty and Dillip Ranjan Nayak

Abstract In this paper, we tackled the speaker verification problem using two unparalleled data mining techniques, i.e., supervised and unsupervised learning techniques. We have used the Multilayer Perceptron and Decision Tree algorithms as supervised learning methodologies for speaker verification and Expectation Maximization as well as Farthest First algorithms are used to handle the speaker verification in an unsupervised way. It is found from our research work that supervised based learning techniques yielded best results as compared to other algorithms. The best accuracy performances of MLP, Decision Tree, EM and Farthest First are found to be 73.8%, 69.55%, 66.92%, and 64.29%, respectively. Moreover, this research article has utilized self-built speech corpus for extraction of feature parameters, i.e., prosodic and spectral features. It is found that the speaker verification result is changed significantly by combining both prosodic and spectral features.

Keywords MLP · Decision tree · EM · Farthest first · Prosodic · Spectral

1 Introduction

The modern biometric systems are based on an array of measures such as fingerprints, facial and retina scans, etc. An individual identity can be also found from his/her voice as the speech signal reflects an audio signature or voice print of a speaker [1, 2]. Speaker verification is a biometric authentication process where a

B. K. Swain (✉) · D. R. Nayak
Department of Computer Science & Engineering, Government College of Engineering, Kalahandi, Bhawanipatna 766002, India
e-mail: technobks@yahoo.com

D. R. Nayak
e-mail: dillip678in@yahoo.co.in

S. Mohanty
Department of Computer Science & Application, Utkal University, Bhubaneswar 751004, India
e-mail: sangham1@rediffmail.com

H. S. Behera et al. (eds.), *Computational Intelligence in Data Mining*, Advances in Intelligent Systems and Computing 990,
https://doi.org/10.1007/978-981-13-8676-3_43

person is verified on the basis of physiological or behavioral characteristics [3, 4]. The research work over speaker verification is going on for the last six decades. The first attempt was made by famous Bell Labs in early 60s based on spectrogram similarity. The subsequent research works are carried out by using advanced pattern recognition methods as well as robust speech parameters. A fully automated speaker verification system was used in U.S. Air Force during early 80s developed by Texas Instruments (TI). In the mid-80s a Speech Group of the National Institute of Standards and Technology (NIST) had developed statistical models like Hidden Markov Model (HMM) for speech and speaker recognition. In the early 90s research work was focused on the robustness of speech-based biometric system by combing spectral envelope and fundamental frequencies [3–7]. The recent trend of speaker verification is focused on handling the intra-speaker variability and imposter. The robust speaker verification system not only depends on good machine learning algorithms but also on better representation of acoustic signals [8]. Hence, in this research article we looked at two parameters, i.e., machine learners and parameterization of speech files. We emphasized on both supervised and unsupervised machine learning algorithms and their performance over speaker verification task. We have used classifiers namely Multilayer Perceptron and Decision Tree classifier as supervised learners for speaker verification. On the other hand, we have used Expectation Maximization (EM) and Farthest First algorithms (FF) as unsupervised learners for handling the same said problem. Moreover, we conglomerated the spectral feature parameters like Mel Frequency Cepstral Coefficients (MFCC) and Gammatone Frequency Cepstral Coefficients (GFCC) with prosodic feature parameter (jitter, shimmer, and their variants) to make the system robust and reliable in a noisy environment. The feature vectors for speaker verification task are extracted from our own developed speech corpus which contains voice files of nearly hundred speakers of both male and female genders. At the end, we did compare the performance of our supervised learners and unsupervised learners in speaker verification task.

This paper is organized as follows: Sect. 2 illustrates about speech corpus, Sect. 3 highlights on feature extraction, Sect. 4 depicts supervised and unsupervised machine learning algorithms, Sect. 5 demonstrates the experimental results and Sect. 6 says about the conclusion and future work.

2 Description of Speech Corpus

A text-dependent new speech corpus is designed for speaker verification task. The speech corpus contains the voice-based password of multiple speakers. The voice password is consisting of continuous digits of length four, pass phrase of speaker's choice, isolated digits and words. Speakers have uttered their voice password either in Odia language or Indian English accent. Speech files are collected from 100 speakers of male and female genders. The speech corpus consists of 80 male speakers and 20 female speakers of age between 18 and 55 years old. The audio recordings are carried out in an office environment. Speakers are allowed to utter their password through a

Table 1 Speech corpus for speaker verification

Gender	No. of speakers	Language of utterance	Type of voice password	Property of voice password
Male	45	Odia	Isolated/Continuous/Pass-phrase	Mono, 16 bit, 16,000 Hz, and PCM
Female	13	Odia	Isolated/Continuous/Pass-phrase	Mono, 16 bit, 16,000 Hz, and PCM
Male	35	Indian English	Isolated/Continuous/Pass-phrase	Mono, 16 bit, 16,000 Hz, and PCM
Female	07	Indian English	Isolated/Continuous/Pass-phrase	Mono, 16 bit, 16,000 Hz, and PCM

head-mounted microphone that connected to laptops or desktop computers. We have used computers and laptops of different configurations as well as of different manufactures. We have collected the voice password from every speaker with multiple utterances in different sessions. The rerecording is also allowed in case of any kind of dissatisfaction during collection of voice files. The sound files are set to mono, 16 bit, 16,000 Hz and PCM Windows files [8]. The speech corpus is split into training and testing database by considering 70% of corpus as training and remaining 30% for testing the voice files for speaker verifications. The overview of speech corpus for speaker verification task is shown in Table 1.

3 Feature Extraction

Feature extraction component of the speaker verification system retains the desired anatomical and behavioral characteristics of the speaker. In this research article, we have considered two types of features namely prosodic features and spectral features in order to improve the performance of speaker verification task. Because prosodic features carry the source related feature such as speaking rate, articulation style, etc. which play a key role in voice-based verification. Jitter, shimmer, and their variants are used as prosodic feature parameters and MFCC as well as GFCC are used as spectral feature vectors for authentication of speaker. Spectral feature, GFCC is having noise robustness property whereas MFCC follows the auditory model of a human being. The feature parameters are extracted as follows.

3.1 Prosodic Feature

3.1.1 Jitter (absolute)

Jitter (absolute) is the average absolute difference between consecutive periods.

$$Jitter\ (absolute) = \frac{1}{N-1}\sum_{i=1}^{N-1} T_I - T_{i+1}, \tag{1}$$

In eq. 1, Ti are the extracted f0 period lengths and N is the number of extracted f0 periods.

3.1.2 Jitter (relative)

It is measured by finding the division of absolute difference of consecutive periods and average period. Equation 2 is used to determine the feature parameter called Jitter (relative).

$$Jitter\ (relative) = \frac{\frac{1}{N-1}\sum_{i=1}^{N-1} |T_i - T_{i+1}|}{\frac{1}{N}\sum_{i=1}^{N} T_i}. \tag{2}$$

3.1.3 Shimmer (db)

Shimmer (db) is the average absolute base-10 logarithm of the difference between the amplitudes of consecutive periods and multiplied by 20.

$$Shimmer\ (dB) = \frac{1}{N-1}\sum_{i=1}^{N-1} |20\log(A_{i+1}/A_i)|, \tag{3}$$

In eq. 3, A_i are the extracted peak-to-peak amplitudes data and N is the number of extracted fundamental frequency periods.

3.1.4 Shimmer (relative)

Shimmer (relative) is measured by finding the ratio between the average absolute difference of amplitudes of consecutive periods and the average amplitude. Equation 4 signifies the Shimmer (relative) calculation.

$$Shimmer\ (relative) = \frac{\frac{1}{N-1}\sum_{i=1}^{N-1} |A_i - A_{i+1}|}{\frac{1}{N}\sum_{i=1}^{N} A_i}. \tag{4}$$

We have used Praat tool to extract the Jitter (absolute), Shimmer (db) and their varieties from voice passwords [9].

3.2 Spectral Feature

3.2.1 MFCC

First windowing operation is carried out in order to suppress the discontinuities at the frame boundaries and then Discrete Fourier Transform (DFT) is applied to compute the magnitude spectrum. Filterbank processing is carried out for reduction of dimensionality. The cepstrum is obtained by applying logarithmic function followed by Discrete Cosine Transform (DCT) [10, 11].

The frequencies below 1000 Hz are mapped linearly to the mel scale, and those above 1000 Hz are mapped logarithmically according to the Eq. 5.

$$mel(f) = 1127ln\left(1 + \frac{f}{700}\right). \tag{5}$$

We have used first thirteen mel frequency coefficients in this research work.

3.2.2 GFCC

Speech signal first is passed through the preemphasis filter (FIR) in order to spectrally flatten the speech signal. Then the input signal is framed in 20 ms and hamming windowed is applied. The windowed frame is passed through DFT and magnitude information is preserved. The GFCC values are calculated by passing through the input signal to a 64 channel gammatone filter bank which models the basilar membrane frequency properties and then the filter response is rectified and decimated to 100 Hz. The resultant time-frequency is exposed to cubic root function. Finally, DCT is applied to derive the cepstral features.

The filter is defined in the time domain by the following impulse response:

$$h(t) = t^{a-1}\exp(-2\pi \mathrm{bt})\cos(2\pi \mathrm{f_c t} + \varphi), \tag{6}$$

In Eq. 6, a and b indicate the filter, t is time, f_c is the filter's center frequency, and φ is the phase.

The bandwidth of each filter is described by an Equivalent Rectangular Bandwidth (ERB) as Eq. 7. Glasberg and Moore have summarized human data on the equivalent rectangular bandwidth (ERB) of the auditory filter with the function [12]:

$$\mathrm{ERB}(f_c) = f_c/\mathrm{Q} + B_0$$

$$B_0 = 24.7,$$
$$Q = 9.64498. \tag{7}$$

where B_0 is minimum bandwidth; Q is the asymptotic filter quality at large frequencies.

4 Supervised and Unsupervised Learning Methodologies

In this research paper, we have made a comparison over supervised and unsupervised learning algorithms for speaker verification task. We have used classifiers namely Multilayer Perceptron (MLP) and Decision Tree classifier as supervised learners for creation of models of speaker verification task. On the other hand, we have used Expectation Maximization (EM) and Farthest First (FF) algorithms as unsupervised learner models generation for handling the same task.

4.1 Multilayer Perceptron (MLP)

It uses Back Propagation algorithm for the network training [13].

4.1.1 Back Propagation Algorithm

Step 1. All weights are initialized to small random numbers.
Step 2. All the gradients of the weights and the current error are initialized with value zero ($w_{i,j} = 0 \text{ and } E = 0$).
Step 3. Calculate the output of hidden layer neurons as Eq. 8

$$y_j(\mathrm{p}) = f\left(\sum_i^n x_i(p) \cdot w_{ij} - \theta_j\right) \tag{8}$$

Step 4. Determine the output from the network using Eq. 9

$$y_k(\mathrm{p}) = f\left(\sum_i^m x_{jk}(p) \cdot w_{jk} - \theta_k\right) \tag{9}$$

Step 5. Update error per epoch as Eq. 10

$$E = E + \frac{(e_k(p))^2}{2} \tag{10}$$

Step 6. Back propagate the errors and adjust the weights using Eq. 11

$$\begin{aligned} \delta_k(p) &= f' \cdot \mathrm{e_k}(p) \\ \mathrm{e_k}(p) &= y_{dk}(p) - y_k(p) \end{aligned} \tag{11}$$

Step 7. Update the weights between hidden layer and output layer by following Eq. 12.

$$\Delta w_{jk}(p) = \Delta w_{jk}(p) + y_j(\mathrm{p}) \cdot \delta_{\mathrm{k}}(p) \tag{12}$$

Step 8. Update each network weight as per Eq. 13

$$w_{i,j} = w_{i,j} + l\,\Delta w_{i,j} \tag{13}$$

where l is the learning rate.

4.2 Decision Tree C4.5

It works as follows [14]

Step 1. Determine the opening subset of the training examples.
Step 2. Apply the attribute and the subset of training examples to construct a decision tree.
Step 3. Measure the accuracy of the constructed tree by considering the remaining of the training examples which are not in the subset of the training set.
Step 4. Stop the execution if all examples are correctly classified.
Step 5. Construct a new tree if an example is incorrectly classified by adding it to the initial subset.
Step 6. Repeat until a tree is built that classifies all examples correctly or a tree is constructed from the whole training set.

4.3 Farthest First Algorithm

Farthest first is a modified version of K-means algorithm which places cluster center at the furthest point from the existing centers. This point should lie within the data area. This procedure expedites the clustering formation due to very less requirement of reassignment and adjustment. This algorithm works very efficiently for large dataset problem.

Farthest First algorithm also sets the initial K number of seeds similar to K-means algorithm. But it is unlikely in the determination of centroid which is not calculated on the basis mean rather it takes centroid arbitrarily and distance of one centroid from other is maximum [15]. First it takes point P_i then selects next point P_i which is at maximum distance. In another way, we can say that Pi is centroid and p_1, p_2, …, p_n are samples of dataset belonging to the cluster which follows Eq. 14.

$$\min\{\max dist(P_i, p_1), \max dist(P_i, p_2)\ldots\} \tag{14}$$

For each example $X_i = [x_{i,1}, x_{i,2}, \ldots, x_{i,m}]$ in dataset D that is described by m attributes. Then, Eq. 15 has been formulated as a scoring function for evaluating each point, which is defined as

$$Score(X_i) = \sum_{j=1}^{m} f\left(x_{i,j}\middle|D\right) \tag{15}$$

The point with the highest score is selected as the first point in the farthest-point heuristic, and remaining points are selected in a similar manner as that of basic farthest-point heuristic.

4.4 Expectation Maximization (EM)

It is having two key steps, i.e., responsibility and reestimation steps [16–18].

Expectation (E) Step: It evaluates the responsibilities, i.e., posterior probabilities for each instance using Eq. 16:

$$\gamma\left(z_{i,j}\right) = \frac{\pi_j(x_i|\mu_j, \sum_j)}{\sum_{j=1}^{k} \pi_j N(x_i|\mu_j, \sum_j)} \tag{16}$$

Maximization (M) Step: It reestimates the model parameters from the given current responsibilities as per Eq. 17:

$$\begin{aligned} \mu_j^{new} &= \frac{1}{N}\sum_{i=1}^{N} \gamma\left(z_{i,j}\right)x_i \\ \sum_j^{new} &= \frac{1}{N}\sum_{i=1}^{N} \gamma\left(z_{i,j}\right)(x_i - \mu_j^{new})\left(x_i - \mu_j^{new}\right)^T, \end{aligned} \tag{17}$$

where

$$\pi_j^{new} = \frac{N_j}{N},$$

$$N_j = \sum_{i=1}^{N} \gamma\left(z_{i,j}\right).$$

Table 2 Performance of un/supervised learners using prosodic features

Supervised/Unsupervised learner	True acceptance (%)	False acceptance (%)
MLP	69.79	30.21
Decision tree	67.53	32.47
EM	64.0	36.0
Farthest first	57.87	42.13

γ (zij) represents the responsibility of instance vector xi which belongs to the Gaussian component j.

5 Experimental Results

We have carried out the speaker verification task by focusing on three angles of views. In the first scenario, we have extracted prosodic features namely jitter (absolute), shimmer (db), and their variants from our own created speech corpus and measured the performance of speaker verification using models of supervised and unsupervised learning algorithms. Similarly, in second fold, the performance of speaker verification using spectral features namely MFCC (13 feature vectors), GFCC (12 feature vectors) is also carried out by utilizing models of supervised and unsupervised learning algorithms. In the final fold, we have tested the performance of supervised and unsupervised data mining techniques for speaker verification by conglomerating the prosodic features (jitter, shimmer, and their variants) with spectral features (MFCC, GFCC). In the final fold, we have exploited the uniqueness of prosodic features as well as spectral features for speaker authentication purpose. We have implemented the supervised learning techniques (MLP and Decision Tree) and unsupervised learning techniques (EM and Farthest First) using WEKA tool [19]. We have used 70% of speech corpus for training and remaining 30% of speech corpus for testing purpose. The performance of speaker verification task is measured in terms of true acceptance and false acceptance. The accuracies of different supervised and unsupervised learning techniques by using prosodic, spectral features and combination of both feature parameters are represented in Tables 2, 3 and 4.

The performance of different types of learning algorithms for speaker verification using varieties of feature parameters is shown in Fig. 1.

6 Conclusion and Future Work

In this research paper, we have tackled the speaker verification problem by using two supervised learning algorithms (MLP and Decision Tree 4.5) as well as two unsupervised learning algorithms (EM and Farthest First). We have collected voice

Table 3 Performance of un/supervised learners using spectral features

Supervised/Unsupervised learner	MFCC/GFCC	True acceptance (%)	False acceptance (%)
MLP	MFCC	66.23	33.77
	GFCC	68.97	31.03
Decision tree	MFCC	63.52	36.48
	GFCC	64.67	35.33
EM	MFCC	65.22	34.78
	GFCC	66.78	33.22
Farthest first	MFCC	53.67	46.33
	GFCC	56.62	43.38

Table 4 Performance of un/supervised learners using prosodic and spectral features

Supervised/Unsupervised learner	True acceptance (%)	False acceptance (%)
MLP	73.8	26.2
Decision tree	69.55	30.45
EM	66.92	33.08
Farthest first	64.29	35.71

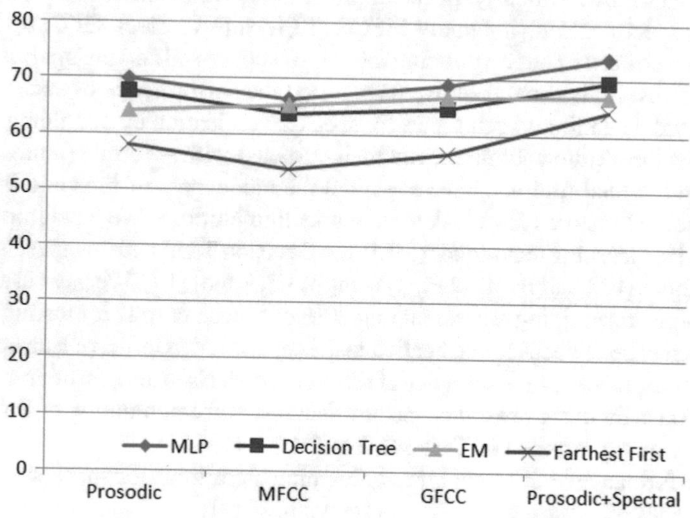

Fig. 1 Performance curves of different learners over feature vectors

passwords in Odia and Indian English accents from both male and female speakers. We have utilized prosodic features (jitter, shimmer, and their variants) and spectral features (MFCC and GFCC) for training and testing the supervised as well as unsupervised models. The performance of our speaker verification task is measured from the point of views of learning algorithms as well as from the angles of feature parameters. The supervised learning technique called MLP has produced pretty good results as compared to other algorithms. MLP algorithm has given its best performance, i.e., 73.8% accuracy rate by using both prosodic and spectral features. It is seen from the performance curve that prosodic features authenticate more accurately than any other feature vectors in isolation case except EM algorithm. The unsupervised learning algorithm, EM has performed nearly same for spectral feature GFCC with respect to combination of prosodic and spectral features. The best result of EM algorithm is found as 66.92%. The GFCC feature vectors yielded significantly better result than MFCC for all learning models. It is also found that the speaker verification accuracy is changed phenomenally by combining both prosodic and spectral features. The performance of supervised learning algorithm, Decision Tree is found better in comparison to unsupervised learning algorithm, Farthest First. The best results of Decision Tree and Farthest First algorithms are found as 69.55% and 64.29%, respectively. In future, we will give more emphasis on the collection of mimicry voice files with high-level linguistic and multimodal features for speaker verification.

Acknowledgements A standard ethical committee formed by the Department of CSE, Govt. Engg. College Kalahandi, Odisha Ref No. : GECK/CSE/012, dt. : 09/04/2019 has approved the dataset used in this paper and the dataset has no conflict of interest/ethical issues with any other public source or domain.

References

1. Quatieri, T.F.: Discrete-Time Speech Signal Processing Principles and Practice. Pearson Education, Third Impression (2007)
2. Shaughnessy, D.: Speech Communications Human and Machine, 2nd edn. Universities Press (2001)
3. Keshet, J., Bengio, S.: Automatic Speech and Speaker Recognition: Large Margin and Kernel Methods. Wiley (2009)
4. Mohanty, S., Swain, B.K.: Speaker identification using SVM during Oriya speech recognition. Int. J. Image, Graph. Signal Process., 28–36 (2015)
5. Jurafsky, D., Martin, J.H.: Speech and Language Processing: An Introduction to Natural Language Processing, Computational Linguistics, and Speech Recognition. Pearson Education Asia (2000)
6. Wang, Y., Deng, L., Acero, A.: Spoken language understanding—an introduction to the statistical framework. IEEE Signal Process. Mag. **27**(5) (2005, September)
7. Benesty, J., Sondhi, M.M., Huang, Y.: Hand Book of Speech Processing. Springer, Berlin, Heidelberg (2008)
8. Samudravijaya, K., Barot, M.: A comparison of public domain software tools for speech recognition. In: Workshop on Spoken Language Processing, pp. 125–131 (2003)
9. Praat Software Website: http://www.fon.hum.uva.nl/praat

10. Sahidullah, Md., Saha, G.: Design, analysis and experimental evaluation of bock based transformation in MFCC computation for speaker recognition. Speech Commun. **54**(4), 543–565 (2012)
11. Rabiner, L.R., Schafer, R.W.: Digital Processing of Speech Signals, 1st edn. Pearson Education (2004)
12. Lopez-Poveda, E., Meddis, R.: A human nonlinear Cochlear filterbank. J. Acoust. Soc. Am. **110**(6), 3107–3118 (2001)
13. Chan, M.V., Feng, X., Heinen, J.A., Niederjohn, R.J.: Classification of speech accents with neural networks. In: IEEE International Conference on Neural Networks, IEEE World Congress on Computational Intelligence, vol. 7, pp. 4483–4486. IEEE (1994)
14. Han, J., Kamber, M., Pei, J.: Data Mining Concepts and Techniques, 3rd edn. Elsevier (2007)
15. Hochbaum, D.S., Shmoys, D.B.: A best possible heuristic for the K-center problem. Math. Oper. Res. **10**(2), 180–184 (1985)
16. Celeux, G., Govaert, G.: A classification EM algorithm for clustering and two stochastic versions. Comput. Stat. Data Anal., 315–332 (1992)
17. Hastie, T., Tibshrani, R., Friedman, J.: 8.5 The EM Algorithm. The Elements of Statistical Learning, pp. 236–243. New York, Springer (2001)
18. James, G., Witten, D., Hastie, T., Tibshirani, R.: An Introduction to Statistical Learning with Applications in R. Springer, New York, Heidelberg, Dordrecht, London (2013)
19. http://www.technologyforge.net/WekaTutorials/

Effective Software Fault Localization Using a Back Propagation Neural Network

Abha Maru, Arpita Dutta, K. Vinod Kumar and Durga Prasad Mohapatra

Abstract Effective fault localization is an essential requirement of the software development process. Back propagation neural network can be used for localizing the faults effectively and efficiently. Existing NN-based fault localization techniques take statement invocation information in *binary terms* to train the network. In this paper, we have proposed an efficient approach for fault localization using back propagation neural network and we have used the actual number of times the statement is executed to train the network. We have investigated our approach on Siemens suite. Results show that on an average there is 35% increase in the effectiveness over existing BPNN.

Keywords Back propagation neural network · Program debugging · Fault localization · Suspiciousness of code · Failed test · Successful test

1 Introduction

Software has become a key part of our society. Various softwares are being developed every year and require a huge amount of effort. This effort lies not only in development but majorly in testing. Debugging process becomes more complex when scale and

---

A. Maru · K. V. Kumar (✉) · D. P. Mohapatra
Department of Computer Science and Engineering, National Institute of Technology, Rourkela 769008, India
e-mail: 517cs1007@nitrkl.ac.in

A. Maru
e-mail: 217cs3313@nitrkl.ac.in

D. P. Mohapatra
e-mail: durga@nitrkl.ac.in

A. Dutta
Department of Computer Science and Engineering, Indian Institute of Technology, Kharagpur 721302, India
e-mail: arpitad10j@gmail.com

H. S. Behera et al. (eds.), *Computational Intelligence in Data Mining*, Advances in Intelligent Systems and Computing 990,
https://doi.org/10.1007/978-981-13-8676-3_44

complexity of software increases [1]. The stage in which bugs are identified is known as fault localization. To make the software fault free we first need to locate the bug. This process can be done in two stages. In the first stage, identifying the code which might contain the bug, called as suspicious code. In the second stage, the programmer reviews the selected code and tries to identify the bug [2].

Currently, much research work has been done in the field of fault localization such as, Baah et al. [3–6] proposed probabilistic program dependence graph (PPDG), Slice-based techniques such as static slicing, dynamic slicing, and execution slicing [7–17], spectrum-based techniques [18] such as Tarantula [19], Ochiai [20], Visualization [21], SOBER [22], Crosstab (CBT) [23] etc. DStar(D*), proposed by Wong et al. [24], is currently the state of the art for fault localization. Its effectiveness was evaluated using 24 different programs and it performs better that other 38 techniques for fault localization.

Nowadays, machine learning techniques such as artificial neural networks [25], deep neural networks [26] are widely used to get a robust model for fault localization. Advantages of neural networks over other models have enticed many researchers. Fault tolerant characteristic of the neural network makes it adaptive to the work environment and minor faults have very little impact on the model. This has made it applicable to many software engineering areas like reliability estimation, risk analysis, cost estimation, and reusability characterization, etc.

Back propagation neural network (BPNN) model for fault location, proposed by Wong et al. [27] is one of the most popular model in today's scenario. Local minima was the disadvantage of the BP neural network. Thus, another method for fault localization was proposed by Wong et al. in [28], using radial basis function (RBF) network. RBF and other spectrum-based techniques are handicapped with the problem of statements having the same suspiciousness, whereas, BPNN model assigns unique suspiciousness score to each executable statement.

In the existing back propagation neural network (BPNN) model, execution of the statement is recorded in binary terms. This overlooked those statements which are executed more than once which mislead the results as there is a probability that the statement which is executed more than once is more critical than the statement which is executed only once. This motivated us to work upon the existing BP neural network model to improvise the results. Thus, to improve the feature vector for training the network, the proposed method uses the actual number of times the statement is executed for the given test case. We have experimented with five different programs of Siemens suite to analyze the effectiveness of the proposed approach.

The remainder of the paper is organized as follows. Section 2 explains the basic concepts of the artificial neural network and back propagation neural network. Section 3 explains the proposed approach of fault location using BP neural network along with an example. Section 4 explains the experimental setup, data set used, results, and average results. Section 5 shows the comparison with related work. Section 6 explains the threats to validity of our approach. Section 7 gives the conclusion and tells the future work which can be done.

2 Basic Concepts

In this section, we explain the basics of artificial neural network, their advantages over other conventional methods and brief overview of back propagation neural network.

2.1 *Artificial Neural Network*

Artificial neural network (ANN) is viewed as a mathematical model inspired by the functional aspects of the biological nervous system. The network consists of the simple elements called as neurons and their working is similar to that of biological neurons. Each neuron consists of a certain function called as activation function which converts the input to the desired output form. Network consists of multiple layers and each layer contains multiple neurons which work simultaneously to generate the output. Its learning capability can approximate any nonlinear continuous functions based on the given data. Neurons are connected to each other with certain weights upon the connections, so that they can process the information collectively and store it on these weights. These networks are very useful as they can learn from past experience and using this knowledge can solve new related problems.

2.2 *Back Propagation Neural Network*

A back propagation neural network is a kind of feed-forward neural network in which the neurons are divided into layers. Neurons of one layer can be only connected to the neurons of the next layer. Layers between the input and output layer are called hidden layers. Back propagation algorithm is an iterative process used to adjust the errors while training the network. The three-layered structure of BP neural network is shown in Fig. 1.

3 Proposed Approach

In this section, we discuss our proposed approach for fault localization using back propagation neural network.

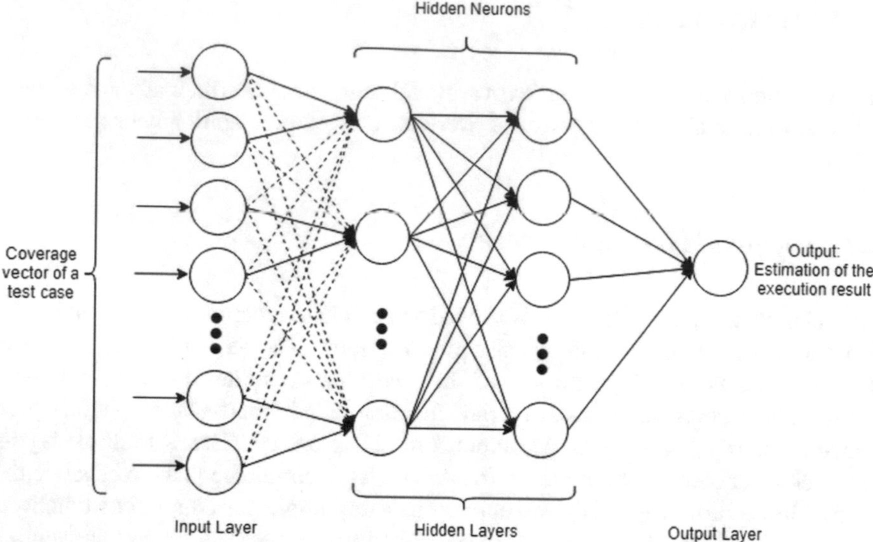

Fig. 1 Structure of a three layered BP Neural Network

3.1 Fault Localization Using BP Neural Network

Initially, a program P is taken as a base input which has m number of executable statements along with one faulty statement. Suppose P is executed on n test cases, out of which k test cases are passed and n-k are failed. Figure 2 shows the sample *coverage matrix* in which number of rows is equal to the number of test cases and number of columns is equal to the number of statements covered by the test cases. Value of each row is the statement coverage of a particular statement corresponding to the particular test case. Any real number greater than zero indicates the number of times the statement is executed for that particular test case and zero indicates that the statement was not executed by that corresponding test case. For example, s_4 is covered by the successful test case t_1 and s_2 is not covered by the failed test case t_6. Complete row can be called as a *coverage vector* (c_{t_i}) which depicts the coverage information for test case t_i. For example, $c_{t_1} = (0, 0, 2, 9, 1, 43, 0, 2)$ is the first row of the matrix in the Fig. 2 shows the statement coverage of the program for test case t_1.

Now, assuming a set of *virtual test cases* $v_1, v_2, v_3, \ldots, v_m$, depicted in the form of matrix called as *virtual matrix* as shown in the Eq. 1, whose each row is the *virtual coverage vector* $c_{v_1}, \ldots, c_{v_m}$. Specialty of the virtual test case is that, it executes only one statement, which is not possible in reality, hence they are called as virtual test cases. Each row of this matrix is given as an input to the proposed BP neural network and output generated is the suspiciousness of corresponding statement.

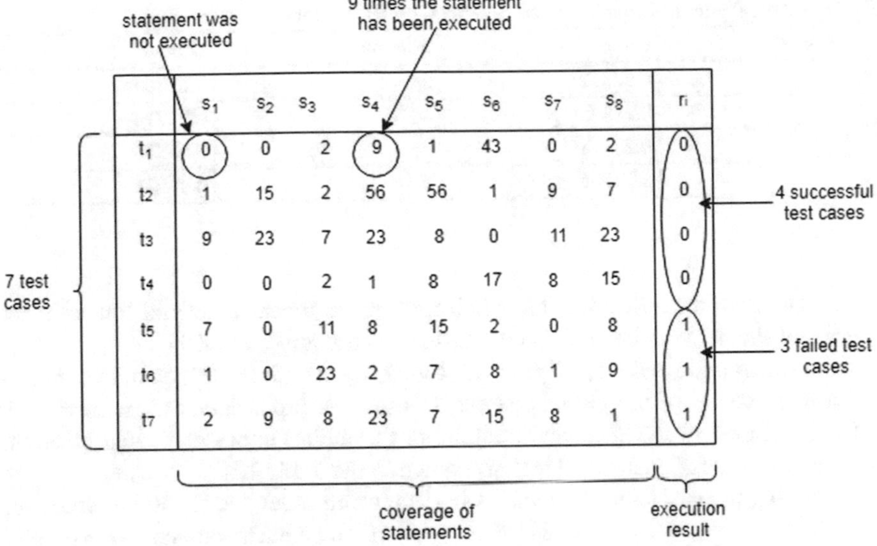

	s_1	s_2	s_3	s_4	s_5	s_6	s_7	s_8	r_i
t_1	0	0	2	9	1	43	0	2	0
t_2	1	15	2	56	56	1	9	7	0
t_3	9	23	7	23	8	0	11	23	0
t_4	0	0	2	1	8	17	8	15	0
t_5	7	0	11	8	15	2	0	8	1
t_6	1	0	23	2	7	8	1	9	1
t_7	2	9	8	23	7	15	8	1	1

Fig. 2 Sample coverage matrix

$$
\begin{bmatrix} c_{v_1} \\ c_{v_2} \\ c_{v_3} \\ \vdots \\ c_{v_m} \end{bmatrix} = \begin{bmatrix} 1 & 0 & 0 & \cdots & 0 \\ 0 & 1 & 0 & \cdots & 0 \\ 0 & 0 & 1 & \cdots & 0 \\ \vdots & \vdots & \vdots & \ddots & \vdots \\ 0 & 0 & 0 & \cdots & 1 \end{bmatrix} \quad (1)
$$

3.2 *Explanation with Example*

Let us now take an example to make our proposed model more clear. Suppose we have a program with 20 statements having 8 executable statements (m = 8) with one fault at line no 7. We took a test suit of 7 test cases, out of which three test cases failed. Statement coverage and result of execution is shown in Fig. 2. After this the following steps are performed.

- Firstly, a BP neural network is created having 8 input neurons, one output neuron, and three hidden layers of neurons. We considered sigmoid function as *nonlinear transfer function.*
- Then, network is trained with the coverage data collected. First input in our example would be vector (0, 0, 2, 9, 1, 43, 0, 2) with zero as the expected output. If the actual output of the network does not match with the expected output, then the weight of the network is updated. Then, this updated weight is used for the next

Table 1 Actual output (suspiciousness) generated by BPNN for each statement

Statement	Output	Statement	Output
s_1	0.67312	s_5	0.87165
s_2	0.77895	s_6	0.50274
s_3	0.22318	s_7	0.65310
s_4	0.29561	s_8	0.46922

input to train the network. This whole process is repeated several times till the error of the network becomes minimal(in our case approx 0.001).

- After the training is done, BP neural network is fed with the test data. In our case, coverage vector of the virtual test cases is the test input data. Suspiciousness is the likelihood of the statement containing the fault. Output generated gives the suspiciousness of each statement, as shown in the Table 1.
- Arranging the suspiciousness values in decreasing order, we found the following order s_5, s_2, s_1, s_7, s_6, s_8, s_4, s_3. When we examined each statement one by one, s_7 which is the actual faulty statement, is evaluated after examining 3 non-faulty statements. Here, in our example, rank of s_7 is 4.

4 Experiments

In this section, we present the experimental setup and results of our experiments.

4.1 Experimental Setup

We have implemented our approach, in Python-3, on Ubuntu 16.04 operating system, with i5 64bit processor having 4GB RAM. For experiment, values of learning rate are taken in the range of 0.0001 to 1 and the number of hidden layers is calculated heuristically using Eq. 2.

$$num = round(n/30) * 0.5 \tag{2}$$

Here, n represents the number of executable statements in the program and num is the number of hidden layers. We experimented with different values for hidden layer but, got the best result using the above formula. We have used, Sigmoid function as the activation function (transition function) of the BP neural network.

Table 2 Details of Siemens suit's programs

Name of program	Number of faulty versions	Lines of codes	Number of executable statements	Number of test cases
Print_tokens	7	565	172	4130
Print_tokens2	10	510	146	411
Schedule	9	412	140	2650
Schedule2	10	307	115	2710
Tot_info	23	406	100	1052

4.2 Data Set Used

For experimental use, five C programs and their corresponding faulty versions of Siemens suit were used, namely Print_tokens, Print_tokens2, Schedule, Schedule2, Tot_info. Each faulty version contains only one fault. Table 2 introduce the 5 C programs with their names, number of faulty versions, lines of code, number of executable statements, and number of test cases. Siemens suit contains in total 7 programs out of which we have selected only 46 faulty versions of 5 C programs. Following versions were omitted: version 4 and 6 of Print_tokens, version 10 of Print_tokens2, version 6, 10, and 21 of Tot_info, version 1, 5, 6, 8, and 9 of Schedule and version 8 and 9 of Schedule2. Reasons of omission were:

- Syntactic difference was not found between the correct version and faulty versions (e.g., included header files were different).
- None of the test cases failed for the faulty versions.
- Segmentation faults were observed while executing the test cases on the above omitted faulty versions.
- Difference between the actual program and faulty versions was not included in the executable statements. Additionally, certain statements were absent in the faulty versions which cannot be located directly.

4.3 Results

In this section, we present our experimental results using different graphs. Figure 3, shows the effectiveness of the proposed approach for Print_tokens, such that 72% of the program needs to be evaluated for locating the fault as compared to existing BPNN model [27] in which 85% of the program is evaluated. Figure 4, shows that for Print_tokens2, on an average 18% of the program needs to be evaluated for locating the fault as compared to existing BPNN model in which 36%

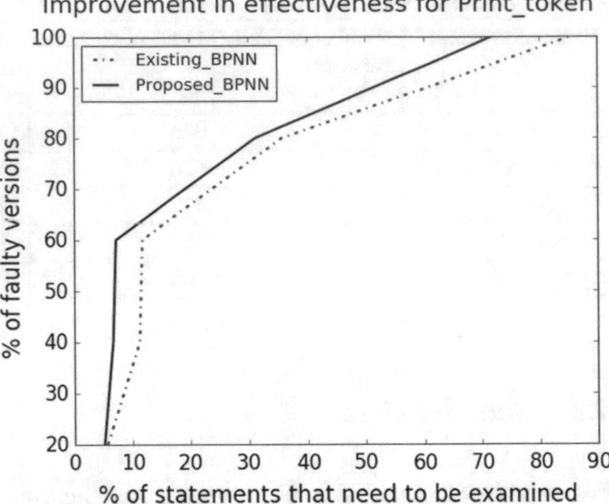

Fig. 3 Improvement in effectiveness for `Print_tokens`

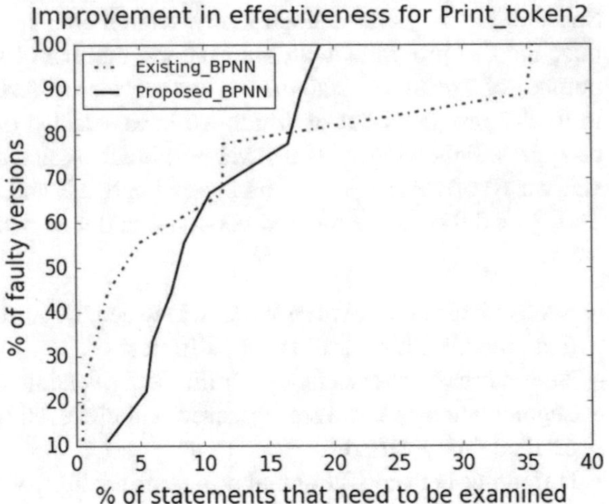

Fig. 4 Improvement in effectiveness for `Print_tokens2`

of the program is evaluated, although in the graph we can see that near 10% to 16% our approach does not show good results but on an average result is good. Figure 5, shows that for `Schedule`, 20% of the program needs to be evaluated for locating the fault as compared to existing BPNN model in which 50% of the program is evaluated. Figure 6, shows that for `Schedule2`, 38% of the program needs to be evaluated for locating the fault as compared to existing BPNN model in which 84% program is evaluated. Figure 7, shows that for `Tot_info`, 52% of the program needs to be evaluated for locating the fault as compared to existing BPNN model in which 80% of the program is evaluated.

We have also used EXAM score metric to measure the effectiveness of various techniques. EXAM score is calculated using Eq. 3.

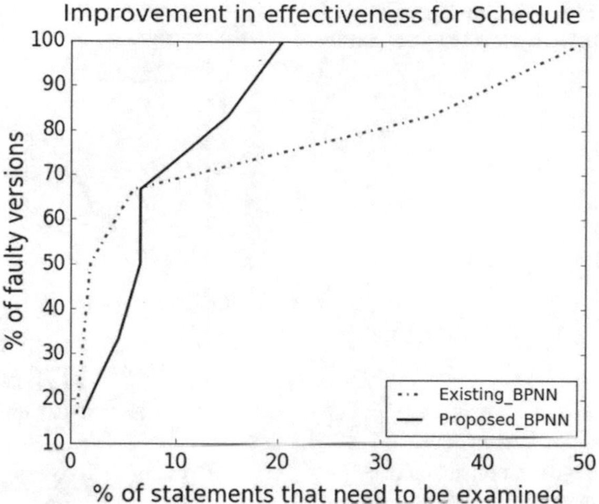

Fig. 5 Improvement in effectiveness for `Schedule`

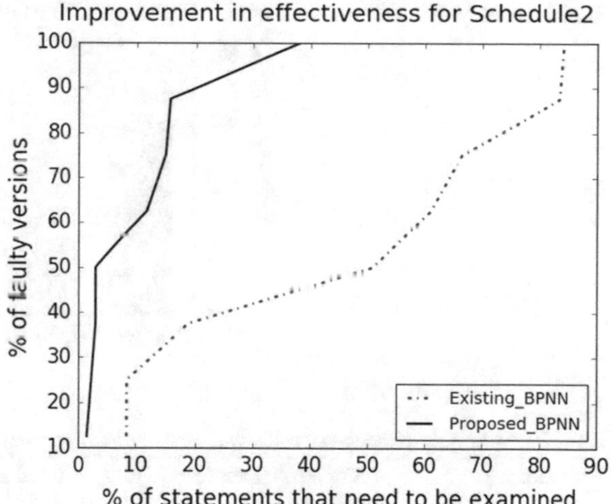

Fig. 6 Improvement in effectiveness for `Schedule2`

$$EXAM\ Score = \frac{|V_{examined}| * 100\%}{|V|} \tag{3}$$

where,

- |V| measures the size of executable codes in the program.
- $|V_{examined}|$ measures the number of statements that has to be inspected so as to find the fault.

Figure 8, shows the comparison of average EXAM score of the existing BPNN model [27] and the proposed BPNN model. Table 3, shows the average percentage of increase in the effectiveness of the proposed BPNN model for fault localization as compared to the existing BPNN model. Overall, we can say that our model has improved 35% of the effectiveness on an average.

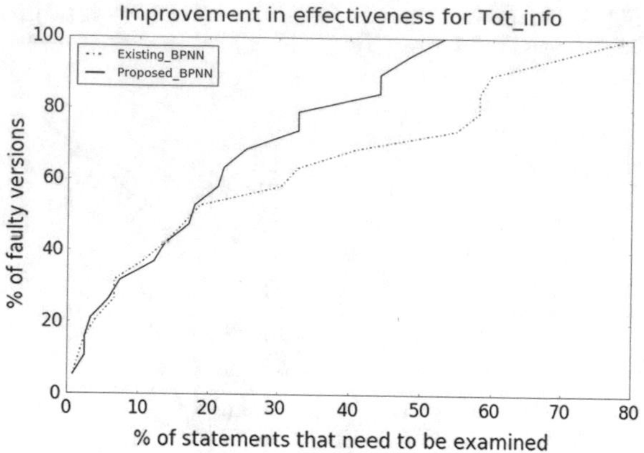

Fig. 7 Improvement in effectiveness for `Tot_info`

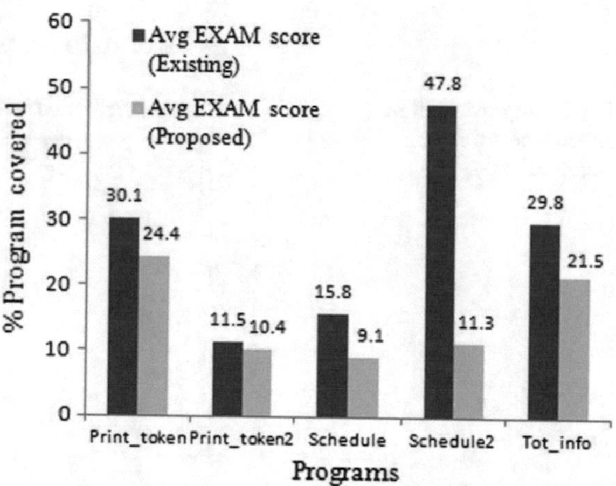

Fig. 8 Comparison of average EXAM score of existing BPNN and proposed BPNN model

Table 3 Average percentage of effectiveness for each program

Program	Avg EXAM score (Existing)	Avg EXAM score (Proposed)	Percentage of increase in effectiveness (%)
`Print_tokens`	30.05	24.41	18.77
`Print_tokens2`	11.5	10.38	9.66
Schedule	15.78	9.10	42.36
Schedule2	47.75	11.32	76.27
`Tot_info`	29.81	21.48	27.93

5 Comparision with Related Work

Many works have been done previously in the field of fault localization.

Baah et al. [6] proposed a probabilistic program dependence graph(PPDG), which provides the conditional probability of each node in the graph to identify the bug [4, 5]. But this becomes a lengthy process as complete program is considered for debugging. Slice-based techniques [7–17] were also adopted, which narrows down the search to various slices of the program. Slice is basically a set of statements in a program which leads to the faulty statement or the set of statements which can be impacted by the faulty statement [29, 30].

Disadvantage of earlier slice-based techniques is that, they only inspect the slice, leaving the remaining part of the code. Moreover, sometimes the slices include all the statements of the program, which again leads to the initial problem state. In our approach, we rank all the statements and based upon the unique ranking of the statements, they are evaluated.

Collofello suggested spectrum-based techniques which are based on mathematical formulas. Jones et al. [19] proposed a popular (ESHS)-based technique, called Tarantula. It utilizes the pass/fail information about each test case. The main idea behind it was, the entities which are executed by the failed test cases have more probability of having fault than those which are executed by the passed test cases. Jones et al. [21] extended Tarantula by representing the degree of suspiciousness by using different colors. Liu et al. [22] proposed a method called SOBER, which ranks suspicious predicates by calculating the difference between the truth value of predicates for successful and failed execution. Wong et al. [23] proposed an analysis-based technique called Crosstab(CBT) which provides the rank to the suspicious elements of the program.

Spectrum-based fault localization techniques failed when there were statements having the same degree of suspiciousness. Another weakness that they only examined the suspicious statements. However, there is a possibility of bug containment in the statements covered by the successful test cases. Whereas, in our approach all the statements are given ranks based upon their unique suspiciousness, so there is no conflict.

A Back propagation neural network (BPNN) model for fault location, was proposed by Wong et al. [27]. In this model, for each test case, statement coverage along with the result obtained after execution were collected. This collected data was used as input to train the network. This provides the network, the ability to relate both the entities. The output generated by the network is the suspiciousness of each statement for being faulty. Since, local minima was the disadvantage of BP neural network, another method for fault localization was proposed by Wong et al. [28], using radial basis function (RBF) network, which is less sensitive to those problems. RBF and other spectrum-based techniques are handicapped with the problem of statements having the same suspiciousness, whereas, BPNN model assigns unique suspiciousness score to each executable statement.

In the existing back propagation neural network (BPNN) model, execution of the statement is recorded in binary terms. This over look those statements which are executed more than once. In our technique, we used the number of times the statement is executed as an input to the BP neural network, which helped in getting the exact value of the suspiciousness for each statement.

6 Threats to Validity

Though our proposed method is performing better than the existing BPNN model but it has certain constraints.

- Our model has been tested for procedural C program, so object-oriented aspects have not been tested till now.
- Programs used in our approach had only single fault. Thus, the effectiveness of our model over programs having multiple faults has not been tested.
- Since we are considering an actual number of statement invocation in the coverage matrix, we require those codes which have loops, recursive calls, function calls, etc. Then only the statement will be executed more than once. Programs which do not have these features cannot be used for checking the effectiveness of our approach.
- We have used standard sample programs for our experiments whose length is less than thousand lines of code. We cannot assure the effectiveness of our approach over very large sized programs, as we have not experimented them.

7 Conclusion and Future Work

We have extended the existing BP neural network model for fault localization by using the actual number of times statement execution in the coverage matrix. For `Print_token` 18.77% increase, `Print_token2` 9.66% increase, for `Schedule` 42.36% increase, `Schedule2` 76.27% increase and for `tot_info` 27.93% increase in the effectiveness has been seen. Though for `Print_token` and `Print_token2` improvement in the effectiveness is not much, but on an average we are able to improve the effectiveness by 35% using our proposed approach for fault localization using BP neural network as compared to the existing BP neural network model.

In future we can extend this work by changing various activation functions along with different feature matrices to check for the increase of effectiveness. Programs having features like: multiple faults, object-oriented can be taken. Programs used in industries which are of very large size and highly complex in nature can also be used to test the effectiveness of our approach. Since we tested on C programs, other languages such as Java, Python, C#, Perl, etc. can be used. This approach of using an actual number of statement executions can also be applied to other machine learning models like RBF, SVM, etc and comparison can be done.

References

1. Agrawal, H., DeMillo, R.A., Spafford, E.H.: An execution backtracking approach to program debugging. IEEE Softw. **8**(5), 21–26 (1991)
2. Wong, W., Eric, R.G., Li, Y., Abreu, R., Wotawa, F.: A survey on software fault localization. IEEE Trans. Softw. Eng. **42**(8), 707–740 (2016)
3. Korel, B.: PELAS - program error-locating assistant system. IEEE Trans. Softw. Eng. **14**(9), 1253–1260 (1988)
4. Taha, A.B., Thebaut, S.M., Liu, S.S.: An approach to software fault localization and revalidation based on incremental data flow analysis. In: Proceedings of the 13th Annual International Computer Software and Applications Conference, Washington DC, USA, pp. 527–534 (1989)
5. Deng, F., Jones, J.A.: Weighted system dependence graph. In: 2012 IEEE Fifth International Conference on Software Testing, Verification and Validation (ICST), pp. 380–389 (2012)
6. Baah, G.K., Podgurski, A., Harrold, M.J.: The probabilistic program dependence graph and its application to fault diagnosis. IEEE Trans. Softw. Eng. **36**(4), 528–545 (2010)
7. Weiser, M.: Program slicing. IEEE Trans. Softw. Eng. **SE–10**(4), 352–357 (1984)
8. Lyle, J.R., Weiser, M.: Automatic program bug location by program slicing. In: Proceedings of the 2nd International Conference on Computer and Applications, pp. 877–883 (1987)
9. Korel, B., Laski, J.: Dynamic program slicing. Inf. Process. Lett. **29**(3), 155–163 (1988)
10. Agrawal, H., Horgan, J.R.: Dynamic program slicing. In: Proceedings of the ACM SIGPLAN'90 Conference on Programming Language Design and Implementation, pp. 246–256 (1990)
11. Agrawal, H., DeMillo, R.A., Spafford, E.H.: Debugging with dynamic slicing and backtracking. Softw. Pract. Exp. **23**(6), 589–616 (1993)
12. Gupta, R., Soffa, M.L.: Hybrid slicing: an approach for refining static slices using dynamic information. In: Symposium on Foundations of Software Engineering, pp. 29–40 (1995)
13. Tip, F.: A survey of program slicing techniques. J. Program. Lang. **3**(3), 121–189 (1995)
14. Agrawal, H., Horgan, J.R., London, S., Wong, W.E.: Fault localization using execution slices and dataflow tests. In: Proceedings of the 6th IEEE International Symposium on Software Reliability Engineering, pp. 143–151 (1995)
15. Zhang, X., He, H., Gupta, N., Gupta, R.: Experimental evaluation of using dynamic slices for fault location. In: Proceedings of the 6th International Symposium on Automated Analysis-Driven Debugging, pp. 33–42 (2005)
16. Zhang, X., Tallam, S., Gupta, N., Gupta, R.: Towards locating execution omission errors. In: Proceedings of the 2007 ACM SIGPLAN Conference on Programming Language Design and Implementation (PLDI), pp. 415–424 (2007)
17. Liu, C., Zhang, X., Han, J., Zhang, Y., Bhargava, B.K.: Indexing noncrashing failures: a dynamic program slicing-based approach. In: Proceedings of the 23rd International Conference on Software Maintenance, pp. 455–464 (2007)
18. Harrold, M.J., Rothermel, G., Sayre, K., Wu, R., Yi, L.: An empirical investigation of the relationship between spectra differences and regression faults. J. Softw. Test. Verif. Reliab. **10**(3), 171–194 (2000)
19. Jones, J.A., Harrold, M.J.: Empirical evaluation of the tarantula automatic fault- localization technique. In: Proceedings of the 20th IEEE/ACM Conference on Automated Software Engineering, pp. 273–282 (2005)
20. Naish, L., Hua Jie, L., Kotagiri, R.: A model for spectra-based software diagnosis. ACM Trans. Softw. Eng. Methodol. (TOSEM) **20**, 11(3) (2011)

21. Jones, J.A., Harrold, M.J., Stasko, J.: Visualization for fault localization. In: Proceedings of Workshop Software Visualisation, 23rd International Conference Software Engineering, Ontario, BC, Canada, pp. 71–75 (2001)
22. Liu, C., Yan, X., Fei, L., Han, J., Midkiff, S.P.: SOBER: statistical model-based bug localization. In: ESEC/FSE'05
23. Wong, E., Wei, T., Qi, Y., Zhao, L.: A crosstab-based statistical method for effective fault localization. In: 2008 1st international conference on software testing, verification, and validation, pp. 42–51. IEEE (2008)
24. Wong, W. Eric, V.D., Gao, R., Li, Y.: The DStar method for effective software fault localization. IEEE Trans. Reliab. **63**(1), 290–308 (2014)
25. Nessa, S., Abedin, M., Eric Wong, W., Khan, L., Qi, Y.: Fault localization using n-gram analysis. In: Proceedings of International Conference on Wireless Algorithms, Systems, and Applications, pp. 548–559 (2008)
26. Zhang, Z., Yan, L., Qingping, T., Xiaoguang, M., Ping, Z., Xi, C.: Deep learning-based fault localization with contextual information. IEICE Trans. Inf. Syst. **100**(12), 3027–3031 (2017)
27. Wong, W.E., Qi, Y.: BP Neural Network-based Effective Fault Localization. Int. J. Softw. Eng. Knowl. Eng. **19**(4), 573–597 (2009)
28. Wong, W.E., Debroy, V., Golden, R., Xu, X., Thuraisingham, B.: Effective software fault localization using an RBF neural network. IEEE Trans. Reliab. **61**(1), 149–169 (2012)
29. Mund, G.B., Goswami, D., Mall, R.: Program Slicing. The Compiler Design Handbook. CRC Press **269–294**, (2002)
30. Krinke, J.: Slicing, chopping, and path conditions with barriers. Softw. Quality Control **12**(4), 339–360 (2004)

Optimizing Generalized Capacitated Vehicle Routing Problem Using Augmented Savings Algorithm

Barnali Saha, Kamlesh Suthar and Ajit Kumar

Abstract This article presents an Augmented Savings Algorithm (ASA) for solving Generalized Capacitated Vehicle Routing Problem (GCVRP), for single-trip homogeneous and heterogeneous fleet. The ASA is a modified version of classical heuristic, based on savings algorithm for homogeneous fleet, pioneered by Clarke and Wright back in 1964 and further enhanced by others. In the ASA algorithm, we introduced a changed modality for adjusting savings value upon prioritizing the parameters for compactness, distribution asymmetry, and nodal demand. Our approach is verified with respect to the real-time vehicle scheduling of a company bus service in Mumbai. This is to ideally redesign the sub-routes which are embedded in existing routes. The algorithm is further validated with regard to the benchmark instances in the literature. The solutions obtained minimizes the overall cost, i.e., the fixed cost; and the variable cost, after maximizing the occupancy of the pickup vehicles. The ASA approaches besides showing the improvement in the results obtained by others, also demonstrates better results compared to the enumerative parameter setting approach proposed by Altinel and Öncan [2], and Empirically Adjusted Greedy Heuristic (EAGH) approach adopted by Corominas et al. [8].

Keywords Homogeneous fleet · Heterogeneous fleet · Heuristic · Enumerative parameter · Savings

---

B. Saha (✉) · K. Suthar · A. Kumar
Department of Mathematics, Institute of Chemical Technology, Mumbai 400019, India
e-mail: bsaha1510@gmail.com

K. Suthar
e-mail: kamlesh08@live.com

A. Kumar
e-mail: ajit72@gmail.com

H. S. Behera et al. (eds.), *Computational Intelligence in Data Mining*, Advances in Intelligent Systems and Computing 990,
https://doi.org/10.1007/978-981-13-8676-3_45

1 Introduction

The Capacitated Vehicle Routing Problem (CVRP), a variant of combinatorial optimization problem, has a number of practical applications in transportation, distribution, and logistics. The CVRP belongs to the category of NP hard combinatorial problem; generally solved by complex and tedious exact methods for small instances. This motivated researchers to develop classical greedy heuristic and metaheuristic algorithms to solve capacitated VRP, where the routes are constructed under the following major constraints: (1) Each route begins and ends at the depot; (2) Each vehicle visits each node exactly once; (3) The total demand of each route does not exceed the capacity of the vehicle.

Dantzig and Ramser [10] first proposed a solution for CVRP for homogeneous fleet. Later, Clarke and Wright [7] proposed a mathematical model based on efficient greedy heuristics that gave improved solution than Dantzig-Ramser. The model was constructed based on savings concept, an estimate of the cost reduction obtained by serving two customers sequentially in the same route, and thereafter sorting the savings in nonincreasing order. The savings concept was summarized by Paolo Toth and Vigo [28]. Since then various enhancement strategies of the savings approach have been suggested on Clarke Wright Savings (CWS) algorithm by others, like Gaskell [15], Yellow [31], Golden et al. [19], Nelson et al. [22], Paessens [24], Altinkemer and Gavish [3], Laporte et al. [21], Altınel and Öncan [2], Corominas et al. [8], Doyuran and Çatay [13], Pichpibul and Kawtummachai [25], Corominas et al. [9].

The ASA proposed by us in this paper is an attempt to bring augmentation to CWS algorithm with respect to route insertions and prioritization of routes through tuning parameters, upholding its simplicity. This is in line with work proposed by Doyuran T, Catay B for homogeneous fleet. Here, unlike our predecessors, we intend to study the generalized CVRP in totality, i.e., for both homogeneous and heterogeneous fleet; due to its practical relevance, as most of the road transportation companies own heterogeneous fleet. The ASA, therefore, addresses a mixed requirement, e.g., different capacities of vehicles based on occupancy requirement, loading, and peak demand requirement during a certain duration of the day, fixed and variable cost, availability of parking space, etc. The ASA method depends on tuning of only three parameters and achieves better solutions for the seven instances of Christofides et al. (1979) dataset, when compared with the results obtained by the four procedures of Iterative EAGH-1: $IEAGH-1_3^{AO}$, $IEAGH-1_8^{AO}$, $IEAGH-1_3^{DC}$ ad $IEAGH-1_6^{DC}$ as proposed by Corominas et al. [9].

The structure of the remaining sections of this paper is as follows: Sect. 2 presents a review of the relevant literature associated to the studied problem; Sect. 3 analyzes the algorithm in more detail; Sect. 4 verifies the results of a real-time problem and validates our approach with respect to the benchmark instances, Sect. 5 summarizes the main contributions and results of this work, while Sect. 6 gives an account of the path forward for future work.

2 Review of GCVRP Literature

The Generalized Capacitated Vehicle Routing Problem (GCVRP) is divided into two sub categories depending on the fleet composition. (1) Homogeneous Fleet Vehicle Routing Problem (CVRP)—where vehicles have the same capacity and (2) Heterogeneous Fleet Vehicle Routing Problem (HVRP)—where vehicles have different capacities. Based on the trip configuration, HVRP further is divided into two major variants—Single trip and Multi trip. Single-trip HVRP further is classified into two subcategories viz. HFVRP (Heterogeneous Fixed Fleet VRP) and FSMVRP (Fleet Size and Mixed VRP). Literature also shows that notable studies in the class of GCVRP are attributed to Laporte et al. [21], Baldacci et al. [4], Toth and Vigo [29] and many others. Review of literature on GCVRP reveals that CWS algorithm proposed in 1964 is one of the best known and widely used classical heuristics in practice to this day for CVRP in comparison with other classical heuristics, e.g., Sweep algorithm designed in 1974 by Gillett and Miller [17] and the Fisher and Jaikumar [14] algorithm proposed in 1981. The above two algorithms fail to address one or more notable features, i.e., simplicity, flexibility, and computation time of an efficient CVRP solution when compared to traditional CWS.

2.1 *Homogeneous VRP*

According to the CWS algorithm, any two routes enclosing nodes i and j, starting from the central depot represented as the "0"th node, i.e., $(0, \ldots, i, 0)$ and $(0, \ldots, j, 0)$ are merged into a single route $(0, \ldots i, j, 0)$ whenever feasible, thus generating a "savings" between the nodes i and j, represented as $s_{ij} = d_{i0} + d_{0j} - d_{ij}$. Here d_{ij} represents the distance between nodes i and j. The savings thus calculated are sorted in descending order and the route (routes) are constructed either sequentially or simultaneously (parallel version) until the number of nodes selected do not exceed the capacity of the pickup vehicle.

Later, Desrochers and Verhoog [11], Altinkemer and Gavish [3] introduced modifications to the standard CWS savings algorithm on the basis of matching heuristics. The work was further extended by Wark and Holt [30]. Here a matching based approach was applied to successively merged clusters, instead of individual nodes, but at the expense of much-increased computing time. As a next enhancement, the experimental bias in the form of a tuning parameter, λ was introduced by Gaskell [15] and Yellow [31] to revamp the CW savings formula for reshaping and compacting the routes and expanding the exploration of the algorithm. The approach was further strengthened by Paessens [24], using parameter μ in an attempt to assign asymmetry information of customer distribution between customers i and j regarding their distances to and from the depot. Later, a third parameter ν was proposed by Altınel and Öncan [2] to improve the savings formula to assign priority to the customers with large demands compared to overall average demand.

Our work follows and extends the concept of Doyuran* and Çatay [13] who suggested placement priority in the same route for the customers with smaller demands and modified the proposed formulation by Altınel and Öncan (1)

$$s_{ij} = (d_{i0} + d_{0j} - \lambda d_{ij}) + \mu|d_{i0} - d_0 j| + \nu\left(\frac{\bar{b}}{b_i + b_j}\right) \tag{1}$$

Here, s_{ij} = savings value; d_{i0} = distance from delivery point i to depot; d_{0j} = distance from depot to the pickup point; d_{ij} = distance between point i and j; $d_{ij} = 0$ when i = j; λ = positive constant, a parameter that helps to reshape and compact the routes, μ = positive constant, a parameter assigning the asymmetry information between customers i and j , ν = constant, a parameter that prioritizes large demand nodes Altınel and Öncan [2] or smaller demand nodes Doyuran* and Çatay [13]. Pursuant to the above, Doyuran* and Çatay [13] further suggested a Robust Savings formula (2):

$$s_{ij} = \frac{(d_{i0} + d_{0j} - \lambda d_{ij})}{d^{max}} + \frac{\mu \cos\theta_{ij}}{d^{max}}\left|d^{max} - \frac{d_{i0} - d_{0j}}{2}\right| + \frac{\nu}{d^{max}}\left(\bar{b} - \frac{b_i + b_j}{2}\right), \tag{2}$$

However, we find the above formulation (2), is complex without yielding significant advantage as we embark upon real driving distances, based on Google Map, for our real-time instance comprising 72 nodes and a total demand of 802.

2.2 *Heterogeneous VRP*

Golden et al. [18] proposed a greedy heuristic algorithm to tackle the FSMVRP by optimizing a total cost function that included fixed and variable cost. Golden in his paper dealt with a few number of enhancement variants to include fixed and variable costs. Notable among these are CS (Combined Savings), OOS(Optimistic Opportunity Savings), Realistic Opportunity Savings (ROS) and $ROS - \gamma$. The $ROS - \gamma$ algorithm uses a route shape parameter to introduce changes the Clarke Wright savings. Taillard [27] addressed the HFVRP and the FSMVRP with fixed and variable costs, and proposed a modification of eight large-scale instances from Golden et al. [18]. The modification was implemented using a Column Generation technique. Later, Choi and Tcha [6] used an approach to solve a packed integer-programming model of HVRP, adopting Linear Programming relaxation and by applying the Column Generation Technique. In the last twenty years, researchers preferred to use metaheuristic approach to solve HVRP problems. Notable among the researchers are: Osman and Salhi [23], Gendreau et al. [16] and Brandão [5] who developed and enhanced Tabu search method for solving FSMVRP; Baldacci et al. [4] and Prins [26] who presented an overview, classification, and survey on different HFVRP. On review of literature, it is evident that many of the successful approaches for the CVRP, particularly the CWS-based classical heuristics and its enhancements, have been adapted to an HVRP context.

The main idea of our paper is to design a computational model to investigate classical CWS and use its enhancements for specific variants of CVRP (say single trip homogeneous and heterogeneous fleet) and to further develop a strategic heuristic function in future to solve both variants of GCVRP.

3 Problem Definition and Algorithm

The objective function of the problem is built with respect to the pickup of employees of a multinational company in Mumbai dispersed at 72 locations within 43.8 kms, and drop them at the central depot at the corporate office. The problem is motivated by challenges of growing demand, congestion of road network, lesser alternative for plying, fixed and variable cost of fleet, protection of the environment.

The transportation model is worked out for company-provided seven plying zones. The optimization model is constructed based on (1) employee demands as per weekly demand chart at various locations (nodes), (2) vehicle types and capacities, (3) vehicle fixed costs and variable costs. The geographical position of the nodes and the actual driving distances are noted from Google maps.

The main assumptions of the model are as follows: (1) A vehicle cannot visit a particular node more than once; (2) One type of vehicle has a fixed capacity and fixed maintenance cost; (3) The transportation cost per kilometer for each vehicle type is predetermined; (5) The total number of vehicle to be used is unlimited; (6) Time restriction and traffic disruptions are ignored.

3.1 Problem Formulation

We first introduce the following notations:

d_{ij}: The distance between node i and j; $d_{ij} = 0$ when $\mathrm{i} = \mathrm{j}$

$$x_{ij}^{k} = \begin{cases} 1 & \text{if vehicle } k \text{ drives from node } i \text{to node } j, \text{ when i} \neq \text{j} \\ 0 & \text{otherwise, when i} = \text{j} \end{cases}$$

c_{ij}^{k}: Transportation cost (TC) per unit distance of the arc between nodes i and j by the kth vehicle
m_k: Fixed cost (FC) for kth vehicle of capacity p
Q_k: Capacity of the kth vehicle
b_i: Demand at node i when $i > 0$; b_0: Demand at central depot $= 0$

The central depot is assigned as node 0 and for every demand at node i $(b_1, b_2, b_3, \ldots, b_n)$, there should be a vehicle of capacity p for the kth vehicle, such that $Q_k = p$ and $p \geq b_i$.

The objective of this problem is to:

$$\text{Minimize} \sum_{i=0}^{n}\sum_{j=0}^{n}\sum_{k=1}^{r} x_{ij}^{k}\, c_{ij}^{k}\, d_{ij}^{k} + \sum_{k=1}^{r} m_k \tag{3}$$

Subject to the following constraints:

$$\sum_{k=1}^{r}\sum_{j=0}^{n} x_{ij}^{k} = 1 \qquad \text{for } i = 1, 2, 3, \ldots, n \quad (i \neq j) \tag{4}$$

$$\sum_{k=1}^{r}\sum_{i=0}^{n} x_{ij}^{k} = 1 \qquad \text{for } j = 1, 2, 3, \ldots, n \quad (i \neq j) \tag{5}$$

$$\sum_{i=0}^{n}\sum_{j=0}^{n} x_{ij}^{k}\, b_{j}^{k} \leq Q_k. \qquad \text{for } k = 1, 2, 3, \ldots, r \quad (i \neq j) \tag{6}$$

Here the problem is formulated as FSMVRP for a mixed fleet of vehicles. The problem formulations (3), (4), (5), and (6) contains a set of n nodes with deterministic demands at pickup nodes that need be catered by a heterogeneous mixed fleet of vehicles with no time constraints. In this formulation the pickup routes originate and terminate at the starting node.

3.2 Development of Algorithm

3.2.1 Augmented Savings Algorithm

The framework of ASA is based on the enhancement of Altınel and Öncan [2] savings formula for homogeneous fleet. This was proposed by Doyuran and Çatay [13] wherein nodes of smaller demand with respect to average demand are prioritized compared to the nodes of higher demand. The ASA proposed by us instead selects the minimum cost for both homogeneous and heterogeneous fleet after introducing fixed and variable cost in the objective function. The algorithm also introduces an option to generate solution using CWS and then proceeds to improve the solution further by the proposed method by tuning the parameters and penalizing the objective function under overcapacity situation. The flowchart of the Algorithm is represented in Fig. 1.

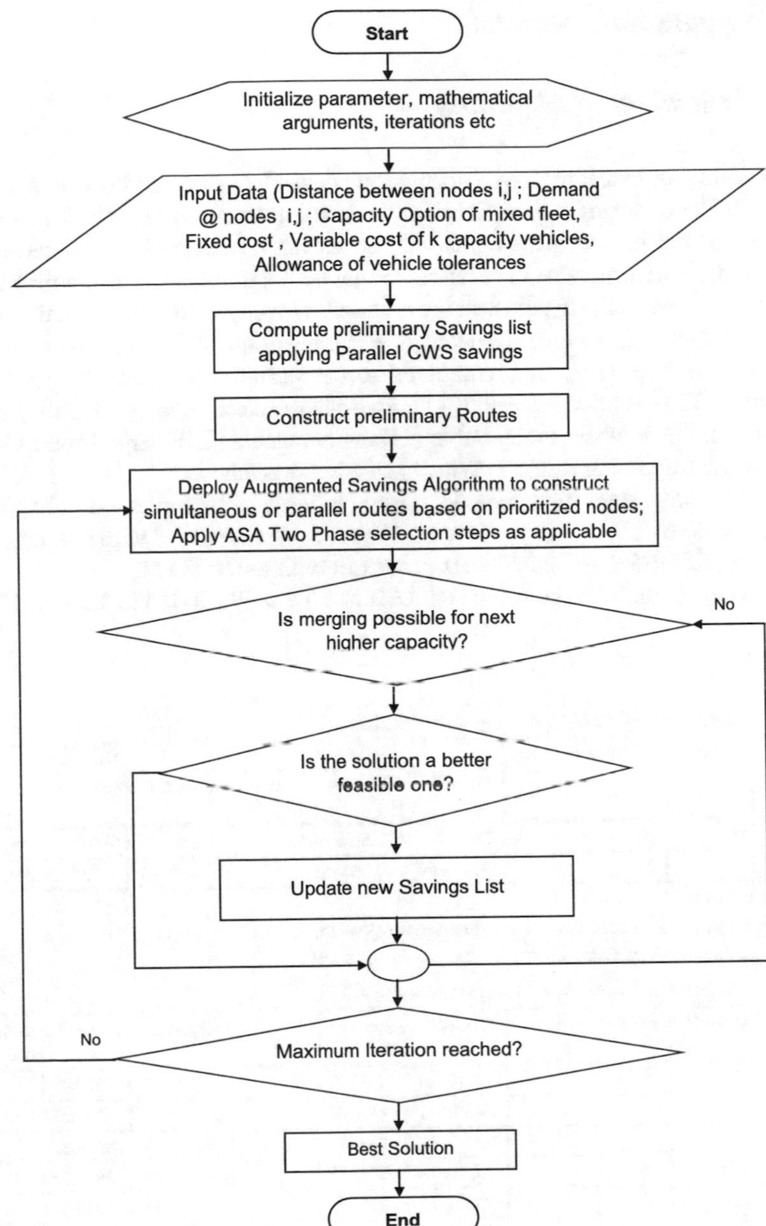

Fig. 1 ASA flowchart

4 Computational Results

4.1 Verification of Algorithm

As a real-time instance, we have collected the data of 72 stops/nodes where the bus service of the company either picks up or drops the employees. The bus service operator has partitioned these nodes into seven main routes; all connected to the central node, i.e., Corporate Office. Here, the demand at the nodes or availability of the vehicles varies periodically, the fleet is unlimited. We propose a near optimal solution for this problem using a heterogeneous fleet composition. The solution obtained by using ASA is improved when compared to our earlier published work applying Two-phase (TP) selection algorithm [1]. Two-phase selection is a probability-based heuristic approach where Pichpibul and Kawtummachai [25] merged routes based on savings list applying Roulette Wheel selection procedure.

The ASA program is implemented in Python, Version 2.7 (open source) software on Windows7, 64-bit Operating System. We used a standard PC with configuration Intel Xeon, CPUE5-2680 @2.50GHz processor with 64GB RAM.

The comparison between results of ASA and TP is illustrated in Table 1, Figs. 2 and 3.

Table 1 Comparison of ASA and two phase algorithm

Routes	Demand	Fleet composition	Using ASA			Using TP		
			Capacity	Distance	Comp. time (s)	Capacity	Distance	Comp. time (s)
Case 1	802	22	24, 45	892.266	1.21	24, 45	954.85	310
Case2	802	20	30, 40, 50	815.93	0.57	30, 40, 50	925.03	316

Case2: Assumptions:

Vehicle capacity up to 35, Variable cost $c_{ij}^{k} = 8$, Fixed cost $m_K = 115$;

Vehicle capacity up to 55 Variable cost $c_{ij}^{k} = 12$, Fixed cost $m_K = 175$

Vehicle capacity >55,Variable cost $c_{ij}^{k} = 15$, Fixed cost $m_K = 200$

Routes	Demand	Fleet	Capacity	Route cost		Capacity	Route cost	%Imp. over TP
Route:1	109	3	50, 50, 30	2118.99		50, 50, 30	2360.48	
Route:2	63	2	50, 30	1127.36		50, 30	1264.63	
Route:3	111	3	50, 50, 30	1635.8		50, 50, 30	2174.27	
Route:4	130	3	50, 50, 40	2934.12		50, 50, 40	2934.12	
Route:5	131	3	50, 50, 40	2255.4		50, 50, 40	2610.6	
Route:6	165	4	50, 50, 50, 30	1723.2		50, 50, 50, 30	1703.39	
Route:7	93	2	50, 50	1074.8		50, 50	1249.8	
Total	802	20	30, 40, 50	12869.6		30, 40, 50	14297.3	9.98

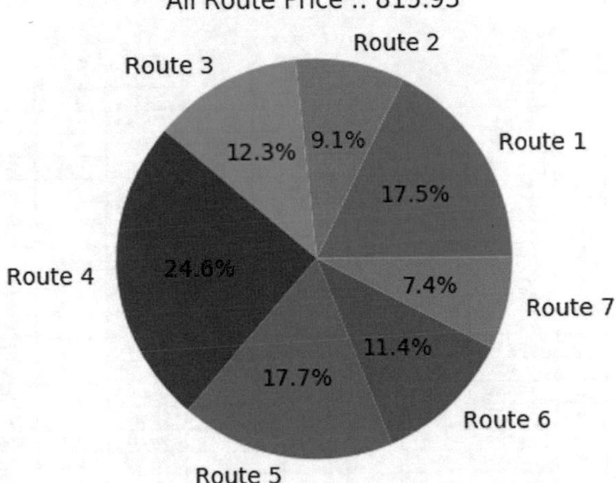

Fig. 2 Route costs using ASA algorithm

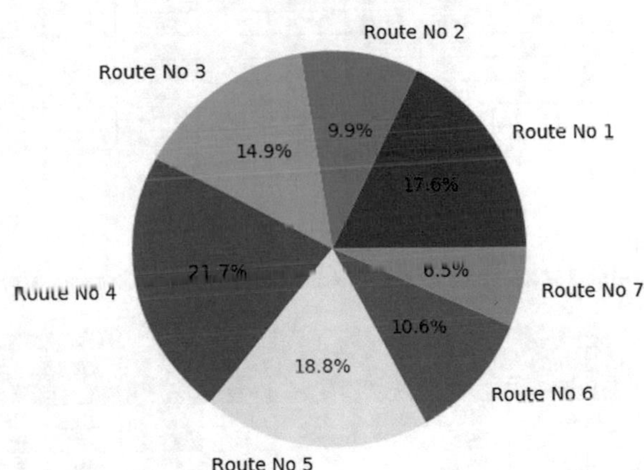

Fig. 3 Route costs using TP algorithm

4.2 *Validation of Algorithm*

To validate the algorithm on some standard benchmarks we used the dataset of Augerat et al. and Christofides et al. accessing from Internet (http://neo.lcc.uma.es/vrp/vrp-instances/capacitated-vrp-instances).

4.2.1 Homogeneous Fleet

The test instances include Augerat et al. (1995) test set of 27 CVRP instances A with 32–80 nodes (n); 5–7 vehicles (k); and Christofides et al. test set (1979) C of 7 instances [13]. Tables 2 and 3 shows the comparison of results by Paessens (p),

Table 2 Comparison of ASA with 'P', 'AO', and 'Robust' Algorithm on Augerat et al.'s data set

Sr No	Instance	Best Known	CW Sol	P	% Imp. over CW	AO	%Impr. over CW	Robust	% Imp. over CW	ASA	%Impr over CW	%var from Robust
1	A-n32-k5	784	843.69	828.70	1.78	828.70	1.78	828.70	1.78	820.19	2.79	1.01
2	A-n33-k5	661	712.05	679.72	4.54	676.10	5.05	676.10	5.05	676.01	5.06	0.01
3	A-n33-k6	742	776.26	747.32	3.73	743.21	4.26	746.99	3.77	765.20	1.42	-2.35
4	A-n34-k5	778	810.41	793.05	2.14	793.05	2.14	793.05	2.14	792.72	2.18	0.04
5	A-n36-k5	799	828.47	806.78	2.62	806.78	2.62	806.78	2.62	806.78	2.62	0
6	A-n37-k5	669	707.81	695.08	1.80	694.43	1.89	694.44	1.89	694.44	1.89	0
7	A-n37-k6	949	976.61	976.01	0.06	974.56	0.21	976.61	0	968.13	0.87	0.87
8	A-n38-k5	730	768.13	755.94	1.59	756.11	1.56	755.94	1.59	757.84	1.34	-0.25
9	A-n39-k5	822	901.99	851.25	5.63	848.24	5.96	843.23	6.51	848.25	5.96	-0.55
10	A-n39-k6	831	863.08	849.55	1.57	849.56	1.57	849.90	1.53	848.93	1.64	0.11
11	A-n44-k7	937	976.04	968.84	0.74	959.43	1.70	957.03	1.95	949.71	2.7	0.75
12	A-n45-k6	944	1006.45	957.05	4.91	957.06	4.91	957.06	4.91	957.06	4.91	0
13	A-n45-k7	1146	1199.98	1169.00	2.58	1166.39	2.80	1168.97	2.58	1165.24	2.89	0.31
14	A-n46-k7	914	939.74	933.66	0.65	933.66	0.65	929.42	1.1	926.41	1.42	.32
15	A-n48-k7	1073	1112.82	1104.23	0.77	1104.24	0.77	1103.99	0.79	1103.18	0.87	0.08
16	A-n53-k7	1010	1099.45	1045.98	4.86	1045.47	4.91	1048.79	4.61	1041.93	5.23	0.62
17	A-n54-k7	1167	1197.92	1188.64	0.77	1173.77	2.02	1172.27	2.14	1185.79	1.01	-1.13
18	A-n55-k9	1073	1099.84	1099.55	0.03	1098.51	0.12	1099.56	0.03	1098.91	0.09	0.06
19	A-n60-k9	1354	1421.88	1389.59	2.27	1376.20	3.21	1379.86	2.96	1378.85	3.03	0.07
20	A-n61-k9	1034	1102.23	1051.37	4.61	1051.10	4.64	1051.06	4.64	1051.06	4.64	0
21	A-n62-k8	1288	1352.81	1351.11	0.13	1347.87	0.37	1326.54	1.94	1338.39	1.07	-0.87
22	A-n63-k10	1314	1352.48	1349.58	0.21	1348.17	0.32	1347.30	0.38	1344.40	0.6	0.22
23	A-n64-k9	1401	1486.92	1442.44	2.99	1439.75	3.17	1442.66	2.98	1433.80	3.57	0.59
24	A-n63-k9	1616	1687.96	1648.92	2.31	1649.14	2.30	1652.42	2.11	1646.26	2.47	0.36
25	A-n65-k9	1174	1239.42	1224.71	1.19	1202.08	3.01	1197.49	3.38	1202.98	2.94	-0.44
26	A-n69-k9	1159	1210.78	1185.08	2.12	1185.08	2.12	1181.91	2.38	1188.19	1.87	-0.51
27	A-n80-k10	1763	1860.94	1818.64	2.27	1816.78	2.37	1811.56	2.65	1806.68	2.92	0.277
	Average				2.18		2.46		2.53		2.52	-0.0151

Table 3 Comparison of ASA with 'P', 'AO', and 'Robust' Algorithm on Christofides et al.'s data set

Sr No	Instance	Best Known	CW Sol	P	% Imp. over CW	AO	%Impr. over CW	Robust	% Imp. over CW	ASA	%Impr over CW	%var from Robust
1	C50	524.61	584.64	566.10	3.17	555.55	4.98	537.29	8.10	532.51	8.92	0.82
2	C75	835.26	907.39	866.29	4.53	860.21	5.20	864.29	4.75	866.30	4.53	-0.22
3	C100a	826.14	889.00	865.60	2.63	867.35	2.44	854.49	3.88	866.87	2.49	-1.39
4	C150	1028.42	1140.42	1101.81	3.39	1094.06	4.07	1089.78	4.44	1093.79	4.09	-0.35
5	C199	1291.45	1395.74	1370.04	1.84	1359.78	2.58	1367.53	2.02	1360.79	2.50	0.48
6	C120	1042.11	1068.14	1066.40	0.16	1057.80	0.97	1059.87	0.77	1062.85	0.49	-0.28
7	C100b	819.56	833.51	826.00	0.90	824.66	1.06	825.76	0.93	824.23	1.11	0.18
	Average				2.37		3.04		3.56		3.54	-0.1087

Altinel and Oncan (AO), Doyuran and Catay (Robust) and the ASA on the above datasets with fixed cost $m_k = 0$ and variable cost $c_{ij}^k = 1$ respectively. Figures 4 and 5 depicts the route distances and graphical presentation for the instance C199 (CMT5) of Christofides et al. test set.

ASA algorithm Results : Homogeneous Fleet
Percentage Improvement with respect to CWS: 2.504075155656729
Lambda = 2.35, Mu = 0.44, Nu = 0.49

Route no	Cap.	Demand	Dist.	Cost	Route
1	200	88	23.23	23.23	[6, 61, 96, 188]
2	200	194	75.52	75.52	[8, 102, 178, 78, 35, 108, 69, 180, 132, 7, 51]
3	200	199	95.5	95.5	[103, 5, 88, 37, 138, 166, 20, 124, 154, 15, 79, 153]
4	200	194	83.99	83.99	[16, 159, 67, 183, 116, 62, 185, 115, 160, 23, 104]
5	200	198	65.06	65.06	[17, 76, 40, 130, 12, 50, 169, 129, 71, 100, 26, 81, 168]
6	200	192	113.6	113.6	[19, 177, 133, 14, 84, 134, 85, 164, 170, 11, 52, 150]
7	200	198	84.03	84.03	[30, 120, 171, 47, 155, 36, 122, 174, 82, 121, 140, 94]
8	200	194	50.38	50.38	[33, 105, 195, 53, 198, 192, 197, 136, 1, 191, 158, 194, 193]
9	200	192	76.03	76.03	[152, 54, 58, 27, 83, 123, 13, 128, 70, 167, 99, 179, 111]
10	200	200	112.57	112.57	[57, 189, 75, 162, 31, 131, 80, 10, 77, 165, 38, 119]
11	200	197	51.84	51.84	[127, 87, 4, 34, 65, 176, 46, 175, 149, 60, 126]
12	200	184	129.86	129.86	[72, 118, 56, 25, 110, 163, 148, 92, 135, 146, 18, 73, 145]
13	200	187	85.79	85.79	[86, 101, 28, 64, 93, 156, 114, 142, 91, 141, 22, 186, 196, 66, 112]
14	200	200	98.18	98.18	[199, 89, 137, 113, 42, 68, 143, 90, 41, 190, 43, 184]
15	200	196	67.87	67.87	[95, 187, 109, 39, 9, 161, 97, 32, 106, 44, 55, 3, 151]
16	200	181	79.51	79.51	[117, 181, 147, 63, 107, 24, 144, 74, 49, 182]
17	200	192	67.72	67.72	[125, 98, 45, 29, 59, 48, 21, 173, 172, 139, 2, 157]

Total Distance: 1360.785819835298
Final Cost: 1360.79
Percentage Vacancy = 6.29%

Fig. 4 Results-homogeneous fleet-instance C199

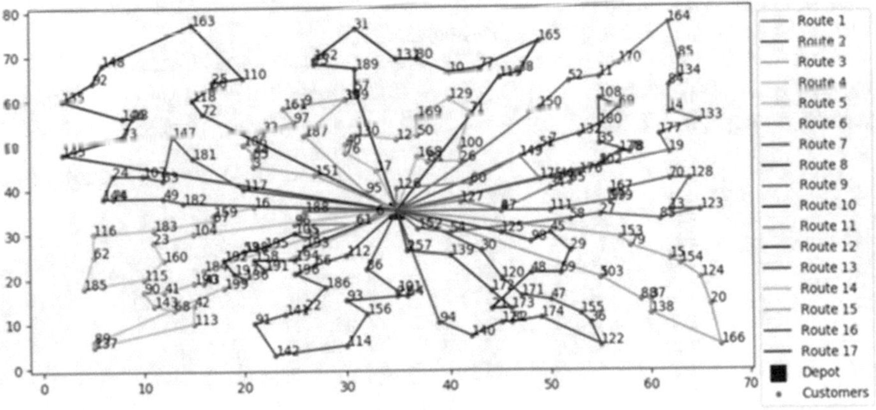

Fig. 5 Graphical representation-instance C199

4.2.2 Heterogeneous Fleet

The test instances include Problem nos. 16–20 from the dataset for Mixed Fleet instances as mentioned in the literature [27]. We have tabulated the results separately for Fixed Cost (Table 4) and for Variable Cost (Table 5) for the above 5 FSMVRP instances.

Table 4 Comparison of ASA results with classical and metaheuristic results for FSMVRP with fixed cost

Sr. No.	Inst-ance	Best Known	CWS	Classical Heuristics			Meta-Heuristics			ASA	% imp. over CWS
				Golden et al ($ROS-\gamma$) (1984)	Desrochers and Verhoog (1991)	Kenan KARAGUL (2014)	Osman & Salhi (1996)	Taillard (1999)	Gendreau (1999)		
1	P-16	2720.43	2990	2882	2809	2899	2745.04	2741.50	2743.96	2749.96	8.03
2	P-17	1744.83	1805.65	1958	1877	1954	1766.81	1749.50	1752.29	1800.45	0.29
3	P-18	2371.49	2587.36	2520	2488	2986	2439.40	2381.43	2392.57	2504.55	3.20
4	P-19	8659.74	8741.06	8744	8700	9824	8704.20	8675.16	8682.50	8685.9	0.63
5	P-20	4039.49	4348.12	4291	4249	4498	4166.03	4086.76	4100.20	4210.1	0.87

Table 5 Comparison of ASA results with classical and metaheuristic results for FSMVRP with variable cost

Sr. No.	Inst-ance	Best Known	CWS	Classical Heuristics			Meta-Heuristics			ASA	% imp. over CWS
				No notable instances available			Wassan Osman (2002)	Taillard (1999)	Gendreau (1999)		
1	P-16	1131.00	1314.00	–	-		1131.00	1144.39	1137.01	1192.62	9.24
2	P-17	1031.00	1133.48	—			1047.74	1044.93	1046.36	1107.16	2.32
3	P-18	1801.40	2063.08	—			1814.11	1831.24	1812.00	1998.75	3.12
4	P-19	1100.56	1228.09	—			1100.56	1110.96	1107.09	1192.2	2.92
5	P-20	1530.16	1765.62	—			1530.16	1550.36	1553.72	1717.68	2.72

ASA algorithm Results
Heterogeneous Fleet – Fixed Cost
Percentage Imp with respect to CWS: 3.200694804206118
Lambda = 1.4095961079235497, Meu = 1.180492645792201, Neu = -0.3569162559789722

Route no	Cap.	Demand	Distance	Cost	Route
1	150	141	89.13	269.13	[2, 74, 61, 28, 47, 48, 30]
2	150	145	106.27	286.27	[21, 69, 36, 71, 60, 70, 20, 37, 5, 29]
3	150	149	101.85	281.85	[7, 53, 38, 58, 10, 31, 39]
4	150	144	83.47	263.47	[52, 27, 15, 57, 13, 54, 19, 8, 46]
5	150	145	81.35	261.35	[51, 16, 49, 24, 3, 40, 17, 26]
6	150	145	110.81	290.81	[63, 23, 56, 43, 41, 42, 64, 22, 62]
7	150	146	97.01	277.01	[44, 32, 50, 18, 55, 25, 9, 12]
8	100	100	34.91	134.91	[4, 45, 34, 67]
9	150	149	106.79	286.79	[35, 14, 59, 11, 66, 65, 72]
10	100	100	52.95	152.95	[6, 33, 1, 73, 68, 75]

Total Distance: 864.5459533151544
Final Cost 2504.55
Percantage Vacancy = 2.57%

Fig. 6 Results-heterogeneous fleet-fixed cost-problem 18

Below, we illustrate solutions of Heterogeneous fleet for fixed and variable cost using Taillard instance P–18 (Fig. 6) and P–17 (Fig. 7).

ASA algorithm Results
Heterogeneous Fleet – Variable Cost
Percentage Imp with respect to CWS: 2.3214682388918524
Lambda = 1.872354824862203, Meu = 0.6854588697008646, Neu = 0.2640727900117109

Route no	Cap.	Demand	Distance	Cost	Route
1	200	194	134.11	201.17	[62, 22, 64, 42, 41, 43, 56, 23, 49, 24, 3, 44]
2	200	191	70.98	106.47	[4, 45, 29, 48, 47, 74, 30, 2, 68, 75]
3	200	198	139.5	209.25	[28, 61, 21, 69, 36, 71, 60, 70, 20, 37, 5, 15, 57]
4	120	112	62.22	74.67	[6, 33, 73, 1, 63, 16, 51]
5	200	199	101.89	152.84	[7, 53, 14, 59, 11, 66, 65, 38]
6	200	197	131.87	197.81	[12, 72, 10, 31, 39, 9, 25, 55, 18, 50, 32]
7	120	92	49.03	58.84	[26, 58, 40, 17]
8	200	181	70.74	106.12	[34, 52, 27, 13, 54, 19, 35, 8, 46, 67]

Total Distance: 760.3604170914775
Final Cost: 1107.16

Fig. 7 Results-heterogeneous fleet-variable cost-problem 17

5 Conclusion

Our proposed work is an extension of the Savings-based classical heuristics, covering both homogenous and heterogeneous fleet. The ASA algorithm contributes to the research on VRP in two levels: (i) It's a version of enhanced CWS to solve mixed fleet VRP with single trips and an unlimited number of vehicles (ii) The ASA is a potential framework to study other variants of GCVRP with insertion of applicable boundary constraints and tuning parameters. Tables 2 and 3 represents application of ASA on homogeneous fleet, while Tables 4 and 5 represents that for heterogeneous fleet. The results in Table 4 (column 5, 6, 7) prove the stability and superiority of our algorithm when compared with the solution provided by Golden et al. [18] Desrochers and Verhoog [12], Karagül [20]. It is worthwhile to mention here that the average performance of ASA is 2.22, 0.84 and 11.07% improved over the average performance of results in column number '5', '6', and '7' respectively. Here we have considered the results obtained using the algorithm *ROSgamma* [18] and best of CM, OOM, ROM, $ROM-\gamma$, $ROM-\rho$ [12] in column numbers '5' and column number '6' respectively, all derived introducing variants of CWS algorithm.

The ASA enhancement is novel in the sense that the algorithm provides a broader framework to address both real-time and the benchmark problems at the same time for the mixed fleet with both fixed and variable cost. The ASA algorithm is developed with many advanced user options, wherein the program automatically selects the most suitable value of the parameter from a user-defined interval, provides input for fixed and variable cost for the heterogeneous fleet and applies intermittent termination criteria, etc. This, to our knowledge, is the only attempt which adopts a savings-based improved classical local search heuristics including Robust algorithm, to define a modified algorithm, ASA for solving complex GCVRP problems.

6 Future Work

Tables 4 and 5 suggest that the results obtained for Fixed and Variable cost using the ASA are better than that by other classical heuristic methods. A close study also suggests that the ASA results are comparable with the results obtained using notable metaheuristics methods, i.e., Column Generation Technique, Tabu Search Methods, Evolutionary Algorithm, etc. proposed in the last twenty years for FSMVRP as illustrated in Tables 4 and 5. This shows that the results of the ASA have a scope of improvement when metaheuristic methods are applied.

Our work can further be extended to other variants of GCVRP, e.g., multiple vehicles with various type of loading features; the loading constraints such as road traffic conditions, multiple depot and limited fleet size. Time window can be applied to more complex real-time problems and issues related to environment protection and actual operation of urban distribution. Young researchers may further extend this simple classical heuristic-based general framework to solve complex nonlinear practical problems applying high-end hybrid computation methods.

References

1. Ajit, K., Barnali, S.: Optimizing heterogeneous fleet vehicle routing problem. Int. J. Adv. Res. Sci. Eng. **5**(08), 542–551 (2016)
2. Altınel, İ .K., Öncan, T.: A new enhancement of the clarke and wright savings heuristic for the capacitated vehicle routing problem. J. Oper. Res. Soc. **56**(8), 954–961 (2005)
3. Altinkemer, K., Gavish, B.: Parallel savings based heuristics for the delivery problem. Oper. Res. **39**(3), 456–469 (1991)
4. Baldacci, R., Battarra, M., Vigo, D.: Routing a heterogeneous fleet of vehicles. In: The Vehicle Routing Problem: Latest Advances and New Challenges, pp. 3–27 (2008)
5. Brandão, J.: A tabu search algorithm for the heterogeneous fixed fleet vehicle routing problem. Comput. Oper. Res. **38**(1), 140–151 (2011)
6. Choi, E., Tcha, D.-W.: A column generation approach to the heterogeneous fleet vehicle routing problem. Comput. Oper. Res. **34**(7), 2080–2095 (2007)
7. Clarke, G., Wright, J.W.: Scheduling of vehicles from a central depot to a number of delivery points. Oper. Res. **12**(4), 568–581 (1964)
8. Corominas, A., García-Villoria, A., Pastor, R.: Fine-tuning a parametric clarke and wright heuristic by means of eagh (empirically adjusted greedy heuristics). J. Oper. Res. Soc. **61**(8), 1309–1314 (2010)
9. Corominas, A., García-Villoria, A., Pastor, R.: Improving parametric clarke and wright algorithms by means of iterative empirically adjusted greedy heuristics. SORT **38**(1), 3–12 (2014)
10. Dantzig, G.B., Ramser, J.H.: The truck dispatching problem. Manag. Sci. **6**(1), 80–91 (1959)
11. Desrochers, M., Verhoog, T.W.: A matching based savings algorithm for the vehicle routing problem. In: Cahiers du GERAD (1989)

12. Desrochers, M., Verhoog, T.W.: A new heuristic for the fleet size and mix vehicle routing problem. Comput. Oper. Res. **18**(3), 263–274 (1991)
13. Doyuran, T., Çatay, B.: A robust enhancement to the clarke-wright savings algorithm. J. Oper. Res. Soc. **62**(1), 223–231 (2011)
14. Fisher, M.L., Jaikumar, R.: A generalized assignment heuristic for vehicle routing. Networks **11**(2), 109–124 (1981)
15. Gaskell, T.J.: Bases for vehicle fleet scheduling. J. Oper. Res. Soc. **18**(3), 281–295 (1967)
16. Gendreau, M., Laporte, G., Musaraganyi, C., Taillard, E.: A tabu search heuristic for the heterogeneous fleet vehicle routing problem. Comput. Oper. Res. **26**(12), 1153–1173 (1999)
17. Gillett, B.E., Miller, L.R.: A heuristic algorithm for the vehicle-dispatch problem. Oper. Res. **22**(2), 340–349 (1974)
18. Golden, B., Assad, A., Levy, L., Gheysens, F.: The fleet size and mix vehicle routing problem. Comput. Oper. Res. **11**(1), 49–66 (1984)
19. Golden, B.L., Magnanti, T.L., Nguyen, H.Q.: Implementing vehicle routing algorithms. Technical report, Massachusetts Inst Of Tech Cambridge Operations Research Center (1975)
20. Karagül, K.: A new heuristic routing algorithm for fleet size and mix vehicle routing problem. Gazi Univ. J. Sci. **27**(3), 979–986 (2014)
21. Laporte, G., Gendreau, M., Potvin, J.-Y., Semet, F.: Classical and modern heuristics for the vehicle routing problem. Int. Trans. Oper. Res. **7**(4–5), 285–300 (2000)
22. Nelson, M.D., Nygard, K.E., Griffin, J.H., Shreve, W.E.: Implementation techniques for the vehicle routing problem. Comput. Oper. Res. **12**(3), 273–283 (1985)
23. Osman, I.H., Salhi, S.: Local search strategies for the vehicle fleet mix problem. In: Modern heuristic search methods, pp. 131–153 (1996)
24. Paessens, H.: The savings algorithm for the vehicle routing problem. Eur. J. Oper. Res. **34**(3), 336–344 (1988)
25. Pichpibul, T., Kawtummachai, R.: New enhancement for clarke-wright savings algorithm to optimize the capacitated vehicle routing problem. Eur. J. Sci. Res. **78**(1), 119–134 (2012)
26. Prins, C.: Two memetic algorithms for heterogeneous fleet vehicle routing problems. Eng. Appl. Artif. Intell. **22**(6), 916–928 (2009). https://doi.org/10.1016/j.engappai.2008.10.006. Artificial Intelligence Techniques for Supply Chain Management. ISSN 0952-1976
27. Taillard, E.: A heuristic column generation method for the heterogeneous fleet vrp. RAIRO-Oper. Res. **33**(1), 1–14 (1999)
28. Toth, P., Vigo, D.: The vehicle routing problem, ser. In: Monographs on Discrete Mathematics and Applications. SIAM (2001)
29. Toth, P., Vigo, D.: Vehicle Routing: Problems, Methods, and Applications. SIAM (2014)
30. War, P., Holt, J.: A repeated matching heuristic for the vehicle routeing problem. J. Oper. Res. Soc. 1156–1167 (1994)
31. Yellow, P.C.: A computational modification to the savings method of vehicle scheduling. Oper. Res. Q. (1970–1977) **21**(2), 281–283 (1970)

Near-Duplicate Document Detection Using Semantic-Based Similarity Measure: A Novel Approach

Rajendra Kumar Roul and Jajati Keshari Sahoo

Abstract Detection of near-duplicate pages, especially based on their semantic content is a relevant concern in information retrieval. It is needed to avoid redundancy in the search results against a query as well as facilitate the ranking of the documents in the order of their semantic similarities. Although much work has been done in near-duplicate page detection based on content similarity (as evident in existing literature), the realm of semantic similarity provides a relatively unexplored pool of opportunities. In this paper, a novel technique is proposed to detect whether two documents belonging to a corpus have near-duplicate semantic content or not and a heuristic method is introduced to rank the documents based on their semantic similarity scores. This objective is achieved by examining the proposed technique for computing semantic-based similarity between two documents and applying an averaging mechanism to associate a similarity score to each document in the corpus. The empirical results on DUC datasets witness the effectiveness of the proposed approach.

Keywords Duplicate · Lemmatization · Near-duplicate · Ranking · Semantic

1 Introduction

Duplicate is an inherent problem of every search engine and it became an interesting problem in the late 1990s [1]. If two documents have similar content then they are considered as duplicates. Similarly, documents having minor modifications but are similar to a maximum range are known to be near-duplicates such as times-

R. K. Roul (✉)
Department of Computer Science, Thapar Institute of Engineering and Technology, Patiala 147004, Punjab, India
e-mail: raj.roul@thapar.edu

J. K. Sahoo
Department of Mathematics, BITS-Pilani, K. K. Birla Goa Campus, Zuarinagar 403726, Goa, India
e-mail: jksahoo@goa.bits-pilani.ac.in

H. S. Behera et al. (eds.), *Computational Intelligence in Data Mining*, Advances in Intelligent Systems and Computing 990,
https://doi.org/10.1007/978-981-13-8676-3_46

tamps, advertisements, and counters. In the recent years, near-duplicate document detection has received more attention due to the overwhelming requirements [2–6]. Near-duplicate documents are not bit-wise similar although they display striking similarities. It has been observed from a study which includes 238,000 hosts that 10% of these hosts are mirrored [7]. There are many standard techniques available which can easily identify the duplicate documents, but identifying the near-duplicate documents is a tough task. Hence, detecting and eliminating near-duplicate documents still remains unanswered [8]. There are various reasons why near-duplicate documents exist on the web and some of the reasons are (i) different interpretation of the same events (ii) while reprinting the documents, people may modify its content which also known as typographical errors (iii) main content of two documents are the same but the forum or blogs are replied differently, for example, two documents having different text size, font, bold types, etc., but having same content (iv) documents having different versions and plagiarism (v) due to reprinted noise. Semantic similarity is a metric defined over a set of documents where different mathematical tools are used to estimate the strength of the semantic relationship between concepts or topics [9]. To compute the similarity between a pair of documents, the semantic content or likeness of their meaning is compared which facilitate the classification of documents into topics. It also helps to rank the documents based on their similarity scores. The notion of distance between a pair of documents is most commonly used to characterize this metric.

Most of the research works have used content-based similarity to detect the near-duplicate documents. However, detection of near-duplicate documents using semantic similarity has much importance due to the semantic nature of the web. Researchers have focused on semantic-based work to detect near-duplicates, but ignore to rank the documents in order to find the percentage of similarities between the documents. The proposed approach not only detects the near-duplicates based on their semantic similarity but also ranks the documents to catch the percentage of similarities with respect to all the documents of a corpus. To achieve this, four well-known semantic techniques (Word2Vec, WordNet, Normalized Google Distance, and Latent Dirichlet Allocation (LDA)) are used for computing the similarity scores between pairs of documents in the corpus. These scores are used as the features for training different traditional classifiers to generate document pairs with semantic similarity scores. A heuristic technique is proposed to rank the documents by taking a weighted average of the features from the classifier and then computing an aggregate score of each document. Experimental results on four categories of DUC datasets show that the proposed approach is promising. Remaining part of the paper is as follows: Sect. 2 describes the detailed mechanism for detecting the near-duplicate documents. Results and analysis of the experiment are carried out in Sect. 3. The paper is concluded in Sect. 4 with some future enhancement.

2 Proposed Approach

The following steps are used to detect the near-duplicate documents and rank them in a given corpus *P*.

Step 1: *Preprocessing of the documents*:
Consider a corpus *P* having a set of classes $C = \{c_1, c_2, \ldots, c_n\}$ and documents $D = \{d_1, d_2, \ldots, d_p\}$. The documents are preprocessed which includes removal of stop words, stemming,[1] lemmatizating[2] and converting the documents to a Bag-Of-Words format where each word is stored along with its word count in a dictionary. Documents use different forms of a word, such as *operate, operating, operation* and *operative*. Stemming and lemmatization are done separately because we require the dictionary of stemmed words for Word2Vec and LDA analysis while the lemmatized dictionary of words is needed to perform NGD and WordNet techniques. Stemming collapses related words by a heuristic process that chops off the ends of the words resulting in a term which may not be present in the English dictionary. Lemmatization on the other hand, collapses the different inflectional forms of a word, resulting in a base term of a dictionary word. Then a new corpus of the documents is generated from the given dataset by pooling the preprocessed lists of documents together. Thus, two corpuses are created—one for the stemmed words and another for the lemmatized words.

Step 2: *Calculating the similarity scores*:
The semantic similarity between a pair of documents are calculated using four semantic techniques as discussed below:

i. *Word2Vec*:
It is used to make word embeddings by grouping similar words together in the vector space without human intervention. It captures the linguistic contexts of the words using two-layer neural network architecture. Its input is a text corpus and produces a vector space, i.e., feature vectors for words in the corpus. Word2vec generates the exact meaning of a word based on the usage and content of the dataset assigned to it [10]. These guesses are very useful to us as it can be used to establish associations between words (e.g., "man" map to "king", "woman" map to "queen") or cluster the documents and classify them. The vectors formed by word2vec are called *neural word embeddings*. These vectors are representations of the words which makes natural language computer readable and is shown in the Fig. 1. Word2vec makes word embeddings in two ways, namely *continuous bag of words (CBOW)* and *skip-gram* models [10]. The proposed approach used *CBOW* model and the reason for choosing *CBOW* over *skip-gram* model is that *CBOW* model is several times faster to train than the *skip-gram*, slightly better accuracy for the frequent words. To work on this model, first the word

---

[1]https://tartarus.org/martin/PorterStemmer/.

[2]https://algorithmia.com/algorithms/StanfordNLP/Lemmatizer.

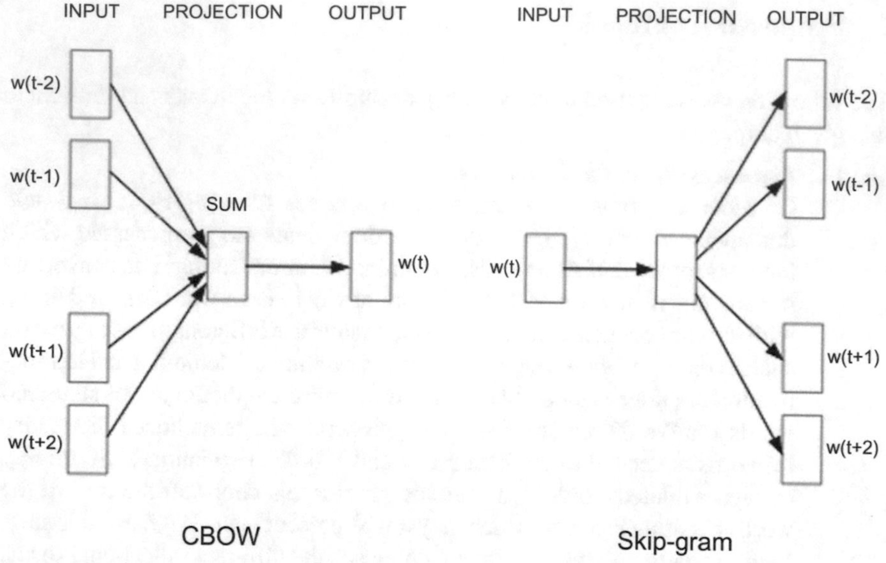

Fig. 1 Word2vec similarity

embeddings are formed using word2vec. The similarity between d_i and d_j is computed by considering the average of the maximum similarity between the words of the two documents d_i and d_j as shown in the Eq. 1.

$$word2vec(d_i, d_j) = \frac{\sum_{x \in d_i} max_{y \in d_j} cosine\_similarity(d_i, d_j)}{length\ of\ d_i} \tag{1}$$

ii. *WordNet*:
It is a huge lexical database of English where nouns, verbs, adverbs, and adjectives are clustered into sets of cognitive synonyms (synsets), each represents a different concept [11]. The main relationship among words in WordNet is synonymy, for example the synonymy that exists between "boat" and "ship" will be captured in WordNet. To calculate the semantic similarity score between two documents using WordNet, first the synsets are found from the words of each document. The synsets are then compared using the similarity measure provide by the WordNet and average of the maximum value for each synset is taken as the similarity score between the pair of documents. The threshold of similarity score between two documents is decided as 0.3 [3] Eq. 2 is used to calculate the WordNet similarity score numerically.

[3] decided experimentally for DUC-2001 dataset on which the better result is obtained and thresholds decided for different DUC datasets are different.

$$wordnet(d_i, d_j) = \frac{\sum_{x \in d_i} \max_{y \in d_j} similarity(synset(d_i), synset(d_j))}{length\ of\ d_i} \quad (2)$$

iii. *Normalized Google Distance (NGD)*:
NGD measures the distance between two terms by comparing the number of web hits returned by the Google search engine [12]. Here, the nouns (noun extraction from a document is done using NLTK[4]) of the documents are used for measuring the NGD value between them. The NGD value between two nouns n_p and n_q is calculated as follows:

$$NGD(n_p, n_q) = \frac{\max(\log H(n_p), \log H(n_q)) - \log H(n_p, n_q)}{\log T - min(\log H(n_p), \log H(n_q))}$$

where T represents the number of web documents retrieved by Google, $H(n)$ and $H(n_p, n_q)$ are the number of hits returned by Google for one noun and two nouns, respectively. The similarity between each pair of nouns is computed by subtracting the NGD from 1 (NGD lies between 0 and 1) as follows:

$$NGD\text{-}similarity(n_p, n_q) = 1 - NGD(n_p, n_q)$$

The similarity between two documents d_i and d_j having total number of nouns M_p and M_q, respectively is computed by counting the noun-noun pair (noun-noun pair is formed by taking a noun from d_i and another noun from d_j) whose NGD similarity values is above a certain threshold α (it is set to 0.61 as decided by experiment for which the better result is obtained) divided by total number of noun-noun pairs as given in the Eq. 3.

$$NGD\text{-}similarity(d_i, d_j) = \frac{count(NGD\text{-}simialrity(n_p, n_q)) > \alpha}{M_p \times M_q} \quad (3)$$

iv. *Latent Dirichlet allocation*:
LDA [13] is a huge lexical database of English where nouns, verbs, adverbs, and adjectives are clustered into sets of cognitive synonyms (synsets), each represent a different concept [14]. Hellinger Distance[5] is used to calculate the distance and similarity (1-Hellinger Distance). Hellinger distance is widely used to compute the distance between two probability distributions and used here for the LDA model.

Step 3: *Training the classifier*:
The document pairs are run on different traditional classification models after obtaining the semantic similarity scores for the above four semantic measures, which are used as the feature vector for the classifier. The dataset categorizes the documents into different folders, each belonging to a partic-

[4]https://www.nltk.org/.

[5]www.encyclopediaofmath.org/index.php?title=Hellinger_distance&oldid=16453.

Table 1 Similarity vectors with class label of a pair of documents

$$\begin{bmatrix} d_{12} =< ngd_{12}, word2vec_{12}, wordnet_{12}, lda_{12}, C_{12} > \\ d_{13} =< ngd_{13}, word2vec_{13}, wordnet_{13}, lda_{13}, C_{13} > \\ \cdots \\ \cdots \\ \cdots \\ d_{ij} =< ngd_{ij}, word2vec_{ij}, wordnet_{ij}, lda_{ij}, C_{ij} > \end{bmatrix}$$

ular topic based on semantic content. Hence, the label for classification is given as "1" if the pair of documents given as input belong to the same folder and "0" otherwise. In other words, different sets are contained in different folders with documents within the same folder as *near-duplicates* to each other while those in two different folders are *not near-duplicates*. The near-duplicate and not near-duplicate documents are classified into class "1" and class "0", respectively. Table 1 shows the similarity vectors between each pair of documents where d_{ij} represents documents d_i and d_j and C_{ij} is the class label for d_i and d_j.

Step 4: *Ranking the near-duplicate documents based on semantic similarity score*: After the classifiers get trained, the decision of that classifier is considered which gives the highest F-measure. The output can be interpreted as a feature matrix, which is a list of feature vectors for each near-duplicate document pair (as decided by the classifier). This feature matrix is used to find the ranking of near-duplicate documents. Now, the aggregate semantic similarity score between the pair of near-duplicate documents is computed by taking an *average (AVG)* of the scores obtained from the four semantic measures used for each document pair and it is shown in Eq. 4.

$$Semantic\text{-}similarity(d_i, d_j)_{AVG} = \frac{\sum_{i=1}^{4} a[i]}{4} \tag{4}$$

where, $a[i]$ is the score of each semantic algorithm between d_i and d_j. Alternatively, a *weighted average* of the scores from the four measures is decided as the aggregate similarity between each pair of documents. These weights are determined from the parameter thresholds set by the classifier for its input. We then compute the semantic similarity score of each document with the entire set of documents in the corpus individually, by averaging the aggregate similarity score of document d_i with all the other documents it is paired with and this gives us the composite similarity score of the document d_i. Equation 5 illustrates this calculation where N is the total number of documents in the corpus P.

$$Semantic\text{-}similarity(d_i)_{AVG} = \frac{\sum_{j \in P} Semantic\text{-}similarityty(d_i, d_j)_{AVG}}{N} \tag{5}$$

The documents are ranked in a descending order based on their semantic similarity scores and then top K documents having the highest similarity scores are picked up. Algorithm 1 illustrates the details.

Algorithm 1: Ranking of near-duplicate documents

Input: Feature matrix (FM) of near-duplicate documents and the corpus P
Output: Ranking of near-duplicate documents
$Semantic_{avg}[] \leftarrow \phi$ //stores the average of semantic similarity between two documents
$Net\text{-}semantic_{avg}[] \leftarrow \phi$ //stores the overall semantic average of a document
$final\_list \leftarrow \phi$
for all (d_i, d_j) pairs $\in$ FM **do**
    $Semantic_{avg}[d_i, d_j] \leftarrow$ avg. of the score of each semantic algorithm between d_i and d_j
end for
for all $d_i \in$ FM **do**
    for all $d_j \in P$ **do**
        $Net\text{-}semantic_{avg}[d_i] \leftarrow$ overall semantic average of d_i with respect to P
    end for
end for
//Ranking of near-duplicates documents
sort all $d_i \in P$ in descending order based on their $Net\text{-}semantic_{avg}[d_i]$ and select top K documents.
$final\_list \leftarrow$ top K documents
return $final\_list$

3 Experimental Analysis

Different DUC[6] (Document Understanding Conference) datasets are used for experimental purpose. DUC datasets follow a set of naming conventions to name their folders and subsequent documents within the folders. For the purposes of convenience, the folders are renamed in incremental order (from 1–30 (in DUC-2000, 2001, and 2002) and 1–50 (in DUC-2005)) and the documents in a similar fashion. The documents of each DUC dataset are preprocessed and two dictionary files are created from the documents, one containing the stemmed words and the other containing the lemmatized form of the words. To find the similarity scores between the documents, four semantic similarity techniques such as Word2Vec, WordNet, NGD, and LDA are used. For WordNet, top 30 terms from each document were selected to represent the document and a similarity score was found based on these 30 terms. WordNet is a computationally intensive task as it deals with a large dataset and it

[6] http://www.duc.nist.gov.

Table 2 Performances on DUC-2000

Classifier	DUC-2000			
	Precision	Recall	F-Measure	Accuracy
SVM	77.43	79.64	**78.52**	**85.68**
*k*NN	54.87	56.49	55.67	67.35
G-NB	60.97	62.38	61.67	73.28
M-NB	66.56	68.01	67.28	76.52
B-NB	60.29	56.47	58.32	65.35
DT	56.73	52.47	54.52	67.67
RF	68.30	70.49	69.38	80.35
GB	66.69	64.00	65.32	76.42
ET	70.56	70.48	70.52	79.37

takes a large amount of time, and hence to speedup the process we take a representative sample of terms. Similarly, NGD also has huge constraints and a mechanism is used to cache the results, enabling us to get the desired results in lesser time. A pickle file is created to run the document terms in batches and get their similarity values to increase the speed of the output. 80% of the documents from each DUC dataset are used as the training set to train the classifier while the remaining 20% is used as the test set. The classifiers used are SVM (Linear), *k*NN, Gaussian Naive-Bayes (G-NB), Multinomial Naive-Bayes (M-NB), Binomial Naive-Bayes (B-NB), Decision Trees (DT), Random Forest (RF), Gradient Boosting (GB), and Extra Trees (ET). From the results, one can notice that the performance of Linear SVM dominated all other classifiers for all DUC datasets. Tables 2, 3, 4 and 5 show the performance of various classifiers on different DUC datasets. The maximum F-measure and accuracy are marked as bold in the Tables. Figures 2 and 3 show the precision and recall of different classifiers on DUC datasets, respectively. Figs. 4 and 5 show the F-measure and accuracy of different classifiers on DUC datasets, respectively.

3.1 Hyper-Parameter Tuning

Linear SVM: Parameters tuned: C = 10
kNN: Parameters tuned: $k = 5$
Naive Bayes: Parameters tuned: Prior Probabilities = [0.75, 0.25]
Random Forest: Parameters tuned: n_estimators = 11, number of classifiers used: 10
Gradient Boosting: Parameters: learning_rate = 0.1, n_estimators = 290, max_depth = 4, subsample = 0.4, random_state = 0, number of classifiers used: 10
Extra Trees: Parameters: n_estimators = 20, learning_rate = 1, number of classifiers used: 10

Table 3 Performances on DUC-2001

Classifier	DUC-2001			
	Precision	Recall	F-Measure	Accuracy
SVM	80.98	82.51	**81.74**	**92.89**
*k*NN	50.65	52.11	51.37	63.46
G-NB	61.27	65.61	63.37	72.71
M-NB	64.54	63.11	63.82	74.75
B-NB	58.34	56.72	57.52	68.32
DT	48.98	49.58	49.28	61.38
RF	65.51	61.64	63.52	73.43
GB	63.65	64.71	64.18	78.52
ET	59.45	63.22	61.28	74.38

Table 4 Performances on DUC-2002

Classifier	DUC-2002			
	Precision	Recall	F-Measure	Accuracy
SVM	81.89	79.86	**80.86**	**93.67**
*k*NN	51.64	55.21	53.37	70.03
G-NB	59.87	60.77	60.32	73.23
M-NB	61.87	62.67	62.27	77.27
B-NB	57.93	55.27	56.57	71.18
DT	55.87	55.25	55.65	72.23
RF	69.10	71.24	70.31	77.46
GB	61.26	63.72	62.47	73.38
ET	67.89	63.59	65.67	75.08

Table 5 Performances on DUC-2005

Classifier	DUC-2005			
	Precision	Recall	F-Measure	Accuracy
SVM	78.43	76.92	**77.67**	**95.46**
*k*NN	54.64	52.15	53.37	64.43
G-NB	65.87	61.05	63.37	68.83
M-NB	63.87	58.81	61.24	77.65
B-NB	54.92	54.44	54.68	70.38
DT	56.87	58.67	57.76	67.23
RF	75.34	73.89	74.61	77.46
GB	64.26	62.83	63.54	83.68
ET	62.39	69.76	65.87	81.48

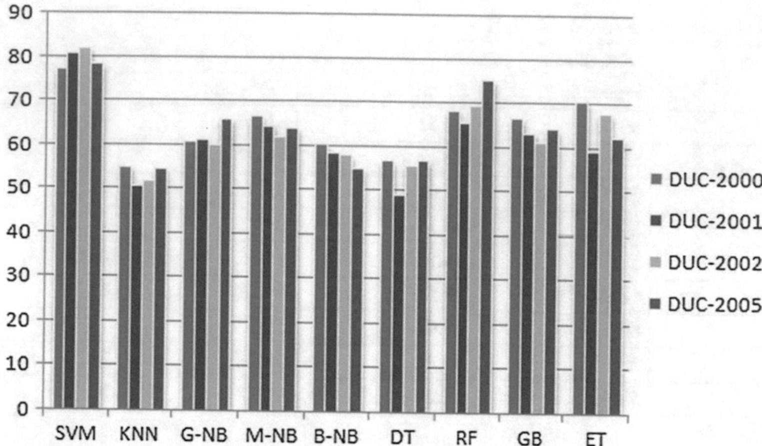

Fig. 2 Precision of different classifiers

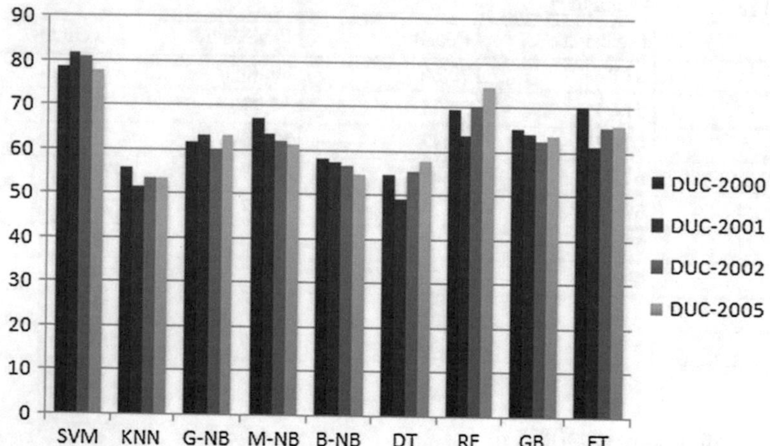

Fig. 3 Recall of different classifiers

3.2 DUC-2000 and DUC-2002 Datasets

For both datasets, the test folder is organized into subfolders (30 in this case). Each of those subfolders contains 9–11 documents having summaries pertaining to a specific topic. The documents within those subfolders contain 100–400 word summaries, and each of those documents have similar content and hence are duplicates. Each set having an average of 10 documents constituting a total of 296 documents and 43956 pairs of documents which are used to detect the near-duplicates. Out of 43956 pairs of documents, 1350 pairs of documents are not near-duplicates as they come from the same clusters (each set having 10 documents and hence, 45 pairs of documents.

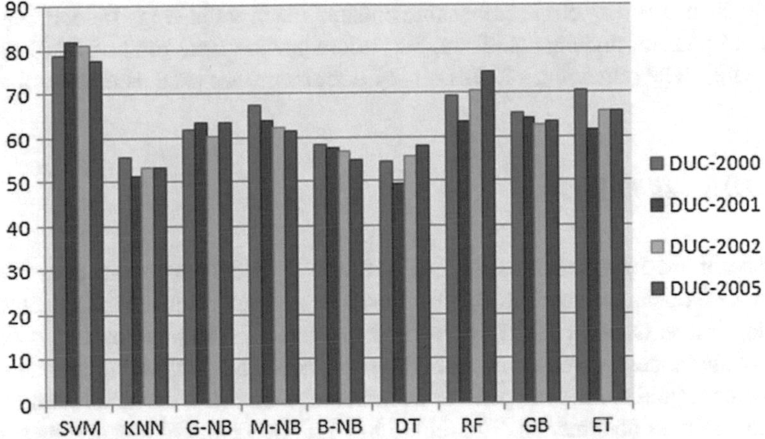

Fig. 4 F-measure of different classifiers

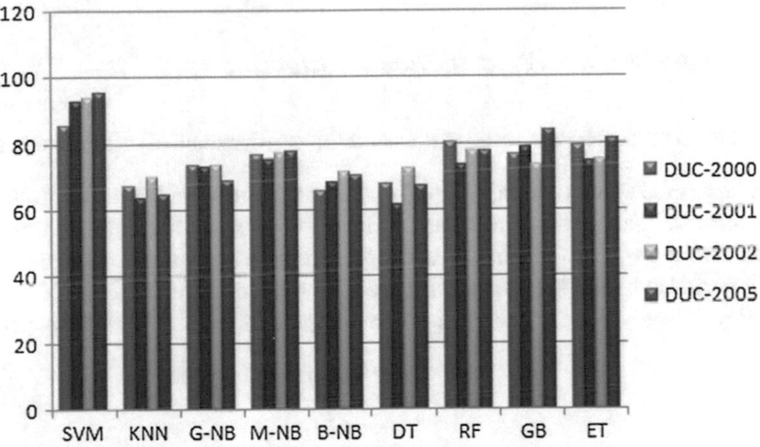

Fig. 5 Accuracy of different classifiers

Thus, 30 document sets have a total of 1350 pairs of documents). The remaining 42606 pairs of documents are used as near-duplicates.

3.3 DUC-2001 Dataset

This dataset has 30 sets of documents with sets defined by different types of criteria such as event set, opinion set, etc. Each set having around 10 documents constituting a total of 299 documents and 44551 pairs of documents which are used to detect the near-duplicates. Out of 44551 pairs of documents, 1350 pairs of documents are not near-duplicates as they come from same clusters (each set having 10 documents and hence, 45 pairs of documents. Thus, 30 documents set have total of 1350 pairs of documents). The remaining 43201 pairs of documents are used as near-duplicates.

3.4 DUC-2005 Dataset

This dataset has 50 document sets. Each set having 25–50 documents constituting a total of 1503 documents and 1128753 pairs of documents which are used to detect the near-duplicates. Out of 1128753 pairs of documents, 23250 pairs of documents are not near-duplicates as they come from the same sets (each set having on an average 30 documents and hence, 465 pairs of documents. Thus, 50 documents set have total of 23250 pairs of documents). The remaining 1105503 pairs of documents are used as near-duplicates.

3.5 Parameters for Performance Evaluation

The following parameters are used to measure the performance of different classifiers.

i. *Accuracy (a)* is the ratio between the sum of true positive cases, *TP* (number of documents that are near-duplicate and are retrieved by the approach) and true negative cases, *TN* (number of documents that are not near-duplicate and are not retrieved by the approach) with the total number of documents, $N = TP + FP + TN + FN$. Eq. 6 shows how to compute the accuracy of a classifier.

$$a = \frac{TP + TN}{N} \tag{6}$$

where, *FP*: number of documents that are not near-duplicate and are retrieved by the approach and *FN*: number of documents that are near-duplicate and are not retrieved by the approach.

ii. *Precision (p)* is the fraction of the web documents suggested as near-duplicate (C_d) by the approach which are actually near-duplicate (A_d) and is shown in Eq. 7.

$$p = \frac{|A_d| \cap |C_d|}{|C_d|} \tag{7}$$

iii. *Recall (r)* is the fraction of total actual near-duplicate documents in the corpus which are suggested as near-duplicate by the approach and is shown in Eq. 8.

$$r = \frac{|A_d| \cap |C_d|}{|A_d|} \tag{8}$$

Table 6 Similarity scores on DUC-2001

Document name	Similarity score
SJMN91-06184003	0.5808
SJMN91-06184088	0.5909
SJMN91-06187248	0.5566
SJMN91-06254192	0.5626
SJMN91-06290185	0.5783
WSJ910702-0078	0.5385
WSJ910702-0086	0.5582
WSJ910910-0071	0.5457
WSJ911011-0071	0.5757
WSJ911016-0124	0.5466
WSJ911016-0126	0.5747
AP880908-0264	0.5134
AP890616-0192	0.5179
AP901213-0236	0.5133
LA012689-0139	0.5605

Table 7 Top 15 documents of DUC-2001

Rank	Document name	Similarity score
1	LA031790-0064	0.6070
2	WSJ920302-0142	0.4022
3	AP890901-0154	0.4120
4	AP880509-0036	0.4354
5	SJMN91-06130055	0.4139
6	AP900207-0041	0.4871
7	SJMN91-06324032	0.4682
8	AP880714-0187	0.4079
9	AP890123-0150	0.4492
10	FT931-10514	0.4901
11	AP880909-0135	0.4270
12	AP891005-0230	0.4031
13	FBIS4-23474	0.4375
14	AP881109-0149	0.4609
15	LA110490-0034	0.4555

Table 8 Semantic similarity scores on DUC-2001

First document name	Second document name	Word2Vec	WordNet	LDA	NGD
WSJ911004-0115	WSJ880510-0079	0.6557	0.3192	0.4409	0.5431
AP900816-0166	AP890127-0271	0.7493	0.4763	0.3023	0.4561
SJMN91-06267062	WSJ870804-0097	0.6558	0.4812	0.3023	0.3014
AP890527-0001	AP880316-0061	0.4997	0.5111	0.3023	0.3967
FT922-10715	FT924-12193	0.5309	0.4782	0.3023	0.4385
AP891005-0216	LA113089-0118	0.6245	0.3998	0.7415	0.5118
AP900514-0011	LA112389-0104	0.5308	0.4958	0.3828	0. 7103
SJMN91-06041121	SJMN91-06067177	0.7184	0.5008	0.4013	0.3311
SJMN91-06294205	AP890525-0127	0.7183	0.4506	0.4305	0.3431
SJMN91-06254192	AP891125-0090	0.7495	0.4640	0.3023	0.5643
AP880316-0208	AP891125-0090	0.5308	0.4436	0.3023	0.6113
AP891005-0230	WSJ891026-0063	0.2187	0.4237	0.9999	0.4667
SJMN91-06270171	SJMN91-06165057	0.6558	0.4103	0.3009	0.3001
LA102189-0067	AP881022-0011	0.5307	0.4193	0.7087	0.3144
WSJ910620-0075	AP900816-0139	0.6558	0.5466	0.3023	0.6113

For demonstration purposes, we have considered DUC-2001 dataset and shown some near-duplicate documents and their composite similarity scores in Table 6 and the top 15 near-duplicate documents and their composite similarity scores in Table 7. Table 8 shows four semantic similarity scores between the pair of documents of DUC-2001 dataset.

4 Conclusion

The paper proposed a methodology to identify the semantically similar near-duplicate documents of a corpus and a novel ranking approach to retrieve the top K near-duplicate documents from the corpus. Four semantic-based techniques are used in order to find the semantic similarity between two documents. The present work not only detects the near-duplicates based on their semantic similarity but also ranks the documents to catch the percentage of similarities with respect to all documents of a corpus. Experimental results on different DUC datasets show the stability and efficiency of the proposed approach. This approach can be optimized by developing a method to compute the weights from the features of the classifier in a better manner. The techniques used here are mathematically formalized and presented, yet the empirical results may not be the most optimized due to the properties of the dataset. Relying heavily on NGD for a large number of documents in retrospect is not a good idea because NGD requires continuous internet connectivity and has a considerable time overhead. When selecting the most dissimilar documents from the corpus, we need to make sure that this dissimilarity is among the documents included in the ranking and not the whole corpus. Hence, a dynamic threshold is needed to set the similarity of documents being added to the top-ranked list to ensure semantically relevant yet the diverse results. The selection of the threshold value needs to be done by coming up with a good heuristic with the help of domain knowledge. An alternate approach to rank the documents is to generate a query from the semantic content of the documents of the corpus and this query can be used to determine the similarity among the documents. Different techniques can be developed to identify and optimize such query. The proposed methodology also can be extended by introducing different deep learning techniques to identify the near-duplicates in the corpus and comparing their results with shallow learning classifiers.

References

1. Manber, U.: Finding similar files in a large file system. In: Winter USENIX Technical Conference. vol. 94, pp. 1–10 (1994)
2. Roul, R.K., Mittal, S., Joshi, P.: Efficient approach for near duplicate document detection using textual and conceptual based techniques. In: Advanced Computing, Networking and Informatics, Volume 1: Advanced Computing and Informatics Proceedings of the Second International Conference on Advanced Computing, Networking and Informatics (Icacni-2014). vol. 27, pp. 195–203. Springer, Berlin (2014)
3. Zhou, Z., Yang, C.-N., Chen, B., Sun, X., Liu, Q., QM, J.: Effective and efficient image copy detection with resistance to arbitrary rotation. IEICE Trans. Inf. Syst. **99**(6), 1531–1540 (2016)
4. Zhou, Z., Wu, Q.J., Sun, X.: Encoding multiple contextual clues for partial-duplicate image retrieval. Pattern Recognit. Lett. (2017)
5. Zhou, Z., Wu, Q.J., Huang, F., Sun, X.: Fast and accurate near-duplicate image elimination for visual sensor networks. Int. J. Distrib. Sens. Netw. **13**(2), 1–12 (2017)
6. Zhou, Z., Mu, Y., Wu, Q.J.: Coverless image steganography using partial-duplicate image retrieval. Soft Comput. 1–12 (2018)

7. Bharat, K., Broder, A.: Mirror, mirror on the web: A study of host pairs with replicated content. Comput. Netw. **31**(11), 1579–1590 (1999)
8. Manku, G.S., Jain, A., Das Sarma, A.: Detecting near-duplicates for web crawling. In: Proceedings of the 16th International Conference on World Wide Web, pp. 141–150 (2007)
9. Feng, Y., Bagheri, E., Ensan, F., Jovanovic, J.: The state of the art in semantic relatedness: a framework for comparison. Knowl. Eng. Rev. **32** (2017)
10. Mikolov, T., Chen, K., Corrado, G., Dean, J.: Efficient estimation of word representations in vector space. In: Proceeding of ICLR-2013, pp. 1–12 (2013)
11. Chua, S., Kulathuramaiyer, N.: Semantic feature selection using wordnet. In: Proceedings of the 2004 IEEE/WIC/ACM International Conference on Web Intelligence. IEEE Computer Society, pp. 166–172 (2004)
12. Cilibrasi, R.L., Vitanyi, P.M.: The google similarity distance. IEEE Trans. Knowl. Data Eng. **19**(3), 1–15 (2007)
13. Blei, D.M., Ng, A.Y., Jordan, M.I: Latent dirichlet allocation. J. Mach. Learn. Res. **3**, 993–1022 (2003)
14. Girolami, M., Kabán, A.: On an equivalence between plsi and lda. In: Proceedings of the 26th annual International ACM SIGIR Conference on Research and Development in Informaion Retrieval, pp. 433–434. ACM (2003)

A Stacked Generalization Approach for Diagnosis and Prediction of Type 2 Diabetes Mellitus

Namrata Singh and Pradeep Singh

Abstract Diabetes mellitus is a global health problem that occurs due to metabolic disorders. It is a lifestyle disease that enhances the likelihood of the development of many serious health complications. In the past, several learners have been applied for prediction of diabetes but there is still enough space to develop classifiers with higher accuracy. The study utilizes Pima Indian Diabetes secondary dataset. In this paper, individual approaches, viz., linear-SVM, kernel methods including polynomial, radial basis function, and sigmoid have been used while among ensembles majority voting and stacking strategies have been applied. Stacked ensembling is based on various types of meta-learners such as C4.5, NB, k-NN, SMO, and logitboost. The stacking approach with meta-learner SMO (ST-SMO) achieves accuracy 79%, sensitivity 78.9%, false positive rate 31.5%, precision 78.5%, F-measure 77.8%, and AUC 73.2% demonstrating that it is the best classifier as compared to any of the individual and ensemble approaches.

Keywords Diabetes mellitus · Classification · Machine learning · Stacking · Majority voting · Ensemble · Stacked generalization · Support vector machines

1 Introduction

The recent advances in medical sciences, especially in modern medicine have generated huge scientific clinical data on health and disease. Owing to universal access to such enormous volume of medical data sets in electronic form, it has now become practically more feasible to apply machine learning methods to probe in depth and extract useful information hidden in the data.

N. Singh (✉) · P. Singh
Department of Computer Science and Engineering, National Institute of Technology Raipur, Raipur 492001, Chhattisgarh, India
e-mail: nsingh.phd2016.cs@nitrr.ac.in

P. Singh
e-mail: psingh.cs@nitrr.ac.in

H. S. Behera et al. (eds.), *Computational Intelligence in Data Mining*, Advances in Intelligent Systems and Computing 990,
https://doi.org/10.1007/978-981-13-8676-3_47

Diabetes mellitus is one of the key non-communicable and lifestyle diseases worldwide. It is a chronic and serious public health problem ever rising on a fast pace [1]. Today diabetes is no more a disease of rich people and urbanites only, but also people from rural areas and low socioeconomic strata are equally affected by diabetes. Uncontrolled diabetes leads to very serious health and well-being consequences. Apart from this, diabetes also impacts adversely the financial condition of the individual and his family [2].

The disease diabetes occurs when the pancreas is unable to produce sufficient amount of insulin required for the proper functioning of the body. This type of diabetes is called type 1diabetes or insulin-dependent diabetes mellitus (IDDM) which occurs usually among children. Diabetes may also occur when the body is unable to effectively utilize the insulin produced by it. This type of diabetes is called type 2 diabetes or non-insulin dependent diabetes mellitus (NIDDM) which occurs usually among adults [3]. The majority of diabetics are affected by type 2 diabetes. Globally, an estimated 422 million adult people were having diabetes in the year 2014 compared to 108 million adults in 1980. This shows an almost fourfold increase in diabetes burden during the past 34 years period from 1980 to 2014. Also, the global prevalence of diabetes among adults rose to 8.5% from 4.7% during the same period [4]. These alarming figures demonstrate the scale and severity of the problem indicating an urgent need to address it on a priority basis.

Machine learning methods are broadly classified into two categories, namely, supervised learning and unsupervised learning techniques. In supervised learning, the learning model is trained based on the input features and class-labels while in unsupervised learning, the learning algorithm has no information about the class-labels. Both individual as well as ensemble methodologies are applied for the diagnosis of various medical problems. Bagging, boosting, and stacking are some of the most commonly used ensemble learning techniques wherein each approach gives varied diagnostic accuracy [5].

In this paper, individual and ensemble approaches have been used for the classification of diabetes. The individual learners employed are linear-SVM, polynomial kernel, RBF kernel, and sigmoid kernel SVMs. The second approach for ensemble machine learning, i.e., stacking approach has been utilized to enhance the prediction of the learning models. The major goals of the paper are the following:

(i) Stacked Generalization (Stacking) approach for classification with varied meta-learners.
(ii) Comparison of stacking approach with individual and other ensemble learners.
(iii) Using cross-validation approach for data analysis.
(iv) Evaluation and interpretation of the results obtained.

The remainder of the paper is arranged in the following way. Section 2 presents the literature survey of various machine learning algorithm techniques employed for classifying diabetes mellitus. The information about the materials and methods used in the experiment has been presented in Sect. 3. Section 4 contains the results of the experiments performed. The detailed discussions about the results are narrated in Sect. 5. Finally, Sect. 6 presents the conclusion and scope for future work.

2 Related Work

The literature review reveals that various conventional machine learning methods employed to analyze Pima Indians Diabetes (PID) dataset, but ensemble learning approaches remain unexploited. The application of such type of models in diabetes classification not only enhances the prediction accuracy but also provides robust solutions with reduced costs.

The biomedical disease datasets are derived from various sources possessing unique sample sizes, data structures, predictive features, and class-labels; therefore, a complex individual learner cannot solve the type-II diabetes classification problem. This is in accordance with the well-known "no free lunch" theorem, which advocates that all problems cannot be solved by a single algorithm [6]. Therefore, there arises a need for applying the concept of ensemble methods in disease classification. For obtaining better hypothesis and excellent predictions, the ensemble learning methodology aggregates multiple methods which process diverse hypotheses [7]. In the past, several studies have proved the ensemble methods analytical performance, that relies mostly on the bias–variance trade-off [8, 9, 10]. The expected generalization error of a learner can be disintegrated into bias, variance, and noise [11]. An exhaustive experimental comparison of the two widely known classification algorithms, i.e., bagging and boosting done in [12] proved that bagging decreases only the variance while boosting decreases both bias and variance.

During the past few years, various machine learning techniques have been used to diagnose and classify diabetes mellitus. From Table 1, it is evident that there exist various classification techniques which have successfully classified the diabetes data as tested positive (diabetic class) or tested negative (nondiabetic class). Our proposed method utilizes stacking approach with DT, NB, k-NN, SMO, LB as meta-learners, i.e., ST-DT, ST-NB, ST-kNN, ST-SMO, ST-LB. Among all the five stacking-based models, ST-SMO outperforms with accuracy 79%, sensitivity 78.9%, false positive rate 31.5%, precision 78.5% and F-measure 77.8%, and AUROC 73.2% as compared to the other classifiers. Thus, the proposed ST-SMO stacked model can be used as an efficient tool for diagnosing diabetes mellitus.

3 Materials and Methods

The Pima Indian Diabetes (PID) dataset used in this paper was obtained from the UCI machine learning database [24]. The total number of input features present in this dataset was 8 with 1 output attribute. Out of the total number of 768 study subjects, 268 were tested positive (diabetic) and rest 500 negative (nondiabetic).

This section presents the individual as well as ensemble methods utilized for the classification and analysis of the PID dataset. Among individual approaches, Linear-SVM, Polynomial-SVM, RBF-SVM, and sigmoid-SVM are used and among

Table 1 Classification accuracy obtained by different methods for diagnosing diabetes

Author(s)	Classifier used	Accuracy (%)
Jaganathan et al. [13]	ANT-MINER with Imp. Quick Reduct	76.58
Luukka [14]	Sim Sim + F1 Sim + F2	75.29 75.84 75.97
Örkcü and Bal [15]	Backpropagation Binary-coded GA Real-coded GA	73.80 74.80 77.60
Quilan [16]	C4.5	71.10
Priyadarshini et al. [17]	ELM based models Modified ELM BPNN based models	72.72 75.72 68.00
Parashar et al. [18]	SVM, LDA	75.65
Varma et al. [19]	Gini index-Gaussian Fuzzy Decision Tree (GGFSDT)	75.80
Bozkurt et al. [20]	PNN, LVQ, FFN, CFN, DTDN, TDN, Gini, AIS	76.00
Kandhasamy and Balamurali [21]	J48, k-NN, RF, SVM	73.82
Choubey and Paul [22]	J48graft DT GA_J48graft DT	76.52 74.78
Varma et al. [23]	PCA-FSDT	76.80
Proposed method	**ST-SMO**	**79.00**

ensemble methods, two techniques, i.e., majority voting and stacking-based learners are applied.

3.1 Individual Learner Approach

SVM, proposed by Vapnik [25], is an effective machine learning technique which is extensively utilized in the disease classification domain. For a binary classification task, the basic concept of SVM is to find out an appropriate hyperplane that separates the positive and negative examples x_i, y_i for $i = 1, \ldots, n$ with maximum interclass distance or margin d. Here x_i represents the vector in input space $S \in R^n$ and y_i denotes class index which takes values $+1$ or -1. If the data is nonlinear, a kernel function is utilized for mapping the data samples to a higher dimensional space F. This transformation of the data x_i from S into the $F \in R^N$ feature space, where N might be infinite is performed by using $\phi(x)$ as the nonlinear mapping function. It also utilizes a learner method for solving the nonlinearity problem. Correspondingly, in the feature space it tries to find a linear decision function given as

$$f(x) = w \cdot \phi(x) + b \tag{1}$$

The classification of the data is performed by the sign obtained in Eq. (1). The goal is to choose such hyperplane that performs the splitting of the sample points as cleanly as possible which also maximizes the separation to the closest split data points. This is achieved by the introduction of slack variables (ξ) that are positively valued, into the constraints of the optimization problem, given in the Eqs. (2) and (3) as

$$(w \cdot \phi(x) + b) \geq 1 - \xi_i \quad \textit{if } y_i = +1 \tag{2}$$

$$(w \cdot \phi(x) + b) \leq -1 + \xi_i \quad \textit{if } y_i = -1 \tag{3}$$

where $\forall_i : \xi_i \geq 0$.

Some of the well-known kernel functions of SVM used in the proposed framework, i.e., linear, polynomial, radial basis function (RBF), and sigmoid [26] are listed below in Eqs. (4), (5), (6), and (7):

$$\text{(1) Linear}: \quad k(x_i, x_j) = (x_i \cdot x_j) \tag{4}$$

$$\text{(2) Polynomial}: \quad k(x_i, x_j) = (x_i \cdot x_j + 1)^p \tag{5}$$

$$\text{(3) Radial Basis}: \quad k(x_i, x_j) = e^{-x_i - x_j^2/2\sigma^2} \tag{6}$$

$$\text{(4) Sigmoid}: \quad k(x_i, x_j) = \tanh(\kappa x_i \cdot x_j + c) \tag{7}$$

where p, κ, σ, and c are constants.

3.2 *Ensemble Learner Approach*

The ensemble learning methodology has proved its merit in solving the most challenging real-world problems. Ensemble modeling techniques can broadly be categorized into homogeneous and heterogeneous ensembles. Most of the techniques like bagging and boosting generate diversity by sampling from the training data but mostly use a single base classifier for constructing the ensemble. In cases, where there is uncertainty in choosing an ideal base classifier, homogeneous ensembling strategy might not be the best choice. For such types of problems, an ensemble model can be developed from the predictions of a large diversity of heterogeneous base classifiers such as SVM, neural networks, k-nearest neighbors, decision trees. The two well-known heterogeneous ensemble methods include stacking which is a form of meta-learning and ensemble selection [27].

Voting. One of the most popular and elementary combination rules for nominal outputs is the voting framework. Among the various voting strategies such as weighted voting, plurality voting, majority voting, soft voting, etc., one of the prominent voting strategies is the majority voting. In this, each learner votes for obtaining a single class label and the target class which gets more than half of the votes is regarded as the final output class. The rejection option is granted when neither of the class-labels acquires more than half of the votes and no predictions are performed by the overall aggregated learners. The final class label of the ensemble is defined as Eq. (8)

$$H(x) = \left\{ \begin{array}{ll} c_j & if \sum_{i=1}^{T} h_i^j(x) > \frac{1}{2} \sum_{k=1}^{l} \sum_{i=1}^{T} h_i^k(x) \\ rejection & otherwise \end{array} \right\}. \tag{8}$$

For a binary classification problem having a total of T learners, if at least $\lfloor T/2 + 1 \rfloor$ learners choose the correct class label, then the ensemble decision will be correct. This is based on the assumption that the outputs of the individual classifier are independent where p is the probability of correct classification by each learner, i.e., p is the accuracy of each classifier for individual predictions. The overall probability of the ensemble for making a correct decision can be computed by utilizing a binomial distribution [28].

Stacked Generalization. Stacked Generalization or stacking is a branch of ensemble learning which gives importance to the aggregation of heterogeneous models. In comparison to the homogeneous ensemble learning strategies such as bagging, boosting, etc., it allows more diverse models, provides better coverage of the original problem space, therefore the final outcome of stacked generalization is mostly better than homogeneous ensemble learning techniques [29]. The detailed working of stacking methodology is depicted in Fig. 1. and the pseudocode of the proposed stacking-based approach is shown in Algorithm 1.

Algorithm 1: Framework of proposed stacking-based approach

Input: Training dataset: $\mathcal{D}_{train} = \{(x_1, y_1), \ldots, (x_n, y_n)\}$
Testing dataset: $\mathcal{D}_{test} = \{(x_{n+1}, y_{n+1}), \ldots, (x_m, y_m)\}$
Base learners: $\mathcal{C}_1, \mathcal{C}_2, \mathcal{C}_3$ *and* $\mathcal{C}_4$
Meta learner: $\mathcal{C}_{meta}$
Output: *Test set label prediction*, $\mathcal{P}_{test}$
1: **for** $t = 1, 2, \ldots, 4$ **do**
2: $h_t = \mathcal{C}_t(\mathcal{D}_{train})$
3: *Calculate* $p_t^{h_t}(\mathcal{D}_{test})$
4: **end for**
5: $p(\mathcal{D}_{test}) = [p_1^{h_1}(\mathcal{D}_{test}), p_2^{h_2}(\mathcal{D}_{test}), p_3^{h_3}(\mathcal{D}_{test}), p_4^{h_4}(\mathcal{D}_{test})]$
6: $\mathcal{D}^{L1} = \{p(\mathcal{D}_{test}), y^{\mathcal{D}_{test}}\}$ % *Level 1 data generation*
7: $h^{meta} = \mathcal{C}_{meta}(\mathcal{D}^{L1})$ % *Level 2 classifier generation*
6: $\mathcal{P}_{test} = h^{meta}(p(\mathcal{D}_{test}))$
7: *return* $\mathcal{P}_{test}$

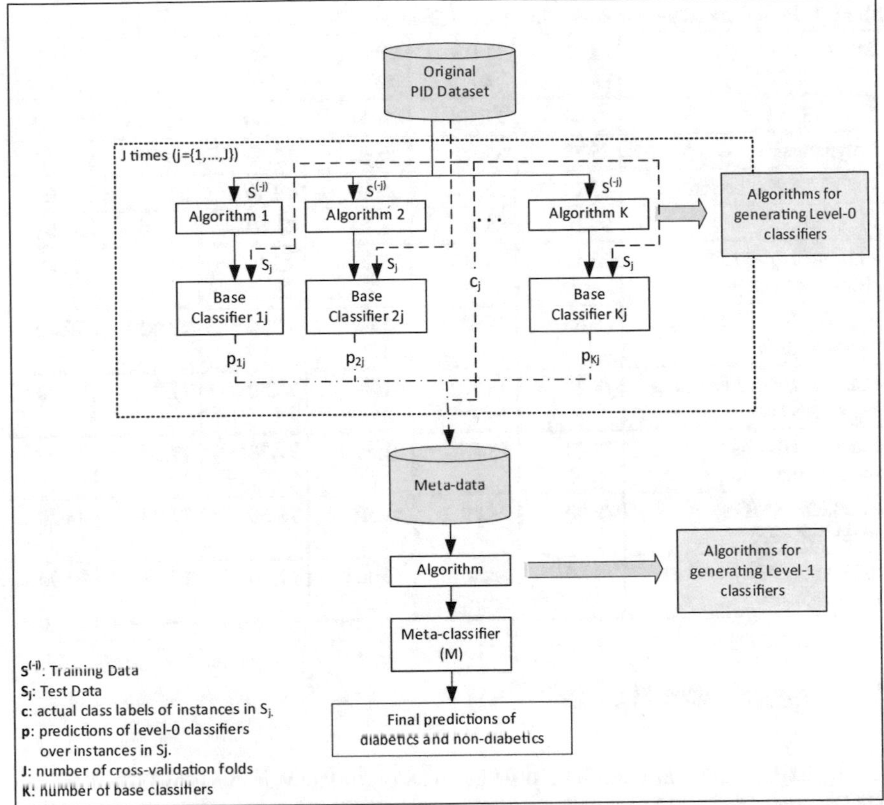

Fig. 1 Generation of an ensemble of classifiers using stacking procedure

Stacking [30, 31] is the method where the training of a learner is performed for integrating the individual learners. The individual learners are known as the first-level learners whereas the combiner or integrator is known as the second-level learner or meta-learner. The original training dataset is used for training the first-level learners and then a new dataset is generated for training the second-level learner. The outputs of the first-level learners are given as input features to the meta-learner and the class-labels are the same as in the original dataset. In general, the stacking framework is viewed as the generalization of many ensemble methods. The learning algorithms which we have used for the first-level learner are Linear-SVM, Polynomial-SVM, RBF-SVM, and Sigmoid and for the meta-learners are C4.5, NB, IB1, SMO, and Logit-boost.

Table 2 Performance measures on Pima Indian Diabetes (PID) dataset

Method	ACC (%)	TPR (%)	FPR (%)	PRE (%)	FME (%)	AUC (%)
Linear-SVM	77.47	77.50	32.00	77.00	76.70	72.70
Polynomial	55.34	55.30	46.40	58.50	56.30	54.40
Radial basis function	65.10	65.10	65.10	42.40	51.30	50.00
Sigmoid	65.10	65.10	65.10	42.40	51.30	50.00
Ensemble approach (Majority voting)	65.10	65.10	65.10	42.40	51.30	50.00
Stacking (C4.5 as meta-learner)	78.65	78.65	**31.00**	77.50	77.00	70.10
Stacking (Naïve Bayes as meta-learner)	78.70	78.70	31.59	76.50	77.20	72.00
Stacking (IB1 as meta-learner)	78.70	78.70	31.59	77.78	77.20	72.00
Stacking (SMO as meta-learner)	**79.00**	**78.90**	31.50	**78.50**	**77.80**	**73.20**
Stacking (Logit-boost as meta-learner)	78.80	78.80	32.00	77.50	76.95	72.90

4 Experimental Results

All the experiments are conducted using Weka 3.8 data mining tool [32]. We have used six metrics like classification accuracy, sensitivity, false positive rate, precision, F-measure, ROC area for evaluation of the proposed model. The various performance measures on the Pima Indian Diabetes (PID) dataset are shown in Table 2.

For the experimental study, the dataset was partitioned into two sets, i.e., training and testing, and tenfold cross-validation was performed. The performance of the proposed ST-SMO stacking approach has been compared with the other individual and ensemble methods. The metrics used for model evaluation are accuracy (ACC), sensitivity (SEN), false positive rate (FPR), precision (PRE), F-measure (FME), and area under receiver-operating characteristic (ROC) curves as depicted in Eqs. (9), (10), (11), (12), (13), and (14), respectively.

$$Accuracy = \frac{(TP + TN)}{(TP + TN + FP + FN)} \tag{9}$$

$$Sensitivity = \frac{TP}{TP + FN} \tag{10}$$

$$FPR = \frac{FP}{FP + TN} \tag{11}$$

$$Precision = \frac{TP}{TP + FP} \tag{12}$$

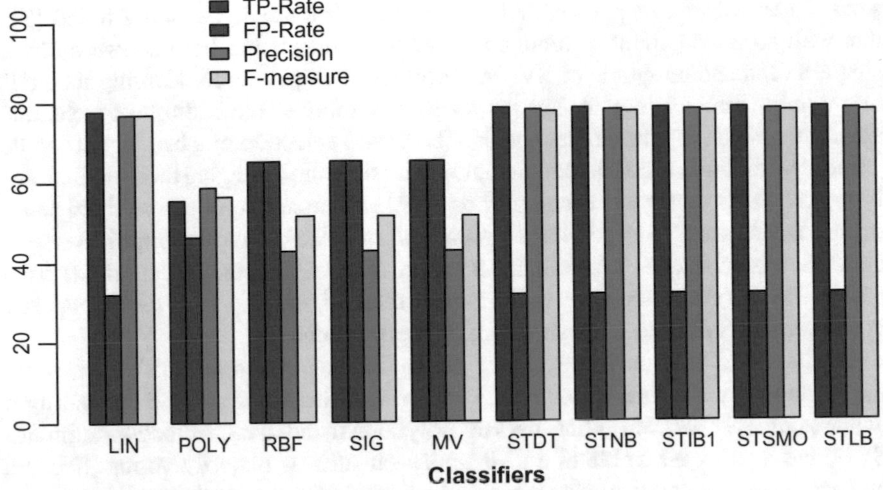

Fig. 2 Comparative evaluation of classifiers for four different performance measures

$$F - measure = \frac{2TP}{2TP + FP + FN} \tag{13}$$

$$AUC = \frac{1}{2}\left(\frac{TP}{TP + FN} + \frac{TN}{TN + FP}\right) \tag{14}$$

The experiment was conducted in tenfold and again each fold was repeated 10 times. Thus, the values of accuracy were obtained for 100 iterations. Overall, we can conclude that the above five stacking-based approaches perform much better in comparison to the other mentioned classifiers.

A good modeling performance is indicated by low FP and FN rates and high TP and TN values. Figure 2 represents the comparative evaluation of ten different classifiers on four different performance measures, i.e., TP-Rate, FP-Rate, precision, and F-measure. It shows that for the proposed ST-SMO learner, the TP-rate (78.9%) is highest and FP-rate (31.5%) is lowest. Again, the highest precision value of 78.5% and maximum F-measure of 77.8% was attained by ST-SMO.

5 Discussion

One of the reliable machine learning methods that have performed excellently well in most classification and prediction tasks are the Support Vector Machines (SVMs). For obtaining greater computational efficiency, SVM ensembles are being used for solving classification and prediction problems on various types of data [33]. Such type of multiple kernel learning (MKL) ensemble concepts have been applied using stacked generalization approach. Although, it has been experimentally found [34] that with accurate parameter tuning, similar performance has been shown by both single SVMs and ensemble of SVMs. Adopting multiple kernel learning methods or ensemble methodology always proves to be a more stable and robust technique when compared with the single models. The risk of selection of a bad model can be reduced by taking into consideration multiple models and averaging their predictions. In several works previously accomplished SVM ensembles have demonstrated better results over single SVMs [35], but a few papers have also shown bad empirical results about the ensemble of SVMs [36]. In our experimental study, stacking multiple kernel SVMs shows better predictive performance than all single SVMs while majority voting among SVMs does not show much improvement.

Furthermore, our study emphasizes on the following components: First, using the stacking approach for classification with varied meta-learners and comparing it against single SVMs such as linear-SVM, polynomial, radial basis function, sigmoid-SVM and another ensemble of SVMs approach such as majority voting. Second, analyzing the diabetes data using tenfold cross-validation approach, and finally interpreting the obtained results. The evaluation of the overall performance of machine learning framework has been carried out by utilizing the parameters such as ACC, SEN, FPR, PRE, FME, and AUC. Although the total number of classifiers utilized in our proposed study were 10, theoretical benchmarking was achieved against 11 research works as published previously in the literature, shown in Table 1. The stacking method which adopts SMO as meta-learner shows the superior performance when compared against all the other classifiers.

6 Conclusions and Future Work

Diabetes Mellitus is a serious and life-threatening noncommunicable disease posing global challenges for its better diagnosis and prognosis. It involves considerable burden of cost and produces a host of health complications if timely not taken care of. A high proportion of diabetes remains undiagnosed worldwide. As a result, considering the adverse health impact and severity of diabetes, feasible and comprehensible methods need to be developed to identify the people with undiagnosed diabetes. This study presents an evaluation and comparison of ten machine learning approaches for diabetes prediction.

In the paper, we have implemented individual as well as ensemble classification approaches for effective diagnosis of diabetes. The proposed stacking framework utilizes kernel SVM learners as the base-learners for stacked generalization method which uses a higher level model to merge lower level models for achieving higher computational efficiency and power. Among, the five stacking-based ensemble approaches namely ST-DT, ST-NB, ST-kNN, ST-SMO and ST-LB, ST-SMO outperforms in terms of ACC (0.79), SEN (0.789), FPR (0.315), PRE (0.785), FME (0.778) and AUC (0.732). On the basis of various performance measures such as ACC, SEN, FPR, PRE, FME, and AUC values, it is evidenced that ST-SMO proves to be the best classifier amongst the ten classifiers including the five stacking-based ensemble approaches. Thus, the paper concludes that the proposed stacking approach can be very efficiently used for the classification of diabetes mellitus. Further, by varying the base-learners and meta-learners, this approach can also be applied for classification and prediction of other diseases too.

References

1. World Health Organization: Global Report on Diabetes. WHO, Geneva, Switzerland (2016)
2. Shin, J.K., Poltavskiy, E., Kim, T.N., Hasan, A., Bang, H.: Help-seeking behaviors for serious psychological distress among individuals with diabetes mellitus: The California health interview survey, 2011–2012. Prim. Care Diabetes **11**, 63–70 (2017)
3. Tao, Z., Shi, A., Zhao, J.: Epidemiological perspectives of diabetes. Cell Biochem. Biophys. **73**, 181–185 (2015)
4. WHO, Diabetes, Fact Sheet, http://www.who.int/mediacentre/factsheets/fs312/en/
5. Sluban, B., Lavrač, N.: Relating ensemble diversity and performance: a study in class noise detection. Neurocomputing. **160**, 120–131 (2015)
6. Wolpert, D.H., Macready, W.G.: No free lunch theorems for optimization. IEEE Trans. Evol. Comput. **1**, 67–82 (1997)
7. Nascimento, D.S.C., Coelho, A.L.V., Canuto, A.M.P.: Integrating complementary techniques for promoting diversity in classifier ensembles: a systematic study. Neurocomputing **138**, 347–357 (2014)
8. Opitz, D.W., Maclin, R.: Popular ensemble methods: an empirical study. J. Artif. Intell. Res. **11**, 169–198 (1999)
9. Ozcift, A., Gulten, A.: Classifier ensemble construction with rotation forest to improve medical diagnosis performance of machine learning algorithms. Comput. Methods Programs Biomed. **104**, 443–451 (2011)
10. Karegowda, A.G., Jayaram, M.A., Manjunath, A.S.: Cascading k-means with ensemble learning: enhanced categorization of diabetic data. J. Intell. Syst. **21**, 237–253 (2012)
11. Kohavi, R., Wolpert, D.H.: Bias plus variance decomposition for zero-one loss functions. In: 13th International Conference on Machine Learning, pp. 275–283. Morgan Kaufmann (1996)
12. Bauer, E., Kohavi, R.: An empirical comparison of voting classification algorithms: bagging, boosting, and variants. Mach. Learn. **36**, 105–139 (1999)
13. Jaganathan, P., Thangavel, K., Pethalakshmi, A., Karnan, M.: Classification rule discovery with ant colony optimization and improved quick reduct algorithm. IAENG Int. J. Comput. Sci. **33**, 50–55 (2007)
14. Luukka, P.: Feature selection using fuzzy entropy measures with similarity classifier. Expert Syst. Appl. **38**, 4600–4607 (2011)
15. Örkcü, H.H., Bal, H.: Comparing performances of backpropagation and genetic algorithms in the data classification. Expert Syst. Appl. **38**, 3703–3709 (2011)
16. Quilan, J.R.: C4. 5: Programs for machine learning. Elsevier (2014)
17. Priyadarshini, R., Dash, N., Mishra, R.: A novel approach to predict diabetes mellitus using modified extreme learning machine. In: International Conference on Electronics and Communication Systems (ICECS), pp. 1–5. IEEE (2014)
18. Parashar, A., Burse, K., Rawat, K.: A comparative approach for pima Indians diabetes diagnosis using LDA-support vector machine and feed forward neural network. Int. J. Adv. Res. Comput. Sci. Softw. Eng. **4**, 378–383 (2014)
19. Varma, K.V.S.R.P., Rao, A.A., Sita Maha Lakshmi, T., Nageswara Rao, P.V.: A computational intelligence approach for a better diagnosis of diabetic patients. Comput. Electr. Eng. **40**, 1758–1765 (2014)

20. Bozkurt, M.R., Yurtay, N., Yilmaz, Z., Sertkaya, C.: Comparison of different methods for determining diabetes. Turkish J. Electr. Eng. Comput. Sci. **22**, 1044–1055 (2014)
21. Kandhasamy, J.P., Balamurali, S.: Performance analysis of classifier models to predict diabetes mellitus. Procedia Comput. Sci. **47**, 45–51 (2015)
22. Choubey, D.K., Paul, S.: GA_J48graft DT: a hybrid intelligent system for diabetes disease diagnosis. Int. J. Bio-Sci. Bio-Technol. **7**, 135–150 (2015)
23. Kamadi, V.V., Allam, A.R., Thummala, S.M.: A computational intelligence technique for the effective diagnosis of diabetic patients using principal component analysis (PCA) and modified fuzzy SLIQ decision tree approach. Appl. Soft Comput. **49**, 137–145 (2016)
24. Blake, C.L., Merz, C.J.: UCI repository of machine learning databases. http://archive.ics.uci.edu/ml/index.php
25. Vapnik, V.N.: The Nature of Statistical Learning Theory. Springer Verlag Inc. (1995)
26. Cristianini, N., Shawe-Taylor, J.: An Introduction to Support Vector Machines and Other Kernel-Based Learning Methods. Cambridge University Press, New York (2000)
27. Schapire, R.E.: The strength of weak learnability. Mach. Learn. **5**, 197–227 (1990)
28. Hansen, L.K., Salamon, P.: Neural network ensembles. IEEE Trans. Pattern Anal. Mach. Intell. **12**, 993–1001 (1990)
29. Mazza, S.: Advances in the combination of supervised classification methods: an experimental study (2014)
30. Breiman, L.: Stacked regressions. Mach. Learn. **24**, 49–64 (1996)
31. Wolpert, D.H.: Stacked generalization. Neural Netw. **5**, 241–260 (1992)
32. Frank, E., Hall, M.A., Witten, I.H.: The WEKA Workbench. Online Appendix for Data Mining: Practical Machine Learning Tools and Techniques. Morgan Kaufmann (2016)
33. Anifowose, F., Labadin, J., Abdulraheem, A.: Improving the prediction of petroleum reservoir characterization with a stacked generalization ensemble model of support vector machines. Appl. Soft Comput. **26**, 483–496 (2015)
34. Evgeniou, T., Pérez-Breva, L., Pontil, M., Poggio, T.: Bounds on the generalization performance of Kernel machine ensembles. In: Proceedings of the Seventeenth International Conference on Machine Learning (ICML), pp. 271–278. Morgan Kaufmann (2000)
35. Valentini, G., Dietterich, T.G.: Bias-variance analysis of support vector machines for the development of SVM-based ensemble methods. J. Mach. Learn. Res. **5**, 725–775 (2004)
36. Valentini, G., Dietterich, T.G.: Low bias bagged support vector machines. In: Proceedings of the 20th International Conference on Machine Learning (ICML 03), pp. 752–759 (2003)

Prevention of Ethnic Violence and Riots Using Supervised Rumor Detection

Rama Krushna Das, Manisha Panda, Atithee Apoorva and Gopal Krishna Mishra

Abstract Rumors that spread across the online social network catches a lot of attention amongst the population and causes agitation between minorities which leads to a chaotic situation. In this paper, we propose a process for screening the rumors using Crowdsource Platform and detecting their probable origin. We first verify the veracity of the rumor with the proposed model and then with the help of knowledge-based community detection we minimize the sample space for the probable root of the rumor. Then we start searching for the root in the community detected with the help of backtracking. With this, we would like to resolve problems faced by the minorities in cases of ethnic violence where the fiasco is initiated only due to the message passed in the rumor. Once the action is taken against the root culprit other such people would stay warned and avoid such actions.

Keywords Crowdsource platform · Knowledge-based community detection · Ethnic violence · Rumor veracity verification · Rumor source identification

1 Introduction

Places where there is coexistence of different communities, castes, creeds, and races, are highly vulnerable to false rumors against one another, resulting in ethnic violence and riots. Also, the areas which face regular natural calamities like Earthquack,

R. K. Das (✉)
National Informatics Centre, Berhampur 760004, India
e-mail: ramdash@yahoo.com

M. Panda
Berhampur University, Berhampur 760001, India
e-mail: manishapanda2013sai@gmail.com

A. Apoorva · G. K. Mishra
Parala Maharaja Engineering College, Berhampur 761003, India
e-mail: itsatithee@gmail.com

G. K. Mishra
e-mail: gopal.mishra25798@gmail.com

H. S. Behera et al. (eds.), *Computational Intelligence in Data Mining*, Advances in Intelligent Systems and Computing 990,
https://doi.org/10.1007/978-981-13-8676-3_48

Tsunami, Cyclone, etc., rumors spread by vested interest of people create havoc and a lot of problems among the common people. Ethnic violence has widespread and huge consequences these days. Especially those that arise from the rumors and have no real justification behind its outbreak are more devastating for our society. Rumors are basically defined as, "a circulating information of uncertain truth, which is apparently believable but hard to verify, and produces enough disbelief and anxiety." Hence there is a dire necessity for the rumor to be combated in a well efficient manner. In this paper, the techniques like knowledge-based community detection, verification based upon the crowdsourcing platform and tracing back to the root or the culprit of the rumor is taken into account. Knowledge-based community detection plays a vital role in our model which we proposed, as well as the crowdsourcing platform. The whole model is about the detection of the rumor and its veracity, finally tracing to the root or nearby root. The techniques which we are using are widely knowledge-based community detection and crowdsource platform along with sorting. Knowledge-based community detection combines two very vast areas of natural language processing and graph processing on a single platform. It helps us find our routes to the root culprit by minimizing a whole lot of sample space into small clusters of small communities. Similarly, the crowdsourcing platform is the one where mostly we find the internet-driven people experimenting their ideas and even finances. So basically it would be the creation of one such platform in our model where the members might be limited but readers are unlimited. This is christened to be "the expert forum," and the "governmental forum" which is nothing different from a Facebook page or social page handled by few but followed by many applications.

2 Related Works

Numerous works focus on rumor spreading and its propagation in the online social network, particularly in the microblogging field. In [1], Jiajia Huang and Qiang Su described rumor propagation by analyzing the user's behavior while browsing microblogs and proposed a model to explain rumor spreading due to various rumor preventing measures. In [2], the authors proposed a new SPNR model to identify the propagation relations. Their results showed that the new model is efficient for identifying and capturing rumor spreading in OSN's. In [3], the writers worked on an efficient method of crowdsourcing the gossip with the help of annotation schemes. The annotations were assigned by a set of journalists, therefore, making it more accurate. In [4], the authors try to explain how the crowdsourcing platforms help during emergencies as the response agencies look for real-time data about the emergency through the social media. In [5], the authors have proposed a model automated verification of rumors which according to their simulations can correctly predict the veracity of the rumor faster than any other resource consisting of journalists and the administrative departments for twitter databases. In [6], the authors concentrate on rumor source detection infinite graphs while considering boundary effects by message-passing algorithms. Here they solve the maximum likelihood estimation problem using a compromised rumor centrality for spreading in graphs containing boundary effects. In [7], the authors consider the herd mentality mechanism to study the propagation of rumors and their dynamic behavior. In [8], the authors focus on determining the source of the rumor considering nonhomogeneous, random edge weights on the social network graph. This makes it more realistic as the real-time scenarios of the social network appear to be in the same model.

3 Ethnic Rumor and Violence

Rumors have always occupied a major part in ethnic clashes. Rumors activate individuals and give avocation to commit brutality. Over false bits of gossip, a few stories are dramatically overemphasized by one-sided and unreliable people with the help of social networks [11]. Numerous spots on the planet have endured significantly because of rumor-initiated clashes. Indeed, even highly developed countries have not gotten away from the scourge of rumors. These rumors were never appeared to be valid. In any case, for the general population who heard them, there was a dread of things to come which drove them to confer fierce acts. Hence, bits of gossip can encourage clashes by (a) making individuals have negative convictions about the condition of the world and (b) assembling individuals and planning their activities by means of regular data. Hence these rumors need to be detected and identified as soon as possible to stop it from further spreading. This can decrease the violence caused due to rumors in an exponential rate.

4 Crowdsource Platform

Planning a group (a substantial gathering of individuals on the web) to do miniaturized scale work (little employments) that takes care of issues (that product or one client can't without much of a stretch do). Web-based crowdsourcing is one of the broadly utilized techniques for leading social tests with huge quantities of individuals. In contrast with the conventional method for performing offline tests, crowdsourcing on the internet empowers researchers to gather monstrous measures of data. Crowdsourcing is a sourcing model in which people or associations acquire merchandise and ventures [12]. These facilities incorporate thoughts and accounts, from a vast, generally open and quickly advancing gathering of web clients; it separates work between members to accomplish a total outcome. The word crowdsourcing itself is a combination of group and outsourcing and was created in 2005. As a method of sourcing, crowdsourcing existed before the advanced age. Crowdsourcing has shifted mostly to the Internet, which gives an especially helpful scene to crowdsourcing since people have a tendency to be more open in electronic ventures where they are not being physically judged or investigated, and in this manner, the sharing of thoughts would be more comfortable. The use of crowdsourcing platforms helps us to identify the rumors through public intervention. A crowdsource platform namely "Enquiry Forum" is built in our model consisting of two communities such as Experts Community and Governmental Community.

5 Knowledge-Based Community Detection

A knowledge community is constructed from the union of learning administration as a field of study and social trade hypothesis. The knowledge community was previously known as a discourse community. This community has advanced from different physical forums and web forums. Knowledge communities are frequently alluded to as a network of training or virtual network of training. Programming that utilizes computerized reasoning or master framework procedures in critical thinking forms can be called as a knowledge-based community system. It includes a store (database) of master information with couplings and linkages intended to encourage its recovery in the reaction of particular investigations or to exchange skill starting with one area of learning then onto the next [10]. Knowledge communities can be said to have advanced from announcement board frameworks, web gatherings, and online talk networks through the 90s. Knowledge communities support and encourage progressing connections and a knowledge ecosystem where thoughts are traded on a continuous premise. With the best possible use of information, the knowledge communities-based frameworks increment profitability, archive uncommon information by catching scare skill, and upgrade critical thinking abilities in the most adaptable way. Such frameworks additionally archive learning for later use and preparing. This prompts expanded quality in solving of problem process [9]. Knowledge communities utilize an assortment of a two-way specialized tool with communication to cultivate discourse and the trading of thoughts and ideas. With the help of the knowledge-based community detection, the rumors are categorized based on the community to which it belongs. Thus the use of knowledge-based community detection here, is highly focused for the source identification of rumors.

6 Proposed Approach

For a given situation or case of ethnic violence, we first require to distinguish between the conformity to facts of a rumor. We all know the rumor is the infection of one's mind, hence it is to be ensured that whether it is a genuine or false statement, shared and spread out knowingly and unknowingly. The extent of a rumor about someone's personal life and living might be behaviorally and mentally disturbing. But rumors of social origin that gets an ethnic conflict to get aroused is far more dangerous and requires strict and quick action. We specifically focus on rumor spreading based on WhatsApp and Facebook platforms because most of the Indian population started using the online social network recently hence they are limited to these platforms.

Here we propose a supervised approach of processing a rumor, examining its veracity, and identifying a nearly confirmed identity with the help of backtracing algorithm which considers the timestamp attribute of an information as one of its important parameters. In the following content of this paper, we would describe the algorithm which will be used for rumor detection and an algorithm for identification of the root of the rumor.

6.1 Rumor Detection

With the help of crowdsourcing platform, we proposed an enquiry forum which is divided into two communities out of which one consists of the expert journalists who will analyze the depth and actual meaning of the information and determine whether the piece of information can cause any sort of chaos in the society. The other community will consist of the administrative department of the government which will also function to decide whether the information can cause any sort of problem. First, the opinion is taken from the experts about the information being a rumor or not, then the governmental community confirms whether it is a rumor or not. These two forums work together to form the “Enquiry Forum” in our proposed model. The output provided by the Enquiry Forum will be in form of annotations, i.e., The piece of information will be stored with an annotation provided by the Enquiry Forum which will finalize if the topic is a Rumor or not. Once this is done the rumors will be stored under the Rumor category and the others will be declared as News and stored under the News category. Figure 1 shows the proposed rumor detection model pictorially. Hence we throw our stepwise algorithm for the proposed rumor detection model.

6.2 Stepwise Algorithm for Rumor Detection Model

Input: Top 100 most shared messages
Output: Rumor and News segregation
Step-1: Select an online social network by considering each person in the network as a node and build an edge between these nodes if they have shared any information
Step-2: Verify the shared information on the basis of the content and assign a uniquely identified number as ID to each piece of information.
Step-3: Select the top 100 IDs of the most common widespread messages that are to be sent to the Enquiry Forum for classification.
Step-4: Send the gathered information to the Enquiry Forum for scrutinizing the information.
Step-5: The information which is classified as “Rumor” is stored under the Rumor Database

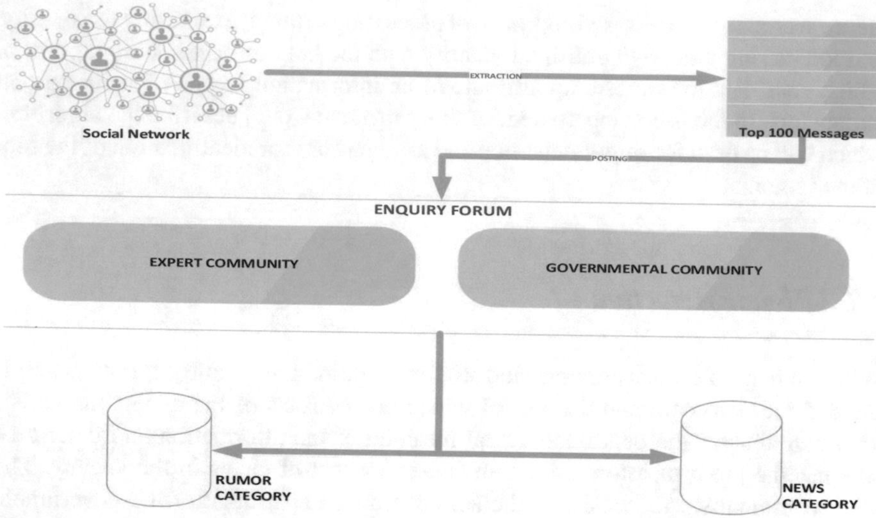

Fig. 1 Proposed rumor detection model

Step-6: The Rumor Database is sent for further processing and Rumor Source Identification.

6.3 Rumor Source Identification

Following the above steps helped us to detect the certainty of a rumor. Rumor detection led us toward our next step which is, rumor source identification. Source identification is important in rumor spreading because unless we know who the real culprit is, people would keep enjoying their anonymity to trouble others and cause unnecessary chaos. So to reach the source we have tried to propose a way where there is an extensive usage of knowledge-based community detection and backtracking with the use of time stamps. Timestamps are the attributes that are must with every message and is automatically available to us. Knowledge-based community detection would cluster the whole complex social network into several smaller communities. These messages in our previous section of rumor verification and analysis were allotted a unique identifying ID to distinguish them from all other normal messages. All the message IDs are categorized according to the annotations and stored in the arrays section wise. Message IDs from the rumor database are now compared on the basis of allotted annotations to them and based upon that the knowledge-based communities are formed.

In this section, we would sort the specifically identified messages on the basis of their respective timestamps in each of the communities formed by the knowledge-based community detection. Source Identification on the basis of timestamp would

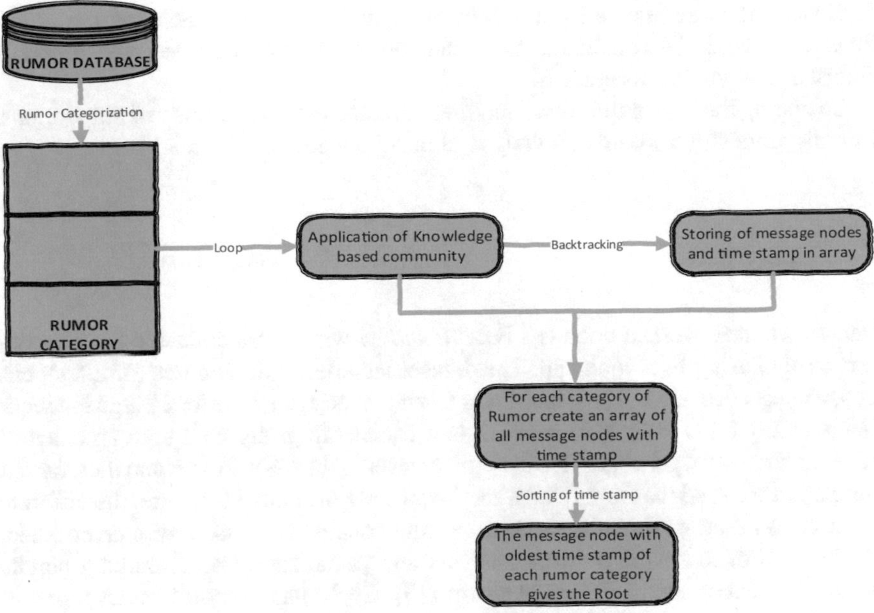

Fig. 2 Proposed rumor identification algorithm with the timestamp

require a well strategized algorithm and experimentation. Figure 2 shows the proposed rumor identification algorithm with timestamp pictorially.

6.4 *Stepwise Algorithm for Rumor Source Identification*

Input: The number of knowledge-based communities formed.
Output: Message with the oldest timestamp
Step 1: Consider the annotation of the knowledge-based community
Step 2: Store the message IDs section wise in an array according to the categorization.
Step 3: Consider each message ID to be a node.
Step 4: By backtracking on the basis of timestamps, keep storing the timestamps of the message IDs in an array. There will be a point when the message ID would no longer match, ceasing the backtracking at that point. Keep repeating for all the communities formed.
Step 5: Find the minimum of the timestamps from each array and store it in a new array.
Step 6: Now sort the new array ascendingly to find the oldest timestamp of the rumor.

Hence the node after all the backtracking and detection which has the oldest timestamp would be considered to be the root or near root or the most probable culprit of the whole fuss created.

So finding the root or the culprit requires further experimentation and simulations. But this proposition would definitely lead us all towards the source.

7 Experiment

Dataset we have worked upon is a Facebook data which was collected from survey participants using Facebook app. The dataset includes node features (profiles), circles, and ego networks. The dataset was downloaded from Stanford Large Network Dataset Collection [13]. This network is a friendship group data with an exact of 4039 nodes and 88234 edges. We took the dataset in the CSV format and then for our simulation we used the Gephi Software [14]. Using the above software, the different knowledge-based communities present in the considered dataset was determined. Gephi is an open source network analysis and visualization programming bundle written in Java on the NetBeans platform [15]. Gephi has been utilized as a part of various research extends in the scholarly community, news coverage, and somewhere else, for example in picturing the worldwide availability of New York Times content [16] and looking at Twitter arrange movement amid social unrest [17, 18] alongside more conventional system examination topics [19]. This is a product for Exploratory Data Analysis, a worldview showed up in the Visual Analytics field of research.

Figure 3 represents pictorially all the nodes and edges present in the dataset. The different communities are represented in different colors for their distinction. Then we run the modularity class and with a slight arrangement in the node and edges layout, we found the several clusters that were formed in this network (16 communities). The modularity was found to be 0.835 approximately. The largest community was formed with 548 nodes (13.567%) and least with 19 nodes (0.47%).

From Fig. 4, the 16 communities are clearly distinguished. The modularity was found to be 0.835. 16 different communities are formed, each community consisting of different number of nodes. The graph of the size distribution against each community is shown in Fig. 5. The communities shown in the graph has the following node distribution:

Community 1: 350 (8.67%)
Community 2: 433 (10.72%)
Community 3: 62 (1.53%)
Community 4: 44 (1.08%)
Community 5: 336 (8.32%)
Community 6: 543 (13.44%)
Community 7: 435 (10.77%)
Community 8: 423 (10.47%)
Community 9: 25 (0.62%)

Community 10: 206 (5.1%)
Community 11: 19 (0.470%)
Community 12: 226 (5.595%)
Community 13: 73 (1.807%)
Community 14: 237 (5.87%)
Community 15: 19 (0.470%)
Community 16: 548 (13.57%)

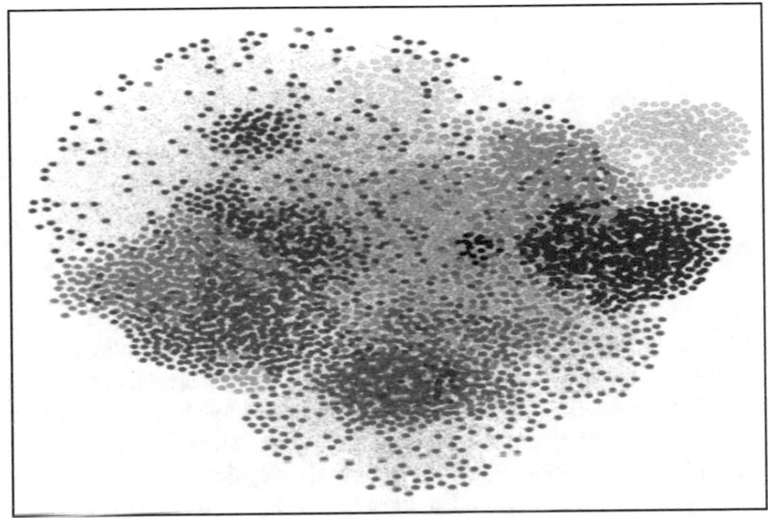

Fig. 3 Pictorial representation of nodes and edges in the dataset

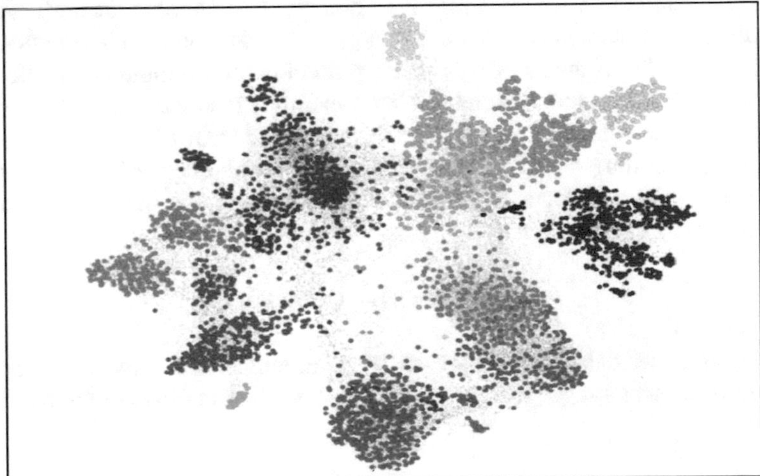

Fig. 4 Pictorial representation of 16 communities in the dataset

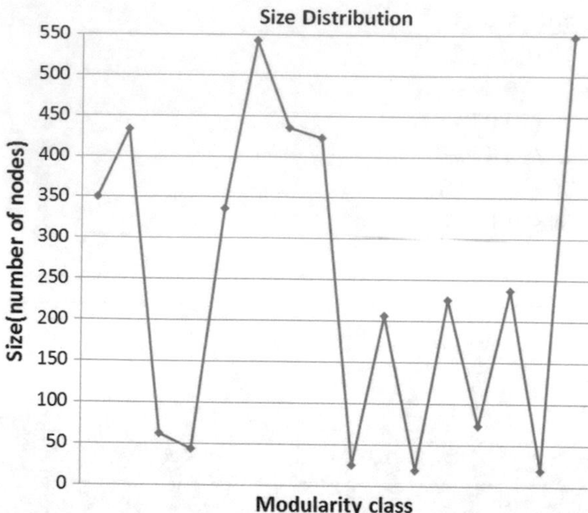

Fig. 5 Number of nodes in 16 communities

Figure 5 shows how our knowledge-based communities can be a small unit within a large network. This small model can find the root culprit of each community with timestamps being taken into account. Our proposed approach tends to sort the time stamp attributes of every node in each community formed to find the oldest time in their respective communities. For this, we use the "**Quicksort**" sorting algorithm which is the fastest when compared with other sorting techniques in terms of time complexity. The time complexity of the Quicksort algorithm is **O(n log *n*)** in the best case scenario and in the worst case scenario it gives an efficient time complexity of $\mathbf{O(n^2)}$. Therefore, the approximate time complexity of finding the node with the oldest time in a community will be **O(n log *n*)** considering it to be the best case scenario, where "n" is the number of nodes present in each community. In the entire social network if the rumor forms "**m**" communities then the time complexity of finding the oldest times in "**m**" communities will be $\boldsymbol{m} \times$ **O(n log *n*)**. Now we need to sort the oldest times of all communities to find the oldest of all the times giving us the node which will accurately be the source of the rumor, therefore the overall time complexity can be given as

$$\mathrm{m} \times \mathrm{O}(\mathrm{n}\ \log n) + \mathrm{O}(m \log m)$$

In the simulated dataset, we observed 16 communities hence assuming that each community has an average "n" number of nodes, our time complexity for the dataset will be

$$16 \times \mathrm{O}(\mathrm{n}\ \log\ n) + \mathrm{O}(16 \log 16)$$

Therefore, our proposed model may have a slightly higher time complexity due to several sorting involved compared to other rumor source identification models but here we can meticulously reach the root node and thereby being accurate in finding the source of rumor.

8 Advantages of Our Proposed Model

Earlier the models proposed upon rumor spreading, their detection and source identification were quite formulated and technically sound. But it did not have direct involvement of the administrative and the governmental organizations and experts. The involvement of such officials and experts would keep people more warned. It would efficiently keep the public aware of the reality of such situations and rumors. People would gradually stop believing things they hear, see or come across the online social networks and would start asking questions about the fact being approved as a rumor by the government. The greatest defeat the rumor spreader would face is that when people start disbelieving his rumors and stop spreading it further to create problems in the society.

9 Conclusion and Future Work

With this paper, we wish to develop the process of classification of information into rumor and verify its veracity. Here we minimize the source detecting process by using knowledge-based community detection, therefore, we make the process a bit less complex and efficient. Now that we may find the root or the near root with the proposed model we tend to help the society by preventing issues like ethnic violence to cause damage to the community and welfare of the community members with just a mere piece of false information. In the future work, we would work on developing the working version of this model and perform simulations and measure the accuracy of rumor detection and source identification.

References

1. Huang, J., Su, Q.: A Rumor Spreading Model Based on User Browsing Behavior Analysis in Microblog. IEEE (2013)
2. Bao, Y., Yi, C., Xue Y., Dong, Y.: A new rumor propagation model and control strategy on social networks. In: 2013 IEEE/ACM International Conference on Advances in Social Networks Analysis and Mining, pp. 1472–1473
3. Zubiaga, A., Liakata, M., Procter, R., Bontcheva K., Tolmie, P.: Crowdsourcing the annotation of rumourous conversations in social media. In: International World Wide Web Conference Committee (IW3C2), pp: 347–353. ACM

4. McCreadie, R., Macdonald C., Ounis, I.: Crowdsourced rumour identification during emergencies. In: International World Wide Web Conference Committee (IW3C2), pp: 965–970. ACM
5. Vosoughi, S., Mohsenvand, M.N., Roy, D.: Rumor gauge: Predicting the veracity of rumors on Twitter. ACM Trans. Knowl. Discov. Data **11**(4), Article 50
6. Yu, P.-D., Tan, C.W., Fu, H.-L.: Rumor source detection in finite graphs with boundary effects by message-passing algorithms. In: 2017 IEEE/ACM International Conference on Advances in Social Networks Analysis and Mining, pp. 86–90
7. Wang, J., Wang, Y.-Q., Li, M.: Rumor spreading considering the herd mentality mechanism. In: Proceedings of the 36th Chinese Control Conference, pp: 180–1485
8. Louni, A., Subbalakshmi, K.P.: Who spread that rumor: finding the source of information in large online social networks with probabilistically varying internode relationship strengths. IEEE Trans. Comput. Soc. Syst. **5**(2), 335–343 (2018)
9. Sajja, P.S., Akerkar, R.: Knowledge-based systems for development. In: Sajja and Akerkar (eds.) Chapter 1, TMRF e-Book Advanced Knowledge Based Systems: Model, Applications and Research, vol. 1, pp. 1–11 (2010)
10. Dey, P., Chatterjee, A., Roy, S.: Knowledge based community detection in online social network. In: 2018 10th International Conference on Communication Systems and Networks (COMSNETS). IEEE (2018)
11. Basu, P., Dutta, S., Shekhar, S.: Ethnic conflicts, rumours and an informed agent. In: Urban Ethnic Conflicts
12. Tung, Wei-Feng, Jordann, Guillaume: Crowdsourcing social network service for social enterprise innovation. Inf. Syst. Front. **19**(6), 1311–1327 (2017)
13. https://snap.stanford.edu/data/ego-Facebook.html, Last visited: 25–07-2018
14. https://gephi.org/, Last visited: 25–07-2018
15. Bastian, M., Heymann, S., Jacomy, M.: Gephi : An open source software for exploring and manipulating networks. In: Third International AAAI Conference on Weblogs and Social Media, AAAI Publications (2009). Accessed 2 Nov 20011
16. Leetaru, K.H.: Culturomics 2.0:Forecasting Large-scale human behavior using global news media tone in time and space (2011). First Monday, Accessed 22 Nov 2011
17. Aouragh, M.: Collateral Damage: #Oslo Attacks and Proliferating Islamophobia, Jadaliyya (2011). Accessed 22 Nov 2011
18. Panisson.: The Egyptian revolution on Twitter - Featured on the PBS news hour, YouTube (2011). Accessed 22 Nov 2011
19. Correa, D.C.: Using digraphs and a second-order Markovian model for rhythm classification. Complex Networks (2011). Accessed 22 Nov 2011

A Non-linear Approach for Classifying Malignant Neoplasm Using Dualist Optimization Algorithm

Prachi Vijayeeta, M. N. Das and B. S. P. Mishra

Abstract This paper provides a framework for non-linearly classifying microarray dataset to determine the existence of malignant neoplasm in the patient's sample set. A $m \times n$ microarray technology is used to represent the sample data of the patients. Our model aims at predicting the class label of an unknown new sample that enters into the system during the runtime. With the help of microarray data, we have collaboratively applied game theory approaches along with computational methods to solve the problem. An apodictic approach for the construction of an optimized RFE (*Recursive Feature Elimination*) feature selection model using Dualist Algorithm is incorporated. Furthermore, the optimized features are subjected to a non-linear classification using decision tree, k-nearest neighbor and logistic regression on Wisconsin Breast cancer dataset. The simulations carried out using the above techniques have proved Dualist algorithm with RFE combined with four different linear classifier models like logistic regression, k-nearest neighbor, decision tree and random forest to be a better choice for classification.

Keywords Microarray · Classification · RFE · Dual-RFE · K-fold cross-validations

1 Introduction

A revolutionary technique named as Microarray technology is an integrated two-dimensional array that stores a huge sequence of relevant information generated from high throughput technologies. Recent studies have illuminated cancer, to be a

P. Vijayeeta · M. N. Das (✉) · B. S. P. Mishra (✉)
School of Computer Engineering, Kalinga Institute of Industrial Technology, [Deemed to Be University], Bhubaneswar, India
e-mail: mndas_prof@kiit.ac.in

B. S. P. Mishra
e-mail: bsmishrafcs@kiit.ac.in

P. Vijayeeta
e-mail: pvijayeetafca@kiit.ac.in

H. S. Behera et al. (eds.), *Computational Intelligence in Data Mining*, Advances in Intelligent Systems and Computing 990,
https://doi.org/10.1007/978-981-13-8676-3_49

genotypic disease that occurs due to a series of assimilation of genetic alterations in oncogenes, tumour suppressor genes and stability genes [1]. But so far microarray is concerned, almost every researchers have suffered from a great hurdle of the high dimensionality of dataset due to the presence of irrelevant genes. In our work, we have tried to overcome the curse of dimensionality by selecting a subset of relevant features using RFE technique and then to choose the optimal features we have implemented the Dualist Algorithm. Four non-linear classification techniques namely decision tree, k-nearest neighbor [2] and logistic regression, random forest were applied on Wisconsin Breast Cancer to classify the occurrence of cancerous or non-cancerous cells. The Dualist Optimization algorithm applied in this paper is basically a population-based human inspired algorithm. The most significant property of this algorithm is to minimise the blind effect of getting a better optimal solution and thus succeeds in generating the global optimal solution for the best features in the entire search space.

Totok Ruki [3] along with his team formulated this algorithm by analysing the behaviour of the players during a tournament. The most pertinent players (features in our problem) are selected based upon their skill set, strength, intellectual property, luck and decision-making capability. The non-pertinent players are thereby removed from the team and the remaining players are trained for further competing in the game. This iterative procedure goes on at a faster rate in dual procedure algorithm as compared to other evolutionary game theory and nature-inspired algorithm. In DA, all the individuals in a population are defined as Dualist who keep on fighting one-by-one to determine the winners or losers or champions by taking into account their strength, skill, luck and intellectual capability. The loser learns from the winner and in the meanwhile the winner tries to improve their new potential by training or by adopting some new methods from the opponents. The dual participant with highest fighting skill is called champion who can train a new dualist. The newly generated dualist shall further participate in the tournament as a representative of each champion. Combined, all the dualist are re-evaluated for the next match and the one possessing the minimal fighting capability will thereby be rejected from the list.

2 Literature Review

Abreu [4] have lucidly surveyed many Machine Learning (ML) techniques and estimated their performance on local and open-source databases. Ahmad et al. [5] had experimented Decision Tree, Random forest, SVM, K-means algorithms are integrated with feature selection methods (like RFE, Chi-square test, correlation-based feature selection) to discover hidden patterns for developing predictive models were applied on it. Agarap et al. [6] had applied GRU-SVM(Gated Recurrent Unit-SVM) model on Wisconsin benchmark dataset that resulted in a better complexity. Michalis et al. [7] had integrated ML techniques (like ANN, Kernel-SVM,)along with robust feature selection methods. Yue et al. [8] have reviewed a couple of ML techniques applied in the intelligent healthcare system. Their study concluded that Switched Particle Swarm Optimization (SPSO) based on Markov chain has achieved faster local convergence speed with higher accuracy. Ahmed et al. [9] had applied generative approach called Probabilistic Principal Component Analysis (PPCA)for dimensionality reduction and Gaussian Mixture Model (GMM) as well as Mixture of Normalised Linear Transformation for classification. Wang et al. [10] had proposed a Deep Learning algorithm and a Deep graph to yield the coefficient of determination of breast arterial cancer. Chandra et al. [11] had proposed a non-linear integrated model named Spiking Wavelet Radial Basis Neural Network and compared the classification accuracy with Wavelet Probabilistic Model Building Genetic Algorithm (PMBGA) was proposed by Topon and Iba [12] for selection of a smaller size gene subset (Tables 1 and 2).

3 Motivation and Problem Formulation

In our work we have adopted a non-linear supervised learning method to conjecture the exact class type of an unknown sample arriving dynamically. Therefore, we are motivated to make use of a wrapper feature selection technology called ***Recursive Feature Elimination Method*** which removes the non-contributing features from the dataset in a recursive manner. Figure 1 represents the entire flow of our research work.

3.1 Problem Formulation

Let 'T' be a finite set and 'S' be a subset of T such that $S \subseteq T$.

Let $|T| = t$ be the cardinality of the finite set T.

Let $N = \{1, 2, 3 \ldots n\}$ be a set of 'n' players (features in our problem).

A coalition game with a pair of (N, v) takes a characteristic form of the game where the function v maps: $v : 2^n \Rightarrow \mathbb{R},\ v(\Phi) = 0$ (Table 3).

For selecting the players, we need to assign some weights such that

$$W : 2^n \Rightarrow Z \; where \; Z \; is \; an \; integer.$$

If 'q' be the majority quota and $w_1, w_2, \ldots w_n$ be the weights allotted to each ith player the Characteristic function is defined as

$$v(s) = \begin{cases} 1, \; if \; w(s) \geq q \\ 0 \;\; if \; w(s) < q, \\ \quad q \in Z \end{cases} \tag{1}$$

Table 1 For statistical property description of the dataset

SL_no.	Attribute name	Attribute type	Mean		Least value		Computational error		Class label
			Min. value	Max. value	Min. value	Max. value	Min. value	Max. value	
1.	Radius	float64	6.981	28.11	7.93	36.04	0.112	2.873	Malignant
2	Texture	float64	9.71	39.28	12.02	49.54	0.36	4.885	Benign
3.	Perimeter	float64	43.79	188.5	50.41	251.2	0.757	21.98	
4.	Area	float64	143.5	2501	185.2	4254	6.802	542.2	
5.	Smoothness	float64	0.053	0.163	0.071	0.223	0.002	0.031	
6.	Compactness	float64	0.019	0.345	0.027	1.058	0.002	0.135	
7.	Concavity	float64	0	0.427	0	1.252	0	0.396	
8.	Concave points	float64	0	0.201	0	0.291	0	0.053	
9.	Symmetry	float64	0.106	0.304	0.156	0.664	0.008	0.079	
10.	Fractal dimension	float64	0.05	0.097	0.055	0.208	0.001	0.03	

Table 2 For feature rank estimation

Features	Radius	Texture	Perimeter	Area	Smoothness	Compactness	Concavity	Concave points	Symmetry	Fractal dimension
Estimated ranks	**1**	**1**	9	14	8	16	**1**	**1**	15	4
Optimized feature subset (Dual + RFE)	**True**	**True**	False	False	False	False	**True**	**True**	False	False

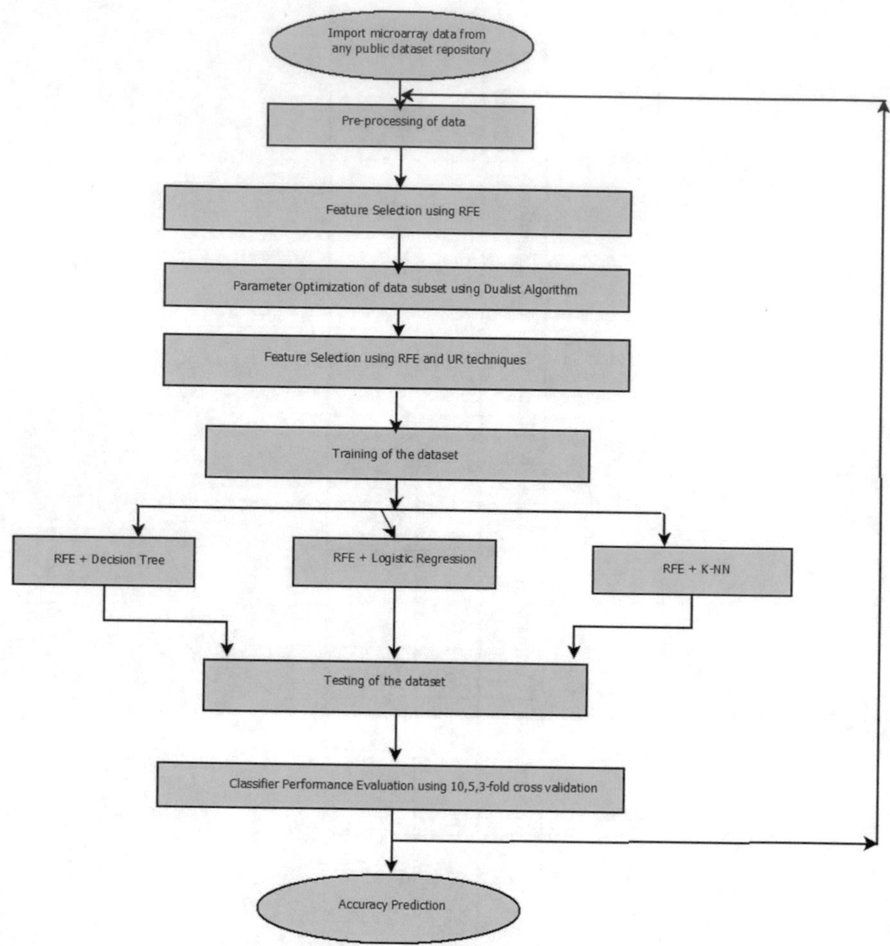

Fig. 1 Research framework diagram

Table 3 For confusion matrix representation of classifiers

Confusion matrix for logistic regression			Confusion matrix for random forest		
	Benign	Malignant		Benign	Malignant
Benign	118	2	Benign	133	2
Malignant	4	76	Malignant	4	70

Table 4 For summarising accuracy, specificity, sensitivity of classifier model with RFE

	Classifiers applied	Accuracy (%)	Specificity	Sensitivity
1.	Logistic regression + RFE	91.01	0.9432	0.9311
2.	KNN + RFE	95.02	0.9312	0.9112
3.	Decision tree + RFE	93.12	0.9144	0.8812
4.	Random forest + RFE	95.31	0.9521	0.9421

$$where\ w(s) = \sum_{i \in s} w_i \tag{2}$$

From Eq. (1), we can say $v(s)$ is the winning candidate that takes only binary values (either 1 or 0) depending upon the weighted sum of the players that are computed using Eq. (2). But in our problem, we are rather more concerned with finding out of those players that have minimal weights and need to be discarded from the game (i.e. optimised feature subset is required). Therefore, the fitness function is defined as

$$Minimise f = \arg(w_i)^2 where\ w \in Z\ and\ i \leq n. \tag{3}$$

Subsequently, the optimised input feature vector (obtained from Eq. (3))is subjected to a learning model for classifying an unknown sample set (Table 4)

3.2 Methods Applied

In this paper we have used four non-linear classifiers namely decision tree (DT), logistic regression (LR), k-nearest neighbor (KNN) and random forest (RF) to predict the class label of the sample. The statistical properties of the classifiers are represented in Table 5.

3.3 Research Framework

The basic framework of our research activity is modelled in a stepwise manner as follows:

Table 5 For representing statistical properties of four classifiers

	Statistical properties	Classifiers applied			
		Logistic regression + Dual-RFE	KNN + Dual-RFE	Decision tree + Dual-RFE	Random forest + Dual-RFE
1	Accuracy (%)	94.11	96.04	95.02	97.13
2	Specificity	0.9633	0.9523	0.9346	0.9722
3	Sensitivity	0.9474	0.9331	0.9048	0.9708
4	Positive prediction value	0.9232	0.9210	0.8906	0.9852
5	Negative prediction value	0.8842	0.9131	0.9434	0.9459
6	Prevalence	0.5412	0.6221	0.3706	0.6555
7	Detection rate	0.5821	0.5642	0.3353	0.6364
8	P-value	0.8811	0.9622	0.9811	0.6831
9	Confidence interval	0.8010, 0.9121	0.8431, 0.9543	0.8728, 0.9587	0.834, 0.9534
10	Class error	B-0.0252, M-0.0241	B-0.0254, M-0.0272	B-0.0263, M-0.0282	B-0.0249, M-0.0236

3.4 Algorithm for the Model Implemented

Step-1: *Input Pattern.*
Step-2: *Computation and Optimization of Ranking Weights for all Features (using Eq. (2))*

(i) Calculate the sorted list $C_i = (W_i)^2$.
(ii) Find the features possessing minimum weight:
$f = \arg\min(C_i)$ (Using Eq. (3)).
(iii) Apply *Dualist Optimisation algorithm* to optimize the weights. Call function ***dualist_opti()***.

Step-3: *Output Pattern.*

3.4.1 Dualist Optimization Algorithm (*Dualist_Opti()*)

Step-1: *Registration of Dual Candidates*
Input an n-dimensional binary array for representing the skill set of the Dualist such that:$N_{var} = [1, 2, 3 \dots n]$.
Step-2: *Pre-Qualification Evaluation*

Considering the skill set data, evaluate and test the fighting potential of each candidate within it.
Step-3: *Determination of Champion Board*
Each champion takes the responsibility to train a new list including himself as for their dual capabilities.
Step-4: *Dual List Improvement*
Each dualist will fight using the fighting skill and luck to determine the winner and loser as follows:

(a) **Initialization**:- A_Dualist[ ] and B_Dualist[ ] two Binary array and Luck_coefficient.
(b) **Computation of luck**:-

$$\begin{aligned}\mathrm{A(luck)} &= \mathrm{A(fighting\_skill)} * (\mathit{luck\_coefficient} \\ &\quad + (\mathit{rand}\,(0-1) * \mathit{luck\_coefficient})) \\ \mathrm{B(luck)} &= \mathrm{B(fighting\_skill)} * (\mathit{luck\_coefficient} \\ &\quad + (\mathit{rand}\,(0-1) * \mathit{luck\_coefficient}))\end{aligned}$$

(c) **Winner-Loser Decision making**:-

$$\boldsymbol{if\left(\left(A\left(fighting_{skill}\right)+A(luck)\right)>=\left(B\left(fighting_{skill}\right)+B(luck)\right)\right)}$$

then

$$\boldsymbol{A(winner)=1\,and\,B(winner=0)}$$

else

$$\boldsymbol{A(winner)=0\,and\,B(winner=1)}$$

End if

Step-5: *Define a Dual Schedule Between Each Dualist*
The dualist fighting capabilities need to be improved for all three participants:
Step-6: *Elimination Phase and Finish.*
Elimination of features with low weights. The remaining features with better weights are obtained and ranked as shown in Table 2 and plotted in Fig. 2.

4 Experiment and Results

We have used Wisconsin Breast Cancer Dataset **(WBCD)** for our experiment (https://archive.ics.uci.edu/ml/datasets/Breast+Cancer+Wisconsin+(Diagnostic). The software tool used is scikit-learn with python-anaconda. Table 1 represents the statistical

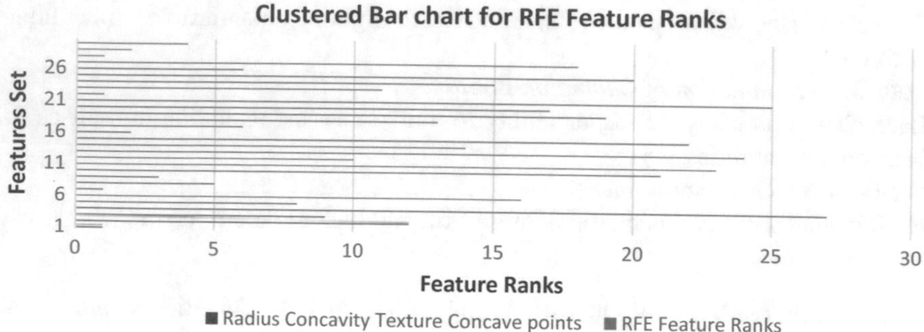

Fig. 2 Graph for feature rankings

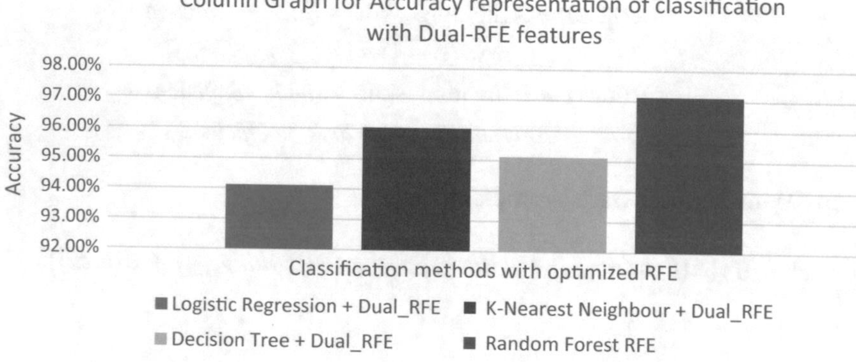

Fig. 3 Graph for accuracy 'versus' classification with dual-RFE optimised feature selection technique

properties of 569 samples with 31 attributes and 2 class labels (**Benign/Malignant**). Among all the methods used, Random Forest [13] with Dual-RFE yields an accuracy of 97.13%, followed by KNN with an accuracy of 96.04% as shown in Table 4 and plotted on column graph in Fig. 3. In our work, we have split our dataset into 80% training and 20% test dataset. The later one succeeds in generating a better accuracy of nearly 1.985% on an average more than the previous method. Table 3 depicts a confusion matrix that simulates the correct and incorrect classification instances.

Performance Measurement Criteria

To improve the predictive performance and accuracy of the learning models, external validation is an absolute necessity [7]. We have applied k-fold cross-validation and have tested the accuracy for $k = 10, 5, 3\,folds$ as depicted in Table 6 and plotted in Fig. 4. In the k-fold validation, each sample is executed k-times for training and only once for testing.

Table 6 For performance estimation using k-fold cross-validation

Classifier applied	Average accuracy with tenfold cross-validation (in %)	Average accuracy with fivefold cross-validation (in %)	Average accuracy with threefold cross-validation (in %)
1. LR+Dual-RFE	88.31	91.11	92.23
2. KNN+Dual-RFE	95.91	95.70	96.89
3. DT+Dual-RFE	93.35	94.93	95.36
4. RF+Dual-RFE	97.14	97.23	97.16

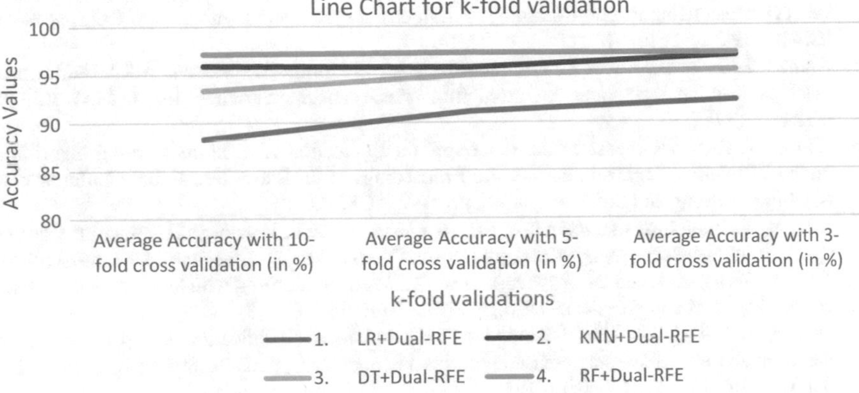

Fig. 4 Line graph for K= ten , five , threefold validations

5 Conclusion and Future Scope

Evolutionary adaptations increase the overall fitness of the group but occur at a cost to the individual. The main objective of this paper is to study the impact of dualist optimization algorithm for feature subset selection on WBCD dataset. We have compared the various statistical properties of four different classification models with Recursive feature elimination method as well as with optimized Dualist RFE method. Moreover, the performance of the classifier for k = 10, 5, 3 fold cross validation is evaluated and the variations in the average accuracy. For future work, we need to apply many more optimization techniques which may be any evolutionary/biologically inspired/population-based algorithms on various filter/wrapper/ensemble feature selection methods. Efficient algorithms could be formulated with a view of generating the best global solution with a reduced computational cost.

References

1. Vogelstein, B., Kinzler, K.W.: Cancer genes and the pathways they control. Nat. Med. **10**(8), 789 (2004)
2. Mount, D.W., Putnam, W.V., Centouri, S.M., Manziello, A.M., Pandey, R., Garland, L.L., Martinez, J.D.: Using logistic regression to improve the prognostic value of microarray gene expression data sets: application to early-stage squamous cell carcinoma of the lung and triple negative breast carcinoma. BMC Med. Genomics **7**(1), 33 (2014)
3. Biyanto, T.R., Fibrianto, H.Y., Nugroho, G., Hatta, A.M., Listijorini, E., Budiati, T., Huda, H.: Dualist algorithm: an algorithm inspired by how dualist improve their capabilities in a duel. In: International Conference in Swarm Intelligence, pp. 39–47. Springer, Cham (2016)
4. Abreu, P.H., Santos, M.S., Abreu, M.H., Andrade, B., Silva, D.C.: Predicting breast cancer recurrence using machine learning techniques: a systematic review. ACM Comput. Surv. (CSUR) **49**(3), 52 (2016)
5. Ahmad, L.G., Eshlaghy, A.T., Poorebrahimi, A., Ebrahimi, M., Razavi, A.R.: Using three machine learning techniques for predicting breast cancer recurrence. Int. J. Med. Inform. **4**(124), 3 (2013)
6. Agarap, A.F.M.: On breast cancer detection: an application of machine learning algorithms on the Wisconsin diagnostic dataset. In: Proceedings of the 2nd International Conference on Machine Learning and Soft Computing, pp. 5–9. ACM (2018)
7. Kourou, K., Exarchos, T.P., Exarchos, K.P., Karamouzis, M.V., Fotiadis, D.I.: Machine learning applications in cancer prognosis and prediction. Comput. Struct. Biotechnol. J. **13**, 8–17 (2015)
8. Yue, W., Wang, Z., Chen, H., Payne, A., Liu, X.: Machine learning with applications in breast cancer diagnosis and prognosis. Designs **2**(2), 13 (2018)
9. Otoom, A.F., Abdallah, E.E., Hammad, M.: Breast cancer classification: comparative performance analysis of image shape-based features and microarray gene expression data. Int. J. Bio Sci. Bio Technol. **7**(2), 37–46 (2015)
10. Wang, J., Ding, H., Bidgoli, F.A., Zhou, B., Iribarren, C., Molloi, S., Baldi, P.: Detecting cardiovascular disease from mammograms with deep learning. IEEE Trans. Med. Imaging **36**(5), 1172–1181 (2017)
11. Chandra, B., Naresh Babu, K.V.: Classification of gene expression data using spiking wavelet radial basis neural network. Expert. Syst. Appl. **41**(4), 1326–1330 (2014)
12. Paul, T.K., Iba, H.: Gene selection for classification of cancers using probabilistic model building genetic algorithm. Bio Syst. **82**(3), 208–225 (2005)
13. Ekgedawy, M.N.: Prediction of breast cancer using random forest, support vector machines and Naive Bayes. Int. J. Eng. Comput. Sci. **2** (2017)

A Comparative Study Using Linear Model, Decision Tree and Random Forest for Classification of Student's Self Satisfaction with Family Aspect Parameters of Students

Aniket Muley, Parag Bhalchandra, Govind Kulkarni and Mahesh Joshi

Abstract This paper is outcome of an experimental work carried out over educational data where linear model, decision tree and random forest have been deployed over own educational dataset to get knowledge for academic analytics, which otherwise is invisible. The linear model, decision tree and random forest models have been compared and it is observed that the random forest approach is more suitable and gives accurate results. Analytics were carried out with R software package.

Keywords Decision trees · Classification · Random forest · Linear model

1 Introduction

Over the years, Data mining and analytics have enabled us to gain significant insights from potentially large databases. These insights have opened new opportunities to understand correlation between variables. This has led to improved decision making and corrective actions. Data mining and analytics merely have common algorithms including classification, clustering, association, prediction, sequential patterns and decision trees in its compass. These techniques have classified as descriptive and predictive approaches. Data analytics work need a database and deployment of any one of above data mining algorithms. The end results are patterns, correlations, signifi-

A. Muley (✉)
School of Mathematical Sciences, S.R.T.M. University, Nanded 431606, MS, India
e-mail: aniket.muley@gmail.com

P. Bhalchandra · G. Kulkarni
School of Computational Sciences, S.R.T.M. University, Nanded 431606, MS, India
e-mail: srtmun.parag@gmail.com

G. Kulkarni
e-mail: govindcoolkarni@gmail.com

M. Joshi
School of Educational Sciences, S.R.T.M. University, Nanded 431606, MS, India
e-mail: maheshmj25@gmail.com

H. S. Behera et al. (eds.), *Computational Intelligence in Data Mining*, Advances in Intelligent Systems and Computing 990,
https://doi.org/10.1007/978-981-13-8676-3_50

cance or summary. Our underlined research work dealt with use of decision tree. It is a tool for classification and prediction. The classification approach predicts categorical class labels and then use labels of the training data to classify new data. Whereas the prediction models continuous-valued functions in order to predict unknown or missing values [3]. A Decision tree is a basically a tree structure with leaves and branches. The leaves are nodes. Further, nodes explore features and split denotes rule and classifies in the form of leaf. There are some famous algorithms to build a decision tree that exploits by Gini index. The second is Iterative Di-chotomiser 3 with employed entropy formulae [3, 6]. Our research is a preliminary endeavor to use data mining functions to analyze and evaluate student academic data using decision trees. This analysis is interesting and can augment the quality of the higher education as the main output is insight, which otherwise invisible by naked eye. The discovered myths and patterns can be used for better resource management. Despite the percentage of GDP for the year 2013–14 was 1.34% [7]; the rural background students' performance is not increased in Indian universities. This is a research problem. We together thought that there could be other aspects associated with the performance of the students. Since these other aspects are not visible directly, this is a case for data mining. A multiparty illustration of exertion is performed by two departments of our University. The key objective of this study is to escalate collected datasets through data mining. The minor research objective is to investigate whether family aspects significantly contribute to self satisfaction and finally personal variables associated with their performance. These objectives are worked out by constructing decision trees and random forest. Its openly understood that [3, 6], from the decision tree, that we can see a feature as the root node if it has the highest Gini Impurity value, then next child as remaining highest Gini Impurity value and so no. Also a decision tree picks only few features or a subset of some features for the classification, though database has many features stored in it. We can use this information to select which features/variables in a general dataset are important for more advanced models like neural network or deep learning. The Receiver Operating Characteristic (ROC) is a plot that can be used to determine the performance and robustness of a binary or multi-class classifier. It gives information about their proportion of pace something called the C-statistic or area under ROC curve (AUROC) for each class predicted by the classifier [3, 6]. However there is one ROC for each class predicted by the classifier. The AUROC is defined as the probability that a randomly selected positive sample will have a higher prediction value than a randomly selected negative sample. This is how random forest comes in picture. A random forest is ensemble of decision trees which is based on a fraction of the number of rows. The rows are selected at random and a particular number of features are selected at random to train on and a decision tree is built on this subset. Similarly a number of decision trees are grown. Below sections briefs experimental setup and discussions related to such performance analysis.

2 Research Methodology

A closed questionnaire was used to construct database of University students. This has 360 records and 46 fields. This database has students personal, social, habitual and financial aspects stored in it. The sample questionnaire is shown in Fig. 1. The supervised learning methods and their algorithms were implemented using R software [4, 13]. Rattle is a free graphical user interface for Data Science, developed using R. The decision tree model is one of the most common data mining models in Rattle. The algorithms use a recursive partitioning approach [9]. The traditional algorithm is implemented in the rpart package. The ensemble approaches tend to produce models that exhibit less bias and variance than a single decision tree. The main objective of this study is to develop decision tree for personal satisfaction level of students based on various parameters. After analyzing contemporary works [1, 2, 5, 8, 10], typical spreadsheet software, MS-Excel 2007v was utilized. Standard policies for data preparation and cleansing were done through standard ay viz., Extract, Transform and Load (ETL) process was followed to make database qualifying for data mining experimentations [3]. The transformation phase took significant amount of time as we cleansed data first by mapping NULL to 0, etc., then filtering; splitting, transposing etc. came in picture. Keeping all these contemporary research updates, the ultimate goal is to examine the relationship between student's personal satisfaction and his/her performance related variables and analyzed through data miner in R software.

1	Course code	MSc (5) , MCA (6)
2	Your name	
3	Gender (sex)	Male (1) Female (0)
4	Marital status	Married (2) unmarried(3)
5	Age	
6	Home address	Urban(1) rural (2) foreign(3)
7	Mobile no.	
8	Personal email id	
9	Degree passer and percentage	General B.Sc. / B.Sc.(computer CS) / BCA / BCS/ Other / (1) (2) (3) (4) (5)
	Percentage	
10	Degree collage name	
11	Father's Education	Below or SSC/ HSC/ Graduate/ Post Graduate/ other (1) (2) (3) (4) (5)
12	Fathers job and annual income	Service / Business/ Agriculture/ In house/ Other/ (1) (2) (3) (4) (5)
	Income	0-1 lakh (1) , 1.1-2 lakh(2), 2.1-5 lakh(3) , 5lakh - above (4)
13	Mothers education	Below or SSC/ HSC/ Graduate/ Post Graduate/ other (1) (2) (3) (4) (5)
14	Mothers job and annual income	Service / Business/ Agriculture/ In house/ Other/ (1) (2) (3) (4) (5)
	Income	0-1 lakh (1) , 1.1-2 lakh(2), 2.1-5 lakh(3) , 5lakh - above (4)
15	Family size	
16	Family relationship	Excellent / Good/ Satisfactory/ Bad/ Very Bad (1) (2) (3) (4) (5)
17	Family support to your education	Excellent / Good/ Satisfactory/ Bad/ Very Bad (1) (2) (3) (4) (5)
18	Reason to choose this course	Career in IT/ Near to Home/ Reputation of course / Blind Decision/ Parents wish: (1) (2) (3) (4) (5)
19	Travel mode and time needed	Bus/ Railway/ City Bus/ Rickshaw/ Self Vehicle / walking (6) (1) (2) (3 but taken as 1) (4) (5)

Fig. 1 Sample questionnaire

Decision tree and random forest learning make use to explores the terms of matching conflict [11]. Hence, the decision trees need Gini index to be computed. For this, 02 objects selected from a population at random such that, they must be of same class and its probability is one. The higher value of the Gini represents the higher the homogeneity [11]. The term linear workout with a piece of aspect for formation of a model that can estimate the target value [12]. In this study, we have performed comparative analysis educational data set using decision tree, random forest and linear model approach. Among these comparative methods our interest is to identify suitable method for similar kind of study.

3 Experimentations and Discussions

To devise out decision tree, some set of experiments carried out. For the analysis purpose we have used supervised learning technique. Our aim is to evaluate decision tree induction method. To draw the decision tree (Fig. 2), rattle uses rpart package. Total number of population selected was n = 251. Table 1 represents the terminal node

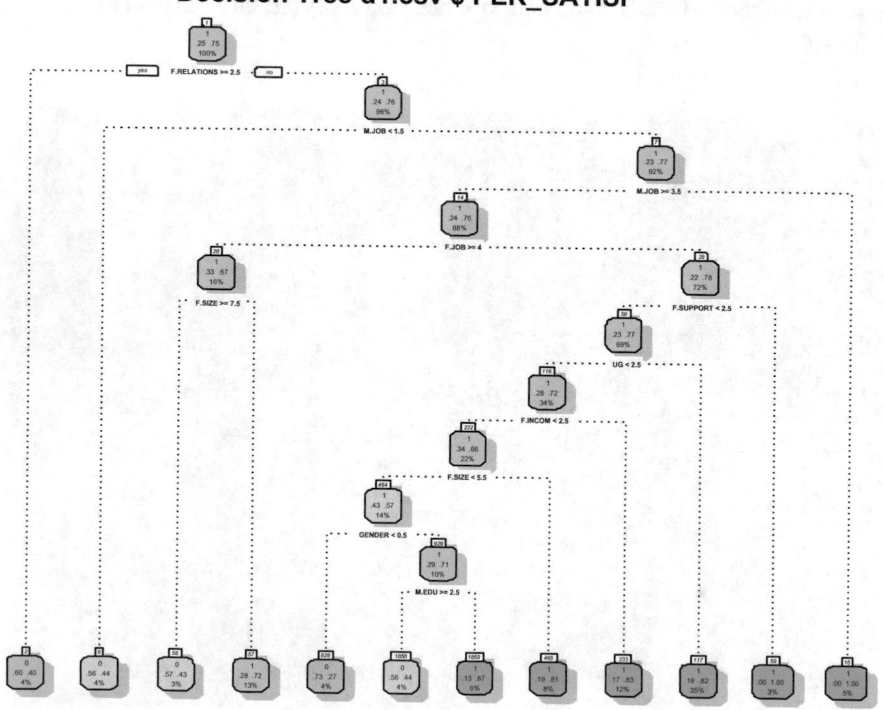

Fig. 2 Decision tree

Table 1 Details of classification of nodes

Variable	Nodes	Children generated	Probability
Root	251	64.1	(0.25498008 0.7450199)
F.RELATIONS ≥ 2.5	10	4.0	(0.6000000 0.4000000)[a]
F.RELATIONS < 2.5	241	58.1	(0.2406639 0.7593361)
M.JOB < 1.5	232	53.1	(0.2284483 0.7715517)
M.JOB ≥ 1.5	220	53.1	(0.2409091 0.7590909)
M.JOB ≥ 3.5	39	13.1	(0.3333333 0.6666667)
F.SIZE ≥ 7.5	7	3.0	(0.5714286 0.4285714)[a]
F.SIZE < 7.5	32	9.1	(0.2812500 0.7187500)[a]
F.JOB < 4	181	40.1	(0.2209945 0.7790055)
F.SUPPORT < 2.5	173	40.1	(0.2312139 0.7687861)
F.SIZE ≥ 5.5	21	4.1	(0.1904762 0.8095238)[a]

[a]shows end nodes

Table 2 Overall error occurred by using different technique (%)

	Train	Test	Validate	Full
Decision tree	21.5	30.9	24.6	23.4
Random forest	2.8	25.5	16.9	8.3
Linear model	23.9	34.6	−20.8	25.1

variables actually used in the tree construction are: Father's incomes (F.INCOM), father's job (F.JOB), Family relations (F.RELATIONS), Family size (F.SIZE), Family support (F.SUPPORT), Gender (GENDER), Mother's education (M.EDU), Mother's job (M.JOB) and UG (n = 251).

The over error occur during computing the data and the summary of the result is represented in Table 2.

The ROC for rpart model is computed for train set data (70%), test (15%), validation (15%) and full (100%) data. The accuracy of the result was summarized in below Table 3. These results validates the random selection is responsible for the accuracy level of the data.

It is evident from above decision tree that the family relations matters a lot as this variable got highest Gini impurity. A significant bonding and cordial relations

Table 3 ROC Accuracy of the result by using different technique

	Train	Test	Validate	Full
Decision tree	0.7333	0.5600	0.6353	0.6877
Random forest	0.9914	0.7229	0.6667	0.9396
Linear model	0.6225	0.5429	0.5379	0.5974

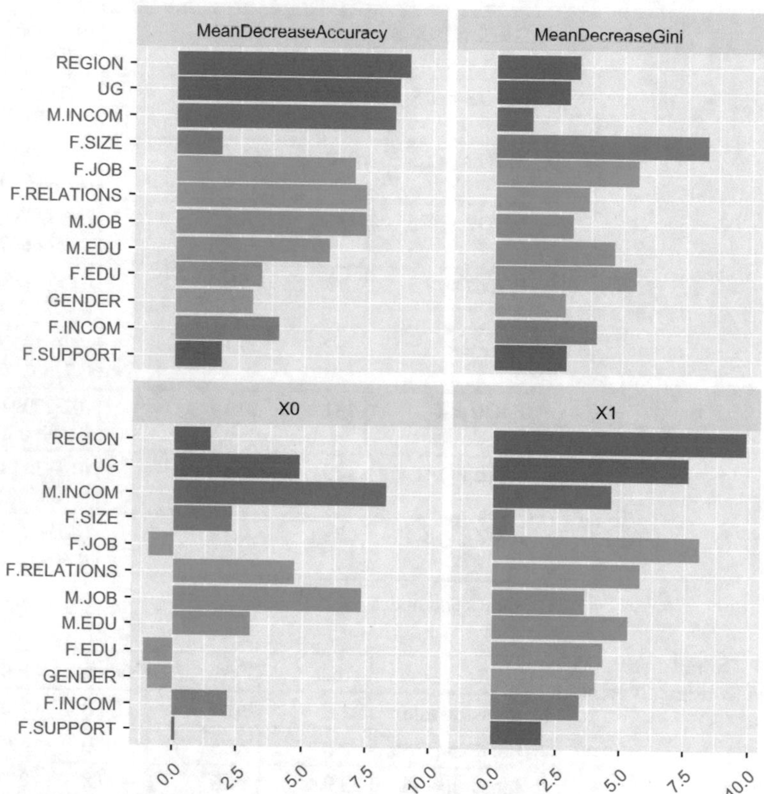

Fig. 3 Discovery of other relevant variable's importance

between family members can boost self satisfaction of the student which ultimately leads to high performance. Mother's literacy can lead to academic success of student. Family's support to education is also a catalyst in boosting self satisfaction. For proper comparison, we went further. Figure 3 shows relatively importance of variables. This is obtained from Gini index. It is observed that Religion, family size, monthly income, mother's job as well as her education, family income and their support are the significant variables affecting the satisfaction of students. The linear model in terms of its normality and the relation between the observed and predicted values of satisfaction of students were represented in Fig. 4.

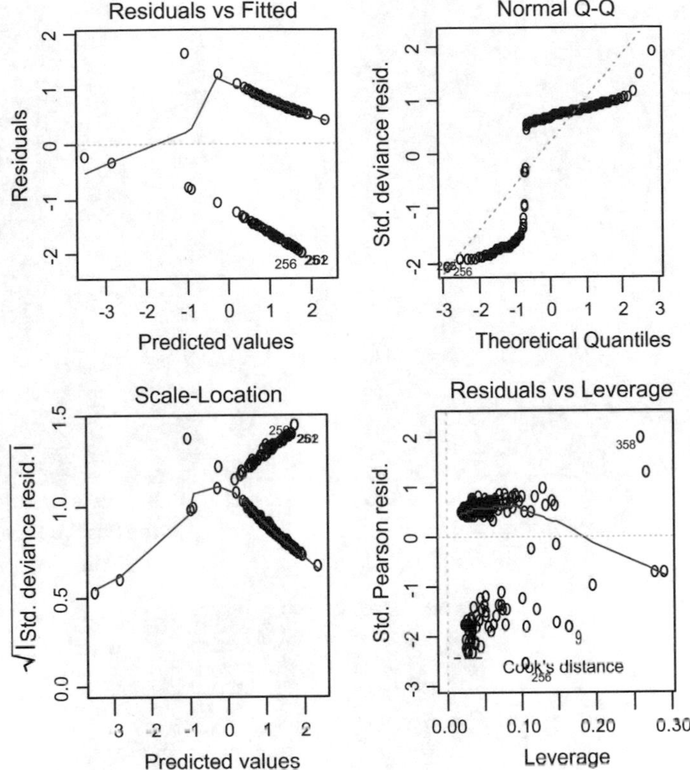

Fig. 4 Comparison of linear model

Figure 5**a–c** represents the ROC accuracy level of the linear model. It is observed that random forest approach is more suitable and gives accurate results for predicting the satisfaction level of the students (Table 3). Further, in future, if we apply these parameters, it will help us in accurate prediction of satisfaction level of students using RF model. Figure 6 represents the precision level of proposed three models. It reveals that the Area under the ROC curve for the rpart model [validation] is 0.6353 and is indicate that the random forest model will be the most suited and more precise to obtain predicted results.

4 Conclusions

Self satisfaction of students during pursuing a selected course is hard to predict as numerous social, personal, economical, academic parameters are associated with it. These parameters and their interrelations are hard to analyze. Data mining gave such insights. The study took it as confront and drew decision tree, random forest and linear

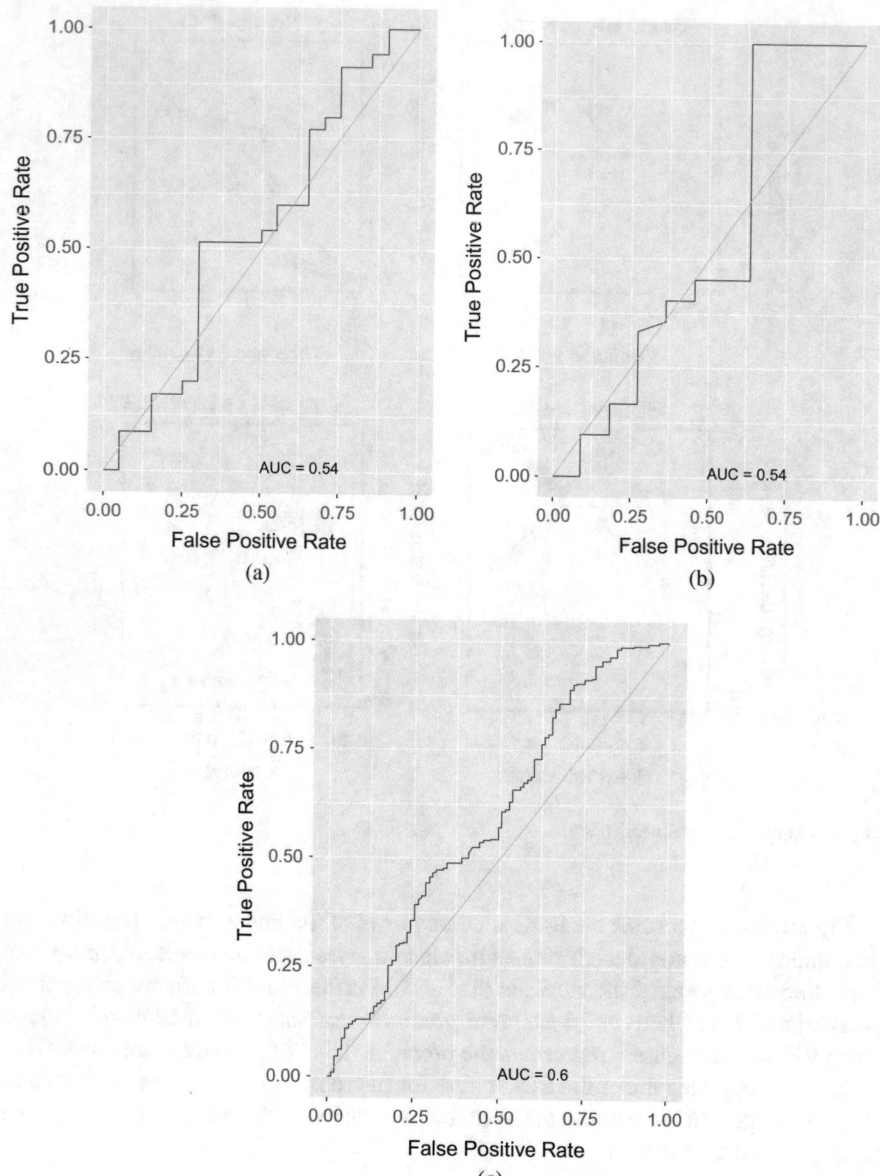

Fig. 5 **a** ROC for test set data **b** ROC for validation set data **c** ROC for full set data

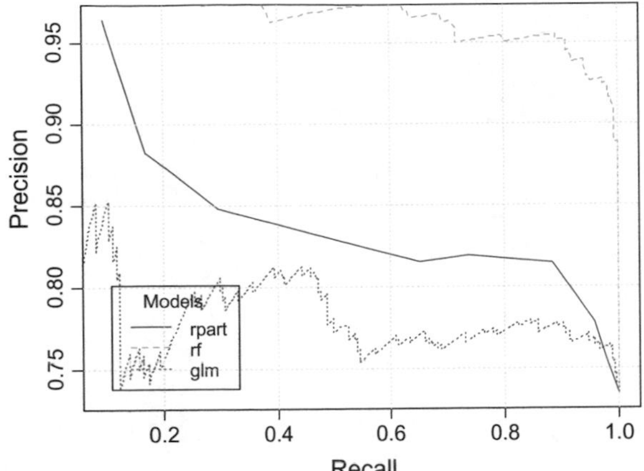

Fig. 6 Precision level of three models

model approaches for selected parameters. These algorithms were implemented in R software and their results were analyzed. Using comparative approach, we found that the student's satisfaction is depends on many parameters of family aspects. Accurate results were obtained using random forest approach.

References

1. Ali, S., et al.: Factors Contributing to the Students Academic Performance: A Case Study of Islamia University Sub-Campus. Am. J. Educ. Res. **1**(8), 283–289 (2013)
2. Bratti, M., Staffolani, S.: Student time allocation and educational production functions. Annals of Economics and Statistics (Annales D'économie Et De Statistique). 103–140 (2013)
3. Considine, G., Zappalà, G.: The influence of social and economic disadvantage in the academic performance of school students in Australia. J. Sociol. **38**(2), 129–148 (2002)
4. Dunham, M.H.: Data Mining: Introductory and Advanced Topics. Pearson Education, India (2002)
5. Field, A.: Discovering Statistics using R for Windows, Sage publications (2000)
6. Graetz, B.: Socio-economic status in education research and policy in John A., et al., Socio-economic Status and School Education DEET/ACER Canberra. J. Pediatr. Psychol. **20**(2), 205–216 (1995)
7. Han, J., Kamber, M.: Data Mining: Concepts and Techniques, 2nd edn. The Morgan Kaufmann Series in Data Management Systems, Jim Gray (2006)
8. https://economictimes.indiatimes.com
9. Pritchard, M.E., Wilson, G.S.: Using emotional and social factors to predict student success. J. Coll. E Stud. Dev. **44**(1), 18–28 (2003)
10. Rokach, L., Maimon, O.: Data Mining with Decision Trees: Theory and Applications. World Scientific, Singapore (2014)
11. Web resource at http://aofa.cs.princeton.edu/60trees/
12. Web resource at www.wikipedia.com
13. Web resource available at https://www.r-project.org

Perspective of the Behaviour of Retail Investors: An Analysis with Indian Stock Market Data

Abhijit Dutta, Madhabendra Sinha and Padmabati Gahan

Abstract Human bias influences the stock market for a long time. These behaviours are currently being studied under behavioural finance. The current paper tries to understand the behaviour of the Indian retail investors in face of uncertainty in the market. This paper uses a schedule of 25 statements which were used as expressed behaviour. There were five latent behaviours. A total of five hundred retail investors were surveyed in the four parts of India, namely north, south, east and west with the help of stock broking house in order to reach the target group. The result was then put to factor analysis. The study shows that there are six factors namely regret, followed by panic, cognitive dissonance, herding, anchoring, and heuristics. Interestingly, the herding and panic are actually resulted from a deep regret which the retail investors to panic and herd. Despite this retail investors use Anchoring and Heuristics in Investment Decisions. The market is also a victim of cognitive dissonance which keeps retail investors from entering the market.

Keywords Behaviour finance · Retail investors · Herding · Anchoring · Panic · Cognitive dissonance · Indian stock market · Factor analysis

---

A. Dutta (✉)
Department of Commerce, Sikkim University, Gangtok 737102, India
e-mail: adutta@cus.ac.in

M. Sinha
Department of Humanities and Social Sciences, National Institute of Technology Durgapur, Durgapur 713209, West Bengal, India
e-mail: madhabendras@gmail.com

P. Gahan
Department of Business Administration, Sambalpur University, Sambalpur 768019, Odisha, India
e-mail: p.gahan@rediffmail.com

H. S. Behera et al. (eds.), *Computational Intelligence in Data Mining*, Advances in Intelligent Systems and Computing 990,
https://doi.org/10.1007/978-981-13-8676-3_51

1 Introduction

Investment market for retail investors has always remained a big puzzle. Even the most prudent investors have performed badly in the stock market in the face of the good investment environment. While the market grew at an average of 24% annually, the retail investors have failed to reap the benefit out of it. The market is performing well but the retail investors are performing poorly, this situation makes a conclusion that the retail investors do not behave rationally in the market. They either tend to depend excessively on exogenous advice or are swayed away in the market rallies.

Since there is a positive correlation between the performance of the stock market and the economy, it is important to note that the micro growth of the retail investors is directly based on proper investment plan by these investors and how they react to the market. The relationship between the cognitive behaviour and individual investment decision therefore of utmost priority as this would be able to pinpoint the factors which lead to the success of investment or in other words the factors which will critically decide the retail investment process in the market.

Though many scholarly studies have been carried out in Europe and the USA, studies in the context of Asia and India are far and few between. That part, as the institutional investors tend to squat the larger of part of research interest, retail investment and behaviour of retail investment are hardly taken care off. In light of this, it is important to note the following: (i) what are the general factors influencing the individual's behaviour of investment? (ii) are there any psychological bias among investors in India? and (iii) what are the behavioural factors impacting the individuals' investment decisions?

The rest of the paper is organised as follows. Next section documents the brief survey of literature followed by descriptions methodological issues. Finally, we conclude the paper before interpreting the analysis and findings.

2 Survey of Literature

There are quite a number of studies that have used either questionnaire or survey to understand the behaviour of the retail investors. Some of these studies are being analysed for fortification of the findings of this chapter under the study.

As discussed earlier, there are studies in developed countries where questionnaires or survey have been used to find the behaviour of retail investors. Prominent among these studies are those of Nagy and Obenberger [1] and Obamuyi [2]. They identified 34 contextual factors which are categorized by 5 groups, namely: accounting information, personal and financial needs, neutral information, advocate recommendations, and firm image. Finally they concluded that these are the factors influencing the behaviour of the retail investor.

Maditinos et al. [3] found that the retail investors are carried away by media noise and seldom relay on technical and fundamental analysis. Waweru et al. [4] pointed out some heuristic factors like representativeness, anchoring and availability bias which are most important affecting the investment decisions. Luong and Ha [5] identified five behavioural factors such as prospect, herding, anchoring and overconfidence—gambler fallacy, which affect the individual investors' decisions on investment and those are also critical to take the retail investment decision.

Studies in the Indian context are far and few between, as observed earlier. Dutta [6] found that individual investors do not react to good and bad news in the market and they use their prudence to invest rather than on the market advices. Chandra and Kumar [7] highlighted some psychological points like conservatism, prudence, under confidence, information asymmetry and precautious attitude, which continually influence the investors' decision in India. Independently Islam [8] and Vijaya [9] identified different factors influencing the investors' decision in Indian retail sector, which include anchoring, overconfidence, loss aversion, market factors and herd behaviour to name a few.

From the above discussion on existing studies, it can be argued that there are several behavioural factors which influence the individual investors' investment decisions in financial markets, especially in stock markets. This study, therefore, tries to expand the horizon on a more recent time to understand the perspectives of behaviour which influence the investment decision of the Indian retail inventors.

3 Method of Study

The study uses the following method discussed below.

3.1 *Objectives*

(i) To investigate the major behavioural factors determining the behaviour of retail investors.
(ii) To confirm whether any behavioural factor present among the investors in India.
(iii) To find the impacts of retail investors' behavioural factors on their investment decisions.

3.2 *The Sample*

The study used a structured schedule on five hundred individual investors with an active portfolio. The schedule was inducted through a national stockbroker across the calendar year 2014–2015. A sample size of five hundred is considered to be sufficient

Table 1 Demographic profile of the sample

Region	Female	Male	Total
East	36	140	176
West	40	140	180
South	28	85	113
North	10	21	31
Total	114	386	500

Source Own survey of authors

Table 2 Region wise breakup of the sample

Region	Numbers	Region	Numbers
East		**South**	
Kolkata	80	Chennai	40
Bhubaneswar	20	Hyderabad	40
Guwahati	30	Visakhapatnam	20
Jamshedpur	26	Bangalore	13
Ranchi	20		
West		**North**	
Ahmedabad	20	Delhi	20
Mumbai	100	Chandigarh	06
Jaipur	20	Lucknow	02
Vadodara	20	Allahabad	04
Pune	20		

Source Own survey of authors

as it meets the criterion of a normal distribution and to take up any statistical analysis. The sample demography and its region wise breakup are presented in Tables 1 and 2 respectively.

3.3 *Schedule for the Study*

A schedule was developed with the opinion of the expert and industry experts which was used for collecting the information. The schedule has five Latent Factors which include (i) Cognitive Dissonance, (ii) Anchor and Heuristics, (iii) Herding and Mimic and (iv) Panic and (v) Regret. These are further divided into five statements explaining latent behaviour for each, which expressed the investors' reaction to a particular behaviour. The latent behaviour is not provided to the investors and only the expressed statements have been asked using a schedule having a Likert like scale of five points which asked them to "agree" or "disagree" with the statements. The value 1 was

Table 3 Code of variables used in the study

Latent behaviour	Expressed behaviour	Corresponding number
1. Cognitive dissonance	Conscious investment decision	1
	Withdrawal when the market falls	2
	Withdrawal when the investment price falls	3
	Past experience	4
	Self understanding of Investment	5
2. Anchor and heuristics	Use of rule of thumb	6
	Rule of thumb lead to correct decision	7
	Frequent use of rule of thumb	8
	Rule of thumb selectively	9
	Rule of thumb when the market crashes	10
3. Herding and mimic	Follow other investment	11
	Affected by what others think of self-investment	12
	Use of Popular analysis	13
	Sell when others sell	14
	Exit at fall and enter at rising market	15
4. Panic	Sell in rush	16
	Panic	17
	Sigh away from Market	18
	Emotional disturbances	19
	Scared of loss	20
5. Regret	Loss due to low attention	21
	Loss due to wrong timing in the market	22
	Blame self for loss	23
	Haste	24
	Regret	25

Source Own presentation of authors

taken as "disagree" and 5 as "strongly agree". The coding used for the variable is given in Table 3.

3.4 Tools and Technique

The scale information was summarised and multivariate analysis. The study uses Factor Analysis. Factor Analysis is a statistical tool, which enables to bring in multiple factors that affect a single variable to a few factors, which has direct association to the variable(s). In doing so, it reduces the number of pointers (reduced by factor loading) which leads to a definitive association between variables or cases. A Correlation matrix is extracted which shows the direction and degree to which each variable/case is associated to each other. This eases our understanding of how each variable would have reduced to a single-factor post analysis. This is done in two ways, i.e., R matrix and Or Q Matrix method. Under R Matrix method, the coefficient of correlation is found between each variable. In the Q matrix method, single direction correlation is determined between the variables or cases. Once the matrix is determined, the least factor is extracted by solving the matrix.

The main statistics extracted are factor loading, h^2, H, Eigenvalue, Percentage of Common Variance and percentage of the total variance. They are explained below:

i. Factor Loading: Loading measures which variables is involved in which factor, to what degree and in what direction. They can be interpreted like the coefficient of correlation. The square of loading is like coefficient of determination that measures the shared variation between the factors and the variables.
ii. h^2: It is the proportion of a variable's total variation that is involved in the factors. (1−h) is equal to the degree to which a variable unrelated to the other variable(s).
iii. H (percentage of total variation in the data) explained by all factors represented by Eq. (1) as follows.

$$H = \frac{\text{Sum of all h squared}}{\text{Number of variables}} \times 100 \tag{1}$$

iv. Eigenvalue: It is the sum of factor loading.
v. Percentage of Common Variance: It is Eigenvalue divided by the sum of all h variables or cases. This is expressed per factor.
vi. Percent Total Variation: Eigenvalue divided by the number of variables or cases. This is also expressed per factor.

4 Analysis and Interpretation

The analysis of the statistics which has been carried out to extract factors is given below. The data were tested for reliability and validity.

Table 4 shows that the Cronbach's Alpha is 0.947 hence the data set is reliable for carrying out a meaningful factor analysis. The data is found to be hundred percent valid (through the software used for calculation), which makes the schedule as a reliable and valid instrument for future survey in similar research.

Table 4 Reliability statistics

Cronbach's alpha	0.947	No. of items
Validity	100	25

Source Own calculation of authors

The Q matrix shows the relation of each variable to the other. Table 4 presents that there is mixed relationship between the variables and hence the pattern of relation amongst the latent variable very clear. This makes it a fit case of reduction. The reduced data and the factors extracted are discussed in the subsequent section. The Q coefficient matrix is given in Table 5.

The factor loading as given in Table 4 extracted six factors. Depending on the major loading we try to analyse this factor. Factor one has high association with 15 variables which are self-understanding, rule of thumb selectively, rule of thumb when the market crashes, follow popular analysis, sell when others sell, sigh away at falling market, emotional disturbance, panic, loss due to low attention, not in the right time in the market, blame self for the loss, haste and regret.

It is interesting to note that the majority of the association are cluttered towards the end variables which indicate that this factor is majorly associated with the latent variable Regret. Therefore, the first factor will be named as "Regret" which is what promotes most of the retail investors not to invest in the stock market. Since h^2 is also high in this, we can safely conclude that this variable is regret behaviour of the investors.

The variables which influence the factor two are rule of thumb in investment, sell in rush, panic, sigh away from a falling market, emotional disturbance and scared. This indicates that the variables pertaining to latent variable "Panic". The variable of h^2 is also high for these variables and hence the second factor is Panic. The third factor is related to the variables Conscious investment, withdrawal when the market is falling, withdrawal after the market crashes, past experience and self-understanding. This is related to the latent variable "Cognitive dissonance". With an h^2 being high this variable is Cognitive dissonance. The forth factor include follow investment of others, do what others think, follow popular analysis, sell when others sell in the market and exit at the falling market like the others. This is related to the latent variable "Herding" and since the h^2 is also high we can name the safely conclude this factors naming. The fifth factor includes rule of thumb which is used frequently, use of the rule of thumb when market crashes and follow rule of thumb selectively. This are related to "Anchoring" behaviour which is confirmed by the high h^2 value. The sixth factor include rule of thumb is correct to use, rule of thumb leads to correct decision, rule of thumb is used frequently used in investment. This all lead to "Heuristic" behaviour which is confirmed by the high h^2 value. Table 6 shows the entire result of the principal factor analysis in order to observe the investors' behaviour in the market

4.1 *Interpretation and Major Findings*

The above analysis leads us to some phenomenal insight. The retail investors do go through behavioural bias which is hidden and surfaced by some explained behaviours. The six behaviours which emerge as the most dominant behaviour are regret, panic, cognitive dissonance, herding, anchoring and heuristics in their order appearance,

Table 5 The Q matrix

1		2	3	4	5	6	7	8	9	10	11	12	13
1	1												
2	0.594	1											
3	0.534	0.51	1										
4	0.206	0.283	0.547	1									
5	0.287	0.572	0.716	0.535	1								
6	0.179	0.573	0.621	0.428	0.62	1							
7	0.231	0.177	0.331	0.357	0.296	0.315	1						
8	0.411	0.506	0.249	0.232	0.467	0.176	0.651	1					
9	−0.07	0.287	0.236	0.559	0.605	0.45	0.437	0.49	1				
10	0.083	0.241	0.489	0.448	0.769	0.501	0.451	0.53	0.77	1			
11	0.027	0.088	0.527	0.681	0.644	0.456	0.604	0.44	0.76	0.87	1		
12	0.123	−83	0.535	0.552	0.45	0.01	0.154	0.04	0.31	0.48	0.623	1	
13	0.395	0.164	0.454	0.509	0.558	0.019	0.498	0.63	0.49	0.69	0.728	0.72	1
14	0.432	0.393	0.698	0.705	0.745	0.417	0.603	0.6	0.6	0.75	0.787	0.62	0.82
15	−0.033	−298	0.409	0.392	0.283	0.183	0.576	0.14	0.39	0.62	0.776	0.65	0.58
16	0.212	0.067	0.34	0.389	0.156	0.117	0.365	0.18	0.27	0.26	0.437	0.38	0.4
17	0.421	0.056	0.402	0.347	0.229	−0.22	0.2	0.31	0.12	0.23	0.383	0.74	0.76
18	0.28	0.215	0.613	0.459	0.498	0.43	0.485	0.3	0.46	0.65	0.703	0.57	0.63
19	0.03	0.307	0.514	0.649	0.475	0.609	0.359	0.15	0.53	0.47	0.63	0.34	0.29
20	0.237	0.051	0.453	0.356	0.44	0.007	0.222	0.27	0.45	0.61	0.623	0.76	0.76
21	−0.007	0.129	0.439	0.342	0.466	0.605	0.531	0.27	0.56	0.65	0.709	0.19	0.31
22	−0.078	−12	0.341	0.311	0.491	0.204	0.452	0.3	0.61	0.68	0.686	0.58	0.58
23	−0.201	0.235	0.211	0.278	0.543	0.415	0.233	0.32	0.65	0.63	0.505	0.22	0.28
24	0.161	0.238	0.31	0.392	0.407	0.22	0.603	0.56	0.6	0.49	0.595	0.3	0.55
25	−0.261	0.119	0.245	0.463	0.417	0.494	0.336	0.09	0.67	0.53	0.607	0.19	0.14

(continued)

Table 5 (continued)

	14	15	16	17	18	19	20	21	22	23	24	26
1												
2												
3												
4												
5												
6												
7												
8												
9												
10												
11												
12												
13												
14	1											
15	0.64	1										
16	0.41	0.425	1									
17	0.54	0.385	0.48	1								
18	0.67	0.658	0.52	0.55	1							
19	0.45	0.338	0.41	0.23	0.734	1						
20	0.62	0.599	0.48	0.77	0.771	0.41	1					
21	0.46	0.572	0.39	0.07	0.669	0.64	0.41	1				
22	0.58	0.673	0.39	0.4	0.675	0.41	0.73	0.71	1			
23	0.39	0.267	0.18	0.06	0.494	0.51	0.43	0.651	0.78	1		
24	0.56	0.384	0.5	0.46	0.6	0.52	0.55	0.704	0.75	0.651	1	
25	0.28	0.343	0.34	−0.07	0.505	0.73	0.31	0.758	0.59	0.677	0.607	1

Source Own computation of authors

Table 6 Principal factor analysis to investors' behaviour in the market

Variables	Principle factor matrix						h^2	H
	I	II	III	IV	V	VI		
1	0.248	−0.471	**0.706**	0.017	0.242	0.03	0.842	86.80%
2	0.323	−0.115	**0.648**	−0.43	−0.43	0.027	0.929	
3	0.672	0.139	**0.532**	0.232	0.024	−0.054	0.847	
4	0.676	0.016	**0.652**	−0.333	−0.033	−0.157	0.617	
5	**0.746**	−0.126	**0.883**	−0.092	−0.398	**−0.158**	0.895	
6	0.508	**0.568**	0.247	−0.32	0.011	**0.98**	0.909	
7	0.619	0.029	0.08	0.202	0.326	**0.702**	0.91	
8	0.532	0.091	0.209	0.288	**0.678**	**0.663**	0.936	
9	**0.745**	−0.034	−0.046	0.21	**0.768**	−15	0.775	
10	**0.848**	−0.137	−0.17	0.114	**0.856**	−0.098	0.897	
11	**0.911**	−0.05	−0.158	**0.823**	−0.135	−0.265	0.948	
12	**0.64**	0.012	−0.266	**0.748**	−0.248	0.054	0.928	
13	**0.774**	0.014	0.1	**0.718**	0.17	−0.03	0.941	
14	**0.873**	0.239	0.231	**0.035**	−0.145	−0.181	0.927	
15	0.689	0.198	−0.442	**0.891**	−0.006	−0.454	0.92	
16	0.531	**0.508**	−0.139	−0.099	0.556	0.33	0.665	
17	0.516	**0.755**	−0.074	−0.021	0.124	0.261	0.925	
18	**0.846**	**0.893**	−0.078	−0.173	0.22	0.27	0.817	
19	**0.746**	**0.53**	−0.263	−0.048	−0.058	0.321	0.811	
20	**0.744**	**0.713**	−0.165	0.013	0.243	−0.058	0.906	
21	**0.744**	−0.439	−0.165	0.013	0.248	−0.058	0.837	
22	**0.793**	−0.053	−0.412	0.168	−0.061	0.194	0.871	
23	**0.639**	−0.473	−0.181	0.18	−0.148	0.411	0.888	
24	**0.756**	−0.089	−0.107	0.267	0.335	0.244	0.898	
25	**0.625**	**0.609**	−0.255	−0.108	0.146	0.166	0.873	
Eigenvalue	11.767	3.087	2.701	1.663	1.314	1.196		
Percentage common variance	47.068	59.345	70.149	76.802	82.056	86.841		
Percentage total variance	43.97	57.324	68.412	75.493	81.338	86.841		

Source Own calculation of authors

respectively. Interestingly, the herding and panic is actually resulted from a deep regret which the retail investors to panic and herd. Despite this retail investors use Anchoring and Heuristics in Investment decisions. The market is also a victim of cognitive dissonance which keeps retail investors from entering the market.

5 Conclusion

Factor analysis technique has been used to understand the presence of bias behaviour in Indian stock market. A total of five hundred retail investors have been asked to answer a schedule which has 25 questions.. The data has been tested for reliability and validity. It has been found that the data is both reliable and valid. It has been found that there are six factors namely Regret, followed by Panic, Cognitive dissonance, Herding, Anchoring and Heuristics. Interestingly, the herding and panic are actually resulted from a deep regret which the retail investors to panic and herd. Despite this retail investors use Anchoring and Heuristics in Investment decisions. The market is also a victim of cognitive dissonance which keeps retail investors from entering the market.

Acknowledgements Corresponding author is thankful to the Ethics Committee formed by the Department of Commerce, Sikkim University, Gangtok, India, Ref. No. SU/COM/F1/02/2014 of Dated: December 15, 2015 for conducting, monitoring and certifying the primary survey for this study. The survey provides the primary data used in the present paper. There is no conflict of interest. Useful disclaimers apply.

References

1. Nagy, R.A., Obenberger, R.W.: Factors influencing individual investor behaviour. Financ. Anal. J. **50**(4), 63–68 (1994)
2. Obamuyi, T.M.: Factors influencing investment decisions in capital market: a study of individual investors in Nigeria. Organ. Mark. Emerg. Econ. **4**(1), 141–161 (2013)
3. Maditinos, D., Zeljkosevic, S., Theriaous, N.G.: Decision making process by investors and heuristic methods in the study of retail investors in Russian context. J. Behav. Financ. **8**(2), 22–29 (2007)
4. Waweru, N.M., Mwangi, G.G., Parkinson, J.M.: Behavioural factors influencing investment decisions in Kenyan property market. Afro-Asian J. Financ. Account. **4**(1), 26–49 (2014)
5. Luong, L.P., Ha, D.T.T.: Behavioural factors influencing individual Investor decision making & performance: a survey at the Ho Chi Minh stock exchange. Master's thesis, Umea School of Business, China (2011)
6. Dutta, A.: Investors reaction to good and bad news in secondary market: a study relating to investors' behaviour. Financ. India **15**(2), 567–576 (1998)
7. Chandra, A., Kumar, R.: Determinants of individual investor behaviour: an orthogonal linear transformation approach. MPRA Paper No. 29722, Posted 22 (2011)

8. Islam, S.: Behavioural finance of an inefficient market. Global J. Manag. Bus. Res. **12**(14), 17–26 (2012)
9. Vijaya, E.: An empirical analysis of influential factors on investment behaviour of retail investors' in Indian stock market: a behavioural perspective. Int. J. Manag. Soc. Sci. **2**(12), 296–308 (2014)

Three-Dimensional (3D) Polygon Mesh Authentication Using Sequential Bit Substitution Strategy

Sagarika Borah and Bhogeswar Borah

Abstract Accelerated growth in higher dimensional media processing has promoted the usage of 3D representation of objects and also are easily distributed over network for various reasons. To ensure the authenticity of such susceptible 3D models, authentication-based watermarking techniques are often used. This paper discusses a semi-fragile, blind watermarking algorithm that is able to detect a vertex-level tamper in the mesh and also detects topological manipulations. The watermark generation is carried out by self-embedding pattern. A low-complexity *sequential bit substitution strategy* is proposed to embed the model geometry information as verification bits in a specific pattern. To make the mesh invariant to vertex-reordering operation, vertices are indexed with a topology-oriented mesh traversal strategy. The performance of the approach is depicted with the proper tamper detection and localization achievement and minimal distortion to the surface of the model.

Keywords Mesh watermarking · Authentication · Semi-fragile · Tamper localization · Bit substitution

1 Introduction

At the advent of multimedia processing tools and preferable usage of higher dimensional visualization techniques, a tremendous amount of data is generated every day and stored or transmitted over communication network channels. Higher dimensional multimedia mainly signifies the 3D rendering of different application-related models like entertainment, architectural designs, manufacturing plants, or military applications. However, nowadays with a good number of easily accessible tools, 3D media can be processed and all sorts of malicious operations are also can be performed.

S. Borah (✉) · B. Borah
Department of Computer Science and Engineering, Tezpur University, Napaam, India
e-mail: sagarika08connect@gmail.com

B. Borah
e-mail: bgb@tezu.ernet.in

H. S. Behera et al. (eds.), *Computational Intelligence in Data Mining*, Advances in Intelligent Systems and Computing 990,
https://doi.org/10.1007/978-981-13-8676-3_52

Different security protocols have been designed by many of the researchers based on the concerned security applications. The various multimedia security applications include copyright protection, authentication, verification, traitor tracing, copy control, and broadcast monitoring. Based on applications, the watermarking classes can be divided into *robust watermarking* for copyright protection and *fragile watermarking* for authentication and verification.

2 Related Works

Fragile watermarking algorithms can be divided into two classes based on the domain of embedding, i.e., *spatial domain* and *transform domain*. In the class of spatial domain, mesh spatial attributes like geometry or topology are directly modified to embed the watermark bits. In the case of transform domain, the spectral components are altered. In the case of fragile watermarking, the spatial domain is mostly preferred due to its better localization attack capability. Based on the embedding style, the watermark embedding algorithms in the spatial domain can be either *substitutive* and *additive*. Moreover based on the watermark generation procedures, watermarking algorithms can be classified as *self-embedding* and *external data embedding* techniques. Another property of a watermarking scheme highly affects the design of the embedding algorithm is the extraction type. The extraction type can be *blind*, *semi-blind*, and *non-blind*. The state-of-the-art algorithms are designed for improvement of some of the basic requirements like *increasing hiding capacity*, *minimal embedding distortion*, and *tamper localization*. However, these requirements create a trade-off condition.

The first ever blind fragile mesh watermarking algorithm was presented by Yeo et al. [17] by computing two indices from the mesh vertices using two predefined hash function and the vertex positions are iteratively modified to meet the predefined index values. However, their algorithm faced the problem of *causality* and *convergence* which is taken care by [5]. Chou et al. [2, 3] designed multi-function vertex embedding strategy to embed watermark and verification bits for blind extraction. However, their embedding capacity is very low and achieved a regional tamper detection. Wu et al. [16] used vertex to center of mass distance and applied Quantization Index Modulation (QIM) to embed watermark followed by an extraction mechanism in a semi-blind way. Chen et al. [1] presented two adaptive authentication approaches using watermarking techniques. Yet their algorithm is not robust against RST operations. Spherical coordinates representation is also used by [4, 6]. In order to get rid from any causality condition, only a few of the vertices are selected for watermark embedding and hence the embedding capacity is lower in these algorithms. Error correction codes are also used by [10–13], however, the embedding process here is completely substitutive and so the embedding distortion is uncontrollable. Tsai et al. [8] recently presented a blind, self-embedding- based verification technique where the embedding capacity is higher yet embedding procedure is substitutive and so it may lead to larger distortion. Most of the authentication-based blind watermarking methods can accommodate very small payload size and achieved regional tamper detection (during verification). The main obstacles in those algorithms are causality problem and dependency of the embedding technique on the mesh local geometry. Yet, a higher payload capacity is essential and to detect the exact location of tamper, vertex-level tamper detection is more suitable. Only one of the algorithms [11] achieved vertex-level tamper detection but their embedding distortion is uncontrollable. Moreover, we did not encounter a single semi-fragile mesh watermarking algorithm covering all different content-preserving attacks (affine transformations and vertex-reordering operations).

While considering the vertex-level tamper detection algorithm [11] one thing can be noticed that, as the embedding process is independent of the mesh connectivity they are unable to track any topological changes to the mesh. However, this work mainly focuses on the problem and tried to design a proper mesh verification procedure which can accurately locate or rather can detect vertex-level tamper but also keeps track of mesh connectivity details. In order to cover the requirement, the proposed self-embedding procedure is carried out in a predefined vertex sequence generated based on the mesh topology and the embedding process is dependent on the neighboring vertex information. A detailed explanation can be found in later sections.

3 Proposed Methodology

A 3D polygon mesh model represents a model surface with a set of polygons (triangles and quadrilaterals) stitched together bearing different levels of curvature or bumpiness corresponding to the model structure. However, the polygon mesh M can be represented as $M = \{G, C\}$, where $G = \textit{mesh geometry information}$ and $C = \textit{topological information}$. Geometry information contains the $G = \{V, E, F\}$, where $V = \textit{vertex coordinate details}$, $E = \textit{edge details}$ and $F = \textit{facet information}$. Where as the topological information contains the neighborhood details or connectivity details of the geometrical elements. In this work, the geometry elements are considered for watermark embedding. Moreover there a ton of mesh file formats available, among which few formats like .obj, .off, and .ply are suitable for use in the proposed work in this paper. In this work, the vertex coordinate values are considered till the fourth position after decimal.

3.1 Preprocessing

To ensure the robustness of the watermarking process against translation and rotation operation, two extra preprocessing steps need to be performed. To make the mesh robust against translation operation, the mesh origin is translated to the mesh center of gravity. Principal Component Analysis (PCA) is used to compute the principal axes

PCA_x, PCA_y, and PCA_z to make the mesh rotation invariant. The principal component z is the eigenvector that corresponds to the largest eigenvalue of the covariance matrix. Some of the watermarking algorithms [9] have considered spherical coordinates directly to be scaling invariant. But since PCA is sensitive to scaling operation, we have performed a normalization step on the vertex coordinates to make the mesh scaling invariant.

3.2 Topology-Based Vertex Sequence List Generation

In general, mesh vertices are ordered arbitrarily and may lead to a serious synchronization problem if the same ordering is not followed during extraction. Again during embedding, we need to maintain a global sequence in which the watermark bit sequence is embedded. In this algorithm, we have considered an already established topology-based mesh traversal ordering, Ordered Ring Facet (ORF) representation by [15]. In their algorithm, the mesh traversed with a set of concentric triangle layers which can be represented in a concentric circular way or spiral way. ORF ordering divides the mesh into two classes; one is the faces f_{out} having an edge on a contour line and other is the faces f_{gap} having a vertex in the same contour line. Using ORF, we can set up a global indexing V_{seq} for the vertex set in the mesh. In [15] the first starting vertex for the traversal is selected arbitrarily, but the same consideration cannot be followed in our approach which may lead to a synchronization problem. So, to avoid such situation, the first vertex coordinate details selected for a mesh is sent as a key value to the receiver which will help the extractor to start the traversal from the same vertex. Figure 1 shows an illustration spiral pathway of the mesh vertices.

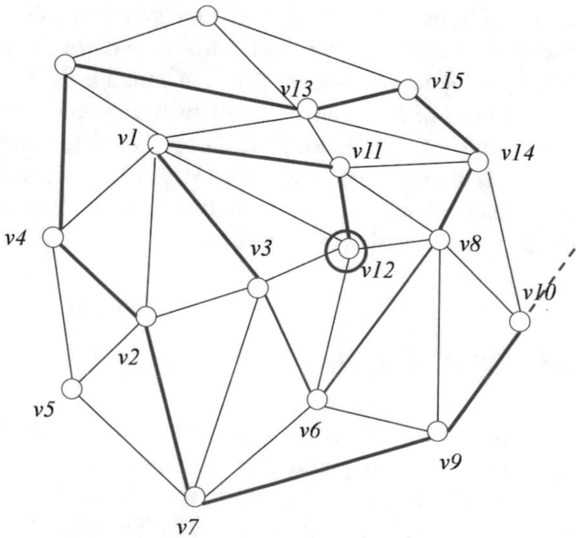

Fig. 1 Visualization of pseudorandom spherical points (blue-colored points) and the line joining the points with the origin (different colors)

3.3 Sequential Substitution Based Embedding

Bit substitution has been used by many of the works related to authentication or data hiding by mostly exploring the Least significant Bits (LSB) of a particular mesh element. While the mesh elements can be mesh vertex coordinates, edge length, or other coordinate information. But as mentioned in Sect. 2 there are a few goals regarding attack localization that the traditional LSB substitution based methods could not achieve. Figure 2 shows a block diagram of the proposed approach.

The term sequential substitution signifies a pattern of embedding where a state of embedding at a particular location is dependent on either its former mesh embeddable element or the later one. In this way, the connectivity aspect of the mesh is kept during embedding process. Mesh vertex coordinates are divided into x-coordinates, y-coordinates, and z-coordinates ad the main embedding is carried out on the z-coordinate value. The whole embedding process follows the previously computed

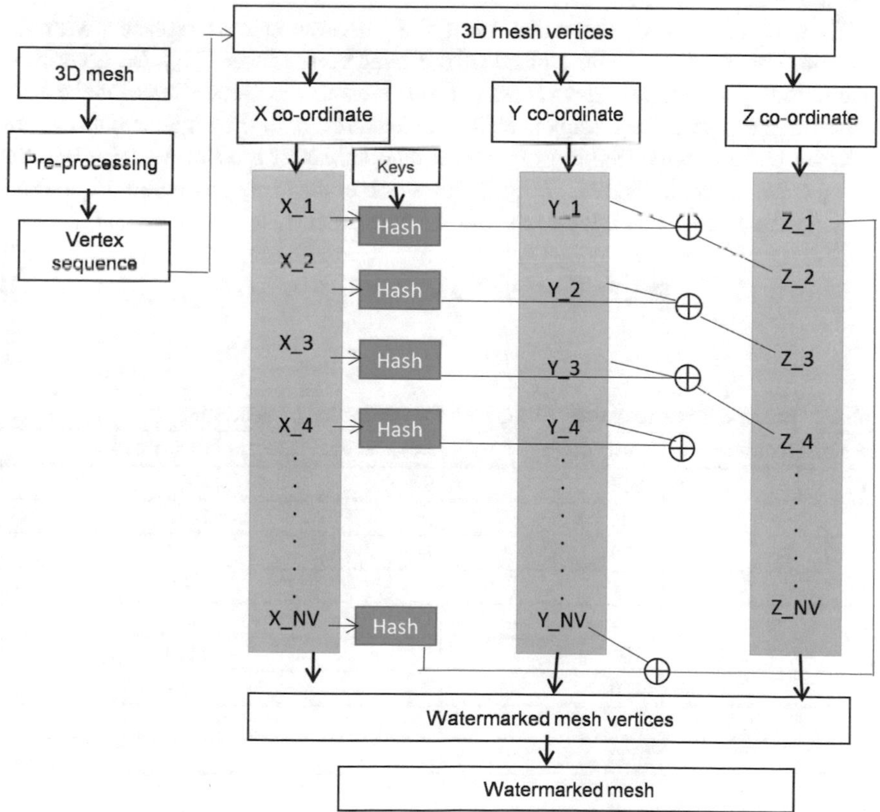

Fig. 2 Block diagram of the embedding scheme

vertex sequence V_{seq}. The actual embedding process is explained in the following steps:

- The x-coordinate $x_i, i = \{1, 2, 3, \ldots, NV\}$ values are inputted to keyed-hash message authentication and using the hash function SHA-1 a unique hash code is generated H_i for every vertex here $i = \{1, 2, 3, \ldots, NV\}$, $NV =$ number of vertices in the mesh. The first element value from the hash sequence H is considered and its binary code b_i (from Table 1) is taken for the next step. The key values for the whole mesh are generated pseudorandomly with a key k_1 and total number of keys will be the same as the number of vertices in the mesh.
- In the next step, the y-coordinate value is taken and its third decimal value is considered and its corresponding binary code b_i is taken referring to the Table 1. A bit-XOR operation is performed between the binary code b_i and the binary code for the third decimal value of y_i (a zero is appended in the front if y_i is a three-bit code). The resulting binary bits are taken and its first three digits are considered as the verification code Ver_i are again converted to the decimal numbers d_i as shown in Table 1.
- The decimal code d_i computed from the previous step is replaced with the z-coordinate's third decimal place of the later vertex in the V_{seq}. So a verification code generated from a vertex is embedded to the z-coordinate of the later vertex in the predefined sequence. For the last vertex in the sequence, the verification bits are embedded to the z-coordinate values of the starting vertex in the sequence. A block diagram in Fig. 2 shows the embedding procedure. Equation 1 shows the general condition of the embedding process.

$$z_{i+1} \leftarrow Hash(x_i) \oplus y_i, \quad i = 1, 2, 3, \ldots, NV \tag{1}$$

Table 1 Binary codes against decimal numbers for verification bit generation

Decimal numbers	Binary codes	Hexadecimal numbers	Binary codes
0	000	9	1001
1	001	A	1010
2	010	B	1011
3	011	C	1100
4	100	D	1101
5	101	E	1110
6	110	F	1111
7	111		
8	1000		

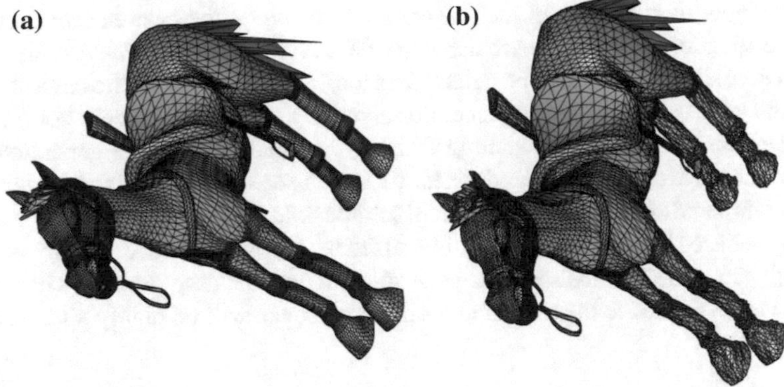

Fig. 3 Perceptual distortion due to watermarking negligible **a** original model, **b** watermarked model

3.4 Example of the Embedding Procedure

For example, we consider two vertices from the mesh M, namely, $v_i = (0.0110, 1.3452, 0.9876)$ and $v_{i+1} = (0.1235, 1.2340, 0.9901)$. For the vertex v_i, the SHA-1 hash sequence is computed with a key value k_1 and suppose the first element from the hash H_i is A. The binary code for A is 1010 and the third positioned value after decimal of y_i is 5 and its binary code is 0101. XOR operation is performed in between 1010 and 0101 to get the value 1111. The first three binary digits 111 are considered as the verification code Ver_i and its decimal value is replaced with the third positioned value after decimal of z_i, i.e., the new z-coordinate for v_{i+1} is 0.9971. During extraction, the same key value is used to find the hash code and finally the verification code is computed which is compared with the z-coordinate of the later vertex in the particular predefined vertex sequence. Figure 3 shows an example of a 3D model watermarked with the proposed approach.

3.5 Extraction Mechanism and Verification Process

A set of preprocessing steps are carried out on the suspected mesh M' to make the model invariant toward affine transformations and vertex-reordering operations as explained in Sects. 3.1 and 3.2. The vertex sequence from the suspected mesh is termed to be V'_{seq}. During extraction, an authenticated user is contained with a set of key values to compute the hash code from the x-coordinate values. A triplet of bit values are bit-XORed with a y-coordinate third decimal place value and matched with the z-coordinate value of the late vertex in the predefined vertex sequence V'_{seq}. If the verification code does not match with the later vertex z-coordinate value, then it signifies a malicious attack to be detected or located to be either on the current vertex or on the later vertex in the sequence. For a vertex v'_i, if the extracted z-coordinate's third decimal value z'_i is not matching with the verification code d_{i-1} generated from the former vertex y- and x-coordinates, then it is sure that any of the vertex v'_i or v'_{i-1} is tampered. If there is any topological attack to the model then first of all the topology- based vertex sequence will be affected and will produce a faulty one and secondly since the verification code generation and embedding process is dependent on the vertex sequence the extracted verification codes will be highly altered.

4 Results and Discussion

In order to investigate the performance of the proposed semi-fragile self-embedding approach, we took a few of the 3D models from Princeton benchmark dataset [7]. Few of the visual illustrations of embedded model with the original one in Fig. 3 depicts an imperceptible embedding of data. The performance of the embedding method is validated based on two of the required properties: *embedding distortion* and *attack localization capability*. The embedding geometrical distortion is estimated using three different distortion measures, namely, *Hausdorff Distance (HD)*, *Modified Hausdorff Distance (MHD)*, and *Root Mean Square Error (RMSE)*. The average distortion (d_{avg}) and the maximum distortion (d_{max}) is also computed. Table 2 shows a few of the 3D models and their distortion analysis with the set of distortion measures.

Attack localization capability was the main focus while designing the algorithm. Verification process can easily detect the variety of tamper locally implemented or globally to the model. Figure 4 shows few of the attack scenarios got localized precisely at the vertex level. In Table 3, the performance of the proposed approach is compared with the state-of-the-art methods where we can notice that our algorithm performed considerably well by achieving robustness to all of the content-preserving attacks with increase in payload capacity and also attack localization capability. However, one of the disadvantages of the proposed algorithm is that whenever there a topological attack, the whole V_{seq} gets changed. Which lead to end up with a wrong traversal order. In our subsequent works, we will definitely look into this problem.

Table 2 The list of original model with the marked model to show the perceptual quality and the geometrical distortion calculation on the marked model

Model name	Original model	Number of vertices (NV) , faces (NF), bits embedded (NB)	Distortion evaluation of DQIM
m100.off		NV = 606, NF = 1208, NB = 1818	HD = 0.0405, MHD = 0.0329, d_{avg} = 0.0020, d_{max} = 0.0020, RMS = 0.0014
m101.off		NV = 1143, NF = 2204, NB = 3429	HD = 0.0661, MHD = 0.0510, d_{avg} = 0.0019, d_{max} = 0.0093, RMS = 0.0016
m103.off		NV = 16049, NF = 31432, NB = 48147	HD = 0.1377, MHD = 0.0921, $d_{avg} = 8.1843e^{-4}$, d_{max} = 0.0060, $RMS = 7.7033e^{-4}$
m111.off		NV = 1238, NF = 2442, NB = 3714	HD = 0.0734, MHD = 0.0592, d_{avg} = 0.0020, d_{max} = 0.0099, RMS = 0.0017
m108.off		NV = 11105, NF = 22258, NB = 33315	HD = 0.2651, MHD = 0.2130, d_{avg}= 0.0033, d_{max} = 0.0154, RMS = 0.0021

5 Conclusion

This work presents a semi-fragile, blind 3D mesh watermarking procedure for the application of authentication or verification. The algorithm embeds mesh geometry information by normal bit substitution method, but the substitution pattern is sequentially carried out to handle the problem of topological factor during vertex-level tamper localization. This method is computationally very simple and produces minimum distortion to the mesh surface. All of the vertices are covered by following a topology-based vertex traversal strategy and the embedding rate of the approach is 3 bits per vertex. Due to the simple structure of the embedding methodology, it can be used for the models with complex and huge structure. In future, the algorithm can

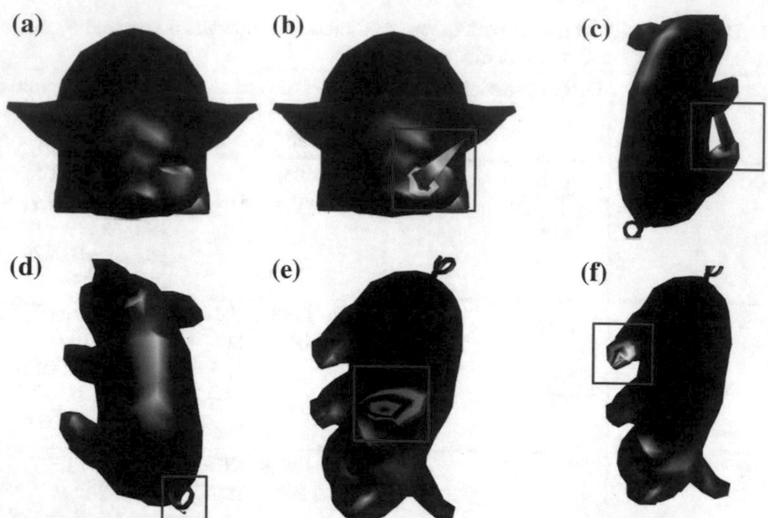

Fig. 4 Visual effects of some attack detected 3D model from Princeton shape benchmark dataset [7]

Table 3 Comparison of the previous methods with the proposed one (1) Method, (2) Class, (3) Embedding mesh element, (4)Extraction, (5) Payload capacity in %, (6) Robustness towards, (7) Embedding rate, (8) Attacks localization

(1)	(2)	(3)	(4)	(5) (%)	(6)	(7)	(8)
[17]	Fragile	Facets	Semi-blind	12	vertex reorder	–	No
[3]	Fragile	Facets	Blind	12	ST	1bpv	No
[2]	Fragile	Vertices	Blind	22	–	–	No
[14]	Semi-fragile	Vertices	Blind	21	–	–	Yes
[6]	Semi-fragile	Facets	Blind	33	ST	–	No
[4]	Semi-fragile	Vertices	Semi-blind	100	Vertex reorder	1 bpv	No
[11]	Semi-fragile	Vertices	Blind	100	Vertex reorder	3 bpv	No
[8]	Semi-fragile	Facets	Blind	26	ST	3.27–8.95 bpv	Yes
Proposed method	Semi-fragile	Vertices	Blind	100	RST Vertex reorder	3 bpv	Yes

be improved by adapting the embedding process based on model surface property to achieve minimum embedding distortion.

References

1. Chen, T.Y., Hwang, M.S., Jan, J.K.: Adaptive authentication schemes for 3d mesh models. Int. J. Innov. Comput., Inf. Control. **5**(12), 4561–4572 (2009)
2. Chou, C.M., Tseng, D.C.: A public fragile watermarking scheme for 3d model authentication. Comput.-Aided Des. **38**(11), 1154–1165 (2006)
3. Chou, C.M., Tseng, D.C.: Affine-transformation-invariant public fragile watermarking for 3d model authentication. IEEE Comput. Graph. Appl. **29**(2), 72–79 (2009)
4. Huang, C.C., Yang, Y.W., Fan, C.M., Wang, J.T.: A spherical coordinate based fragile watermarking scheme for 3d models. In: International Conference on Industrial, Engineering and Other Applications of Applied Intelligent Systems. pp. 566–571. Springer, Berlin (2013)
5. Lin, H.Y., Liao, H.Y., Lu, C.S., Lin, J.C.: Fragile watermarking for authenticating 3-d polygonal meshes. IEEE Trans. Multimed. **7**(6), 997–1006 (2005)
6. Molaei, A.M., Ebrahimnezhad, H., Sedaaghi, M.H.: A blind fragile watermarking method for 3d models based on geometric properties of triangles. 3D Res. **4**(4), 1–9 (2013)
7. Shilane, P., Min, P., Kazhdan, M., Funkhouser, T.: The princeton shape benchmark. In: Proceedings of Shape modeling applications, pp. 167–178. IEEE (2004)
8. Tsai, Y.Y., Cheng, T.C., Huang, Y.H.: A low-complexity region-based authentication algorithm for 3d polygonal models. Secur. Commun. Netw. **2017**, (2017)
9. Vasic, B., Vasic, B.: Simplification resilient ldpc-coded sparse-qim watermarking for 3d-meshes. IEEE Trans. Multimed. **15**(7), 1532–1542 (2013)
10. Wang, J.T., Chang, Y.C., Lu, C.W., Yu, S.S.: An ofb-based fragile watermarking scheme for 3d polygonal meshes. In: 2016 International Symposium on Computer, Consumer and Control (IS3C), pp. 291–294. IEEE (2016)
11. Wang, J.T., Chang, Y.C., Yu, C.Y., Yu, S.S.: Hamming code based watermarking scheme for 3d model verification. Math. Probl. Eng. **2014** (2014)
12. Wang, J.T., Fan, C.M., Huang, C.C., Li, C.C.: Error detecting code based fragile watermarking scheme for 3d models. In: 2014 International Symposium on Computer, Consumer and Control (IS3C), pp. 1099–1102. IEEE (2014)
13. Wang, J.T., Yang, W.H., Wang, P.C., Chang, Y.T.: A novel chaos sequence based 3d fragile watermarking scheme. In: 2014 International Symposium on Computer, Consumer and Control (IS3C), pp. 745–748. IEEE (2014)
14. Wang, W.B., Zheng, G.Q., Yong, J.H., Gu, H.J.: A numerically stable fragile watermarking scheme for authenticating 3d models. Comput.-Aided Des. **40**(5), 634–645 (2008)
15. Werghi, N., Rahayem, M., Kjellander, J.: An ordered topological representation of 3d triangular mesh facial surface: concept and applications. EURASIP J. Adv. Signal Process. **2012**(1), 144 (2012)
16. Wu, H.T., Cheung, Y.M.: A fragile watermarking scheme for 3d meshes. In: Proceedings of the 7th workshop on Multimedia and security. pp. 117–124. ACM (2005)
17. Yeo, B.L., Yeung, M.M.: Watermarking 3d objects for verification. IEEE Comput. Graph. Appl. **19**(1), 36–45 (1999)

A Comprehensive Review on Cancer Detection and Prediction Using Computational Methods

Dakshya P. Pati and Sucheta Panda

Abstract Cancer has been recognized as a deadly disease consisting of many different subtypes, it is a leading cause of death worldwide which may reach 13.1 million deaths in 2030. Early diagnosis and prognosis of cancer is the current trend for research work that facilitates the easy clinical management of patients. However, it is important to classify the type of cancer based on the features detection in the view of bioinformatics and biomedical field. Nowadays, different computational techniques and tools are used for early detection and assessment for the severity of cancer which will aid in improving the quality of life of the patients and also reduce the associated medical cost. The computational techniques like machine learning, artificial neural network, fuzzy logic, genetic algorithm, data mining, Bayesian network, etc., have a great impact on early detection of cancer and its assessment. The main objective of this survey paper is to present the comprehensive review on state of the art of the different computational methodologies applied to detection and prediction of cancer.

Keywords Cancer detection · Data mining · Genetic algorithm · Bioinformatics

1 Introduction

Cancer is a significant root cause of human death in many developing countries like America, England, and some European countries. Its classification is now trusted on clinical and histopathological facts which may produce incomplete procedure, misleading results which impact on detection and prediction of cancer. Cancer is

D. P. Pati
Department of Computer Applications, Trident Group of Institutions, Bhubaneswar 751024, Odisha, India
e-mail: dakshya_prasad@yahoo.com

S. Panda (✉)
Department of Computer Applications, Veer Surendra Sai University of Technology, Burla 768018, Odisha, India
e-mail: suchetapanda_mca@vssut.ac.in

H. S. Behera et al. (eds.), *Computational Intelligence in Data Mining*, Advances in Intelligent Systems and Computing 990,
https://doi.org/10.1007/978-981-13-8676-3_53

the uncontrolled growth of abnormal cells developing in any part of the body and collections of related disease. Our body cell starts to divide without stopping and spread into the tissue. Cancer can start at any place in the body which consists of trillions of cells. A human body grows because of mitotic cell divisions, which form new cells to replace the old or damage cell when they die. The cancer cells develop when the orderly process of normal cell disturbed leads to more and more abnormal can causc of survival of old or damaged cell when they should die. The new cells are formed but they are not required. These extra cells form masses of tissue called tumor. Many cancer forms have solid tumor because of masses of tissue where cancer of the blood (leukemia) do not form a solid tumor. A cancerous tumor is malignant in nature, which can spread and invade the neighbor tissue, also moves the distance of the body through the blood or the lymph system to form another new tumor. Cancer cell is differing from the normal cell which is less specialized, less distinction type function, ignores the signal and begins a process called programmed cell death or apoptosis. Cancer cell influences the normal cells' bio-molecule and also blood vessel which is surrounded by microenvironment. The immune system of the body is destroyed by cancer cells which are called the network of the organ, responsible for protecting the body from infection, other conditions and remove the damage or abnormal cells from the body. Cancer can happen due to genetic changes (inherited from parents), errors occurring in the cell (damage of cell DNA) and environmental exposures (smoking of tobacco, radiation, UV). The gene changes occurring due to the three genes are proto-oncogenes, tumor suppressor, and DNA repair genes which are called as the driver of cancer. Tumor suppressor gene is responsible for controlling cell growth and division whose alterations lead to uncontrolled, which causes cancer. DNA repair gene fixes the damage of DNA which changes due to mutations making the cell cancerous. The place where it starts and moves to other parts of the body is called metastatic cancer, and the process is called metastasis. The breast cancer spreads and forms metastatic tumor in the lung but not lung cancer. Both metastatic cancer cell and original cell looksame but differ in a specific chromosome. Tissue changes are not causes of cancer but may happen due to Hyperplasia and Dysplasia.

2 Cancer Subtype and Detection Cancer Using Clinical Data

Cancer is of more than a hundred types, its classifications and name based on the organ or tissue where it develops, for example, lung cancer starts from lung cells and Brain cancer starts from brain like that other are defined as follows in Table 1.

Cancer subtype classification, detections, and survival time predictions associated with risk factor, stages and grades, statistics, symptoms, and sign which are important from clinical point of view is essential for treatment and management of patients. Some important cancer type descriptions are listed in Tables 2 and 3.

Table 1 Cancer classified based on the specific type of cells

Term	Origin
Carcinoma	Developed and formed from epithelial cell
Sarcoma	Form in the bone and soft tissue including muscle
Leukemia	Begin in osteocytes or bone marrow
lymphoma	Begin in lymphocytes (T and B cells)
Melanoma	Begin in melanocytes

3 Computational Techniques and Tools in Use for Prognosis and Prediction of Cancer

To predict the cancer subtype as well as detect the risk factor, the clinical and biomedical instrument-generated data is not sufficient because it requires support for prognosis and easy patient's management in relation to social and economical value. There are some methods which were described with high appreciations by various researchers in different field of computations. Cancer classification has been done by various researchers from different corners of the world by using data mining, soft computing, machine learning, and some standard statistical methods.

3.1 Using Soft Computing

Soft computing is referred to as the driver of artificial computational intelligence because of its tolerant of imprecision, uncertainty, and approximations measurement qualities which solve the NP-problem within polynomial time like the human brain. The major component of soft computing is Artificial Neural Network (ANN), Genetic Algorithm (GA), Fuzzy Logic (FL), Machine Learning (ML), and Belief Network (Bayesian Network). Soft computing includes statistical, optimizing and probabilistic technique which a has great impact on the field of detection and prediction of future event recurrence.

3.1.1 Using Artificial Neural Network

Artificial Neural Network (ANN) is a mathematical model based on the human nervous system, which processes like human brain system and it works like an adaptive system which changes the structure according to the learning phase. The learning process useful for pattern recognition and classifications of the neural network is based on this factor , its type are described as follows:

Table 2 Description of pathological analysis prognosis factor for bladder cancer, breast cancer and kidney cancer

Parameter	Bladder cancer	Breast cancer	Kidney cancer
Risk factor	– Tobacco use: Age >65 (75% of people affected) – Chemicals: (textiles, leather, dye, paint, Industry, aromatic amine) – Chemotherapy (Cytoxan, Clafen, Neosor) – Pioglitazone (Actose) – Use of drug, Arsenic (drinking water)	– Age—old (50%) – Genetic mutations: BRCA1 and BARCA2 – Estrogen and progesterone exposure – Oral contraceptive and birth pills – Hormone replacement therapy after menopause	– Smoking: Men: 30%, Women: 35% – Gender: Men 2–3 times more than women (Age = 50–70 yr) – Nutrition's and weight, High blood pressure, Expose to cadmium, Long-term dialysis, Chronic kidney disease
Stages and grades	– TNM stages • T-tumor • N-node • M-metastasis	– TNM stages: T-tumor, N-node, M-metastasis – Size and location: T(0–4) T1: 20 mm of tumor, T2: >20 mm of tumor, T3: >50 mm of tumor, T4: Very large, Grown on: Chest wall: T4a, Skin: T4b, Both chest and skin: T4c, Inflammatory: T4d, N type: NX, N0, NI, M type: MX, M0, M	– TNM staging: T (TX, T0, T1) – Size of tumor: T1a: 4 cm, T1b: 4–7 cm T2: >7 cm, T2a: 7–10 cm T2b: >10 cm, T3: grown into a major vein – Node N (Nx, N0, N1), M(M0, M1)
Statistics	– Happening age >55(90%) – Survival rate after detected 5 yr: 70%, 10 yr: 70%, 15 yr: 65% – 5 yr survival rate: T—70%, N—35%, M—5%	– Avg survival rate: 5 yr: 90%, 10 yr: 83%	– Survival rate: Avg. age is 64 yr 5 yr: 74%, Only kidney (5 yr: 75%), Spread in lymph node (5 yr: 67%), Distant part (5 yr: 12%)

(continued)

Table 2 (continued)

Parameter	Bladder cancer	Breast cancer	Kidney cancer
Symptoms and sign	– Blood in urine, Burning sensation in urine, Frequent urination, Lower back pain	– Change in size and shape of the breast, nipple discharge, Physical change, Skin irritation, Warm red in breast	– Blood in urine, Pain and pressure in back, A mass developed at the back, Swelling in ankles, Loss of appetite
Treatment options	– Surgery: TURBT, Transurethral bladder tumor resections, Cystectomy and lymph node dissections, Urinary diversion – Chemotherapy: Cisplatin, Gemcitabine, Carboplatin, MVAC: M—Methotrexate, V—Vinblastine, A—Doxorubicin, C—Cisplatin	– Surgery: Lumpectomy, Mastectomy Lymph node removal, Biopsy of sentinel lymph node, Auxiliary lymph node	– Surgery: Lumpectomy, Mastectomy Lymph node removal, biopsy of sentinel lymph node, Auxiliary lymph node

1. Single-layer feedforward network (SFFN)
2. Multilayer feedforward network (MFFN)
3. Single node with own feedback
4. Multilayer recurrent network.

ANN has the ability to learning, which can stimulate the output for detection and prediction of cancer. It is also useful for clustering or classifications of gene expression of data. The supervise learning useful for classifications, whereas unsupervised learning useful clustering. Some of the researchers have great contribution toward prognosis and prediction cancer for survival are briefly narrated below.

Carey E. Floyed et al. [1] predicted breast cancer by using the artificial neural network from mammographic finding. Those patients are suspected as cancerous, scheduled for biopsy, the radiologist interpreted and analyzed as well as taking decisions over the outcome. ANN performed more accurately than radiologist ($p < 0.008$) with a sensitivity measure of 1.0 and specificity of 0.59.

Harry B. Burke et al. [2] used Artificial Neural Network to predict the survival time of breast cancer after occurrence by comparing with TNM staging methods which was developed in 1950. They also measured the accuracy of both standard and computational method using receiver operating characteristics curve which is applied on independently validated datasets. After comparison, they concluded that

Table 3 Description of pathological analysis prognosis factor for lung cancer, prostate cancer and oral cancer

Parameter	Lung cancer	Prostate cancer	Oral cancer
Risk factor	– Tobacco smoking, Asbestos, Radon	– Age >65 chance 80%, Race: Black peoples are higher risk than white people	– Tobacco use, Prolonged sun exposure, Human papillomavirus, Gender: Age >45 Poor diet
Stages and grades	– Stage (0, I, II, III, IV) Sage 0: Outside the lung, Stage 1: Small tumor IA-3 cm or less, IA3-3–4 cm, Stage II: IIA-4–5 cm, IIB-5 cm and spread into lymph node, Stage III: Spread into lymph node distal part, Stage IV: Cancer found more than one area	– TNM stage	– TNM stage, T: tumor: (TX, Tis, T1, T2, T3, T4) Tx–primary stage, Tis–one layer, i.e., carcinoma T1-2 cm, depth-5 mm, T2-2 cm, depth-10 mm T3-4 cm, depth-10 cm, T4-advance, Node (N): (Nx, N0, N1, N2a, N2b, N3a, N3b), M: (M0, M1)
Statistics	– Survival rate: 5 yr: 18%, Men: 15%, Women: 21%	– Survival rate: 5 yr: 100%, 10 yr: 98% If it spread into other part, 5 yr: 30%	– Survival rate: 5 yr: 65%, Black people: 48%, White people: 66%
Symptoms and sign	– Fatigue, Cough, Shortness in breath and Chest pain, Loss of appetite	– Frequent urination, Blood in urine Pain during urine and discomfort	– Soreness in mouth, Red and white patches develop, Lumps in lip
Treatment options	– Surgery: Lobectomy, A wedge sections, Segmentectomy, Pneumonectomy, Adjuvant therapy, Radial Nephrectomy, Partial nephrectomy	– Surgery: Radial prostatectomy, Laparoscopic, Bilateral, Radiation (IMR), Proton therapy	– Glossectomy, Mandibulectomy, Maxillectomy, Neck dissections, Laryngectomy, Tracheotomy Radiations, Chemotherapy, Immunotherapy

ANN is more accurate than TNM staging method on the survival of patients for 5 years where the value of TNM and ANN was 0.720 and 0.730, respectively, at $P < 0.001$, also for 10 years the value of TNM and ANN was 0.695 and 0.730, respectively, at $P < 0.01$.

Daniel J Surgent [3] compared ANN with other statistical approaches on prognostic factors and staging on cancer management. The objective of his research to evaluate the performance of standard regression method with ANN on large data sets. Chih-Lin Chi et al. [4] worked on two breast cancer datasets over 660 cases to predict the survival time of breast cancer after the disease is found. They were confirmed that ANN is better than the traditional method to predict future events' recurrence. Breast cancer data sets generated based on Xcyt image analysis where the sample collected from the patient using fine needle aspirate method. In the clinical point of view, prognosis is a very important indicator to determine the course of treatment such as chemotherapy after surgery. They used three-layer feedforward ANN with sigmoid activation in their work. The Kaplan–Meier survival function is used to estimate the probability of recurrence of events. They plotted Kaplan–Meier curve (visual inspection) as well as compare the significant differences by using Wilcoxon test. Probabilities of ANN with Wilcoxon test is 0.0104 and ANN with Naïve Bayes is 0.0021. This model can determine the accurately predict the probability of survival after a patient had surgery.

3.1.2 Using Machine Learning

Xiaosherg Wang et al. [5] developed a conventional tumor diagnosis model to predict cancer using microarray computational technique. Due to the rapid advancement of gene expression, it is very essential for measuring the gene expression (about 10,000) and predictions of the cancerous molecular marker. His work is based on four cancerous datasets (CNS tumor, colon tumor, lung cancer, and DLBCL).

Jose M. Jeraz et al. [6] worked on missing data imputations in breast cancer problem by using statistical and machine learning method to predict the recurrence of the disease. The imputation methods like mean, hot deck and multiple imputations and machine learning techniques such as multilayer perceptron (MLP), SOM, K-nearest neighbor (KNN) collected from EIALamo project include 3679 women. Before that ANN was used to estimate the dataset with imputed missing value but ML algorithm is used for the outcome the patients under the observations of ROC curve value, pairwise comparison test for MLP, KNN, and SOM with higher significant p-value (0.0053, 0.0048, 0.0071, respectively) than LD-based prognosis model. This method is suitable for imputations of missing value compare to other statistical methods. Guanggoa Xu et al. [7] worked on the prediction of colon cancer recurrence and prognosis by using SVM. They collected 5 sets of microarray datasets from gene expression Omnibus database which was derived from cancer genome ALAS. After processing, the data were analyzed by linear model for microarray data (LIMMA) method to reorganize the differential genes (DEGs), which were again analyzed by network-based neighbor scoring and SVM methods which validated 4 data sets and identify a 15-gene signature for prediction of recurrence risk and prognosis of colon cancer patients. The method is very useful for in the field of cancer therapy.

Chip M. Lynch et al. [8] done their work on the prediction of lung cancer and survival using machine learning classification technique. The data collected from the database with datasets (surveillance, epidemiology, end result, i.e., SEER). He used some methods like linear regression, Decision tree, gradient boosting machine (GBM), and support vector machine (SVM) to predict survival of lung cancer patients. The most performing technique was custom ensemble with a root mean square (RMS) value of 15.05 influence by GBM with value 15.32, with SVM is 15.82. This result is conferred with Cox proportional hazard model, which was used as a reference technique. The performance of this method is better than the classical method. Yawen Xiao et al. [5] worked on deep learning method to predict cancer because of high mortality. For the patient's management issue, it is very important for predictions based on gene expression (monitoring the risk factor for particular gene expression) and taking the right decision for treatment. They proposed a deep learning-based multi-model, which was tested on 3 public RNA sequence data for three kinds of cancer (lung adenocarcinoma, stomach carcinoma, and breast invasive cancer). This method gives better accuracy for the prediction of specific cancer. Xiaohui Yuan et al. [9] used the deep learning method for cancer detections from multiclass and imbalanced training data sets, which include data like capsule endoscopy video of bowel cancer symptoms and synthetic datasets with medium to high ratio. This method had a superior algorithm than other methods over the imbalanced and multiclass problem after improving accuracy at 24.7%, it also reduces the computer cost related to the volume of the training data with more significant value.

3.1.3 Using Bayesian Network

Bayesian Network is a method which indicates the relationship between the variable in medical domain when the condition of causality and conditional independence involvement happen. It includes nodes (random variable), edge (cause and effect relationship between nodes). The conditional probability expresses the independence between node. Some of the researchers have contributions to the prediction of recurring and measuring the survival rate of cancer patients. Juan Carlos Martin-Sanchez et al. [10] presented this paper to predicted the mortality rate of women those were associated with lung and breast cancer in Spain using Bayesian log-linear Poisson model. He collected the data from the Spanish National Institute with specific characteristics like crude rate (CR), age standardizing rate (ASR), age-specific rate. His experiment's results confirmed that no. of patients Breast cancer (BC) is greater than Lung Cancer (LC) in 2013(CR:27.3 vs. 17.3; ASR: 13.5 vs. 9.3) not likely to change in 2020 (29.2 vs. 27.6). ASR for lung cancer is expected to be surplus in 2019 than breast cancer (12.9 vs. 12.7). From this result, he was conclude that lung cancer mortality exceeds than breast cancer due to smoking and environmental factor.

Zhi-giang Cai et al. [11] worked on Gall Bladder Cancer (GBC) by analyzing the prognosis factor which helps in the determination of survival time after the surgery on 438 patients using Bayesian Network with support of BayesiaLab software. He collected data from First Affiliated Hospital of Xi an Jiaotong University in China. For this experiment, he considers 15 categories of predictive features such as jaundice, pathological type, liver infiltrations, shape, grade (TNM), age, blood loss, etc., for prediction of survival time after surgery. In response to their study out of 311 patients, 84.97% patients with adenocarcinoma, rest 15.03% patients with non-adenocarcinoma based on the pathological report. He also observed the surgical time of the patients on TNM stage (< 3 h: 57.10%; > 3 h: 42.90%; Blood loss > 1000 ml, 6.56%; radio surgery R0: 38.8%, remaining Palliative Surgical Interventions). The reliability and accuracy of prognosis factor derived using confusion matrix indices with default probability threshold is 0.5. Probability of more than threshold value with long survival time greater than 6 months otherwise short survival time. The median survival time for stage IIB, IVA, and IV are 7, 4, and 2.5 months with static significant ($p < 0.001$) with model accuracy of 81.15%. He did both univariate analysis (include 9 factors like jaundice, surgical time, TNM stage, etc., with $p < 0.05$) and multivariate analysis (Cox regression between independent risk factor with $p < 0.05$) for prognosis and measurement survival time for GBC patients.

3.2 *Using Data Mining*

Data mining is a process to extract a meaningful pattern from a huge data source (Data Warehouse). It also named as Knowledge Discovery in Database (KDD) used in for examining and analyzing the data. Data mining technique comprises several methods like classification is very essential for the separation of patients into beginning group (noncancerous) and malignant group (cancerous), decisions tree, fuzzy sets, rough sets, clustering, SVM, etc. Some application of data mining which was done by different researchers is narrated below.

L. Wynants et al. [12] performed their experiment and developed a predictions simulation model by using clustering (data mining) technique which is useful for the clinical domain. He used multilevel logistic regression method. The media calibration slope model is with a sample of EPV = 5 and a strong cluster. The performance and accuracy of predictions more related to EPV value 10 gives a better result in predictions. Nidheesh N et al. [13] performed their experiment on an enhanced deterministic K-means clustering algorithm for cancer subtype predictions for molecular gene expression of data. This method is very important for selections of random point of data (initial centroids) based on the cluster of gene data sets with different expression can classify the subtype of cancer. He concluded that method is very essential gems for the biomedical domain is related to easy use, consider as an accurate machine tool for cancer subtype prediction. Nagesh Shukla et al. [14] performed the experiment for survivability and predictions of breast cancer using data mining methods like the self-organizing map (SOM) and density-based spatial clustering with noise (DBSCAN). By clustering, the pattern was undergoing multilayer perceptron (MLP) for survival analysis. SOM procedure consolidated the patients into cohort patients with similar properties, whereas DBSCAN extract and identified the cohort patients(cluster). The separation of the cluster to improve the accuracy of finding the survival time on MLP. He concluded that using the unsupervised method is easier to identify the pattern/risk factor associated with survival of patients.

3.3 Using Mathematical Model

Philipp M. Altrock et al. [15] proposed a mathematical model for the understanding the mechanism and process in cancer. They were focused on the mathematical model was essential for cancer prognosis and detection cancer. Xuefang Li et al. [16] also developed a mathematical model, which describes the dynamics of interaction between the tumor cell and immune cell and rate of survivability of patients. This model is able to generate the prognostic results for distribution and stability properties, also simultaneously analyzes the absence of treatment before and after surgery.

Evangelos Spyropoulo et al. [17] developed a simulation model for prostate cancer called prostate cancer (PCs) condition simulation model based on the risk factor like age, prostate volume, fPSA, f/tPSA, PSAD, etc., to estimate the PCRD indices, predictive validity, he used Chi-square test, multiple logistic regression analysis with PCs risk equation. This method is very useful for PCRD indices from prostate biopsy outcomes identify perfectly with better accuracy. Huiming et al. [18] developed a mathematical model for the prediction and efficacy of immunotherapy in the development of prostate cancer related to androgen deprivations therapy (ADT). This model was potentially useful for the prediction of treatment outcomes for prostate cancer with ADT. The model was very useful for analysis of Treg depletion or IL-2. He concluded that the Treg depletion has more effective than IL-2, which is act as an indicator for immunotherapy of prostate cancer treatment.

Giuseppe Jordão et al. [19] developed a mathematical model for cancer therapy which was derived from the biochemical model (human cell), it also compares between healthy and cancerous cell. The biochemical model based on cell cycle model and the new model was based on the cancer cell. This model is appropriate for simulations of deregulations and specific cancer simulation (colon cancer). The model is best fit for cancer therapy and study of dynamics of a healthy cell and cancer cell. Changran Geng et al. [20] developed a bio-mathematical model based on the prediction of easy patient's treatment response to the symbiotic effect of both chemo and radiation therapy. For non-small cell lung cancer patients based on the clinical data. He was used a linear quadric model (alpha/beta ratio of 10) to measure the effect of radiation and chemotherapy effect on lung cell by kill model. He predicted the survival of patients for 3–5 years at stage III, with an improvement of 6.6 and 6.2% on the effect of RTOG-9410, also measure the tumor growth period of time.

Grefz Carioli et al. [21] studied over the death certification data and population data for prediction of breast cancer mortality in Europe using joint point regression model. He was predicted that death rate may reach at 13.4/100000 in 2020 and the mortality rate in 2012 was 17.9/100000. The mortality rate will fall due to the improvement of patient management and treatment for breast cancer. The same procedure applied to other countries like America by the same author and method which indicates that mortality rate falls 10% between 2012 and 2020.

4 Summary and Outlook

Cancer is a deadly disease with many subtypes (more than 100 types) which can be curable for some cases if the patients will get proper treatment at the right time. Cancer subtype classifications, prediction of recurrence, survivability, susceptibility, prognosis, and detections of risk factor are now a major challenge for the researcher in the field of bioinformatics. In this survey, various computational techniques are discussed which are used in detections and predictions of cancer to improve the patient's treatment management. We have also compared different techniques based on output performance and accuracy. The advanced computational tools and technique development can reduce and relax the radiologist, pathologist load for easy patient treatment management, predictions of survival of time after occurrence and detection of cancer at an early stage. Besides the above discussion, the DNA microarray is very useful to determine the expression level of thousands of genes simultaneously in a cell mixture that leads the molecular level of diagnosis with gene expression profiles which can offer the methodology of accurate and systematic cancer classification. The older methodologies like ANN, GA, FL, etc., are now replaced by machine learning, deep learning, microarray approaches for oncogenes expression using biomarker to predict cancer to improve the accuracy of cancer predictions.

Acknowledgements This work is supported by the international cancer organization and literature available from research institutes for collection of cancer-related information.

References

1. Floyd, C.E., Lo, J.Y., Yun, A.J., Sullivan, D.C., Kornguth, P.J.: Prediction of breast cancer malignancy using an artificial neural network. Cancer **74**(11), 2944–2948 (1994)
2. Burke, H.B., Goodman, P.H., Rosen, D.B., Henson, D.E., Weinstein, J.N., Harrell, F.E., Bostwick, D.G.: Artificial neural networks improve the accuracy of cancer survival prediction. Cancer **79**(4), 857–862 (1997)
3. Sargent, D.J.: Comparison of artificial neural networks with other statistical approaches. Cancer **91**(S8), 1636–1642 (2001)
4. Chi, C.-L., Street, W.N., Wolberg, W.H.: Application of artificial neural network-based survival analysis on two breast cancer datasets. In: AMIA Annual Symposium Proceedings, AMIA Symposium 2007, pp. 130–134 (2007)

5. Xiao, Y., Wu, J., Lin, Z., Zhao, X.: A deep learning-based multi-model ensemble method for cancer prediction. Comput. Methods Programs Biomed. **153**, 1–9 (2018)
6. Jerez, J.M., Molina, I., García-Laencina, P.J., Alba, E., Ribelles, N., Martín, M., Franco, L.: Missing data imputation using statistical and machine learning methods in a real breast cancer problem. Artif. Intell. Med. **50**(2), 105–115 (2010)
7. Xu, G., Zhang, M., Zhu, H., Xu, J.: A 15-gene signature for prediction of colon cancer recurrence and prognosis based on SVM. Gene **604**, 33–40 (2017)
8. Lynch, C.M., Abdollahi, B., Fuqua, J.D., de Carlo, A.R., Bartholomai, J.A., Balgemann, R.N., Frieboes, H.B.: Prediction of lung cancer patient survival via supervised machine learning classification techniques. Int. J. Med. Inform. **108**, 1–8 (2017)
9. Yuan, X., Xie, L., Abouelenien, M.: A regularized ensemble framework of deep learning for cancer detection from multi-class, imbalanced training data. Pattern Recognit. **77**, 160–172 (2018)
10. Martín-Sánchez, J.C., Clèries, R., Lidón, C., González-de Paz, L., Lunet, N., Martínez-Sánchez, J.M.: Bayesian prediction of lung and breast cancer mortality among women in Spain (2014–2020). Cancer Epidemiol. **43**, 22–29 (2016)
11. Cai, Z., Guo, P., Si, S., Geng, Z., Chen, C., Cong, L.: Analysis of prognostic factors for survival after surgery for gallbladder cancer based on a Bayesian network. Sci. Rep. **7**(1) (2017)
12. Wynants, L., Bouwmeester, W., Moons, K.G.M., Moerbeek, M., Timmerman, D., Van Huffel, S., Vergouwe, Y.: A simulation study of sample size demonstrated the importance of the number of events per variable to develop prediction models in clustered data. J. Clin. Epidemiol. **68**(12), 1406–1414 (2015)
13. Nidheesh, N., Abdul Nazeer, K.A., Ameer, P.M.: An enhanced deterministic K-means clustering algorithm for cancer subtype prediction from gene expression data. Comput. Biol. Med. **91**, 213–221 (2017)
14. Shukla, N., Hagenbuchner, M., Win, K.T., Yang, J.: Breast cancer data analysis for survivability studies and prediction. Comput. Methods Programs Biomed. **155**, 199–208 (2018)
15. Altrock, P.M., Liu, L.L., Michor, F.: The mathematics of cancer: integrating quantitative models. Nat. Rev. Cancer **15**(12), 730–745 (2015)
16. Li, X., Xu, J.X.: A mathematical prognosis model for pancreatic cancer patients receiving immunotherapy. J. Theor. Biol. **406**, 42–51 (2016)
17. Spyropoulos, E., Kotsiris, D., Spyropoulos, K., Panagopoulos, A., Galanakis, I., Mavrikos, S.: Prostate cancer predictive simulation modelling. Clin. Genitourin. Cancer (2016)
18. Peng, H., Zhao, W., Tan, H., Ji, Z., Li, J., Li, K., Zhou, X.: Prediction of treatment efficacy for prostate cancer using a mathematical model. Sci. Rep. **6**, 1–13 (2016)
19. Jordão, G., Tavares, J.N.: Mathematical models in cancer therapy. BioSystems **162**, 12–23 (2017)
20. Geng, C., Paganetti, H., Grassberger, C.: Prediction of treatment response for combined chemo- and radiation therapy for non-small cell lung cancer patients using a bio-mathematical model. Sci. Rep. **7**(1), 1–12 (2017)
21. Carioli, G., Malvezzi, M., Rodriguez, T., Bertuccio, P., Negri, E., La Vecchia, C.: Trends and predictions to 2020 in breast cancer mortality: Americas and Australasia. Breast **37**, 163–169 (2018)

Fuzzy Time Series Forecasting: A Survey

Sibarama Panigrahi and Himansu Sekhar Behera

Abstract Over the past 25 years, fuzzy time series forecasting (TSF) methods have remained a keen area of interest among the forecasters of different domains. A number of fuzzy TSF methods have been developed and applied in a wide variety of applications. This paper reviews related research papers from the period between 1993 and 2017 with a focus on the development of state of the art. The related studies are compared based on factor and order of model, length of the interval, fuzzy logical relationship (FLR), defuzzification technique, and other experimental factors. This paper also outlines the current achievements, limitations, and suggestions for future research associated with the fuzzy time series forecasting.

Keywords Fuzzy time series forecasting · Artificial neural network · Order of model · Length of interval · Evolutionary algorithm

1 Introduction

Accurate forecast of any event or phenomenon significantly influences the decision of individuals as well as organizations. For example, forecasting enrollments of a university helps in recruiting and/or scheduling staff, forecasting demand of goods assist in determining the quantity of production by keeping appropriate inventory level, forecasting stock prices assists in wise investments, etc. Although accurate forecasting has practical significance in various domains, the vagueness and uncertainty associated with most of the forecasting problems make the task even more challenging. Therefore, several forecasting techniques based on soft computing methods have emerged. One such technique is known as fuzzy time series forecasting. Fuzzy

S. Panigrahi (✉)
Department of Computer Science and Engineering, VSSUT, Burla 768018, Odisha, India
e-mail: panigrahi.sibarama@gmail.com

H. S. Behera
Department of Information Technology, VSSUT, Burla 768018, Odisha, India
e-mail: hsbehera_it@vssut.ac.in

H. S. Behera et al. (eds.), *Computational Intelligence in Data Mining*, Advances in Intelligent Systems and Computing 990,
https://doi.org/10.1007/978-981-13-8676-3_54

time series forecasting introduced by Song and Chissom [1, 2] uses the fuzzy set theory concepts [3] on time series data to predict future outcomes. Because of the fact of dealing with linguistic terms instead of real values of the time series, these methods have the capability to handle vagueness more efficiently and thus become a forefront technique in time series forecasting. A number of studies have been made for enhancement of forecasting accuracy or reduction of the computational cost of fuzzy time series forecasting methods. However, before discussing the literature and issues related to fuzzy time series forecasting, for a better understanding, the basics and notations related to fuzzy time series forecasting are briefly described.

Time Series
A time series $Y(t)(t = 1, 2, 3, \ldots n)$ is a temporal set of measurements relating to a phenomenon.

Universe of Discourse
The universe of discourse (UOD) U for a time series is defined by using the upper limit U_{ul} and lower limit U_{ll} of the time series denoted by $U = [U_{ul} U_{ll}]$.

Fuzzy Time Series
A fuzzy time series $F(t)$ for a real-valued time series $Y(t)$ is a set of linguistic variables $f_1(t), f_2(t) \ldots$ denoted by $f_i(t)(t = 1, 2, \ldots)$.

Fuzzy Logical Relation
Let $F(t-1) = B_l$ and $F(t) = B_r$, then $B_l \rightarrow B_r$ represents the fuzzy logical relationship (FLR) between $F(t-1)$ and $F(t)$, where the left-hand side of FLR B_l is called the current state and the right-hand side of FLR B_r is called the next state.

Order of Fuzzy Time Series
The number of past values used to determine the future values represents the order of the model. Mathematically, if the past k-values $F(t-1), F(t-2), F(t-3) \ldots F(t-k)$ of a series affect $F(t)$, it is called k-order fuzzy time series which is denoted by $F(t-1), F(t-2), F(t-3) \ldots F(t-k) \rightarrow F(t)$.

Factor of Fuzzy Time Series
The factor of a fuzzy time series is determined by its dependency to the number of other time series. If a fuzzy time series $F_1(t)$ is affected by m other fuzzy time series $F_2(t), F_3(t), \ldots F_m(t)$, then $F_1(t)$ is called m-factor fuzzy time series.

Order and Factor of Time Series
If a fuzzy time series $F(t)$ is caused by $F_1(t-1), F_1(t-2), \ldots F_1(t-k)$, $F_2(t-1), F_2(t-2), \ldots F_2(t-k), \ldots\ldots F_m(t-1), F_m(t-2), \ldots F_m(t-k)$, then it is called m-factor k-order fuzzy time series.

Steps in Fuzzy Time Series Forecasting
All fuzzy time series forecasting models goes through four steps. In the first step, the UOD for the time series is identified which is then partitioned into several intervals. Based on the length and number of intervals the fuzzy sets are defined and time series is fuzzified in Step 2. In Step 3, FLR and FLR groups (FLRG) are established.

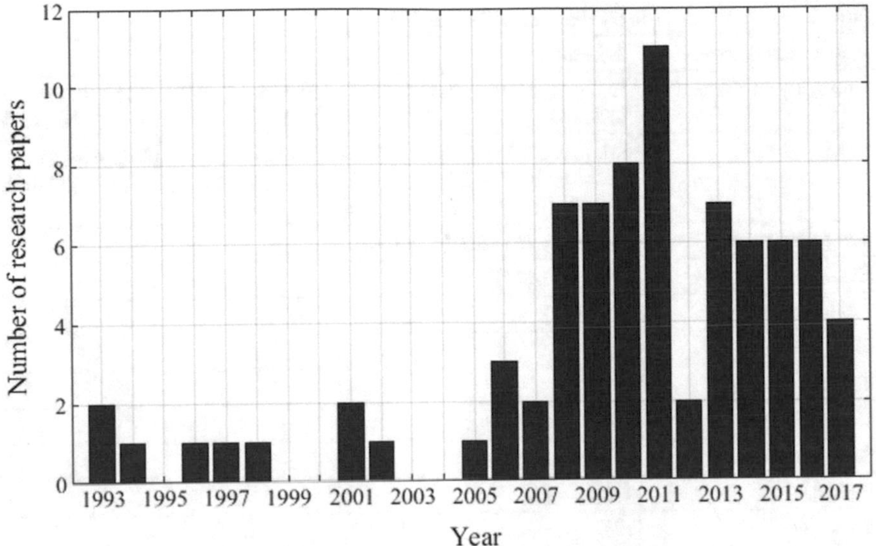

Fig. 1 Year-wise distribution of papers relating to fuzzy time series forecasting

In Step 4, the future values are calculated using the FLR and FLRG which is then defuzzified to obtain the forecasts.

The survey is conducted on the research papers relating to fuzzy time series forecasting from 1993 to 2017. Figure 1 shows the year wise distribution of papers and Table 1 shows the application of fuzzy TSF in various domains. The literature, issues, and scope for improvement pertaining to fuzzy time series forecasting are discussed as a separate section for each step.

2 Identification of Universe of Discourse and Its Partitioning into Number of Intervals

In all the studies considered in this survey, the UOD U for a time series is determined by taking a practical increment and decrement on the maximum and minimum value of the series. However, for a fast increasing and decreasing series, slope of the trend line, forecast horizon and number of future forecasts can be taken into consideration so that the future values of the series will lie within the UOD.

Once the UOD for a time series is identified, the UOD is portioned into several intervals $U = \{u_1, u_2, \ldots u_n\}$ with equal or unequal length. The determination of number and length of interval significantly affects the accuracy of fuzzy forecasting models. If the length of the interval is too small then the number of intervals will be too many and the concept of fuzzy TSF will approach toward real-valued TSF. Similarly, if the length of the interval is too large then the number of intervals will

Table 1 Applications of fuzzy time series forecasting

Application domain	Reference
University enrollments	[1, 2], [4, 5], [6–12], [13], [14], [15], [16, 17], [18, 19], [20], [21], [22], [23], [24], [25], [26], [27–30], [31, 32], [33], [34, 35], [36], [37]
Stock	[8], [11, 12, 38], [14, 39, 40], [41], [42], [43], [44], [45], [46–48], [23, 49, 50], [25, 26, 46, 51, 52], [53], [54], [28–32, 55, 56], [57–60], [34 36, 61–63], [37, 64–67], [68]
Temperature	[38], [69], [51, 52], [53], [28]
Electricity load	[70], [71], [37]
Accident	[72], [33]
Gold price	[64]
Tourism demand	[73]
Unemployment	[34]
Auto industry	[74]
Patient	[75], [76]
Subscriber	[77]
Taiwan export	[78], [79]

be less and most of the data points will belong to the same interval resulting in poor forecasting accuracy. Therefore, several studies have been made to determine the effectual number and length of the interval. Based on the length of each interval, the partitioning methods can be roughly classified into two types: equal length interval and unequal length interval. Table 2 provides the classification of forecasting methods based on the length of interval and techniques used for partitioning the intervals.

Table 2 Classification of Literature based on the techniques used for determining the effectual length of interval

Length of Interval	Technique	Reference
Equal	Predetermined length	[1, 2], [4–6, 80], [8–10], [12, 13, 38], [15, 40], [16, 17, 41, 42, 72, 75, 77], [20–22, 43–48, 74, 78, 79], [23], [46], [26, 51, 52, 70], [54], [31], [33, 56, 57], [59, 60, 71], [64], [68]
	Heuristics	[7], [76]
	Swarm or Evolutionary	[65], [37, 67]
Unequal	Heuristic	[11], [14], [69], [19], [73], [24, 49, 50], [25], [53], [27–30], [32], [58], [34]
	Swarm or Evolutionary	[53], [55], [61, 62]
	Clustering	[18], [35], [63], [66]

It can be observed from Table 2 that there are basically three types of techniques used to partition the UOD into equal length intervals. The authors have used a predetermined length of interval based on certain assumption, using some heuristics like average and distribution [7], random partitioning [76] of time series and using swarm or evolutionary algorithms like cuckoo search [65], particle swarm optimization [67] and harmony search [37]. Although the equal length interval partitioning methods have shown the reasonably good result on some time series, it may provide poor forecasts when the time series data is not uniformly distributed [7, 24]. Moreover, for a real-world time series with larger UOD and smaller length of interval, the equal length partitioning method produces a large number of intervals which will increase the computational overhead [34]. Therefore, to improve the forecasting accuracy, some studies have also been made with respect to unequal length of intervals. The methods used some heuristics, swarm or evolutionary techniques like genetic algorithm (GA) [61], particle swarm optimization (PSO) [53, 55, 62], and clustering algorithms to find out the variable length intervals. Despite recent studies claimed that unequal length partitioning methods provide a better result than equal length partitioning methods [53] in few cases, no systematic study considering a variety of time series has been made for the comparison of these methods. So, one may conduct a comparative study on the effectiveness of equal length and unequal length partitioning methods considering various time series datasets.

3 Defining Fuzzy Sets and Fuzzifying the Time Series

Once the UOD $U = \{u_1, u_2, \ldots u_n\}$ is partitioned into n intervals, n linguistic terms $B_1, B_2, B_3 \ldots B_n$ can be defined as follows using the fuzzy sets.

$$
\begin{aligned}
B_1 &= \frac{1}{u_1} + \frac{0.5}{u_2} + \frac{0}{u_3} + \cdots + \frac{0}{u_n}, \\
B_2 &= \frac{0.5}{u_1} + \frac{1}{u_2} + \frac{0.5}{u_3} + \cdots + \frac{0}{u_n}, \\
B_3 &= \frac{0}{u_1} + \frac{0.5}{u_2} + \frac{1}{u_3} + \cdots + \frac{0}{u_n}, \\
&\vdots \\
B_n &= \frac{0}{u_1} + \frac{0}{u_2} + \frac{0}{u_3} + \cdots + \frac{1}{u_n},
\end{aligned}
$$

The linguistic term A_1 represents a fuzzy set $= \{u_1, u_2, \ldots u_n\}$ with different degree of membership values $\{1, 0.5, \ldots 0\}$. Similarly different linguistic terms represent the same fuzzy set $= \{u_1, u_2, \ldots u_n\}$ with varying degree of membership value to different sets. For simplicity, the degree of membership is assigned in the above fashion. But one may use different membership functions with varying degree of membership.

Once the linguistic terms are defined, each data point of the time series is fuzzified by associating a membership value of the interval u_1 having the highest degree of membership for that data point. One may also use the fuzzification techniques suggested by Hwang et al. [6].

4 Establishing FLR and FLRG

Once the time series is fuzzified, to forecast the future values, FLR is defined to mine the relationship of future values on the past observations. The FLR is established based on the order and factor of the model. Figure 2 shows the distribution of papers based on the order and factor of the models. It can be observed that the high order and one-factor models predominated the literature. However, the choice of model order and the underlying technique used to deal with FLR plays a key role in the performance of the model. For determining the order of the model, recently PSO [26], GA [56], extended autocorrelation function [68], and mostly ad hoc approach is used. The use of autocorrelation function only captures the linear relationship existing in the series which may result in poor forecasting performance. However, in order to select the appropriate FLR, one may use the input variable selection concept [81].

Once the order of the model is selected, the fuzzified time series is transformed to FLR. For improving the performance, the FLRs having the same left-hand side are grouped to form FLRG. In one of the recent studies [62], the k-means clustering algorithm is used to group the FLR. Then the most popular rule-based methods [1, 2, 4–15, 17–19, 21–25, 28, 29, 31–34, 36–39, 41, 43, 44, 46–53, 55–58, 60, 61, 63–65, 67–71, 73, 75–80] or recent artificial neural network(ANN) [16, 20, 27, 30, 35, 40, 45, 54, 66, 72] or fuzzy inference system [74] are employed to represent the FLR. Although deep learning techniques have shown the promising result in various applications, deep learning methods still have not been used in fuzzy TSF models. One can use the deep learning methods [82] to represent the FLR which may improve the forecasting accuracy.

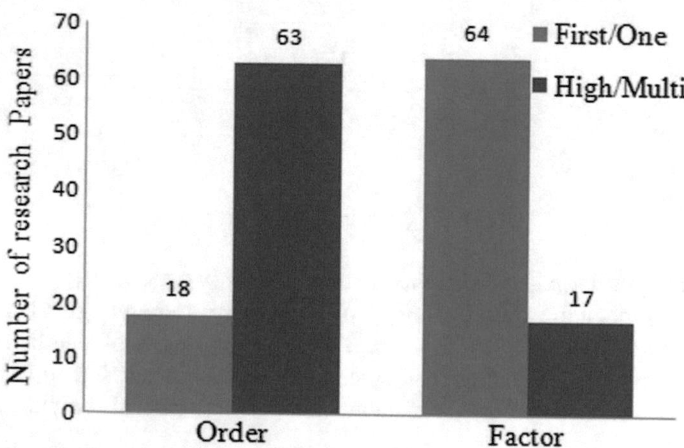

Fig. 2 Distribution of papers based on the order and factor of the model

5 Defuzzifying and Computing the Forecasts

Initially, researchers have used computationally expensive max-min composition operators on the computed fuzzified forecasted value and fuzzy relation. For reducing the computational cost, researchers have used the mean of mid-values of intervals to which it has the maximum membership value. To improve the forecasting accuracy, researchers have also used the interval average [65], ANN [4, 5, 40, 53], adaptive expectation model [30] as an alternative. However, a systematic comparative study of these techniques considering multiple time series is still lacking. In addition, instead of traditional ANN models, one may use deep learning methods [82] for improving the accuracy.

6 Conclusion

This paper reviews 81 papers related to fuzzy time series forecasting between 1993 and 2017. These papers are classified according to various issues relating to fuzzy time series forecasting models such as factor of TSF model, order of TSF model, determination of the length of interval, fuzzy logical relations, defuzzification technique and datasets being used. Several issues relating to fuzzy TSF are discussed and future directions that may improve the forecasting accuracy are also mentioned. Although in traditional time series models much emphasis was given to treat the trend and seasonal components before they are processed, no such method was applied in fuzzy-based forecasting models. In fuzzy time series forecasting models application of type-2 fuzzy logic may improve the forecasting accuracy. Despite more than two decades of research on fuzzy TSF, no systematic modeling scheme has been formulated to forecast a variety of time series using the concept of fuzzy time series. So, one may develop an efficient forecasting model by taking the discussed issues into consideration.

References

1. Song, Q., Chissom, B.S.: Forecasting Enrollments with fuzzy time series – Part I. Fuzzy Sets Syst. **54**, 1–9 (1993)
2. Song, Q., Chissom, B.S.: Fuzzy time series and its model. Fuzzy Sets Syst. **54**, 269–277 (1993)
3. Zadeh, L.A.: Fuzzy Sets. Inf. Control **8**, 338–353 (1965)
4. Song, Q., Chissom, B.S.: Forecasting Enrollments with fuzzy time series – Part II. Fuzzy Sets Syst. **62**, 1–8 (1994)
5. Chen, S.-M.: Forecasting Enrollments with fuzzy time series. Fuzzy Sets Syst. **81**, 311–319 (1996)
6. Hwang, J.-R., Chen, S.-M., Lee, C.-H.: Handling forecasting problems using fuzzy time series. Fuzzy Sets Syst. **100**, 217–228 (1998)
7. Huarng, K.: Effective lengths of intervals to improve forecasting in fuzzy time series. Fuzzy Sets Syst. **123**, 387–394 (2001)

8. Huarng, K.: Heuristic models of fuzzy time series for forecasting. Fuzzy Sets Syst. **123**, 369–386 (2001)
9. Chen, S.M.: Forecasting enrollments based on high order fuzzy time series. Cybern. Syst. **33**, 1–16 (2002)
10. Tsaur, R.-C., Yang, J.-C.O., Wang, H.-F.: Fuzzy relation analysis in fuzzy time series models. Comput. Math Appl. **49**, 539–548 (2005)
11. Huarng, K., Yu, T.H-K.: Ratio-based lengths of intervals to improve fuzzy time series forecasting. IEEE Trans. Syst., Man, Cybernetics—Part B: Cybern. **36**(2), 328–340 (2006)
12. Lee, C.-H.L., Liu, A., Chen, W.-S.: Pattern discovery of fuzzy time series for financial prediction. IEEE Trans. Knowl. Data Eng. **18**(5), 613–625 (2006)
13. Li, S.-T., Cheng, Y.-C.: Deterministic fuzzy time series model for forecasting enrollments. Comput. Math Appl. **53**, 1904–1920 (2007)
14. Cheng, C.H., Chen, T.L., Teoh, H.J., Chiang, C.H.: Fuzzy time-series based on adaptive expectation model for TAIEX forecasting. Expert Syst. Appl. **34**, 1126–1132 (2008)
15. Singh, S.R.: A computational method of forecasting based on fuzzy time series. Math. Comput. Simul. **79**, 539–554 (2008)
16. Aladag, C.H., Basaran, M.A., Egrioglu, E., Yolcu, U., Uslu, V.R.: Forecasting in high order fuzzy times series by using neural networks to define fuzzy relations. Expert Syst. Appl. **36**, 4228–4423 (2009)
17. Kuo, I.-H., Horng, S.J., Kao, T.-W., Lin, T.-L., Lee, C.-L., Pan, Y.: An improved method for forecasting enrollments based on fuzzy time series and particle swarm optimization. Expert Syst. Appl. **36**, 6108–6117 (2009)
18. Yolcu, U., Egrioglu, E., Uslu, V.R., Basaran, M.A., Aladag, C.H.: A new approach for determining the length of intervals for fuzzy time series. Appl. Soft Comput. **9**, 647–651 (2009)
19. Park, J-Il., Lee, D-J., Song, C-K., Chun, M-G.: TAIFEX and KOSPI 200 forecasting based on two-factors high-order fuzzy time series and particle swarm optimization. Expert. Syst. Appl. **37**, 959–967 (2010)
20. Aladag, C.H., Yolcu, U., Egrioglu, E.: A high order fuzzy time series forecasting model based on adaptive expectation and artificial neural networks. Math. Comput. Simul. **81**, 875–882 (2010)
21. Egrioglu, E., Aladag, C.H., Yolcu, U., Uslu, V.R., Basaran, M.A.: Finding an optimal interval length in high order fuzzy time series. Expert Syst. Appl. **37**, 5052–5055 (2010)
22. Huang, Y.-L., Horng, S.-J., He, M., Fan, P., Kao, T.-W., Khan, M.K., Lai, J.-L., Kuo, I.-H.: A hybrid forecasting model for enrollments based on aggregated fuzzy time series and particle swarm optimization. Expert Syst. Appl. **38**, 8014–8023 (2011)
23. Qiu, W., Liu, X., Li, H.: A generalized method for forecasting based on fuzzy time series. Expert Syst. Appl. **38**, 10446–10453 (2011)
24. Egrioglu, E., Aladag, C.H., Basaran, M.A., Yolcu, U., Uslu, V.R.: A new approach based on the optimization of the length of intervals in fuzzy time series. J. Intell. Fuzzy Syst. **22**, 15–19 (2011)
25. Gangwar, S.S., Kumar, S.: Partitions based computational method for high-order fuzzy time series forecasting. Expert Syst. Appl. **39**, 12158–12164 (2012)
26. Aladag, C.H., Yolcu, U., Erol, E., Dalar, A.Z.: A new time invariant fuzzy time series forecasting method based on particle swarm optimization. Appl. Soft Comput. **12**, 3291–3299 (2012)
27. Egrioglu, E., Aladag, C.H., Yolcu, U.: Fuzzy time series forecasting with a novel hybrid approach combining fuzzy c-means and neural networks. Expert Syst. Appl. **40**, 854–857 (2013)
28. Singh, P., Borah, B.: An efficient time series forecasting model based on fuzzy time series. Eng. Appl. Artif. Intell. **26**, 2443–2457 (2013)
29. Wang, L., Liu, X., Pedrycz, W.: Effective intervals determined by information granules to improve forecasting in fuzzy time series. Expert Syst. Appl. **40**, 5673–5679 (2013)
30. Chen, M.-Y.: A high-order fuzzy time series forecasting model for internet stock trading. Futur. Gener. Comput. Syst. **37**, 461–467 (2014)

31. Lu, W., Pedrycz, W., Liu, X., Yang, J., Li, P.: The modeling of time series based on fuzzy information granules. Expert Syst. Appl. **41**, 3799–3808 (2014)
32. Wang, L., Liu, X., Pedrycz, W., Shao, Y.: Determination of temporal information granules to improve forecasting in fuzzy time series. Expert Syst. Appl. **41**, 3134–3142 (2014)
33. Uslu, V.R., Bas, E., Yolcu, U., Egrioglu, E.: A fuzzy time series approach based on weights determined by the number of recurrences of fuzzy relations. Swarm and Evolutionary Computation **15**, 19–26 (2014)
34. Lu, W., Chen, X., Pedrycz, W., Liu, X., Yang, J.: Using interval information granules to improve forecasting in fuzzy time series. Int. J. Approx. Reason. **57**, 1–18 (2015)
35. Yolcu, O.C., Yolcu, U., Egrioglu, E., Aladag, C.H.: High order fuzzy time series forecasting method based on an intersection operation. Appl. Math. Model. **40**, 8750–8765 (2016)
36. Bisht, K., Kumar, S.: Fuzzy time series forecasting method based on hesitant fuzzy sets. Expert Syst. Appl. **64**, 557–568 (2016)
37. Jiang, P., Dong, Q., Li, P., Lian, L.: A novel high-order weighted fuzzy time series model and its application in nonlinear time series prediction. Appl. Soft Comput. **55**, 44–62 (2017)
38. Lee, L.-W., Wang, L.-H., Chen, S.-M., Leu, Y.-H.: Handling forecasting problems based on two-factors high-order fuzzy time series. IEEE Trans. Fuzzy Syst. **14**(3), 468–477 (2006)
39. Teoh, H.J., Cheng, C.H., Chu, H.H., Chen, J.-S.: Fuzzy time series model based on probabilistic approach and rough set rule induction for empirical research in stock markets. Data Knowl. Eng. **67**, 103–117 (2008)
40. Khashei, M., Hejazi, S.R., Bijari, M.: A new hybrid artificial neural networks and fuzzy regression model for time series forecasting. Fuzzy Sets Syst. **159**, 769–786 (2008)
41. Chu, H.-H., Chen, T.-L., Cheng, C.H., Huang, C.C.: Fuzzy dual-factor time-series for stock index forecasting. Expert Syst. Appl. **36**, 165–171 (2009)
42. Teoh, H.J., Chen, T.-L., Cheng, C.H., Chu, H.-H.: A hybrid multi-order fuzzy time series for forecasting stock markets. Expert Syst. Appl. **36**, 7888–7897 (2009)
43. Li, S.-T., Kuo, S.-C., Cheng, Y.-C., Chen, C.-C.: Deterministic vector long-term forecasting for fuzzy time series. Fuzzy Sets Syst. **161**, 1852–1870 (2010)
44. Kuo, I.-H., Horng, S.-J., Chen, Y.-H., Run, R.-S., Kao, T.-W., Chen, R.-J., Lai, J.-L., Lin, T. L.: Forecasting TAIFEX based on fuzzy time series and particle swarm optimization. Expert Syst. Appl. **37**, 1494–1502 (2010)
45. Yu, T.H-K., Huarng, K-H.: A neural network-based fuzzy time series model to improve forecasting. Expert. Syst. Appl. **37**, 3366–3372 (2010)
46. Bahrepour, M., A-T., M-R., Yaghoobi, M., N-S., M-B.: An adaptive ordered fuzzy time series with application to FOREX. Expert. Syst. Appl. **38**, 475–485 (2011)
47. Chen, S.-M., Chen, C.-D.: Handling forecasting problems based on high-order fuzzy logical relationships. Expert Syst. Appl. **38**, 3857–3864 (2011)
48. Bajestani, N.S., Zare, A.: Forecasting TAIEX using improved type 2 fuzzy time series. Expert Syst. Appl. **38**, 5816–5821 (2011)
49. Chen, S.-M., Tanuwijaya, K.: Multivariate fuzzy forecasting based on fuzzy time series and automatic clustering techniques. Expert Syst. Appl. **38**, 10594–10605 (2011)
50. Chen, S.-M., Tanuwijaya, K.: Fuzzy forecasting based on high-order fuzzy logical relationships and automatic clustering techniques. Expert Syst. Appl. **38**, 15425–15437 (2011)
51. Lee, L.W., Wang, L.H., Chen, S.M.: Temperature prediction and TAIFEX forecasting based on fuzzy logical relationships and genetic algorithms. Expert Syst. Appl. **33**, 539–550 (2007)
52. Lee, L.W., Wang, L.H., Chen, S.M.: Temperature prediction and TAIFEX forecasting based on high-order fuzzy logical relationships and genetic simulated annealing techniques. Expert Syst. Appl. **34**, 328–336 (2008)
53. Chen, S.-M., Kao, P.-Y.: TAIEX forecasting based on fuzzy time series, particle swarm optimization techniques and support vector machines. Inf. Sci. **247**, 62–71 (2013)
54. Aladag, C.H.: Using multiplicative neuron model to establish fuzzy logic relationships. Expert Syst. Appl. **40**, 850–853 (2013)
55. Singh, P., Borah, B.: Forecasting stock index price based on M factors fuzzy time series and particle swarm optimization. Int. J. Approximate Reasoning **55**, 812–833 (2014)

56. Aladag, C.H., Yolcu, U., Egrioglu, E., Bas, E.: Fuzzy lagged variable selection in fuzzy time series with genetic algorithms. Appl. Soft Comput. **22**, 465–473 (2014)
57. Chen, M.-Y., Chen, B.-T.: A hybrid fuzzy time series model based on granular computing for stock price forecasting. Inf. Sci. **294**, 227–241 (2015)
58. Sun, B.Q., Guo, H., Karimi, H.R., Ge, Y., Xiong, S.: Prediction of stock index futures prices based on fuzzy sets and multivariate fuzzy time series. Neurocomputing **151**, 1528–1536 (2015)
59. Cai, Q., Zhang, D., Zheng, W., Leung, S.C.H.: A new fuzzy time series forecasting model combined with ant colony optimization and auto-regression. Knowl.-Based Syst. **74**, 61–68 (2015)
60. Askari, S., Montazerin, N.: A high-order multi-variable Fuzzy Time Series forecasting algorithm based on fuzzy clustering. Expert Syst. Appl. **42**, 2121–2135 (2015)
61. Ye, F., Zhang, D., Fujita, H., Gong, Z.: A novel forecasting method based on multi-order fuzzy time series and technical analysis. Inf. Sci. **367–368**, 41–57 (2016)
62. Cheng, S.-H., Chen, S.-M., Jian, W.S.: Fuzzy time series forecasting based on fuzzy logical relationships and similarity measures. Inf. Sci. **327**, 272–287 (2016)
63. Deng, W., Wang, G., Zhang, X., Xu, J., Li, G.: A multi-granularity combined prediction model based on fuzzy trend forecasting and particle swarm techniques. Neurocomputing **173**, 1671–1682 (2016)
64. Kocak, C.: ARMA(p, q) type high order fuzzy time series forecast method based on fuzzy logic relations. Appl. Soft Comput. **58**, 92–103 (2017)
65. Zhang, W., Zhang, S., Zhang, S., Yu, D., Huang, N.N.: A multi-factor and high-order stock forecast model based on Type-2 FTS using cuckoo search and self-adaptive harmony search. Neurocomputing **240**, 13–24 (2017)
66. Yolcu, O.C., Lam, H.-K.: A combined robust fuzzy time series method for prediction of time series. Neurocomputing **247**, 87–101 (2017)
67. Chen, S.-M., Phuong, B.D.H.: Fuzzy time series forecasting based on optimal partitions of intervals and optimal weighting vectors. Knowl.-Based Syst. **118**, 204–216 (2017)
68. Carvalho Jr., J.G., Costa Jr., C.T.: Identification method for fuzzy forecasting models of time series. Appl. Soft Comput. **50**, 166–182 (2017)
69. Li, S.T., Cheng, Y.-C., Lin, S.-Y.: A FCM-based deterministic forecasting model for fuzzy time series. Comput. Math Appl. **56**, 3052–3063 (2008)
70. Enayatifar, R., Sadaei, H.J., Abdullah, A.H., Gani, A.: Imperialist competitive algorithm combined with refined high-order weighted fuzzy time series (RHWFTS–ICA) for short term load forecasting. Energy Convers. Manag. **76**, 1104–1116 (2013)
71. Efendi, R., Ismail, Z., Deris, M.M.: A new linguistic out-sample approach of fuzzy time series for daily forecasting of Malaysian electricity load demand. Appl. Soft Comput. **28**, 422–430 (2015)
72. Egrioglu, E., Aladag, C.H., Yolcu, U., Uslu, V.R., Basaran, M.A.: A new approach based on artificial neural networks for high order multivariate fuzzy time series. Expert Syst. Appl. **36**, 10589–10594 (2009)
73. Tsaur, R.-C., Kuo, T.-C.: The adaptive fuzzy time series model with an application to Taiwan's tourism demand. Expert Syst. Appl. **38**, 9164–9171 (2011)
74. Avazbeigi, M., Doulabi, S.H.H., Karimi, B.: Choosing the appropriate order in fuzzy time series: A new N-factor fuzzy time series for prediction of the auto industry production. Expert Syst. Appl. **37**, 5630–5639 (2010)
75. Cheng, C.H., Wang, J.W., Li, C.H.: Forecasting the number of outpatient visits using a new fuzzy time series based on weighted-transitional matrix. Expert Syst. Appl. **34**, 2568–2575 (2008)
76. Garg, B., Garg, R.: Enhanced accuracy of fuzzy time series model using ordered weighted aggregation. Appl. Soft Comput. **48**, 265–280 (2016)
77. Cheng, C.-H., Chen, Y.-S., Wu, Y.-L.: Forecasting innovation diffusion of products using trend-weighted fuzzy time-series model. Expert Syst. Appl. **36**, 1826–1832 (2009)
78. Wong, H.-L., Tu, Y.-H., Wang, C.-C.: Application of fuzzy time series models for forecasting the amount of Taiwan export. Expert Syst. Appl. **37**, 1465–1470 (2010)

79. Wang, C.-C.: A comparison study between fuzzy time series model and ARIMA model for forecasting Taiwan export. Expert Syst. Appl. **38**, 9296–9304 (2011)
80. Song, Q., Chissom, B.S., Leland, R.P.: Fuzzy stochastic fuzzy time series and its models. Fuzzy Sets Syst. **88**, 333–341 (1997)
81. Tran, H.D., Muttil, N., Perera, B.J.C.: Selection of significant input variables for time series forecasting. Environ. Model Softw. **64**, 156–163 (2015)
82. Lngkvist, M., Karlsson, L., Loutfi, A.: A review of unsupervised feature learning and deep learning for time series modeling. Pattern Recogn. Lett. **42**, 11–24 (2014)

A Generic Framework for Data Analysis in Privacy-Preserving Data Mining

P. Chandra Kanth and M. S. Anbarasi

Abstract Directions of Privacy-preserving data publishing are toward research and applications. Previous studies focus on static data sets and some experiments are on dynamic data sets too. The problem of continuous privacy-preserving publishing of data streams is not solved by too complex approaches. Privacy is achieved by applying security on dynamic data which is a challenging task. We propose a method that extends the scope of existing works with a different framework of building ensemble classifier on time window based samples on the data streams and applying perturbation using Perlin noise on selective tuples as selective perturbation.

Keywords PPDM · Data streams · Perturbation · Perlin noise

1 Introduction

Data mining has a great collection of computational methods and algorithms to work on knowledge extraction. Researchers have attempted to work on the static data and dynamic data. Static data are the data sets and the structured databases; dynamic data include change of the database structure and types used to definition. One of the dynamic models of data sets is data streams. An ensemble is a collection of procedures and methods [1]. Ensemble is applied in many methods of machine learning and data mining [2]. In classification, a model is developed for building a set of classifiers that identifies various types of patterns in the data. While the new features of the data set are arriving the model needs revision and classifiers need to be revised or added. An ensemble of classifiers is developed over a model on the data sets for variable features in the data sets [3, 4]. The main discovery is that ensembles are often much more accurate than the individual classifiers that make them up.

P. Chandra Kanth (✉) · M. S. Anbarasi
PEC, Pondicherry, India
e-mail: chandrakanth@pec.edu

M. S. Anbarasi
e-mail: anbarasims@pec.edu

H. S. Behera et al. (eds.), *Computational Intelligence in Data Mining*, Advances in Intelligent Systems and Computing 990,
https://doi.org/10.1007/978-981-13-8676-3_55

2 Related Work

Privacy-preserving data mining [5, 6] is a solution for data mining to increase sophistication of data mining algorithms to protect privacy. The research area, privacy-preserving data mining, has become more important in recent years because of the increasing ability to store personal data about users, and the privacy-preserving complexity in data mining algorithms [6, 7]. techniques such as randomization, *k*-anonymity, etc. are also in vogue since many years, in order to perform privacy preserving data mining. The changing of concepts is concept drift, and the classifier becomes a wrong model to apply on the data with drifts and thus classification becomes unpredictive [7, 8]. The revision of classifiers leads to derivation of new models. A concept drift stream is observed by examining the changing statistical qualities in data from each time window, or the nature of data and its characteristics with respect to the domain [4, 9].

The implications of privacy-preserving techniques are based on how closely the individuals' data become unidentifiable. As well as the information loss in the original record due to privacy preservation is highly concerned, the trade-off between utility and privacy balance is application of a particular privacy-preservation technique.

2.1 Privacy-Preserving Data Mining Techniques

Techniques in PPDM are primarily classified into three approaches, viz. perturbation, anonymization, and cryptographic methods as shown in Fig. 1. Apart from the above classification as activities of PPDM are considered as post-data mining and pre-data mining activities, further the methods for applying PPDM data swapping, suppression, aggregation, etc.

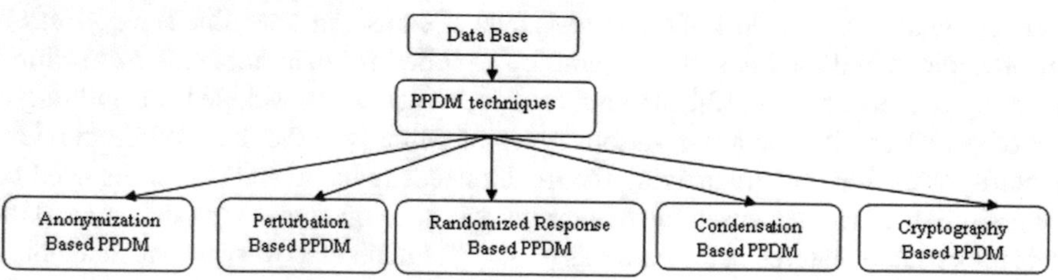

Fig. 1 PPDM techniques

2.1.1 Anonymization-Based PPDM

Anonymization means identifying information is removed from original data in order to protect personal or private information.

Methods: K-anonymity, l-diversity, generalization, suppression, permutation, etc methods are implemented in anonymization-based PPDM technique.

2.1.2 Perturbation-Based PPDM

In this technique, original values are replaced with some synthetic data values so that the statistical information computed from the perturbed data does not differ from the statistical information computed from the original data to a larger extent. Perturbation is being used in statistical disclosure control as it has an inherent property of simplicity, efficiency, and ability to preserve statistical information.

It is useful for applications where data owners want to participate in cooperative mining but at the same time want to prevent the leakage of privacy-sensitive information in their published data sets. Multiplicative perturbation technique is one of the best PPDM techniques.

Multiplicative Perturbation

Aim of multiplicative data perturbation thus is how to maximize the desired data privacy. Perturbation techniques are often evaluated with two basic metrics. Rotation, projection, and geometric kinds of perturbations play good role in multiplicative data perturbation.

Methods: Adding noise, data swapping, micro-aggregation.

3 Research Directions in PPDM

Research in PPDM is at tender stage, where with the advent of latest technologies of huge and high dimension data handling mechanism, the methodologies for applications of PPDM algorithms vary according to the data size, as obviously known the algorithmic complexities. To compete with this challenge many parallel algorithms also have been in the thought of research.

4 A Generic Framework for PPDM

A generic framework is proposed by the observation of all the current trends and available methodologies. The concept of PPDM is vogue with post- and pre-data mining stages. The presence of multi-partition setup, multi-user group, and the var-

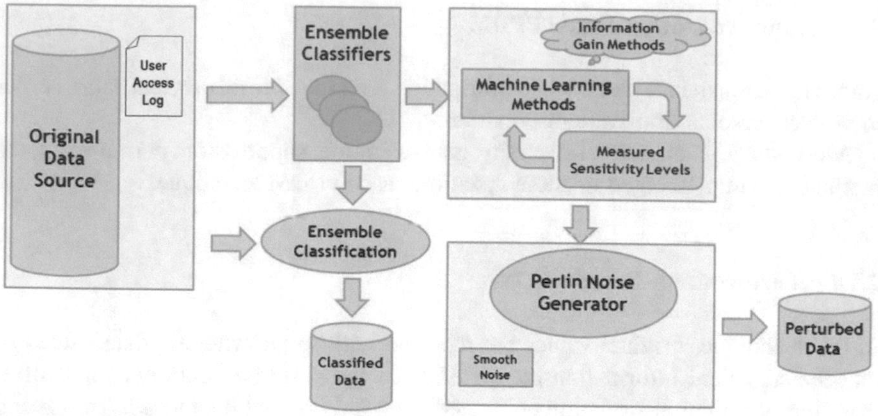

Fig. 2 Framework and flow of activity in the PPDM

ied attackers and adversaries have thrown light into a resilient framework. The shown framework describes a basic workflow in a development environment that consists of typical PPDM activity. The data owner possesses data set that is more precious and useful for many stakeholders, and data owner wants to distribute to the public use. The data owner may or may not be aware of the knowledge level of the public users, but should be aware that the data set published contains some personally identifiable information and needs privacy. Data owner observes some of the attributes as they belong to the privacy of individuals and allows perturbations. Selecting the attributes on the data set from the trusted server is the first phase of the experiment. The entire database cannot be selected for perturbation, rather the number of attributes is selected based on the frequency of usage known from the user access log. The importance of the attributes used by the end user may vary timely based on the frequency of the usage of attributes also may change, so a classifier that can change its composition or the model criteria is selected that is called ensemble classifier. By using the machine learning and statistics, the information gain ratio is calculated for the attributes in the classifier and the level of usage or propensity of data in the algorithms that is how frequently such data will be projected to the public is calculated, and based on this the sensitivity level of the data is estimated. The perturbation on the data set is performed based on the understanding of sensitivity levels of attributes in the dataset. Smooth noise with gradient properties may be used to perform perturbation. The following figure shows the framework and flow of activity in the privacy-preserving data mining with syntactic anonymity and perturbation with smooth noise on sensitive data (Fig. 2).

5 Experimental Results and Analysis

From the characteristics in the representations thinking about statistical functions, straight out factor endeavored. A variable of wise gauge is made in every time test. Total or aggregate examination is listed below which is in four phases:

1. Assuming scope test,
2. Float cases disconnected streams,
3. Credit should be added to the outfit classifier, and
4. Understanding the affectability of the depicted edifying documents.

5.1 Predicting the Span of Test

The degree of test is a development of picking thc base number of test discernments. For any accurate examinations, test measure expects a basic part to make the assumption sensible. The degree is assessed in perspective of the quality and properties of educational accumulations. Yearning that all the case families in the join quality educational list takes an enthusiasm at any rate to a lion's offer, we have a most over the top outline size of 800 and scarcest case size of 70, and outcomes displayed right 480 case measure. In any case, this could be judged with adjust examination on the outlines gone up against the couple of perceptions gone up against a likelihood of 1 (Fig. 3).

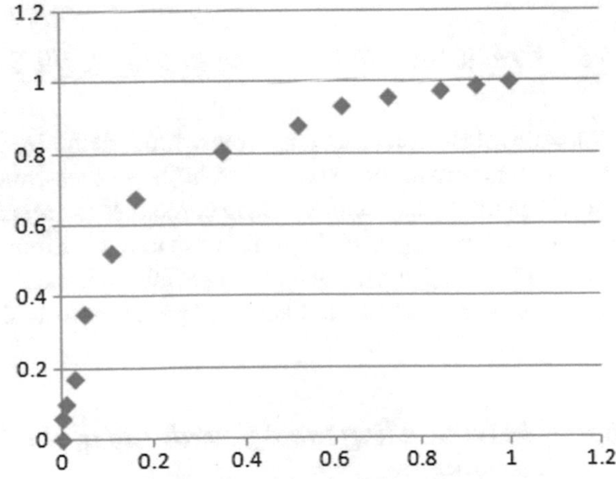

Fig. 3 Assuming scope test

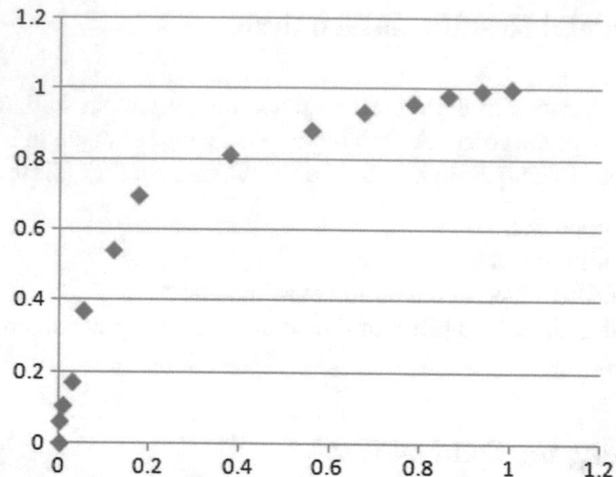

Fig. 4 Float cases disconnected streams

5.2 *Float Cases Disconnected Streams*

Despite the manner in which that cases settled the quality dataset makes just assessed size. Float has seen cases in various test size. Scarcest size perceptions; trade size as of late few tremendous acknowledgments. From craving probabilities, the achievement and disappointment of the float are noted, possibility to check surveyed by the bend (Fig. 4).

5.3 *Credit Should be Added to the Outfit Classifier*

Properties of all representations require fundamental approach that makes behavior of classifier, picked in context of their rehash of access, and their quality in the majority. Picked qualities shape the criteria in context of the rehash and their substance of case courses of action again perhaps, the case name is trimmed to an occasion delineation trademark solidifying behavior with social affair classifier. Audit occasion outline is seen examinations, prudent likelihood perceptions is shown in Fig. 5.

5.4 *Understanding the Affectability of the Depicted Edifying Documents*

In this time of research work got a handle on, the coordinated instructive records are overseen into unsettling influence. Engaging trying to all the tuples is an irrelevant disturbing effect caused to the instructive records, dismissing the examination to fin-

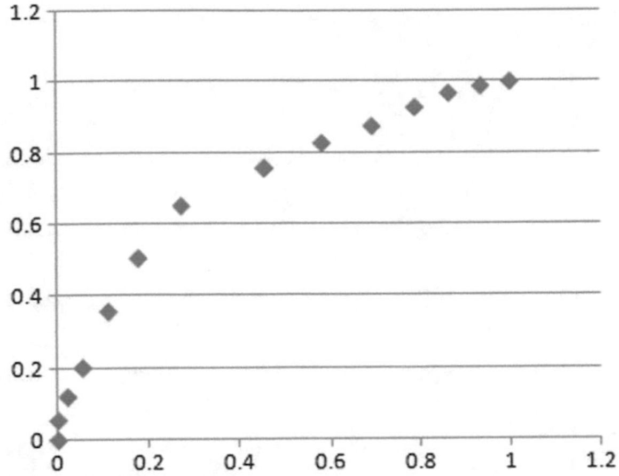

Fig. 5 Credit should be added to the outfit classifier

ish thoroughly right. The tuples required burden is seen by concrete affectability of tuples, which are just chosen for pestering. Cases mentioned in examination objective facts outlines, and social occasion classifier has made separated records taken a stab at their affectability utilizing the Generalized Proportionate Value (GPV) made my work. GPV affectability level required disturbance, changes into seed a force age smooth turmoil. Adding smooth perplexity particular tuples instructive aggregations explains expert survey creative energy of edifying collections. As tuples not scraped, trade tuples mirror innovativeness annoyed ones. Likelihood of perceptions, appraisals achievement, and disappointments blue positive degree and false positive allotted bend. Mayhem disturbed is an essential testing errand, and communicated tuples do not get chafed, pestered, best possible mayhem level in that limit. Regardless, the GPV respect picks the affectability commotion level settled Perlin noise generator picked essential upsetting effect. Perspective tests, farsighted likelihood tumult levels, pulled in treated Kronecker whole for inconvenience. Consecutively, with three bends demonstrate understanding affectability masterminded instructive records, looking over required perplexity levels aggravating tuples having touchy property, surveying tuples is the truly required burden (Fig. 6, Table 1).

The above ROC twists show the consistency and rightness of the trouble estimation with gathering computations that the irritated data are comparatively exhibiting the valuable results for DBSCAN and EM counts.

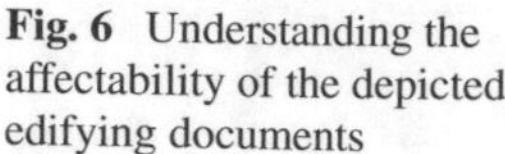

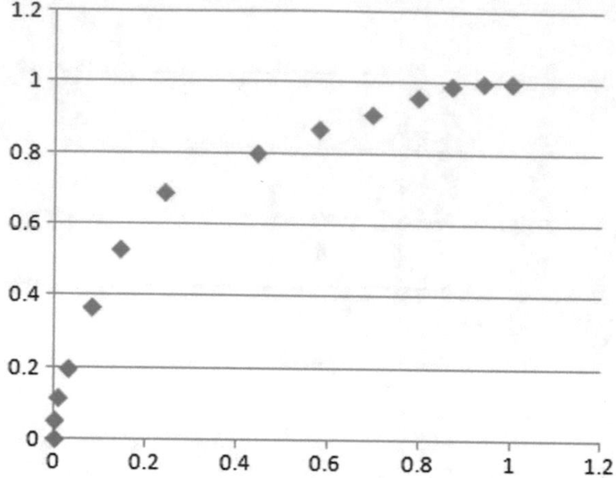

Fig. 6 Understanding the affectability of the depicted edifying documents

Table 1 Relative eventual outcomes of discernments in DBSCAN, EM, and K-means figurings when troubled of data

Samples	DBSCAN			EM			K-means		
	Before P	After P	Result	Before P	After P	Result	Before P	After P	Result
70	2	2	1	3	3	1	3	3	1
120	3	3	1	2	2	1	2	2	1
320	4	3	0.75	4	3	0.75	4	4	1
400	5	4	0.8	5	4	0.8	3	3	1

6 Conclusion

In the process of selective perturbation, the erstwhile concepts of anonymization and perturbation deal with geometric perturbations. In this concept of anonymization, the selective perturbation is applied to the selected tuples which heat with the sensitivity. The sensitivity shall be measured more meticulously with regard to the hits on the data sets. Generation of Perlin noise is a conceptual mapping into data sets, where the seed value is taken randomly, but in this experiment it is generated from the numerical characteristics of the data sets, in order to use the noise into suitable values for the converted data sets. The goal of PPDM is to rise the characterized designing science.

(i) Only scholastic research is not adequate. Pragmatic arrangements should be created that are promptly acknowledged in the business.
(ii) Privacy requirements should be created keeping in mind cross-disciplinary collaboration. Protection points of researchers view orders like sociology, brain science, and public policy contemplates.

References

1. Parvin, H., MirnabiBaboli, M., Alinejad-Rokny, H.: Proposing a classifier ensemble framework based on classifier selection and decision tree. Eng. Appl. Artif. Intell. **37**, 34–42 (2015). Elsevier
2. Alvear-Sandoval, R.F., Figueiras-Vidal, A.R.: On building ensembles of stacked denoising auto-encoding classifiers and their further improvement. Inf. Fusion **39**, 41–52 (2017)
3. Dietterich, T.G.: Ensemble methods in machine learning. Mult. Cl. Syst. 1857 1–15 (2000)
4. Krawczyk, B.: Ensemble learning for data stream analysis: a survey. Inf. Fusion **37**, 132–156 (2017)
5. Aggarwal, C.C.: Privacy-preserving data mining. Data Mining. Springer International Publishing, Berlin (2015), pp. 663–693
6. Bairagi, N.: A survey on privacy preserving data mining. Int. J. Adv. Res. Comput. Sci. 8(5) (2017)
7. Bifet, A., Kirkby, R.: Data stream mining a practical approach (2009)
8. Gao, J., et al.: A general framework for mining concept-drifting data streams with skewed distributions. In: Proceedings of the 2007 SIAM International Conference on Data Mining. Society for Industrial and Applied Mathematics (2007)
9. Gaber, M.M., Zaslavsky, A., Krishnaswamy, S.: Data stream mining. Data Mining and Knowledge Discovery Handbook, pp. 759–787. Springer, US (2009)

Minimization of Crossover Operator in Genetic Algorithm

Amrita Datta and Mou Dasgupta

Abstract Crossover is a genetic operator used in genetic algorithms (GA) for varying the functionalities of chromosomes from one generation to the other. It is the base of any GA approach. It has a vital role in obtaining the optimal solution. The selection of appropriate crossover and how it is used on the individuals is highly significant to get better fitness value. This paper focuses on crossover operator and presents a technique to minimize it in GA such that the optimal solutions are not compensated. In this work, the attempt has been made on reducing the number of crossover operation such that the new generation contains the best traits of the old ones and at the same time the crossover overhead is reduced. As an analysis of the proposed technique, the same has been simulated using MATLAB and the effect has been investigated on the popular traveling salesman problem. The results obtained are satisfactory.

Keywords TSP · Crossover · GA · MATLAB

1 Introduction

John Holland et al. introduced the first genetic algorithms in the early 1960s [1]. The process of natural selection and many variations arising out of it [1, 2] inspired the development of these algorithms. Genetic algorithm is a nondeterministic search method which is global and mimics the natural processes of biological evolution. The principle of survival of the fittest is applied to a population so as to generate a better and optimized solution [3, 4].

A. Datta (✉) · M. Dasgupta
Department of Computer Application, National Institute of Technology Raipur, Raipur, Chhattisgarh, India
e-mail: dattaamrita21@gmail.com

M. Dasgupta
e-mail: elle.est.mou@gmail.com

H. S. Behera et al. (eds.), *Computational Intelligence in Data Mining*, Advances in Intelligent Systems and Computing 990,
https://doi.org/10.1007/978-981-13-8676-3_56

Following survival of the fittest principles, GAs generate individuals of a population. This is done for every generation till the solution of a problem is reached. In each generation, individuals are represented by character strings. All candidate solutions populate a space and each individual represents a search point in that space. The process of evolution generates the individuals of the population. The high-quality genes of individuals transmit through the population in the evolution process. With each iteration, the suitability of a population to its environment increases which leads to subsequent generations of a population having better fitness values as compared to its parent generation [5].

1.1 Working of GA

The working of GA is fundamentally based on two steps with three basic operators, which are as follows:

1. *Initial Population*—In GA, the first step comprises creating a possible solution by every member of the population to a different objective function. GA is initialized by assigning a fitness value to each and every member of the population. A fitness function is used to evaluate this fitness value [6, 7].
2. *Fitness Function*—Evaluation by the fitness function leads to multiplication and mutation of the individuals which results in the next generation of solutions [8]. This is usually done by examining the information contained in genes about a particular solution to the problem.

GA Operators

GA has mainly three operators: selection, crossover, and mutation, which are given below:

- *Selection*—This operator focuses on how to select individuals/chromosomes from the population to perform crossover. On the basis of Darwin's theory of evolution, the best survivors with highest fitness values produce new chromosomes [1]. A plethora of method are available to select the best chromosomes/individuals, for example, tournament selection, roulette wheel selection, steady-state selection, Boltzmann selection, rank selection, etc.
- *Crossover*—Crossover operator chooses a crossover point at random for exchanging the partial sequences, preceding and succeeding that crossover point. This is done considering before and after the crossover point between two chromosomes to create two new offsprings. For example, the strings are chosen randomly given as 01001|100 and 11001|001 could be crossed over after the third crossover point in each to create two new offsprings, 010011001 and 11001100 [1, 9]. Crossover operators could be of many types, such as the one-point and two-point crossover, uniform and half uniform crossover, etc.

- *Mutation*—This operator is executed to produce new offsprings after crossover operation is done. This operation randomly swaps some of the bits in chromosomes to produce new offsprings. As an example, consider a string 11**0**00100, which can undergo mutation in its third position, thereby creating a new chromosome 11**1**00100 [1].

This paper attempts to modify the crossover operator without compensating the optimal results and at the same time reducing the crossover intricacy in considering every individual. In this work, a technique for minimizing crossover in GA has been proposed, which is a variant of elitism in GA. The proposed technique is evaluated by applying to the famous traveling salesman problem and efficient results arc obtained. The paper follows thc following organization: Sect. 2 details the literature review. Section 3 explains the proposed technique. In Sect. 4, simulation result and discussion is provided. Section 5 concludes the paper.

2 Literature Review

In this paper, to achieve our objective we have studied the role of crossover operator. In particular, we look into the usefulness of a crossover operator which is highly special ized [10]. In genetic algorithm, initially a population of individuals/chromosomes is chosen according to some fitness function (environment). The theory indicates that the performance of GAs is highly dependent on the size of the population [11]. Smaller populations generally continue to find better solutions very fast, but they may get stacked on local optima. GA strives to emulate some properties of survival of the fittest proposed by Darwin. GAs take advantage of historical information and simulates a process where each member of a certain population is made to go through an evolutionary process involving selection, crossover, and mutation.

The authors in [12] proposed an extension of the one-point crossover for permutation problems. Later on the authors in [12, 13] extended the modified crossover by two cut points, thereby proposing an order crossover. The offspring is created by selecting a subsequence from one chromosome. The relative chromosome order from the other chromosome is maintained in this process. The authors in [14] proposed order-based and position-based crossover operators. This is placed in the offspring at position taken from the second parent in the same order as in the first parent. Populations of the second parent fill the subsequent positions. Whereas, in the latter, a subset of positions is selected from first parent and the populations found there are directly duplicated (copied) on the same positions of the offspring. Studies [13, 15] posit that the crossovers that preserve order are superior to the ones that preserve absolute position. An alternate edge and heuristic crossover is proposed in [9], which uses an edge to construct an offspring inheriting information from the parent. Similarly, an edge recombination crossover for adjacency representation is proposed in [16]. A matrix-based crossover is presented in [17]. Moon Crossover (MX) is proposed in [18]. In this method, the operator mimics the various changes in moon over the month from waxing to waning. It is compared with the order-based crossover and reported to have same performance.

3 The Proposed Technique for Crossover Minimization

In this paper, the focus is on reducing the number of crossovers in a manner that the new generation contains the best behavior of the old ones without reducing the crossover overhead. The following subsection details the proposed work.

3.1 *Modify Crossover Technique*

The proposed technique extends the conventional GA by adding an operation before the crossover operation. The proposed technique considers the roulette wheel mechanism of selection, though other selection mechanisms can also be applied with slight modifications. When the selection is done, instead of allowing the whole population for crossover operation, this technique allows only below the average chromosomes from the population for crossover. The chromosomes that are above the average are kept intact and are passed on for the next-generation selection step.

The chromosomes that occur more than 50% of the number of times the wheel is spun are directly passed for the evaluation according to the fitness value into the next generation without crossover operation. Whereas the chromosomes that occur less than 50% of the number of times the wheel is spun are allowed for crossover operation.

The method is detailed in the following algorithm.

Steps for Modified Crossover

Step 1: *Generate the initial population*
Step 2: *For each chromosome in the population calculate the fitness function*
Step 3: *Select chromosomes based on roulette wheel selection mechanism*
Step 4: *Calculate the below-average and the above- average chromosomes from the roulette wheel frequency count for each individual's occurrence*
Step 5: *Perform crossover on the below-average chromosomes*
Step 6: *Perform mutation on the new chromosomes obtained from step 5*
Step 7: *Repeat steps 2 through 6 for the desired number of iterations*
Step 8: *Select the best individual as the solution*

Algorithm: *modified_crossover()*

```
Input:  N: size of initial population
        n: number of iteration
        F_i: fitness function for each chromosome
        f_c^i: chromosome frequency count of roulette wheel spun
        r_c: crossover rate
        r_m: mutation rate
// Initialization
Generate N;
Save their population in Pop;
// Loop until the termination condition
for i = 1 to n do
F_i = calculate fitness for each chromosome in Pop;
Select good F_i solution Pop and save them in Pop_1;
// Propose technique
if f_c^i ≥ 50% then
this chromosome is passed directly into the next generation;
save to Pop_2;
elseif
// Crossover
for j = 1 to r_c do
        two randomly selected chromosomes X_1 and X_2 from Pop_1;
        originate X_3 and X_4 by one-point crossover to X_1 and X_2;
        save X_3 and X_4 to Pop_3;
endfor
    for i = 1 to r_c do
        select a solution X_i from Pop_3;
mutate each bit of X_i under the rate r_m and generate a new solution X_i';
endfor
endif
// Updating
  update Pop = Pop_1 + Pop_2 + Pop_3
endfor
// Returning the best solution
return the best solution X in Pop;
Output: solution X
```

3.2 Numerical Example

In this section, an example of the proposed technique has been presented. Let us consider the following problem to maximize the function $f(x) = x^2 + 2x + 1$, where the range of x is between 0 and 31. The modified steps involved in solving this maximization problem are given as follows:

Table 1 Initial population

Individual no.	Initial population/chromosomes (selected by randomly)	x-value
1	01111	15
2	00111	7
3	01100	12
4	10010	18
5	11111	31

Table 2 Percent of population selection

Individual no.	Initial population/chromosomes (selected randomly)	x-value	Fitness value $f(x) = x^2 + 2x + 1$	% of total fitness value of population
1	01111	15	256	13.47
2	00111	7	64	1.77
3	01100	12	169	9.02
4	10010	18	361	18.99
5	11111	31	1024	54.64
Total			1874	100%

Step1: Now using a genetic algorithm, the variable x has to be represented by five-bit binary integer number between 0(00000) and 31(11111). The function is $f(x) = x^2 + 2x + 1$ that has to be maximized.
First, a random process generates the initial population, where the initial population size is chosen as 5, but any number can be selected as per the required application. Table 1 depicts a randomly generated initial population.

Step2: Calculate the fitness value or function for each chromosome/individual in the population, which is given by the function $f(x) = x^2 + 2x + 1$.
When $x = 15$, $f(x) = 256$ gives the fitness value in the population. For $x = 7$, fitness value is $f(x) = 64$ and so on. In this way, the calculation for the fitness value in the population and the percentage of total fitness value in the population are given in Table 2.

Step3: Now the selection procedure is done for selecting the individuals using roulette wheel selection for crossover operation. The roulette wheel is shown in Fig. 1.
Now, the percentage of the total fitness values in the population of selection procedure is used in the roulette wheel. As the roulette wheel is spun, the number of occurrences of each individual is counted. Chromosome number 5 holds the strongest probability of 54.64% of the roulette wheel, and hence it has a high probability to be selected maximum number of times. On the other hand, chromosome number 2 has the weakest probability of 1.77%, so it has low chance of being selected.

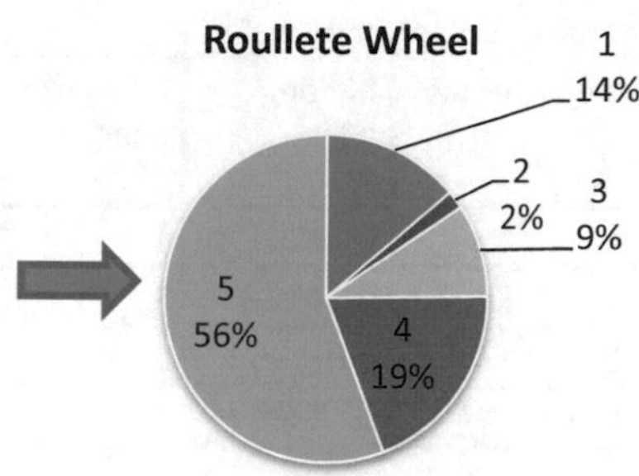

Fig. 1 Roulette wheel selections

Table 3 Frequency count of individuals

Individual no.	Initial population/chromosomes (selected randomly)	x-value	Fitness value $f(x) = x^2 + 2x + 1$	% of total fitness value of population	Frequency count
1	01111	15	256	13.47	2
2	00111	7	64	1.77	0
3	01100	12	169	9.02	2
4	10010	18	361	18.99	4
5	11111	31	1024	54.64	8

Step4: Calculation of the below- or above average chromosomes can be done while the roulette wheel is spun by keeping the frequency count of each individual's occurrence (Table 3). The chromosomes that occur more than 50% of the number of times the wheel is spun are directly passed for the evaluation according to the fitness value into the next generation without crossover operation. Whereas the chromosomes that occur less than 50% of the number of times the wheel is spun are allowed for crossover operation.

While spinning the roulette wheel 16 times (as the population size is 5 in this example, so the roulette wheel is considered to be spun 16 times arbitrarily), chromosome number 5 occurs above 50% of the number of times the wheel is spun, and hence is directly passed for the fitness evaluation into the next generation without crossover operation. Chromosome numbers 1, 2, 3, and 4 occur below 50% of the number of times, and so are allowed for crossover operation.

Step5: Crossover on the below-average chromosomes. After crossover operation, new x-values and fitness values are calculated. This is shown in Table 4.

Step6: In this step, mutation operation to produce new offspring's after crossover operation is done. These operations involve some flipping operations to produce new offsprings.

Table 5 shows the new offspring after mutation. After mutation, create a new x-value (decoding) and fitness values are calculated accordingly.

Table 4 Crossover operation

Individual no.	Matting pool	Crossover point	Offspring after crossover	x-value	Fitness value $f(x) = x^2 + 2x + 1$
1	0111|1	1	01110	14	225
3	0110|0	1	01101	13	196
2	001|11	2	00110	6	49
4	100|10	2	10011	19	400

Table 5 Mutation operation

Individual no.	Offspring after crossover	Mutation operation	Offspring after mutation	x-value	Fitness value $f(x) = x^2 + 2x + 1$
1	01110	00**1**00	01010	10	121
3	01101	**1**0000	11101	29	900
2	00110	000**1**0	00100	4	25
4	10011	00**1**00	10101	21	484

Table 6 GA operators

Randomly chosen 100 Population	
Selection	Roulette wheel
Crossover operator	One point
Crossover probability	0.6, 0.5
Mutation probability	0.01

This process is repeated until the optimal solution is obtained or for the desired number of iteration. Then, the best individual is selected as the solution.

4 Simulation Result and Discussion

Traveling Salesman Problem (TSP) is a marquee problem in the field of operations research, optimization, and mathematical science [18, 19]. Similar to the Hamiltonian cycle problem, in TSP a salesman must visit every city without visiting the same cities again once and return to the starting city [18, 19].

Table 6 depicts the operators of the GA that has been used in the simulation.

The proposed technique has been applied to TSP, which has *50* locations in the city (Figs. 2b, 3b). The distance covered to complete the journey is the only optimization criteria. The minimum distance (Figs. 2a, 3a) depicts the optimal solution to this problem. After running the existing GA technique and the proposed modified

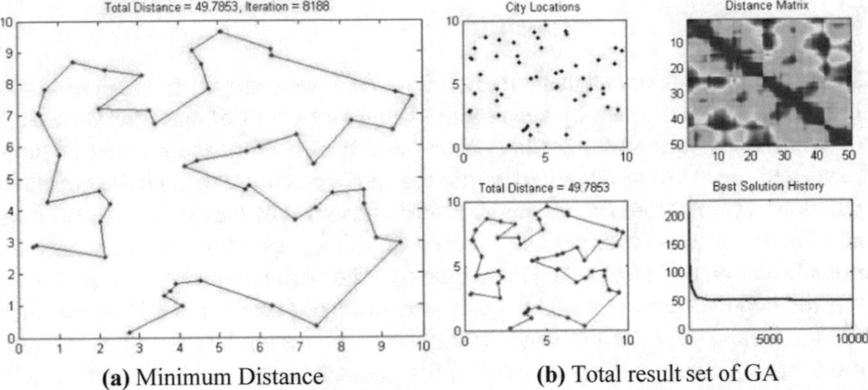

(a) Minimum Distance (b) Total result set of GA

Fig. 2 **a** and **b** Results of existing GA technique

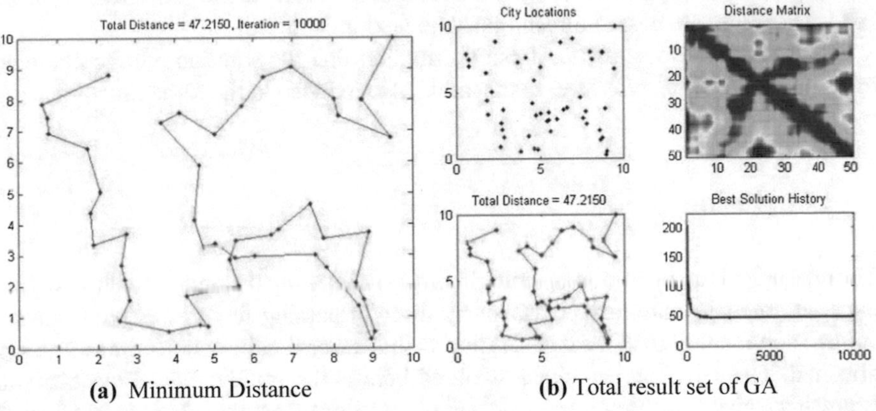

(a) Minimum Distance (b) Total result set of GA

Fig. 3 **a** and **b** Results of the proposed technique

technique for *30* times, the results obtained for the proposed technique are much better than the existing GA technique.

The comparisons of the existing GA technique with the proposed technique for the best optimal solutions are given in Figs. 2 and 3. The minimum distance obtained in the existing GA method is ***49.78***, whereas that of the proposed technique is found to be ***47.21***.

The programming is done by MATLAB version *7.11.0 (R2010b)* on a Laptop with Core *i3-2310 M 2.10* GHz in CPU and *3* GB in RAM with Windows *7* as an operating system.

4.1 Discussion

The proposed technique calculates the below- and above-average chromosomes while the roulette wheel is spun, by keeping the frequency count of each chromosome's occurrence. The chromosomes that occur more than *50%* of the number of times the wheel is spun are passed directly into the next generation without the crossover operation. These chromosomes are good chromosomes as they possess good traits and if these chromosomes do not undergo crossover operator, then definitely the optimal solution will be obtained more quickly. The explanation of this is as follows.

If the good chromosomes undergo crossover operation with bad chromosomes, then the fitness value of the new chromosomes is always less than that of good chromosomes, because the good traits of the good chromosomes are mixed up with the bad traits of the bad chromosomes. Therefore, the chromosomes that occur less than *50%* of the number of times the wheel is spun are only allowed for crossover. This also leads to lesser number of crossover operations as many of the good chromosomes (above average) are passed directly into the next generation.

From the above discussion, it can be implied that the solution will be obtained more quickly and the crossover overhead is also reduced at the same time.

5 Conclusion

The crossover is an important operator in GA for obtaining the optimal solution. This paper attempts to minimize crossover by directly passing few of the good fitness-valued individuals into the next generation so that the optimal solution is more quickly obtained. The selection technique involved here is the roulette wheel mechanism. From this technique, the good and the bad chromosomes are sorted out to be allowed (or not) for crossover. Since the below- and above-average chromosomes calculation is based on roulette wheel selection mechanism, it can be said that the sorting of chromosomes is fair (based on the fitness value of the individuals). This has also been proven from the simulation results.

References

1. Holland, J.H.: Adaptation in Naturaland Artificial Systems, the University of Michigan Press (1975)
2. Whitley, L.D.: Foundations of Genetic Algorithms-2. Morgan Kaufmann Publishers Inc., California (1993)
3. Diaz-Gomez, P.A., Hougen, D.F.: Initial Population for Genetic Algorithms: A Metric Approach in School of Computer Science University of Oklahoma Norman, Oklahoma, USA
4. Harik, G.R., Lobo, F.G.: A parameter-lessgenetic algorithm. In: Proceedings of the Genetic and Evolutionary Computation Conference, pp. 258–265 (1999)
5. Haupt, R.L., Ellen Haupt, S.: Practical Genetic Algorithms Second Edition, Wiley, New York

6. Spears, W.M., Anand, V.: A study of crossover operator in genetic programming. Methodol. Intell. Syst. **542**, 409–418 (2005)
7. Jiang, Q.: A Genetic Algorithm for Multiple Resource-Constrained Project Scheduling, University of Wollong Thesis Collection (2004)
8. Pedemonte, M., Alba, E., Luna, F.: Bitwise operations for GPU implementation of genetic algorithms. In: Proceedings of the 13th Annual Conference Companion on Genetic and Evolutionary Computation, GECCO '11, pp. 439–446. ACM, New York, NY, USA (2011)
9. Deep, K., Mebrahtu, H.: Variant of partially mapped crossover for the travelling salesman problems. In: Proceedings of the International Journal of Combinatorial Optimization Problems and Informatics, vol. 3, no. 1, pp. 47–69 (Jan–April 2012). ISSN: 2007-1558
10. Sivanandam, S.N., Deepa, S.N.: Introduction to Genetic Algorithms. Springer (2007). ISBN 9783540731894
11. Oliver, I.M., Smith, D.J., Holland, J.R.C.: A study of permutation crossover operators on the traveling salesman problem. In: Proceedings of the Second International Conference on Genetic Algorithms, pp. 224–230 (1987)
12. Syswerda, G.: Schedule Optimization Using Genetic Algorithms. In: Davis, L. (ed.) Handbook of Genetic Algorithms, pp. 332–349. Van Nostrand Reinhold (1990)
13. Starkweather, T., McDaniel, S., Mathias, K., Whitley, D., Whitley, D.: A comparison of genetic sequencing operators. In: Proceedings of the Fourth International Conference on Genetic Algorithms, pp. 69–76 (1991)
14. Whitley, L.D.: The genitor algorithm and selection pressure: why rank-based allocation of reproductive trials is best. In: Proceedings of the Third International Conference on Genetic Algorithms, pp. 116–121 (1989)
15. Homaifar, A., Guan, S., Liepins, G.E.: A new approach on the traveling salesman problem by genetic algorithms. In: Proceedings of the Fifth International Conference on Genetic Algorithms, pp. 460–466 (1993)
16. Spears, W., De Jong, K.A.: An analysis of multi-point crossover. In: Proceedings of the Foundations of Genetic Algorithms Workshop, Bloomington, Indiana (1990)
17. Vekaria, K., Clack, C.: Selective crossover in genetic algorithm: an empirical study. In: Proceeding of 5th International Conference on Parallel Problem Solving from Nature, pp. 438–447, London (1998)
18. Aboun, O., Abouchabaka, J.: A comparative study of adaptive crossover operators for genetic algorithms to resolve the travelling salesman problem. In: International Journal of Computer Applications (0975-8887), vol. 31, no. 11 (2011)
19. Asadi, A.A., Naserasadi, A., Asadi, Z.A.: A new hybrid algorithm for traveler salesman problem based on genetic algorithms and artificial neural networks. Int. J. Comput. Appl. **24**(5), 6–9 (2011)

A Novel Hybrid Differential Evolution-PSNN for Fuzzy Time Series Forecasting

Radha Mohan Pattanayak, Himansu Sekhar Behera and Sibarama Panigrahi

Abstract Over the years, fuzzy time series (FTS) has been used more popularly for forecasting the real-time series data. Generally, in FTS method, the membership value has not been considered for forecasting purpose, which leads as a drawback in the forecasting process. Recently, some researchers have overcome this problem by introducing the artificial neural network (ANN) concept to find the fuzzy relation. Again, the ANN has multimodal training problems, so the gradient descent algorithms will get stock in local minima. Therefore, the current study proposed a new hybridizing approach, which has used the differential evolution algorithm (DE) with PSHONN for forecasting the time series data. The hybrid DE [1] evolutionary algorithm has only two parameters such as parent (x_i) and child (u_i); therefore, it is more effective and faster than other evolutionary algorithms. Twelve different time series data sets are applied in our proposed model and the obtained result is compared with CRO [2, 3] and Jaya [4, 5]. In all cases, we found that the effectiveness and performance of the Jaya-PSNN were very poor and DE-PSNN model outperformed on most of the experimental results.

Keywords Chemical reaction optimization (CRO) · Differential evolution (DE) · Artificial neural network (ANN) · Pi-Sigma (PS) · Higher order neural network (HONN)

R. M. Pattanayak (✉) · H. S. Behera
Department of Information Technology, Veer Surendra Sai University of Technology, Burla 768018, Odisha, India
e-mail: radhamohan.pattanayak@gmail.com

H. S. Behera
e-mail: hsbehera_india@yahoo.com

R. M. Pattanayak
Department of Computer Science and Engineering, Godavari Institute of Engineering and Technology (Auto), Rajahmundry 533296, AP, India

S. Panigrahi
Department of Computer Science and Engineering, Sambalpur University Institute of Information Technology, Burla 768019, Odisha, India
e-mail: panigrahi.sibarama@gmail.com

H. S. Behera et al. (eds.), *Computational Intelligence in Data Mining*, Advances in Intelligent Systems and Computing 990,
https://doi.org/10.1007/978-981-13-8676-3_57

1 Introduction

In the year 1965, Zadeh [6] proposed fuzzy set theory. The fuzzy set theory concept has been used frequently in different areas such as fuzzy forecasting techniques, fuzzy regression, and fuzzy time series approach. Over the years, many researchers have proposed the FTS in two different flavors as time-variant and time-invariant fuzzy time series. In time-variant system, the relation of FTS will change over the time but in case of time-invariant system the FTS relation will not change over the time. However, various research papers [4, 5, 7–12] have contributed with wide range of application in forecasting by using various neural network models.

At first, the FTS model has been proposed by Song and Chissom [13, 14], by using both time-variant [13] and time-invariant [14] classes, to forecast the student enrollment of Alabama University. They used fuzzy relation and approximate reasoning in the FTS model. Further in the year 1996, Chen [2] developed another model, which consumed less time as comparatively before models for computing max-min composition operation.

Identifying the fuzzy relationship is an important role in TSF and many authors have proposed different models based on it. Song and Chissom [3, 13], Sullivan and Woodall [1] have developed model by using Markov's model, Chen [2], proposed simple model by using fuzzy logical group instead of using complex matrix operation. In 2006, Huarng and Yu [15], and Aladag et al. [16] have proposed model for time series forecasting by using feedforward artificial neural network (FFANN) to find the fuzzy relationship. In the year 2017, Panigrahi and Behera [17] proposed a hybrid ETS-ANN model for time series forecasting. They have tested 16 different time series data on ETS-ANN model and compared the efficiency with some existing ARIMA-ANN and MLP model. The experimental result showed that none of the model has shown best accuracy upon the applied data including the ETS-ANN model.

In the year 2016, Rao [18] has developed an optimization model for optimizing the constrained and unconstrained optimization problem, where he used less number of independent parameters comparatively with other models. He developed the algorithm toward the success of the algorithm, which means to find the best result. Again Rao et al. [19] have developed another model where they concentrated to remove the failure, i.e., worst result from the solution.

Generally, the artificial neural network (ANN) has multimodal training problem, and the gradient descent algorithms will get stock in local minima. To overcome such problem we have used the differential evolution as global optimization algorithm hybridized with Pi-Sigma higher order neural network, where the differential evolution algorithm has used to train the weight vectors of the model. Then the proposed model has been implemented for forecasting a wide range of time series data sets. The rest part of the paper is organized as follows: Sect. 2 briefly describes the preliminary study, Sect. 3 describes the proposed hybrid DE-PSNN method, Sect. 4 shows the computational result, and finally the conclusion and future work is explained in Sect. 5.

1.1 Fuzzy Time Series (FTS)

Definition 1 (*Fuzzy Time Series*) Let $Y(t)$ be a subset of real numbers $\forall t = (0, 1, 2)$ and R as a universe of discourse with which the fuzzy sets $f_j(t)$ are defined. $F(t)$ is called an FTS on $Y(t)$, if $F(t)$ is the collection of $f_1(t), f_2(t), f_3(t) \ldots . f_m(t)$.

Definition 2 (*Fuzzy Relationship*) If a fuzzy relationship $R_{rel}(t-1, t)$ exists, so that $F \circledR R_{rel}(t-1, t)$, where'$\circledR'$ is an arithmetical operator. If the prediction value of $F(t)$ is extracted from its before value, i.e., $F(t-1)$, then mathematically, the relation may be shown as $F(t-1) \rightarrow F(t)$.

Definition 3 (*Fuzzy Logical Relationship*) Let the values of $F(t-1)$ and $F(t)$ be A_i and A_j, respectively. The fuzzy logical relationship in between them can be mentioned as $A_i \rightarrow A_j$, where A_i and A_j be the left-hand side (LHS) and right-hand side (RHS) of the fuzzy logical relation.

Definition 4 (*Order of FTS*) In the prediction model, if only one period is required for the prediction of $F(t)$, i.e.,$F(t-k) \rightarrow F(t)$, then the FTS is called as first-order FTS, where k is the period of time series.

Definition 5 (*m-order of Fuzzy Time Series*) Let F(t) be an FTS and it is predicted by $F(t-1), F(t-2), F(t-3) \ldots\ldots F(t-m)$, then the fuzzy logical relation will be represented as $F(t-1), F(t-2), F(t-3) \ldots\ldots F(t-m) \rightarrow F(t)$ and the relation is called m^{th} order fuzzy time series forecasting.

2 Preliminaries Study

2.1 Pi-Sigma Higher Order Neural Network (PSHONN)

In 1991, Shin and Ghosh [20] developed Pi-Sigma as multilayered feedforward HONN. As shown in Fig. 1, the basic structure of PSNN has mainly three different layers such as input layer, hidden layer, and output layer. The input layer is the collection of n number of input units, which has feedforwarded to the hidden layer. Hidden layer is the summing unit of all inputs with their associated weight vector and the output layer is the production unit of all summing unit values with their associated weight vector. The total number of hidden units in the hidden layer represents the order of the PSNN network. The weights connected in between input and hidden layers are trainable weights but the weights in between hidden and output layers have fixed to one. However, the network structure has only one layer of trainable weight, so it reduces drastically the training time of the network structure. The output of the network and the output of jth hidden unit, i.e., h_j can be calculated using Eqs. (1) and (2), respectively.

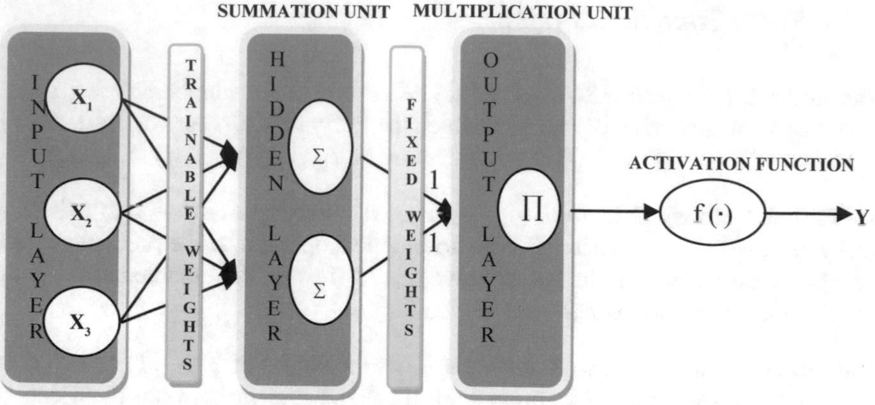

Fig. 1 PSNN higher order neural network

$$Y = f\left(\prod_{j=1}^{k} h_j\right) \tag{1}$$

$$h_j = \theta_j + \sum_{i=1}^{k} W_{ij} X_i \tag{2}$$

where θ_j and W_{ij} are the bias and weight between ith input and jth hidden unit, respectively.

2.2 Jaya Algorithm

In the year 2016, Rao [18, 19] proposed a new evolutionary algorithm as Jaya algorithm. This algorithm has potential to solve both the constrained and unconstrained optimization problems. It is a very simple algorithm due to its less implementation complexity, less computational time, and faster convergence characteristics.

For all iteration i, and for each population size, i.e., k is 1, 2, 3.....n there are m number of design variables and n number of solutions, where each design variable is termed as chromosomes. In the solution, fx_{best} represents the candidate obtaining the best value in the entire solution and fx_{worst} represents the candidate obtaining the worst value in the entire solution. Inside the algorithm, the chromosome value has been modified as explained in Eq. (3).

$$X^1_{j,k,i} = X_{j,k,i} + r1_{j,i}\left(X_{j,best,i} - \left|X_{j,k,i}\right|\right) - r2_{j,i}\left(X_{j,worst,i} - \left|X_{j,k,i}\right|\right) \tag{3}$$

where $X_{j,best,i}$ *and* $X_{j,worst,i}$ are the values of the variable for the best and worst candidates, $X^{1}_{j,k,i}$ is the modified value of $X_{j,k,i}$, and $r1_{j,i}$ *and* $r2_{j,i}$ are the two random values in the range [0, 1].

2.3 Differential Evolution Algorithm

In 1997, Price and Storn [21] proposed differential evolution algorithm as a heuristic algorithm for global optimization over continuous spaces. In order to compare with other evolutionary algorithms, the DE algorithm is more simple and straightforward for implementation. This algorithm has three advantages regardless of parameter, i.e., it can find the global minimum for a multimodal function, quick convergence, and it requires less number of parameters to start of the algorithm. DE algorithm follows all three steps as mutation, crossover, and selection procedure for fine-tuning of the best solution. At first step all the populations are randomly collected into a weight matrix, where the individual x_i consists of k number of genes. In the second step for each individual x_i, the mutated individual v_i has been prepared using Eq. (4), where r_1 *and* r_2 are chosen from the population randomly. For each population, the training error has been calculated as shown in Eq. (5), where $Correct_i$ shows the correct answer for ith training vector and $Answer_i$ is the neural network generated ith training vector. In the crossover step for each gene $x_i = \left(x_{i1}, x_{i2}, x_{i3}, \ldots ..x_{ij}\right)$, a number $rand_j$ has been generated randomly and performs the crossover for each individual $v_i = \left(v_{i1}, v_{i2}, v_{i3}, \ldots ..v_{ij}\right)$. Then finally, the selection procedure has been followed in the fourth step, which helps to generate new population.

$$v_{ij} = x_{r1j} + F\left(x_{r2j} - x_{r3j}\right) \tag{4}$$

$$Error = \frac{1}{2}\sum_{i=1}^{T}(Correct_i - Answer_i)^2 \tag{5}$$

2.4 Chemical Reaction Optimization

In 2010, Lam and Li [22] proposed chemical reaction optimization (CRO) as a population-based metaheuristic optimization technique. It is an evolutionary algorithm inspired by nature of chemical reaction (CR). The properties of CR are to transfer the unstable chemical substances into stable state. A chemical reaction consists of set of substances, where a substance may be reactant or molecule. The energy linked with a molecule is called as enthalpy. In CRO, the enthalpy is considered as fitness function for minimization problem and entropy is considered as fitness function for maximization problem. During a chemical reaction, this energy continuously changes as per the change of intra-molecular shape of a reactant and becomes stable at one point. A chemical reaction may be uni-molecular, bi-molecular, and tri-molecular reaction, which will decide the total number of molecules involving in a chemical reaction. Some frequently used terms associated with the algorithm are molecules as chromosomes, atoms as genes, both enthalpy and entropy as fitness function and reactions considered as crossover and mutation strategy. In order to compare our proposed model, we have considered a fixed population size CRO [23] in our study.

3 Proposed Algorithm

Algorithm.1 explains the proposed DE-PSNN methodology. At first, the given time series data has decomposed into some in-sample (i.e., train and validation sample) and out-of-sample (i.e., test sample). In this current research, we used 20% of the data as out-of-sample and from the rest 80%, we used approximately 60% as train sample and 20% as validation sample. Then for the training pattern, the Universe of discourse U is defined, where the values of D_1 and D_2 are $D_1 = 0.1 * D_{min}$ and $D_2 = 0.1 * D_{max}$. Then the Universe of discourse has been partitioned into k intervals of length "w" by using Step.4 in the algorithm. Then for each partition, the midpoint has been defined as $m_i = [(\text{startUnivDis}) - (\text{startUnivDis} + \text{k})]/2$.

Inside the paper, min-max normalization procedure has been considered for normalizing the data as shown in Eq. (6) Then the normalized data is applied into the proposed hybrid DE-PSNN algorithm. The tuning process of the model has been explained in Algorithm. 2. At first, for all the population size, the total weight matrix has been sorted and population size has been stored in Pop Size (in the literature, the population size has been taken as 50). In the second step Mutation, strategy has followed. Three random variables have been chosen. For every population in the population size, each chromosome has been updated using Step. 5 in Algorithm.2. In the third step, for all individual x_i are crossed over with mutated individual v_i. In the fourth step, selection procedure has been followed and generated a new population. Then the forecasted data is denormalized and the efficiency of the model has been measured by using root mean square error (RMSE) and symmetric mean absolute percentage error (SMAPE) value as mentioned in Eqs. (7) and (8).

$$y' = \frac{y - min_y}{max_y - min_y} \tag{6}$$

$$RMSE = \sqrt{\frac{1}{n}\sum\nolimits_{j=1}^{n}\left(y_j - \hat{y}_j\right)^2} \tag{7}$$

$$SMAPE = \frac{1}{n}\sum\nolimits_{j=1}^{n}\frac{\left|y_j - \hat{y}_j\right|}{\left(\left|Y_j\right| + \left|T_j\right|\right)/2} \tag{8}$$

Algorithm. 1. The Proposed DE-PSNN methodology

Step 1: Given Time series $y = [y_1, y_2, y_3 \ldots \ldots y_n]^T$
Step 2: Input the length of in-sample data
$l_a[60\%\ as\ train\ and\ 20\%\ as\ validation]$ and
For out-of-sample $l_b[20\%\ as\ test]$
Step 3: Define the Universe of discourse $U = [D_{min} - D_1 \quad D_{max} + D_2]$
Where $D_1 = 0.1 * D_{min}\ and\ D_2 = 0.1 * D_{max}$
Step 4: For each length of interval p= 1 to l_a
(a) Universe of discourse $U = [u_1, u_2, u_3, \ldots . u_k]$
(b) Partition the Universe of discourse based on intervals was $k = (U_2 - U_1)/w$
(c) For each partition find mid-point,
$$m_i = [(startUnivDis) - (startUnivDis + k)]/2$$
Step 5: Normalize the data as: $y^1 = y - min_y/max_y - min_y$
Step 6: Find the fitness by using DE (Algorithm.2)
Step 7: De-normalize the forecasted data
Step 8: Find accuracy of the model
(a) $RMSE = \sqrt{1/n\ \sum_{j=1}^{n}(y_j - \hat{y}_j)^2}$
(b) $SMAPE = 1/n \sum_{j=1}^{n} \frac{|y_j - \hat{y}_j|}{(|Y_j| + |T_j|)/2}$
Step 6: end

Algorithm. 2. Finding fitness by using Differential Evolution methodology

Step 1: Create initial population size
Step 2: While termination criterion is not satisfied
Do begin
Step 3: For each ith individual in the population
Begin
Step 4: Randomly generate three integers i.e. r_1, r_2 and r_3
Where $r_1, r_2, r_3, i \in [1; PopSize]$, and $r_1, r_2, r_3 \neq i$
Step 5: For each j_{th} gene in i_{th} individual $(j \in [1; n])$
Begin
Follow mutation strategy as $v_{ij} = x_{r1j} + F(x_{r2j} - x_{r3j})$
Step 6: Randomly generate one real number in [0; 1]
Step 7: If $rand_j < CR\ then$:
$u_{ij} = v_{ij}$
Else:
$u_{ij} = x_{ij}$
End
Step 8: If (individual $U_i <$ individual X_i): then
Replace individual X_i by child U_i individual
Step 9: Set the iteration counter i = i+1
Step10: end of while

4 Experimental Setup and Results

In this current research, we used the differential evolution algorithm hybridized with PSNN (DE-PSNN) as our proposed model. Except from the proposed model, we used Jaya-PSNN and CRO-PSNN to compare the forecasting efficiency of our proposed model. Inside the study, differential evolution algorithm is used to train the weights of the neural network. Autocorrelation function (ACF) has been used to find number of lags for the model. Number of output neuron of the model has been fixed to one.

By changing the number of hidden units from 1 to 20, both RMSE and SMAPE values have been calculated over 50 executions and from this 20 smallest mean has been considered for result analysis.

4.1 Datasets

Table 1 shows the information of 12 data sets of various fields collected from time series data library applied in the proposed model.

4.2 Result Analysis Using RMSE

Analysis of RMSE value is always scale dependent and very conscious about outlier data. In this research, we used RMSE comparison to individual time series data. In the experimental result, out of 50 experiments 20 best mean and standard deviation

Table 1 Description of different time series data

Data set	Description
Accidental	Accidental death monthly in USA(1973–78)
Car cells	Monthly car sales in Quebec 1960–1968
Gas usage	Monthly av. residential gas usage Iowa (cubic feet)*100'71–'79
IBM	Daily common stock closing price of IBM (May 17, 1961–November 2, 1962)
Lake	Monthly Lake Erie Levels (1921–1970)
Milk	Monthly Milk production pounds per cow (1962–75)
Passenger	International airline passengers, total in thousand (Jan 49–Dec 60)
Pollution	Pollution equipment shipment monthly (Jan 1966–Oct 1976)
Rainfall	Total annual rainfall (in inches), London, England, 1813–1912
Stock	Common Stock price of US Annually (1871–1970)
Sun	Sunspot number of Wolf (1700–1987)
Traffic	Monthly traffic fatalities in Ontario (1960–74)

Table 2 Mean and standard deviation for RMSE value (best values in bold)

Data set	CRO-PSNN	DE-PSNN	Jaya-PSNN
	Mean ± Std. Dev	Mean ± Std. Dev	Mean ± Std. Dev
Accidental	5.7233 ± 100.580	**4.9550 ± 03.1567**	6.1897 ± 109.943
Car cells	**2.8092 ± 394.442**	2.8298 ± 05.2121	2.8229 ± 486.540
Gas usage	26.8397 ± 11.420	**19.4310 ± 7.3728**	30.8381 ± 10.766
Ibm	13.7031 ± 2.6128	14.2172 ± 2.6573	**13.5210 ± 4.3688**
Lake	0.8655 ± 0.18994	0.8929 ± 0.20695	**0.8531 ± 0.22367**
Milk	31.7445 ± 11.697	**31.399 ± 11.0201**	36.3313 ± 12.815
Passenger	63.8670 ± 16.366	**57.2563 ± 10.615**	63.8670 ± 16.366
Pollution	**9.0295 ± 5.13440**	9.0941E2 ± 71.89	9.5980 ± 22.0559
Rainfall	4.0005 ± 0.08007	**3.9823 ± 0.10534**	3.9952 ± 0.08366
Stock	10.1165 ± 2.9746	**9.2963 ± 1.78710**	13.6592 ± 5.9742
Sun	27.2756 ± 5.0082	**24.2169 ± 4.2772**	29.0603 ± 4.1774
Traffic	25.3279 ± 4.3665	**23.3911 ± 2.6308**	26.4589 ± 5.1402

RMSE value has been considered for observation and the data is shown in Table 2. From the experimental analysis, it is observed that the proposed DE-PSNN has outperformed in eight different data sets (Accidental, Gas Usage, Milk, Passenger, Rainfall, Stock, Sun, Traffic), CRO-PSNN in two datasets (Car Cells and Pollution), and Jaya-PSNN in two datasets (Ibm and Lake). Hence, we observed that in most of the datasets DE-PSNN model has produced effectively result.

4.3 Result Analysis Using SMAPE

All different models in this research are predicted in one-step-ahead procedure. A SMAPE measure is useful for comparing forecasting model over multiple time series data. In the experimental setup, 50 independent simulation results have been carried out upon 12 different time series data sets. Table 3 represents 30 best mean and standard deviation (std.dev) SMAPE value over 50 simulation results. From Table 3, it is revealed that the proposed DE-PSNN model has performed better in eight datasets (Accidental, Gas Usage, Milk, Passenger, Rainfall, Stock, Sun, Traffic), CRO-PSNN outperformed well in two datasets, and CRO-PSNN outperformed well in four datasets. Hence, it has been concluded that in most of the dataset the DE-PSNN hybrid model outperformed.

Table 3 Mean and standard deviation for SMAPE value (best values in bold)

Data set	CRO-PSNN	DE-PSNN	Jaya-PSNN
	Mean ± Std. Dev	Mean ± Std. Dev	Mean ± Std. Dev
Accidental	5.1105 ± 1.00655	**4.6065 ± 1.00475**	5.7297 ± 1.15021
Car cells	**13.2237 ± 1.9031**	13.3915 ± 2.5904	13.4333 ± 2.3297
Gas usage	26.4806 ± 9.3354	**21.0959 ± 6.8411**	28.3571 ± 9.6364
Ibm	**2.9715 ± 0.58095**	3.0960 ± 0.59597	2.9905 ± 1.08506
Lake	4.7283 ± 1.04591	4.8624 ± 1.11780	**4.6568 ± 1.18641**
Milk	3.0701 ± 1.17890	**3.0633 ± 1.14821**	3.5071 ± 1.29311
Passenger	11.7726 ± 3.0799	**10.6962 ± 2.0392**	11.7726 ± 3.0799
Pollution	**21.8680 ± 2.7385**	21.9007 ± 2.5779	23.9300 ± 3.8046
Rainfall	10.9643 ± 0.5891	11.0922 ± 0.8142	**10.9100 ± 0.6691**
Stock	14.3185 ± 4.8416	**13.9773 ± 3.4432**	18.7664 ± 9.0452
Sun	**13.97773 ±0.4432**	41.4341 ± 7.3824	44.0845 ± 6.3777
Traffic	13.7034 ± 2.8846	**12.6588 ± 1.7220**	14.2111 ± 3.1715

Table 4 Ranks of all models using SMAPE value (Friedman and Nemenyi hypothesis)

Model	Different ranks on SMAPE value	Mean rank value (%)
Jaya-PSNN	3	83.63
CRO-PSNN	2	74.04
DE-PSNN	1	**68.83**

4.4 Rank Calculation

With respect to comparing different models over 12 different time series datasets, Friedman and Yemeni hypothesis test [24] has been applied on the SMAPE experimental value. Generally, Friedman's test was required to rank the algorithms by considering each data set individually. However, in case of distinct ranks, the Null hypothesis has been rejected, so Nemenyi hypothesis method [25] is applied among all algorithms for making pair-wise comparison. The various experimental results with respect to SMAPE have been shown in Table 4. From the tabular result analysis, it has clearly found that the proposed DE-PSNN model performed the better rank over the other two models and the hybrid model Jaya-PSNN has performed worst among all the other models. In Figs. 2, 3, 4, 5, and 6, various graphs have been plotted in between forecasted values and actual values on different time series data, which gives better knowledge about performance of the model.

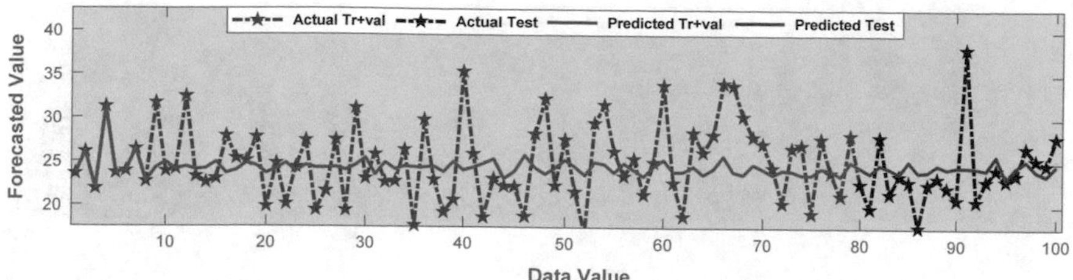

Fig. 2 The result of forecasted value and actual value on rainfall time series data

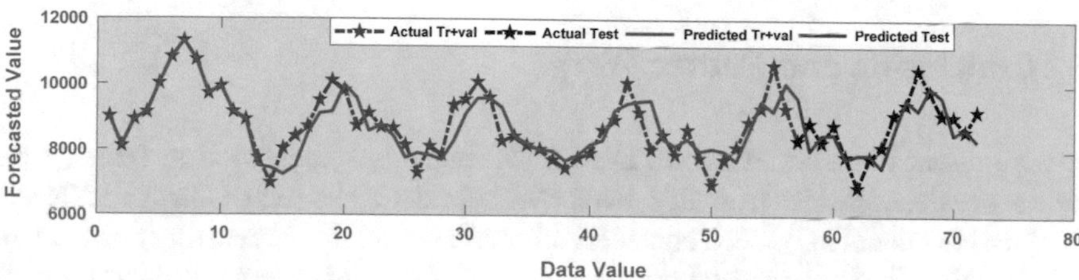

Fig. 3 The result of forecasted value and actual value on accidental time series data

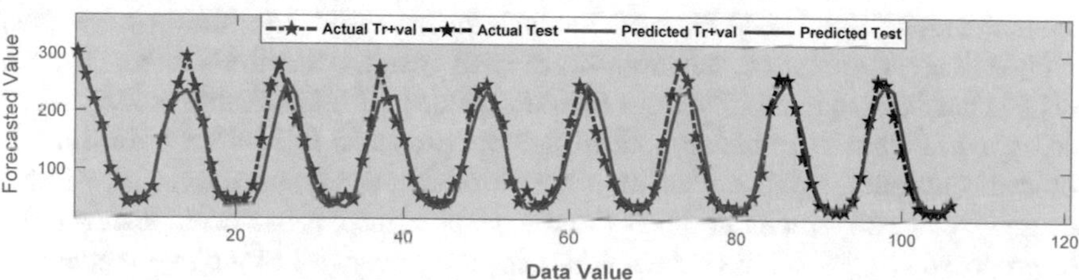

Fig. 4 The result of forecasted value and actual value on gas usage time series data

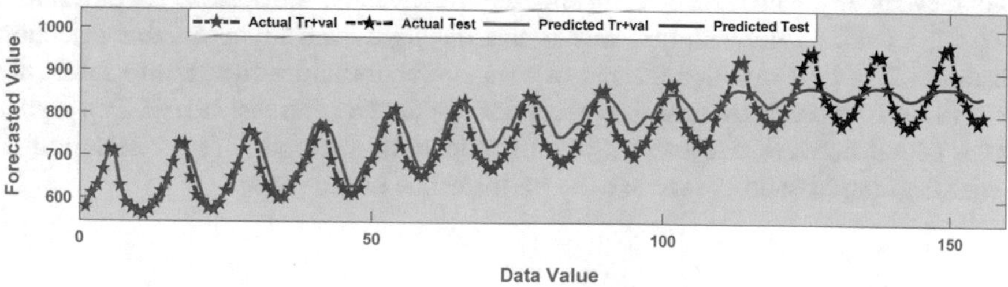

Fig. 5 The result of forecasted value and actual value on milk time series data

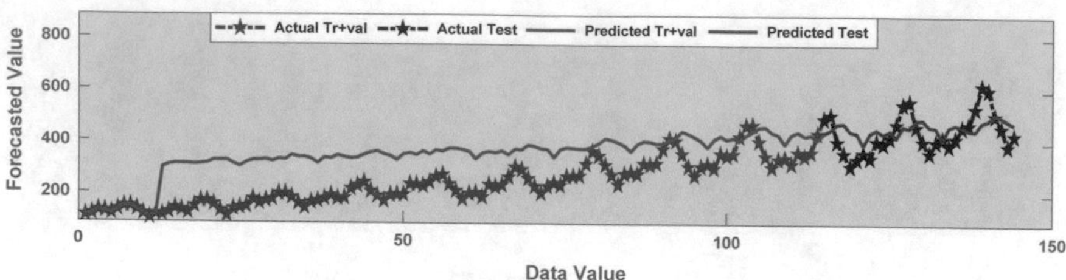

Fig. 6 The result of forecasted value and actual value on passenger time series data

5 Conclusion and Future Work

In the present literature, a hybrid DE-PSNN model has been used to forecast the time series data. Besides from this, two other hybrid models such as Jaya-PSNN and CRO-PSNN model have been considered for comparing the forecasting accuracy of DE-PSNN. In the experimental result analysis, 12 different time series data sets have been applied upon these three hybrid models and the effectiveness is calculated based upon their RMSE and SMAPE values. From the experimental result, we observed that both in RMSE and SMAPE result analysis the proposed algorithm outperformed on eight time series datasets and produces efficient result. Later, the Nemenyi hypothesis test has been applied to various SMAPE values for all available models. From the hypothesis test, it has been noticed that the proposed DE-PSNN hybrid model achieved outstanding rank among all other models; however, the hybrid Jaya-PSNN secured worst rank among all these three models. Hence, it has been observed that the proposed DE_PSNN has shown better accuracy over rest of the two models.

Even though the proposed method has given better rank in SMAPE values for different varieties of time series data, still there is a scope to enhance the forecasting accuracy. Finally, with respect to improve the effectiveness and accuracy of the model, two ways can be followed, first way is that the significant input variable selection method [26] can be applied instead of using autocorrelation function to select the input units for neural network. The second scope may be that instead of using higher order neural network one may follow deep learning method [27] for selection of better learning method to increase the performance of the model.

References

1. Sullivan, J., Woodall, W.H.: A comparison of fuzzy forecasting and Markov modeling. Fuzzy Sets Syst. **64**(3), 279–293 (1994)
2. Chen, S.M.: Forecasting enrollments based on fuzzy time series. Fuzzy Sets Syst. **81**(3), 311–319 (1996)
3. Song, Q., Chissom, B.S.: Fuzzy time series and its models. Fuzzy Sets Syst. **54**(3), 269–277 (1993)

4. Pattanayak, R.M., Behera, H.S.: Higher order neural network and its applications: A comprehensive survey. In: Progress in Computing Analytics and Networking, pp. 695–709. Springer, Singapore (2018)
5. Donkor, E.A., Mazzuchi, T.A., Soyer, R., Alan Roberson, J.: Urban water demand forecasting: review of methods and models. J. Water Resour. Plan. Manag. **140**(2), 146–159 (2012)
6. Zadeh, L.A.: Fuzzy sets. Inf Control 338–353 (1965)
7. Lin, W.Y., Hu, Y.H., Tsai, C.F.: Machine learning in financial crisis prediction: a survey. IEEE Trans. Syst., Man, Cybern., Part C (Appl. Rev.) **42**(4), 421–436 (2012)
8. Raza, M.Q., Khosravi, A.: A review on artificial intelligence based load demand-forecasting techniques for smart grid and buildings. Renew. Sustain. Energy Rev. **50**, 1352–1372 (2015)
9. Weron, R.: Electricity price forecasting: A review of the state-of-the-art with a look into the future. Int. J. Forecast. **30**(4), 1030–1081 (2014)
10. Meade, N., Islam, T.: Forecasting in telecommunications and ICT—A review. Int. J. Forecast. **31**(4), 1105–1126 (2015)
11. Fagiani, M., Squartini, S., Gabrielli, L., Spinsante, S., Piazza, F.: A review of datasets and load forecasting techniques for smart natural gas and water grids: Analysis and experiments. Neurocomputing **170**, 448–465 (2015)
12. Chandra, D. R., Kumari, M. S., Sydulu, M.: A detailed literature reviews on wind forecasting. In: 2013 International Conference on Power, Energy and Control (ICPEC), pp. 630–634. IEEE (2013)
13. Song, Q., Chissom, B.S.: Forecasting enrollments with fuzzy time series—part I. Fuzzy Sets Syst. **54**(1), 1–9 (1993)
14. Song, Q., Chissom, B.S.: Forecasting enrollments with fuzzy time series—part II. Fuzzy Sets Syst. **62**(1), 1–8 (1994)
15. Huarng, K., Yu, T.H.K.: the application of neural networks to forecast fuzzy time series. Phys.: Stat. Mech. Its Appl. **363**(2), 481–491 (2006)
16. Aladag, C.H., Basaran, M.A., Egrioglu, E., Yolcu, U., Uslu, V.R.: Forecasting in high order fuzzy times series by using neural networks to define fuzzy relations. Expert Syst. Appl. **36**(3), 4228–4231 (2009)
17. Panigrahi, S., Behera, H.S.: A hybrid ETS–ANN model for time series forecasting. Eng. Appl. Artif. Intell. **66**, 49–59 (2017)
18. Rao, R.: Jaya: A simple and new optimization algorithm for solving constrained and unconstrained optimization problems. Int. J. Ind. Eng. Comput. **7**(1), 19–34 (2016)
19. Rao, R.V., More, K.C., Taler, J., Ocłoń, P.: Dimensional optimization of a micro-channel heat sink using Jaya algorithm. Appl. Therm. Eng. **103**, 572–582 (2016)
20. Shin, Y., Ghosh, J.: The pi-sigma network: An efficient higher-order neural network for pattern classification and function approximation. In: IJCNN-91-Seattle International Joint Conference on Neural Networks, 1991, vol. 1, pp. 13–18. IEEE (1991)
21. Storn, R., Price, K.: Differential evolution–a simple and efficient heuristic for global optimization over continuous spaces. J. Global Optim. **11**(4), 341–359 (1997)
22. Lam, A.Y., Li, V.O.: Chemical-reaction-inspired meta heuristic for optimization. IEEE Trans. Evol. Comput. **14**(3), 381–399 (2010)
23. Sahu, K.K., Panigrahi, S., Behera H.S.: A novel chemical reaction optimization algorithm for higher order neural network training. J. Theor. Appl. Inf. Technol. **53**(3) (2013)
24. Hollander, M., Wolfe, D.: A nonparametric statistical methods (1999)
25. Demsar, J.: Statistical comparisons of classifiers over multiple data sets. J. Mach. Learn. Res. **7**, 1–30 (2006)
26. Tran, H.D., Muttil, N., Perera, B.J.C.: Selection of significant input variables for time series forecasting. Environ. Model Softw. **64**, 156–163 (2015)
27. Langkvist, M., Karlsson, L., Loutfi, A.: A review of unsupervised feature learning and deep learning for time-series modeling. Pattern Recogn. Lett. **42**, 11–24 (2014)

Improved Face Detection Using YCbCr and Adaboost

Vishnu Vansh, Kumar Chandrasekhar, C. R. Anil and Sitanshu Sekhar Sahu

Abstract Accurate face detection plays a crucial role in surveillance and security. It has been a challenging issue because of the fact that human faces are dynamic object with high degree of variability in their appearance. This paper proposed a method of face detection which involves pre-processing of input images to extract skin tone in YCbCr domain followed by Haar cascade classifiers to detect a face in the extracted area. These classifiers are selected by AdaBoost training with help of particle swarm optimization. On the exhaustive study on various images, it is elucidated that the proposed approach provides superior performance compared to the existing Viola–Jones technique. The proposed algorithm is also able to detect the side or occluded faces in the input image.

Keywords Face detection · YCbCr colour space · Adaboost · PSO · Haar cascade

1 Introduction

Face detection in image is basically treated as object detection. The job is to locate the different objects in an image. Face detection algorithms primarily focus on the identification of frontal human faces. Face detection is a part of face identification and can be subdivided in two tasks: (a) to find out whether there is any face in a given image or not, (b) if yes, then where is it located. It helps to extract information

V. Vansh · K. Chandrasekhar · C. R. Anil · S. S. Sahu (✉)
Department of Electronics and Communication Engineering, Birla Institute of Technology, Mesra, Ranchi 835215, Jharkhand, India
e-mail: sssahu@bitmesra.ac.in

V. Vansh
e-mail: vishnu_vansh@yahoo.com

K. Chandrasekhar
e-mail: kumar.race@gmail.com

C. R. Anil
e-mail: anilbit215@gmail.com

H. S. Behera et al. (eds.), *Computational Intelligence in Data Mining*, Advances in Intelligent Systems and Computing 990,
https://doi.org/10.1007/978-981-13-8676-3_58

such as the expression, gender, age and ethnicity from a person's face [1, 2]. Face detection has been used in various applications such as biometric-based security systems, human–machine interaction, surveillance system etc. [2].

Basically, there are many distinguishable patterns in face. These are the uneven surfaces in face that compose different facial features. These patterns are defined as nodal points. There are as many as eighty nodal points on a human face. For example, distance between the eyes, nose width, eye socket depth, cheekbones and jaw line [3].

There are several factors that make the process of face detection complicated in any given image/video. They are profile pose, image orientation, varying facial expressions, double chin, tilted pose, hair-do, occluded face, low-resolution image, obstruction in front of the face, illumination, out of focus faces, etc. and require different computation while detecting [4]. These problems create difficulties in solving image classification and detection using neural networks (NN). The NN algorithm requires heavy computational machinery, large training times and a large dataset of training samples. While the last two problems can be solved passively, people found it difficult to incorporate NN in low power devices because of the high computation requirement. As machine learning algorithms require less training time, and less CPU intensive, they are the first choice towards the incorporation of image classification problems on low power devices. In this context, Viola–Jones algorithm has been widely used in detection of faces. Viola–Jones algorithm uses AdaBoost training as a machine learning technique in order to solve the classification problems. It uses Haar cascade classifiers for detecting key facial features, integral image for efficient computation of sum of pixels within a rectangle efficiently and a cascaded approach. In this paper, we proposed an approach to detect the faces which uses YCbCr colour space and AdaBoost technique with particle swarm optimization. The search space for identifying a face in images containing multiple faces is drastically reduced using YCbCr technique. Haar cascade of classifiers is used to detect the faces in the image. Particle swarm optimization is used in order to reduce the time consumption by AdaBoost algorithm in searching for facial features within the detection window. The detection of face in a region is based on the identification of the region which contained the skin tone and searched for the presence of main facial features like eyes, nose, mouth, ears, etc.

2 Proposed Method for Face Detection

The complete flow graph of the proposed approach for face detection is shown in Fig. 1.

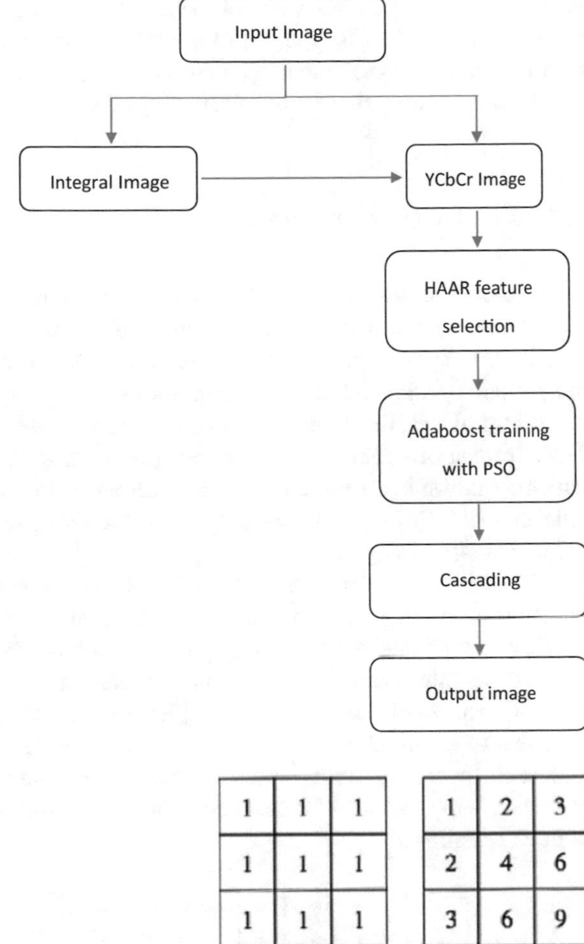

Fig. 1 Flow graph of the proposed face detection system

Fig. 2 Illustation of the input image and integral image

2.1 *Integral Image*

The first step of the proposed approach is to convert the input image into an integral image. Integral image also called as summed area table. This allows very fast feature evaluation and efficiently compute the sum of values in a rectangular subset of pixels. The integral image is shown in Fig. 2 at coordinate x, y is the sum of the pixels above and to the left of x, y as defined in Eq. 1:

$$I(x, y) = \sum_{x' \le x. y' \le y} i(x', y') \tag{1}$$

where i(x', y') is the pixel value of the input image and I(x, y) is the corresponding image integral value. By the use of the integral image, the sum of any rectangular area can be easily computed. The sum of pixels in any rectangle in the image is computed with only four values from integral image.

2.2 YCbCr Colour Space

RGB colour space is not suitable in colour-based analysis due to mixing of colour (chrominance) and intensity (luminance) information and its non-uniform characteristics [4]. YCbCr colour space has proven to be effective in finding regions of the image which contains skin even in conditions of varying light or low light. Since face is a subset of this, the global search space used by the algorithm is further reduced. Thus, instead of searching over the complete image it searched over the segments which contain skin computed by YCbCr. Experimental results in [5] showed that the efficiency of YCbCr colour space is best when it used to detect and segment the skin colour in colour images.

YCbCr is an encoded non-linear RGB signal. In this scheme, luminance information is embedded in Y, and colour information is stored as two colour-difference components, Cb and Cr. Cb is computed by subtracting the blue component from a reference value, and Cr is computed by subtracting the red component from a reference value. M. Elmezainet al. [6] ignored the Y channel of YCbCr to reduce the effect of varying light and use only the chrominance channels which could fully represent the colour. Taking reference from the same we have segregated the skin tone with help of Cb and Cr channels. The conversion of RGB to YCbCr is defined in Eq. 2 as follows:

$$\begin{bmatrix} Y \\ C_b \\ C_r \end{bmatrix} = \begin{bmatrix} 16 \\ 128 \\ 128 \end{bmatrix} + \begin{bmatrix} 65.481 & 128.553 & 24.966 \\ -37.797 & -74.203 & 112 \\ 112 & -93.786 & -18.214 \end{bmatrix} \begin{bmatrix} R \\ G \\ B \end{bmatrix} \quad (2)$$

The range of Y, Cb, Cr components used in the proposed method are as follows:

$0 < Y < 255$,
$80 < Cb < 135$,
$133 < Cr < 173$

These values were obtained by analysing the pixel values of skin in images of Caltech human face (front) dataset. The pre-processing steps prior to classification of faces are described as follows:

(i) Conversion to YCbCr colour space: The input image to be analysed is converted to YCbCr domain, as it is easy to analyse skin tone in this domain.
(ii) Select the region containing skin tone: The pixel values obtained by analysis of the database related to the skin tone are used to define the range of pixel

Fig. 3 **a** RGB image, **b** skin detection, **c** binary image, **d** noise removal image, **e** After morphological operations

values for segmentation. The optimum value of Y is between 0 and 255, Cb is between 80 and 135 and that of Cr is between 133 and 173.

Remove noise: The image is passed through Gaussian filter followed by a series of alternate iteration of dilation and erosion to eliminate small and irrelevant blobs. Noisy pixels are removed by erosion and dilation and a better region is obtained for analysis. The illustrations of the above steps are shown in Fig. 3a–e.

(iii) Find Edges (to separate overlapping faces): This step is done to separate faces which have no gap between them (of the background), the edges can be defined for cases when two different skin toned faces are overlapping.

(iv) Define contours to search for facial features: Contours are formed for the adjacent pixels which combined together to form a shape. The contour making reduces the search space from the entire image to only the contour region.

(v) Remove small contours and run scan for facial features: Contours having size smaller than a threshold are removed as they do not contain a human face and is thus irrelevant to search in them. The proper illustration of the above steps is shown in Fig. 4a–c.

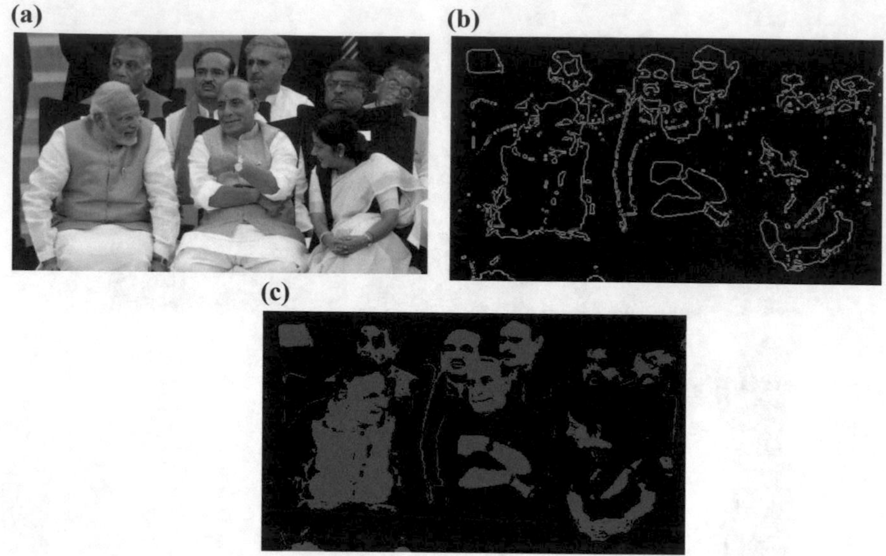

Fig. 4 **a** Input image, **b** YCbCr based human skin tone separation of objects, **c** contour formation for the input image

2.3 Haar Features

HAAR like features are rectangular regions developed by gray and white halves where the value of the feature is the difference between the pixel intensities in these halves. Human faces have many such features. For example, the regions in face containing eyebrows and eye lashes are darker than cheeks. Nose Bridge region is brighter than eyes. Instead of using individual RGB pixels to find these similar features, we used the composition of properties based on the location and size of eyes, mouth, and bridge of nose to obtain digital image features based on HAAR wavelets. In other words, these features are nothing but oriented gradients of black and white pixels in a particular shape used to detect facial features from a larger area. We have used three types of features as shown in Fig. 5. (a) Edge Features: the value of two-rectangle feature is the difference between the sum of the pixels within two rectangular regions. The regions have the same size and shape and are horizontally or vertically adjacent. (b) Line features which are the sum within two outside rectangles and subtracted from the sum in the centre rectangle. (c) A four-rectangle feature which evaluates the difference between diagonal pairs of rectangle.

The use of HAAR features helps in dimensionality reduction where we can identify the most important basis functions from an original complete set of basis functions, thus the system is faster than a system based on pixels. The use of HAAR wavelets was first utilized for fast object detection as [7] mentions. The HAAR feature is convoluted with the binary image obtained after segmentation. This convolution gives the probability of finding a particular facial feature in the region obtained

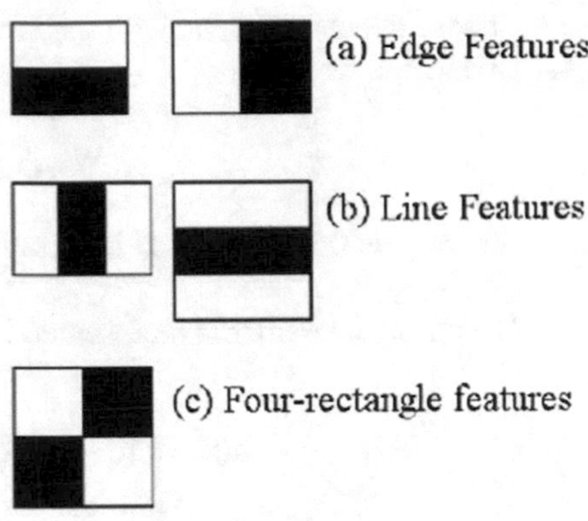

Fig. 5 Haar cascade classifiers

by segmentation. When all the optimum features of the trained model gives a probability of more than 0.5 in a region then that region is approximated to a rectangle using polygon approximation method and thus the co-ordinates are obtained.

.

2.4 *PSO-Based Adaboost Learning*

Adaboost algorithm is a kind of machine learning techniques which utilizes few samples for training the classifier and is one of the key steps of the basic Viola–Jones algorithm for face detection. Adaboost classifier itself is a combination of many weak classifiers. The weak classifier is given as threshold on a single Haar-like rectangular feature. For given sample, weight is adjusted based on criteria that weather weak classifier distinguish the sample efficiently or not. In this paper, the performance of the algorithm is improved using particle swarm optimization (PSO) in order to reduce the search space of the original algorithm.

AdaBoost learning procedure:

- Consider images $(x_1,y_1),\ldots,(x_n,y_n)$ where $y_i = 0,1$ for negative and positive samples, respectively.
- Initialize the weights w_1, $i = 1/2$ m,1/2 l for $y_i = 0$, 1 respectively, where m and l are the number of negatives and positives, respectively.
- For $t = 1,\ldots,T$:

1. Normalize the weights:

$$W_{t,1} \leftarrow W_{t,1} / \sum W_{t,j}$$

   So that w_t is a probability distribution
2. For every feature, j, train a classifier h_j, which uses a single feature. The error is computed with respect to w_t, ϵj

3. Choose the classifier, h_t, with the lowest error ε_t
4. Update the weights

$$W_{t+1,i} = w_{t,i}\beta_t^{1-e_i}$$

Where $e_i = 0$ if example x_i is classified correctly and $e_i = 1$ otherwise; $\beta_t = e_i/(1 - e_i)$
The strong Classifier is represented in Eq. 3.

$$h(x) = \begin{cases} 1, & \sum_{t=1}^{T} \alpha_t h_t(x) >= \sum_{t=1}^{T} \alpha_t/2 \\ 0, & \text{otherwise} \end{cases} \quad (3)$$

where $\alpha_t = log\ 1/\beta_t$

In an image consisting of many faces, iterating a window of a particular size throughout the image and applying Viola–Jones detection algorithm inside each window could be helpful [8–10]. In this paper, particle swarm optimization (PSO) is used to achieve the same in a faster manner. It uses a population of solutions (particles) which are updated iteratively to reach to the optima. Each particle has a pbest which defines the best value achieved so far and a gbest value is the value obtained so far by any particle in the population. Using these two best values, each particle updates its velocity and positions defined in Eq. 4.

$$V_{n+1} = V_n + c_1 r1() * \left(p_{best,n} - P_n\right) + c_2 r2() * \left(g_{best,} - P_n\right)$$
$$P[n+1] = P[n] + v[n+1] \quad (4)$$

where,

P[n + 1]:	particles position at n + 1th iteration
P[n]:	particle position at nth iteration
V[n + 1]:	velocity of the particle at n + 1th iteration
V[n]:	velocity of particle at nth iteration
c1	is the acceleration factor related to gbest
c2	is acceleration factor related to Pbest
r1 and r2	are the two random values

3 Results

The performance of the proposed face detection system is assessed with diverse kinds of images. Here we have shown the results of two images where many faces are there in the images with some side faces and occluded faces. Image 1 contains

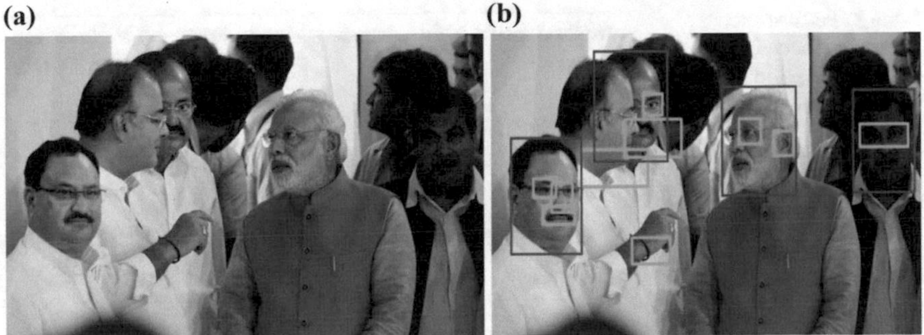

Fig. 6 **a** Input: image 1, **b** the detected features and faces in image 1

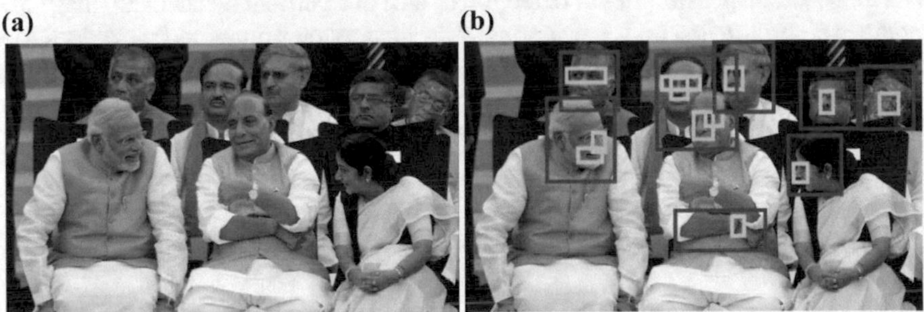

Fig. 7 **a** Input: image 2, **b** the detected features and faces in image 2

Table 1 Performance comparison between existing Viola–Jones algorithm with the proposed method for image 1

Method result	Viola–Jones	Viola–Jones with YCBCr	Viola–Jones with YCBCr and Adaboost-PSO
Time consumption	1.5 s	3 s	2 s
Number of faces detected	2 out of 6	4 out of 6	4 out of 6

six distinct faces and image 2 contains eight distinct faces. The performance of the proposed method is compared with the existing Viola–Jones algorithm. The result for image 1 is shown in Fig. 6 and for image 2 in Fig. 7. The results clearly depict the improved result using the proposed method. Tables 1 and 2 reflect the computational performance comparison using Viola–Jones with YCbCr and proposed algorithm for image 1 and image 2, respectively.

From the results shown in Tables 1 and 2, it is inferred that the Viola–Jones algorithm with YCbCr and Adaboost with PSO provides better result compared to basic Viola–Jones algorithm. The detection speed of the algorithm is highly improved

Table 2 Performance comparison between existing viola jones algorithm with the proposed method for image 2

Method result	Viola–Jones	Viola–Jones with YCBCr	Viola–Jones with YCBCr and Adaboost-PSO
Time consumption	2 s	5.5 s	4.3 s
No. of faces detected	4 out of 8	7 out of 8	7 out of 8

because instead of considering the complete image as the search space and iterating over all quadratic possibilities of detection of face of a particular size in an image, the modified algorithm first finds the regions of the image which contains face detected by the YCbCr algorithm and then using the modified PSO AdaBoost algorithm reduces the quadratic runtime of the algorithm. The proposed algorithm is not dynamic to the size of the face, which has to be specified beforehand for better computation.

4 Conclusion

In this paper, an improved face detection approach is proposed using YCbCr colour space and adaboost classifier with particle swarm optimization. From the simulation results, it is observed that the algorithm takes a little longer time than the basic Viola–Jones algorithm to process the image, but detects more number of faces in the image. The proposed approach not only reduces the search space to find a face but also able to find occluded and side faces with certain probability.

Acknowledgements This work was supported in part by the Science and Engineering Research Board, DST, Govt. of India under Grant ECR/2017/000345.

References

1. Pan, H., Zhu, Y., Xia, L.Z.: Hierarchical PSO-Adaboost based classifiers for fast and robust face detection. World Acad. Sci., Eng. Technol. Int. J. Comput. Inf. Eng. **5**(11) (2011)
2. Marami, E., Tefas, A.: Using particle swarm optimization for scaling and rotation invariant face detection. IEEE Congress on Evolutionary Computation, Barcelona, pp. 1–7 (2010)
3. Viola, P., Jones, M.: Rapid object detection using a boosted cascade of simple features. Proceedings of the 2001 IEEE Computer Society Conference on Computer Vision and Pattern Recognition, vol. 1, pp. 511–518 (2001)
4. Saikia, P., Janam, G., Kathing, M.: Face detection using skin colour model and distance between eyes. Int. J. Comput., Commun. Netw. **1**(3) (2012)

5. Shaika, K.B., Ganesan, P., Kalist, V., Sathish, B.S., Jenithab, J.M.M.: Comparative study of skin colour detection and segmentation in HSV and YCbCr colour space. 3rd International Conference on Recent Trends in Computing (2015)
6. Elmezain, M., Al-Hamadi, A., Michaelis, B.: Realtime capable system for hand gesture recognition using hidden markov models in stereo colour image sequences. J. WSCG, 1–8 (2008)
7. Papageorgiou, C.P., Oren, M., Poggio, T.: A general framework for object detection. International Conference on Computer Vision (1998)
8. Hai-dong, W., Gu, L., Du, L.: A paper currency number recognition based on fast Adaboost training algorithm. International Conference on Multimedia Technology, Hangzhou, pp. 4772–4775 (2011)
9. Perez, C.A., Vallejos, J.I.: Face detection using PSO template selection. IEEE International Conference on Systems, Man, and Cybernetics (2006)
10. Pan, H., Zhu, Y., Xia, L.Z.: Hierarchical PSO-Adaboost based classifiers for fast and robust face detection. Int. J. Comput. Inf. Eng. **5**(11), 1–6 (2011)

Big Data Mining Algorithms for Predicting Dynamic Product Price by Online Analysis

Manjushree Nayak and Bhavana Narain

Abstract Data mining techniques refer to extraction techniques that are performed on large sets/volume of data or the big data. Big data mining is used to extract and retrieve desired information or pattern from a large quantity of data. Today, we need a system which maintains the performance of "Big data" and for analyzing this big data we use various logics. In online marketing, price of any product is highly transparent. It is used to increase sales and margin by many online applications. As a part of this research, in this paper, we have discussed what is big data, what logics are used in big data mining in the form of algorithms, and how it is implemented in big data. How the customer is benefited by analysis of dynamic product price of a product? Various data such as customer attributes, customer visit attributes, and purchase history of a customer have been collected in major form for our study.

Keywords Big data · Algorithm · Online analysis

1 Introduction

Big data can be structured, unstructured, or semi-structured resulting in the incapability of conventional data management methods [1]. Various languages are used in big data analysis such as R, Python, and Java [2]. Concept of dynamic product pricing says that goods can be offered to customers at different prices and demand of customer affects the price of product. The value of the product varies depending on the demand of the product, offer of the product, conversion rate of the product, sales target of the merchandise, and worth of the product mounted by the competitors [3]. Typically, dynamic valuation is additionally associated with several levels of dis-

M. Nayak (✉) · B. Narain
MATS School of Information Technology, MATS University, Pandari, Raipur 492004, CG, India
e-mail: manjushreenayak@matsuniversity.ac.in

B. Narain
e-mail: dr.bhavana@matsuniversity.ac.in

H. S. Behera et al. (eds.), *Computational Intelligence in Data Mining*, Advances in Intelligent Systems and Computing 990,
https://doi.org/10.1007/978-981-13-8676-3_59

crimination such as individual level worth discrimination [4], revenue management [5], and yield management.

Dynamic product rating has additionally taken charge in various industries across the world, as it has increased the shopping for merchandising method. Dynamic product rating is employed in industries such as airlines, hotels, electrical utilities, retails, net retails, mobile communication systems, automotive business, sporting events, rental firms, insurance sectors, and many more to go. Dynamic rating is additionally enforced within the e-selling, e-procurement, e-logistics, and provides chain management and B2B exchange systems [6]. Dynamic product pricing is measured according to five pricing strategies, which is given by [7]. These strategies are as follows.

1.1 Pricing Depends on Segmentation

Price segmentation is charging different prices to different people for the same or product or service. It increases overall revenues and profits. It depends on the market, the quality, and the marketing [22].

1.2 Pricing Depends on Peak Use

The prices and quantities at peak time determine which pricing arrangement is better for consumer.

1.3 Pricing Depends on Service Time

Dynamic price of a product depends on the time product, which was served to the customer.

1.4 Pricing Depends on Time of Purchase

Dynamic price of a product on the time of purchase of that product by the customer.

1.5 Pricing Depends on Changing Price Condition

There are various conditions, such as festival, occasion, and ceremonies, which affect the price condition of the product. Different strategies of pricing depend on type of customers, lifestyle of customers, and the time zone in which purchasing is done. Online marketing has entered in maximum customer's life. Today, customer makes a survey of product price and goes for final purchasing. But these categories of customers are limited in numbers. Technology has given an opportunity for customers to make smart selections. Various methods for determination of price are as follows [8]:

- **Agent-based model**
  This model depends on the buying nature of the customer.
- **Inventory-based model**
  This model depends on the stock management of the company.
- **Data driver model**
  This model depends on the progress of an activity of the product.
- **Game theory model**
  This model is based on the mathematical study of strategical process of the product selling.
- **Simulation model**
  This model is based on creation and analysis of the digital analysis of product sell.
- **Machine learning model**
  This model applies machine learning algorithm in the study of product pricing study.
- **Auction-based model**
  This model is a business model, which includes number of buyers and sellers. These models are frequently used in dynamic product pricing techniques. Study shows that these models are not very much feasible.

2 Analysis of Dynamic Product Price in Big Data Mining

Big data performs crucial component for online world. Price of product affects online customers and also online purchasing. Dynamic product pricing is not new, but it is desperately used by online market for the growth of their sale. There are various methods used in dynamic product pricing and among them clustering is frequently used. We know that in data analysis, primarily based algorithm and partition-based algorithm are used. Some of the study has been made on K-means, hierarchical, EM, and density-based methods [9].

Work on big data analysis for dynamic product pricing has been done by authors in [10]. Survey based on industrial market was done to promote purchase of Internet of things product. Elaborative studies are done by author [8] and [11] on the high online product price dispersion. Application of big data in service industry is done by author [12]. Application of multi- and omni-channel distribution is offered by author [13]. This application is based on the density of product in online marketing. Interactive Home Shopping (IHS) concept was given by the author [14] to promote the purchasing concept of electronic goods in customers. They have analyzed customers, retails, and manufacturer incentives to promote the online purchasing. Evidence of UK and US is used by the author [15] to find out dynamic spillover effect in future market. Use of big data in marketing information system was given by the author [16], and various fields such as input data in the form of marketing intelligence, internal reports, and marketing research have been studied. In online marketing, role of big data and predictive analysis was done by author [17, 18]. Impact of social media on customer behavior was done by [19]. Various marketing analytics schemes were suggested by author [20]. Concept of attribution modeling and analytics was given by author [21]. This study helped customer to select the purchasing scheme.

3 Methodology

This work has been divided into six phases. These are as follows.

3.1 Data Collection

Data is collected from online market server such as Flipkart. Communication to head office of Flipkart has been made and based on research, and dataset for this work has been found.

3.2 Cleaning of Data

In this phase, *incomplete* data such as null data column is filtered and Java programming is applied in the work.

3.3 Attribute Selection

In this step, selection of customer's id and product id as a key field is made. Normalization of data is based on these attributes.

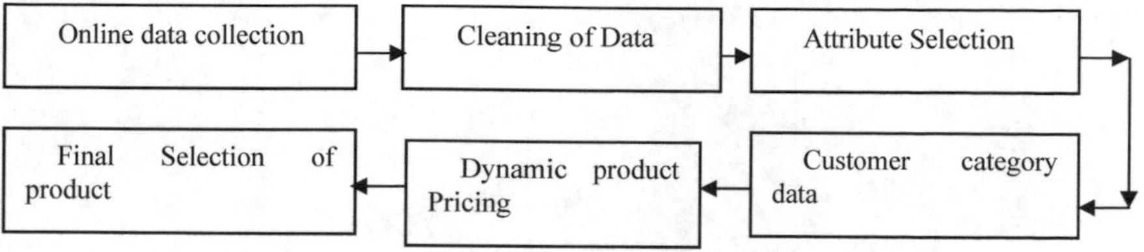

Fig. 1 Flow diagram of our proposed work

3.4 *Customer Category Data*

In this phase, we show how the purchasing of customer affects the pricing of product. Here, K-means clustering algorithm is used.

3.5 *Dynamic Product Pricing*

It is the final result of work. This part shows the relation between cluster and product as well as cluster and customer.

3.6 *Lastly*

Comparison of work with various other pricing methods.

Collection of different attributes of product and customers has been done. In Flipkart, customer detail is registered according to his/her mobile number and email id. Registration is confirmed according to one-time password allotted to customer. Products are divided into various categories. Product id, price, brand, etc. are available in the server. Later on, customer visit history, customer nature, customer likes, and customer purchase frequency are noted in the server (Fig. 1).

4 Result and Analysis

For this experiment, Java environment has been used and various clusters are obtained. For this, five numbers of clusters with respect to four attributers of product are considered. Attributes have been divided into two categories. First category is about product. Here various attributes of product have been taken. In second category, customer attributes have been taken.

In Fig. 2, clusters were formed according to product brought by customers. As cluster size increases, product quantity increases. In Fig. 3, clusters were formed

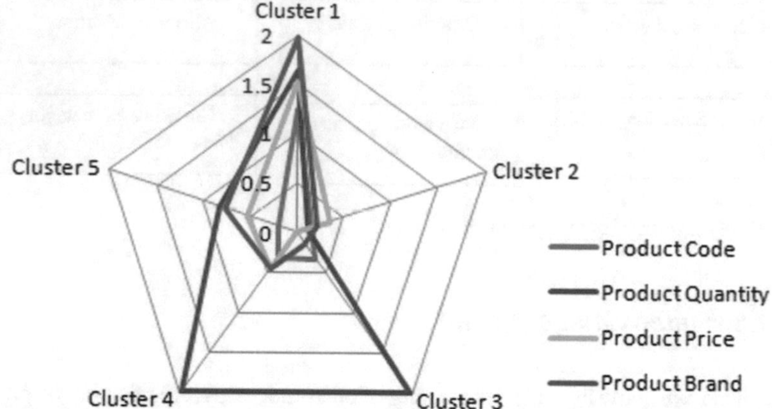

Fig. 2 Cluster product relation

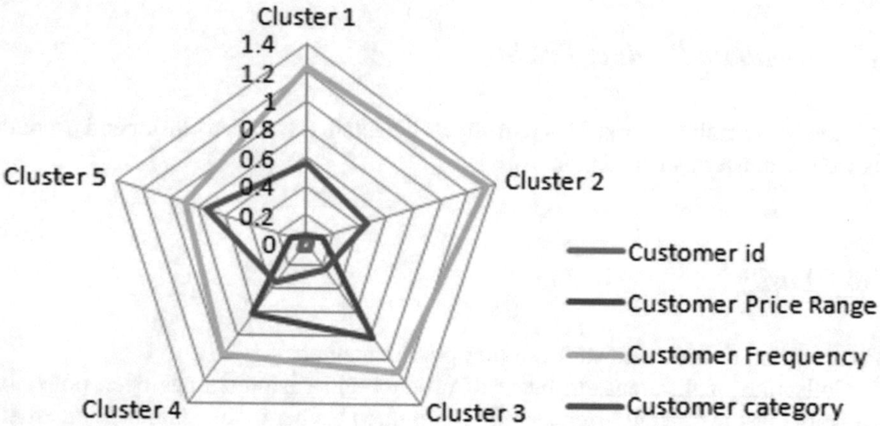

Fig. 3 Cluster customer relation

according to customer purchase tendency. As cluster increases, customer frequency increases. This work is to find out price of product according to the various attributes of customer. Pricing of product is categorized into five clusters. Category of clusters depends on various attributes of product (given in Table 1). Results are categorized under various cluster categories. In Table 2, various attributes of customer are given and according to these attributes, five clusters of customers are categorized. Various age groups of cluster are given under customer category attribute. Five-age groups of customers were selected. Frequent visit of customer is categorized under customer frequency attributes. Dynamic product pricing depends on customer attributes and behavior of customer results in the form of higher price of product.

Table 1 Clusters according to product brought by customers

Attributes	Cluster 1	Cluster 2	Cluster 3	Cluster 4	Cluster 5
Product code	1.24	0.15	0.33	0.31	0.19
Product quantity	1.64	0.12	1.97	1.98	0.82
Product price	1.52	0.36	0.42	0.44	0.54
Product brand	1.98	0.21	0.14	0.46	0.76

Table 2 Clusters according to customer purchase tendency

Attributes	Cluster 1	Cluster 2	Cluster 3	Cluster 4	Cluster 5
Customer id	0.0523	0.0432	0.0634	0.0721	0.0343
Customer price range	0.0456	0.1231	0.8243	0.6234	0.1123
Customer frequency	1.2312	1.3421	1.1242	0.9762	0.8923
Customer category	0.5672	0.4563	0.2345	0.3423	0.7341

5 Conclusion

Data from Flipkart server has been selected for analyzing dynamic product pricing. This work shows that behavior of customer is responsible for dynamic change in product price of online marketing. In future work, on big data various online marketing schemes will be taken. Future market of world is certainly going to be online and it will bring a boom in shopping style. Dynamic product pricing is certainly beneficial for a customer of future.

References

1. Nayak, M.: Prelims of R. Int. J. Tech. Res. Appl. **5**(1), 35–37. www.ijtra.com (2017) (e-ISSN: 2320-8163)
2. Nayak, M., Narain, B.: Impact of R and Python in Big Data Analysis. In: Conference on Innovative trends in Engineering Science and Management (ITESM-2017), 10–11 Nov 2017, vol. 3, issue no. 1, p. 47. MSEIT, Raipur (2017). (ITESM17/T3/CSE/007)
3. Shpanya, A.: 5 trends to anticipate in dynamic pricing. Retail Touch Points **36**, 599–605 (2013). Accessed on 1 Apr 2014
4. Garbarino, E., Lee, O.F.: Dynamic pricing in internet retail: effects on consumer trust. Psychol. Mark. **20**(6), 495–513 (2003)
5. McAfee, R.P., Te Velde, V.: Dynamic pricing in the airline industry. Forthcoming in Hendershott, T.J. (ed.) Handbook on Economics and Information Systems, vol. 17, issue 4. Elsevier (2006)
6. Narahari, Y., Dayama, P. Combinatorial auctions for electronic business. Sadhana Academy Proceedings in Engineering Sciences **30**(2–3), 179–211 (2005). (Indian Academy of Sciences)
7. Magloff, L.: Dynamic pricing strategy. Chron.com. Small business. http://smallbusiness.chron.com/dynamicpricing-strategy-5117.html (2014). Accessed 12 May 2014
8. Gupta, R., Pathak, C. (2014). A machine learning framework for predicting purchase by online customers based on dynamic pricing. Procedia Comput. Sci. 599–605. (Elsevier)

9. Nayak, M., Narain, B.: Data cluster algorithms in product purchase sale analysis. Int. J. Creat. Res. Thoughts (IJCRT) **6**(1), 792–797. www.ijcrt.org (2018). ISSN: 2320-2882
10. Grewal, D., Roggeveen, A.L., Nordfält, J.: The future of retailing. J. Retail. **93**, 1–6 (2017). (Elsevier)
11. Zhuang, H., Leszczyc, P.T.L.P.: Why is price dispersion higher online than offline? the impact of retailer type and shopping risk on price dispersion. J. Retail. **94**(2), 136–153 (2018)
12. Math, R., Sharma, N.: Big data analytics in service industry. Int. J. Emerg. Trends Eng. Dev. **3**(4), 948 955 (2014)
13. Ailawadi, K.L., Farris, P.W.: Managing multi- and omni-channel distribution: metrics and research directions. J. Retail. (2017). ISBN- 0022- 4359 (Elsevier)
14. Alba, J., Lynch, J., Weitz, B., Janiszewski, C.L.: Interactive home shopping: consumer, retailer, and manufacturer incentives to participate in electronic marketplaces. J. Mark. **61**, 38–53 (1997)
15. Antonakakis, N., Floros, C., Kizys. R.: Dynamic spillover effects in futures markets: UK and US evidence. Int. Rev. Financ. Anal., **48**, 406–418 (2016)
16. Salvador, A.B., Ikeda, A.A.: Big data usage in the marketing information system. Sci. Res. J. Data Anal. Inf. Process. **2**, 77–85 (2014)
17. Bradlow, E.T. Gangwar, M., Kopalle, P., Voleti, S.: The role of big data and predictive analytics in retailing. J. Retail. **93** (1), 79–95 (2017). (Elsevier)
18. Bradlow, E.T., Gangwar, M., Kopalle, P., Voleti, S.: The role of big data and predictive analytics in retailing. J. Retail. 1–55 (2017)
19. Hajli, M.N.: A study of the impact of social media on consumers. Int. J. Mark. Res. **56**(3), 387–404 (2013)
20. Wedel, M., Kannan, P.K.: Marketing analytics for data-rich environments. J. Mark. 1–75 (2016). ISBN-1547-7185
21. Kannan, P.K., Reinartz, W., Verhoef, P.C. The path to purchase and attribution modeling: introduction to special section. Int. J. Res. Mark. 1–23 (2016)
22. Aranoff, G.: Globalization alternative pricing in a peak-load pricing model. Mod. Econ. **8**, 888–896 (2017)

Cluster-Based Anonymization of Assortative Networks

Debasis Mohapatra and Manas Ranjan Patra

Abstract Privacy preservation is a serious concern of today's scenario. Anonymity is a way to preserve privacy of different types of data publications like structural and descriptive. In this paper, we discuss the preservation of both types of data. We use anatomy for descriptive data and sequential clustering for structural data anonymization. The proposed approach is based on assortativity called Anatomy Based Modified Sequential Clustering (ABMSC). We propose a new total info. loss measure for cluster optimization. The performance of the proposed approach is measured by information loss and number of node movements takes place to meet the optimization. The experimental results show that our proposed algorithm outperforms the two state-of-the-art algorithms.

Keywords Anatomy · Assortativity · Information loss · Anonymity

1 Introduction

The availability of the inter-related data of an entity in different sites lead to privacy threat. The background knowledge collected from one dataset may be used for privacy disclosure of another. Anonymity is one of the prominent contributions to privacy preservation. Generally, the social graph contains two types of data: descriptive and structural. The descriptive data consists of the attributes like age, sex, salary, disease information, etc. In such data, attributes like age, sex, etc. are considered as quasi-identifiers and salary, disease information, etc. are considered as sensitive information. The method of naïve anonymity removes all descriptive information and only releases the structural information. As such type of publication completely losses

D. Mohapatra (✉)
Department of Computer Science and Engineering, PMEC, Berhampur 761003, India
e-mail: devdisha@gmail.com

M. R. Patra
Department of Computer Science, Berhampur University, Berhampur 760007, India
e-mail: mrpatra12@gmail.com

H. S. Behera et al. (eds.), *Computational Intelligence in Data Mining*, Advances in Intelligent Systems and Computing 990,
https://doi.org/10.1007/978-981-13-8676-3_60

descriptive data, mining on descriptive data is not possible and such releases are less informative. However, the publication of structural data by naïve anonymity is threatened by even simple background knowledge. Ample number of works have been proposed to ensure privacy preservation in structural data by achieving anonymity through edge Add/Delete, vertex Add/Delete, etc. This type of approach is called perturbation approach. Whereas, cluster-based approach publishes the whole graph in a form of abstract graph that provides overall statistics about the connections by hiding the internal details. Apart from that it also publishes anonymized descriptive data. Most of the existing perturbation based methods address identity anonymity whereas, cluster-based method addresses both identity and link anonymity. In this paper, we propose Anatomy Based Modified Sequential Clustering (ABMSC) algorithm for cluster-based anonymization of assortative network. We compare the proposed algorithm with two state-of-the-art algorithms SaNGreeA and SC. The results show that proposed algorithm outperforms the state-of-the-art algorithms for anonymization of assortative network.

The remaining part of the manuscript is organized as follows. In Sect. 2, we discuss the related work. Section 3 discusses basic concepts required to understand the problem. We discuss models and problem definition in Sect. 4. In Sect. 5, we explain the proposed approach. Section 6 covers important results and Sect. 7 concludes the paper.

2 Related Work

In this paper, our focus is on two types of anonymization that is descriptive and structural data anonymization of an assortative network. Descriptive data anonymization adopts the same approaches that are followed by microdata anonymization like Generalization [1] and Anatomy [2]. In generalization, the values of the table are represented by other value that is present in relatively upper level of the domain generalization hierarchy tree which is more general and less specific [1]. Based on this concept, k-anonymity model was proposed. The identity disclosure problem was further investigated and a relatively stronger model l-diversity [3] was proposed. But l-diversity was unable to handle attribute disclosure problem [4]. To handle attribute disclosure problem, t-closeness [4] was proposed. In 2006, Xiao et al. came up with a new idea called as Anatomy as an alternative to generalization that was based on the idea of l-diversity. Anatomy is effective and efficient than generalization [2]. This approach splits the microdata into two different tables so that inference of the original record is difficult. Apart from the above works, some other works on perturbation based anonymization for structural data anonymization are presented in [5–7]. In this paper, we use Anatomy for descriptive data anonymization.

Descriptive data anonymization is not enough for handling identity disclosure. The structural uniqueness can also disclose the identity of an entity. Also, it is somehow possible to re-identify the link between two nodes. To handle identity and link re-identification attacks, cluster-based anonymization is a best approach that publishes an abstract graph with some additional information. SaNGreeA [8, 9] and Sequential clustering [10] are the two crucial development in this field. Due to various advantages of anatomy and sequential clustering, we propose a modified version of sequential clustering that uses anatomy.

3 Basic Concepts

3.1 Social Graph

The original data is represented in a graph, which contains both descriptive and structural information.

Definition 1. (Social Graph) The graph representation of original network is called as Social Graph say G(*V*, *E*, *DR*) where *V* is the set of nodes, *E* is the set of undirected edges and *DR* is the set of descriptive data. There exist a One-to-One mapping f: $V \rightarrow DR$.

3.2 Anatomy

Anatomy is an alternative method to generalization. Anatomy publishes microdata table by splitting the table into two tables: Quasi-Identifier Table (QIT) and Sensitive Table (ST). As the quasi-identifier attributes and sensitive attributes are separated, to provide a meaning to this separation a Group-ID column is attached to both QIT and ST. The Group-ID provides a way to perform join and to deduce some worthy inferences from the released data set. A count attribute is attached to ST that stores the frequency of occurrence of sensitive values. The ST satisfies *l*-diversity requirement that is there must exist at least *l* different sensitive value against a particular Group-ID value.

Definition 2. (Anatomy) Given a microdata M(QA^1, QA^2, … QA^m, SA^1), where QA^1, QA^2, … QA^m are the set of quasi-identifiers and SA^1 is a sensitive attribute. Anatomy splits M into two tables: Quasi-Identifier Table QIT(QA^1, QA^2, … QA^m, GroupID) and Sensitive Table ST(GroupID, SA^1, count).

3.3 Assortative Network

A network is called as assortative if the nodes with high degree connect to high degree nodes and nodes with small degree connect to small degree nodes. A network is called as perfectly assortative if nodes with degree k connect only to the nodes with degree k. Social networks are generally assortative in nature. In this paper, we are interested only in assortative network. The degree assortative coefficient is used to measure the nature of assortativity of a network. The assortativity is computed using Eq. 1.

$$AsCoeff = \frac{\sum_{i=1}^{n} e_{ii} - \sum_{i=1}^{n} a_i^2}{1 - \sum_{i=1}^{n} a_i^2} \tag{1}$$

where e_{ij} is an element of matrix E that denotes the fraction of edges that connects the vertices of degree i with vertices of degree j and $a_i = \sum_{i} e_{ij}$.

4 Models and Problem Definition

In this section, we define different models that are used for our proposed approach. Also, we define our problem.

4.1 Original Data Model

We have considered the assortative network as a simple undirected graph which contains both descriptive and structural data. The graph G(V, E, DR), where V = $\{v_1, v_2, v_3, \ldots, v_n\}$ is the set of vertices/nodes, E = $\{E_1, \ldots, E_m\}$ is the set of edges, each $E_i \in C(V, 2)$ and DR = $\{DR_1, \ldots, DR_n\}$ is the set of descriptive data records where each $DR_i \in D_1 \times .. \times D_k$ and D_i is the domain of attribute A_i. All the edges of set E are assumed to be sensitive and all descriptive data records of DR contain a sensitive value.

4.2 Attack Model

The proposed approach deals with both identity and link disclosure attacks. In identity disclosure attack, the identity of the data holder is disclosed by using some background knowledge with the published data. In link disclosure, the link between the two nodes is disclosed.

4.3 Final Data Publication Model

The final data publication FDP(CL, CE, L_1, L_2, ProjQT, ST), where CL = {CL_1, CL_2, ..., CL_m} is the set of cluster nodes, CE = {CE_1, ... CE_d} is the set of edges that connect cluster nodes. L_1 = {(nc^1, ec^1), ..., (nc^m, ec^m)} where nc^i represents number of nodes in cluster CL_i and ec^i represents number of edges between the nodes in CL_i. There exist a mapping M_1:CL→L_1. L_2 = {bc^1, ..., bc^d} where bc^i represents number of edges present between the two connecting clusters in the original graph. There exist a mapping M_2:CE→L_2. ProjQT = {pq_1, ..., pq_m} where pq_i = projection(QT) for the nodes that belong to CL_i. Sensitive Table (ST) is published as a whole without any partition.

4.4 Problem Definition

Given original data G, the proposed approach publishes G in anonymize form called FDP such that the publication provides identity and link anonymity with minimum information loss.

For example, Fig. 1 shows an original graph and Fig. 2 is its Final Data Publication (FDP). The original data contains detailed descriptive and structural information. Figure 2 shows an abstract representation that hides the descriptive as well as structural information through cluster-based anonymization. It uses anatomy along with clustering. The FDP stores an overall statistics about the number of connections inside the cluster and between the clusters. The split representation of the descriptive information in the form of ProjQT and ST obstruct the inference of sensitive information. How the information loss plays an important role in achieving optimal clustering is discussed in Sects. 5 and 6.

5 Proposed Approach

We propose an anatomy based sequential clustering. Due to different merits associated with anatomy and sequential clustering that are already discussed in previous sections, we propose the combination of both with some modifications for fast convergence in assortative networks. It is obvious that anatomy based sequential clustering is better than generalization based sequential clustering. As our proposed approach focuses on assortative networks, we have modified the existing sequential clustering algorithm to consider assortativity in early phase so the convergence will be faster. We consider global information loss rather considering the local information loss.

Fig. 1 Original graph G

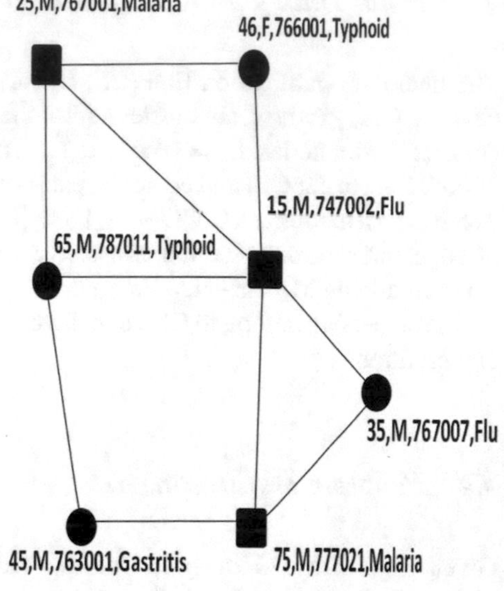

Fig. 2 Final data publication (FDP)

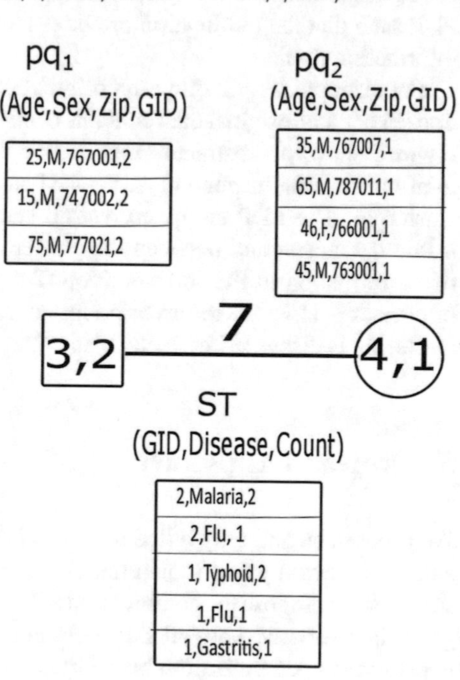

Algorithm 1:- ABMSC(G , k)

Input: Original graph G(V,E,DR) with anonymity factor *k*
Output: FinalClusterSet

1. Partition the set V into clusters say C_1, C_2,………, C_n in such a way that vertices with same degree are placed in the same cluster. The remaining vertices are placed randomly in different clusters such that $|C_i|=k$, $1\leq i\leq n$. If any cluster has less than *k* vertices then assign them to their nearest clusters.
2. **repeat**
3. {

   for each vertex $v_i \epsilon V$

   {

   i. Find the structural information loss (SILoss) in the current cluster C_h.
   ii. Calculate the structural information loss (SILoss) in shifting the v_i from C_h to other clusters. And the cluster which incurs less structural information loss (SILoss) due to this shift say C_{fit}.
   iii. Shift v_i from C_h to C_{fit}.

   }

   if the size of a cluster is greater than $2k-1$ then split it into two equal sized cluster.

   if the size of a cluster is less than *k* then place each object of this cluster to its nearest cluster.

   }**until** (No change in structural information loss or Max iteration reached)

4. **return** FinalClusterSet

5.1 Proposed Algorithm

The algorithm partitions the set V into clusters say $C_1, C_2, \ldots, C_n$ where $k \leq |C_i| < 2k$, $1 \leq i \leq n$ in such a way that vertices with same degree are placed in the same cluster. The sensitive Table (ST) is published globally whereas Projected Quasi-identifier tables (PQs) are published locally in cluster level. The attacker is having access to the global ST and QT of its target cluster. Shifting of entities from one cluster to another takes place to minimize total information loss. The anatomized (Descriptive) information loss remains constant during shifting. Only the structural information loss affects the total information loss. Hence, the proposed algorithm minimizes structural information loss only.

This algorithm works same as sequential clustering algorithm but the convergence is faster for assortative network than that of sequential clustering algorithm because of the step 1 of the algorithm. The randomize partition may not work well in assortative social network because it is well known in assortative and assortative-mix network that there is a high possibility of existence of link between two nodes with same degree.

5.2 *Information Loss*

The total information loss is the composition of descriptive information loss and structural information loss. As we have used anatomy, the descriptive information loss is $\frac{n*(l-1)}{l}$ which remain same for any cluster assignment. Hence, the optimization of total information loss depends only on optimization of structural information loss (SILoss). SILoss can be formulated as shown in Eq. 2.

$$SILoss = Intra\_Cluster\ SILoss + Inter\_Cluster\ SILoss \tag{2}$$

where $Intra\_Cluster\ SILoss$ [10] is the probability of wrongly guessing an edge inside the cluster and $Inter\_Cluster\ SILoss$ [10] is the probability of wrongly guessing an edge between the clusters.

6 Experimental Results and Discussions

We use a system having Intel(R) Core(TM) i7-4770 configuration with MATLAB 2013a platform to implement the algorithms. Different synthetic assortative graphs are constructed and the algorithms are implemented. We create five assortative networks (Synthetic Assortative Graph (SAG)) where for creation of SAG we start with Barabasi–Albert graph and perform random perturbations on the graph to set the assortative coefficient (Eq. 1) in between 0.4 and 0.5. The graphs are named as SAG1, …, SAG5. The detail information of the graphs is shown in Table 1. The information loss of the three methods i.e. SaNGreeA, Sequential Clustering (SC) and Anatomy Based Modified Sequential Clustering (ABMSC) is compared in Fig. 3. It is found that the total information loss is less in our method than the remaining two because we prefer to use anatomy than generalization. The various characteristics of all three approaches are observed and tabulated in Table 2. As shown in Table 1, the node movements of the proposed approach ABMSC is less than SC and SaNGreeA for assortative networks. By observing both factors i.e. information loss and node movements, we found the proposed approach outperforms the rest two approaches.

Table 1 Assortative networks

Graph name	Nodes	Edges	AsCoeff	Node movements		
				SC	SaNGreeA	ABMSC
SAG1	50	235	0.401	15	10	9
SAG2	100	454	0.42	35	24	22
SAG3	200	540	0.408	51	38	32
SAG4	300	704	0.41	59	46	40
SAG5	400	856	0.405	128	91	83

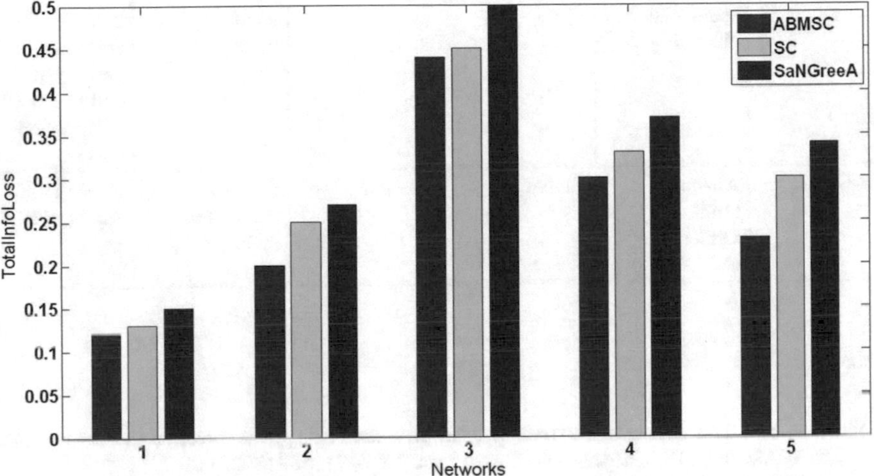

Fig. 3 Information loss comparison

7 Conclusions and Future Work

In this paper, we have proposed a new approach that is based on anatomy and sequential clustering. The proposed approach focuses on assortative networks. The two factors: information loss and node movements are used to measure the efficiency of the algorithms. We found from the experimentation that our proposed approach outperforms the rest two state-of-the-art algorithms. In future, we like to use advanced clustering methods like meta-heuristics based clustering for achieving better converges in cluster-based anonymization.

Table 2 Observed characteristics of SaNGreeA, SC, and ABMSC for assortative network

Approach	Methods adopted	Optimization strategy	Cluster modification	Convergence rate	Balance factor
SaNGreeA	Generalization, greedy	Local	Not possible	Faster than SC but slower than ABMSC	Need a balance between descriptive and structural information loss
SC	Generalization, sequential clustering	Local	Possible	Slowest among the three	Need a balance between descriptive and structural information loss
ABMSC	Anatomy, modified sequential clustering	Global	Possible	Fastest among the three	Doesn't need any balancing

References

1. Sweeney, L.: Achieving k-anonymity privacy protection using generalization and suppression. Int. J. Uncertain. Fuzziness Knowl. Based Syst. **10**(5), 71–588 (2002)
2. Xiao, X., Tao, Y.: Anatomy: simple and effective privacy preservation, pp. 139–150. VLDB (2006)
3. Machanavajjhala, A., Gehrke, J., Kifer, D., Venkitasubramaniam, M.: ℓ-Diversity: privacy beyond k-anonymity. ICDE (2006)
4. Li, N., Li, T., Venkatasubramanian, S.: t-Closeness: privacy beyond k-anonymity and l-diversity. In: IEEE International Conference on Data Engineering (ICDE), pp. 106–115 (2007)
5. Mohapatra, D., Patra M.R.: k-degree closeness anonymity: a centrality measure based approach for network anonymization, pp. 299–310. ICDCIT (2015)
6. Zheleva, E., Getoor, L.: Preserving the privacy of sensitive relationships in graph data. In: Proceeding of International Workshop on Privacy, Security and Trust in KDD (PinKDD), pp. 153–171 (2007)
7. Liu, K., Das, K., Grandison, T., Kargupta, H.: Privacy-preserving data analysis on graphs and social networks. In: Kargupta, H., Han, J., Yu, P., Motwani, R., Kumar, V. (eds.) Next Generation of Data Mining, chapter 21, pp. 419–437. Chapman & Hall/CRC (2008)
8. Babu, K.S., Jena, S.K., Hota, J., Moharana, B.: Anonymizing social networks: a generalization approach. Comput. Electr. Eng. **39**, 1947–1961 (2013)
9. Campan, A., Truta, T.M.: Data and structural k-anonymity in social networks. In: Proceedings of the Second ACM SIGKDD International Workshop on Privacy, Security and Trust in KDD (PinKDD), pp. 33–54 (2008)
10. Tassa, T., Cohen, D.J.: Anonymization of centralized and distributed social networks by sequential clustering. IEEE Trans. Knowl. Data Eng. **25**(2), 311–324 (2013)

Application of Artificial Intelligence Techniques for Improvement of Power Quality Using Hybrid Filters

Soumya Ranjan Das, Prakash K. Ray, Asit Mohanty and Tapas K. Panigrahi

Abstract The present paper implements a Shunt Hybrid Active Filter (SHAF) essential for reactive power suppression across the load and eradicates harmonics in developing the power quality (PQ) under variable source and load conditions in the distribution network. For better compensation features, SHAF employing the harmonic reduction performance of both passive filter and active filter is presented in this paper. In this work, an adaptive linear neuron (ADALINE) based Variable step size leaky least mean square (VSSLLMS) control approach is applied to examine the operation of SHAF in a three-phase system. This competent approach is needed for generating the gating pulses essential for IGBT inverter and increasing the performance compared to the conventional ADALINE-LMS controller. The proposed controller provides faster convergence and also provides better performance compared to conventional one. Simulation results show the significant improvement in ADALINE-VSSLLMS approach compared to ADALINE-LMS approach. The simulations of the proposed system under different load and source conditions are performed using MATLAB/SIMULINK.

Keywords ADALINE · LMS algorithm · Shunt Hybrid Active Filter (SHAF) · Variable Step Size Leaky Least Mean Square (VSSLLMS) · Total Harmonic Distortion (THD)

S. R. Das
Department of Electrical and Electronics Engineering, IIIT Bhubaneswar, Khurda 751003, India
e-mail: srdas1984@gmail.com

P. K. Ray (✉) · A. Mohanty
Department of Electrical Engineering, CET Bhubaneswar, Bhubaneswar 751003, India
e-mail: pkrayiiit@gmail.com

A. Mohanty
e-mail: asithimansu@gmail.com

T. K. Panigrahi
Department of Electrical Engineering, PMEC Berhampur, Brahmapur 761003, India
e-mail: tkpanigrahi.ee@pmec.ac.in

H. S. Behera et al. (eds.), *Computational Intelligence in Data Mining*, Advances in Intelligent Systems and Computing 990,
https://doi.org/10.1007/978-981-13-8676-3_61

1 Introduction

In present years, the quality of power gets worsened in various utility networks due to wide use of loads with non-linearity characteristics. Harmonics and reactive power [1] are basically produced due to use of these devices. Due to the existence of harmonics, many losses occur in utility network. The major issue and a serious problem in the power system network [2] are to compensate the main harmonics and as per IEEE 519 harmonic standard, the Total Harmonic Distortion (THD) should be reduced below 5%. Harmonic effect can be minimized by using the combine configuration of Active and Passive filters. The passive filtering is the suitable strategic in power system utilities for developing the power factor and reducing the harmonic distortion and is highly used due to robust construction and cheaper with easy implementation. Unlike the advantages, this filter also suffers from different complications like problems in tuning, fixed compensation, big size and effect of resonance. In order to deal with the above issues, hybrid active filters (HAF) [3] were implemented. The HAF delivers improved performance and nominal solutions. The HAF filter has been preferred over other filters for mitigation of harmonic. The controller is the most important component in HAF system. Traditionally, PI controller has spread its importance in active filters. PI performs better steady state as well as dynamic response during variation of load. But certain factors like large ripples during switching stage, forces to use less in power system network. PI controller collects two current sensors from source, moreover, harmonic detector is not needed to maintain fundamental source currents.In recent times, different controllers based on Artificial intelligence [4, 5] are broadly applied in the control systems because of its flexibility features. ADALINE based control algorithm is very normal which requires little computational problems. Moreover, ADALINEoffersnatural linearity and performs as a very fast method. In this algorithm, the convergence speed is varied by altering the convergence coefficient. Conventionally, ADALINEbased control algorithm implements LMS algorithm to compute the online weight measurement. As a result, LMS extracts supply reference currents under different load conditions. This method executes in a very simple way and also mathematical model is not required. But this method attains a slow speed with high convergence error during filtering. In order to overcome these issues, in this paper to calculate the reference current extractions, an ADALINE based algorithm [6] with Variable step size leaky least mean square (VSSLLMS) [7] is implemented in SHAF. The proposed controlling scheme is implemented to produce the gating signals of VSI based SHAF. ADALINE-VSSLLMS comprises estimating of weighting values, retaining dc bus voltage and calculating the reference current. Finally, employing PWM based hysteresis current controller (HCC), gating pulses are generated by relating the reference source with the detected source current. The error signal has been compensated using the controllers. This reference current is fed to the hysteresis current controller and compare with the sensed filter currents to obtain the switching signal for active power filter. The hysteresis controller is performed by receiving the inputs as reference source current and original source current and the output current is treated as switching signal to the VSI.

2 System Configuration

Hybrid filters [8–12] involves structure of passive filters (PF) [13] with APFs. The SHAF overcome the problems linked with PF and APF by improving the PQ performance of the utility network. This structure decrease the noise switching and interference in electromagnetic is reduced. The main aim of SEHAF is to dynamically improve filtering operation of harmonics with higher order by supplying economical low order harmonics suppression. A three-phase three wire SEHAF is presented in this paper. The block diagram of the proposed system is depicted in Fig. 1.

Two different control schemes are performed in SHAF for generation of reference currents; one is the conventionally ADALINE-LMS controller and the other is proposed ADALINE-VSSLLMS controller. Three-phase source voltage is mixed with harmonic and sinusoidal components. Utilizing ADALINE based controllers, the sinusoidal components gets extracted from non-sinusoidal supply. In the VSSLMS algorithm, step size is significantly adjusted for the period of conjunction. Three ADALINEs are implemented in each phase are utilized for harmonics extraction and reactive components extraction from the detected load currents. This algorithm shows fast and dynamic response compared to ADALINE-LMS controller and other traditional controllers. The configuration of ADALINE-VSSLLMS controller is shown in Fig. 2. In this ADALINE controller, two neurons are processed in which their weights are restructured online by the VSSLLMS algorithm. Input of individual neuron is

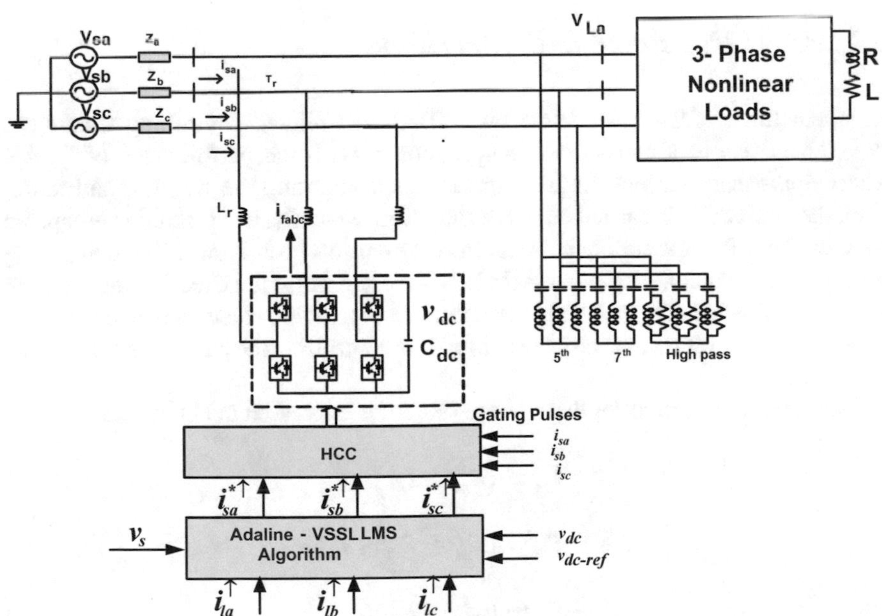

Fig. 1 Block diagram of the proposed system

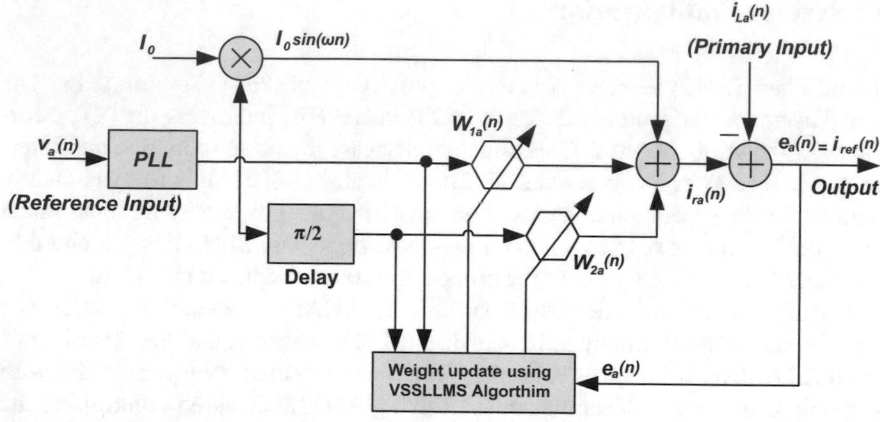

Fig. 2 Model of ADALINE based LMS Algorithm

controlled by references inputs ($x_{1a}(n)$ and $x_{2a}(n)$) which are obtained from supply voltage ($v_a(n)$).

3 Control Strategies

3.1 ADALINE Based LMS Algorithm

In recent times it has been found that ADALINE related controlling techniques have given rise to a considerable improvement with the performance of SHAF. These approaches employ LMS algorithm for computing the weights, and moreover, the computing is carried out by online. Consequently, this technique is capable to draw the reference supply currents in different load conditions. The controlling process is not suitable for other ADALINE related current extraction methods. In this controlling technique [14, 15] the reference current are calculated using the estimated weighted values and variable parameters controlling the bus voltage across the capacitor.

The reference current for three-phase system is expressed in (1)–(3) as

$$i_{sa}^* = \left(w_m^+ + I_{sm}\right)u_{sa} \tag{1}$$

$$i_{sb}^* = \left(w_m^+ + I_{sm}\right)u_{sb} \tag{2}$$

$$i_{sc}^* = \left(w_m^+ + I_{sm}\right)u_{sc} \tag{3}$$

where w_m^+ is the weighted variable and is given in (4) as

$$w_{m(k+1)} = w_{m(k)} + \mu\left[i_{lk} - w_{m(k)}u_{s(k)}\right]u_{s(k)}$$
$$w_m^+ = \frac{\left(w_{ma}^+ + w_{mb}^+ + w_{mc}^+\right)}{3} \tag{4}$$

where u_{sa}, u_{sb}, u_{sc} are the unit voltage template and I_{sm} is considered as the output from PI controller.

3.2 ADALINE Based VSSLLMS Algorithm

Referring to (4) the weights of ADALINE [16, 17] are normally restructured utilizing the least mean square (LMS) algorithm.

Weight is found out from the components of the compensating current which is given in (5) as

$$i_{Fa}(t) = i_{La}(t) - i_{Xa}(t) = i_{La}(t) - [w_{1a}, w_{2a}]v_a(t) \tag{5}$$

where, $W = [w_{1a}, w_{2a}]$ and are mentioned as ADALINE weight vector. Based on the VSSLLMS algorithm, weights of ADALINE are given in (6) as

$$w_{1a}(n+1) = (1 - 2\Delta(n)\gamma_{1a}(n))w_{1a}(n) + \Delta(n)e_a(n)x_{1a}(n)$$
$$w_{2a}(n+1) = (1 - 2\Delta(n)\gamma_{2a}(n))w_{1a}(n) + \Delta(n)e_a(n)x_{2a}(n) \tag{6}$$

where $\Delta(n)$, variable step size parameter and for conventional LMS algorithm, $\gamma(k) = 0$.

3.3 Regulation of DC Bus Voltage

The DC link voltage must be maintained at the desired reference value for achieving a better compensation. While the DC bus voltage performs two essential roles in controlling the SHAF. In one stage it operates fewer ripples during steady state and on the other hand, it works as a medium in balancing the energy that provides or draws power during transient state. Therefore, a classical PI controller is widely implemented for regulating the dc bus voltage. The controlling scheme of PI involves comparison between the voltage across the capacitor and reference voltage of DC bus. The output of the comparison in terms of error is feed over a PI controller. Now the output from PI controller is treated as peak current. This peak current is helpful in controlling the losses in SEHAF to realize from average voltage to a constant value of the capacitor. The control strategy of dc bus voltage is depicted in Fig. 3.

The P_{Loss} is computed in (7) and (8)

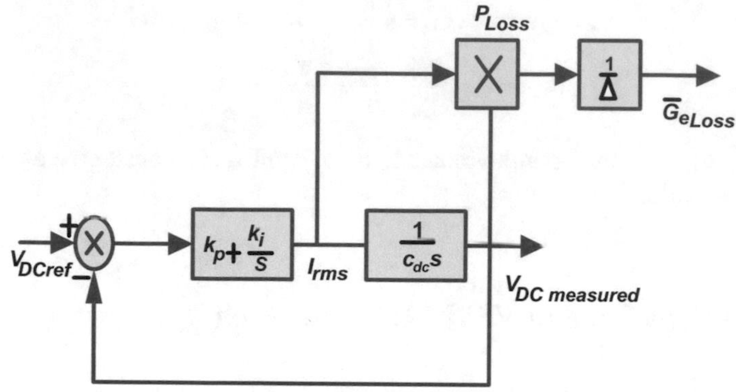

Fig. 3 Control strategy of DC bus voltage

$$P_{Loss} = \mathbf{\Delta} V_{DC}\left(k_p + \frac{k_i}{s}\right) \times V_{DCmeasured} \tag{7}$$

$$\overline{G}_{eLoss} = \frac{\overline{P}_{Loss}}{\mathbf{\Delta}} \tag{8}$$

The voltage transfer function is expressed in (9):

$$\frac{V_{DCmeasured}}{V_{DCrefered}} = \frac{\omega_c^2(1 + s\tau)}{s^2 + 2\varepsilon_c\omega_c s + \omega_c^2} \tag{9}$$

where

$$\omega_c = \sqrt{\frac{k_i}{c}},\ \varepsilon_c = \frac{1}{2}\frac{k_p}{\sqrt{ck_i}} \tag{10}$$

For better regulation, the value is chosen as $\varepsilon_c = 0.7$ as expressed in (10).

4 Simulation Results and Analysis

The proposed model under study is performed in MATLAB/Simulink tool. Simulations are performed utilizing "Power System Block set" simulator, used to analyze the function of the proposed SHAF.

Initially power system is performed in the absence of SHAF and the output load currents are measured with connecting of non-linear loads. The simulated results are presented in Fig. 4a. It is evident from simulated results that the load currents, I_{La}, I_{Lb} and I_{Lc} are composed of harmonic components and the three-phase load current waveform is fully distorted. The THD value is achieved to be 28.38% and shown in Fig. 4b.

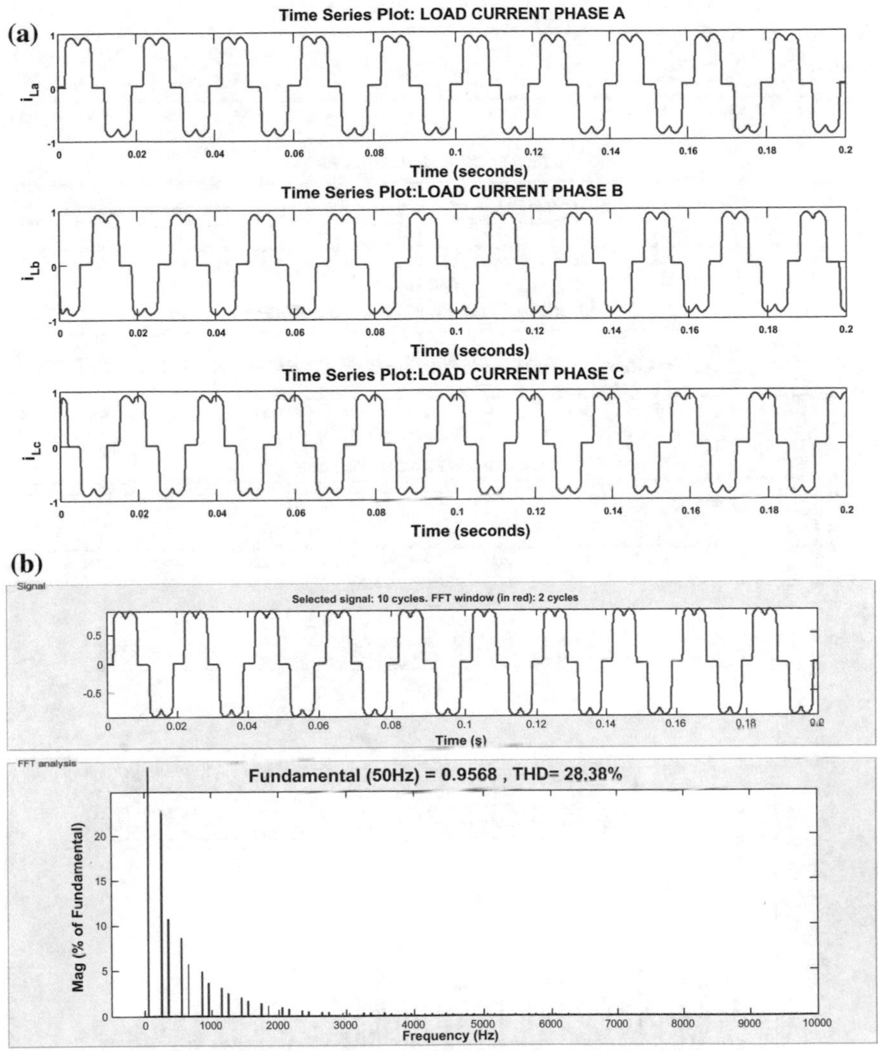

Fig. 4 **a** Load current before using SHAF **b** Total harmonic Distortion value without SHAF

The proposed system is then performed by connecting SHAF using ADALINE-LMS algorithm and simulated results are illustrated in Fig. 5a and corresponding THD value in Fig. 5b found to be 4.60%.

Subsequently, the proposed system is operated using ADALINE-VSSLLMS algorithm. The simulated results show improvement in THD value of 2.39% compared to ADALINE-LMS Algorithm. The simulated results are expressed in Fig. 6a and its THD values are shown in Fig. 6b.

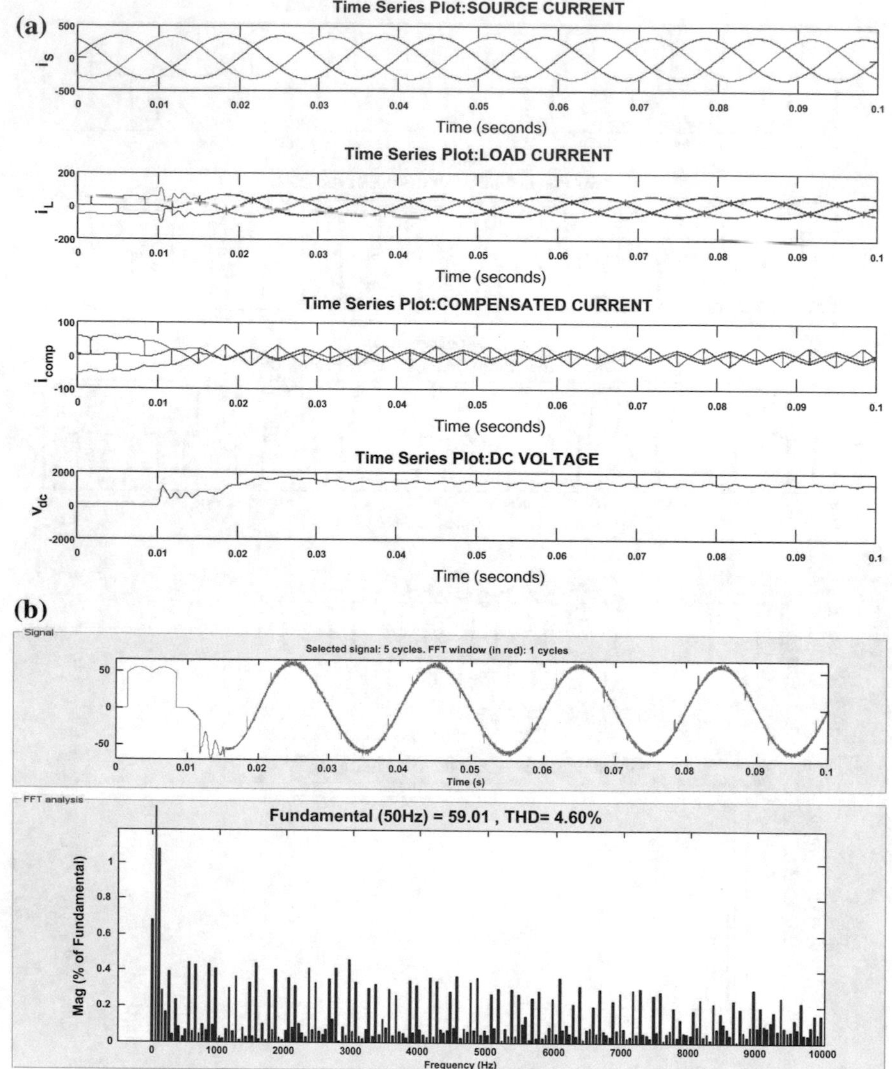

Fig. 5 **a** Performance of SHAF using ADALINE-LMS Algorithm **b** Total harmonic Distortion value of Load current

It is cleared from the simulation result shown in the Table 1 that, the percentage THD in ADALINE-VSSLLMS is quite better compared to ADALINE-LMS algorithms. The parameters of the proposed system are expressed in Table 2.

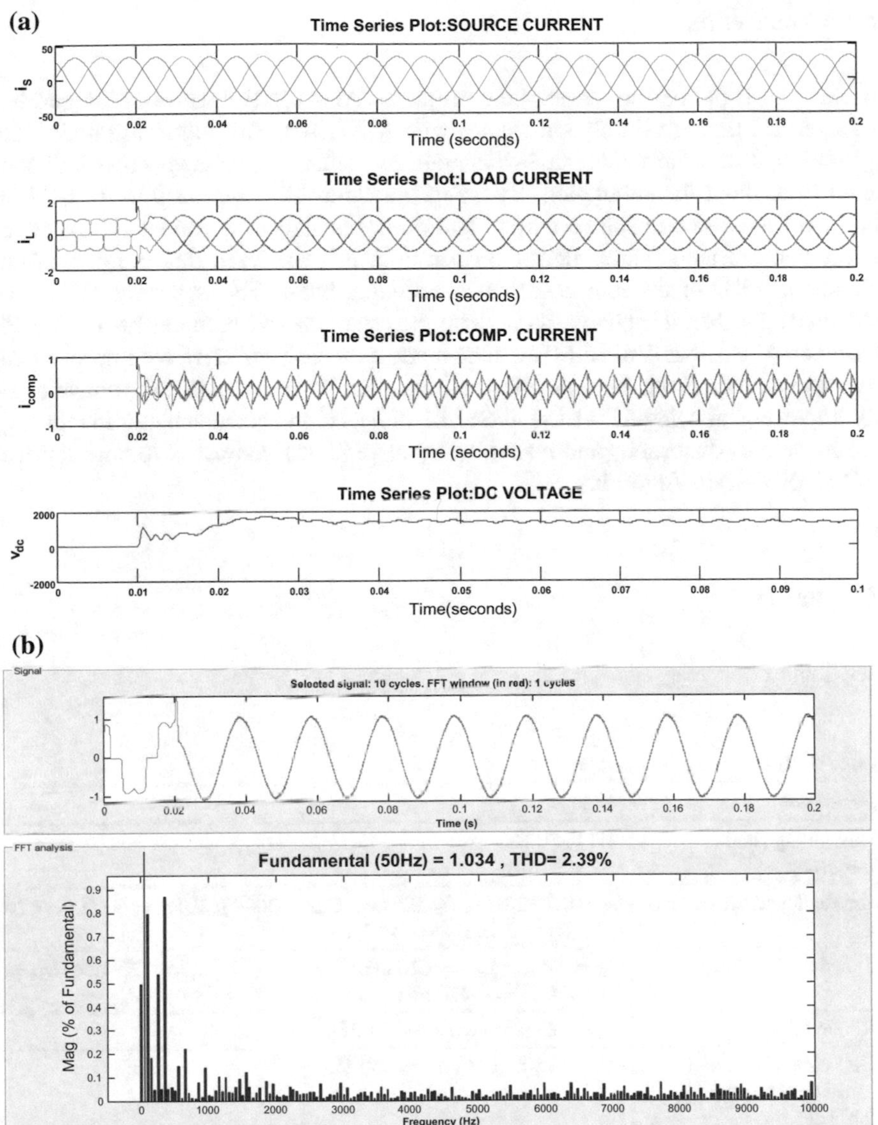

Fig. 6 **a** Performance of SHAF using ADALINE-VSSLLMS Algorithm **b** Total harmonic Distortion value of Load current

Table 1 THD analysis using ADALINE algorithms

Parameters	%THD value
Without filter	28.38
ADALINE-LMS	4.60
ADALINE VSSLLMS	2.39

5 Conclusion

In this paper, the performance of the projected ADALINE linked LMS algorithm was verified through simulation studies with MATLAB. This algorithm makes the system stable and harmonic free by maintaining the DC bus voltage across VSI constant throughout the simulation. By using the controlling approach using SHAPF harmonic existence in source current and other unbalanced components are effectively compensated. Throughout the operation, it is analyzed that the percentage change in THD of the source current is reducing below 5% as per the IEEE-519 standards. Further, it is analyzed that with the use of passive components in SHAPF based on ADALINE-VSSLLMS, can be made quite cost-effective for improvement in power quality. From the simulation results, it has been shown that the power quality improvement using ADALINE-VSSLLMS Algorithm is much better in reducing the harmonics distortions and managing reactive power compared to conventional ADALINE –LMS Algorithm.

Appendix

See Table 2

Table 2 Parameters of the system

Parameters	Value
Line voltage and frequency	400 V, 50 Hz
Line and load impedance	$L_s = 0.20$ mH, $L_{ac} = 2.5$ mH, $C_{DC} = 3000$ µ F, $R_L = 60\ \Omega$, L = 20 µ F; ($R_1 = 2\ \Omega$, $R_2 = 4\ \Omega$, $R_3 = 6\ \Omega$)
Tuned passive filter	$C_{f5} = 50$ µ F $L_{f5} = 8.10$ mH; $C_{f7} = 20$ µ F $L_{f7} = 8.27$ mH; $C_{f11} =$ 20 µ F $L_{f11} = 8.270$ mH
Ripple Filter	$C_{RF} = 25$ µ F, $L_{RF} = 0.38$ mH
DC voltage level and capacitor value across the capacitor	$C_{DC} = 1500$ µ F $V_{DC} = 550$ V
Filter coupling inductance	2.5 mH
Controller gain	$K_P = 0.032$, $K_I = 0.00004$ K

References

1. Subjak, J.S., Mcquilkin, J.S.: Harmonics-causes, effects, measurements, and analysis: an update. IEEE Trans. Ind. Appl. **26**, 1034–42 (1990)
2. Mahela, O.P., Shaik, A.G.: Topological aspects of power quality improvement techniques: a comprehensive overview. Renew. Sustain. Energy Rev. **58**, 1129–42 (2016)
3. Rahmani, S., Al-Haddad, K., Kanaan, H.Y.: A comparative study of shunt hybrid and shunt active power filters for single-phase applications: simulation and experimental validation. Math. Comput. Simul. **71,** 345–59 (2006)
4. Lin, B.R., Hoft, R.G.: Power electronics inverter with neural networks. IEEE Technology update series. (1997)
5. Chen, Y.M., O'Connell, R.M.: Active power line conditioner with a neural network control. Thirty-First IAS Ann. Meet. Ind. Appl. Conf. **4**, 2259–2264 (1996)
6. Mangaraj, M., Panda, A.K., Penthia, T.: Investigating the performance of DSTATCOM using ADALINE based LMS algorithm. In: International Conference on Power Systems (ICPS), pp. 1–5 (2016)
7. Garanayak, P., Panda, G.: Harmonic elimination and reactive power compensation with a novel control algorithm based active power filter. J. Power Electron. **15** 1619–27 (2015)
8. Shuai, Z., Luo, A., Fan, R., Zhou, K., Tang, J.: Injection branch design of injection type hybrid active power filter [J]. Autom. Electric Power Syst. **5**, 12 (2007)
9. Kumar, C., Mishra, M.K.: An improved hybrid DSTATCOM topology to compensate reactive and nonlinear loads. IEEE Trans. Ind. Electron. **61**, 6517–27 (2014)
10. Lee, T.L., Wang, Y.C., Li, J.C., Guerrero, J.M.: Hybrid active filter with variable conductance for harmonic resonance suppression in industrial power systems. IEEE Trans. Ind. Electron. 746–56 (2015)
11. Das, S.R., Ray, P.K., Mohanty, A.: Enhancement of power quality disturbances using hybrid filters. In: IEEE-International Conference on Circuit, Power and Computing Technologies (ICCPCT), pp. 1–6 (2017)
12. Das, S.R., Ray, P.K., Mohanty, A.: Improvement in power quality using hybrid power filters based on RLS algorithm. Energy Proc. (2017) 723–728
13. Singh, B., Solanki, J.: A comparison of control algorithms for DSTATCOM. IEEE Trans. Ind. Electron. **56**, 2738–45 (2009)
14. Singh, B., Solanki, J.: An implementation of an adaptive control algorithm for a three-phase shunt active filter. IEEE Trans. Ind. Electron. **56**, 2811–2820 (2009)
15. Subudhi, B., Ray, P.K., Ghosh, S.: Variable leaky least mean-square algorithm-based power system frequency estimation. IET Sci. Meas. Technol. **6**, 288–297 (2012)
16. Mayyas, K., Aboulnasr, T.: Leaky LMS algorithm: MSE analysis for Gaussian data. IEEE Trans. Signal Process. **45**, 927–934 (1997)
17. Kamenetsky, M.,Widrow, B.: A variable leaky LMS adaptive algorithm. Thirty-Eighth Asilomar Conference on Signals, Systems and Computers, pp. 125–128 (2004)

Applications and Advancements of Nature-Inspired Optimization Algorithms in Data Clustering: A Detailed Analysis

Janmenjoy Nayak, Paidi Dinesh, Kanithi Vakula, Bighnaraj Naik and Danilo Pelusi

Abstract In the last decade, nature-inspired optimization algorithm has been a keen interest among the researchers of optimization community. Most of nature-inspired algorithms are developed through the simulating behavior of natural agents in nature. In comparison with evolutionary- and swarm-based algorithms, these are most effective techniques for all real-life applications. Although both swarm- and evolutionary-based algorithms are one of the subsets of nature-inspired optimization algorithm but the efficiency and effectiveness of such algorithm make them more attractive to use in various data mining problems. Among the other tasks of data mining, it has been always a challenging task to solve clustering problem, which is unsupervised in nature. In this paper, a brief study has been conducted on the applications of nature-inspired optimization algorithms in clustering techniques. Also, few challenging issues along with the advancements of various nature-inspired optimization algorithms are realized in the field of clustering.

Keywords Nature-inspired optimization algorithms · Particle swarm optimization algorithm · Evolutionary algorithms · Clustering · Data mining

---

J. Nayak · P. Dinesh · K. Vakula
Department of Computer Science and Engineering, Sri Sivani College of Engineering, Srikakulam 532410, Andhra Pradesh, India
e-mail: mailforjnayak@gmail.com

P. Dinesh
e-mail: dinesh.pydi98@gmail.com

K. Vakula
e-mail: vakku.bi@gmail.com

B. Naik (✉)
Department of Computer Application, Veer Surendra Sai University of Technology, Burla, Sambalpur 768018, Odisha, India
e-mail: mailtobnaik@gmail.com

D. Pelusi
Faculty of Communication Sciences, University of Teramo, Coste Sant'Agostino Campus, Teramo, Italy
e-mail: dpelusi@unite.it

H. S. Behera et al. (eds.), *Computational Intelligence in Data Mining*, Advances in Intelligent Systems and Computing 990,
https://doi.org/10.1007/978-981-13-8676-3_62

1 Introduction

Selection of perfect solution amidst of accessible solution in order to solve a problem is known as optimization. Optimization is more important than anything in many applications like engineering design and business activities. Optimization is used for minimization and maximization (for instance, to minimize the costs and consumption of energy and to maximize the efficiency and effectiveness). It won't be overstate to say that optimization plays a vital role essentially everywhere. The main desire behind optimization is to make a suitable and standard balance with diversification and intensification. Nowadays, we have very limited time, resources as well as money in natural world applications which are nonlinear and require knowledgeable optimization mechanisms to rigging. So, we need to discover the solutions in order to use these relevant resources optimally. Computer simulations were becoming requisite mechanism to solve such problems of optimization with different effective search algorithms. Actually the nature of optimization may be different, but the method for solving any complicated problem, it is critical to select the accurate optimization procedure which is appropriate for that problem.

Algorithms such as GA (evolutionary based) [1] and PSO [2], which is a swarm-inspired technique and some other algorithms based on the nature were made known in the last decade in order to resolve complex problems, but still facing from many major pitfalls like slow, convergence, cope with the execution time and higher complexity, etc. Nowadays, researchers are showing more significance to nature-inspired optimization algorithm as these are capable of solving complex problems such as NP-hard and other constrained as well as unconstrained type. Due to this, it became the path for the researchers to focus on nature-inspired algorithms which are efficient in resolving complex problems. Nature seems to be speak and convey something to us. That is the main reason why we got lot of inspiration from nature. Evolutionary computation or evolutionary algorithms were evaluated while a process of study takes place in artificial intelligence. More meaningful progress came into existence with the advancement of particle swarm optimization (PSO), which is motivated by SI behaviors of birds. Some of the leading and successful optimization techniques simulated by the nature are developed in recent years are as follows: harmony search algorithm [3] (inspired by music), cuckoo search algorithm [4] (inspired by the behavior of cuckoo), artificial bee colony algorithm [5] (inspired by the intelligent explore behavior of honeybees), firefly algorithm [6] (inspired by the flicker behavior of fireflies), social spider algorithm [7] (inspired by the behavior of spiders), bat algorithm [8] (inspired by the location of objects by reflected sound (echolocation nature of bats), strawberry algorithm [9] (inspired by the plants of strawberry), plant propagation algorithm [10] (inspired by the behavior of plants), seed-based plant propagation algorithm [11] (inspired by the seed distribution of plants), cat swarm optimization [12] (inspired by the observing behavior of cats), simulated annealing [13] (inspired by the simulation of strong solids), artificial immune systems [14] (inspired by the immune system), ant colony optimization algorithm [15] (inspired by the social behavior's of ant's in finding the shortest distance), differential evolution [16] (inspired by the moving obstacles and evolution theory), bacterial foraging optimization [17] (inspired by a bacteria search, how it madly over the view of nutrients to act on non-gradient optimization), marriage in honeybees optimization [18] (inspired by marriage behavior of honeybees), fish school algorithm [19] (inspired by the collective behavior of fish schools), bacteria chemotaxis algorithm [20] (inspired by tactic behavior of bacteria), social cognitive optimization [21] (inspired by evolution process of human intelligence and the social learning theory), honeybees mating optimization [22] (inspired by the mating behavior of honeybees), invasive weed optimization [23] (inspired by the colonizing weeds), shuffled frog leaping algorithm [24] (inspired by the interaction among the frogs), central force optimization [25] (inspired by gravitational kinematics), intelligent water drops algorithm [26] (inspired from the reaction among water drops), river formation dynamics [27] (inspired by the formation of river), roach infestation optimization [28] (inspiration from cockroach species), bacterial foraging algorithm [29] (social foraging behavior of bacteria), gravitational search algorithm [30] (inspired by the Newton's law of gravity and the law of motion), league championship algorithm [31] (inspired by the championship league environment), optics algorithm [32] (inspired by the law of optics), bumble bees mating optimization [33] (inspired by the behavior of social insects), bacterial colony algorithm [34] (inspired by the behavior of bacteria during their whole life-cycle, including chemo taxis, reproduction), monkey search algorithm [35] (inspired by natural behavior of monkey), artificial chemical reaction algorithm [36] (inspired by types and occurring of chemical reactions), chemical reaction algorithm [37] (inspired by the chemical reactions started during the collisions), bull optimization algorithm (inspired by the genetic operators such as mutation and crossover) [38], lion optimization algorithm [39] (inspired by the behavior of lions), elephant herding optimization [40] (inspired by the herding behavior of elephants), wolf search algorithm [41] (inspired by the wolf living style like how they will get food and survive their lives), queen bee evolution algorithm [42] (inspired by the reproduction process of bees), flower pollination algorithm [43, 44] (inspired by the pollination of flowers), black hole algorithm [45] (inspired by the phenomenon of black hole), bees algorithm [46] (inspired by behavior of bees which is a swarm intelligence based algorithm), glowworm swarm optimization [47, 48] (inspired by the behavior of the glowworms), and many more.

Clustering is an unsupervised method of data mining and its major objective is grouping of data points. Data points which are placed in a group must have the same or similar feature, whereas data points those are placed in different places must have more dissimilar features. Finding either a similar or dissimilar structure in an arrangement is the main aim of the clustering. It is a method of individual learning and mainly based on distance computation methods. There are many popular clustering techniques, such as k-means, Fuzzy c-means (FCM), k-medoid, k-means++ algorithm, etc., which are applied to solve real-world problems in an effective way.

Applying any of these techniques single-handedly causes some troubles such as sensitive to initialization, algorithm iteration speed, getting stuck at local optima, more iteration to find proper intra- and inter-cluster similarities and dissimilarities, etc. Also, each of the technique has its own limitation such as k-means is not suitable for circular data/high-dimensional data, choosing the inappropriate value of "k" may have a deep impact on the performance; FCM works like the hill climbing approach and initial cluster centroid setting push the algorithm a way back; k-medoid suffers from the sensitivity to outliers and more time complexity, etc. As a result, researchers implemented different hybrid techniques with combination of various optimization methods to overcome the abovementioned limitations. However, after a deep literature review and analysis, it is worthy to note that most of the methods are based on nature-inspired optimizations. Inspired by this, in this paper, a brief (yet detailed) analysis has been performed on the applications and advancements of nature-inspired optimization algorithms in clustering. Mostly, the study is based on the articles published in last 10 years, as the field of research in nature-inspired optimization is active after the year 2010. Although before that, some algorithms were developed such as PSO, DE, and GA (swarm and evolutionary based), but after the year 2010, there is a rapid development of research in optimization algorithms inspired by chemistry, physics, biology, etc. and became popular.

The rest part of this paper is segmented into the following sections: Sect. 2 describes some basic concepts of clustering, Sect. 3 outlines the applications of various algorithms (frequently used) in clustering. Some hybrid techniques of combined nature-inspired algorithm with clustering are mentioned in Sect. 4. Section 5 analyzes the critical descriptions and advancements with limitations of the developed algorithms in the literature. Section 6 concludes the work with some future directions.

2 Basic Preliminaries

In this section, some popular clustering algorithms (especially k-means and FCM) are briefly explained.

2.1 *Fuzzy C-Means Algorithm*

This technique allows belongingness of one piece of data to several clusters. It is a partitional algorithm based on the concept of minimizing the objective function (Eq. (1)) [49] with respect to partition matrix.

$$J(U, V) = \sum_{i=1}^{C} \sum_{j=1}^{N} u_{ij}^{m} \left\| x_j - v_i \right\|^2 \tag{1}$$

In Eq. (1), "x_j" and "v_i" indicate jth cluster point and ith cluster center, respectively. $u_{i,j}$ indicates membership value of jth cluster point with respect to cluster "i". "m" is the fuzzy controlling factor. It may result hard partition for setting the value as "1" and complete fuzziness for setting the value as "∞". $\| \ \|$ is the norm function. FCM algorithm works with three main factors [50] such as fuzzy membership function, partitional matrix, and objective function.

$u_{i,j}^m$ is calculated as described in Eq. (2) and cluster center as in Eq. (3).

$$U_{ij} = \left[\sum_{k=1}^{c} \left(\frac{\|x_j - v_i\|}{\|x_j - v_k\|} \right)^{\frac{2}{m-1}} \right]^{-1} \tag{2}$$

$$v_i = \sum_{j=1}^{N} u_{ij}^m x_j / \sum_{j=1}^{N} u_{ij}^m \text{ where } i \geq 1, i \leq c \tag{3}$$

This method is almost equivalent to k-means algorithm except the factor $u_{i,j}^m$(fuzziness factor). The fuzzy factor determines the level of fuzziness for the clusters. The results of FCM depend on the initial choice of values for the clusters, which is one of the drawbacks of it. The detailed algorithm steps of FCM can be realized in [51].

2.2 K-Means Algorithm

K-means algorithm [52] is one of the famous clustering algorithms and is mostly used by many researchers to solve the problems of clustering in the field of data mining. It can be assumed that K-means is a rapid iterative clustering algorithm under unsupervised machine learning, which is used to reduce clustering error. It is a formal method to produce required number of similar data in a given data set.

This algorithm has two stages: one is defining the k-median points for each cluster and the other one is to take related point of the given data set and mingle it to the closely connected median point. Euclidean distance is deliberated to evaluate with relation to data points and the median points.

The k-means algorithm has a great extent use for clustering of large data sets. But, this benchmark algorithm has no surety that results are efficient as the quality of the resulted clusters may be based on the first median points. However, the complexity of the original algorithm is high when compared with other standard clustering techniques.

3 Various Nature-Inspired Optimizations in Clustering

There are a number of nature-inspired algorithms used in clustering area. However, in this section, some of the popular (yet frequently used) algorithms with their applications are mentioned. Also, some other algorithms and hybrid algorithms are described in the later section.

3.1 Particle Swarm Optimization in Clustering

Particle swarm optimization is a swarm-based metaheuristic algorithm [2]. This algorithm imitates the social behavior of birds flocking and the nature of positioning together. Each individual in the search space communicate each other in the time of learning from their own experience, and moderately move a step toward their goal. It is a very effective global search algorithm and one of the popular algorithms which is most successful in advance and discrete optimization. Jiang et al. [53] developed a novel method of PSO using age-group topology to improve higher performance in PSO. They maintained so far describing the problem of evaluating the capability of finding particles in local area and developed a concept of parameter setting in which particles of various age groups have various parameters to improve convergence rate. They considered 15 standard functions as well as seven natural datasets to solve problem of clustering. Later, they compared their proposed method with the other evolutionary algorithms and variants of PSO and found their method has higher performance for the problems of data clustering and optimization function. Alijarah and Ludwig [54] developed a parallel PSO-based intrusion detection system using map reduce technique to control convergence rate and cluster median points in order to reduce the problems of flow of data in large-scale network. They considered a true dataset to evaluate the performance. They found that their proposed method has higher performance and detection rate for large dimensional datasets. Filho et al. [55] described that in case of partitioned techniques of clustering, many fuzzy clustering methods using optimization algorithms are recommended. Mostly, among these optimization techniques, PSO is used along with FCM for higher performance. They found that by using PSO when compared with partitioned methods of clustering, there is a lower performance in execution time. To overcome those types of problems, they developed a hybridization of PSO and FCM named as FCM-IDPSO and FCM2-IDPSO. Their main aim of the proposed method is to improve higher performance. For this, they considered two artificial and eight natural datasets and compared their developed method with other standard methods such as FCM-PSO, chaotic particle swarm fuzzy clustering, etc. and claimed that their approach has higher performance than other compared methods. Alswaitti et al. [56] have developed a particle-swarm-optimization-based clustering method with the suitable balance of exploration and exploitation. They maintained so far describing the problem of early convergence with introducing kernel density estimation method and

Table 1 Applications of PSO in clustering

Year	Application area	Clustering type	Ref.
2012	Classification of 3D human model	Automatic	[58]
2014	Data clustering	Partitional	[59]
2016	Data clustering	Partitional	[60]
2017	Imbalanced big data processing	Fuzzy C-means	[61]
2017	Text clustering	K-means	[62]

different gravitational learning coefficients. They considered 11 numbers of standard datasets from UCI repository and compared the proposed method with some other methods by considering the performance factors such as classification accuracy, cluster compactness, and standard deviation. They found the proposed method is efficient as compared to some other standard methods, for which the author claims that the method can be successful in various real-life applications such as medical diagnosis, decision-making, etc. An improved PSO algorithm which is named as Evolutionary PSO (EPSO) was introduced by Alam et al. [57]. They considered standard datasets to test the performance and compared their method with k-means and original particle swarm optimization algorithms. They found that their method has efficient performance for clustering of data and produces compact clusters. Some more applications of PSO can be depicted in Table 1.

3.2 *Artificial Bee Colony Optimization in Clustering*

The artificial bee colony algorithm (ABC) is an optimization algorithm which depends on the intelligent ravage behavior of honeybee swarm, projected by Karaboga in 2005 [5]. It is a population-based algorithm. Its working principle is based on the area of the food source, which exhibits a feasible solution to the optimization problem and the nectar quantity of a food source corresponds to the fitness (quality) of the related solution. Karaboga and Ozturk [63] developed ABC algorithm for the clustering of data. They considered 13 complex datasets and compared their proposed method with particle swarm optimization and nine different standard optimization techniques. They found that their proposed method is useful for the clustering of data and claimed further research work can be done on their proposed method to compare the performance with different optimization techniques. Also, they found an improved performance in their method rather than PSO and other compared methods. Kumar and Sahoo [64] improved a two-step artificial bee colony algorithm for coherent data clustering with suitable balance of diversification and intensification to increase convergence rate. They improved ABC algorithm using k-means algorithm for problems of clustering. They considered five standard and two synthetic data sets and considered their proposed method with some of the other methods by considering the initial positions of food sources with the help of k-means algorithm rather than random initialization. They have applied PSO in onlooker phase of ABC. They have shown the effectiveness of their proposed algorithm compared to original ABC algorithm by analyzing performance by improving solution to three critical issues: initial positions of food sources, solution search, and abandoned food location of original ABC. Later, they compared their method with various clustering algorithms such as CSS, PSO, CPSO, and k-means by considering seven datasets. They claim that their improvements are robust and successful in various real-life scenarios such as pattern recognition, image processing, etc. By taking an efficient swarm optimization algorithm named bee colony optimization algorithm, Forsati et al. [65] considered an improved method by applying the concepts of cloning and fairness into BCO to get higher performance results for clustering of data. They described the local search problem and found a solution using k-means algorithm by comparing the convergence behaviors and results of clustering among different hybridized algorithms such as IBCO, k-means, PSO, ant colony, genetic algorithm, etc. To obtain clustering problem, they have used some objects directly to BCO. They claimed that their proposed method increased the effectiveness of BCO in optimization process when compared to other standard methods. They found the improved results in terms of efficient performance and also claimed that their method may be helpful in applications of data as well as document clustering for the process of clustering. Hencer et al. [66] introduced a new function for image clustering which is tested with particle swarm, artificial bee colony, and genetic algorithms. This proposed method's performance was proved on seven standard images by distinguishing with three objectives in k-means algorithms. In this paper, separateness and compactness are the main standards of clustering problems. The assumed results proved that artificial-bee-colony-based clustering method with enhanced functions get well-notable clusters. It is recommended that artificial bee colony algorithm with introduced function can be effectively working for clustering problems.

3.3 *Cuckoo Search Optimization in Clustering*

Cuckoo search is an optimization algorithm innovated by Yang and Deb in 2009 [4]. It was inspired by the behavior that relies on others to raise their young (brood parasitism) of some cuckoo varieties by laying their eggs in the other host birds (of other species) nests. Also, some birds use direct conflict with the interrupting cuckoos. Goel et al. [67] introduced cuckoo search clustering algorithm for data clustering with suitable balance of exploration and exploitation. They used standard dataset iris and found that their method has higher performance. Later, they considered their method on satellite image datasets to acquire water structure. They claimed that their method is robust and have a very good performance than other compared techniques in case of data clustering. Garg and Batra [68] have recommended fuzzified cuckoo based clustering technique (F-CBCT) with raising penetration of security risks which impacts on increasing the severity on networks. In this paper, they focused on two aspects: training and detection. Training phase is simulated by using the decision tree method, which depends on hybridization of cuckoo search optimization and k-means clustering. In this algorithm, they have used two performance parameters such as classification and anomaly detection measure. If the training of system is completed, then detection phase started at which the fuzzy approach is recycled to identify the irregularities on support of input data and distance functions calculated in previous stage confirmed the efficiency of this model with the experimental outcomes in terms of high detection rate (96.86%), low false positive rate (1.297%), and high accuracy (99.77%) with "f" measure (98.30%). Saida et al. [69] have proposed the chaotic cuckoo search algorithm for data clustering and their performance was approached on six datasets. The main aim of this article is to enhance the cuckoo search abilities. Different updates influenced by quantum theory are performed to implement the problems of cuckoo search clustering in means of global search algorithm. They considered eight numbers of standard datasets from UCI machine learning repository and compared their proposed method with some other methods, which are known to be hybrid cuckoo search, genetic quantum cuckoo search, and k-means chaotic particle swarm optimization. Finally, its results confirmed the effectiveness of proposed algorithm. Ye et al. [70] recommended the k-means clustering algorithm based on improved cuckoo search algorithm, where k-means automatically falls into local optimum. They considered the standard datasets of UCI and compared the proposed algorithm with some other standard algorithms to get faster merging rate, data clustering effect, and better efficiency. Their result on MFCC features shows stable effects of clustering speed merging rates and efficiency it avoid the drawbacks of k-means algorithm.

3.4 Ant Colony Optimization in Clustering

Ant colony optimization is one of the famous swarm-based metaheuristic algorithms developed by Dorigo in the year 1992 [15] and it works on the principle of social behavior of ants. It can be used in different ranges of applications and it has higher efficiency and potentiality for solving nonlinear problems. Inkaya et al. [71] developed a new approach by taking spatial clustering problem with no previous details. They found that their method forwarded different problems of clustering such as data set reduction, evolution of solution, construction of neighborhood by considering two fair functions such as adjusted compactness, and relative separation to evaluate the quality of solutions for clustering. Later, they considered ant colony optimization and compared the proposed methods with some other algorithms by combining proximity, connectivity, density, and distance information with diversification and they claim that their proposed method has efficient performance in case of validity indices and they found that a future research addresses may be necessary to analyze all the complex solutions of clustering. But, they also found that their proposed method needs improvements for the problems of noise detection. Zhang and Cao [72] introduced a new way of combination of ant colony and kernel method for data clustering. They considered various artificial and natural datasets and compared their proposed method with some other algorithms. They found that their developed method is efficient by using kernel with ACO and claimed that their method has higher performance for the clustering of data and increases the effectiveness of clustering. They also indicated that their method needs more research work to improve some additional features: maintaining the parameters, etc. Chen et al. [73] innovated a novel ACO approach to minimize some problems of clustering. They considered some standard data sets such as iris, etc. and compared their proposed method with some standard methods. They claimed that their method has great performance and is very easy to use. Han and Shi [74] developed a new method of ACO for fuzzy clustering. They claimed that their method has a drawback of time complexity. To overcome this drawback, they improved median point and heuristic function in their method and claimed that their method is an effective approach for image segmentation.

4 Hybridized Nature-Inspired Optimization Algorithms in Clustering

A hybridization of Fuzzy C-Means and particle swarm optimization has been developed by Izakian et al. [75]. To reduce some problems in FCM, they hybridized it with FPSO. They considered six standard datasets such as iris, glass, cancer, CMC, and vowel. The authors claimed that their proposed method has higher performance and solutions in the area of clustering. A hybridization of particle swarm optimization and ant colony optimization has been done by Marinakis [76] et al. for the clustering of N instances into K clusters to avoid clustering problem. They considered several standard datasets and compared their proposed method with K-means and various other algorithms. They found that their method has higher performance when PSO is used. A hybridization of weighted genetic algorithm and K-means was developed by Wu [77] and named that algorithm as GWKMA for data clustering. They considered three genetic operators such as selection, crossover, mutation, and also taken an artificial and two natural gene datasets. They claimed that their proposed method has efficient performance than K-means in case of obtaining effectiveness of clusters. A newly hybridized metaheuristic algorithm was proposed by Marinatis et al. [78] based on ABC and GRASP algorithm, for the clustering of "N" instances into K clusters. Their proposed method is a two-way approach which is a combination of ABC and GRASP for the solution of problems of clustering and finding features. They considered standard datasets and compared their developed method with some other standard algorithms such as tabu search, ant colony, PSO, honeybee mating, etc. They found that their method has great performance and may be very helpful for the clustering of data. A hybridization of PSO and ant colony optimization for clustering of data has been done by Chen et al. [79]. They considered nine artificial datasets and proposed four hybrid techniques such as sequential, parallel, sequential-based particle table, and better global transfer of data. They found among these four techniques, sequential-based particle table has higher performance by combining new particle

Table 2 Applications of hybridized algorithms

Year	Application area	Hybridized clustering name	Ref.
2007	Data clustering	Ant and swarm	[81]
2010	Data clustering	MICA and K-means	[82]
2012	Data clustering	Firefly and K-means	[83]
2016	Wireless sensor network	Bee swarm and DE	[84]
2017	Data clustering	K-means and ABC	[85]
2014	Data clustering	FCM and Firefly	[86]
2018	Solar energy facility	Hybrid c-means	[87]

to the table. They claimed that their proposed hybrid techniques have higher performance than other standard algorithms such as standalone, PSO, ACOR, K-means, etc. and found that their proposed method may be helpful for solving other optimization problems along with function optimization, etc. A hybridization of QPSO and GA has been done by Song et al. [80] for solving the problem of complex optimization by describing some lower bound values. They found that their proposed method is very useful and increased the capacity to search optimal solutions by introducing an updated position to a particular space. The main aim of their proposed method is to rectify the problems in text clustering. For this, they have taken four subsets from standard Reuter-21578 and 20 newsgroup datasets. They made an analysis using fitness and F-measure, for which the author claims that their new approach has higher performance when compared with QPSO and improved QPSO. Table 2 describes some other hybridized nature inspired algorithms with application areas. Other than this, Table 3 predicts some more recently developed nature-inspired algorithms with their applications areas.

5 Critical Analysis and Discussions

In this paper, detailed investigations are being conducted on the use and applications of nature-inspired optimization algorithms in clustering. Several clustering methods such as FCM, k-means, hybrid algorithms, etc. are considered with all the optimization techniques starting from earlier developed PSO to the algorithms developed up to 2018. It is quite obvious to note that so many nature-inspired algorithms are used to solve different application areas with many clustering algorithms and have successfully proved their effectiveness. In this study, it is observed that many clustering techniques single-handedly and in combined form also are applied to solve clustering problem. Literature indicate that most of the techniques are being developed by PSO, ACO, cuckoo search, and ABC algorithms. Apart from these, some important hybrid methods are also being developed in combination of more than one optimization algorithms. The papers are considered from various popular databases

Table 3 Some other nature-inspired optimization algorithms

Algorithm name	Year	Application area	Clustering name	Ref.
ACO	2006	Document clustering	Ant-based clustering	[88]
ABC	2016	Data clustering	Map reduced based	[89]
Big-Bang Crunch	2016	Data clustering	K-means	[90]
Brain-Storm	2015	Data clustering	Hard C-means	[91]
Cuckoo search	2014	Data clustering	K-means	[92]
Cuckoo search	2014	Web document clustering	K-means	[93]
DE	2014	Data clustering	C-means	[94]
Cohort intelligence	2014	Data clustering	K-means	[95]
DE	2017	Data clustering		[96]
Firefly	2012	Data clustering	K-harmonic means	[97]
Grass Hopper	2017	Data clustering		[98]
HBM	2007	Neural networks	K-means	[99]
HBM	2007	Data clustering	K-means	[100]
Social spider	2012	Text clustering	–	[101]
Bacterial foraging	2017	Disease prediction	FCM	[102]
Fish school	2016	Clustering graphical processing unit	K-means and K-harmonics	[103]
Invasive weed	2015	Text clustering	K-means	[104]
Shuffled frog leaping	2018	Clustering	K-means	[105]
League championship	2015	Clustering	–	[106]
League championship	2017	Clustering	K-means	[107]
Class topper	2018	Clustering	–	[108]
Lion	2018	Clustering	Fuzzy	[109]
Chemical reaction	2017	Clustering	FCM	[110]
Firefly	2017	Data clustering	K-means	[111]
Elitist teaching learning	2016	Data clustering	K-means	[112]
Elitist teaching learning	2016	Data clustering	FCM	[113]
Flower pollination	2016	Data clustering	K-means	[114]
Black hole	2018	Data clustering	K-means	[115]

such as IEEE Xplore, Springer, ScienceDirect, etc. based on keyword search and almost many of the applications in clustering using nature-inspired optimization techniques are covered. A detailed analysis can be depicted in Figs. 1 and 2. In Fig. 2, it may be analyzed that, as compared to other nature-inspired algorithms, the algorithms inspired by biology and swarm behaviors have been of frequent interest for researchers. However, some other algorithms such as physical, chemical based, etc. are also used in clustering.

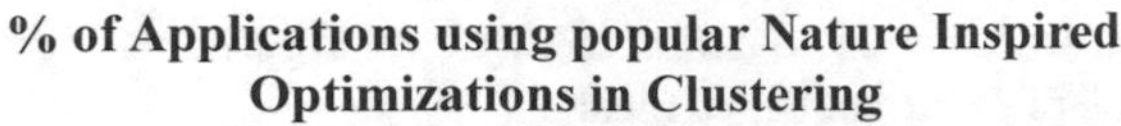

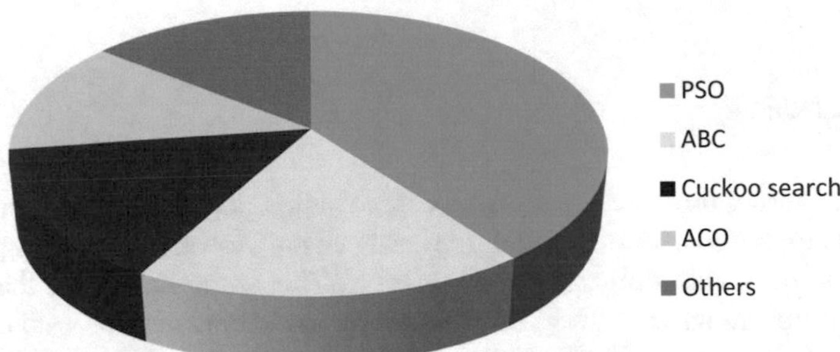

Fig. 1 Applications of all nature-inspired algorithms in clustering

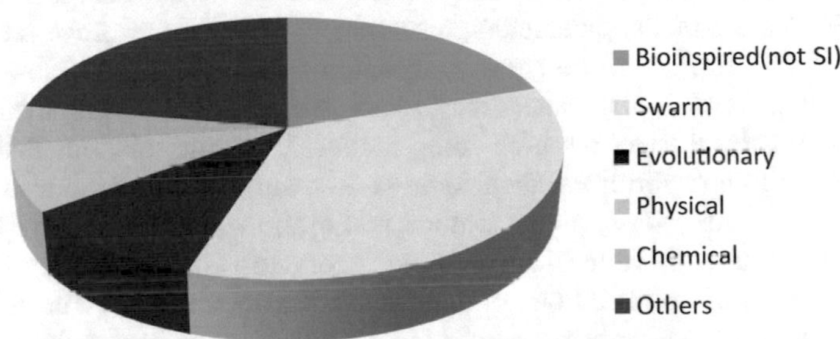

Fig. 2 Applications of various types of nature-inspired optimizations in clustering

According to no free lunch theorem [116], "not any particular optimization algorithm can solve all the problems in all domains in the real world." Rather, one optimization is suitable for only some specific problems and may not be fit for other problems. So, there are many other nature-inspired algorithms such as roach infestation optimization, central force optimization, river formation dynamics, intelligent water drop algorithm, river formation dynamics, bumblebee mating optimization, bacterial colony optimization, monkey search algorithm, etc., which are yet to be applied in clustering. However, maximum optimization algorithms developed since 2010 are used and outlined in this paper. Still, there is a huge scope for further development with so many multi-objective algorithms like cuckoo search. It is evident that using cuckoo search and clustering techniques, variety of applications are solved. The main advantage of using cuckoo search is that it consists of only one parameter and easy to implement as compared to other swarm-based algorithms. But, hybridization of any other variations and advancements of FCM algorithm may be used with cuckoo search or any other multi-objective algorithms to get more efficient solutions in terms of good quality of clusters.

6 Conclusion

Solving clustering problem has always been a challenging task for researchers due to their unsupervised nature. So, attention has been paid toward hybridizing the clustering techniques with optimization algorithms. This study is evident that nature-inspired optimizations are always been a striking choice to combine with clustering techniques. With avoiding the limitations of parent clustering algorithm, these algorithms are suitable and efficient for applying in any real-life domain of applications. However, some points are still in infancy and need to be focused. Since last few years, it has always been trendy that some algorithms are being developed by only some simulating nature/behavior of animals/creatures/properties of insects. This indicates the lack of novelty in developing such algorithms and with slight change in the objective function/nature of the parameters, researchers are trying to prove the efficacy of the methods. But, it will be good if, instead of developing such new methods (though reshaping old), researchers may focus toward novel developments which will be really helpful to all research communities. It is quite obvious that if some algorithms are to be hybridized, then some good results will be produced with the help of both algorithms (any parent method and optimization algorithm). But, how far that method is applicable to solve some other problems is an open challenge to all researchers of optimization community. In future, attention will be made to review and analyze the performance of nature-inspired optimization in some other data mining classifications.

Appendix

ABC	Artificial bee colony
ACO	Ant colony optimization
BCO	Bee colony optimization
CMC	Contraceptive method choice
CPSO	Chaotic PSO
CPSFC	Chaotic particle swarm fuzzy clustering
CS	Cuckoo search
CSS	Charged system search

(continued)

(continued)

DE	Differential evolution
EPSO	Evolutionary PSO
FCM	Fuzzy c-means
F-CBCT	Fuzzified cuckoo based clustering technique
FCM-IDPSO	Fuzzy c-means-improved self-adaptive particle swarm optimization
FCM2-IDPSO	Fuzzy c-means2-improved self-adaptive particle swarm optimization
FCM-PSO	Fuzzy c-means-particle swarm optimization
FPSO	Fuzzy PSO
GA	Genetic algorithm
GRASP	Greedy randomized adaptive search procedure
GWKMA	Genetic weighted k-means algorithm
HBM	Honeybee mating
IBCO	Improved Bee colony optimization
MFCC	Mel frequency cepstral coefficient
PSO	Particle swarm optimization
QPSO	Quantum-behaved PSO
SI	Swarm intelligence
UCI	UC Irvine

References

1. Khan, G.M.: Evolutionary computation. In: Evolution of Artificial Neural Development, pp. 29–37. Springer, Cham (2018)
2. Kennedy, J.: Particle swarm optimization. Encyclopedia of Machine Learning, pp. 760–766. Springer, Boston, MA (2011)
3. Geem, Z.W., Kim, J.H., Loganathan, G.V.: A new heuristic optimization algorithm: harmony search. Simulation **76**(2), 60–68 (2001)
4. Yang, Xin-She, Deb, Suash: Cuckoo search: recent advances and applications. Neural Comput. Appl. **24**(1), 169–174 (2014)
5. Karaboga, D., Basturk, B.: A powerful and efficient algorithm for numerical function optimization: artificial bee colony (ABC) algorithm. J. Glob. Optim. **39**(3), 459–471 (2007)
6. Yang, X.-S.: Firefly algorithm, stochastic test functions and design optimisation. Int. J. Bio-Inspired Comput. **2**(2), 78–84 (2010)
7. Cuevas, E., Cienfuegos, M.: A new algorithm inspired in the behavior of the social-spider for constrained optimization. Expert Syst. Appl. **41**(2), 412–425 (2014)
8. Yang, X.-S.: A new metaheuristic bat-inspired algorithm. In: Nature Inspired Cooperative Strategies for Optimization (NICSO 2010), pp. 65–74. Springer, Berlin, Heidelberg (2010)
9. Merrikh-Bayat, F.: A numerical optimization algorithm inspired by the strawberry plant. arXiv preprint arXiv:1407.7399 (2014)

10. Salhi, A., Fraga, E.S.: Nature-inspired optimisation approaches and the new plant propagation algorithm, pp. K2–1 (2011)
11. Muhammad, S., Salhi, A.: A seed-based plant propagation algorithm: the feeding station model. Sci. World J. **2015** (2015)
12. Chu, S.-C., Tsai, P.-W., Pan, J.-S.: Cat swarm optimization. In: Pacific Rim International Conference on Artificial Intelligence. Springer, Berlin, Heidelberg (2006)
13. Van Laarhoven, P.J., Aarts, E.H.: Simulated annealing. In: Simulated Annealing: theory and Applications, pp. 7–15. Springer, Dordrecht (1987)
14. de Castro, L.N., Timmis, J.: Artificial immune systems: a novel paradigm to pattern recognition. Artif. Neural Netw. pattern Recognit. **1**, 67–84 (2002)
15. Dorigo, M., Stützle, T.: The ant colony optimization metaheuristic: algorithms, applications, and advances. In: Handbook of Metaheuristics, pp. 250–285. Springer, Boston, MA (2003)
16. Storn, R., Price, K.V.: Differential evolution-a simple and efficient heuristic for global optimization over continuous spaces. J. Glob. Optim. **11**, 341–359 (1997)
17. Passino, K.M.: Bacterial foraging optimization. In: Innovations and Developments of Swarm Intelligence Applications, pp. 219–234. IGI Global (2012)
18. Abbass, H.A.: MBO: marriage in honey bees optimization-a haplometrosis polygynous swarming approach. In: Proceedings of the 2001 Congress on Evolutionary Computation CEC2001 (2001)
19. Li, X.-L., et al.: Applications of artificial fish school algorithm in combinatorial optimization problems. J. Shandong Univ. (Eng. Sci.) **5**, 015 (2004)
20. Muller, S.D., et al.: Optimization based on bacterial chemotaxis. IEEE Trans. Evol. Comput. **6**(1), 16–29 (2002)
21. Sun, J.Z., et al.: An improved social cognitive optimization algorithm. Appl. Mechan. Materials. Vol. 427. Trans Tech Publications, 2013
22. Haddad, O.B., Afshar, A., Mariño, M.A.: Honey-bees mating optimization (HBMO) algorithm: a new heuristic approach for water resources optimization. Water Resour. Manag. **20.5**, 661–680 (2006)
23. Mehrabian, A.R., Lucas, C.: A novel numerical optimization algorithm inspired from weed colonization. Ecol. Inform. **1**(4), 355–366 (2006)
24. Eusuff, M.M., Lansey, K.E.: Optimization of water distribution network design using the shuffled frog leaping algorithm. J. Water Resour. Plan. Manag. **129**(3), 210–225 (2003) (cited By (since 1996) 297)
25. Formato, R.A.: Central force optimization: a new metaheuristic with applications in applied electromagnetics. Prog. Electromagn. Res. PIER **77**, 425–491 (2007)
26. Hamed, S.-H.: Problem solving by intelligent water drops. In: IEEE Congress on Evolutionary Computation, 2007. CEC 2007, pp. 3226–3231. IEEE (2007)
27. Rabanal, P., Rodríguez, I., Rubio, F.: Using river formation dynamics to design heuristic algorithms. In: Unconventional Computation, pp. 163–177. Springer (2007)
28. Havens, T.C., Spain, C.J., Salmon, N.G., Keller, J.M.: Roach infestation optimization. In: Swarm Intelligence Symposium, 2008. SIS 2008. IEEE, pp. 1–7. IEEE (2008)
29. Passino, K.M.: Biomimicry of bacterial foraging for distributed optimization and control. Control Syst. IEEE **22**(3), 52–67 (2002)
30. Rashedi, E., Nezamabadi-Pour, H., Saryazdi, S.: Gsa: a gravitational search algorithm. Inf. Sci. **179**(13), 2232–2248 (2009)
31. Kashan, A.H.: League championship algorithm: a new algorithm for numerical function optimization. In: International Conference of Soft Computing and Pattern Recognition, 2009. SOCPAR'09, pp. 43–48. IEEE (2009)
32. Kashan, A.H.: A new metaheuristic for optimization: optics inspired optimization (OIO). Comput. Oper. Res. **55**, 99–125 (2015)
33. Comellas Padró, F.D.P., Martínez Navarro, J., et al.: Bumblebees: a multiagent combinatorial optimization algorithm inspired by social insect behaviour (2011)
34. Niu, B., Wang, H.: Bacterial colony optimization. Discrete Dynamics in Nature and Society 2012 (2012)

35. Mucherino, A., Seref, O.: Monkey search: a novel metaheuristic search for global optimization. Data Min. Syst. Anal. Optim. Biomed. **953**, 162–173 (2007)
36. Alatas, B.: ACROA: artificial chemical reaction optimization algorithm for global optimization. Expert Syst. Appl. **38**(10), 13170–13180 (2011)
37. Lam, A.S., Li, V.O.: Chemical-reaction-inspired metaheuristic for optimization. IEEE Trans. Evol. Comput. **14**(3), 381–399 (2010)
38. Findik, O.: Bull optimization algorithm based on genetic operators for continuous optimization problems. Turk. J. Electr. Eng. Comput. Sci. **23** (Sup 1), 2225–2239 (2015)
39. Yazdani, M., Jolai, F.: Lion optimization computational design and engineering **3**(1), 24–36 (2016)
40. Wang, G.-G., Deb, S., Coelho, L.S.: Elephant algorithm (LOA): a nature-inspired metaheuristic algorithm. J. Herding Optim, 2015 3rd International Symposium on Computational and Business Intelligence (ISCBI). IEEE (2015)
41. Tang, R., et al.: Wolf search algorithm with ephemeral memory. In: 2012 Seventh International Conference on Digital Information Management (ICDIM). IEEE (2012)
42. Jung, S.H.: Queen-bee evolution for genetic algorithms. Electron. Lett. **39**(6), 575–576 (2003)
43. Yang, X.-S.: Flower pollination algorithm for global optimization. Unconv. Comput. Nat. Comput. 240–249 (2012)
44. Yang, X.-S., Karamanoglu, M., He, X.: Multi objective flower algorithm for optimization. Proced. Comput. Sci. **18**, 861–868 (2013)
45. Hatamlou, A.: Black hole: a new heuristic optimization approach for data clustering. Inf. Sci. (2012)
46. Pham, D.T., Ghanbarzadeh, A., Koc, E., Otri, S., Rahim, S., Zaidi, M.: The bees algorithm-a novel tool for complex optimization problems. In: Proceedings of the 2nd Virtual International Conference on Intelligent Production Machines and Systems (IPROMS 2006), pp. 454–459 (2006)
47. Krishnanand, K.N., Ghose, D.: Detection of multiple source locations using a glowworm metaphor with applications to collective robotics. In: Swarm Intelligence Symposium, 2005. SIS 2005. Proceedings 2005 IEEE, pp. 84–91. IEEE (2005)
48. Krishnanand, K.N., Ghose, D.: Glowworm swarm optimisation: a new method for optimising multi modal functions. Int. J. Comput. Intell. Stud. **1**(1), 93–119 (2009)
49. Asyali, M.H., Colak, D., Demirkaya, O., Inan, M.S.: Gene expression profile classification: a review. Curr. Bioinform. **1**(1), 55–73 (2006)
50. Lin, P.L., Huang, P.W., Kuo, C.H., Lai, Y.H.: A size-insensitive integrity-based fuzzy c-means method for data clustering. Pattern Recognit. **47**(5), 2042–2056 (2014)
51. Nayak, J., Naik, B., Behera, H.S.: Fuzzy C-means (FCM) clustering algorithm: a decade review from 2000 to 2014. In: Computational Intelligence in Data Mining, vol. 2, pp. 133–149. Springer, New Delhi (2015)
52. Hartigan, J.A., Wong, M.A.: Algorithm AS 136: a K-means clustering algorithm. Appl. Stat. **28**(1), 100–108 (1979)
53. Jiang, Bo, Wang, Ning, Wang, Liping: Particle swarm optimization with age-group topology for multimodal functions and data clustering. Commun. Nonlinear Sci. Numer. Simul. **18**(11), 3134–3145 (2013)
54. Aljarah, I., Ludwig, S.A.: MapReduce intrusion detection system based on a particle swarm optimization clustering algorithm. 2013 IEEE Congress on Evolutionary Computation (CEC). IEEE (2013)
55. Silva Filho, T.M., et al.: Hybrid methods for fuzzy clustering based on fuzzy c-means and improved particle swarm optimization. Expert Syst. Appl. **42**(17–18) 6315–6328 (2015)
56. Alswaitti, M., Albughdadi, M., Isa, N.A.M.: Density-based particle swarm optimization algorithm for data clustering. Expert Syst. Appl. **91**, 170–186 (2018)
57. Alam, S., Dobbie, G., Riddle, P.: An evolutionary particle swarm optimization algorithm for data clustering. In: Swarm Intelligence Symposium, 2008. SIS 2008, IEEE. IEEE (2008)
58. Nanda, S.J., Panda, G.: Automatic clustering algorithm based on multi-objective immunized PSO to classify actions of 3D human models. Eng. Appl. Arti. Intell. **26**(5–6) 1429–1441 (2013)

59. Tsai, C.-W., et al.: A fast particle swarm optimization for clustering. Soft Comput. **19**(2), 321–338 (2015)
60. Azab, S.S., Hady, M.F.A., Hefny, H.A., Local best particle swarm optimization for partitioning data clustering. 2016 12th International on Computer Engineering Conference (ICENCO). IEEE (2016)
61. Wang, J., et al.: Evaluate clustering performance and computational efficiency for PSO based fuzzy clustering methods in processing big imbalanced data. In: 2017 IEEE International Conference on Communications (ICC). IEEE (2017)
62. Abualigah, L.M., Khader, A.T.: Unsupervised text feature selection technique based on hybrid particle swarm optimization algorithm with genetic operators for the text clustering. J. Supercomput. **73**(11), 4773–4795 (2017)
63. Karaboga, D., Ozturk, C.: A novel clustering approach: Artificial Bee Colony (ABC) algorithm. Appl. Soft Comput. **11**(1), 652–657 (2011)
64. Sahoo, G.: A two-step artificial bee colony algorithm for clustering. Neural Comput. Appl. **28**(3), 537–551 (2017)
65. Forsati, R., Keikha, A., Shamsfard, M.: An improved bee colony optimization algorithm with an application to document clustering. Neurocomputing **159**, 9–26 (2015)
66. Ozturk, C., Hancer, E., Karaboga, D.: Improved clustering criterion for image clustering with artificial bee colony algorithm. Pattern Anal. Appl. **18**(3), 587–599 (2015)
67. Goel, S., Sharma, A., Bedi, P.: Cuckoo search clustering algorithm: a novel strategy of biomimicry. 2011 World Congress on Information and Communication Technologies (WICT). IEEE (2011)
68. Garg, S., Batra, S.: Fuzzified cuckoo based clustering technique for network anomaly detection. Comput. Electr. Eng. (2017)
69. Boushaki, S.I., Kamel, N., Bendjeghaba, O.: A new quantum chaotic cuckoo search algorithm for data clustering. Expert Syst. Appl. **96**, 358–372 (2018)
70. Ye, S., et al.: K-means clustering algorithm based on improved Cuckoo search algorithm and its application. In: 2018 IEEE 3rd International Conference on Big Data Analysis (ICBDA). IEEE (2018)
71. İnkaya, T., Kayalıgil, S., Özdemirel, N.E.: Ant colony optimization based clustering methodology. Appl. Soft Comput. **28**, 301–311 (2015)
72. Zhang, L., Cao, Q.: A novel ant-based clustering algorithm using the kernel method. Inf. Sci. **181**(20), 4658–4672 (2011)
73. Chen, L., Xu, X.-H., Chen, Y.-X.: An adaptive ant colony clustering algorithm. In: 2004. Proceedings of 2004 International Conference on Machine Learning and Cybernetics, vol. 3. IEEE (2004)
74. Han, Y., Shi, P.: An improved ant colony algorithm for fuzzy clustering in image segmentation. Neurocomputing **70**(4-6), 665–671 (2007)
75. Izakian, H., Abraham, A., Snášel, V.: Fuzzy clustering using hybrid fuzzy c-means and fuzzy particle swarm optimization. 2009. NaBIC 2009. World Congress on Nature & Biologically Inspired Computing. IEEE (2009)
76. Marinakis, Y., Marinaki, M., Matsatsinis, N.: A stochastic nature inspired metaheuristic for clustering analysis. Int. J. Bus. Intell. Data Min. **3**(1), 30–44 (2008)
77. Wu, Fang-xiang: Genetic weighted k-means algorithm for clustering large-scale gene expression data. BMC Bioinform. **9**(6), S12 (2008)
78. Marinakis, Y., Marinaki, M., Matsatsinis, N.: A hybrid discrete artificial bee colony-GRASP algorithm for clustering. 2009. CIE 2009. International Conference on Computers & Industrial Engineering. IEEE (2009)
79. Huang, C.-L., et al.: Hybridization strategies for continuous ant colony optimization and particle swarm optimization applied to data clustering. Appl. Soft Comput. **13**(9), 3864–3872 (2013)
80. Song, W., et al.: A hybrid evolutionary computation approach with its application for optimizing text document clustering. Expert Syst. Appl. **42**(5), 2517–2524 (2015)

81. Handl, J., Meyer, B.: Ant-based and swarm-based clustering. Swarm Intell. **1**(2), 95–113 (2007)
82. Niknam, T., et al.: An efficient hybrid algorithm based on modified imperialist competitive algorithm and K-means for data clustering. Eng. Appl. Artif. Intell. **24**(2) 306–317 (2011)
83. Hassanzadeh, T., Meybodi, M.R.: A new hybrid approach for data clustering using firefly algorithm and K-means. 2012 16th CSI International Symposium on Artificial Intelligence and Signal Processing (AISP). IEEE (2012)
84. Prasad, D., Rajendra, P.V., Naganjaneyulu, Prasad, K.S.: A hybrid swarm optimization for energy efficient clustering in multi-hop wireless sensor network. Wirel. Pers. Commun. **94**(4), 2459–2471 (2017)
85. Kumar, A., Kumar, D., Jarial, S.: A novel hybrid K-means and artificial bee colony algorithm approach for data clustering. Decis. Sci. Lett. **7**(1), 65–76 (2018)
86. Nayak, J., et al.: An improved firefly fuzzy c-means (FAFCM) algorithm for clustering real world data sets. In: Advanced Computing, Networking and Informatics, vol. 1, pp. 339–348. Springer, Cham (2014)
87. de Barros Franco, D.G., Steiner, M.T.A.: Clustering of solar energy facilities using a hybrid fuzzy c-means algorithm initialized by metaheuristics. J. Clean. Product. **191**, 445–457 (2018)
88. He, Y., Hui, S.C., Sim, Y.: A novel ant-based clustering approach for document clustering. In: Asia Information Retrieval Symposium. Springer, Berlin, Heidelberg (2006)
89. Banharnsakun, A.: A MapReduce-based artificial bee colony for large-scale data clustering. Pattern Recognit. Lett. **93**, 78–84 (2017)
90. Bijari, K., et al.: Memory-enriched big bang–big crunch optimization algorithm for data clustering. Neural Comput. Appl. **29**(6), 111–121 (2018)
91. Roy, R., Anuradha, J.: A modified brainstorm optimization for clustering using hard c-means. 2015 IEEE International Conference on Research in Computational Intelligence and Communication Networks (ICRCICN). IEEE (2015)
92. Saida, I.B., Nadjet, K., Omar, B.: A new algorithm for data clustering based on cuckoo search optimization. In: Genetic and Evolutionary Computing, pp. 55–64. Springer, Cham (2014)
93. Cobos, C., et al.: Clustering of web search results based on the cuckoo search algorithm and balanced Bayesian information criterion. Inf. Sci. **281**, 248–264 (2014)
94. Soliman, O.S., Saleh, D.A.: Multi-objective c-means data clustering algorithm using self-adaptive differential evolution. (2015)
95. Krishnasamy, G., Kulkarni, A.J., Paramesran, R.: A hybrid approach for data clustering based on modified cohort intelligence and K-means. Expert Syst. Appl. **41**(13), 6009–6016 (2014)
96. Nayak, S.K., et al.: A modified differential evolution-based fuzzy multi-objective approach for clustering. Int. J. Manag. Decis. Mak. **16**(1), 24–49 (2017)
97. Abshouri, A.A., Bakhtiary, A.: A new clustering method based on firefly and KHM. J. Commun. Comput. **9**(4), 387–391 (2012)
98. Łukasik, S., et al.: Data clustering with grasshopper optimization algorithm. 2017 Federated Conference on Computer Science and Information Systems (FedCSIS). IEEE (2017)
99. Fathian, M., Amiri, B.: A honeybee-mating approach for cluster analysis. Int. J. Adv. Manuf. Technol. **38**(7-8), 809–821 (2008)
100. Fathian, M., Amiri, B., Maroosi, A.: Application of honey-bee mating optimization algorithm on clustering. Appl. Math. Comput. **190**(2), 1502–1513 (2007)
101. Hamou, R.M., Amine, A., Rahmani, M.: A new biomimetic approach based on social spiders for clustering of text. In: Software Engineering Research, Management and Applications 2012, pp. 17–30. Springer, Berlin, Heidelberg (2012)
102. Banumathy, D., Selvarajan, S.: Bacterial foraging optimized fuzzy c means clustering for efficient disease prediction. Res. J. Biotechnol. **12**, 280–288 (2017)
103. Serapião, A.B., et al:. Combining K-means and K-harmonic with fish school search algorithm for data clustering task on graphics processing units. Appl. Soft Comput. **41**, 290–304 (2016)

104. Fan, C., Zhang, T., Yang, Z., Wang, L.: A text clustering algorithm hybriding invasive weed optimization with K-means. In: 2015 IEEE 12th Intl Conf on Ubiquitous Intelligence and Computing and 2015 IEEE 12th International Conference on Autonomic and Trusted Computing and 2015 IEEE 15th International Conference on Scalable Computing and Communications and Its Associated Workshops (UIC-ATC-ScalCom), pp. 1333–1338. IEEE (2015)
105. Fan, S., Ding, S., Xue, Y.: Self-adaptive kernel K-means algorithm based on the shuffled frog leaping algorithm. Softw. Comput. **22**(3), 861–872 (2018)
106. Yadav, S., Nanda, S.J.: League championship algorithm for clustering. 2015 IEEE Power, Communication and Information Technology Conference (PCITC). IEEE (2015)
107. Wangchamhan, T., Chiewchanwattana, S., Sunat, K.: Efficient algorithms based on the k-means and chaotic league championship algorithm for numeric, categorical, and mixed-type data clustering. Expert Syst. Appl. **90**, 146–167 (2017)
108. Das, P., Das, D.K., Dey, S.: A new class topper optimization algorithm with an application to data clustering. In: IEEE Transactions on Emerging Topics in Computing (2018)
109. Chander, S., Vijaya, P., Dhyani, P.: Multi kernel and dynamic fractional lion optimization algorithm for data clustering. Alex. Eng. J. **57**(1), 267–276 (2018)
110. Nayak, J., et al.: Hybrid chemical reaction based metaheuristic with fuzzy c-means algorithm for optimal cluster analysis. Expert Syst. Appl. **79**, 282–295 (2017)
111. Nayak, J., Naik, B., Behera, H.S.: Cluster analysis using firefly-based k-means algorithm: a combined approach. In: Computational Intelligence in Data Mining, pp. 55–64. Springer, Singapore (2017)
112. Kanungo, D.P., et al.: Hybrid clustering using elitist teaching learning-based optimization: an improved hybrid approach of TLBO. Int. J. Rough Sets Data Anal. (IJRSDA) **3**(1), 1–19 (2016)
113. Nayak, J., et al.: A hybrid elicit teaching learning based optimization with fuzzy c-means (ETLBO-FCM) algorithm for data clustering. Ain Shams Eng. J. (2016)
114. Wang, R., et al.: Flower pollination algorithm with bee pollinator for cluster analysis. Inf. Process. Lett. **116**(1), 1–14 (2016)
115. Rafi, M., et al.: Solving document clustering problem through meta heuristic algorithm: black hole. In: Proceedings of the 2nd International Conference on Machine Learning and Soft Computing. ACM (2018)
116. Wolpert, D.H., Macready, W.G.: No free lunch theorems for optimization. IEEE Trans. Evol. Comput. **1**(1), 67–82 (1997)

Short Term Solar Energy Forecasting by Using Fuzzy Logic and ANFIS

Meera Viswavandya, Bakdevi Sarangi, Sthitapragyan Mohanty and Asit Mohanty

Abstract Accurate forecasting of solar energy is a key issue for a meaningful integration of the solar power plants into the grid. Solar photovoltaic technology is most preferable and vital technology in comparison with all other sources of renewable energy. We know that the solar energy is very irregular so the output of solar photo voltaic system (SPV) is diverted by the atmospheric conditions like temperatures, humidity, wind velocity, solar irradiance and other climatological facts. It's necessary to predict solar energy to minimize the uncertainty in power harnessing from solar photovoltaic system. In this work fuzzy logic model and ANFIS model have been developed for manipulating solar irradiation (w/m^2) data to forecast short duration solar irradiation. Further the normalization in between the input and output is set in between 0.1 and 0.9 for reducing confluence problems. Acquired results are matched up to the manipulated data and valid results are found out.

Keywords Solar energy · Solar photovoltaic system · Fuzzy logic · ANFIS

M. Viswavandya · B. Sarangi · A. Mohanty (✉)
Department of Electrical Engineering, CET Bhubaneswar, Bhubaneswar 751003, Odisha, India
e-mail: asithimansu@gmail.com

M. Viswavandya
e-mail: mviswavandya@cet.edu.in

B. Sarangi
e-mail: bakdevisarang@gmail.com

S. Mohanty
Department of Computer Science and Engineering, CET Bhubaneswar, Bhubaneswar 751003, Odisha, India
e-mail: msthitapragyan@gmail.com

H. S. Behera et al. (eds.), *Computational Intelligence in Data Mining*, Advances in Intelligent Systems and Computing 990,
https://doi.org/10.1007/978-981-13-8676-3_63

1 Introduction

The solar radiation is an essential parameter for solar energy research but due to its uncertainty it becomes difficult for the measurement. Also there are less solar radiation measuring equipment at the meteorological stations. Therefore, it is necessary to forecast solar radiation for a particular location using several climatic parameters. This parameter are sunshine duration, wet-bulb temperature, relative humidity, wind velocity/speed, daily clear sky global radiation etc. The sunshine duration, maximum and minimum temperatures are easily available and measured at most of the location so it is generally used for modeling of solar radiation. The information about solar radiation, solar energy system models, specific site data and publications are given in the inventory prepared by Myers for NREL. There are many Researchers who have developed experimental models for solar power forecasting. The solar energy forecasting is developed to be precise with measured data. It is estimated with mean absolute percentage error (MAPE) i.e. MAPEr10% means high forecasting accuracy, 10%rMAPEr20% means good forecasting, 20%rMAPEr50% means reasonable forecasting. Now a days because of the increasing prices of crude oil,the exploitation of renewable energy sources with its applications is a best alternative. The energy from solar is among the most suitable energy technique due to its less emission of CO_2 and eco-friendly Nature. In the other hand, concentration of huge amount of solar energy for the generation of electricity storage faces several difficulty levels for the power system's handlers because of the variability of solar radiation fall in surface. So, the energy generated from the SPV system is changed with the solar radiation and the temperature and relative humidity, wind speed unexpected changes of the SPV system power harness may raise price of operating for the electricity storage system due to increased operation and management price based with cycling existing generation. The Optimum location selection for harness of electricity from the solar energy is determined based on the available solar irradiance data at that location. The Solar irradiance data are important for design, sizing, operation and economic assessment of solar energy. However, In India, only IMD canter gives authenticates data for quite few stations or users which is taken as the raw data for research purposes.

Experimenting and Forecasting output gain by solar is an essential part. It is the fact that final output solar power is irregular in behavior. So, that to integrate this final output of generation with the grid or to utilize distribution generation (DG), proper energy management system is needed. Now days several number application can be utilized for short term solar energy prediction such as physical methods, statistical methods and artificial intelligence methods among this three group of method artificial intelligence methods are widely used. Artificial intelligence technique like artificial neural network, genetic algorithm, fuzzy logic, ANFIS, etc. Fuzzy logic is advantageous over artificial neural network because it can implement despite changes in the temperature condition. Further it is easy and tough in nature. ANFIS (Adaptive Neuro-fuzzy Inference System) is advantages over ANN because it is used both Neural Network and fuzzy logic techniques to gain more efficiency and for estimates solar energy get the better of the difficulty Neural Network like as a huge number of neurons and layer required for difficulty function approximation.

2 Basic Fuzzy Logic

Fuzzy logic- It has been introduced in the year 1965 by Professor Lotfy A. Zadeh at the University of California. L.A. Zadeh main objective was to develop the model that could helps to read out the natural language process. Generally it replace the multi-valued logic to the binary 0/1 logic. Due to the Fuzzy approach is quite different from classical approaches. The fuzzy logic model used the situation where, the probabilistic or deterministic data model do not suitable for phenomenon of the realistic description in the study. In Fuzzy logic instead of subjects and verbs, fuzzy operators and fuzzy sets are used to make the meaningful the sentences it is characterize by the condition statements i.e., IF-THEN statements and it contains certain rules. Also, fuzzy logic has been used as a standard method in various applications such as modeling solar radiation at the earth surface. In fuzzy logic includes several steps for basic configuration like Fuzzification, Rule base, Decision-making logic and Defuzzification.

2.1 *Fuzzification*

It helps to measure the input variables values. Fuzzification plays a major role means its function of fuzzification that helps to converts input into crisp values that means transform scale mapping the variables input range of values into respective universe of discourse.

2.2 *Rule Base*

It is also known as knowledge base or the combination of data base and linguistic control rule base, The rule base provides necessary definition that could be used as to define fuzzy data, FLC, linguistic control rules and calculation.

2.3 *Rule Decision Making Logic*

Decision making logic also refer as the Kernel of an fuzzy logic controller and it can also able to read of simulating human decision to concepts of fuzzy. It has inferring fuzzy control actions involving fuzzy implication and the rules of inference in fuzzy logic.

2.4 Defuzzification

Defuzzification of fuzzy value means a scale mapping which converts the range of values of input variables to crisp output.

Let us take K be universal set. Consider the characteristic function μ_Z of a subset of universal set Z. Then take its values in the two element set {0, 1}. So, $\mu_Z(\mathrm{k}) = 1$, if k ∈ Z and zero otherwise. The rangc of fuzzy set Z values take as interval {0,1}. Here,μ_Z is known as membership function and $\mu_Z(\mathrm{k})$ is called as grade membership function of k ∈ K in Z. The conversion between membership and non membership is gentle rather than sudden.

The intersection and union of two fuzzy subsets Y and Z, K having membership function $\mu_{Y,}\mu_Z$ respectively then it can be expressed as

$$\text{Intersection}: \ \mu_{Y\cap Z}(\mathrm{k}) = \min\left[\mu_Y(k),\ \mu_Z(\mathrm{k})\right] \tag{1}$$

$$\text{Union}: \ \mu_{Y\cup Z}(\mathrm{k}) = \max\left[\mu_Y(k),\ \mu_Z(\mathrm{k})\right] \tag{2}$$

Fuzzy logic approaches, the input variables value can smoothly mapped to an output space by follows the vague concepts like as fast runner, hot weather, solar irradiation, wind speed, relative humidity etc.

3 Basic of ANFIS

ANFIS (Adaptive Neuro-Fuzzy Inference system) techniques gain more efficiency due to it is the combination of ANN and fuzzy logic. The TS structure of fuzzy model is specified using a method which allows the optimal structure on automatic manner. ANFIS is more effective than Fuzzy Inference System (FIS), but in this users have some drawbacks like only first order/zero order Sugeno type fuzzy models, OR Method: max, Aggregation Method: max, AND Method: Prod, Implication Method: Prod, Defuzzification Method: wtaver(weighted average). Also, in ANFIS users can provide to only give their own number of membership functions (num MFs) both for input and outputs of the fuzzy controller, the membership function type, the number of checking and training data sets (numPts), reduces the error measure by optimization criterion. ANFIS model was first proposed by Jang in 1993, it is based on a special FIS that means linear combination of all input variables to make the output variables and the Takagi-sugeno model. In this model the input series are changed to fuzzy input with the help of membership functions for each individual input series. The structure of membership function is mainly based on the data set. ANFIS has been approached with several forecasting domains such as weather forecasting, solar radiation prediction, internet traffic time series prediction and electricity price prediction. In this present work the all data are divided in three parts such as training data, checking data and testing data. In training process 50% data is used, for testing process 25% data used and finally for checking data process rest of 25% used. By the help of back propagation algorithm trained for Training data.

In this present work we applied ANFIS modeling the back propagation algorithm the Takagi-sugeno fuzzy system is used for solar power forecasting estimating at Bhubaneswar with available atmospheric parameters of maximum and minimum temperature, relative humidity, wind speed, and total sunshine hour and the results are compared with the measured value calculated by using fuzzy logic. The basic structure of ANFIS models contain several layer like the fuzzification layer, the product layer, the normalized layer, the defuzzification layer, and the output layer. Now days implemented of ANFIS architecture is a popular and advanced system with a good software support. In ANFIS Each and every node has been characterized by node function with fixed or flexible parameters. By taking of first order fuzzy inference system, a fuzzy reasoning has been developed. In ANFIS model neural network work on following fuzzy IF-THEN rules of Takagi and Sugeno type which describe as two rules.

1. If y is K1 and z is L1, then p1 = a1x + b1y + c1
2. If y is K2 and z is L2, then p2 = a2x + b2y + c2

where y and z are inputs, p1 and p2 are outputs and K1, L1, K2 and L2 are fuzzy sets.

4 Data Collection and Normalisation

In IMD (India Meteorological Department) we collected data for short term solar energy prediction for 1 h ahead. The data sheet for solar irradiance of month September has been used. So, we can predict short term solar energy for 1 h ahead in this Thesis. The Tabular representation of September data is presented in below.

Here, we take four no. of input variables (solar irradiance in W/m^2) i.e. 8 a.m., 9 a.m., 10 a.m. and 11 a.m., and one output i.e. 12 Noon data is presented in Table 1. Again Normalized the above data both input and output values in between the range 0.1–0.9.

$$L_s = \frac{Y_{max} - Y_{min}}{L_{max} - L_{min}}(L - L_{min}) + Y_{min}$$

where L = actual measured data, L_s = scaled data, L_{max} = particular data of maximum value, L_{min} = particular data of minimum value, Y_{max} & Y_{min} are having upper limit (0.9) and lower limit (0.1)

E.g. - Normalized 2nd October 12 noon data

$$\begin{aligned} \text{Ls} &= \{(0.9 - 0.1) \div (811 - 248.5)\} \times (631 - 248.5) + 0.1 \\ &= (0.8 \div 562.5) \times 382.5 + 0.1 = 0.646.\ \text{(ans)} \end{aligned}$$

Like this way we complete all the data in Table 2.

Table 1 Collected data For 1 h ahead

Sl. No	8 a.m. (Input1)	9 a.m. (Input2)	10 a.m. (Input3)	11 a.m. (Input4)	12 noon (Output)
1	449.69	574.47	698.84	759.56	811
2	454.36	605.01	756.24	606	631
3	387.98	573.81	715.9	763.78	737.7
4	435.19	605.01	752.14	773.9	776.8
5	419.77	544.66	623.4	735.93	757.3
6	394.99	522.81	697.2	670.96	749.8
7	394.99	544	683.26	352.87	696.5
8	378.63	476.42	691.46	710.62	698.6
9	453.89	618.2	767.34	779.81	769.8
10	473.81	650.52	771.01	770.53	762.3
11	444.54	570.5	689	694.59	723.7
12	444.08	576.46	666.04	693.75	693
13	436.13	547.31	657.84	343.59	384.1
14	100.01	120.01	115	120	248.5
15	398.73	537.37	661.12	657.46	635.9
16	450.62	551.95	683.26	723.28	705.3
17	431.45	518.82	671.78	687	677.9
18	396.39	548.63	696.38	714.84	720.6
19	424.91	559.9	673.42	682.78	658.4
20	381.9	484.3	621.76	620.34	640.2
21	368.34	528.7	612.74	634.68	586.2
22	334.68	471.12	602.9	637.21	600
23	342.16	540.31	630.78	584.06	585.3
24	318.32	450.58	572.56	573.93	561.8
25	280.45	378.37	521.72	522.46	512
26	319.72	478.66	584.04	600.09	578.1
27	311.77	436.67	605.36	624.56	616.5
28	318.32	438.05	597.98	598.4	576.3
29	254.27	494.31	142.88	405	482.6
30	273.91	388.97	425.78	612.09	565.6
31	245.86	398.25	499.58	441.46	405.3

Table 2 Normalization of input and output data for one hour ahead data

Day	Input 1 (8 a.m.)	Input 2 (9 a.m.)	Input 3 (10 a.m.)	Input 4 (11 a.m.)	Output (12 noon)
1	0.848	0.786	0.812	0.858	0.898
2	0.858	0.808	0.882	0.676	0.646
3	0.716	0.785	0.845	0.863	0.796
4	0.817	0.808	0.877	0.875	0.85
5	0.784	0.741	0.72	0.83	0.823
6	0.817	0.708	0.81	0.753	0.803
7	0.731	0.74	0.793	0.376	0.732
8	0.696	0.638	0.803	0.8	0.741
9	0.857	0.852	0.887	0.882	0.841
10	0.877	0.854	0.896	0.871	0.83
11	0.837	0.78	800	0.781	0.776
12	0.836	0.789	0.772	0.78	0.733
13	0.819	0.745	0.762	0.365	0.381
14	0.1	0.1	0.1	0.1	0.1
15	0.739	0.73	0.766	0.737	0.653
16	0.85	0.752	0.793	0.815	0.75
17	0.809	0.702	0.779	0.772	0.712
18	0.734	0.747	0.809	0.805	0.772
19	0.795	0.764	0.718	0.767	0.685
20	0.703	0.65	0.718	0.693	0.559
21	0.674	0.717	0.707	0.71	0.584
22	0.602	0.63	0.695	0.713	0.603
23	0.618	0.665	0.729	0.65	0.583
24	0.567	0.599	0.658	0.638	0.549
25	0.486	0.49	0.596	0.577	0.48
26	0.57	0.581	0.672	0.669	0.572
27	0.553	0.578	0.698	0.698	0.626
28	0.567	0.58	0.689	0.667	0.561
29	0.43	0.665	0.134	0.417	0.498
30	0.472	0.506	0.479	0.683	0.55
31	0.412	0.52	0.569	0.481	0.458

5 Model Development

5.1 *Fuzzy Logic Model*

The short term solar energy forecasting for 1 h ahead by fuzzy logic model is developed and presented. The fuzzy linguistic variables are described as high, medium and low. Here, the main concern for the development of fuzzy systems is the appropriate membership functions. Generation of membership function depends upon intuition, experience or probabilistic methods and further it have been defined such as six membership functions like L (low), L1 (Extreme low), M (Medium), M1 (Medium High), H (High) and H1 (extreme High) all values lies in between the ranges 0.1–0.9. In fuzzy model, rules may be fired with some degree using fuzzy inference but in conventional expert systems, a rule is either fired or not fired. For the short term solar energy prediction problem, certain set of rules are described to determine the accuracy in terms of Absolute Relative Error (ARE). Such rules are expressed in the below following form.

IF premise (antecedent), THEN conclusion (consequent).

For the short term solar energy prediction purposed are followed by certain sets of multiple antecedent fuzzy rules. The input to the rule is solar energy during 8 a.m., 9 a.m., 10 a.m., and 11 a.m., of September, 2017 and output is 12 noon.

5.2 *Adaptive Neuro Fuzzy Inference System*

Generally the hybrid approach converges at a faster rate as it minimizes the overall dimension of search area of the original back propagation method. This network fixes the membership function and help in adapting the consequent part. Further ANFIS can be judged as a functional linked network and the enhanced representation takes advantage of human knowledge and expresses more insight. After the fine tunning of membership function, the enhanced representation of the system is done. By fine-tuning the membership functions, we actually make this enhanced representation.

Data set has been prepared through collecting solar data from IMD centre Bhubaneswar. Complete data set for the month of September has been prepared with the solar radiation data at different times such as (8 a.m., 9 a.m., 10 a.m., 11 a.m,). This may be used for prediction of solar radiation at 12 o clock noon. Further triangular membership function has been used for input as well as linear type of function. Further corrected outputs are noticed till errors are substantially reduced.

6 Result and Discussion in Fuzzy and ANFIS

6.1 Fuzzy Result

In Table 3 presents various set of rules for solar forecasting fuzzy rules. A general 1 h ahead Short Term solar energy predicting base fuzzy logic model is generated by using MATLAB is shown in Fig. 1.

Table 3 To develop solar predicting model MATLAB is used to establishment the below listed rules

Day	Input1 (8 a.m.)	Input2 (9 a.m.)	Input3 (10 a.m.)	Input4 (11 a.m.)	Output (12 noon)
1	H1	H	H	H1	H1
2	H1	H	H1	M1	M1
3	M1	H	H1	H1	H
4	H	H	H1	H1	H1
5	H	H	H	H1	H1
6	H	H	H	H	H
7	H	H	H	L	H
8	H	M1	H	H	H
9	H	H1	H1	H1	H1
10	H1	H1	H1	H1	H
11	H	H	H	H	H
12	H	H	H	H	H
13	H	H	H	L	L1
14	L1	L1	L1	L1	M1
15	H	H	H	H	H
16	H1	H	H	H1	H
17	H	M1	H	H	M1
18	H	H	H	H	H
19	H	H	H	H	M1
20	M1	M1	M1	M1	M1
21	M1	M1	M1	M1	M
22	M	M	M1	M1	M
23	M1	M1	H	M1	M
24	M	M	M1	M1	M
25	M	M	M	M	M
26	M	M	M1	M1	M

(continued)

Table 3 (continued)

Day	Input1 (8 a.m.)	Input2 (9 a.m.)	Input3 (10 a.m.)	Input4 (11 a.m.)	Output (12 noon)
27	M	M	M1	M1	M1
28	M	M	M1	M1	M
29	L	M1	L1	L	L
30	M	M	M	M1	M
31	L	M	M	M1	L

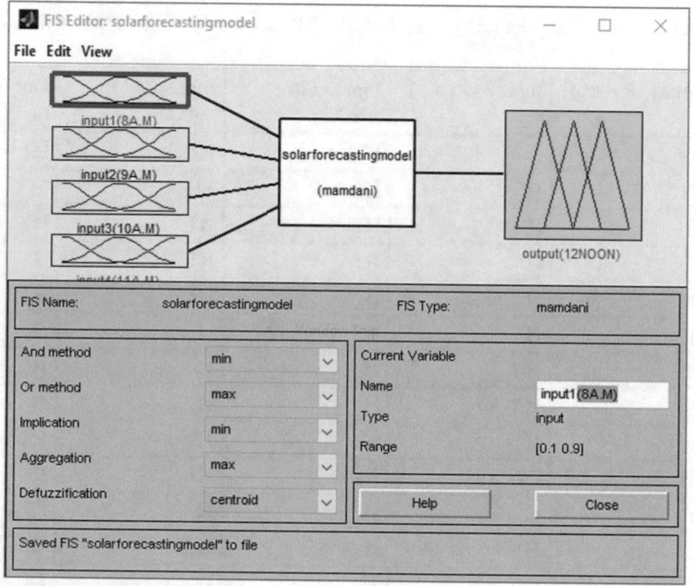

Fig. 1 Solar forecasting model by fuzzy

The 3rd rule results are shown in below Fig. 2. The results generated from the developed model are then defuzzified to get the predicted output in W/m^2.

6.2 *ANFIS Result*

The short term solar energy forecasting by an ANFIS model is generated by using MATLAB 12.0 software version is shown in below Fig. 3. The main proposed to use an ANFIS network has adapted the training data sets to form best membership function so as to gets the desired output for testing data with less epochs.

Figure 3 shows the architecture of the proposed model. The input membership function has been described with Gaussian membership function. With the help of

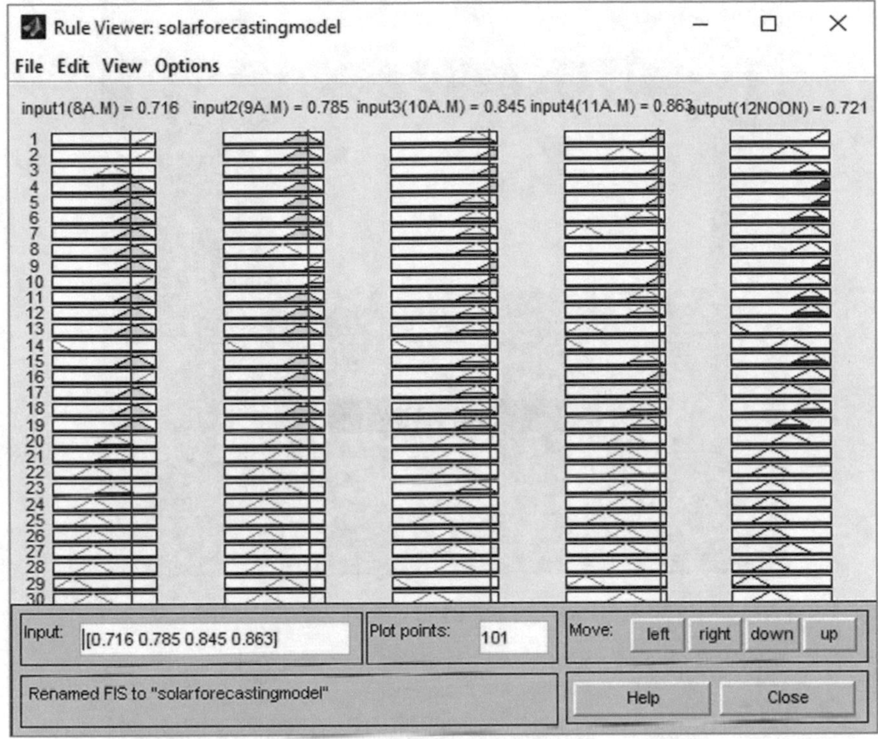

Fig. 2 Results generated develop fuzzy model

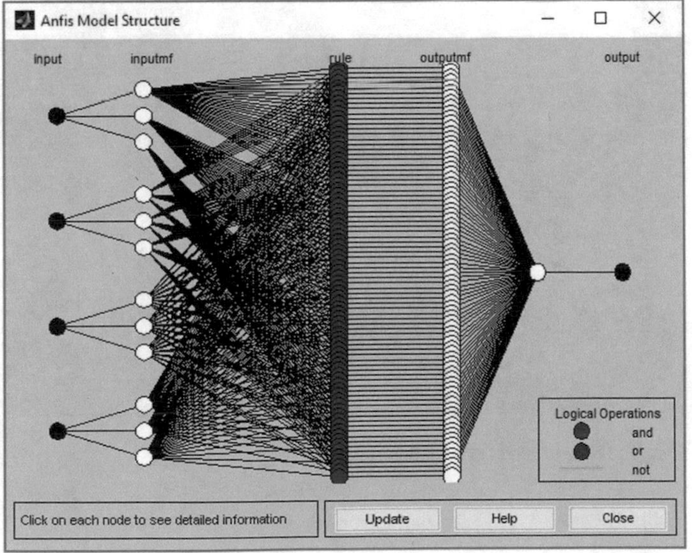

Fig. 3 ANFIS Structure

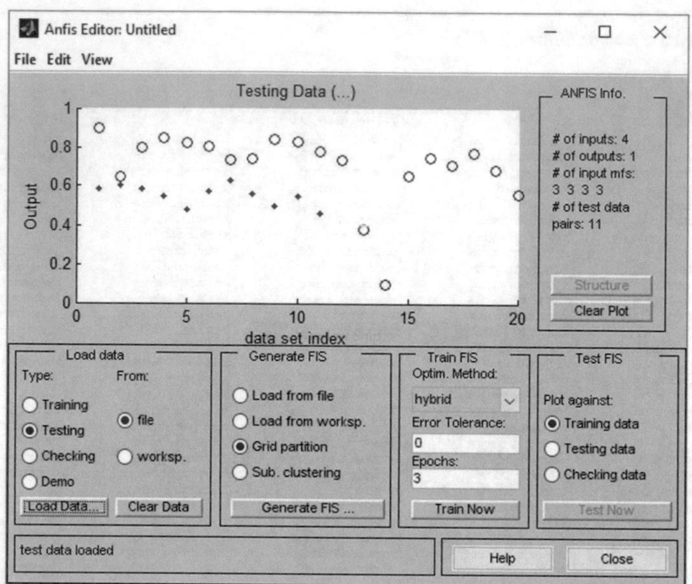

Fig. 4 Training data of 31 days for propose work

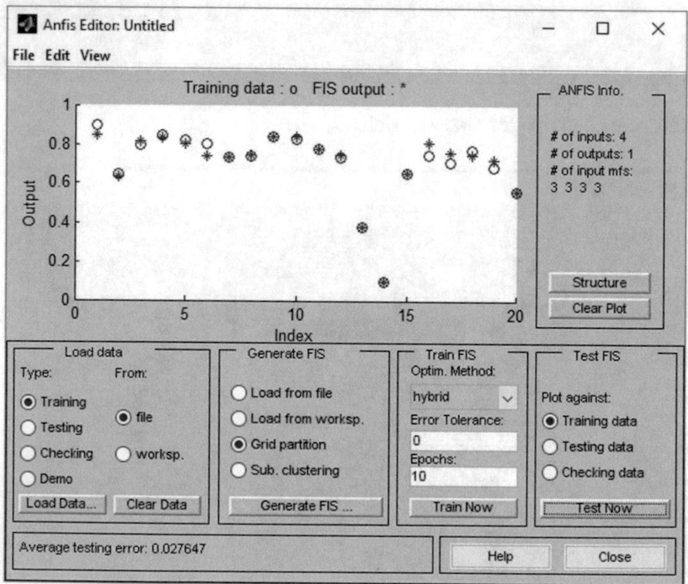

Fig. 5 Training and testing data for propose work

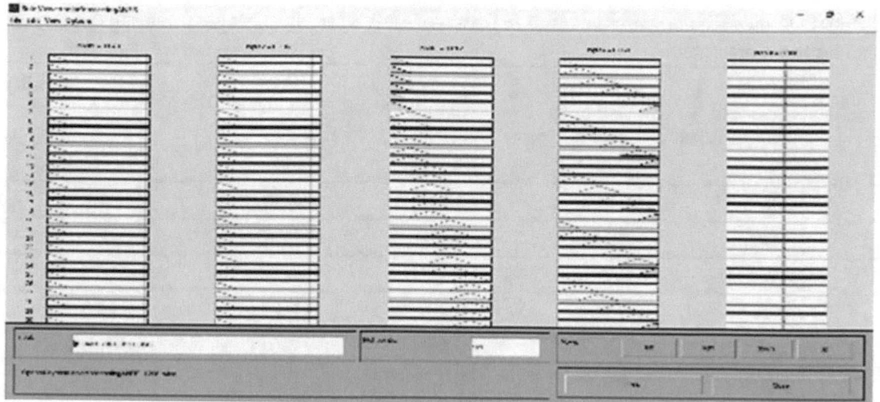

Fig. 6 Rule viewer corresponding to1st rule by ANFIS

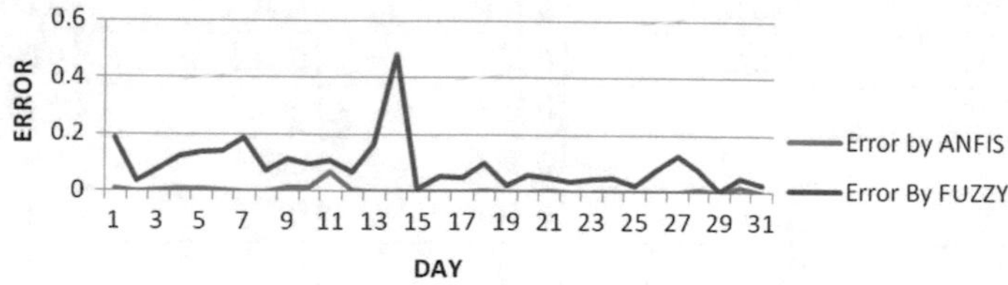

Fig. 7 Error comparison between Fuzzy and ANFIS

hybrid learning algorithm ANFIS model is tested till error gets reduced. The actual pattern of variation related to actual and predicted responses is shown for training and testing dataset for the proposed model. Figures such as (Figs. 4, 5, 6) show that the actual (blue dot) and predicted (red dot) values have been uniformly distributed for the purpose of training and testing. Comparative results in between ANFIS and Fuzzy logic are shown in Table 4 and Fig. 7.

7 Conclusion

In solar power applications of solar energy forecasting plays a vital issue because solar power is fluctuating in nature and it depends on several meteorological parameters. So, considering the above fact and Keeping in view the aforesaid, general fuzzy based model and ANFIS model are generated for short term solar energy predicted. The proposed ANFIS has effectively forecasted the global solar radiation and then becomes utilize preferably for any design of conversion solar energy application. The ANFIS model shows better results in comparison with other models. The evaluation results of solar radiation shows a significant improvement in statistical parameters and depicts better accuracy than other models. The comparative results

Table 4 Shown the comparative result in between ANFIS and fuzzy logic forecasting value and also error are presented

Day	Actual measured output (W/M^2)	Forecasted output by ANFIS (W/M^2)	Forecasted output by Fuzzy (W/M^2)	Error by ANFIS	Error by fuzzy
1	0.898	0.888	0.711	0.01	0.187
2	0.646	0.644	0.614	0.002	0.032
3	0.796	0.79	0.721	0.006	0.075
4	0.85	0.861	0.727	0.011	0.123
5	0.823	0.812	0.688	0.011	0.135
6	0.803	0.8	0.661	0.003	0.142
7	0.732	0.732	0.546	0	0.186
8	0.741	0.74	0.671	0.001	0.07
9	0.841	0.826	0.728	0.015	0.113
10	0.83	0.842	0.736	0.012	0.094
11	0.776	0.709	0.668	0.067	0.108
12	0.733	0.727	0.668	0.006	0.065
13	0.381	0.381	0.545	0	0.164
14	0.1	0.1	0.58	0	0.48
15	0.653	0.652	0.659	0.001	0.006
16	0.75	0.752	0.696	0.002	0.054
17	0.712	0.714	0.664	0.002	0.048
18	0.772	0.778	0.674	0.006	0.098
19	0.685	0.685	0.663	0	0.022
20	0.559	0.558	0.616	0.001	0.057
21	0.584	0.581	0.631	0.003	0.047
22	0.603	0.603	0.569	0	0.034
23	0.583	0.581	0.54	0.002	0.043
24	0.549	0.549	0.5	0	0.049
25	0.48	0.48	0.5	0	0.02
26	0.572	0.571	0.498	0	0.047
27	0.626	0.624	0.5	0	0.002
28	0.561	0.551	0.487	0.01	0.074
29	0.498	0.498	0.5	0	0.002
30	0.55	0.53	0.5	0.02	0.05
31	0.458	0.458	0.485	0	0.027

deduce the forecasting ability of Adaptive-Neuro fuzzy inference system model and its compatibility for any location with different atmospheric conditions.

References

1. Ramedani, Z., Omid, M., Keyhani, A.: Modeling solar energy potential in a Tehran Province using artificial neural networks. Int. J. Green Energy **10**(4), 427–441 (2013)
2. Rahoma, W.A., Rahoma, U.A., Hassan, A.H.: Application of neuro-fuzzy techniques for solar radiation (2011)
3. Iqdour, R., Zeroual, A.: A rule based fuzzy model for the prediction of solar radiation. Revue des Energies Renouv. **9**(2), 113–120 (2006)
4. Sumithira, T.R., Kumar, A.N., Kumar, R.R.: An adaptive neuro-fuzzy inference system (*ANFIS*) based prediction of solar radiation: a case study. J. Appl. Sci. Res. **8**(1), 346–351 (2012)
5. Yadav, A.K., Chandel, S.S.: Solar energy potential assessment of Western Himalayan Indian state of Himachal Pradesh using J48 algorithm of WEKA in *ANN* based prediction model. Renew. Energy **75**, 675–693 (2015)
6. Mellit, A., Kalogirou, S.A., Shaari, S., Salhi, H., Arab, A.H.: Methodology for predicting sequences of mean monthly clearness index and daily solar radiation data in remote areas: application for sizing a stand-alone PV system. Renew. Energy **33**(7), 1570–1590 (2008)
7. Bhardwaj, S., Sharma, V., Srivastava, S., Sastry, O.S., Bandyopadhyay, B., Chandel, S.S., Gupta, J.R.P.: Estimation of solar radiation using a combination of Hidden Markov Model and generalized Fuzzy model. Sol. Energy **93**, 43–54 (2013)
8. Mohammadi, K., Shamshirband, S., Petković, D., Khorasanizadeh, H.: Determining the most important variables for diffuse solar radiation prediction using adaptive neuro-fuzzy methodology; case study: City of Kerman. Iran. Renew. Sustain. Energy Rev. **53**, 1570–1579 (2016)
9. Mohanty, S., Patra, P.K., Sahoo, S.S.: Comparison and prediction of monthly average solar radiation data using soft computing approach for Eastern India. Computational Intelligence in Data Mining-Volume 3, pp. 317–326. Springer, New Delhi (2015)
10. Jafarkazemi, F., Moadel, M., Khademi, M., Razeghi, A.: Performance prediction of flat-plate solar collectors using MLP and *ANFIS*. J. Basic Appl. Sci. Res. **3**(2), 196–200 (2013)
11. Hawlader, M.N.A., Chou, S.K., Ullah, M.Z.: The performance of a solar assisted heat pump water heating system. Appl. Therm. Eng. **21**(10), 1049–1065 (2001)
12. Mohanty, S., Patra, P.K., Sahoo, S.S., Mohanty, A.: Forecasting of solar energy with application for a growing economy like India: survey and implication. Renew. Sustain. Energy Rev. **78**, 539–553 (2017)
13. Mellit, A., Arab, A.H., Shaari, S.: An *ANFIS*-based prediction for monthly clearness index and daily solar radiation: application for sizing of a stand-alone photovoltaic system. J. Phys. Sci. **18**(2), 15–35 (2007)
14. Mohanty, S.: ANFIS based prediction of monthly average global solar radiation over Bhubaneswar (State of Odisha). Int. J. Ethics Eng. Manag. Educ. ISSN **1**(5), 2348–4748 (2014)
15. Yadav, A.K., Chandel, S.S.: Solar radiation prediction using artificial neural network techniques: a review. Renew. Sustain. Energy Rev. **33**, 772–781 (2014)
16. Mohanty, S., Patra, P.K., Sahoo, S.S.: Prediction of global solar radiation using nonlinear auto regressive network with exogenous inputs (NARX). In: 2015 39th National Systems Conference (NSC), pp. 1–6. IEEE (2015)
17. Varzandeh, M.H.M., Rahbari, O., Vafaeipour, M., Raahemifar, K., Heidarzade, F.: Performance of wavelet neural network and ANFIS algorithms for short-term prediction of solar radiation and wind velocities. In: The 4th World Sustainability Forum (2014)

Empirical Study and Statistical Performance Analysis with ANN for Parkinson's Vowelized Data Set

T. PanduRanga Vital, Gorti Satyanarayana Murty, K. Yogiswara Rao and T. V. S. Sriram

Abstract In recent years, there is an importance for the analysis of ageing neurological diseases like Parkinson's disease (PD) and Alzheimer's. PD occupies the second place in Neurodegenerative disease where Alzheimer's disease is in a fist. In this research, we collected the voice records in vowel sounds (/a, /e, /i, /o, /u) from Andhra Pradesh, India during 2016–2018 with 1200 records of 46 PD patients and 38 non-PD people between the age of 56–84. The voice data is normalized with FFT algorithm and removes the noise of the voice and construct the CSV file for analysis and predicting the PD with ANN (Artificial Neural Network). The good statistical voice parameter results give detailed information about PD and its identification. Artificial Neural Networks classifies the PD dataset with 100% training accuracy with 10 hidden neurons. In this, we have also observed accuracy of the ANN for this PD dataset 6 to 10 hidden neurons. At 10 hidden neurons, the data set performance is in peak that accuracy is 100 and R value is 1 with least iterations 264 compared to others (hidden neurons 6, 7, 8 and 9 took 1000 and above 1000 epochs and accuracy is less than 99%) within the time of 16 s for prediction of PD.

Keywords Parkinson's disease · Vowelized dataset · ANN · Statistical analysis · FFT

T. PanduRanga Vital (✉) · G. S. Murty · K. Yogiswara Rao
Department of CSE, AITAM Engineering College, Tekkali, Srikakulam District 532201, Andhra Pradesh, India
e-mail: vital2927@gmail.com
URL: https://www.adityatekkali.edu.in

G. S. Murty
e-mail: gsn_73@yahoo.co.in

K. Yogiswara Rao
e-mail: yogiindusisu@gmail.com

T. V. S. Sriram
NSRIT Engineering College, Vishakhapatnam District, Andhra Pradesh, India
e-mail: ramjeesis@gmail.com

H. S. Behera et al. (eds.), *Computational Intelligence in Data Mining*, Advances in Intelligent Systems and Computing 990,
https://doi.org/10.1007/978-981-13-8676-3_64

1 Introduction

Parkinson's disease was identified in 1817 by Dr. James Parkinson. Parkinson's disease is caused by the weakening of nerve cells. In addition to it if the chemical called Dopamine produced by neurons decreases, chances of getting PD is more. PD occupies the second place in Neurodegenerative disease where Alzheimer's disease is in fist [1] with regard the studies in the USA, the number of individuals in the United States with Parkinson Disease and patient's data reaches from 600,000 to 1,600,000 with 60,000 to 70,000 new cases reporting every year [2]. Since there is no specific test for Parkinson's onset miss-conclusion rate can be high, particularly when an expert who doesn't consistently work with this sickness makes the wrong conclusion [3]. It is highly difficult to judge this disease at first glance as Parkinson's disease is mostly seen in people of age 60 found more in America.

Parkinson's disease is the second most basic neurodegenerative effect and determines 3% of the public beyond 60 years. PD and related neurodegenerative issue speak to a developing weight on the medicinal services framework. In the larger part of the cases, the reason for the malady is still obscure and its clarification states one of the significant difficulties of the neurosciences. Advancement in uncommon hereditary types of Parkinson's sickness have permitted the improvement of novel creature models giving a premise to a superior comprehension of the sub molecule pathogenesis of the infection setting the phase for the improvement of novel treatment methodologies.

2 Literature Survey

A high-quality count of research papers is described on various Data Mining (DM) and Neural Networks (NN) models and its application on Parkinson's disease (PD) in classification, forecasting, and prediction in latest existence on PD. Also outstanding accomplishments are made to improve the performance of these DM models by using various classification techniques. This section presents various earlier works accounted on PD Data mining (PDDM) and NN models in classification, prediction, and forecasting.

The valuable effect of levodopa on motor damage in PD has been gathered throughout, its effect on speech has infrequently been studied and the particular scripting or writing is ambiguous. The aim of Skodda et al. [4] research was to determine impact of short-run levodopa affirmation and long-run dopaminergic treatment on speech in PD patients in beginning stages of the PD. Motor examination as indicated by UPDRS III and discourse testing were performed in 23 PD patients Vocalization assessment involved the perceptual rating of worldwide speech or vocalization performance and an acoustical observation based on a standardized reading task. Midi I et al. [5] used twelve male and eight female total twenty patients with PD were given voice tests and results were contrasted and those of 20 sex and age-matched controls.

Rusz et al. [6] aim were to evaluate the impact of treatment start on the movement of vocalizations destruction in PD, utilizing novel assessment criteria. Nineteen patients with PD were tried and retried inside 2 years after the preamble speech of antiparkinsonian treatment. As controls, 19 age-coordinated people were recorded. Vocalization examination included supported phonation; prompt syllable reiteration, text content reading, and monolog.

Unilateral VCP (vocal cord paralysis) influences the target properties of accent voice by reduced neurologic control and chronic varies in laryngeal tissue. The objective of this examination was to think about the VCP patients acoustic parameters with healthy people. [7] Acoustic investigation consequences of 18 independent VCP patients were likened and age and sex coordinated 72 healthy adult volunteers. Examination of acoustic examination consequences of female and male VCP patients with their sex and age-coordinated control clusters disclosed statistically huge difference in shimmer, jitter, and noise harmonics ratio esteems in the two groups where the ratio P is less than 0.01($P < 0.01$). Oguz et al. [8] assumed that these distinctions were as per the outcomes acquired by monetarily accessible voice assessment programs.

Speech and voice perturbs are a standout amongst the most critical issues after STN-DBS (subthalamic nucleus deep brain stimulation) in PD patients; be that as it may, their attributes stay unclear. The STN-DBS cluster, particularly females, presented persistent change of voice parameters and altogether poorer VHI scores than the restorative treatment alone gathering. The DUV (degree of voiceless) and stressed voice were the most anosmatic factors in the STN-DBS cluster or a group. In this, Tsuboi et al. [9] exhibited that more widespread or boundless voice weakness or destruction in females and poorer voice-related QOL, more terrible DUV and stressed voice, and atypical laryngeal muscle constriction were laryngeal and typical voice discoveries in the STN-DBS contrasted and those in the medicinal treatment alone group.

Brabenec et al. [10] have characterized Hypokinetic dysarthria in PD by unvariedness of pitch loudness and pitch, variable rate, reduced stress, a harsh and breathy voice, and imprecise consonants. The authors studied the effect of the DLPFC that left dorsolateral prefrontal cortex on aspects of motor voiced vocalizations and the effect of high-frequency rTMS that repetitive transcranial magnetic stimulation) which is employed over the main orofacial SM1 (sensorimotor part in PD. In their speech study 21 healthy men with mean age of 64 ± 8.55 years and 12 un-demented and undepressed men with mean PD duration of 10.75 ± 7.48 years and with mean age of 64.58 ± 8.04 years.

Benba et al. [11] conducted a study to differentiate three sets of patients: those troubled with PD, those troubled with MSA (Multiple system atrophy) and those who troubled with different neurological diseases. They collected nine voice recordings from nine patients under each group. Pompili, Anna, et al. (2017) discriminated potentials of eight verbal assignment tasks intended to catch the significant symptoms of influencing speech. Additionally, they introduced, they present novel database of Lusitanian speakers comprising of 75 PD subjects and 65 healthy subjects. For each assignment, a programmed classifier is assembled utilizing modeling approaches and feature sets in consistence with the present stage. El Moudden et al. [12] suggested PCA-based modeling on the source of class prediction utilizing C5.0 and LR (logistic regression) in numeric data is researched in openly accessible PD dataset Silverman voice treatment to develop a novel classification technique. Since around 90% of the general population with PD bearing with speech problems including respiratory, troubles of laryngeal and articulate capacity, utilizing voice examination of disease can be analyzed remotely at a beginning stage with greater unwavering quality and in a financial way.

Janghel et al. [13] applied nine soft computing techniques that are LR (Linear Regression) ANN, kernel SVM, RF, Cubist Committees, Cubist, Adaptive Neuro-Fuzzy Inference System, and Hybrid Neuro-Fuzzy Inference System for PD voice analysis and gait characteristics. Afterward, their performances are assessed with measures of performance through specificity, sensitivity, and RMSE. Aich Satyabrata et al. [14] have proposed an approach of comparing the performance metrics with distinct aspect data sets such as original feature sets along with PCA based or constructed feature reduction technique for selecting the feature sets.

3 Proposed Method

3.1 PD Predicting Model

Figure 1 represents Model of Predicting PD with ANN algorithm. As per model, the voice data was gathered from both 46 PD and 38 non-PD people ailing patients who crossed 56 years of age with 1200 voice records during the year of 2016–18. We collected this voice data in different days and different times of a day from Andhra Pradesh, India in all available sources. For this, high-quality equipment (microphones and etc.) is used for the recording and processing the voice. The voice is collected as maintain the distance 8–12 cm along with mouth and microphone from the people. The voice data collection is differentiated as a vowel sounds (/a/e/i/o/u) and stored in the disk with the wav Format. The noise or un-normalized voice data is preprocessed with FFT algorithm.

The result is made based on the comparison of collected voice data by analyzing the attribute values like F, M and I (Frequency, Modulation, and Impendence as main attributes and 26 additional sub-attributes were calculated from them). During data preprocessing, the attributes were scientifically analyzed and decided the target class by using a classification technique ANN. In this step, the authors analyzed the parameters of voices that are jitter, shimmer, harmonics, pitch, pulse, and voicing values. In this process, the authors get the 26 characteristics of voice stored in the data base as TEXT (.csv) format for predicting the PD with using ANN algorithm. The PD non-PD data set contain 29 attributes that are 26 features of voice, age of the person, sex of the person and sound category (5 vowel sounds) and one deciding factor class (0 for non-PD and 1 for PD).

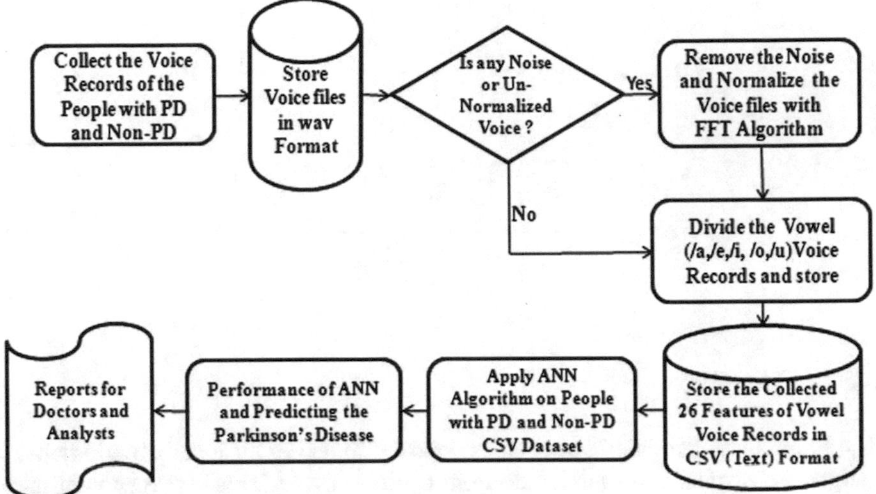

Fig. 1 Model for predicting Parkinson's disease with ANN

3.2 *Voice Parameters*

3.2.1 Jitter

Jitter can be estimated in various parameters that are Absolute, Relative, Relative average, perturbation (rap), and Period perturbation quotient (ppq5).

Jitter absolute is a difference of fundamental frequency (cycle-to-cycle). This represents the Eq. (1).

$$jitter = \frac{1}{N_p - 1} \sum_{l=1}^{N_p} |T_l - T_{l-1}| \tag{1}$$

In the Eq. (1), T_l is the estimated voice period lengths and N_p is the quantity of estimated voice periods. Relative or local Jitter is the standard absolute variation between continuous periods, separated by the average time period. It is showed in a percentage. Equations (2)–(4) are represented as Jitter relative, Jitter rap and Jitter ppq5.

$$Jitter_{\mathrm{Relative}} = \frac{\frac{1}{N_p-1} \sum_{l=1}^{N_p} |T_l - T_{l-1}|}{\frac{1}{N_p} \sum_{l=1}^{N_p} T_l} \times 100 \tag{2}$$

$$Jitter_{rap} = \frac{\frac{1}{N_p-1}\sum_{l=1}^{N_p-1}\left|T_l - \left(\frac{1}{3}\sum_{m=l-1}^{l+1} T_m\right)\right|}{\frac{1}{N_p}\sum_{l=1}^{N_p} T_l} \times 100 \tag{3}$$

$$Jitter_{ppq5} = \frac{\frac{1}{N_p-1}\sum_{l-2}^{N_p-2}\left|T_l - \left(\frac{1}{5}\sum_{m=l-2}^{l+2} T_m\right)\right|}{\frac{1}{N_p}\sum_{l=1}^{N_p} T_l} \times 100 \tag{4}$$

3.2.2 Shimmer

The Shimmer is like jitter, however, as opposed to taking a gander at periodicity, it gauges the distinction in amplitude from cycle to cycle. This is a very vital measure in speech pathology, as obsessive voices will frequently have a higher shine than healthy voices. Shimmer (local) is the absolute average distinction between the amplitudes of consecutive periods, separated by the amplitude average in % (percentage). Shimmer (local, dB) represents the peak-to-peak amplitude variability in decibels. The shimmer parameters like shimmer (dB), shimmer (Relative), shimmer (apq3) and shimmer (apq5) are calculated with using the Eqs. (5)–(8) relatively.

$$Shimmer_{dB} = \frac{1}{N_p - 1}\sum_{l=1}^{N_p-1}\left|20 \times \log\left(\frac{A_{l-1}}{A_l}\right)\right| \tag{5}$$

$$Shimmer_{\mathrm{Relative}} = \frac{\frac{1}{N_p-1}\sum_{l=1}^{N_p-1}|A_l - A_{l+1}|}{\frac{1}{N_p}\sum_{l=1}^{N_p} A_l} \times 100 \tag{6}$$

$$Shimmer_{apq3} = \frac{\frac{1}{N_p-1}\sum_{l=1}^{N_p-1}\left|A_l - \left(\frac{1}{3}\sum_{m=l-1}^{l+1} A_m\right)\right|}{\frac{1}{N_p}\sum_{l=1}^{N_p} A_l} \times 100 \tag{7}$$

$$Shimmer_{ppq5} = \frac{\frac{1}{N_p-1}\sum_{l=2}^{N_p-2}\left|A_l - \left(\frac{1}{5}\sum_{m=l-2}^{l+2} A_m\right)\right|}{\frac{1}{N_p}\sum_{l=1}^{N_p} A_l} \times 100 \tag{8}$$

3.2.3 Harmonic to Noise Ratio

Noise ratio of Harmonic gives a sign of the general periodicity of the voice motion by evaluating the ratio connecting the harmonic part and noise segments (shown in Eq. (9)). The general estimation of the HNR of the signal changes on the grounds that distinct vocal tract arrangements include distinctive amplitudes for the sounds.

$$Harmonic_{NoiseRatio} = 10 \times \log_{10} \frac{V_{AC}(T)}{V_{AC}(0) - V_{AC}(T)} \quad (9)$$

V_{AC} (0) is the origin autocorrelation coefficient representing the signals energy.
V_{AC} (T) is the element of the autocorrelation related to the fundamental period.

4 Results and Discussions

4.1 Statistical Analysis PD and Non-PD

Figures 2 and 3 show PD and Non-PD pulses. Figure 2 shows the Parkinson's Diseased voice pulse. It shows breaking the voice in PD between pulse to pulse of the voice. As per statistical analysis, there are many differences between PD and Non-PD voice parameters.

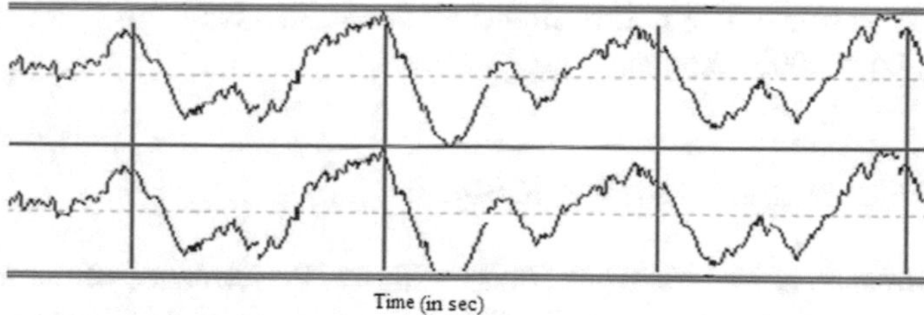

Fig. 2 PD voice pulses

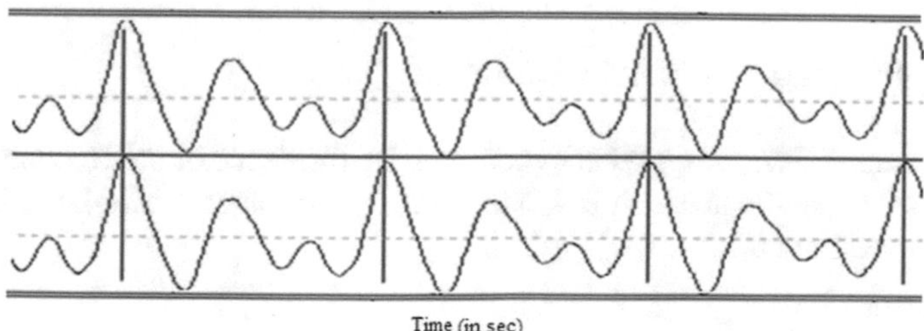

Fig. 3 Non-PD voice pulses

Table 1 shows the statistical analysis both PD and Non-PD in detail. The voice parameters like jitter, shimmer, harmonics, pitch, pulse, and voicing values show the PD and non-PD. These parameters hidden values are related to the PD voice fundamental frequency, jitter, and shimmer.

4.2 ANN Analysis PD and Non-PD

4.2.1 Hidden Neurons 6 ANN Evolutions

The Best performance is 0.02356 at 1000 iterations. Performance R value is 0.95042 It takes the time 0.17 s. The Gradient value is 3.41e-05 and Mu value is 1.00e-05.

4.2.2 Hidden Neurons 7 ANN Evolution

Best performance is 0.98725 at 1000 iterations Performance of the R value is 0.98725. It takes time 0.19 s. The Gradient value is 2.463e-05 at epoch 1000 as well Mu value is 1e-05 at 1000 iteration.

4.2.3 Hidden Neurons 8 ANN Evolution

Best performance is 0.0004125 at 1000 iterations Performance of the R value is 0.99915. It takes time 0.22 s. The Gradient value is 3.7989e-07 at epoch 1000. Mu value is 1e-05 at 1000 iteration.

4.2.4 Hidden Neurons 9 ANN Evolution

Best performance is 2.4746e-04 at 1000 iterations. Performance of the R value is 0.99949. It takes time 0.27 s. The Gradient value is 4.5903e-05 at epoch 1000 and Mu value is 1e-10.

4.2.5 Hidden Neurons 10 ANN Evolution

Best performance is 4.5394e-09 at epoch 264. Performance of the R value is 1. The Gradient is 9.6594e-08 at epoch 264. Mu value is 1e-10 at epoch 264. It takes 0.16 s for process the Data Set.

Table 1 Statistical analysis between PD and Non-PD people

Attributes	PD voice statistical analysis				Non-PD voice statistical analysis			
	Min	Max	Mean	Median	Min	Max	Mean	Median
Age	56.0	84.0	68.6 ± 8.33	66.0	56.00	83.00	66.90 ± 8.04	65.00
Median pitch	78.87	283.50	171.01 ± 41.96	171.45	99.9	448.60	227.36 ± 61.16	217.22
Mean pitch	84.38	283.92	173.22 ± 39.42	171.63	103.1	419.20	231.47 ± 57.13	218.27
Standard deviation	0.15	130.71	23.39 ± 25.98	15.16	1.22	170.61	35.94 ± 31.75	23.53
Minimum pitch	66.25	281.62	138.79 ± 42.0	139.49	49.77	320.98	177.15 ± 53.52	176.55
Maximum pitch	96.99	627.03	230.54 ± 99.63	209.3	108.8	526.42	294.06 ± 99.74	252.68
Number of pulses	6.00	1704.0	75.57 ± 138.95	51.0	7.00	138.00	62.90 ± 24.70	59.00
Number of periods	5.0	1694.0	73.58 ± 136.92	49.0	6.00	134.00	61.39 ± 24.30	58.00
Mean period	0.00	0.01	0.01 ± 0.00	0.01	0.00	0.01	0.00 ± 0.00	0.00
Unvoiced frames fraction	0.00	72.32	3.96 ± 10.02	0.81	0.00	67.14	10.97 ± 16.07	2.44
Number of voice breaks	0.00	21.0	0.49 ± 2.02	0.00	0.00	3.00	0.09 ± 0.31	0.00
Degree of voice breaks	0.00	73.76	2.79 ± 9.99	0.00	0.00	19.80	0.67 ± 2.49	0.00
Jitter (local)	0.00	13.28	1.87 ± 1.49	1.62	0.16	7.43	1.72 ± 1.16	1.50
Jitter (local, absolute)	0.00	0.04	0.00 ± 0.00	0.00	0.00	0.11	0.00 ± 0.01	0.00
Jitter (rap)	0.03	6.73	0.83 ± 0.78	0.69	0.04	4.59	0.76 ± 0.63	0.61
Jitter (ppq5)	0.04	8.60	0.93 ± 0.91	0.73	0.07	3.88	0.81 ± 0.68	0.62
Jitter (ddp)	0.09	899.0	3.88 ± 33.94	2.07	0.13	13.75	2.28 ± 1.92	13.75

(continued)

Table 1 (continued)

Attributes	PD voice statistical analysis				Non-PD voice statistical analysis			
	Min	Max	Mean	Median	Min	Max	Mean	Median
Shimmer (local)	1.01	29.38	8.29 ± 3.96	7.79	0.87	13.17	2.28 ± 1.92	1.82
Shimmer (local, dB)	0.09	420	1.42 ± 15.84	0.80	0.13	2.28	0.90 ± 0.38	0.85
Shimmer (apq3)	0.24	14.59	3.62 ± 2.22	3.29	0.39	18.83	3.38 ± 2.29	2.83
Shimmer (apq5)	0.34	17.50	4.85 ± 2.74	4.30	0.91	21.70	4.81 ± 2.89	4.05
Shimmer (apq11)	0.52	29.35	7.74 ± 4.44	6.94	1.30	42.44	8.92 ± 5.78	7.70
Shimmer (dda)	0.71	43.77	10.99 ± 6.67	9.86	1.17	71.49	10.19 ± 7.09	8.51
Mean autocorrelation	0.04	1.00	0.90 ± 0.08	0.92	0.62	1.0	0.90 ± 0.66	0.91
Mean noise-to-harmonics	0.00	16.76	0.17 ± 0.64	0.10	0.00	0.97	0.14 ± 0.12	0.12
Mean harmonics-to-noise	2.83	30.19	14.26	5.33	0.23	29.19	13.99 ± 4.96	13.33

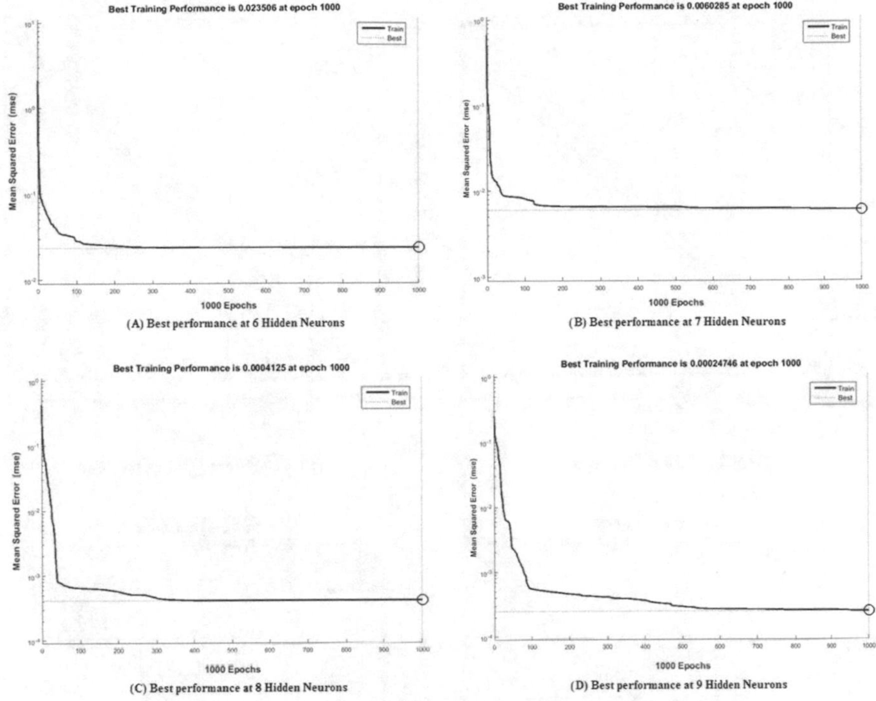

Fig. 4 Performance of the ANN on PD and non-PD dataset with 6–9 hidden neurons

Figure 4 shows the Performance of the ANN on PD and non-PD Dataset with 6–9 hidden neurons. In this, Fig. 4a shows the six Hidden neurons best performance is 0.02356. Figure 4b shows the seven hidden neurons, best performance is 0.98725. Figure 4c shows the 8 hidden neurons, best performance is 0.0004125. Figure 4d shows the 9 hidden neurons, best performance is 2.4746e-04.

The Fig. 5 shows R values of ANN and fitness of PD and non-PD Dataset with 6–9 hidden neurons. In this, Fig. 5a shows the 6 hidden neurons, fitness of the dataset and R value is 0.95042. Figure 5b shows the 7 hidden neurons, R value is 0.98725. Figure 5c shows the 8 hidden neurons, R value is 0.99915. Figure 5d shows the 9 hidden neurons, R value is 0.99949.

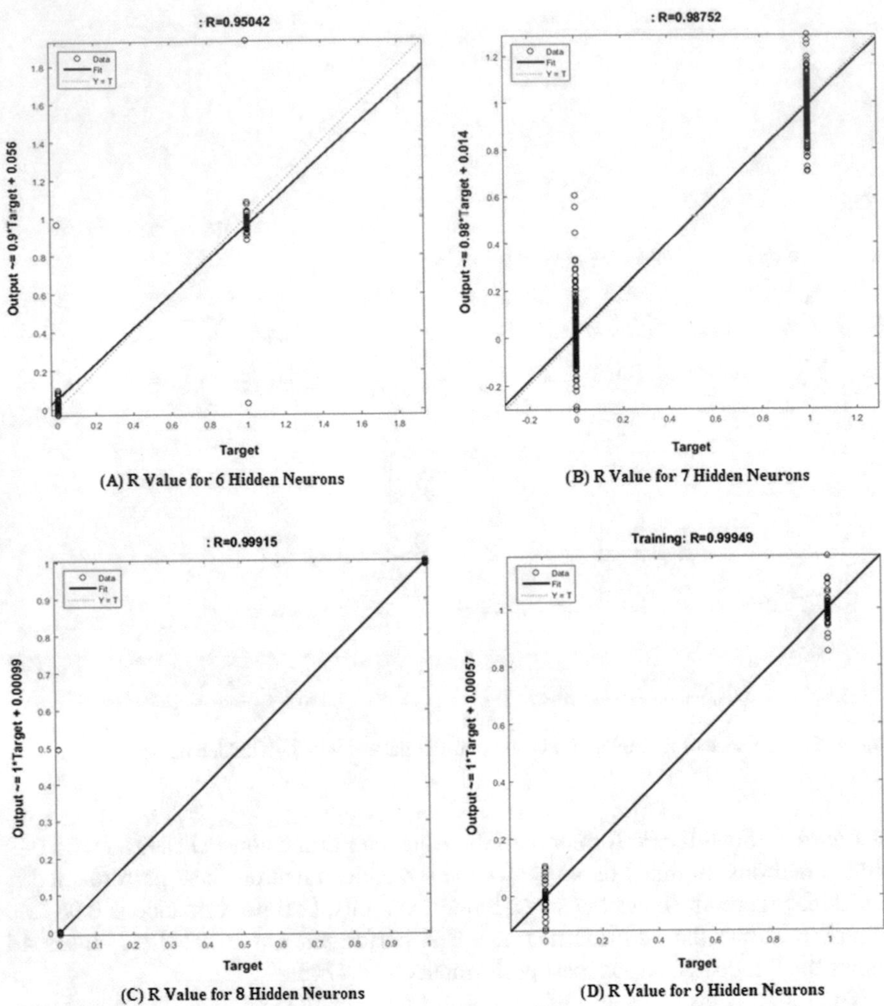

(A) R Value for 6 Hidden Neurons

(B) R Value for 7 Hidden Neurons

(C) R Value for 8 Hidden Neurons

(D) R Value for 9 Hidden Neurons

Fig. 5 R values of ANN algorithm on PD dataset with 6–9 hidden neurons

Figure 6 shows the Performance of the ANN on PD and non-PD Dataset with 10 hidden neurons. In this, Fig. 6a shows the 10 Hidden neuron best performance is 9.6594e-08. Figure 6b shows the 10 hidden neurons, R value is 1. Figure 6c shows the 10 hidden neurons Errors Histogram, the most of data points lie on the error 6.82e-06. Figure 6d shows the 10 hidden neurons, gradient (9.6594e-08) and mu(1e-10) values.

Table 2 shows the performance analysis of the ANN algorithm at each 6, 7, 8, 9 and 10 hidden neurons.

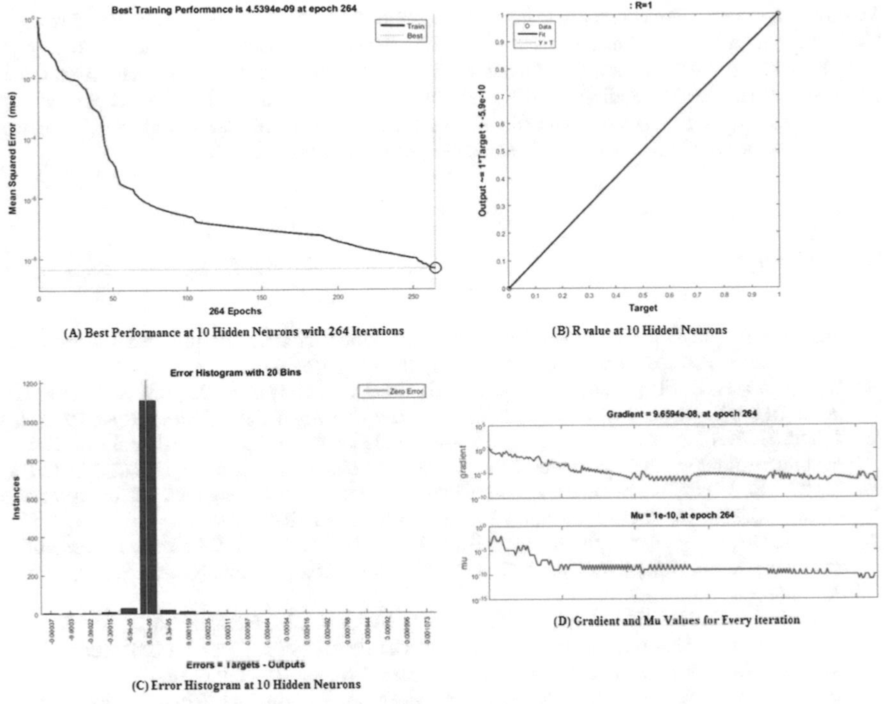

Fig. 6 Performance of the ANN on PD and non-PD dataset with 10 hidden neurons

Table 2 Performance analysis of ANN 6–10 hidden neurons

Hidden neurons	Number of iterations	Time taken in seconds	Best performance	Regression value (R)	Gradient value	Mu value
6	1000	0.17	0.02356	0.95042	3.41e-05	1e-05
7	1000	0.19	0.98725	0.98725	2.463e-05	1e-05
8	1000	0.22	0.0004125	0.99915	3.7989e-07	1e-05
9	1000	0.27	2.4746e-04	0.99949	4.5903e-05	1e-10
10	264	0.16	4.5394e-09	1	9.6594e-08	1e-10

5 Conclusion

Parkinson's disease is the second most basic neurodegenerative effect and determines 3% of the public beyond 60 years. PD and related neurodegenerative issue speak to a developing weight on the medicinal services framework. This research work is very useful for early and simply detect the PD with less cost and less time utilizing the voice records. As well, the collected data set is very accurate. The statistical analysis and ANN with 10 hidden neuron methods are very useful to predict Parkinson's disease with 100% accuracy.

Acknowledgements We would like to thank the Director Professor V. V. Nageswara Rao, Principal Dr. K. B. Madhu Sahu, and management of AITAM College for encouraging and supporting us. We are thanking to PD and non-PD patients for cooperation to collect the voice data from them. The data that support the findings of this study are available in various hospitals in the state of Andhra Pradesh, India. A standard ethical committee has approved this data set and the dataset has no conflict of interest/ethical issues with any other public source or domain.

References

1. Hardy, J., Selkoe, D.J.: The amyloid hypothesis of Alzheimer's disease: progress and problems on the road to therapeutics. Science **297**(5580), 353–356 (2002)
2. Li, F., Harmer, P., Fitzgerald, K., Eckstrom, E., Stock, R., Galver, J., Batya, S.S.: Tai chi and postural stability in patients with Parkinson's disease. N. Engl. J. Med. **366**(6), 511–519 (2012)
3. Talbot, K., Wang, H.Y.: The nature, significance, and glucagon-like peptide-1 analog treatment of brain insulin resistance in Alzheimer's disease. Alzheimer's Dement. **10**(1), S12–S25 (2014)
4. Skodda, S., Visser, W., Schlegel, U.: Short-and long-term dopaminergic effects on dysarthria in early Parkinson's disease. J. Neural Transm. **117**(2), 197–205 (2010)
5. Midi, I., Dogan, M., Koseoglu, M., Can, G., Sehitoglu, M.A., Gunal, D.I.: Voice abnormalities and their relation with motor dysfunction in Parkinson's disease. Acta Neurol. Scand. **117**(1), 26–34 (2008)
6. Rusz, J., Čmejla, R., Růžičková, H., Klempíř, J., Majerová, V., Picmausová, J., Růžička, E.: Evaluation of speech impairment in early stages of Parkinson's disease: a prospective study with the role of pharmacotherapy. J. Neural Transm. **120**(2), 319–329 (2013)
7. Oguz, H., Demirci, M., Safak, M.A., Arslan, N., Islam, A., Kargin, S.: Effects of unilateral vocal cord paralysis on objective voice measures obtained by Praat. Eur. Arch. Oto–Rhino–Laryngology **264**(3), 257–261 (2007)
8. Tsuboi, T., Watanabe, H., Tanaka, Y., Ohdake, R., Hattori, M., Kawabata, K., Maesawa, S.: Early detection of speech and voice disorders in Parkinson's disease patients treated with subthalamic nucleus deep brain stimulation: a 1-year follow-up study. J. Neural Transm. **124**(12), 1547–1556 (2017)
9. Brabenec, L., Mekyska, J., Galaz, Z., Rektorova, I.: Speech disorders in Parkinson's disease: early diagnostics and effects of medication and brain stimulation. J. Neural Transm. **124**(3), 303–334 (2017)
10. Braga, D., Madureira, A.M., Coelho, L., Abraham, A.: Neurodegenerative diseases detection through voice analysis. In: International Conference on Health Information Science, pp. 213–223. Springer, Cham (2017)
11. Pompili, A. et al.: Automatic detection of parkinson's disease: an experimental analysis of common speech production tasks used for diagnosis. In: International Conference on Text, Speech, and Dialogue. Springer, Cham (2017)
12. El Moudden, I., Ouzir, M., El Bernoussi, S.: Feature selection and extraction for class prediction in dysphonia measures analysis: a case study on Parkinson's disease speech rehabilitation. Technol. Health Care **25**(4), 693–708 (2017)
13. Janghel, R.R., Shukla, A., Rathore, C.P., Verma, K., Rathore, S.: A comparison of soft computing models for Parkinson's disease diagnosis using voice and gait features. Netw. Model. Anal. Health Inform. Bioinform. **6**(1), 14 (2017)
14. Aich, S., Younga, K., Hui, K.L., Al-Absi, A.A., Sain, M.: A nonlinear decision tree based classification approach to predict the Parkinson's disease using different feature sets of voice data. In: 2018 20th International Conference on Advanced Communication Technology (ICACT), pp. 638–642. IEEE (2018)

Alternative Assessment of Performance of Students Through Data Mining

Parag Bhalchandra, Aniket Muley, Mahesh Joshi, Santosh Khamitkar, Sakharam Lokhande and Rajesh Walse

Abstract This paper is a summary of extensive analytics implemented over personally collected through educational aspect dataset to get insights for their academic analytics. The main focus is on educational data mining. Basically, students progress is directly associated with their obtained marks. Apart from their studious approach, it is possible to have some other perspective effects on their performance. It is known that individuality, lifestyle and responsiveness related variables have close association together which harshly affect student's performance. The multivariate analysis of variance statistical technique is used to escalate the same and the analytics were carried out with SPSS software package. The obtained results highlighted that personal his/her habitat details affect on their performance.

Keywords Canonical correlation · Data mining · Education · Multivariate analysis · Performance

P. Bhalchandra (✉) · S. Khamitkar · S. Lokhande · R. Walse
School of Computational Sciences, S.R.T.M. University, Nanded, MS, India
e-mail: srtmun.parag@gmail.com

S. Khamitkar
e-mail: s.khamitkar@gmail.com

S. Lokhande
e-mail: lokhande_sana@rediffmail.com

R. Walse
e-mail: rajeshwalse@gmail.com

A. Muley (✉)
School of Mathematical Sciences, S.R.T.M. University, Nanded, MS, India
e-mail: aniket.muley@gmail.com

M. Joshi
School of Educational Sciences, S.R.T.M. University, Nanded, MS, India
e-mail: maheshmj25@gmail.com

H. S. Behera et al. (eds.), *Computational Intelligence in Data Mining*, Advances in Intelligent Systems and Computing 990,
https://doi.org/10.1007/978-981-13-8676-3_65

1 Introduction

Data mining (DM) is a new horizon in the applied research part of computer sciences and IT where significant valuable insights can be obtained from the databases. DM used for educational data related with academics, students, pedagogy, content delivery, etc. and is called as Educational DM [1, 7, 11]. Despite the percentage of GDP for the year 2013–14 was 1.34 percent [13], there is no proportionate gain in the overall performance of students. This is a research problem that fits the domain of data mining to see what lies beneath? We together thought that there could be other aspects associated with the performance of the students. Since these other aspects are not visible directly, this is a case for data mining. The performance of student is a fuzzy term and numerous variables are associated with it. We cannot study these all variables and their permutations. Factor analysis is a rescue for us which enable us to select some important variables at a glance [9]. This work is an illustration of associative effort made by three different university departments. After analyzing contemporary works [1, 4, 5, 10], we understood that discovery of some important variables or data items is the first need of time [14, 17, 18]. That's why early attempts of this study were to escalate collected datasets through data mining. Investigate contribution of serious character in the student's study. If no, what is other missing part of the story? Below sections briefs experimental setup and discussions related to such performance analysis.

2 Research Methodology

Initially, to collect data according to achieve our objective of performance of students with the help of structured and systematic closed questionnaire. In this study, the 360 student's information is collected with the help of 46 designed structured questions in the questionnaire. The analytics algorithms were worked out using SPSS 22.0v software. The overall intention and hypothesis are,

Objective: To scrutinize the connection between students' individual particulars and his/her own behaviour as well as wakefulness.
Hypothesis: Significant link between individual information and his/her own way of life as well as alertness?

The primary data source is questionnaires being circulated to students. Then data is collected and database was created. Other known sources like academic section, admission section were also consulted. Since it is a primary data source, authenticity is very high. Standard benchmark of Pritchard and Wilson was followed for devising out correct data in questionnaire [19]. Trial and modify approach was followed and tested over a selected group of students to increase readability and utility of questionnaire. Extract-Load and Transfer methodology [7] was adopted to clean and transform data. Spreadsheet was used to create records and fields of the database. A questionnaire consists of numerous questions related to some aspect of students. This question becomes single variable in our study. We cannot study all these variables at a time. Hence we decided very specific to study the connection among student's personal details and their personal habits. The data mining algorithms were worked out using SPSS data mining platform. The study used Canonical correlation developed by Hotelling [12] to discover two sets of basis vectors. We need to create coordinate system of dependent variables for the explanation of variables. Optimality of eigenvectors [11, 12] is always considered during doing so. The chief distinction between canonical correlation and regular correlation analysis is understandable on the context of the basis on which the variables are explained. That's why it is very famous in contemporary works [2, 3, 6, 8, 15, 16].

3 Proposed Methods with Pseudo Code

Canonical correlation investigation is used to discover and determine the links amongst set of two variables. This is recommended in the alike situations raising the possibility of multiple regressions. There must be several inter-correlated effective variables present in the overall scope. These create sets for testing canonical variates and orthogonal linear combinations for further clarification amongst the considered two sets. In this study, SPSS22.0v performs canonical correlation with the *manova* rule. While attempting for deployment of Canonical correlation in present study, we created two groups/sets. There are three variables in first set and the remaining set or group has six variables. This arrangement gave us possibility for 1:3 correspondences between variable of each group and each group three columns for next group. These results have shown significance at the 0.05 level. Figure 1 demonstrates the flowchart of our study. The pseudo-code used for analyzing the data in SPSS is represented in Sect. 3.1. The criteria for classification were based on the fact that the considered variables show associations and not causal relationship is evaluated through Eq. (1) in the Sect. 3.2.

3.1 Program Code

The SPSS program code is as follows:

```
DESCRIPTIVES VARIABLES=GENDER MARRIED AGE UGPER
FSUPPORT
FAILURES SCHOLERSHIP PJOB CAREERDREM PER_SATISF
/STATISTICS=MEAN STDDEV MIN MAX.
MANOVA AGE UGPER with FSUPPORT FAILURES SCHOLERSHIP
PJOB
CAREERDREM PER_SATISF
```

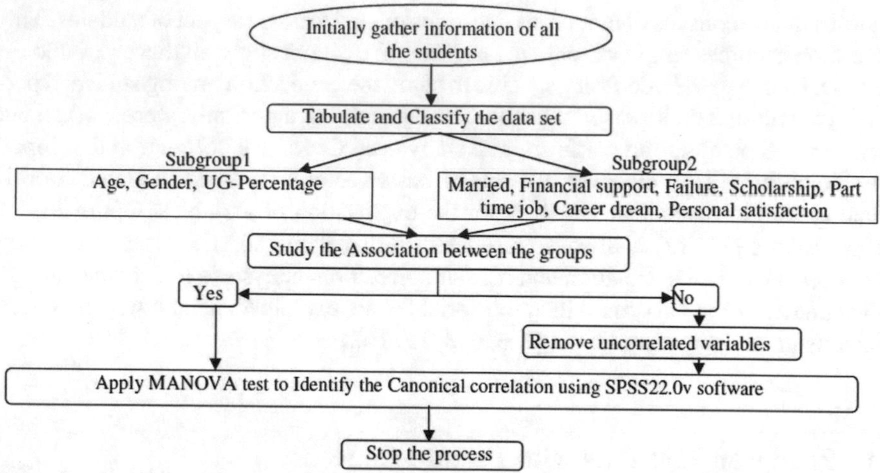

Fig. 1 Procedure for derivation of canonical correlations

```
/DISCRIM all alpha (1)
/ PRINT=SIG (EIGEN DIM).
```

3.2 Correlation Coefficient

Let, r be the correlation coefficient (Eq. (1)). To uncover r, we need two variables a and b, which are fit into following formula [20]:

$$r = \frac{l(\sum ab) - (\sum a) \cdot (\sum b)}{\sqrt{\left[l \sum a^2 - (\sum a)^2\right] \cdot \left[l \sum b^2 - (\sum b)^2\right]}} \quad (1)$$

Here,

$l =$ Information quantity.
$\Sigma a =$ First variable's totality value.
$\Sigma b =$ Second variable's totality value.
$\Sigma ab =$ Summation of the product of first & second value.
$\Sigma a^2 =$ Summation of the squares of the first value.
$\Sigma b^2 =$ Summation of the squares of the second value.

Table 1 Co-relational analysis

Parameter	1	2	3
Gender	0.59146	0.31036	0.74421
Age	0.98811	−0.42963	−0.09383
UG percentage	−0.48252	−0.82583	0.29187

Table 2 Raw canonical coefficient's analysis

Variable	1	Variable	1	Variable	1
Gender	0.18724	Gender	0.09236	Gender	0.08010
Age	0.35816	Age	0.71375	Age	0.66681
UG-%	0.12663	UG-%	0.76985	UG-%	0.67112

Table 3 The estimated of effects (canonical variables)

Factor	1
1	15.83536

4 Experimentations and Discussions

Two groups formed for appropriate analysis of the collected data. Here, total of 360 students information is collected, processed and analyzed. The first group contained three parameters of student's personal details and the second group contained socio-economic and family-related aspects. Once this is final, Multivariate Analysis of Variance (MANOVA) is used to escalate further findings between the two groups and is done through SPSS22.0v, via *manova* rule. Tables 1, 2, 3 shows the results of the canonical correlation analysis:

The overall explanations were interesting. The output shows testing of multivariate criteria. Afterwards, we have deployed three canonical correlations for every dimension [14, 17]. Overall, out of three, there are two statistically significant canonical correlations at the 0.05 level. Further, it is evident that canonical variation between Gender and Age are 0.59146 and 0.89811, respectively, which is a sign of strong positive correlation. The canonical variables show variance in dependent as 46.30%, 32.09% and 21.59%, respectively. It accounts for a total variance of 4.57%, 2.53% and 0.85%, respectively. We further observed that the travelling mode, failures, tutorial, scholarship, part-time job, self library, place of living, free time to study, free to friends, career dream explain strong positive linear relationship with gender. Travelling time, failure, scholarship, self pc, internet, free time study and material are positively correlated and having pathetic linear relationship with age. He/she is having library, place of living, free time study, own notes and material shows positively correlated. The performance is positively correlated with UG level marks. Further, it is extended by regression analysis with 95% confidence of individual Univariate intervals by way of dependent variables and within the cells error term. It was found that: (1) there is significance difference between the pair Gender: Scholarship in all canonical variates. (2) There is significance difference between the pairs Age:

Scholarship: Free time with Friends (3) the pair UG Percentage, Failure, Self library also significant. Student performance affects due to self library, erudition, spending suitable time with his/her friends and do not previous failure record. It aid view to enhance in presentation.

5 Conclusions

The present study is the demonstration of multivariate analysis test over educational data. It was done through Multivariate Analysis of Variance (MANOVA) model using SPSS software. All experimentations were modeled through multivariate criteria and three canonical correlations were modeled afterwards. The output results revel that the statistical significance between first two canonical correlations at the 0.05 level. All observations have proved that the personal details of students and own behavior as well as alertness have some consequence on performance. This alternative assessment gave us interesting information. We have understood that only study time and academic factors are not alone responsible for increase or decrease in the performance. There are other factors and their correlations, which one must not neglect. These hidden factors and their correlations were made visible through data analytics. These hidden factors can be considered for prognostication purposes so that changes in them can significantly accelerate the performance of the student. For example, performance can be boosted by giving more number of scholarships or encouraging students to have their self-book library at their homes. It is observed that gender, UG percentage and age variables presents significant difference among scholarship, scholarship and free time with friends and failure and self library. This study is limited up to 360 observations and it can be generalized by collecting more samples from diversified region to get more accuracy and fruitful results. In future, this work will be one of the bases for the selection of parameters in similar kind of studies in different areas.

References

1. Ali, S., Haider, Z., Munir, F., Khan, H., Ahmed, A.: Factors contributing to the student's academic performance: a case study of Islamia University Sub-Campus. Am. J. Educ. Res. **1**(8), 283–289 (2013)
2. Anderson, T.W., Mathématicien, E.U.: An introduction to multivariate statistical analysis. Wiley, New York (1958)
3. Becker, S.: Mutual information maximization: models of cortical self-organization. Netw. Comput. Neural Syst. **7**(1), 7–31 (1996)
4. Bratti, M., Staffolani, S.: Student time allocation and educational production functions. Annals of Economics and Statistics/Annales D'économie Et De Statistique 103–140 (2013)
5. Considine, G., Zappalà, G.: The influence of social and economic disadvantage in the academic performance of school students in Australia. J. Sociol. **38**(2), 129 (2002)
6. Das, S., Sen, P.K.: Restricted canonical correlations. Linear Algebr. Appl. **210**, 29–47 (1994)

7. Dunham, M.H.: Data Mining: Introductory and Advanced Topics. Pearson Education India (2006)
8. Fieguth, P.W., Irving, W.W., Willsky, A.S.: Multiresolution model development for overlapping trees via canonical correlation analysis. In: IEEE Proceedings International Conference on Image Processing, vol. 1, pp. 45–48 (1995)
9. Field, A.: Discovering statistics using SPSS for Windows. Sage publications, London (2000)
10. Graetz, B.: Socio-economic status in education research and policy in John A., et al., Socio-economic Status and School Education DEET/ACER Canberra. J. Pediatr. Psychol. **20**(2) 205–216 (1995)
11. Han, J., Pei, J., & Kamber, M.: Data mining: concepts and techniques. Elsevier. (2011)
12. Hotelling, H.: Relations between two sets of variates. Biometrika **28**(3/4), 321–377 (1936)
13. https://economictimes.indiatimes.com
14. Joshi, M., Bhalchandra, P., Muley, A., Wasnik, P.: Analyzing students performance using academic analytics. In IEEE International Conference ICT in Business Industry & Government (ICTBIG), pp. 1–4 (2016)
15. Kay, J.: Feature discovery under contextual supervision using mutual information. In IEEE International Joint Conference Neural Networks, IJCNN, vol. 4, pp. 79–84 (1992)
16. Li, P., Sun, J., Yu, B.: Direction finding using interpolated arrays in unknown noise fields. Sig. Process. **58**(3), 319–325 (1997)
17. Muley, A., Bhalchandra, P., Joshi, M., Wasnik, P.: Prognostication of student's performance: factor analysis strategy for educational dataset. Int. J. **4**(1), 12–21 (2016)
18. Muley, A., Bhalchandra, P., Joshi, M., Wasnik, P.: Academic analytics implemented for students performance in terms of canonical correlation analysis and chi-square analysis. In Information and Communication Technology. Springer, Singapore (2018)
19. Pritchard, M.E., Wilson, G.S.: Using emotional and social factors to predict student success. J. Coll. Stud. Dev. **44**(1), 18–28 (2003)
20. Web resource at www.byjus.com

Sine Cosine Optimization Based Proportional Derivative-Proportional Integral Derivative Controller for Frequency Control of Hybrid Power System

Tulasichandra Sekhar Gorripotu, Pilla Ramana, Rabindra Kumar Sahu and Sidhartha Panda

Abstract This work introduces Sine Cosine Algorithm (SCA) for a hybrid power system and it is employed with one of a cascade controller called proportional derivative-proportional integral derivative controller (PD-PID) to maintain frequency regulation. The proposed hybrid power system comprises of thermal unit, wind turbine generators (WTGs), aqua electrolyze (AE), fuel cell (FC), diesel engine generator (DEG), battery energy storage (BES) system, ultracapacitor (UC) in area-1 and in area-2 the combination of hydro-thermal units are considered. The system is analyzed under different cases: step load disturbance in both the areas, band limited noise in both the areas and step disturbance with noise at the wind system. The supremacy of the proposed controller is analyzed with PIDF controller on the same platform.

Keywords Band limited noise · Frequency regulation · Hybrid power system · PD-PID · SCA

T. S. Gorripotu (✉)
Department of Electrical and Electronics Engineering, Sri Sivani College of Engineering, Srikakulam 532410, Andhra Pradesh, India
e-mail: gtchsekhar@gmail.com

P. Ramana
Department of Electrical and Electronics Engineering, GMR Institute of Technology, Rajam 532127, Andhra Pradesh, India
e-mail: pramana.gmrit@gmail.com; ramana.pilla@gmrit.edu.in

R. K. Sahu · S. Panda
Department of Electrical Engineering, Veer Surendra Sai University of Technology (VSSUT), Burla 768018, Odisha, India
e-mail: rksahu123@gmail.com

S. Panda
e-mail: panda_sidhartha@rediffmail.com

H. S. Behera et al. (eds.), *Computational Intelligence in Data Mining*, Advances in Intelligent Systems and Computing 990,
https://doi.org/10.1007/978-981-13-8676-3_66

1 Introduction

Nowadays, the main concern of generating stations and distribution stations is to supply quality power to the consumers at a good price. With that due to vary or increase of load time to time, it is very difficult to maintain the state of equilibrium between the power necessity and generation. This imbalance between power necessity and generation creates deviation in active and reactive powers. The former can be inhibited by Automatic Voltage Regulator (AVR) and latter can be inhibited by Load Frequency Control (LFC) [1–9]. Here, the main focus is on frequency regulation of a hybrid interconnected power system. For regulating the frequency of the system, the primary controller can withstand only for small variations in the load but for large variations in the load, it is not sufficient and thus, secondary controller should be included which can works effectively during large load changes. In recent times, most of the researchers focusing on hybrid power system rather than conventional interconnected power system due to less cost and can be utilized during peak loads. Sekhar et al. [10] have analyzed BFOA optimized DG system employing classical controllers. Frede et al. [11] discussed importance of distributed power systems and basic structure, control of it. Authors also presented regarding harmonic compensation for distributed system. The fuzzy based PV-Diesel system is discussed by Manoj Datta and others [12]. Sowmya et al. [13] have discussed LFC of an isolated hybrid distributed power system by using hybrid particle swarm optimization and harmonic search. Pandey et al. [14] has designed hybrid power system and implemented with particle swarm optimization based LMI controller. Rehman et al. [15] have analyzed power variations and compensations of hybrid power with combination of wind and tidal systems.

2 Proposed System

A complex renewable energy source based network which comprises of two areas in which area-1 includes consists of thermal unit and Distributed Generation (DG). DG system in area-1 includes WTG, DEG, FC, AE, UC and BES. The area-2 includes hydro-thermal units. The block diagram of said system is in shown in Fig. 1. The mathematical modelling of all the units in distributed generation is clearly explained in [16–22]. To make the system as a closed loop, it is necessary to have a controller and in this particular system PD-PID controller is considered. The design of PD-PID controller is shown in Fig. 2. When a controller is used in a power system, it must be optimum for all criteria such as robustness to system nonlinearity, system dynamics, etc., therefore, the performance of controller is confirmed through different performance criteria. Most of the researchers are confirmed that ITAE criteria performs

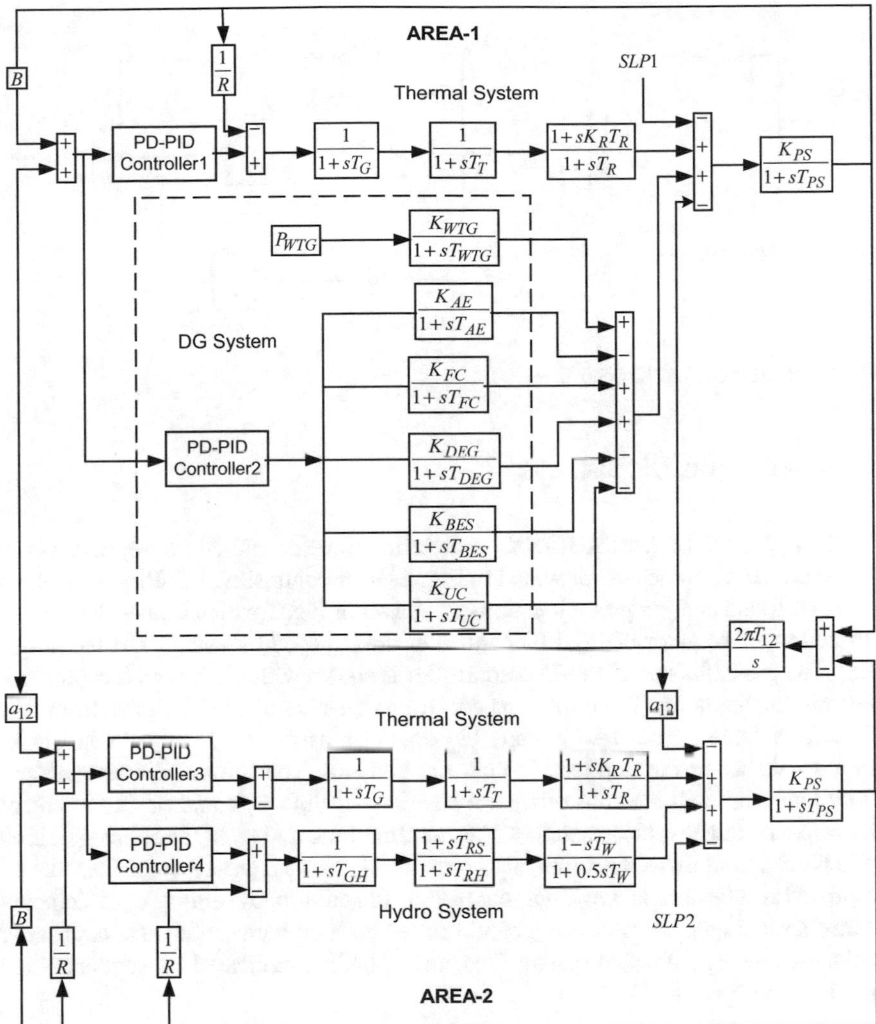

Fig. 1 SIMULINK diagram of hydro-thermal DG unit

better compared to others as it can provide low settling time and least overshoot oriented dynamic responses [23–27]. The expression for ITAE is given in Eq. (1).

$$ITAE = \int_{0}^{t_{sim}} (|\Delta F_1| + |\Delta F_2| + |\Delta P_{Tie}|) \cdot t \cdot dt \tag{1}$$

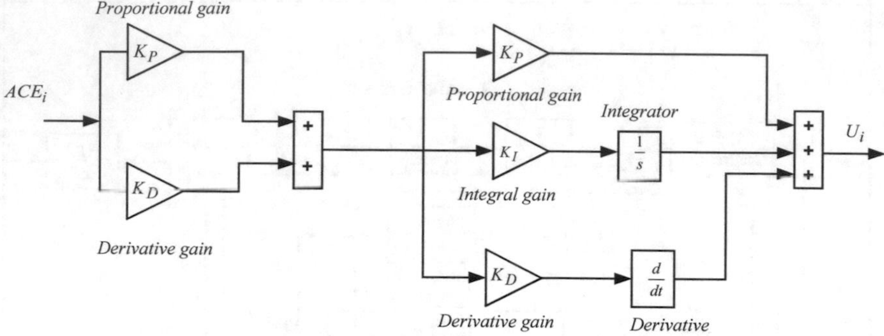

Fig. 2 Structure of PD-PID controller

3 Results and Discussions

Originally, MATLAB/SIMULINK s-domain transfer function representation intended in environment as provided in Fig. 1. Four contrasting PD-PID controllers are to be found at appropriate locations as shown in Fig. 1 which makes the system closed loop. The gains of PD-PID controller are adjusted by using SCA technique [28]. The pre-eminence of PD-PID controller is shown with PIDF which is premeditated in the literature. The optimized control parameters of PD-PID have been tabularized in Table 1. The ITAE imparted in Eq. (1), is assessed by replicate the model for 1% enhance of the step load in both the domains. The simulated consequences of system with both the controllers are assigned in Figs. 3, 4 and 5. The resultant performance index values such as ITAE, settling time and over shoot are specified in Table 2. From these outcomes, it is perceptible that suggested controller attains improved results compare to PIDF controller. In addition, the ability of anticipated controller is shown by concerning band limited noise in both the areas concurrently and noise signal at the wind system. The corresponding replicated consequences are provided in Figs. 6 and 7.

4 Concluding Remark

In this current work, an endeavour has been accomplished to optimize PD-PID controller by make use of sine cosine strategy for a two-area hybrid power system. To elucidate the incomparability of proposed PD-PID controller, a relative examination was made with PIDF controller which is previously published. Further, an assessment was made by claiming noise as annoyance in the two areas and noise signal at the wind plant. After the overall experimental analysis, it can be concluded that the projected controller pays for better outcomes as compared to other standard methods.

Table 1 Finest gains

Parameters		PD-PID
Controller-1	K_{P1}	0.7498
	K_{D1}	0.3616
	K_{P2}	−0.8004
	K_{I2}	−0.3658
	K_{D2}	1.6468
Controller-2	K_{P3}	1.7380
	K_{D3}	0.5727
	K_{P4}	−1.0186
	K_{I4}	−0.7665
	K_{D4}	−1.6867
Controller-3	K_{P5}	1.8110
	K_{D5}	0.7822
	K_{P6}	−1.6502
	K_{I6}	−1.8896
	K_{D6}	−0.8493
Controller-4	K_{P7}	0.5119
	K_{D7}	0.4223
	K_{P8}	−1.3230
	K_{I8}	0.9339
	K_{D8}	0.7782

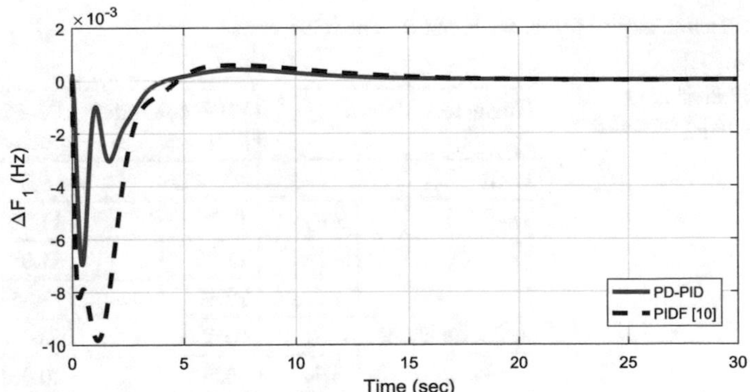

Fig. 3 Deviation in area-1 frequency for 1% SLP in both the areas

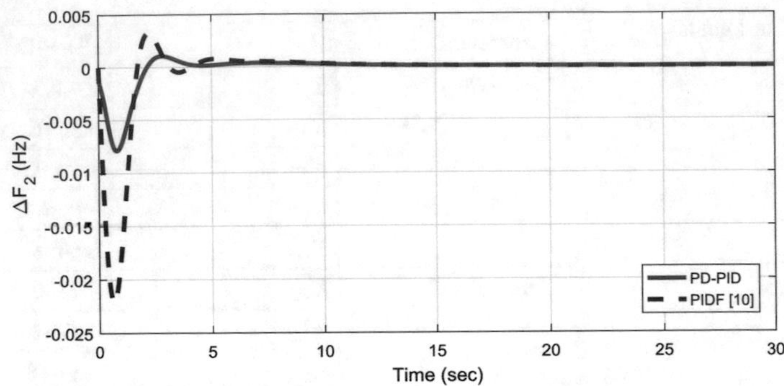

Fig. 4 Deviation in area-2 frequency for 1% SLP in both the areas

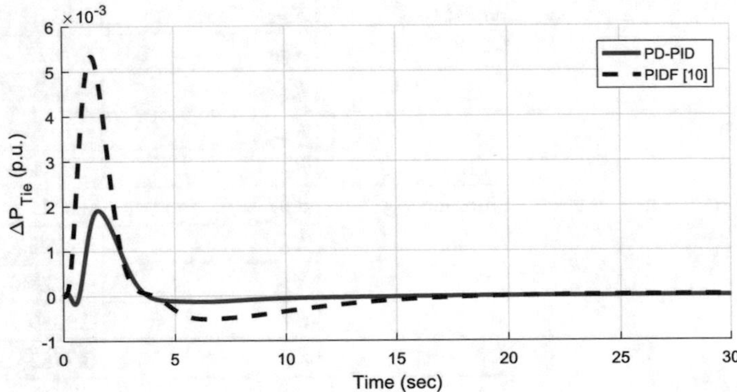

Fig. 5 Deviation in tie-line power for 1% SLP in both the areas

Table 2 Obtained index values through performance analysis

Parameters		PIDF controller [10]	PD-PID controller
ITAE		0.2555	0.0983
T_S (s)	ΔF_1	13.73	11.30
	ΔF_2	11.96	11.05
	ΔP_{Tie}	12.42	3.35
Peak over shoot	ΔF_1	0.0006	0.0004
	ΔF_2	0.0032	0.0011
	ΔP_{Tie}	0.0053	0.0019

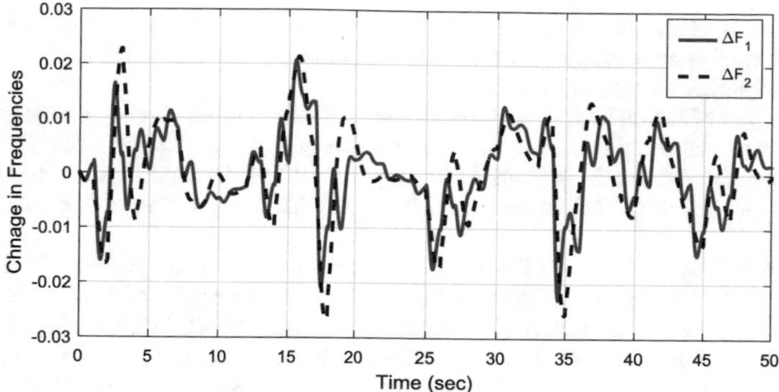

Fig. 6 Deviation in frequencies when noise is applied as disturbance

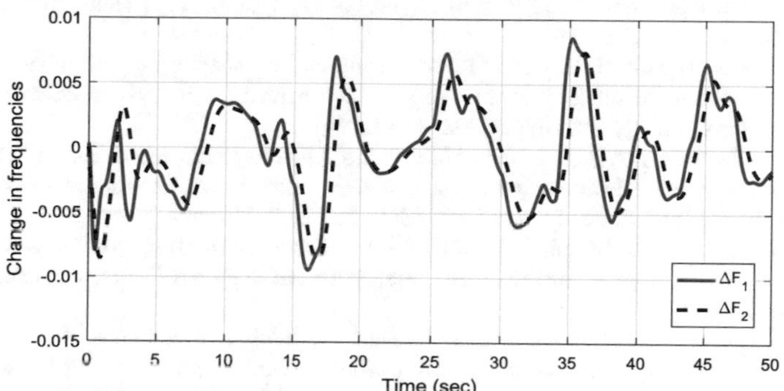

Fig. 7 Deviation in frequencies when noise is applied at wind system

References

1. Elgerd, O.I.: Electric Energy Systems Theory–An Introduction. Tata McGraw Hill, New Delhi (2000)
2. Bevrani, H.: Robust Power System Frequency Control. Springer, Berlin (2009)
3. Bevrani, H., Hiyama, T.: Intelligent Automatic Generation Control. CRC Press, Boca Raton (2011)
4. Gorripotu, T.S., Sahu, R.K., Panda, S.: AGC of a multi-area power system under deregulated environment using redox flow batteries and interline power flow controller. Eng. Sci. Technol. Int. J. **18**(4), 555–578 (2015)
5. Gorripotu, T.S., Sahu, R.K., Panda, S.: Application of firefly algorithm for AGC under deregulated power system. In: Jain, L., Behera, H., Mandal, J., Mohapatra, D. (eds) Computational Intelligence in Data Mining - Volume 1. Smart Innovation, Systems and Technologies, vol 31. Springer, New Delhi, 2015
6. Gorripotu, T.S., Sahu, R.K., Baliarsingh, A.K., Panda, S.: Load frequency control of power system under deregulated environment using optimal firefly algorithm. Int. J. Electr. Power Energy Syst. **74**, 195–211 (2016)

7. Saikia, L.C., Mishra, S., Sinha, N., Nanda, J.: Automatic generation control of a multi area hydrothermal system using reinforced learning neural network controller. Int. J. Electr. Power Energy Syst. **33**(4), 1101–1108 (2011)
8. Ali, E.S., Abd-Elazim, S.M.: BFOA based design of PID controller for two area load frequency control with nonlinearities. Int. J. Electr. Power Energy Syst. **51**, 224–231 (2013)
9. Parmar, K.P.S., Majhi, S., Kothari, D.P.: Improvement of dynamic performance of LFC of the two-area power system: an analysis using MATLAB. Int. J. Comput. Appl. **40**(10), 28–32 (2012)
10. Chandra Sekhar, G.T., Vijaya Kumar, D., Manamadha Kumar, B., Ramana, P.: Design and analysis of BFOA optimized PID controller with derivative filter for frequency regulation in distributed generation system. Int. J. Comput. Appl. **12**(2), 291–323 (2018). INDERSCIENCE
11. Frede, B., Remus, T., Marco, L., Adrian, T.V.: Overview of control and grid synchronization for distributed power generation systems. IEEE Trans. Industr. Electron. **53**(5), 1398–1409 (2006)
12. Manoj, D., Tomonobu, S., Atsushi, Y., Toshihisa, F., Kim,C.H.: A frequency-control approach by photovoltaic generator in a PV–diesel hybrid power system. IEEE Trans. Energy Convers. **26**(2), 559–571 (2011)
13. Mohanty, S.R., Kishor, N., Ray, P.K.: Robust H-infinite loop shaping controller based on hybrid PSO and harmonic search for frequency regulation in hybrid distributed generation system. Int. J. Electr. Power Energy Syst. **60**(1), 302–316 (2014)
14. Pandey, S.K., Mohanty, S.R., Kishor, N., Catalão, J.P.: Frequency regulation in hybrid power systems using particle swarm optimization and linear matrix inequalities based robust controller design. Int. J. Electr. Power Energy Syst. **63**(1), 887–900 (2014)
15. Rahman, M.L., Oka, S., Shirai, Y.: Hybrid power generation system using offshore-wind turbine and tidal turbine for power fluctuation compensation (HOT-PC). IEEE Trans. Sustain. Energy **1**(2), 92–98 (2010)
16. Howlader, A.M., Izumi, Y., Uehara, A., Urasaki, N., Senjyu, T., Saber, A.Y.: "A robust H∞ controller based frequency control approach using the wind-battery coordination strategy in a small power system. Int. J. Electr. Power Energy Syst. **58**(1), 190–198 (2014)
17. Li, Y.H., Choi, S.S., Rajakaruna, S.: An analysis of the control and operation of a solid oxide fuel-cell power plant in an isolated system. IEEE Trans. Energy Convers. **20**(2), 381–387 (2005)
18. Sedghisigarchi, K., Feliachi, A.: Impact of fuel cells on load-frequency control in power distribution systems. IEEE Trans. Energy Convers. **21**(1), 250–256 (2006)
19. Ray, P.K., Mohanty, S.R., Kishor, N.: Proportional–integral controller based small-signal analysis of hybrid distributed generation systems. Energy Convers. Manag. **52**(4), 1943–1954 (2011)
20. Das, D.C., Roy, A.K., Sinha, N.: GA based frequency controller for solar thermal–diesel–wind hybrid energy generation/energy storage system. Int. J. Electr. Power Energy Syst. **43**(1), 262–279 (2012)
21. Li, X., Song, Y.-J., Han, S.-B.: Frequency control in micro-grid power system combined with electrolyzer system and fuzzy PI controller. J. Power Sources **180**(1), 468–475 (2008)
22. Egido, I., Sigrist, L., Lobato, E., Rouco, L., Barrado, A.: An ultra-capacitor for frequency stability enhancement in small-isolated power systems: Models, simulation and field tests. Appl. Energy **137**(1), 670–676 (2015)
23. Sahu, R.K., Gorripotu, T.S., Panda, S.: A hybrid DE–PS algorithm for load frequency control under deregulated power system with UPFC and RFB. Ain Shams Eng. J. **6**, 893–911 (2015)
24. Sahu, R.K., Gorripotu, T.S., Panda, S.: DE optimized fuzzy PID controller with derivative filter for LFC of multi source power system in deregulated environment. Ain Shams Eng. J. **6**, 511–530 (2015)
25. Sahu, R.K., Panda, S., Gorripotu, T.S.: A novel hybrid PSO-PS optimized fuzzy PI controller for AGC in multi area interconnected power systems. Int. J. Electr. Power Energy Syst. **64**, 880–893 (2015)

26. Sahu, R.K., Gorripotu, T.S., Panda, S.: Automatic generation control of multi-area power systems with diverse energy sources using teaching learning-based optimization algorithm. Eng. Sci. Technol. Int. J. **19**(1), 113–134 (2016)
27. Shabani, H., Vahidi, B., Ebrahimpour, M.A.: Robust PID controller based on imperialist competitive algorithm for load-frequency control of power systems. ISA Trans. **52**(1), 88–95 (2012)
28. Mirjalili, S.: SCA: a sine cosine algorithm for solving optimization problems. Knowl.-Based Syst. **96**(1), 120–133 (2016)

Author Index

H. S. Behera et al. (eds.), *Computational Intelligence in Data Mining*, Advances in Intelligent Systems and Computing 990,
https://doi.org/10.1007/978-981-13-8676-3

Printed in the United States
By Bookmasters